Contents

Preface

The Eighth International Symposium of Robotics Research was held in Shonan near Kamakura, Japan from October 4 to 7, 1997, organized by the International Foundation of Robotics Research (IFRR). The goal of the Symposium was to bring together active, leading robotics researchers from academia and industry to assess the state of art of Advanced Robotics and to discuss future research directions.

The Symposium was held in an informal setting with fifty nine participants. Attendance was limited in order to maximize interaction. The selection of participants was made by the Program Committee using a process of nomination and voting. The Program Committee consisted of the seventeen officers of the IFRR:

Ruzena Bajcy	(University of Pennsylvania, USA)
Andrew Blake	(Oxford University, UK)
Robert Bolles	(SRI International, USA)
Paolo Dario	(Scuola Superiore Sant'Anna, Pisa, Italy)
Joris De Schutter	(Katholieke Universiteit, Leuven, Belgium)
Georges Giralt	(LAAS-CNRS, France), Secretary
Gerd Hirzinger	(DLR, Germany)
Shigeo Hirose	(Tokyo Institute of Technology, Japan)
Hirochika Inoue	(University of Tokyo, Japan), President
Ray Jarvis	(Monash University, Australia)
Takeo Kanade	(CMU, USA), Treasurer
Dan Koditschek	(University of Michigan, USA)
Hirofumi Miura	(University of Tokyo, Japan)
Richard Paul	(University of Pennsylvania, USA)
Bernie Roth	(Stanford University, USA)
Yoshiaki Shirai	(Osaka University, Japan)
Tsuneo Yoshikawa	(Kyoto University, Japan)

While a limited number of participants were nominated in each area by the Program Committee members of the area, a call for papers was issued in order to include researchers who had made significant contributions to robotics. All submitted extended abstracts were reviewed by the Program Committee before final paper selection was made. Attendance at the Symposium was limited to the authors of selected papers and the nominees.

During the four days of sessions, forty-four papers were presented in a single track to cover the broad research area of robotics. One of the featured topics of the symposium was robots working in close contact with humans. Manipulators or mobile robots, for example, work in the same environment as humans interacting with humans. Further efforts are being made in medical applications, humanoid robots and human-robot interface technologies. In addition robotics was shown to be useful for human welfare, social infrastructure and entertainment.

Reflecting this trend, a three year project under the name of "tightly-coupled perception-motion behavior" was started two years ago as one of the priority researches by the Japanese Ministry of Education. In one of the evening sessions of this Symposium, on going research results of the project together with another new project on humanoid robots were introduced and discussed. We hope that readers of this book will share the excitement of the attendees.

We would like to express, on behalf of the IFRR, our thanks to the following institutions for funding: New Technology Foundation, the French embassy, Mitsubishi Heavy Industry and Fujitsu Laboratories. Our special thanks go to Mr Kazuhiko Nishi for his financial support. We also wish to thank the Organizing Committee members: H. Miura, H. Inoue, M. Uchiyama and Y. Nakamura, and the staff of the members who helped in organizing and operating the Symposium.

Yoshiaki Shirai and Shigeo Hirose
Co-Chair
The Eighth International Symposium of Robotics Research

List of Authors

PART 1
ADVANCED MANIPULATION
SESSION SUMMARY

D. E. Koditschek
AI Lab and Control Systems Lab
EECS Department, College of Engineering
University of Michigan, Ann Arbor 48109-2110
kod@umich.edu

One of the most fundamental problems in robotics concerns "task encoding" — the translation of abstract human goals into sensory driven force and torque profiles that guarantee our machines achieve them. In the 80's, much fundamental work in the field of robotics was devoted to the development of control algorithms (and proofs of their correctness) that would force general kinematic chains to track asymptotically exactly any sufficiently smooth reference trajectory. Toward the end of that decade, adaptive versions of these algorithms were developed that afford the same results even absent knowledge of dynamical parameters. In the 90's, attention has been focused on tracking reference trajectories in the presence of force requirements as well. More recently, we have even begun to see some of these advanced trajectory tracking ideas make their appearance in real commercial products. Thus, although important work remains to be done in this area, it seems fair to assert that trajectory tracking methods have reached a rather mature point of development in robotics. In other words, while advanced new tools of this kind are very important to the field, they now have a very clearly understood place in the robotics toolbox.

In contrast, this chapter presents a set of three papers offering distinct approaches to one of the most important next steps: the question of "where does the trajectory come from?" in the first place. This question shifts the focus of inquiry in manipulation research from purely controls oriented concentration on stability and transient performance (although these issues, of course, remain at the foundations of the problem) up to a more general mathematical arena – the representation of motions. Said a little more formally, this amounts to the choice of a finite parametrization of an infinite dimensional space. As usual, the issue of what parametrization to choose is tightly bound up with how expressive of human goals one or another parameter space can be as well as how easily the link to sensory data can be used to move parameters around in an effectively reactive manner.

The paper, "Elastic Strips: Real-Time Path Modification for Mobile Manipulation" by Brock and Khatib addresses exactly this latter issue. The problem domain arises with a robotic motion planning scenario where a pre-planned obstacle free path may be partially or totally blocked by unanticipated but accurately sensed execution-time moving obstacles. The desired goal is a perturbation of the path with the original initial and final points that remains free of obstacles. A perturbed path must be efficiently computed since it is to be imposed in real time during motion execution. A central problem surrounds the matter of how to parametrize in a computationally feasible manner the infinite variety of shapes that a perturbed path can take on. A finite dimensional parametrization having been selected, it remains to develop an algorithm for "pushing away" the original path from entanglements detected by sensors. In previous work, the authors have introduced the notion of an "elastic band" – a linked list of "knot" points in the free configuration space that can be "pushed around" by artificial potential forces corresponding to obstacles whose dislocating effects are countered by spring-like restoring forces designed to keep contiguous knot points proximal. In the present paper, the authors introduce an analogous construction, an "elastic strip," in the workspace. They describe a set of preliminary experiments with a nine degree of freedom mobile manipulator platform suggesting the practicability of the idea.

A very different point of view would be to parametrize not the space of reference trajectories or motions itself but, rather, the space of "infinitesimal generators," — i.e., the differential (or difference) equations of motion. Typically, one implements such a parametrization in a working system by manipulation of feedback laws, for such an approach offers off-line planning capability readily expressive of human goals as well as the necessary execution-time link between sensory data and control decisions. Moving around in a parameter space that represents dynamics or transition rules is in some sense a more abstract way to encode tasks than via motions directly, since each parameter value generates an entire family of motions (one corresponding to each initial condition). Note that this is explicitly the point of view taken by both Nakamura and Yamazaki's paper and by this author's own paper in the chapter on emergent motions. It is in some sense the generic rationale for a "reactive systems" point of view in robotics.

In the present chapter, the paper "Modeling and Control for Mobile Manipulation in Everyday Environments," by Feiten *et al.* represents this point of view in a most satisfyingly concrete and practical setting. To date, two prototype floor cleaning machines based upon the principles articulated in this paper have been functioning in working supermarkets operating several times a day throughout the normal hours of business. The authors describe a next step in the exciting evolution of their working service robots. Since these machines are designed to work in human dominated surroundings they must be safe (i.e., react robustly, reliably, and effectively when the environment departs from its expected state) and able to navigate in a human environment with no further engineering such as beacons or special doorknobs or elevator controls. The system operates on the principle of plan execution through layered goal seeking. A high level planner works upon an abstracted location graph, an intermediate planner works within a geometric floor plan, and a safety layer navigates locally with respect to an occupancy grid. The key to the reliability and robustness of this architecture is the strict adherence to a discipline wherein plans are represented as sequences of goal states to be achieved by servoing on sensory phenomena. For example, the task of reaching to grasp a door knob is represented as a visual servoing task requiring the robot to diminish the discrepancy between existing and specified image plane feature locations. Thus, the authors parametrize tasks by manipulating the goal points and the feedback gains of a servoing law. Formally speaking, they are moving around settings that parametrize the space of infinitesimal generators of the resulting motions.

A still higher level question arising from the necessity of relating abstract human tasks and changed environmental circumstances to robot motions concerns the matter of how to "compile" into a working "program" of subgoals and responses whatever parametrization has been adopted. For example, in the paper by Feiten *et al.* just described, such higher level goals are planned by backchaining the sensory benchmarks together in such a fashion that the domain of attraction for the next goal always includes the final state (completed task) of the previous. This idea of backchaining has a long and honorable history in the field dating back to work arising from the students and colleagues of Lozano-Perez in the mid '80s.

A very different approach to this issue of how motion primitives might be effectively joined together to yield higher level procedures is presented in the last paper of this chapter, "Scale Dependent Grasp," by Kaneko. This is a human behavior based inquiry into high level planning of grasp strategies. Tool-handle-like lengths of stock at various thicknesses and cross sections were presented to human subjects whose grasping strategies were contrasted when the objects were covered with a rubber sheet and with paper. A qualitative taxonomy of grasp types was used to construct a strategy histogram from the observed human behaviors. The most popular strategies identified in this fashion were translated into parametrized machine procedures for robotic sensing and grasping. A further set of experiments was performed studying the success of the machine procedures as a function of the parameter settings. Notice, in this approach, that most of the key decisions concerning how to parametrize both the motion primitives as well as the higher level decision procedures regarding their composition are bound up in the taxonomy of grasp types. These "borrowed" strategies, when translated into machine procedures and tuned properly were capable of grasping a wide range of the object types studied and do indeed suggest the efficacy of "reverse engineering" human grasp strategies from direct observation. Generalizing such success, of course, presumes the availability of an effective parametrization with respect to which the human strategies can be represented, the reverse engineering recorded, and the translation to machine effected. There are intriguing hints in this paper about the relatively greater importance of scale over shape in building up such representations. It would be of great interest, then, to understand more about the author's favorable choice of scale sensitive taxonomy in this setting, as well as his ideas about how such clever insight might be generalized to other classes of tasks.

In summary, the present chapter offers three distinct approaches to the central problem of task encoding in robotics. The papers all share the virtue of presenting successful physical laboratory implementations of tasks at the cutting edge of what is presently possible in our field. From this perspective, the reader can be reassured that each of these architectural views confers some evident virtues upon the resulting implementation. Clearly, much more empirical work and theoretical exploration will be required before these exciting ideas settle into their appropriate places in the standard robotics toolbox.

Elastic Strips: Real-Time Path Modification for Mobile Manipulation

Oliver Brock   Oussama Khatib
Robotics Laboratory, Department of Computer Science
Stanford University, Stanford, California 94305
email: {oli, khatib}@CS.Stanford.EDU

Abstract

The elastic strip framework presented in this paper is a new approach to real-time obstacle avoidance for robots with many degrees of freedom. The avoidance behavior can be integrated with task execution, a capability necessary for the application of mobile manipulation to dynamic real-world environments. The elastic strip initially represents a collision-free path generated by a planner. As the environment changes, the elastic strip is incrementally modified to maintain a smooth, collision-free path. Since the elastic strip is entirely represented in workspace, costly computation in a high-dimensional configuration space is avoided. This results in an efficient modification procedure for the elastic strip, even for robots with many degrees of freedom. The general framework of elastic strips is introduced and its application to obstacle avoidance during task execution is discussed.

1 Introduction

The robust execution of planned robot motion in dynamic environments and in the presence of uncertainty is essential for achieving intelligent robot behavior. In mobile manipulation this problem becomes particularly challenging because the execution of a task has to be integrated with reactive collision avoidance for robots with many degrees of freedom.

Recent research in control structures for mobile manipulation has resulted in an effective approach to task description and execution for arm/base systems at the object level (Khatib et al. 1996a). Compliant motion and force control strategies allow multiple arm/base systems to cooperate with a decentralized control structure, accomplishing complex manipulation tasks.

Reactive collision avoidance for manipulators or mobile platforms is a well-studied problem (Krogh 1984; Khatib 1986; Arkin 1987; Latombe 1991). Due to the recent interest in mobile manipulation more advanced methods are being investigated to incorporate obstacle avoidance with limited coordination between manipulator and mobile base (Carriker et al. 1989; Seraji 1993; Yamamoto and Yun 1995; Nassal 1996). These approaches are mainly concerned with positioning the base according to a manipulator-dependent cost function and suspend task-level behavior while performing collision avoidance. Mobile manipulation, however, requires obstacle avoidance during task execution.

In this paper we present a new approach to real-time path modification for robotic systems with many degrees of freedom. It is an extension to the elastic band framework (Quinlan and Khatib 1993). The new approach, called the *elastic strip*, is an efficient framework for collision avoidance for robots with many degrees of freedom. In addition, it allows for the integration of collision avoidance with the execution of manipulation tasks. Furthermore, it maintains the topological properties of the path and therefore is immune to local minima.

Similarly to elastic bands, this novel approach decomposes the task of executing robot motion plans in dynamic environments into a planning and an execution phase. The path resulting from the planning phase is assumed to be valid. During execution of the path, it is incrementally modified. This could be viewed as augmenting a motion plan with a reactive component to tolerate unpredictable changes in the environment.

The efficiency of this approach is based on the fact that all computations are carried out directly in the workspace, avoiding the computational complexity of high-dimensional configuration spaces. We present the general algorithm and illustrate coordination schemes for obstacle avoidance and task execution.

2 Arm/Base Coordination

The joint space dynamics of a manipulator is described by

$$A(\mathbf{q})\ddot{\mathbf{q}} + \mathbf{b}(\mathbf{q}, \dot{\mathbf{q}}) + \mathbf{g}(\mathbf{q}) = \mathbf{\Gamma}; \tag{1}$$

where $\mathbf{q}$ is the n joint coordinates and $A(\mathbf{q})$ is the $n \times n$ kinetic energy matrix. $\mathbf{b}(\mathbf{q}, \dot{\mathbf{q}})$ is the vector of centrifugal and Coriolis joint-forces and $\mathbf{g}(\mathbf{q})$ is the gravity joint-force vector. $\mathbf{\Gamma}$ is the vector of generalized joint-forces.

The operational space equations of motion of a manipulator are (Khatib 1987)

$$\Lambda(\mathbf{x})\ddot{\mathbf{x}} + \mu(\mathbf{x}, \dot{\mathbf{x}}) + \mathbf{p}(\mathbf{x}) = \mathbf{F}, \tag{2}$$

where $\mathbf{x}$ is the vector of the m operational coordinates describing the position and orientation of the effector, and $\Lambda(\mathbf{x})$ is the $m \times m$ kinetic energy matrix associated with the operational space. $\mu(\mathbf{x}, \dot{\mathbf{x}})$, $\mathbf{p}(\mathbf{x})$, and $\mathbf{F}$ are respectively the centrifugal and Coriolis force vector, gravity force vector, and generalized force vector acting in operational space.

These equations of motion describe the dynamic response of a manipulator to the application of an operational force $\mathbf{F}$ at the end effector. For non-redundant manipulators, the relationship between operational forces $\mathbf{F}$ and joint forces $\mathbf{\Gamma}$ is

$$\mathbf{\Gamma} = J^T(\mathbf{q})\mathbf{F}, \tag{3}$$

where $J(\mathbf{q})$ is the Jacobian matrix.

For redundant systems it can be shown that the relationship between joint torques and operational forces is

$$\mathbf{\Gamma} = J^T(\mathbf{q})\mathbf{F} + \left[I - J^T(\mathbf{q})\overline{J}^T(\mathbf{q})\right]\mathbf{\Gamma}_0, \tag{4}$$

with

$$\overline{J}(\mathbf{q}) = A^{-1}(\mathbf{q})J^T(\mathbf{q})\Lambda(\mathbf{q}), \tag{5}$$

where $\overline{J}(\mathbf{q})$ is the dynamically consistent generalized inverse. This relationship provides a decomposition of joint forces into two dynamically decoupled control vectors: joint forces corresponding to forces acting at the end effector ($J^T\mathbf{F}$) and joint forces that only affect internal motions, $\left([I - J^T(\mathbf{q})\overline{J}^T(\mathbf{q})]\mathbf{\Gamma}_0\right)$.

Using this decomposition, the end effector can be controlled by operational forces, whereas internal motions can be independently controlled by joint forces that are guaranteed not to alter the end effector's dynamic behavior. This relationship is the basis for implementing the dynamic coordination strategy for an arm/base system. It can be exploited for collision avoidance during task execution. In equation (4) $\mathbf{F}$ are the commanded forces at the end-effector to accomplish the task and $\mathbf{\Gamma}_0$ are the joint forces mapped into the null space to implement task-independent behavior.

The end-effector equations of motion for a redundant manipulator are obtained by the projection of the joint-space equations of motion (1), by the dynamically consistent generalized inverse $\overline{J}^T(\mathbf{q})$:

$$\begin{aligned} &\overline{J}^T(\mathbf{q})\left[A(\mathbf{q})\ddot{\mathbf{q}} + \mathbf{b}(\mathbf{q}, \dot{\mathbf{q}}) + \mathbf{g}(\mathbf{q}) = \mathbf{\Gamma}\right] \\ &\quad\Longrightarrow\ \Lambda(\mathbf{q})\ddot{\mathbf{x}} + \mu(\mathbf{q}, \dot{\mathbf{q}}) + \mathbf{p}(\mathbf{q}) = \mathbf{F} \end{aligned} \tag{6}$$

This property also applies to non-redundant manipulators, where the matrix $\overline{J}^T(\mathbf{q})$ reduces to $J^{-T}(\mathbf{q})$.

3 Elastic Band Framework

The elastic strips are based on elastic band model (Quinlan and Khatib 1993), this latter is briefly introduced in this section and its limitations are discussed.

3.1 Overview

An elastic band (Quinlan and Khatib 1993) is a discretized curve in the configuration space of a robot, representing a collision-free path. This path is initially obtained from a motion planner and subsequently subjected to artificial forces. Objects in the environment exert external, repulsive forces on the elastic band, ensuring obstacle avoidance. Internal forces act to shorten and smoothen the path, imitating the physical behavior of elastic material.

In the original framework the modification procedure of the elastic band requires that the robot is holonomic. In an extension to the framework bubbles were introduced that incorporate non-holonomic motion constraints (Khatib et al. 1997).

In Figure 1 an example of an elastic band is shown. The initial trajectory that was obtained

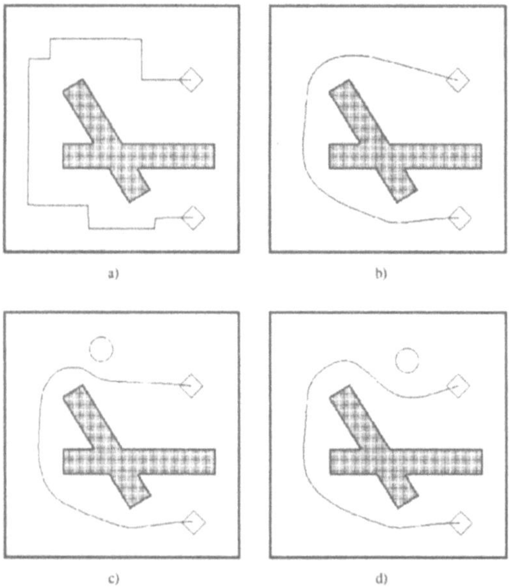

Figure 1: Example of an elastic band

by a planner (a) is changed by external and internal forces to yield a smoothened and shortened one (b). The introduction of a new obstacle deforms the path in real-time, maintaining its global topological properties (c,d). Note that in the example the configuration space coincides with the workspace, since we consider a planar robot that translates in the plane.

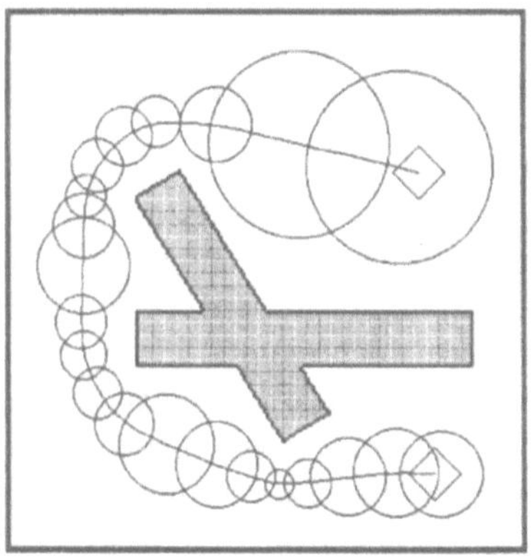

Figure 2: Bubbles on an elastic band

The key ingredient for the efficiency of the modification procedure for the elastic band is the description of local free space around it. With only one distance computation in the workspace a hypersphere of free configuration space can be computed. This hypersphere is called *bubble* (Quinlan 1994). By covering the entire elastic band with overlapping bubbles we can guarantee that the path it represents is collision-free. This is illustrated in Figure 2.

3.2 Limitations

The elastic band framework represents the trajectory of the robot and the local free space around it in the configuration space. This entails disadvantages that make an application of the approach to mobile manipulation difficult.

In mobile manipulation the task is described in terms of end-effector motions. A joint space trajectory is an unnatural representation for this motion. Since the elastic band approach is based on joint space trajectories, it is impractical for application to mobile manipulation.

In the elastic band framework the estimate of the local free space around a configuration becomes excessively conservative, as the number of degrees of freedom of the manipulator increases (Quinlan 1994). As a consequence, an excessive amount of bubbles is needed to cover the elastic band, affecting the real-time performance of the algorithm.

4 Elastic Strip Framework

To overcome the problems of elastic bands and to allow robust execution of motion for robots with many degrees of freedom, we have devised an extension to the elastic band approach, called elastic strip. Using this novel framework, obstacle avoidance can be integrated with task execution behavior.

4.1 Overview

In the elastic strip, a trajectory is incrementally modified by external forces originating from obstacles in the environment and internal forces applied to shorten and smoothen it.

To minimize the impact of the dimensionality of the configuration space on the computational complexity of the algorithm and thereby its real-time performance, we choose to represent the trajectory in the workspace.

Whereas with elastic bands a configuration of the robot could be described as a point in configuration space, a workspace description of that configuration consists of the robot's workspace volume. Similarly, in the elastic band framework a bubble represents a configuration free space hypersphere; this corresponds to an obstacle-free workspace volume in the new approach.

A robot trajectory will then be represented by the workspace volume swept by the robot during the motion along the path, as opposed to a one-dimensional curve in configuration space. This volume can be imagined as a strip of elastic material that is deformed by obstacles and shortens when obstacles are removed.

The behavior of the elastic strip differs in certain aspects from elastic material. Since we are not restrained by the physical laws of elasticity, we choose the effect of forces on the elastic strip to be task dependent, thereby integrating collision avoidance and task execution.

4.2 Rigid Body Representation

During the execution of a trajectory a robot sweeps out a volume in the workspace. The trajectory is collision-free if no obstacle is inside this volume. To warrant collision avoidance the volume has to be computed using a model of the rigid bodies in motion and checked against obstacles in the environment.

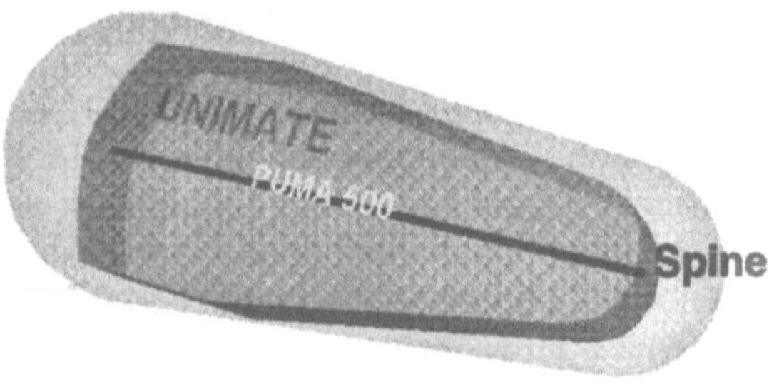

Figure 3: Illustration of a spine

A rigid body can be approximated by a line segment that is parameterized by a varying width. We call that line segment the *spine* of the rigid body. The width specifies the free space required around the line segment for the body to be collision free. In Figure 3 the spine of a transparent PUMA 560 link is shown as the black line along its central axis; the volume associated with the spine encloses the link.

This model is only a very coarse approximation of the rigid body. However, it can be assumed that in most cases the robot will maintain a safe distance to obstacles. The computationally more efficient check against a coarse representation of its volume then suffices to ensure collision avoidance.

As a rigid body comes closer to obstacles, the model described above might result in incorrect collision detection. To address this difficulty the representation of rigid bodies can be generalized to describe rigid bodies at an arbitrary level of accuracy.

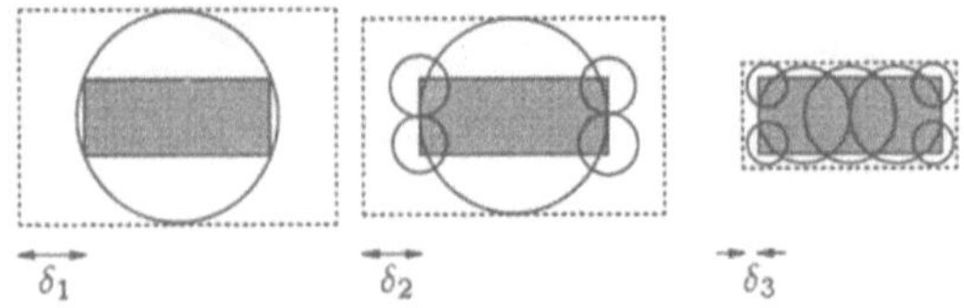

Figure 4: Covering a body with spines

The basic idea of this generalization is to introduce more spines to cover the volume of the rigid body with increasing accuracy as the distance to obstacles decreases. In Figure 4 the rectangular cross section of a body and its circular covering by spines are shown for different resolutions δ_i, indicated by the dotted region.

The three pictures could also be interpreted as the evolution of the volume description as the rigid body approaches an obstacle at distance δ_i. If at resolution δ_1 a collision is detected, the next level of accuracy in representation is used. The successive refinement of the model used for collision detection ends when the maximum resolution of the model is reached.

For the sake of simplicity we assume in the remainder of this paper that the volume of a rigid body is described by a single spine.

4.3 Representing Local Free Space

To compute a bubble in the elastic band approach, a distance measurement to a point in the workspace has to be translated into a hypersphere in configuration space. For elastic strips the free space is represented in the workspace and the distance computation translates directly into a bubble in the workspace. Let $\rho(\mathbf{p})$ be the function that computes the minimum distance from a point $\mathbf{p}$ to any obstacle. The *workspace bubble* of free space around $\mathbf{p}$ is defined as the set of points that are closer to $\mathbf{p}$ than the closest obstacle:

$$\mathcal{B}(\mathbf{p}) = \{\, \mathbf{q} : \|\mathbf{p} - \mathbf{q}\| < \rho(\mathbf{p})\}.$$

An approximation of the local free space around a rigid body b in configuration q can be computed by generating a set of workspace bubbles centered on the spine. This set of bubbles is called *protective hull* $\mathcal{P}_q^b$ of rigid body b in configuration q. The local free space or protective hull

$\mathcal{P}_q^{\mathcal{R}}$ of a robot $\mathcal{R}$ at a configuration q is described by the union of protective hulls of each rigid body of $\mathcal{R}$,

$$\mathcal{P}_q^{\mathcal{R}} = \bigcup_{b \in \mathcal{R}} \mathcal{P}_q^b.$$

Figure 5 shows a protective hull of the *Stanford Mobile Platform*. Note that a single workspace bubble may contain multiple rigid bodies or even the entire robot, implying that for large clearances a simple description of the local free space suffices.

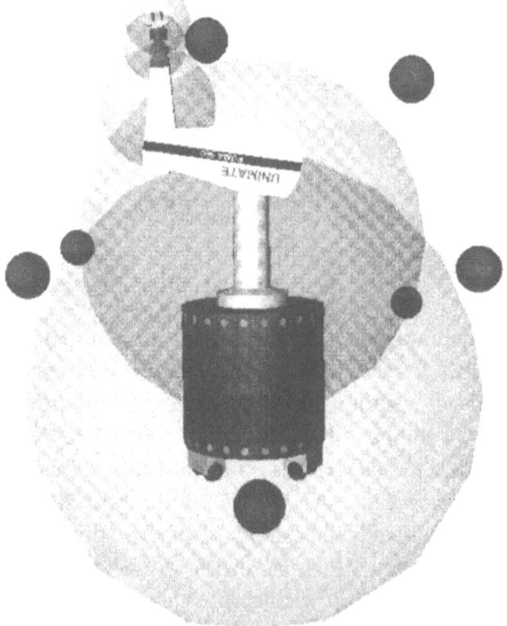

Figure 5: Protective hull of the Stanford Mobile Platform amidst spherical obstacles

4.4 Connectedness of Elastic Strip

An elastic strip $\mathcal{S}_{\mathcal{T}}^{\mathcal{R}} = (q_1, q_2, q_3, \cdots, q_n)$ is a sequence of configurations q_i on the trajectory $\mathcal{T}$ of the robot $\mathcal{R}$. The local free space for a configuration q_i is described by the protective hull $\mathcal{P}_i^{\mathcal{R}}$.

Since each configuration q_i is guaranteed to be free of collisions by the protective hull $\mathcal{P}_i^{\mathcal{R}}$, it remains to be shown that the union of all protective hulls contains the volume $V_{\mathcal{T}}^{\mathcal{R}}$ swept by the robot along the trajectory. The condition of feasibility of trajectory $\mathcal{T}$ described by $\mathcal{S}_{\mathcal{T}}^{\mathcal{R}}$ is

$$V_{\mathcal{T}}^{\mathcal{R}} \subseteq V_{\mathcal{S}}^{\mathcal{R}} = \bigcup_{1 \leq i \leq n} \mathcal{P}_i^{\mathcal{R}}. \quad (7)$$

A trajectory is feasible, if the volume swept by the robot along the trajectory is contained with the volume described the elastic strip. The volume of the elastic strip is defined by the protective hulls. This illustrated in Figure 6, where three consecutive protective hulls cover the trajectory of the robot. The initial and the final configuration are shown by a robot. An obstacle is reducing the size of the intermediate protective hull, indicated by a line.

It suffices to describe a procedure that verifies the existence of a path between two consecutive protective hulls $\mathcal{P}_i^{\mathcal{R}}$ and $\mathcal{P}_{i+1}^{\mathcal{R}}$. By applying this procedure repeatedly the condition of feasibility (7) can be ensured.

We will make the assumption that every point on a rigid body b moves on a straight line as b transitions from q_i to q_{i+1}. This ignores the effect of rotation. However, this effect can be bounded and taken into account at a computational expense, when computing the protective hull of b. The justification for this assumption is that two adjacent configurations will be similar enough for this effect to be insignificant when the robot is close to an obstacle. This is a simplification but not an inherent limitation of the approach.

Using this assumption the path of each rigid body b can be examined independently. If a trajectory between q_i and q_{i+1} exists for all rigid bodies $b \in \mathcal{R}$, one exists for $\mathcal{R}$.

The existence of a trajectory $\mathcal{T}_{i,i+1}$ for a rigid body b from configuration q_i to q_{i+1} is guaranteed if the volume $V_{\mathcal{T}_{i,i+1}}^b$ swept by b along $\mathcal{T}_{i,i+1}$ is contained within the protective hulls of the configuration q_i and q_{i+1},

$$V_{\mathcal{T}_{i,i+1}}^b \subseteq \left(\mathcal{P}_i^b \cup \mathcal{P}_{i+1}^b\right). \quad (8)$$

Let the protective hulls $\mathcal{P}_i^b$ and $\mathcal{P}_{i+1}^b$ be the sets of workspace bubbles $\{s_1, \cdots, s_m\}$ and $\{t_1, \cdots, t_n\}$, respectively.

To verify condition (8) the union $\mathcal{U} = \mathcal{P}_i^b \cup \mathcal{P}_{i+1}^b$ is examined. If b can pass through $\mathcal{U}$ on a straight line trajectory form q_i to q_{i+1} then $V_{\mathcal{T}_{i,i+1}}^b$ is contained within $\mathcal{P}_i^b \cup \mathcal{P}_{i+1}^b$. Hence, condition (8) holds and $\mathcal{T}_{i,i+1}$ must be collision-free.

Due to the symmetry and convexity of the workspace bubbles and taking into account that the rigid body b is only allowed to move on a straight line, it becomes clear that it is sufficient to examine the intersection of two consecutive protective hulls $\mathcal{I} = \mathcal{P}_i^b \cap \mathcal{P}_{i+1}^b$. If b passes through the projection of $\mathcal{I}$ onto a plane orthogonal to the direction of motion without intersection the

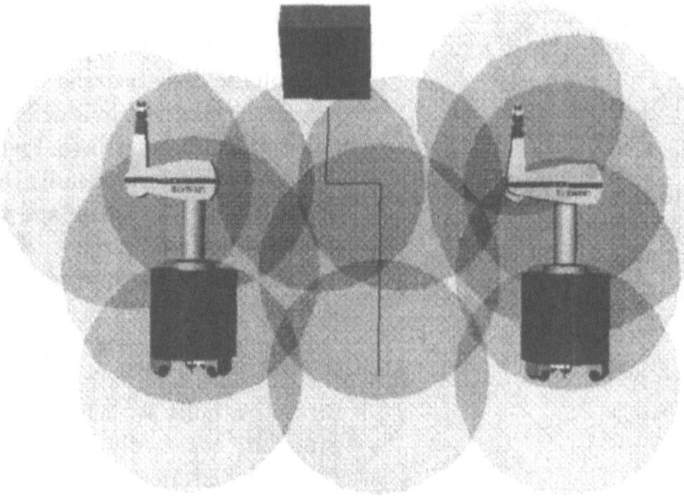

Figure 6: Protective hulls covering a trajectory

boundary of that projection, it can be concluded that the entire trajectory of b must be inside the protective hulls and hence free of collisions.

To determine $\mathcal{I}$, the bubbles in $\mathcal{P}_i^b$ and $\mathcal{P}_{i+1}^b$ are examined in the order they occur along the spine of b. Intuitively, we only need to check the locally narrowest passages of $\mathcal{I}$ and only the extent of these passages orthogonal to the plane p the spine moves on.

We describe this procedure for the general case, when the spine is not parallel to the direction of motion. If that were the case a simpler procedure than the one described below would be sufficient to compute $\mathcal{I}$.

Due to the convexity of bubbles, the locally narrowest passage is defined by the intersection[1] of three bubbles. The intersection of the boundary of two bubbles, one from each spine, defines a circle. Intersecting that circle with the boundary of a third sphere results in two points. These points define a line segment orthogonal to p. The length of this line segment can be checked against the width of the spine. Repeating this procedure for all intersections of three bubbles verifies if the rigid body b can move from q_i to q_{i+1}, given the local free space $\mathcal{P}_i^b \cup \mathcal{P}_{i+1}^b$.

If for all rigid bodies $b \in \mathcal{R}$ the union of their protective hulls $\mathcal{P}_i^b \cup \mathcal{P}_{i+1}^b$ is large enough to allow a straight-line trajectory, we say that two consecutive protective hulls $\mathcal{P}_i^{\mathcal{R}}$ and $\mathcal{P}_{i+1}^{\mathcal{R}}$ are *connected*.

[1]For the ease of presentation, the intersection of any two bubbles s_1 and s_2 is assumed to be non-empty and a true subset of each of the two bubbles. $I = s_1 \cup s_2 \neq \emptyset$, $I \subset s_1$, $I \subset s_2$. If the intersection is empty, no path is possible and if one bubble contains the other, a path must exist.

4.5 Forces Acting on the Strip

Elastic material deforms under the influence of external forces. Once these external forces are removed, internal forces cause it to reassume its original shape. For the elastic strip to behave similarly, these external and internal forces have to be modeled. An elastic strip, however, differs from a physical strip of elastic material in some aspects. Whereas elastic material is homogeneous and its principal physical properties do not vary over its volume, this is not true for an elastic strip. In this framework an elastic strip can be seen as a two-dimensional grid of links and springs. Figure 7 illustrates that for the elastic strip $\mathcal{S} = (q_1, q_2, q_3)$ for an arm mounted on a mobile base. The elastic strip has no elasticity along the links of the arm.

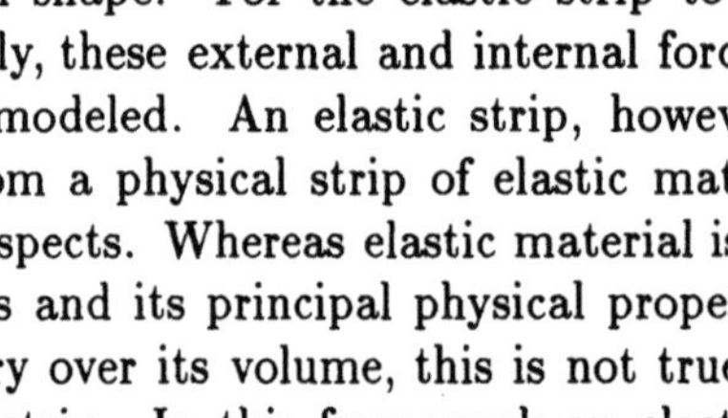

Figure 7: Principal structure of elastic strip

The internal forces acting on the elastic strip are generated by the virtual springs attached to control points in subsequent configurations along the trajectory. Let $\mathbf{p}_j^i$ be the position vector of the origin of the frame attached to the j-th joint of the robot in configuration q_i. We use these points as control points. The internal contraction force $\mathbf{F}_{i,j}^{int}$ caused by the springs attached to joint j is defined as

$$\mathbf{F}_{i,j}^{int} = k_c \left[\frac{d_j^{i-1}}{d_j^{i-1} + d_j^i} (\mathbf{p}_j^{i+1} - \mathbf{p}_j^{i-1}) - (\mathbf{p}_j^i - \mathbf{p}_j^{i-1}) \right]$$

where d_j^i is the distance $\|\mathbf{p}_j^i - \mathbf{p}_j^{i+1}\|$ in the initial, unmodified trajectory and k_c is a constant determining the contraction gain of the elastic strip.

These forces cause the elastic strip to contract, maintaining a constant ratio of distances between every three consecutive configurations. This additional constraint is necessary to accurately model the elasticity in the desired direction along the elastic strip. Note that the force acting on the control points depends only on the local curvature of the elastic strip and not on its elongation.

The external forces are caused by a repulsive potential associated with the obstacles. For a point $\mathbf{p}$ this potential function is defined as

$$V_{ext}(\mathbf{p}) = \begin{cases} \frac{1}{2} k_r (d_0 - d(\mathbf{p}))^2 & \text{if } d(\mathbf{p}) < d_0 \\ 0 & \text{otherwise} \end{cases},$$

where $d(\mathbf{p})$ is the distance from $\mathbf{p}$ to the closest obstacle, d_0 defines the region of influence around obstacles, and k_r is the repulsion gain.

The external force $\mathbf{F}_{\mathbf{p}}^{ext}$ acting at point $\mathbf{p}$ is defined by the gradient of the potential function at that point:

$$\mathbf{F}_{\mathbf{p}}^{ext} = -\nabla V_{ext} = k_r (d_0 - d(\mathbf{p})) \frac{\mathbf{d}}{\|\mathbf{d}\|},$$

where $\mathbf{d}$ is the vector between $\mathbf{p}$ and the closest point on the obstacle.

4.6 Elastic Strip Modification

Let $\mathcal{S} = (q_1, q_2, q_3, \cdots, q_n)$ be an elastic strip. When $\mathcal{S}$ is subjected to the forces described in section 4.5, it is deformed by altering each of the configurations q_i in turn. To change a configuration according to the internal and external forces, these forces have to be mapped to joint torques. The choice of this mapping is task-dependent.

For collision avoidance in the absence of a task requirement, we use the Jacobian $J_{\mathbf{p}}$ associated with the point $\mathbf{p}$ at which the force $\mathbf{F}_{\mathbf{p}}$ is acting for this mapping. The joint torques $\mathbf{\Gamma}$ caused by $\mathbf{F}_{\mathbf{p}}$ are given by

$$\mathbf{\Gamma} = J_{\mathbf{p}}^T \mathbf{F}_{\mathbf{p}}. \tag{9}$$

Joint limits can be avoided using a potential field function (Khatib 1986).

To implement collision avoidance during the execution of a task the joint torques computed in equation (9) have to be mapped into the null space associated with the Jacobian J of the task frame, as shown in equation (4).

The dynamic model of the system can be used to compute the joint displacements caused by the joint torques. The displacements for a configuration q_i define the new protective hull $\mathcal{P}_i'$, resulting in the modified elastic strip $\mathcal{S}' = (\mathcal{P}_1, \cdots, \mathcal{P}_i', \cdots, \mathcal{P}_n)$. $\mathcal{S}'$ represents a valid trajectory, only if the protective hulls $\mathcal{P}_{i-1}$, $\mathcal{P}_i'$, and $\mathcal{P}_{i+1}$ are connected. This is verified using the procedure described in section 4.4.

If $\mathcal{P}_i'$ and $\mathcal{P}_{i+1}$ are not connected, the elastic strip $\mathcal{S}'$ becomes invalid. This means that the trajectory represented by $\mathcal{S}'$ cannot be proven to be collision-free, using the representation of local free space associated with $\mathcal{S}'$. In order to reconnect $\mathcal{P}_i'$ and $\mathcal{P}_{i+1}$ intermediate protective hulls are inserted into the elastic strip. By imposing constraints on the transition from $\mathcal{P}_i$ to $\mathcal{P}_i'$ this procedure can be guaranteed to succeed.

As obstacles recede from the vicinity of the elastic strip, the protective hulls of configurations increase in volume and potentially move closer together. This can result in protective hulls $\mathcal{P}_i$ and $\mathcal{P}_{i+2}$ to be connected. In that case $\mathcal{P}_{i+1}$ is redundant and can be removed.

4.7 Motion Behavior

Given a planned motion, the elastic strip allows a robot to dynamically modify its motion to accommodate changes in the environment. For redundant mechanisms this modification is not uniquely determined and may be chosen depending on the task. A transportation task for a mobile manipulator, for instance, can be described by the motion of the mobile base, while only a nominal posture of the arm and load are specified. For a manipulation task, the description consists of the motion of the end effector and its contact forces, while only a nominal posture of the mobile base and arm is given. In both cases some degrees of freedom are used for task execution, while others can be used to achieve task-independent motion behavior.

The elastic strip also provides an effective framework for executing partially described task. If only those degrees of freedom necessary for execution have been specified, reactive obstacle avoidance combined with an attractive potential to the desired posture can complete the robot control in real-time. With a partial plan, however, the elastic strip can be subjected to local minima.

The framework for combining task behavior and motion behavior relies on the general structure for redundant robot control (equation 4):

$$\mathbf{\Gamma} = J^T(\mathbf{q})\mathbf{F} + \left[I - J^T(\mathbf{q})\overline{J}^T(\mathbf{q})\right]\mathbf{\Gamma}_0, \quad (10)$$

with

$$\overline{J}(\mathbf{q}) = A^{-1}(\mathbf{q})J^T(\mathbf{q})\Lambda(\mathbf{q}), \quad (11)$$

where $\overline{J}(\mathbf{q})$ is the dynamically consistent generalized inverse, $A(\mathbf{q})$ the joint space mass matrix, and $\Lambda(\mathbf{q})$ the kinetic energy matrix associated with the operational space (Khatib 1987; Khatib, Yokoi, Chang, Ruspini, Holmberg, and Casal 1996b).

Equation 4 provides a decomposition of of the joint torques into those caused by forces at the end effector ($J^T\mathbf{F}$) and those that only affect internal motion $\left(\left[I - J^T(\mathbf{q})\overline{J}^T(\mathbf{q})\right]\mathbf{\Gamma}_0\right)$. This decomposition can be exploited to achieve different kinds of task behavior. Simple obstacle avoidance without the incorporation of task behavior can be achieved by using equation 9 to map internal and external forces to joint torques.

To ensure the execution of a task specified in a particular task frame f, the internal and external forces are mapped into the null space of the Jacobian J_f associated with the task frame. This corresponds to the sets of tasks where the end effector is required to move on a certain trajectory and the redundant degrees of freedom can be used for obstacle avoidance.

Simple obstacle avoidance behavior can be easily augmented by specifying a desired posture for the robot. This posture can be chosen according to some optimization criterion, for example, to minimize the torques necessary to support the load.

5 Experimental Results

The results of a preliminary implementation are shown in Figure 8. The framework is applied to the Stanford Mobile Platform, a PUMA 560 robot arm mounted on a holonomic mobile base with a total of nine degrees of freedom. The elastic strip is represented by a set of intermediate configurations, displayed as lines connecting joint frames.

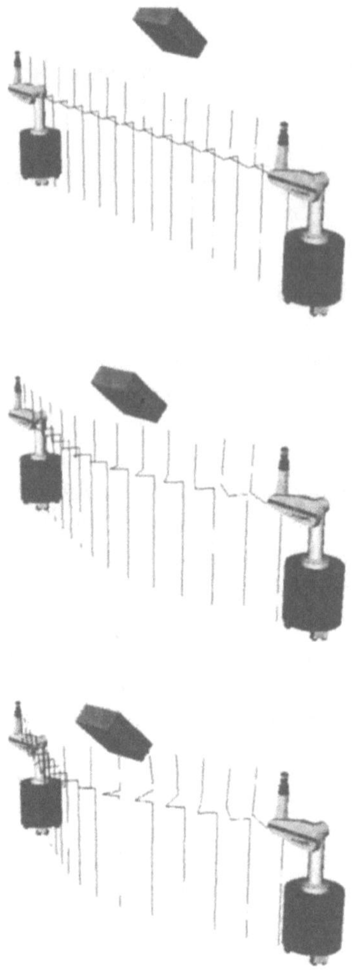

Figure 8: Elastic strip for a 9-dof robot incrementally modified by a moving obstacle

The approaching obstacle deforms the elastic strip to ensure obstacle avoidance. As the obstacle moves away, internal forces cause the elastic strip to assume the straight line trajectory. Our current implementation achieves an update rate for the elastic strip in Figure 8 of 10 to 15 Hz on an SGI Indigo2. In the figure the stationary obstacles of the environment have been removed for clarity.

6 Conclusion

The elastic strip framework is an efficient approach to real-time obstacle avoidance for robots with many degrees of freedom. It is particularly well suited for mobile manipulation, since it allows the integration of task-level behavior with collision avoidance in dynamic and uncertain environments.

An elastic strip represents the workspace volume swept by a robot along a pre-planned trajectory. This representation is incrementally modified by external, repulsive forces originating from obstacles to maintain a collision-free path. Internal forces act on the elastic strip to shorten and smoothen the trajectory. To represent the volume swept by the robot the notion of a protective hull is introduced. A protective hull represents the local free space around a configuration and can be computed efficiently. A sequence of these hulls is used to contain the robot at any point along the trajectory, thus ensuring collision avoidance. The effectiveness of the elastic strip framework is demonstrated in real-time path modification for the Stanford Mobile Platform, a nine degree of freedom manipulator, in a dynamic environment.

Acknowledgments

The authors would like to thank Diego Ruspini and Kyong-Sok Chang for their helpful insights and discussion in preparing this paper. The financial support of Boeing, Nomadic Technologies, General Motors, and NSF (grant IRI-9320017) is gratefully acknowledged.

References

Arkin, R. C. (1987, August). Motor schema-based mobile robot navigation. *International Journal of Robotics Research 8*(4), 92–112.

Carriker, W. F., P. K. Khosla, and B. H. Krogh (1989). An approach for coordinating mobility and manipulation. In *IEEE International Conference on Systems Engineering*, pp. 59–63.

Khatib, M., H. Jaouni, R. Chatila, and J.-P. Laumond (1997). How to implement dynamic paths. In *Proceedings of the International Symposium on Experimental Robotics*, pp. 225–36. Preprints.

Khatib, O. (1986). Real-time obstacle avoidance for manipulators and mobile robots. *International Journal of Robotics Research 5*(1), 90–8.

Khatib, O. (1987). A unified approach to motion and force control of robot manipulators: The operational space formulation. *IEEE Journal of Robotics and Automation 3*(1), 43–53.

Khatib, O., K. Yokoi, K.-S. Chang, D. Ruspini, R. Holmberg, and A. Casal (1996b). Coordination and decentralized cooperation of multiple mobile manipulators. *Journal of Robotic Systems 13*(11), 755–64.

Khatib, O., K. Yokoi, K.-S. Chang, D. Ruspini, R. Holmberg, and A. Casal (1996a). Vehicle/arm coordination and multiple mobile manipulator decentralized cooperation. In *Proceedings of International Conference on Intelligent Robots and Systems. IROS '96*, Volume 2, pp. 546–53.

Krogh, B. H. (1984). A generalized potential field approach to obstacle avoidance control. In *Robotics Research*, pp. MS84-484/1–15.

Latombe, J.-C. (1991). *Robot Motion Planning.* Boston: Kluwer Academic Publishers.

Nassal, U. (1996). Motion coordination and reactive control of autonomous multi-manipulator systems. *Journal of Robotic Systems 13*(11), 737–54.

Quinlan, S. (1994). *Real-Time Modification of Collision-Free Paths.* Ph. D. thesis, Stanford University.

Quinlan, S. and O. Khatib (1993). Elastic bands: Connecting path planning and control. In *Proceedings IEEE International Conference on Robotics and Automation*, Volume 2, pp. 802–7.

Seraji, H. (1993). An on-line approach to coordinated mobility and manipulation. In *Proceedings of International Conference on Robotics and Automation*, Volume 1, pp. 28–35.

Yamamoto, Y. and X. Yun (1995). Coordinated obstacle avoidance of a mobile manipulator. In *Proceedings of IEEE International Conference on Robotics and Automation*, Volume 3, pp. 2255–60.

Modeling and Control for Mobile Manipulation in Everyday Environments

Wendelin Feiten Björn Magnussen
Jochen Bauer
Siemens AG

Gregory D. Hager Kentaro Toyama
Department of Computer Science
Yale University

email: wendelin.feiten@mchp.siemens.de, hager-greg@cs.yale.edu

Abstract

In recent years, mobile navigation research has matured to the point that practical, cost-effective navigation systems are starting to appear in the marketplace. However, today's systems are limited to a carefully chosen set of tasks and environments which do not require manipulation capabilities. Future robotic systems must be able to function in everyday environments containing doors to be opened, elevators to be called, and objects which must be manipulated.

In this article, we describe our first efforts toward building a system capable of both mobility *and* manipulation. In particular, we outline the system hardware and software architecture, paying particular attention to our use of specialized vision processing and visual servoing techniques for manipulation. We include a set of preliminary experiments showing the feasibility of the approach on the problem of opening a door.

1 Introduction

For several decades, mobility and manipulation have remained largely separate areas of robotics research even though, in reality, mobility and manipulation are highly complementary skills. Manipulation provides the means for mobile systems to move beyond passive tasks such as patrolling or floor cleaning and out of environments where the doors and elevators can be remotely or passively activated. Conversely, mobility provides a means for manipulators to expand their workspace and provides redundant degrees of freedom for avoiding obstacles or singularities.

Historically, limited on-board computing resources, engineering complexity and cost have prohibited building effective mobile manipulation systems. We feel that the technological and scientific state of the art have advanced to the point that combining mobility and manipulation is not only feasible, but in a few years is likely to be both a practical and a cost-effective means of enhancing today's mobile systems.

Combining mobility and manipulation also opens up a host of interesting avenues for research. On one hand, mobility forces manipulation research to move out of a pre-engineered, static "workcell" into a changing (and changeable) environment. On the other hand, endowing a mobile system with manipulation challenges us to design systems with much richer sensing capabilities and more flexible execution architectures.

In this article, we focus on how we are working to overcome two hurdles in mobile manipulation. The first is to use rich sensing and smart processing to move from the relatively "coarse" geometry and imprecise location information of a navigation system to the much richer representations needed for manipulation. The second is to design an execution architecture which can react quickly, flexibly and safely to a variety of environment situations, both anticipated and unanticipated.

2 Previous Work

Siemens Corporate Technology has developed a navigation system for autonomous mobile robots. The project started in 1991. The first application of this navigation system is an autonomous cleaning robot developed in cooperation with Siemens Automation and Drives and with the German cleaning machine maker Hefter Cleantech. This prototype is equipped with redundant sensors of different modalities: a laser scanner, 48 ultrasonic transducers, tactile bumpers and a gyroscope.

In August 1996 and February 1997, the first two prototypes went into operation in two supermarkets, cleaning the vegetable department several times a day during ordinary business hours. Since then, these robots have operated autonomously without a single incident, halting operation only under very unfavorable conditions.

The architecture of these autonomous mobile robots is composed of three layers. At the bottom, there is a safety layer that ensures safe operation in all system states. On top of this layer, there are a number of reactive schemes which allow the robot to reach intermediate target locations which effect the task at hand. The behaviors are sufficiently rich to allow the robot to cope with deviations between its internal models and the real environment (Feiten, Bauer, and Lawitzky 1994; Wienkop, Lawitzky, and Feiten 1994). Finally, on top of the reactive layer there is a collection of higher level planners that are able to generate suitable intermediate targets to accomplish a task. Each of these layers utilizes a different modeling scheme, including a dynamically updated certainty grid at the lowest level, a 2-D floor plan for navigation at the intermediate level (Rencken 1993; Rencken 1994) and a graph-like representation at the highest level.

Apart from cleaning in the commercial sector, more applications are to follow. The plan is to introduce the navigation system (comprising obstacle avoidance, localization, application specific reactive behaviors and higher level planners) as a subsystem to be used by original equipment manufacturers in the market in 1998.

3 The Next Generation

With the first generation of autonomous intelligent robots rapidly approaching the product stage, the general expectation is that the next generation will exhibit at least limited capabilities of manipulation in unstructured, unmodified everyday environments. For example, one particularly useful enhancement of an autonomous mobile robot is the ability to open and close doors and to use unmodified elevators[1].

It may seem that combining mobility and manipulation involves solving two independently equally complex problems. However the context supplied by navigation greatly simplifies the manipulation problem in the following aspects:

Figure 1: Prototype of an autonomous cleaning robot for supermarkets

- The locations of the objects to be manipulated are roughly known from the localization of the base and contextual information stored in the environment map.
- The objects to be manipulated are a limited class, for example door handles, elevator operating panels, meal trays or mail boxes.
- The range of uncertainty about the manipulated objects themselves is highly constrained. For example, a door only swings on hinges, elevator buttons are dark or lit, and trays are on (relatively small) bedside tables.

To perform these tasks, the autonomous mobile robot has to be equipped with a manipulator and with additional sensors able to sense objects to be manipulated. Likewise, the execution architecture must be augmented to accommodate the greater range of situations, sensing and flexibility in the system.

3.1 Hardware

We have recently completed the first generation mobile manipulation system. The mobile base is a differential drive mechanism. The base carries

[1] Currently, instrumenting doors and elevators in a typical building rivals or exceeds the cost of the robot itself (Helpmate 1997)

a scara-like robot arm consisting of a prismatic and revolute joint operating on a common vertical axis, a second parallel revolute joint, and a 3 DOF wrist outfitted with a conventional parallel jaw gripper. Each joint is self-contained—it contains a a motor, power amplifier, a 16-bit controller, an encoder, limit switches and a bus interface. Via a bus, the joints are connected to a PC that serves as the arm controller. This kinematic was designed for low energy consumption, good payload characteristics over the workspace and for a reach ranging from floor level to roughly one and a half meters above the floor.

In addition to the current sonar system, our new mobile manipulator is equipped with vision, laser ranging and tactile sensing. Vision is provided by a color CCD camera mounted close to the gripper. The camera, which has its own dedicated processor (a Pentium PC), has computer controlled pan, tilt and zoom. It can be pointed in such a way that the gripper fingers as well as the object to be manipulated are in the image. We have also included active lighting with a laser light plane that can be projected from within the gripper. This arrangement is favorable for visual servoing purposes.

The tactile sensors are placed on the inner sides of the two gripper fingers. Each tactile sensor consists of two plates that move freely, suspended by springs, each having a 3x3 force measurement array on it.

Figure 4 shows the tactile features related to a cylindrical object. The features are not symmetrical because of a slight offset between the two sides of the tactile sensor.

Another set of tactile sensors is planned for collision avoidance of the robot arm. These sensors will be pressure sensitive cushions placed around the arm.

3.2 Software Architecture

We plan to use our existing navigation system unchanged. The system for controlling the manipulation system is currently under development. Our development has been driven by criteria including: ***Robustness*** - the robot must be able to adapt its behavior online to compensate varying environmental conditions, ***Reactivity*** - the robot must be able to respond quickly and safely to unexpected events, ***Flexibility*** - the architecture should allow for the easy incorporation of situation or task specific behaviors, ***Ease of programming*** - it should be easy to specify a task to the system, and finally, *low cost.*

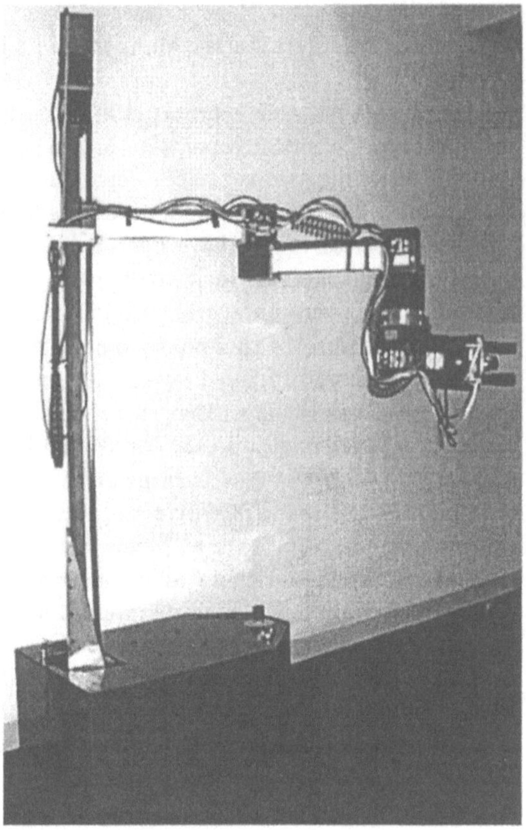

Figure 2: Hardware of the mobile manipulator

Figure 5 gives an outline of our proposed system architecture. The central tenet of this architecture is that an appropriate sequencing of relatively simple sensor-based control loops—visual, force, or position-based servoing—suffice for most of the simple manipulations to be performed by the system. In particular, we plan to rely strongly on the visual tracking and control concepts developed by Hager (Hager 1997; Hager and Toyama 1996; Hager and Toyama 1995).

Briefly, the behavior of the system is driven by the ***system state model*** and the ***target state model.*** Intuitively, the target state model specifies values for specific sensor features which correspond to a correctly performed task or subtask. These values may be physical values or, in some cases, error or quality measures on information. The target state values are supplied in sequence by a program causing the system to "step" through the various configurations which constitute task performance (as with our previously described cleaning robot). The system state model is maintained continuously by monitoring the sensor data

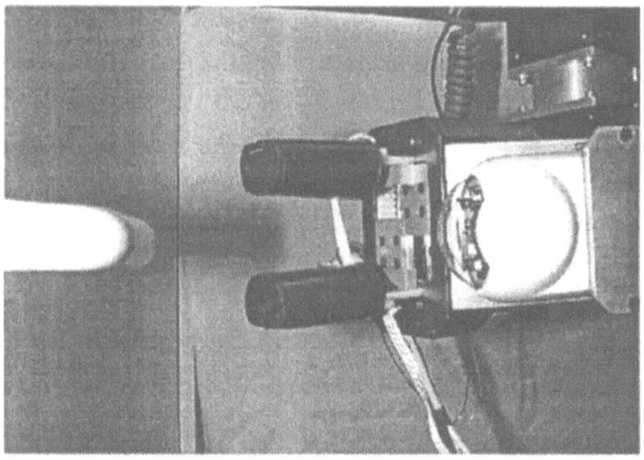

Figure 3: Gripper with camera and tactile sensors

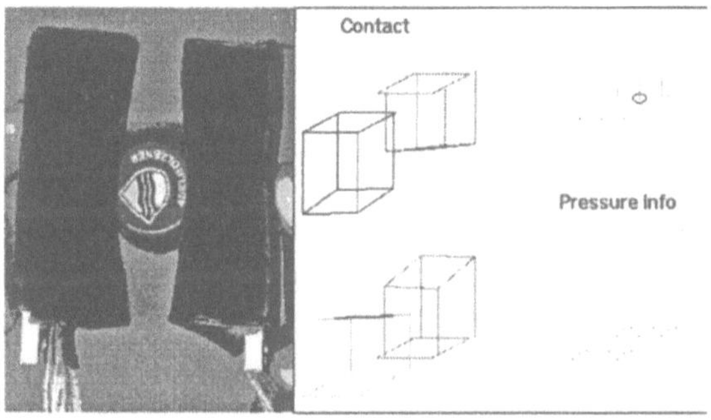

Figure 4: Tactile sensor measurement

and by updating the inferred error measures or quality measures. In particular, in the case of vision, visual tracking is used heavily.

Achieving in the target state is what drives the system. At each time point, the system compares the current and target states and computes mismatches. The types of mismatches include: mismatches which can be reduced by applying a primitive (continuous) motor control routine, mismatches due to missing information requiring information gathering, or mismatches which can be reduced through a series of motor or informational actions. After the matching procedure, actions are prioritized, and the highest priority chosen.

Thus, the behavior of the system involves *continuous control* regimes (e.g. a primitive visual servoing operation), *discrete switching* for choosing a control regime to use, and *sensor control* for acquiring or estimating state quantities. The change to the next target state is triggered when no more actions are required to achieve the current target.

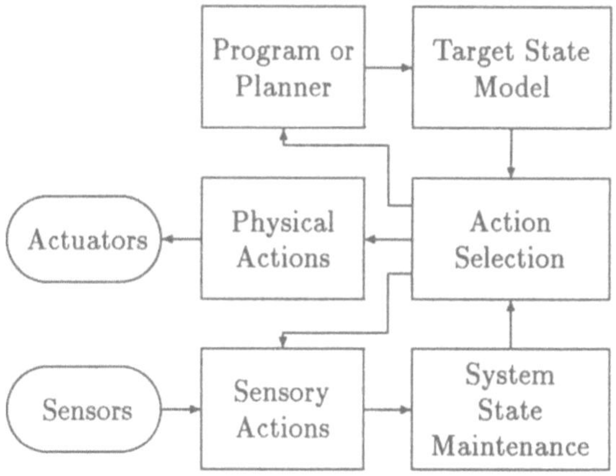

Figure 5: System architecture for the mobile manipulator

3.3 An Example

To illustrate how this architecture functions, consider the problem of driving from the interior of a room into a corridor. In order to make the example interesting, we assume the door to the corridor to be closed. We assume further that information on the environment, including landmarks for navigation, suitable intermediate target points for traveling and the approximate location of doors, is stored in the environmental map for navigation.

Simply specifying the target state by a constraint on the robot position (namely being out in the corridor) is not an option in this case, as we currently have no onboard planning capabilities. Instead, a sequence of intermediate target states has to be specified, each of which can be reached in an obvious way by some primitive action, such that the sequence reliably leads the robot to the final target state. Here we give just the first four of about twenty intermediate target states:

State 1: The robot stands in front of the door handle.

State 2: The handle is identified in the camera image.

State 3: The handle is between the gripper fingers.

State 4: The handle is grasped.

For the control system, these target states have to be formulated in terms of the sensory features. This reformulation will be given below where each step is discussed in more detail.

State 1

The target configuration for the robot is specified in global coordinates, and the robot is controlled to this target configuration. In this control loop, the "system state" is the estimate of the current position of the robot. As long as the robot is not yet at this configuration within a given accuracy, i.e. as long as the corresponding constraint is not yet satisfied, the execution system drives the robot toward the target configuration using the navigation system. Eventually, the mobile base will be at the target configuration and the first intermediate target state will be satisfied. One of the conditions which this state was designed to achieve is to have the camera pointed at the handle of the door, – that is, it is guaranteed that the door handle is in the camera image if the door is closed.

State 2

The second intermediate target state concerns missing information. To satisfy the constraint that the position (in the image) of the door handle is known, a visual search for the upper and lower edges of the door handle in the image is initiated. This visual search is based on features that are characteristic for the handle. The approach is described in more detail in the next section. The result of the visual search will be that the upper and lower edge of the door handle are identified as line features in the image. Once the features are found, their location is actively maintained by tracking (in system state maintenance).

At this point, the robot is standing in front of the door, the camera is pointed at the handle and the upper and lower edge of the handle are identified in the camera image.

State 3

This target state is described in terms of visual features and of metrical information. The visual features are the upper and lower edge of the handle and the lines corresponding to the inner sides of the gripper jaws. These features are constrained so that the door handle edges have to be between the gripper jaws in the image. This automatically implies that the camera is looking through the gripper jaws. The constraint on the metrical information is that the distance from gripper to handle is 5 cm with an accuracy of 1 cm.

At the beginning of this phase this distance is estimated (based on the accuracy of the navigation system) to be 50 cm, with a pessimistic variance of 20 cm. The mismatch in the variance triggers a depth measurement based on motion stereo. A Kalman Filter is started to estimate the depth from motion, and a suitable motion of the camera (and the gripper) is generated. This continues until the estimate has a sufficiently low variance. Now the distance between gripper and handle can be adjusted by moving along the line defined by the camera axis. If this intermediate target state is reached, the handle is between the gripper jaws.

Another alternative to the motion stereo based depth measurement is to use the laser light plane projected from within the gripper. The reflections of this light plane are easily identifiable. Using these features, the target state description of "gripper jaws around the handle" becomes: the door handle edges are between the gripper jaws, one laser feature is between the gripper jaws and between the door handle edges, and all other (if any) laser features are not between the gripper jaws, as shown in Figures 6 and 7.

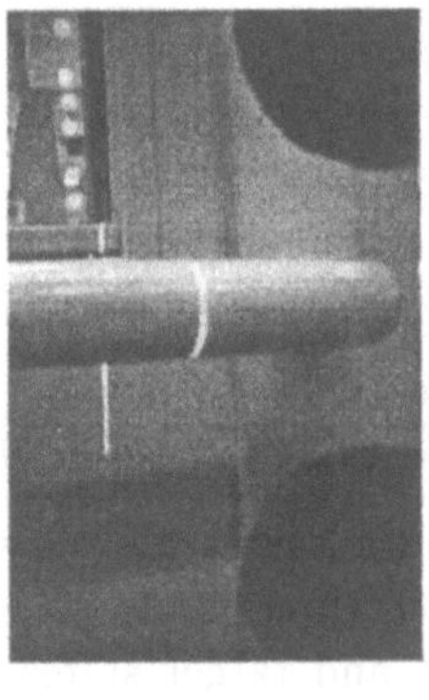

Figure 6: Laser feature at a distance

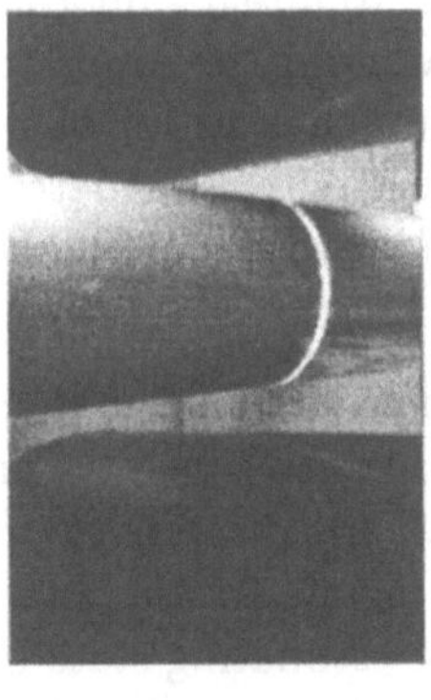

Figure 7: laser feature at target state

State 4

This target state is described in terms of tactile features: on the two opposing sides, the tactile arrays have to show a line, and the two sensor plates in each jaw will have to be tilted towards the jaw in the middle. Also, the tactile features should be symmetrical.

When the control first switches to this intermediate target state, the system state is "no force measurement, no tilt". The action associated with this mismatch is "close the gripper". Any violation of the symmetry constraint results in an additional motion of the gripper base. When this intermediate target state is reached, the robot should have grasped the handle at the desired position, and with the desired force.

In a similar way, the handle will then be turned, the door pushed open, the handle turned back and released and so on.

4 Vision Processing

As discussed at the outset, one key problem in mobile manipulation is to be able to locate objects to be manipulated, and to be able to specify a manipulation relative to the object. We have chosen to use vision and visual servoing for this purpose. One significant challenge is to "package" vision in a way which is flexible, robust, and cost-effective.

4.1 An Overview

The vision processing to be performed by the system can be divided into two different modes of operation: visual tracking and visual search. *Visual tracking*, used primarily for feature-based visual servoing, is performed by the XVision package (Hager and Toyama 1996; Hager and Toyama 1995). XVision is a portable, software-based processing system optimized for local feature tracking on commodity computing hardware. It is organized around the idea of decomposing tracking problems into a small set of highly optimized image-level features. These features include unstructured blobs, edges, and structured regions. Complex objects are defined in terms of these primitives using constraints. For example, tracking the contour of an object is defined in terms of tracking a set of line segments with the constraint that the segments meet at their endpoints. Using XVision, we can track several tens of features at or near frame rate – enough information to attack an interesting class of manipulation tasks.

Although tracking itself is efficient, it requires initialization of the tracked object. Unfortunately, navigation is not itself sufficiently precise to be able to perform initialization "open loop." Thus, we are faced with a problem of *visual search*, in which a specific target object must be found in an image with only coarse prior knowledge about its location or aspect.

The methods we use for visual search are an outgrowth of previous work at Yale on *Incremental Focus of Attention* (henceforth IFA), a framework for robust motion tracking systems (Toyama and Hager 1996). Conceptually, IFA layers different tracking algorithms and heuristics into a control hierarchy which focuses the "attention" of the system onto relevant subregions of the image. The result is an effective method of target re-acquisition which also operates on commodity computing hardware.

For this project, we have modified IFA's layered structure specifically for mobile manipulation. First, more complex "recognition" modules are used instead of simpler tracking algorithms, allowing for greater freedom in selecting objects of visual search. Second, we use non-vision-based estimates of object position, which significantly speeds up visual search. For example, we do not have to search blindly for a door handle in images. Instead, we can drive to a location in front of the door and know within a few centimeters where we expect the handle to be.

4.2 An Example

For obvious reasons, an important class of potential visual targets are door handles. Although a general "door handle finder" is impossible with today's understanding of cognitive processes, one can use environmental constraints as positive design criteria for a *specialized* door handle finder (Horswill 1993).

The particular door handles used in our experiments exhibit the following visual qualities: they occur at waist height, they are yellow, they are relatively edge-rich (*i.e.*, they have a high ratio of pixels on detectable edges to pixels without detectable edges), among detectable edges they have a high proportion of horizontal edges, and they can be modeled approximately by a long rectangle whose longitudinal axis is parallel to the horizontal.

Although many objects might satisfy any one

or more of these qualities, no objects in the environment satisfy all qualities simultaneously. A visual search for door handles therefore proceeds by finding objects in the image which satisfy the conjunction of all qualities. The search is broken down into several steps, where each step narrows the possibilities of the door handle position through heuristic examination of the image. The obvious strategy, which works well in practice, is to perform the steps in an order which eliminates more candidate objects earlier in the search process using less computation. We thus perform the search as follows:

- First, those regions which are not approximately at waist height are eliminated from the search. Since camera height is known, the algorithm is trivial.

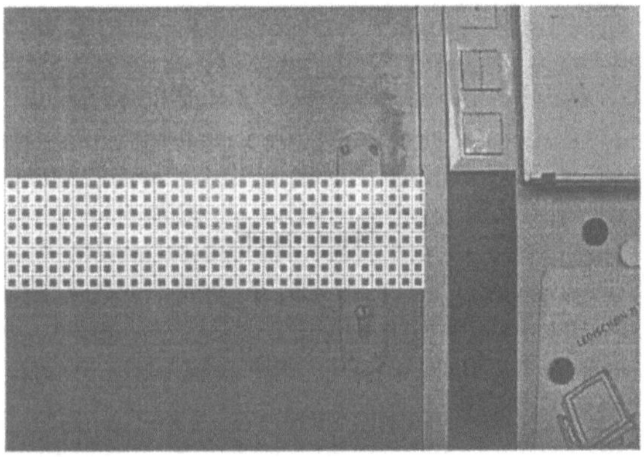

Figure 8: Yellow regions at waist height

- Then, we eliminate regions which are not yellow. Finding yellow pixels requires only a single comparison per pixel, and because the entire door handle with its immediate background (the door) is always yellow, the image can be coarsely subsampled. Figure 8 shows those regions (marked by white squares) which are at waist height and are yellow.

- Among yellow regions at waist height, we find regions which are edge-rich, particularly with horizontal edges. A simple directed edge operator is applied to the regions, and regions with edge-to-non-edge ratios below some threshold are thrown out (see Figure 9). Edge detection is relatively expensive compared to color detection.

- The remaining regions are examined more closely for purposes of tracking initialization.

Figure 9: Edge-rich regions

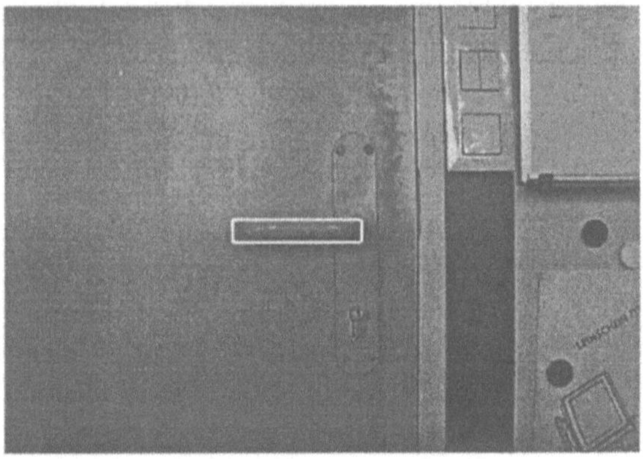

Figure 10: Rectangle model of doorknob

  A further search for a rectangle, whose approximate position was determined from the previous steps, results in the door handle being identified (Figure 10). Optionally, a more accurate model of the door handle can then be initialized.

The total time to find a door handle once the robot is positioned is no more than 0.1 second. The search would be far less efficient were we to begin, for example, by examining the entire image for edge-rich regions. Edge detection is expensive, and in our environment, there may be more edge-rich objects than there are yellow objects at waist height.

In addition to using the rectangle model, a more sophisticated model of the door handle can also be used. This model is formulated as multiple tracked lines from the image and with constraints over the orientation and relative position of the lines. Tracked lines include the horizontal edges of the handle and the vertical edges of the plate behind the handle. In Figure 11 the door handle is shown together with the best instantiation of the model with tracked lines.

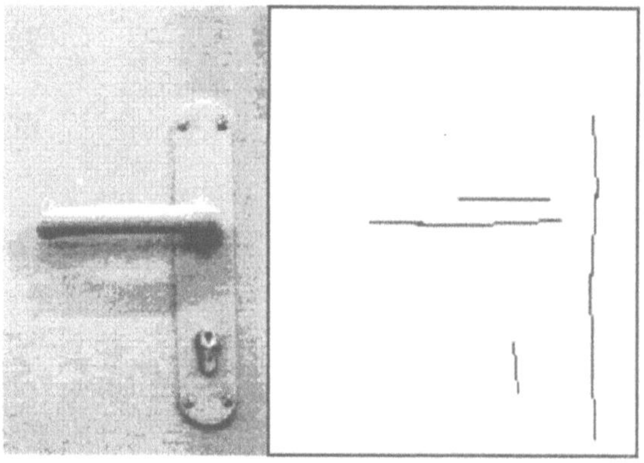

Figure 11: Handle-and-plate model

The flexibility of this particular algorithm, developed for the Siemens laboratory, is illustrated by the fact that it is a simple variation of an algorithm used to find door handles at the Yale lab.

5 Current Status and Future Work

The navigation system for the mobile base (comprising among other modules obstacle avoidance, localization and a set of planners) has reached product maturity. This system provides the basis for our mobile manipulator. Context information can be stored in the environment map for use in the manipulation tasks, and target configurations can be sent down to the platform to drive it to suitable initial configurations.

The implementation of the control system for the arm is under way. The unified framework for servoing based on visual, tactile and kinesthetic features has been implemented. It has been tested for tasks involving single sensor modalities (visual tracking, depth measurement, grasping). The architecture has been shown to be flexible enough to incorporate these different modes of operation. The first implementation of the action selection is based on the priority of the related constraints. Currently, work is under way to integrate the different steps into one coherent action sequence, based on the mechanism of constraint satisfaction and action selection.

Suitable sequences of intermediate target states are generated manually. The programmer is responsible to ensure that they can be subsequently reached based on the reactive behavior.

The parts of our architecture that have been implemented and tested perform well. The unified framework for sensor feature based servoing seems to be suited to take the role of the reactive layer in the architecture of the navigation system of the mobile base. The next step is to collect all the pieces together to make the robot go through the door in the context of a larger transport mission.

Error recovery is an important aspect of reactive robot control. In the layered architecture of the mobile base, error recovery relies on the use of planners. In this scheme, the progress in achieving an (intermediate) target is monitored. If the target cannot be reached within a given time limit, a planner is called to generate a new sequence of intermediate targets. Thus the robot has the robustness with respect to major changes in the environment that results from the use of high level planners, while at the same time having the speed and flexibility with respect to minor changes that result from the reactive layer.

In order to use the same approach for the manipulation, work will be done on the planner. How to predict sensor features from higher level models is basically known. However, to automatically derive suitable elementary actions and priorities is an open problem.

Acknowledgements Greg Hager and Kentaro Toyama were supported by National Science Foundation grant IRI-9420982, a grant from Siemens Corp., and by funds provided by Yale University.

The work at Siemens has in part be supported by the BMBF under the project 01IN601A2.

References

Feiten, W., R. Bauer, and G. Lawitzky (1994). Robust obstacle avoidance in unknown and cramped environments. In *Proceedings of the ICRA '94*, pp. 2412–2417. IEEE Computer Society Press.

Hager, G. D. (1997). A modular system for robust hand-eye coordination. *IEEE Trans. Rob. Automat.*.

Hager, G. D. and K. Toyama (1995). The "X-vision" system: A general purpose substrate for real-time vision applications. DCS RR-1078, Yale University. To Appear in Comp. Vis. Image Understanding.

Hager, G. D. and K. Toyama (1996). XVision: Combining image warping and geometric constraints for fast visual tracking. In *Computer*

Vision-ECCV'96, pp. 507–517. Springer Verlag.

Helpmate (1997). HelpMate Presentation, Yale University.

Horswill, I. (1993). *Specialization of Perceptual Processes.* Ph.d.thesis, MIT.

Rencken, W. D. (1993). Concurrent localization and map building for mobile robots using ultrasonicsensors. In *Proceedings of the IROS '93*, pp. 2192–2197. IEEE Computer Society Press.

Rencken, W. D. (1994). Autonomous sonar navigation in indoor, unknown and unstructured environments. In *Proceedings of the IROS '94*, pp. 431–438. IEEE Computer Society Press.

Toyama, K. and G. D. Hager (1996). Incremental focus of attention for robust visual tracking. In *CVPR '96*, pp. 189–195. IEEE Computer Society Press.

Wienkop, U., G. Lawitzky, and W. Feiten (1994). Intelligent low-cost mobility. In *Proceedings of the IROS '94*, pp. 1708–1715. IEEE Computer Society Press.

Scale-Dependent Grasps

Makoto Kaneko
Industrial and Systems Engineering, Hiroshima University
Higashi-Hiroshima, 739, Japan
email: kaneko@huis.hiroshima-u.ac.jp

Abstract

This paper discusses the scale-dependent grasps. Suppose that an object is initially placed on a table without touching by human hand and, then he (or she) finally achieves an enveloping grasp after an appropriate approach phase. Under such initial and final conditions, human unconsciously changes the grasp strategy according to the size of object, even though they have similar geometry. We call the grasp planning the scale-dependent grasp. Along the grasp patterns observed in human grasping, we present a couple of procedures applicable to multi-fingered robot hands.

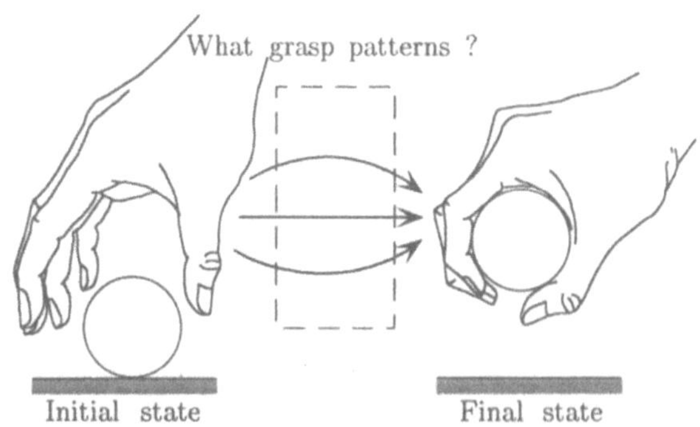

Figure 1: Enveloping grasp for an object placed on a table.

1 Introduction

There have been a number of works concerning multi-fingered robot hands. Most of them address a finger tip grasp, where it is assumed that a part of inner link never makes contact with the object. Enveloping grasp (or power grasp) provides another grasping style, where multiple contacts between one finger and the object are allowed. Such an enveloping grasp can support a large load in nature and is highly stable due to a large number of distributed contact points on the grasped object. While there are still many works discussing enveloping grasps, most of them deal with the grasping phase only, such as contact force analysis, robustness of grasp, contact position sensing, and so forth. Suppose that human eventually achieves an enveloping grasp for an object placed on a table as shown in Fig.1. Actually, such a situation is often observed in a practical environment, for example, in grasping a table knife, an ice pick, a hammer, a wrench, and so on. In many cases, the tool handle can be modeled as a cylindrical shape. For a cylindrical object having a sufficiently large diameter, human wraps it directly without any regrasping process, since the table makes no interference with the finger links at all. As the diameter decreases, human is obliged to utilize a different strategy so that he (or she) may avoid interference caused by the table. By simple experiments, we can easily show that human chooses the grasp planning according to the scale of objects, even though they are geometrically similar. We call the grasp planning the scale-dependent grasp planning. We would note that scale-dependent grasp does not mean the final grasp style but means the change of the grasp patterns existing between the initial and final states according to the size of objects.

In this paper, we first observe the human behavior for grasping column objects with different sizes and surface friction, and then discuss a couple of grasp patterns applicable to a multi-fingered robot hand.

2 Related Work

Human grasping based approach: In robotic hands, there have been a number of papers learnt by human behaviors[1]–[5]. Cutkosky and Wright[1] have analyzed manufacturing grips and correlation with the design of robotic hands by examining grasps used by humans working with tools and metal parts. Stansfield discussed the robotic grasping based on knowledge[4]. These works [1]–[4], and [5] focus on either the final grasp mode or finding an appropriate grasp posture under a set of grasp modes, target geometric characteristics and task description. Jeannerod[6] has shown that during the approaching phase of grasping, the hand preshapes in order to prepare the shape matching with the object to be grasped. Bard and Troccaz[7] introduced such a preshaping motion into a robotic hand and proposed a system for preshaping a planar two-fingered hand by utilizing low-level visual data.

Enveloping grasp or power grasp: Mirrza and Orin[8] applied a linear programming approach to formulate and solve the force distribution problem in power grasps, and showed a significant increase in the maximum weight handling capability for completely enveloping type power grasps. Trinkle[9] analyzed planning techniques for enveloping, and frictionless grasping. Salisbury[10] has proposed the Whole-Arm Manipulation (WAM) capable of treating a big and heavy object by using one arm which allows multiple contacts with an object. Bicchi[11] showed that internal forces in power grasps which allow inner link contacts can be decomposed into active and passive. Omata and Nagata[12] also analyzed the indeterminate grasp force by considering that sliding directions are constrained in power grasps. Zhang et. al.[13] evaluated the robustness of power grasp by utilizing the virtual work rate for all virtual displacements. Kleinmann et.al.[14] showed a couple of approaches for finally achieving power grasp from finger tip grasp. In our previous work[15], we have shown a preliminary work on Scale-Dependent Grasp.

3 Observation of Human Grasping

Fig.2 shows the experimental results for cylindrical objects whose surfaces are covered by (a) drawing papers and (b) rubbers, respectively, where d is the normalized diameter defined by $d = D/L$ (D = diameter of object, L = length from thumb's tip to pointer's tip) and "No." denotes the number of subjects who took the particular grasp pattern. Utilizing d is very convenient since it is non-dimensional and, therefore, suppress the scale effect brought by the hand size. Each pattern in Fig.2 has the following procedure.

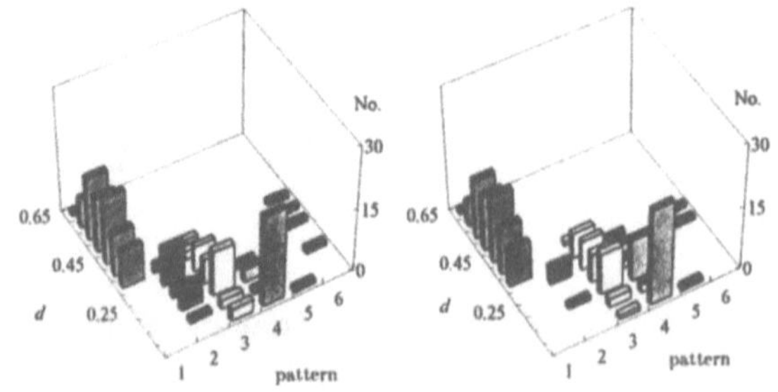

(a) Drawing paper (b) Rubber

Figure 2: Grasp pattern classification map.

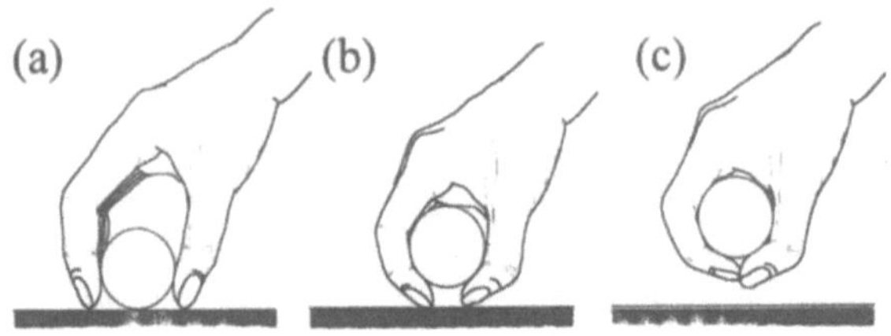

Figure 3: Grasp pattern 2.

Pattern 1: The object is directly grasped (Direct grasp).

Pattern 2: Finger tips are pushed between the bottom part of object and table, such that the object is lifted up automatically, as shown in Fig.3(wedge-effect based grasp).

Pattern 3: The object is first picked up by thumb and the remaining four fingers, and then the object is rolled up over the surface of thumb.

Pattern 4: The object is first picked up by thumb and index (or middle) finger tip. The remaining fingers hook the object and then squeeze it till the finger tip grasp is broken and the object contacts the palm, as shown in Fig.4.

Pattern 5: Four fingers except thumb roll the object forward on the table until the object is fully wrapped by four fingers, and simultaneously thumb dips up the object by pushing the tip to the bottom part of object.

Pattern 6: Palm first makes contact with the top of the object and then each finger is gradually closed to complete enveloping grasp as shown in Fig.5

For a large object ($d \geq 0.35$) Fig.2 shows that human directly grasps irrespective of the surface friction. On the other hand, many grasp patterns appear for the objects in the range of $0.05 \leq d \leq 0.35$, such as wedge-effect based one (pattern 2), grasp-transition based one (pattern 4), rolling motion based one (pattern 5), palm-

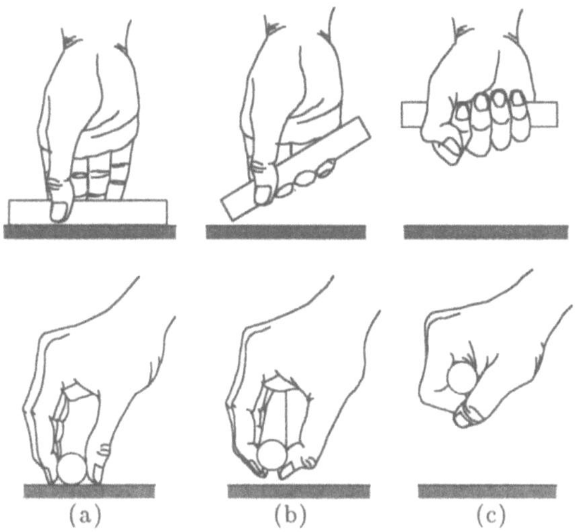

Figure 4: Grasp pattern 4.

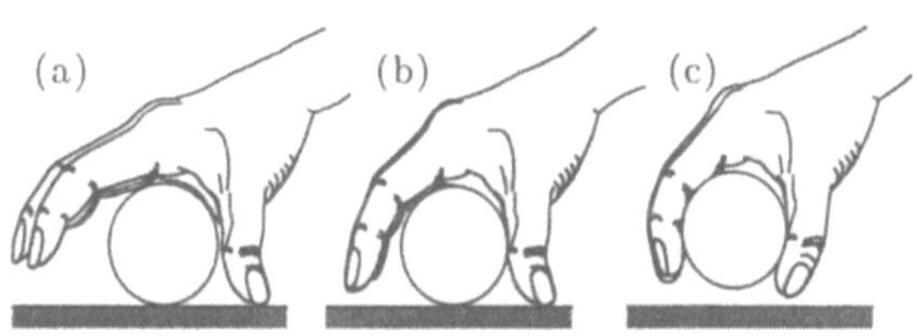

Figure 5: Grasp pattern 6.

contact based one (pattern 6), and so forth. As a general tendency, both grasp patterns 2 and 3 appear just after the grasp pattern 1 disappears when the object size decreases. As the size decreases continuously, these patterns gradually disappear and pattern 4 becomes more dominant. Of course, there are some person-dependent grasp patterns. From Fig.2(a) and(b), we can also observe the friction dependency, where the friction of rubber sheet is much larger than that of drawing paper. For example, the number of subjects who utilize wedge-effect based pattern (pattern 2) drastically decreases in Fig.2(b). This is because the pattern 2 is based on the slip, and therefore it is blocked under significant friction on the object surface. Instead, the number of subjects taking the rolling motion based pattern (pattern 3) increase.

We also observed the grasp pattern for column objects whose cross sections are triangular. For such objects, two main grasp patterns are observed, one is the direct grasp which appears for relatively large objects and the other is the rotating-motion based grasp as shown in Fig.6, where human first rotates the object around one edge and then in-

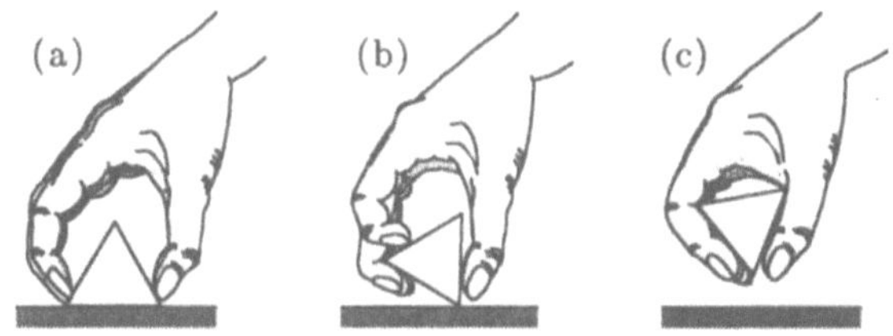

Figure 6: Grasp pattern for a column object whose cross section is triangle.

sert the finger tip to the small gap for wrapping the object. The rotating motion is essential for detaching the object from the table.

4 Application to Robot Hands

While grasp experiments by human provide a number of person-dependent grasp patterns, we choose a couple of grasp patterns applicable to multi-fingered robot hands. Since the direct grasp can be easily realized if the robot hand satisfies the geometrical condition, we do not discuss this particular one here. Instead, we focus on objects with relatively small size compared with that of hand. We start by discussing the wedge-effect based grasp pattern, since it is simple enough to apply a multi-fingered robot hand, and show its effectiveness and limitation. For a while, we assume that the contact friction is sufficiently small to ensure that the object motion is never blocked by the friction between finger link and object. We also assume that a robot hand includes position sensor and torque sensor for each joint.

4.1 Wedge-effect based grasp

To cover a large area, initially each finger is opened and then approaches the table until the finger tip makes contact with it, where the table detection can be easily checked by torque sensor outputs. In the next step, each finger tip follows along the table until a part of finger link makes contact with the object as shown in Fig.7(b). These two phases are what we call the approach phase. Then, each finger tip pushes the object each other, so that we can make the most use of wedge-effect. The object will be automatically lifted up by the slip between the finger tip and the object surface. At the same time each link is closed to remove the degrees of freedom of the object gradually. Constant torque control can

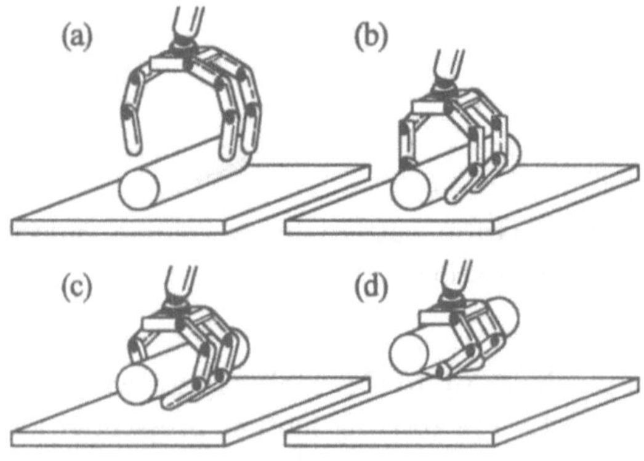

Figure 7: Wedge-effect based grasp.

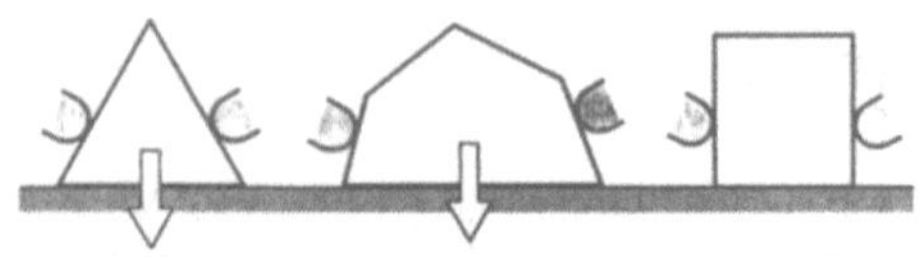

Figure 8: Examples of objects where the upward force is not expected by a simple pushing motion.

be utilized effectively for achieving both lifting and grasping phases. Whether the object really reaches the palm or not, and how firmly the hand grasps the object, strongly depend on how much torque command is imparted to each joint.

For a cylindrical object, there usually exists an enough space to insert a finger tip between the bottom part of the object and the table, unless the object's diameter is smaller than that of finger tip. As a result, the finger tip can easily produce the upward force. For a general column object, however, depending upon the object's shape, the finger tip forces may balance within the object or they may produce the downward force. Under such situations, the lifting force is not produced, even though we increase the contact force. For example, such situations will be observed for the objects shown in Fig.8. Failure in lifting can be easily detected by the joint torque sensor, because their outputs will sharply increase during a pushing motion in the horizontal direction.

4.2 Rotating motion based grasp

As grasp experiments by human suggest, the rotating motion based grasp should be a key for partly detaching an object having rectangular or triangular cross section from a table. However,

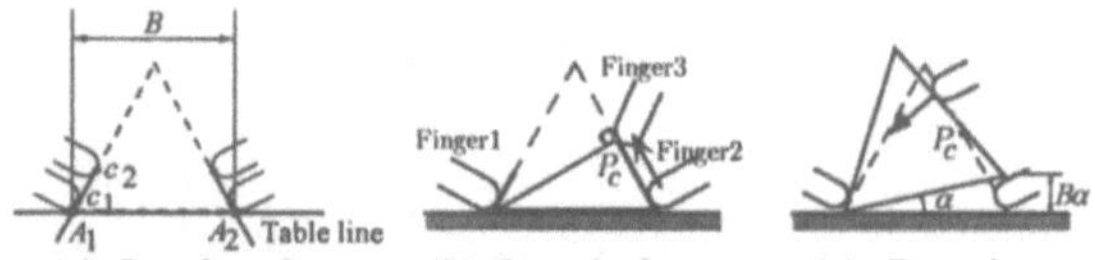

(a) Sensing for B and m (b) Local shape sensing (c) Rotating motion

Figure 9: Sensing procedure for triangular object.

it should be noted that sensors available for the robot hand are joint position and torque sensors only, while numerous tactile sensing organs covering all over human hand can provide useful information quickly. Therefore, it is even more difficult for a robot hand to achieve the task similar to that of human. Active sensing is known as a sensing technique to extend the sensing capability. For example, suppose that we impart an active motion to robot finger so that it can follow the object surface. By continuously connecting tangential line in each finger tip, we can estimate the local shape of object. As a result, we can obtain the tactile information as if there were tactile sensor over each finger tip. We produce such a virtual finger tip sensor by moving finger tip. The information which we intend to estimate by active sensing is the geometrical parameters necessary for the robot to rotate the object. For achieving rotating motion of object, there are four important parameters, base length B, slope of the bottom part of object m, pushing point P_c, and rotating angle α.

(a) Base length B and local slope m:
By making each finger tip contact the object with two slightly different heights, we can obtain one straight line tangential to both finger tips. The local slope m is directly obtained by the tangential line. By extending the tangential line until it passes through the table line, B is obtained as shown in Fig.9(a).

(b) Pushing point P_c :
Suppose an extreme case, where the friction between the finger tip and the object is zero. Even for such an extreme case, in order to produce a rotating moment around one side of the support polygon, we have to impart a pushing force at the upper point than P_c, where P_c is the intersection between the object surface and the normal line from the supporting edge, as shown in Fig.9(b). However, P_c does not always exist over the object surface since it strongly depends on the object's geometry. When P_c is not detected during the active sensing, we assign the top point as P_c in

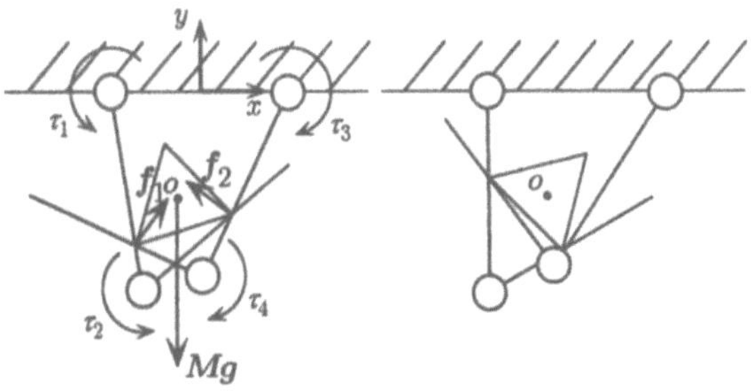

Figure 10: Triangle object enveloped by two jointed fingers

which the finger tip can apply the largest moment under a constant pushing force. When the finger tip placed at the top of the object can not rotate the object due to the slip, the robot hand fails in achieving the rotating motion based grasp.

(c) Rotation angle α :
The rotation angle α is determined according to an approximate vertical displacement $B\alpha$ produced by the rotation. A sufficient condition for inserting a finger tip into the gap produced by the rotating motion is given by the following inequality.

$$B\alpha > e \tag{1}$$

where e is the diameter of a finger tip or an equivalent diameter if the cross section is not exactly circle. An example satisfying the sufficient condition is shown in Fig.9(c).

After a sufficient gap is produced, one finger is removed away from the object's surface to be inserted into the gap. After the finger tip is sufficiently inserted into the gap between the object and the table, we apply the same grasping mode as that taken for cylindrical objects. More precise discussions on sensing are described in [16].

(d) Lifting and Grasping Phase :
After the finger tip is sufficiently inserted into the gap between the object and the table, we apply the same grasping mode as that taken for a cylindrical object[2]. However, the condition for the object to reach the palm is not as simple as that for a cylindrical object. Because, for a general column object, contact points between the object and finger links change according to the orientation of object, even though the center of gravity of object does not change. This is a big difference between a cylindrical object and a general column one. If the resultant force acted by each contact force is greater than the gravitation force, irrespective of both orientation and position, then the lifting motion of object is guaranteed. For

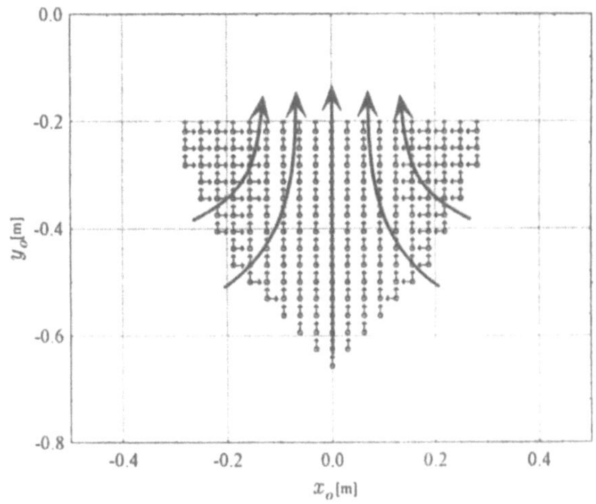

(a) Force flow diagram

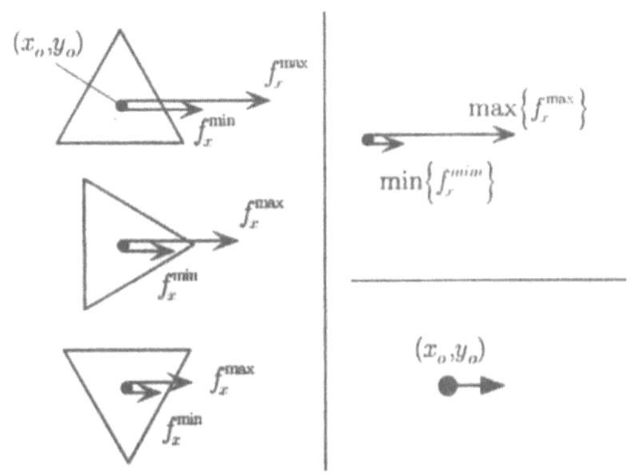

(b) Force flow

Figure 11: Force Flow Diagram

a triangle object as shown in Fig.10, the resultant force $\boldsymbol{f}_o$ of produced by each contact force is given by $\boldsymbol{f}_o = \sum_{i=1}^{4} \boldsymbol{f}_i - M\boldsymbol{g}$. Under the constant torque control, the resultant force $\boldsymbol{f}_o$ is not uniquely determined, because there exist infinite combinations of contact forces that can balance with a set of given torque. By applying the linear programming technique, we can compute both the maximum and the minimum resultant forces. Fig.11(a) shows the force flow diagram where the positive arrow means that both the maximum and the minimum forces are always positive irrespective of object orientation as shown in Fig.11(b). This diagram visually tells us a rough behaviour of object during the lifting phase without solving any complicated differential equations with respect to time. From Fig.11(a) we can see that the object moves upward without moving away from the hand working space, where Mg = -1.0[N], τ_1

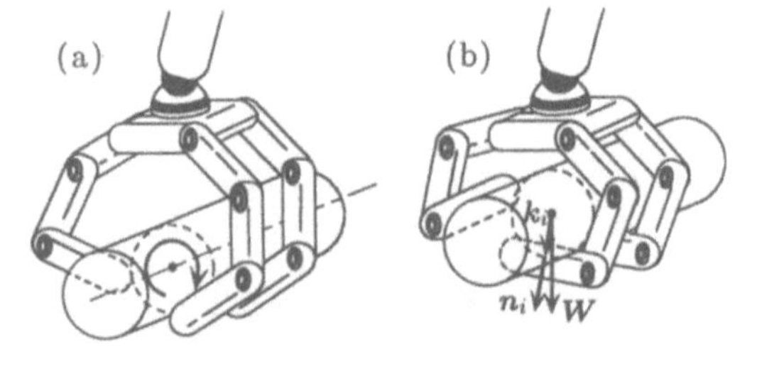

Figure 12: Rolling motion based grasp under significant friction.

$= 2.0$, $\tau_2 = 1.0$, $\tau_3 = 2.0$, $\tau_4 = 1.0$ and the coefficient of friction is 0.18.

4.3 Rolling motion based grasp

For both wedge-effect and rotating motion based grasps, we implicitly assume that the contact friction is small enough to smoothly achieve a slipping motion during the lifting phases. Under significant friction, however, both procedures may eventually fail in enveloping an object, since a slipping motion is not always guaranteed between the robot finger and the object. This failure can be detected by monitoring both position and torque sensors' outputs. At the moment such failure happens, each torque sensor output sharply increases, while finger posture does not change at all. In such a case, we switch from a slip motion based strategy to rolling motion based grasp strategy after the object is put down on the table. Fig.12 shows an example of rolling motion based grasp, where it is assumed that the center of gravity exists between the two right fingers. First, the left finger starts to make the object roll over along the surface of right fingers being in slowly closing until the following conditions are confirmed.

$$\boldsymbol{W} \otimes \boldsymbol{n}_i < 0 \quad \cap \quad \boldsymbol{W} \otimes \boldsymbol{k}_i < 0$$

where $\boldsymbol{W}$, $\boldsymbol{n}_i$ and $\boldsymbol{k}_i$ are the gravitational vector, the inward normal vector of link surface at i-th contact point, and the vector with minimum norm from the gravitational center axis to i-th contact point, respectively and $\otimes$ is a scalar operator computing $\boldsymbol{x} \otimes \boldsymbol{y} = x_1 y_2 - x_2 y_1$ for two vector $\boldsymbol{x} = (x_1, y_1)^{\mathrm{T}}$ and $\boldsymbol{y} = (x_2, y_2)^{\mathrm{T}}$.

The above condition is a sufficient one for not making the object falling down from the right fingers without any support from the left finger. After confirming this condition, the left finger is removed and finally an enveloping grasp is completed.

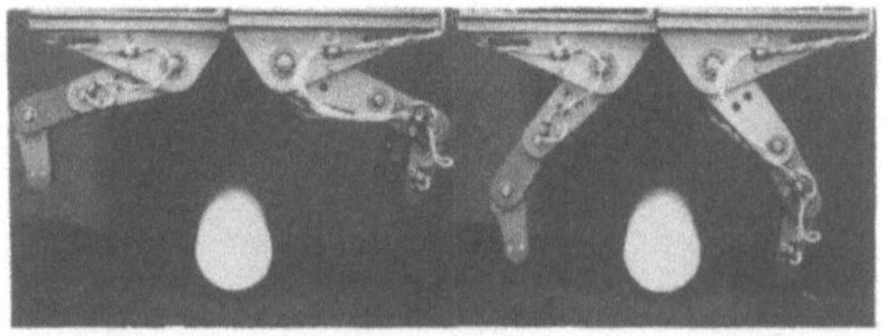

(a)Initial phase (b)Table detection

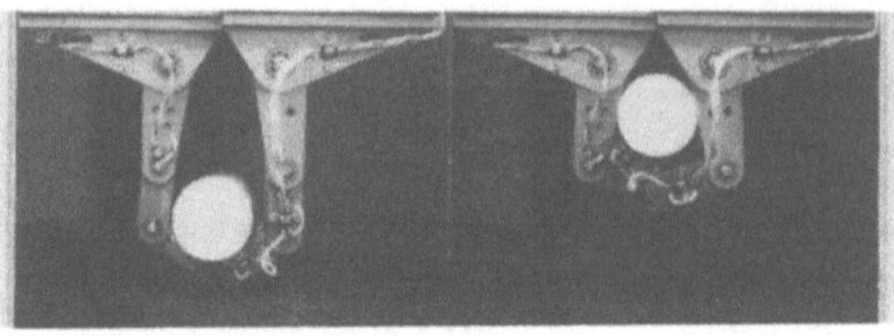

(c)Lifting phase (d)Grasping phase

Figure 13: Wedge-effect based grasp strategy for a cylindrical object (R=15[mm], $\mu = 0.27$).

5 Experiments

The experiments were done by using the Hiroshima Hand[16] which is controlled by pulley-wire driving system. The hand includes a specially designed joint torque sensor in each joint.

Fig.13 shows continuous photos for grasping a cylindrical object, the cylindrical object is covered by a drawing such that we can keep the contact friction small ($\mu = 0.27$). In experiments, we assume that the robot does not know what kind of object is tested. The robot first tries to apply the wedge-effect based strategy[15][16] since it is the simplest one. Under either a column object with the cross section as shown in Fig.8 or a cylindrical object with significant friction, however, the wedge-effect based approach will fail. The robot hand can recognize this failure, because the finger posture does not change while each joint torque sharply increases. Once the robot results in such a situation, it starts to detect the local shape of object, especially the bottom part of object. If the robot judges that the object has the cross section as shown in Fig.8, it starts the rotating motion based strategy as shown in Fig.14. If the robot judges that the cross section is similar to circle, then it applies the rolling motion based strategy as shown in Fig.15.

Fig.15 shows continuous photos for grasping a cylindrical object, where the cylindrical object is

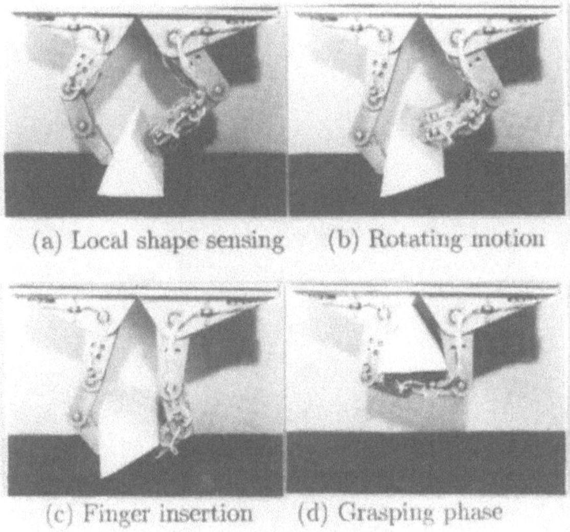

Figure 14: Rotation motion based grasp strategy for a triangular object.

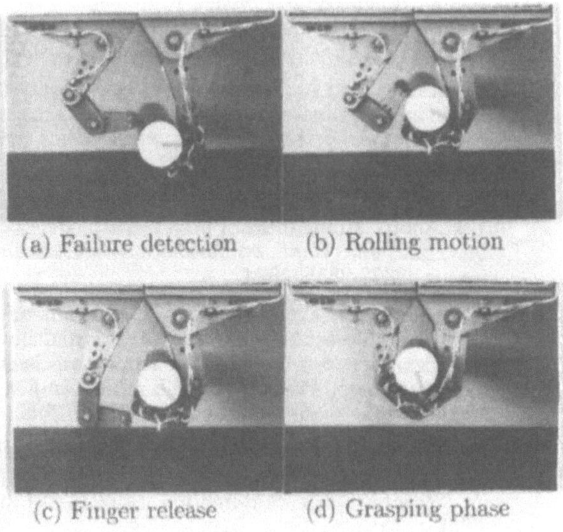

Figure 15: Rolling motion based grasp strategy under significant friction.

covered by rubber such that we can increase the surface friction purposely ($\mu = 1.3$).

6 Conclusion

We discussed the Scale-Dependent Grasp by observing the grasp patterns produced by human. We found that as the size of object decreases, several patterns appear based on personal choice, while the direct grasp is the only feasible pattern if the object is large compared to with human hand. Based on the observation of human grasping, we choose three grasp patterns which are easily applicable for a multi-fingered robot hand. We showed that by carefully choosing one of these patterns, we can achieve an enveloping grasp for most of column object irrespective of their surface friction.

This work is supported by Inter University Robotic Project provided by Ministry of Education, Japan. Finally the author would like to express his sincere thanks to Dr. T. Tsuji, Mr. N. Thaiprasert, Mr. Y. Hino, Mr. M. Higashimori, Mr. Y. Tanaka, Mr. K. Nakagawa, and Mr. K. Furutera for their cooperation for this work.

References

[1] Cutkosky, M.: On grasp choice, grasp models, and the design of hands for manufacturing tasks, *IEEE Trans. on Robotics and Automation*, vol.5, no.5, pp269–279, 1989.

[2] Bekey, G.A., H. Liu, R. Tomovic, and W. Karplus: Knowledge-based control of grasping in robot hands using heuristics from human motor skills, *IEEE Trans. on Robotics and Automation*, vol.9, no.6, pp709–722, 1993.

[3] Kang, S.B., and K. Ikeuchi: Toward automatic robot instruction from perception—Recognizing a grasp from observation, *IEEE Trans. on Robotics and Automation*, vol.9, no.4, pp432–443, 1993.

[4] Iberall, T., J. Jackson, L. Labbe, and R. Zampano: Knowledge-based pretension: Capturing human dexterity, *Proc. of the IEEE Int. Conf. on Robotics and Automation*, pp82–87, 1988.

[5] Stansfield, S.: Robotic grasping of unknown objects: A knowledge based approach, *Int. J. of Robotics Research*, vol.10, pp314–326, 1991.

[6] Jeannerod, M.: Attention and performance, chapter Intersegmental coordination during reaching at natural visual objects, pp153-168, Erlbaum, Hillsdale, 1981.

[7] Bard, C., and J. Troccaz: Automatic preshaping for a dexterous hand from a simple description of objects, *Proc. of the IEEE Int. Workshop on Intelligent Robots and Systems*, pp865-872, 1990.

[8] Mirza, K., and D. E. Orin: Control of force distribution for power grasp in the DIGITS system, Proc. of the IEEE 29th CDC Conf., pp1960-1965, 1990.

[9] Trinkle, J. C., J. M. Abel, and R. P. Paul: Enveloping, frictionless planar grasping, *Proc. of the IEEE Int. Conf. on Robotics and Automation*, 1987.

[10] Salisbury, J. K., Whole-Arm manipulation, *Proc. of the 4th Int. Symp. of Robotics Research, Santa Cruz*, CA, 1987. Published by the MIT Press, Cambridge MA.

[11] Bicchi, A: Force distribution in multiple whole-limb manipulation, *Proc. of the IEEE Int. Conf. on Robotics and Automation*, pp196-201, 1993.

[12] Omata, T., and K. Nagata: Rigid body analysis of the indeterminate grasp force in power grasps, *Proc. of the IEEE Int. Conf. on Robotics and Automation*, pp1787-1794, 1996.

[13] Zhang, X-Y., Y. Nakamura, K. Goda, and K. Yoshimoto: Robustness of power grasp, *Proc. of the IEEE Int. Conf. on Robotics and Automation*, pp2828-2835, 1994.

[14] Kleinmann, K. P., J. Henning, C. Ruhm, and H. Tolle: Object manipulation by a multifingered gripper: On the transition from precision to power grasp, *Proc. of the IEEE Int. Conf. on Robotics and Automation*, pp2761-2766, 1996.

[15] Kaneko, M., Y. Tanaka, and T. Tsuji: Scale-dependent grasp, *Proc. of the IEEE Int. Conf. on Robotics and Automation*, pp2131-2136, 1996.

[16] Kaneko, M., N. Thaiprasert, and T. Tsuji: Experimental Approach on Enveloping Grasp for Column Objects, *Preprint of 5th Int. Symp. on Exp. Robotics*, pp17-29, 1997.

PART 2

DYNAMICS AND CONTROL

SESSION SUMMARY

Tsuneo Yoshikawa
Department of Mechanical Engineering, Kyoto University, Kyoto 606 (Japan)
email: yoshi@mech.kyoto-u.ac.jp

This article presents a summary of the session "Dynamics and Control." Although the title "Dynamics and Control" sounds rather classical in robotics, the theme is still one of the main research fields in more advanced way. In fact, among the three papers presented in the session, the first paper is concerned with a unified formulation of dynamics of a wide class of manipulators and at the same time it is related to space application of manipulators. The second paper is related to deep sea application of robotic vehicles. The last paper is concerned with validity of the task frame concept, a theoretical issue in the formulation of manipulator force control.

"A General Formulation of Under-Actuated Manipulator systems" by K. Yoshida and D. N. Nenchev presents a general formulation of kinematics and dynamics for a wide variety of under-actuated manipulator systems including free-joint manipulators, free-floating manipulators, flexible-base manipulators, and flexible-arm manipulators. A merit of this general formulation is that some concepts originally developed for some particular system are useful for other systems as well. Examples are the generalized Jacobian matrix, generalized inertia tensor, measure of dynamic coupling, kinematic compensability, and reaction null-space. It is shown that a main difference among these four classes of manipulators is what type of nonholonomic constraint they have. It might be true that similarity among the kinematics and dynamics equations for the four classes of manipulators has been well recognized, but the authors are the first to explicitly point this out. Experimental and simulation results of application of reaction null-space have been shown by video.

A new underwater vehicle navigation and control system employing a new commercially available 1,200 Khz doppler sonar is reported in **"Towards Precision Robotic Maneuvering, Survey, and Manipulation in Unstructured Undersea Environments" by L. Whitcomb, D. Yoerger, H. Singh, D. Mindell, and J. Howland**. This system is installed in JASON, a 1,200 Kg 6000 meter underwater vehicle. The performance of this new system was compared with those of conventional 12Khz and 300 Khz long-baseline (LBL) acoustic navigation systems. The experimental results show that a hybrid system incorporating both doppler and LBL provides superior tracking performance over doppler or LBL alone. A field deployment of JASON with this new system in the Tyrrhenian Sea is also reported by a video tape.

"Where Does the Task Frame Go?" by H. Bruyninckx and J. De Schutter discusses the role of "Task Frame" as the central modeling, specification, sensor processing and control concept in the hybrid control paradigm for robot manipulators in contact with its environment. In a previous paper, the authors have tried to formalize Mason's original intuitive ideas of task frame in four rules: geometric compatibility, causal compatibility, time-invariance, and maximal decoupling. These rules cover the "classical" compliant motion tasks, but not tasks in more complex contact situations. In this paper, a specification of the task frame is presented for the task to move an end-effector along the line of intersection of two surfaces maintaining the two contacts between the end-effector and the surfaces. This result could be regarded as giving a particular choice of the generalized coordinates in the modern formulation of hybrid control taking the manipulator dynamics into consideration.

All three papers were very interesting and it seems that there are a lot more to be done in the field of robot dynamics and control.

A General Formulation of Under-Actuated Manipulator Systems

Kazuya Yoshida and Dragomir N. Nenchev
Dept. of Aeronautics and Space Engineering, Tohoku University
Aobayama campus, Sendai 980-77 (Japan)
yoshida@astro.mech.tohoku.ac.jp
nenchev@space.mech.tohoku.ac.jp

Abstract

This paper discusses a general formulation both in kinematics and dynamics for under-actuated manipulator systems, represented by a free-joint manipulator, a free-floating manipulator, a flexible-base manipulator and a flexible-arm manipulator. We highlight the fact that a concept, originally developed for a particular system, is applicable to the others as well because the dynamic equations of all four systems show common characteristics. Examples of such concepts are the generalized Jacobian matrix and the generalized inertia tensor which were originally developed for a free-floating manipulator, and also the reaction null-space concept, which was introduced for a free-floating and a flexible-base manipulator. We show also that the differences of the four systems are essentially due to the integrability conditions.

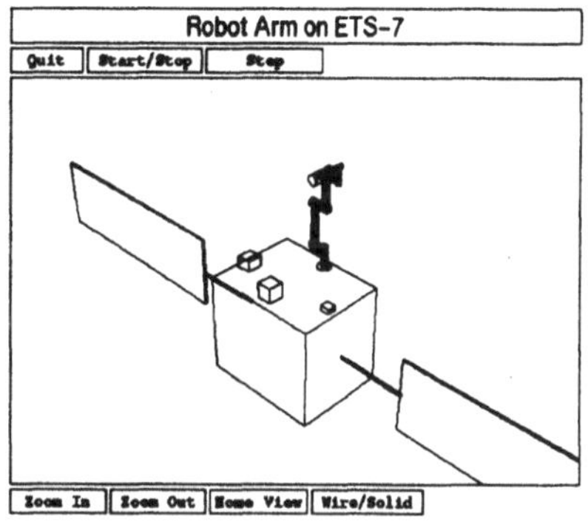

Figure 1 An example of Free-Floating Space Manipulator

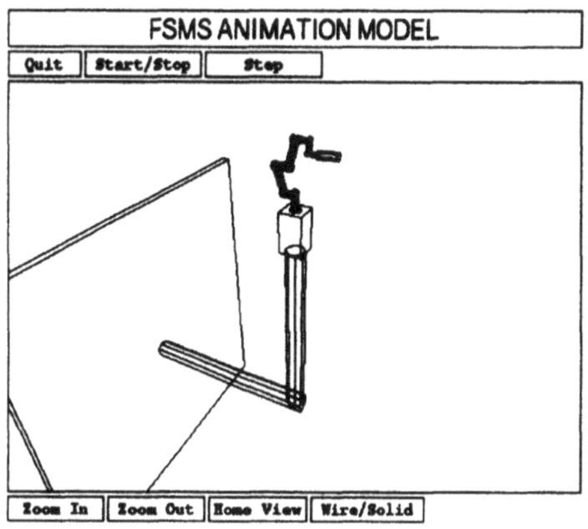

Figure 2 An example of Flexible-Base Manipulator

1 Introduction

The interest toward complex robot systems is expanding for new application areas. A class of such robot systems are under-actuated systems, characterized by the number of control actuators being less than the number of degree of freedom. Typical examples are serial manipulators with passive joints, free-floating space manipulators, flexible-arm manipulators and others. Jain and Rodriguez [1] have shown that the dynamics of various types of under-actuated systems can be formulated within a common framework. Currently, intensive research in the field is going on, using geometrical mechanics and differential geometry as the theoretical base [2][3].

Another example of an under-actuated system is a dextrous manipulator arm mounted on a passive flexible base (**Figure 2**). In literature, such a system is known under the name ***long-reach manipulator*** [4][5], or ***flexible structure mounted manipulator system (FSMS)*** [6]. A major issue concerning long-reach manipulators is dynamic interaction between the motion of the dextrous manipulator and the vibrational motion of the flexible base due to induced reaction. One can find that the problem here is similar to the problem in free-floating manipulators in the sense that the dynamic reaction of the manipulator arm induces the interactive motion in the supporting base. The difference between these two systems is that the manipulator base of a free-floating manipulator is a floating inertia, but that of a long-reach manipulator is an inertia-spring-damper system.

One more example of an under-actuated system is a flexible-arm manipulator, the system DOF of which is a sum of the DOF of controllable joints and the DOF of elastic displacements. The joint motion and elastic displacements are dynamically interacting in some fashion similar to other under-actuated systems. Compared with long-reach manipulators which have a concentrated compliance at the base, flexible-arm manipulators have distributed compliance along the kinematic chain.

Figure 3 shows a family of under-actuated manipulator systems. The top two represent systems with *free* joints, while the bottom two show manipulators with *compliant* or visco-elastic joints. Since a flexible-link manipulator can be modeled as a serial kinematic chain composed of active and elastic joints alternately, we regard a flexible-*arm* manipulator and a flexible-*joint* manipulator as systems of the same type. In the figure, the left column is a group of *separable* systems, which means that free or compliant (passive) joints are separately located from active joints. The right column is a group of *distributed* systems, where passive joints are distributed among other active joints.

The above systems have been studied independently although, they eventually show a lot of similarity in the dynamic formulation. For example, a free-floating manipulator is a system composed by a free base (which can be modeled as a 6 DOF free joint in general) and active manipulator(s). But once the free joints are distributed, the system can be regarded as a serial manipulator with free joints, and both systems basically have common dynamics. A long-reach manipulator is a system composed by a compliant base (which can be modeled as a 6 DOF visco-elastic beam in general) and active manipulator(s). But when the compliant joints are distributed, the system can be regarded as a manipulator with compliant joints or a flexible-arm manipulator.

If we look at differences among these four systems, non-holonomy plays a significant role. A free-floating manipulator is known to be generally a first-order non-holonomic system [7]. On the other hand, a free-joint manipulator can be holonomic or non-holonomic depending on its physical construction. For example, it is known for a 2R horizontal manipulator that the one with the 1st joint free and the 2nd active is holonomic, but the other with the 1st joint active and the 2nd free is non-holonomic [8]. As far as flexible-base and flexible-arm manipulators are concerned, we consider them non-holonomic in general, but here we will discuss some special conditions for their integrability.

The main purpose of this paper is to compare the various types of under-actuated systems, highlighting the fact that a concept originally developed for a particular system is applicable to the others as well. The paper is organized in three parts: In the first part, we focus on a general formulation for under-actuated manipulator systems, paying attention to the fact that all the systems in this class have common characteristics in the dynamic formulation. We use a matrix-based approach and show that the Generalized Jacobian Matrix [9][10] and the Generalized Inertia Tensor [11]–[12], which were originally developed for a free-floating space manipulator, are applicable to other systems as well. In the second part, we examine the characteristics of constraint dynamics for the above four types of under-actuated systems, case by case. Here, the differences of the dynamic character are discussed from the point of view of holonomy or integrability. In the last part, we focus on some further common concepts, such as the measure of dynamic coupling, the kinematic compensability, and the reaction null-space.

2 General Formulation

Let us consider a system whose motion is described by n degrees of freedom of the generalized coordinate $\boldsymbol{q} \in R^n$ for *active* joints and m degrees of freedom of the generalized coordinate $\boldsymbol{p} \in R^m$ for *passive* joints. Define $\mathcal{F}_q$ as active generalized force generated on coordinate $\boldsymbol{q}$, and $\mathcal{F}_p$ as a passive generalized force on coordinate $\boldsymbol{p}$. Also, define $\boldsymbol{x}$ as a coordinate of a point of interest (the operational space coordinate), and let an external wrench $\mathcal{F}_{ex}$ be applied on $\boldsymbol{x}$. This wrench i mapped as $\boldsymbol{J}_q^T \mathcal{F}_{ex}$ and $\boldsymbol{J}_p^T \mathcal{F}_{ex}$ onto each generalized coordinate respectively, using corresponding Jacobian matrices. The equation of motion of

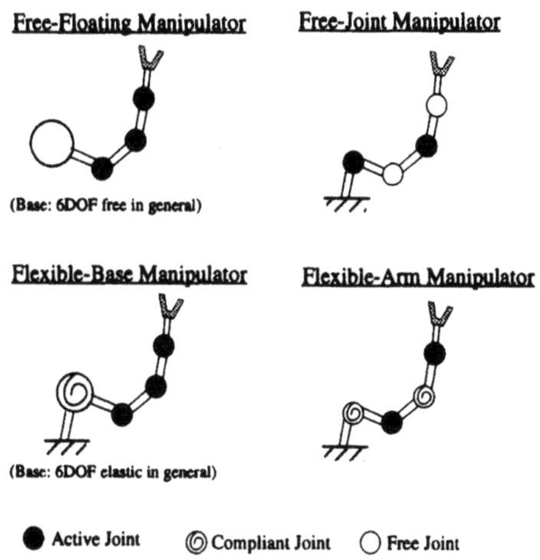

Figure 3 A family of under-actuated manipulator systems

such system is expressed as:

$$\begin{bmatrix} \boldsymbol{H}_p & \boldsymbol{H}_{pq} \\ \boldsymbol{H}_{pq}^T & \boldsymbol{H}_q \end{bmatrix} \begin{bmatrix} \ddot{\boldsymbol{p}} \\ \ddot{\boldsymbol{q}} \end{bmatrix} + \begin{bmatrix} \boldsymbol{c}_p \\ \boldsymbol{c}_q \end{bmatrix} + \begin{bmatrix} \boldsymbol{g}_p \\ \boldsymbol{g}_q \end{bmatrix} = \begin{bmatrix} \mathcal{F}_p \\ \mathcal{F}_q \end{bmatrix} + \begin{bmatrix} \boldsymbol{J}_p^T \\ \boldsymbol{J}_q^T \end{bmatrix} \mathcal{F}_{ex} \quad (1)$$

where $\boldsymbol{H}_p \in R^{m \times m}$, $\boldsymbol{H}_q \in R^{n \times n}$, $\boldsymbol{H}_{pq} \in R^{m \times n}$ are inertia matrices, $\boldsymbol{c}_p$, $\boldsymbol{c}_q$ non-linear Coriolis and centrifugal forces, and $\boldsymbol{g}_p$, $\boldsymbol{g}_q$ gravity terms.

The kinematic relationship among $\boldsymbol{p}$, $\boldsymbol{q}$ and $\boldsymbol{x}$ is expressed as:

$$\dot{\boldsymbol{x}} = \boldsymbol{J}_p \dot{\boldsymbol{p}} + \boldsymbol{J}_q \dot{\boldsymbol{q}}, \quad (2)$$

$$\ddot{\boldsymbol{x}} = \boldsymbol{J}_p \ddot{\boldsymbol{p}} + \dot{\boldsymbol{J}}_p \dot{\boldsymbol{p}} + \boldsymbol{J}_q \ddot{\boldsymbol{q}} + \dot{\boldsymbol{J}}_q \dot{\boldsymbol{q}}. \quad (3)$$

The upper and lower sets of Equation (1) are written as:

$$\boldsymbol{H}_p \ddot{\boldsymbol{p}} + \boldsymbol{H}_{pq} \ddot{\boldsymbol{q}} + \boldsymbol{c}_p + \boldsymbol{g}_p = \mathcal{F}_p + \boldsymbol{J}_p^T \mathcal{F}_{ex}, \quad (4)$$

$$\boldsymbol{H}_q \ddot{\boldsymbol{q}} + \boldsymbol{H}_{pq}^T \ddot{\boldsymbol{p}} + \boldsymbol{c}_q + \boldsymbol{g}_q = \mathcal{F}_q + \boldsymbol{J}_q^T \mathcal{F}_{ex}. \quad (5)$$

Since $\mathcal{F}_q$ is an actively controlled force, while $\mathcal{F}_p$ is a resultant "passive" one, Equation (4) works as a dynamic constraint to the system. If $\boldsymbol{H}_p$ is non-singular, the acceleration $\ddot{\boldsymbol{p}}$ of the passive coordinate is:

$$\ddot{\boldsymbol{p}} = -\boldsymbol{H}_p^{-1}(\boldsymbol{H}_{pq}\ddot{\boldsymbol{q}} + \boldsymbol{c}_p + \boldsymbol{g}_p - \mathcal{F}_p - \boldsymbol{J}_p^T \mathcal{F}_{ex}) \quad (6)$$

By substituting it into Equation (5), we obtain the equation of motion involving the acceleration $\ddot{\boldsymbol{q}}$ of the active joints:

$$\widehat{\boldsymbol{H}}\ddot{\boldsymbol{q}} + \widehat{\boldsymbol{c}} + \widehat{\boldsymbol{g}} = \widehat{\mathcal{F}} + \widehat{\boldsymbol{J}}^T \mathcal{F}_{ex} \quad (7)$$

where

$$\widehat{\boldsymbol{H}} = \boldsymbol{H}_q - \boldsymbol{H}_{pq}^T \boldsymbol{H}_p^{-1} \boldsymbol{H}_{pq}, \quad (8)$$

$$\widehat{\boldsymbol{J}} = \boldsymbol{J}_q - \boldsymbol{J}_p \boldsymbol{H}_p^{-1} \boldsymbol{H}_{pq}, \quad (9)$$

$$\widehat{\boldsymbol{c}} = \boldsymbol{c}_q - \boldsymbol{H}_{pq}^T \boldsymbol{H}_p^{-1} \boldsymbol{c}_p, \quad (10)$$

$$\widehat{\boldsymbol{g}} = \boldsymbol{g}_q - \boldsymbol{H}_{pq}^T \boldsymbol{H}_p^{-1} \boldsymbol{g}_p, \quad (11)$$

$$\widehat{\mathcal{F}} = \mathcal{F}_q - \boldsymbol{H}_{pq}^T \boldsymbol{H}_p^{-1} \mathcal{F}_p. \quad (12)$$

The matrix $\widehat{\boldsymbol{H}}$ is known as the *generalized inertia tensor* [11].

The matrix $\widehat{\boldsymbol{J}}$ is known as the *generalized Jacobian matrix* [9][10]. It was originally defined for a free-floating manipulator under the condition of momentum conservation. But here, we see that the generalized Jacobian exists regardless of this condition. Hence, the generalized Jacobian can be defined for any under-actuated system.

Equations (10)(11)(12) have the same structure: the forces $\boldsymbol{c}_p$, $\boldsymbol{g}_p$ and $\mathcal{F}_p$ related to the passive variables are mapped onto the active variable space through the dynamic coupling expression $\boldsymbol{H}_{pq}^T \boldsymbol{H}_p^{-1}$.

On the other hand, if the passive acceleration (6) is substituted into the kinematic equation (3), the following is obtained:

$$\begin{aligned} \ddot{\boldsymbol{x}} &= (\boldsymbol{J}_q - \boldsymbol{J}_p \boldsymbol{H}_p^{-1} \boldsymbol{H}_{pq})\ddot{\boldsymbol{q}} + \dot{\boldsymbol{J}}_p \dot{\boldsymbol{p}} + \dot{\boldsymbol{J}}_q \dot{\boldsymbol{q}} \\ &\quad - \boldsymbol{J}_p \boldsymbol{H}_p^{-1}(\boldsymbol{c}_p + \boldsymbol{g}_p - \mathcal{F}_p - \boldsymbol{J}_p^T \mathcal{F}_{ex}) \\ &\equiv \widehat{\boldsymbol{J}}\ddot{\boldsymbol{q}} + \ddot{\boldsymbol{\zeta}}. \end{aligned} \quad (13)$$

The non-linear acceleration $\ddot{\boldsymbol{\zeta}}$ can be computed from a model or measurement. Then the inverse kinematic solution is given as:

$$\ddot{\boldsymbol{q}} = \widehat{\boldsymbol{J}}^{-1}(\ddot{\boldsymbol{x}} - \ddot{\boldsymbol{\zeta}}). \quad (14)$$

The effectiveness of the generalized Jacobian matrix in the inverse kinematic problems has been discussed only for a free-floating robot so far, but the above equation suggests the importance of the generalized Jacobian for other classes of under-actuated systems as well.

3 Case Formulation

Based on the general formulation, we derive the equation of motion for the following under-actuated systems: free-joint manipulators, free-floating manipulators, flexible-base manipulators and flexible-arm manipulators.

3.1 Free-Joint Manipulator

A free-joint manipulator is a serial link manipulator composed of active and passive joints. For this system replace the subscript notation in Equation (1) as follows:

p	$\rightarrow$	ϕ_p	:passive joint angle
q	$\rightarrow$	ϕ_a	:active joint angle
x	$\rightarrow$	x_h	:position/orientation of the hand
$\mathcal{F}_p$	$\rightarrow$	O	:passive joint torque
$\mathcal{F}_q$	$\rightarrow$	τ	:active joint torque
$\mathcal{F}_{ex}$	$\rightarrow$	O	:external wrench on the hand

The dynamic equation becomes:

$$\begin{bmatrix} \boldsymbol{H}_p & \boldsymbol{H}_{pa} \\ \boldsymbol{H}_{pa}^T & \boldsymbol{H}_a \end{bmatrix} \begin{bmatrix} \ddot{\boldsymbol{\phi}}_p \\ \ddot{\boldsymbol{\phi}}_a \end{bmatrix} + \begin{bmatrix} \boldsymbol{c}_p \\ \boldsymbol{c}_a \end{bmatrix} + \begin{bmatrix} \boldsymbol{g}_p \\ \boldsymbol{g}_a \end{bmatrix} = \begin{bmatrix} \boldsymbol{O} \\ \boldsymbol{\tau} \end{bmatrix} \tag{15}$$

The characteristics of the dynamic constraint provided by the upper set of (15) has been recently studied for integrability. According to [8], the condition of existence of a first integral is

A-1. the gravity force $\boldsymbol{g}_p$ is constant,

A-2. the passive joint angles $\boldsymbol{\phi}_p$ do not appear in the manipulator inertia matrix.

The condition of existence of a second integral is

B-1. a first integral exists,

B-2. the distribution defined by the null-space of matrix $\begin{bmatrix} \boldsymbol{H}_p & \boldsymbol{H}_{pa} \end{bmatrix}$ is involutive.

For example, a 2R horizontal manipulator with the 1st joint free and the 2nd active is holonomic, but one with the 1st joint active and the 2nd free is non-holonomic.

3.2 Free-Floating Manipulator

A free-floating manipulator is composed of a floating base and a serial articulated arm. Assume no gravity and zero external forces. Then replace the subscript notation in Equation (1) as follows:

p	$\rightarrow$	x_b	:position/orientation of the base
q	$\rightarrow$	ϕ	:joint angle of the arm
x	$\rightarrow$	x_h	:position/orientation of the hand
$\mathcal{F}_p$	$\rightarrow$	O	:generalized force on the base
$\mathcal{F}_q$	$\rightarrow$	τ	:joint torque of the arm
$\mathcal{F}_{ex}$	$\rightarrow$	O	:external wrench on the hand

We obtain the following equations:

$$\begin{bmatrix} \boldsymbol{H}_b & \boldsymbol{H}_{bm} \\ \boldsymbol{H}_{bm}^T & \boldsymbol{H}_m \end{bmatrix} \begin{bmatrix} \ddot{\boldsymbol{x}}_b \\ \ddot{\boldsymbol{\phi}} \end{bmatrix} + \begin{bmatrix} \boldsymbol{c}_b \\ \boldsymbol{c}_m \end{bmatrix} = \begin{bmatrix} \boldsymbol{O} \\ \boldsymbol{\tau} \end{bmatrix} \tag{16}$$

$$\dot{\boldsymbol{x}}_h = \boldsymbol{J}_m\dot{\boldsymbol{\phi}} + \boldsymbol{J}_b\dot{\boldsymbol{x}}_b \tag{17}$$

$$\ddot{\boldsymbol{x}}_h = \boldsymbol{J}_m\ddot{\boldsymbol{\phi}} + \dot{\boldsymbol{J}}_m\dot{\boldsymbol{\phi}} + \boldsymbol{J}_b\ddot{\boldsymbol{x}}_b + \dot{\boldsymbol{J}}_b\dot{\boldsymbol{x}}_b \tag{18}$$

In the equation of motion, the gravity is zero and the inertia matrices $\boldsymbol{H}_b \in R^{6\times 6}$, $\boldsymbol{H}_m \in R^{n\times n}$, $\boldsymbol{H}_{bm} \in R^{6\times n}$ are functions of $\boldsymbol{\phi}$ only. Thus, the conditions A-1 and A-2 are satisfied. Hence, the constraint dynamic equation

$$\boldsymbol{H}_b\ddot{\boldsymbol{x}}_b + \boldsymbol{H}_{bm}\ddot{\boldsymbol{\phi}} + \dot{\boldsymbol{H}}_b\dot{\boldsymbol{x}}_b + \dot{\boldsymbol{H}}_{bm}\dot{\boldsymbol{\phi}} = \boldsymbol{O} \tag{19}$$

is integrated to yield the system momentum:

$$\mathcal{L} = \boldsymbol{H}_b\dot{\boldsymbol{x}}_b + \boldsymbol{H}_{bm}\dot{\boldsymbol{\phi}} = const. \tag{20}$$

$\mathcal{L}$ involves linear and angular momentum: linear momentum is once more integrable to yield the position of the system centroid, while angular momentum is not anymore integrable because the condition B-2 is not satisfied. Therefore, a free-floating manipulator is a first-order non-holonomic system [7].

Solving (20) for $\dot{\boldsymbol{x}}_b$ and substituting into (17), we obtain a useful kinematic equation with the generalized Jacobian matrix.

$$\dot{\boldsymbol{x}}_h = \hat{\boldsymbol{J}}\dot{\boldsymbol{\phi}} + \dot{\boldsymbol{x}}_{h0} \tag{21}$$

where

$$\hat{\boldsymbol{J}} = \boldsymbol{J}_m - \boldsymbol{J}_b\boldsymbol{H}_b^{-1}\boldsymbol{H}_{bm} \tag{22}$$

and

$$\dot{\boldsymbol{x}}_{h0} = \boldsymbol{J}_b\boldsymbol{H}_b^{-1}\mathcal{L} \tag{23}$$

Since $\boldsymbol{H}_b$ is the inertia tensor of a single rigid body (the manipulator base), it is always positive definite and its inverse exists. Note that Equation (22) is the original definition of the generalized Jacobian matrix. For a first-order non-holonomic system, the generalized Jacobian can be derived from the velocity equations.

In the presence of an external wrench $\mathcal{F}_h$ at the manipulator hand, we get the following dynamic equation, by analogy to Equation (7):

$$\widehat{\boldsymbol{H}}\ddot{\boldsymbol{\phi}} + \widehat{\boldsymbol{c}} = \boldsymbol{\tau} + \hat{\boldsymbol{J}}^T\mathcal{F}_h \tag{24}$$

where

$$\widehat{\boldsymbol{H}} = \boldsymbol{H}_m - \boldsymbol{H}_{bm}^T\boldsymbol{H}_b^{-1}\boldsymbol{H}_{bm}. \tag{25}$$

An external wrench at the hand occurs when a free-floating manipulator collides or catches an object, and at this moment, the system momentum changes. If the definition of $\widehat{\boldsymbol{H}}$ and $\hat{\boldsymbol{J}}$ relies on the momentum conservation, we cannot use these matrices for the impact modeling. But as discussed in the previous section, Equation (24) is generally valid regardless of the momentum conservation. This fact justifies the modeling and analysis of impact dynamics [12][13] based on Equation (24).

3.3 Flexible-Base Manipulator

Next, consider a flexible-base manipulator composed by a single manipulator base which is constrained by a flexible-beam or a spring and damper (visco-elastic) system, and a serial manipulator arm at whose endpoint an external wrench may apply. For such a flexible-base manipulator, replace the subscript notation in Equation (1) as follows:

p	$\rightarrow$	x_b	:position/orientation of the base
q	$\rightarrow$	ϕ	:joint angle of the arm
x	$\rightarrow$	x_h	:position/orientation of the endpoint
$\mathcal{F}_p$	$\rightarrow$	$\mathcal{F}_b$	:forces to deflect the flexible base
$\mathcal{F}_q$	$\rightarrow$	τ	:joint torque of the arm
$\mathcal{F}_{ex}$	$\rightarrow$	$\mathcal{F}_h$	:external wrench on the endpoint

Then we obtain the following equations:

$$\begin{bmatrix} \boldsymbol{H}_b & \boldsymbol{H}_{bm} \\ \boldsymbol{H}_{bm}^T & \boldsymbol{H}_m \end{bmatrix} \begin{bmatrix} \ddot{\boldsymbol{x}}_b \\ \ddot{\boldsymbol{\phi}} \end{bmatrix} + \begin{bmatrix} \boldsymbol{c}_b \\ \boldsymbol{c}_m \end{bmatrix} + \begin{bmatrix} \boldsymbol{g}_b \\ \boldsymbol{g}_m \end{bmatrix} = \begin{bmatrix} \mathcal{F}_b \\ \boldsymbol{\tau} \end{bmatrix} + \begin{bmatrix} \boldsymbol{J}_b^T \\ \boldsymbol{J}_m^T \end{bmatrix} \mathcal{F}_h \tag{26}$$

$$\dot{\boldsymbol{x}}_h = \boldsymbol{J}_m \dot{\boldsymbol{\phi}} + \boldsymbol{J}_b \dot{\boldsymbol{x}}_b \tag{27}$$

$$\ddot{\boldsymbol{x}}_h = \boldsymbol{J}_m \ddot{\boldsymbol{\phi}} + \dot{\boldsymbol{J}}_m \dot{\boldsymbol{\phi}} + \boldsymbol{J}_b \ddot{\boldsymbol{x}}_b + \dot{\boldsymbol{J}}_b \dot{\boldsymbol{x}}_b \tag{28}$$

The difference from the previous two classes of under-actuated systems is the existence of the base constraint force $\mathcal{F}_b$. Let $\boldsymbol{D}_b, \boldsymbol{K}_b$ be damping and spring factors of the flexible-base. Then, the constraint force $\mathcal{F}_b$ is expressed as:

$$\mathcal{F}_b = -\boldsymbol{D}_b \dot{\boldsymbol{x}}_b - \boldsymbol{K}_b \Delta \boldsymbol{x}_b \tag{29}$$

where $\Delta \boldsymbol{x}_b$ denotes elastic base displacement from the equilibrium position.

The upper set of Equation (26) is rearranged as:

$$\boldsymbol{H}_b \ddot{\boldsymbol{x}}_b + \boldsymbol{D}_b \dot{\boldsymbol{x}}_b + \boldsymbol{K}_b \Delta \boldsymbol{x}_b = -\boldsymbol{g}_b - \mathcal{F}_m + \boldsymbol{J}_b^T \mathcal{F}_h \tag{30}$$

where

$$\mathcal{F}_m = \boldsymbol{H}_{bm} \ddot{\boldsymbol{\phi}} + \dot{\boldsymbol{H}}_{bm} \dot{\boldsymbol{\phi}} \tag{31}$$

is the dynamic reaction force induced by the manipulator. Equations (26) and (30) are familiar expressions for flexible-base manipulators [14] [15].

The constraint dynamic equation (30) is not integrable, i.e. it imposes a non-holonomic constraint, generally. But let us consider some special cases when we can find an integral.

Suppose that $\mathcal{F}_h = \boldsymbol{O}$ and the gravity $\boldsymbol{g}_b = \boldsymbol{k}_1$ is constant. Then Equation (30) is expressed as:

$$\mathcal{F}_b + \mathcal{F}_m + \boldsymbol{k}_1 = \boldsymbol{O}. \tag{32}$$

Here we also assume that $\mathcal{F}_b$ does not contain $\boldsymbol{\phi}$ and has a constant value, and that $\mathcal{F}_m$ does not contain $\boldsymbol{x}_b$ and has a constant value. Under such condition, Equation (32) can be integrated to yield

$$\mathcal{V}(t) + \mathcal{L}_m + \boldsymbol{k}_1 t + \boldsymbol{k}_2 = \boldsymbol{O} \tag{33}$$

where $\mathcal{V}(t)$ is a function to express the vibration around a dynamic equilibrium given by Equation (32). Also,

$$\mathcal{L}_m = \boldsymbol{H}_{bm} \dot{\boldsymbol{\phi}} = \ell_1 t + \ell_2 \tag{34}$$

is the momentum of the manipulator part, which is termed the *coupling momentum* [16]. If there are enough degrees of freedom in the manipulator, i.e. not less than the number of passive coordinates, $n \geq 6$ for this case, we can find manipulator motions to satisfy the above integrable condition.

A reduced form of the dynamic equation is derived as:

$$\widehat{\boldsymbol{H}} \ddot{\boldsymbol{\phi}} + \widehat{\boldsymbol{c}} + \widehat{\boldsymbol{g}} = \boldsymbol{\tau} + \widehat{\boldsymbol{J}}^T \mathcal{F}_h + \boldsymbol{R} \mathcal{F}_b \tag{35}$$

where $\boldsymbol{R} = \boldsymbol{H}_{bm}^T \boldsymbol{H}_b^{-1}$. This will be useful for force control with explicit measurement of both $\mathcal{F}_h$ and $\mathcal{F}_b$ wrenches.

3.4 Flexible-Arm Manipulator

Finally, consider a manipulator system composed by flexible links. A flexible link has an infinite number of vibrational DOF (modes) actually, but it can be approximately modeled as a successive chain of a finite number of virtual elastic joints [17][18]. The formulation has exactly the same structure as other under-actuated systems. This is apparent by replacing the subscript notation in Equation (1) as follows:

p	$\rightarrow$	e	:elastic deflection of the flexible links
q	$\rightarrow$	ϕ	:active joint angle
x	$\rightarrow$	x_h	:position/orientation of the endpoint
$\mathcal{F}_p$	$\rightarrow$	$\mathcal{F}_e$	:forces to deflect the flexible links
$\mathcal{F}_q$	$\rightarrow$	τ	:joint torque
$\mathcal{F}_{ex}$	$\rightarrow$	$\mathcal{F}_h$	:external wrench on the endpoint

Then we obtain the following equations:

$$\begin{bmatrix} \boldsymbol{H}_e & \boldsymbol{H}_{em} \\ \boldsymbol{H}_{em}^T & \boldsymbol{H}_m \end{bmatrix} \begin{bmatrix} \ddot{\mathbf{e}} \\ \ddot{\boldsymbol{\phi}} \end{bmatrix} + \begin{bmatrix} \boldsymbol{c}_e \\ \boldsymbol{c}_m \end{bmatrix} + \begin{bmatrix} \boldsymbol{g}_e \\ \boldsymbol{g}_m \end{bmatrix} = \begin{bmatrix} \mathcal{F}_e \\ \boldsymbol{\tau} \end{bmatrix} + \begin{bmatrix} \boldsymbol{J}_e^T \\ \boldsymbol{J}_m^T \end{bmatrix} \mathcal{F}_h \tag{36}$$

$$\dot{\boldsymbol{x}}_h = \boldsymbol{J}_m \dot{\boldsymbol{\phi}} + \boldsymbol{J}_e \dot{\mathbf{e}} \tag{37}$$

$$\ddot{\boldsymbol{x}}_h = \boldsymbol{J}_m\ddot{\boldsymbol{\phi}} + \dot{\boldsymbol{J}}_m\dot{\boldsymbol{\phi}} + \boldsymbol{J}_e\ddot{\boldsymbol{e}} + \dot{\boldsymbol{J}}_e\dot{\boldsymbol{e}} \tag{38}$$

The force of elastic deflection is generally expressed with stiffness and damping matrices as:

$$\mathcal{F}_e = -\boldsymbol{D}_e\dot{\boldsymbol{e}} - \boldsymbol{K}_e\Delta\boldsymbol{e} \tag{39}$$

where $\Delta\boldsymbol{e}$ denotes the elastic displacement from the equilibrium position.

Here we can have the same discussion on integrability as for the flexible-base manipulator. But the difference is that the number of elastic coordinates is generally larger than the number of manipulator joints $(m > n)$. Hence, for any given end-effector path, we cannot find a manipulator motion which would satisfy the integrability condition. This non-holonomic nature renders the inverse path tracking control problem difficult.

4 Common Concepts

In the above sections, we showed that the four types of under-actuated systems have a common structure in the equations of motion, and their differences are found in the integrability conditions. In this section, we focus on some concepts which were originally proposed and developed for one class of under-actuated systems, but are applicable to any other class as well. More specifically, we will discuss the measure of dynamic coupling, kinematic compensability, and the reaction null-space concept.

4.1 Measure of Dynamic Coupling

Dynamic coupling between the active coordinate and the passive coordinate is evaluated in some papers [19][20]. Torres [19] derived a simple expression for the displacement of the flexible beam induced by the manipulator reaction under a quasi-statically condition as:

$$\delta\boldsymbol{x}_b = -\boldsymbol{H}_b^{-1}\boldsymbol{H}_{bm}\delta\boldsymbol{\phi}, \tag{40}$$

which lead to the *Coupling Map* of flexible-base manipulators.

Following this expression, and using the general notation of Equation (1), we obtain:

$$\delta\boldsymbol{p} = -\boldsymbol{H}_p^{-1}\boldsymbol{H}_{pq}\delta\boldsymbol{q}. \tag{41}$$

The matrix product $\boldsymbol{H}_p^{-1}\boldsymbol{H}_{pq}$ appears in Equations (7)-(11) to represent the dynamic coupling property. For example, when the expression $\boldsymbol{H}_p^{-1}\boldsymbol{H}_{pq}$ appearing in (7) and (8) is a null matrix (e.g. due to specific inertia and mass properties), then the generalized Jacobian matrix and the generalized inertia tensor coincide with the Jacobian and inertia tensor of a "conventional" manipulator. As the eigenvalues of $\boldsymbol{H}_p^{-1}\boldsymbol{H}_{pq}$ get larger, the dynamic coupling effects become significant. It would be appropriate to have a measure of such coupling. We propose the following one:

$$I_d = \sqrt{\det\left(\boldsymbol{H}_p^{-1}\boldsymbol{H}_{pq}\boldsymbol{H}_{pq}^T\boldsymbol{H}_p^{-T}\right)} \tag{42}$$

4.2 Kinematic Compensability

The idea of compensability is originally proposed for a flexible-arm manipulator [21] and for a system composed by flexible-macro and rigid-micro manipulators [22]. Kinematic compensability can be defined via Equation (37), when the elastic deflections are slightly perturbed from their equilibrium position:

$$\delta\boldsymbol{x}_h = \boldsymbol{J}_m\delta\boldsymbol{\phi} + \boldsymbol{J}_e\delta\boldsymbol{e}. \tag{43}$$

When the elastic deflection $\delta\boldsymbol{e}$ is totally compensated by the joint action $\delta\boldsymbol{\phi}$, we can get $\delta\boldsymbol{x}_h = 0$. The amount of compensation is

$$\delta\boldsymbol{\phi} = -\boldsymbol{J}_m^+\boldsymbol{J}_e\delta\boldsymbol{e} \tag{44}$$

Such a compensation is possible, only when

$$\text{range}(\boldsymbol{J}_m) \supseteq \text{range}(\boldsymbol{J}_e). \tag{45}$$

The degree of compensability can be evaluated through the measure [18]:

$$I_k = \frac{1}{\sqrt{\det\left(\boldsymbol{J}_m^+\boldsymbol{J}_e\boldsymbol{J}_e^T\boldsymbol{J}_m^{+T}\right)}}. \tag{46}$$

The above equations are applicable to any type of under-actuated system, by replacing the subscript notation as:

$$\boldsymbol{e} \to \boldsymbol{p}, \quad \boldsymbol{\phi} \to \boldsymbol{q}, \quad \boldsymbol{x}_h \to \boldsymbol{x}$$

$$\boldsymbol{J}_e \to \boldsymbol{J}_p, \quad \boldsymbol{J}_m \to \boldsymbol{J}_q$$

For example, the application to a flexible-base manipulator is straightforward, considering the situation that the deformation of the flexible-beam is compensated by the joint displacement.

For manipulators with free-joints or a floating-base, an interesting interpretation is obtained in terms of the so-called *manipulator inversion* [23]. The manipulator can be operated to change the angles of the free-joints (or the base), keeping the endpoint displacement $\delta \boldsymbol{x} = 0$.

Note that the fixed-attitude-restricted dexterity proposed in [23] and expressed in our general notation as:

$$I_{FAR} = \sqrt{\det \left[\boldsymbol{J}_q (\boldsymbol{E} - \boldsymbol{H}_{pq}^{+} \boldsymbol{H}_{pq}) \boldsymbol{J}_q^T \right]} \quad (47)$$

is a closely related concept. $\boldsymbol{E} \in R^{n \times n}$ is the identity matrix. This index is derived through a kinematic relationship excluding the passive coordinates:

$$\dot{\boldsymbol{x}} = \boldsymbol{J}_q (\boldsymbol{E} - \boldsymbol{H}_{pq}^{+} \boldsymbol{H}_{pq}) \dot{\boldsymbol{q}}. \quad (48)$$

4.3 Reaction Null-Space

The reaction null-space is a useful idea to discuss the coupling and decoupling of dynamic interaction between the active coordinates and the passive coordinates. This concept has its roots in the earlier work on free-floating space manipulator by Nenchev et al, where the Fixed-Attitude-Restricted (FAR) Jacobian has been proposed as means to plan [23] and control [24] manipulator motion that does not disturb the attitude of the free-floating base. Application of the reaction null-space with relation to impact dynamics can be found in [13]. In a recent study [6] the authors emphasized the fact that the reaction null-space concept is general, and can be applied to a broad class of moving base manipulators.

Now let us recall the definition of the reaction null-space in terms of our general notation. In the system described by Equations (1), or (4) and (5), let us consider the case when the external wrench at endpoint $\mathcal{F}_{ex} = \boldsymbol{O}$, say free-space motion of the manipulator. without any contact with the environment. If the constraint equation (4) is integrable and the coupling momentum $\mathcal{L}_m$ is defined with a constant value, then the manipulator reaction $\mathcal{F}_m = \boldsymbol{O}$ and no reaction force or torque is induced on the passive coordinates. In case the number of DOF of the active coordinates is larger than that of the passive ones $n > m$, the solution for the manipulator operation to satisfy $\bar{\mathcal{L}}_m = const$ is given by:

$$\dot{\boldsymbol{q}} = \boldsymbol{H}_{pq}^{+} \bar{\mathcal{L}}_m + (\boldsymbol{E} - \boldsymbol{H}_{pq}^{+} \boldsymbol{H}_{pq}) \dot{\boldsymbol{\xi}} \quad (49)$$

where $\dot{\boldsymbol{\xi}} \in R^n$ is an arbitrary vector.

The projector $(\boldsymbol{E} - \boldsymbol{H}_{pq}^{+} \boldsymbol{H}_{pq})$ maps vectors onto the null space of the inertia matrix $\boldsymbol{H}_{pq}$. This is an inertial null space, which we termed *reaction null-space* .

In the special case of zero initial condition ($\bar{\mathcal{L}}_m = \boldsymbol{O}$), we obtain:

$$\dot{\boldsymbol{q}}_{ns} = (\boldsymbol{E} - \boldsymbol{H}_{pq}^{+} \boldsymbol{H}_{pq}) \dot{\boldsymbol{\xi}}. \quad (50)$$

As long as we operate the manipulator joints using the joint velocities given by (50), no reaction force or torque is generated on the base, therefore no reactive motion or vibration is observed in the passive coordinates. The integration of (50) yields *reactionless paths* which are joint space trajectories not exciting the motion of the passive coordinates.

On the other hand, the first term in Equation (49) suggests maximum interaction with the base, which is due to the properties of the pseudoinverse. This maximum interaction characteristics can be used, for example, to effectively damp out the motion of the passive coordinates. Using the measurement of the passive displacement $\Delta \boldsymbol{p}$ as a feedback signal and G as a gain matrix, we have a simple, but effective damping control law:

$$\dot{\boldsymbol{q}}_v = -G \boldsymbol{H}_{pq}^{+} \Delta \boldsymbol{p} \quad (51)$$

The above control space is perpendicular to the reaction null-space. Therefore these two operations (50) and (51) can be easily superimposed without interfering each other, just by simple addition.

$$\dot{\boldsymbol{q}}_c = \dot{\boldsymbol{q}}_v + \dot{\boldsymbol{q}}_{ns}$$

$$\dot{\boldsymbol{q}}_c = -G \boldsymbol{H}_{pq}^{+} \Delta \boldsymbol{p} + (\boldsymbol{E} - \boldsymbol{H}_{pq}^{+} \boldsymbol{H}_{pq}) \dot{\boldsymbol{\xi}} \quad (52)$$

The availability and effectiveness of the reaction null-space has been successfully demonstrated by simulations and experiments, as shown in the following section. For a planar 2R flexible-base manipulator, $n = 2$ and $m = 1$ then it is relatively easy to obtain the reaction null-space solutions, yielding the reactionless paths. On the other hand, a spatial flexible-base or floating-base manipulator is characterized with $m = 6$ in general. But the number of passive coordinates can be artificially restricted, for example, considering base position only, or base attitude only (i.e. $m = 3$), then it would be possible to make use of the reaction null-space concept and the reactionless paths. Such a possibility exists also without the artificial restriction, but with a kinematically redundant manipulator (i.e. $n > 6$). On the other hand, for flexible-arm manipulators we have $m > n$ in general and the reaction null space does not exist. Only when the number of elastic coordinates (or vibration modes) is limited and less than the number of active joints, then it would be possible to use the reaction null space.

5 Experiments and Simulations of Reactionless Operation

One of the remarkable outcomes from the discussion in this paper is that there exists a unique operation to yield zero reaction to the passive coordinates in the under-actuated manipulator systems. In this section, we shall highlight such an operation, called *reactionless operation*, evidenced by experiments and numerical simulations.

5.1 Planar 2R Experiments

A planar experimental setup of a flexible-base manipulator, called TREP, is developed in Tohoku University [25]. The setup consists of a small 2R rigid link manipulator attached to the free end of a flexible double beam representing an elastic base (see **Figure 4**). The manipulator is driven by DC servomotors with velocity command input.

Figure 4 Photo of the experimental setup TREP

The TREP setup is modeled according to **Figure 5**. Since the elastic base has been designed as a double beam, the torsion of the beam can be neglected as a disturbance. This is also the case with the reaction force component along the longitudinal axis of the base. Thus, we shall consider just the reaction force along the so-called low stiffness direction, which coincides with the x axis of the elastic-base coordinate frame. This means that $m = 1$. Since the manipulator has two motors ($n = 2$), the reaction null-space is one-dimensional, meaning that there is one nonzero vector in the reaction null space. Recalling the

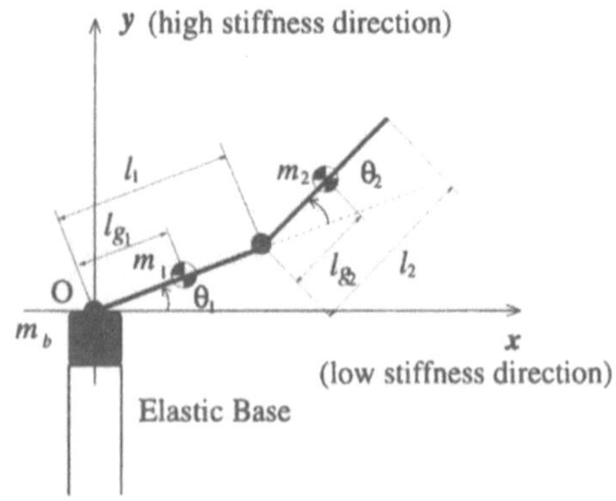

Figure 5 The model of TREP

definition of the coupling momentum (34), we obtain as equation for our experimental setup.

$$\mathcal{L}_m = \boldsymbol{h}_{bm} \begin{bmatrix} \dot{\theta}_1 \\ \dot{\theta}_2 \end{bmatrix} \tag{53}$$

where $\boldsymbol{h}_{bm} = \begin{bmatrix} h_{bm1} & h_{bm2} \end{bmatrix}$ denotes the coupling matrix, with

$$h_{bm1} = -(m_1 l_{g1} + m_2 l_1) sin(\theta_1) - m_2 l_{g2} \sin(\theta_1 + \theta_2),$$

$$h_{bm2} = -m_2 l_{g2} \sin(\theta_1 + \theta_2).$$

The reaction null space vector becomes then

$$\boldsymbol{n} = \begin{bmatrix} h_{bm2} & -h_{bm1} \end{bmatrix}^T. \tag{54}$$

A zero initial coupling momentum will be conserved with any joint velocity parallel to the reaction null space vector. This vector induces a one-dimensional distribution in joint space, which, with well-conditioned coupling, is integrable. Consequently, the set of reactionless paths of the system can be obtained. This set is displayed in **Figure 6**.

We have conducted experiments of reactionless operation by tracking a reactionless path. **Figure 7** shows the experimental results, obtained with the initial configuration $\theta_1 = \theta_2 = 0$ where the arm was extended and aligned with the flexible base. The trace of the end-point is depicted in the upper part of the figure. This operation is generated by Equation (50) with the corresponding coupling inertia shown above. The (scalar) velocity on the path was determined from the variable $\dot{\xi}$, which was designed as a fifth-order spline function of time. In order to verify the possibility for an arbitrary choice of $\dot{\xi}$, we performed the motion on the same path twice, with different velocities

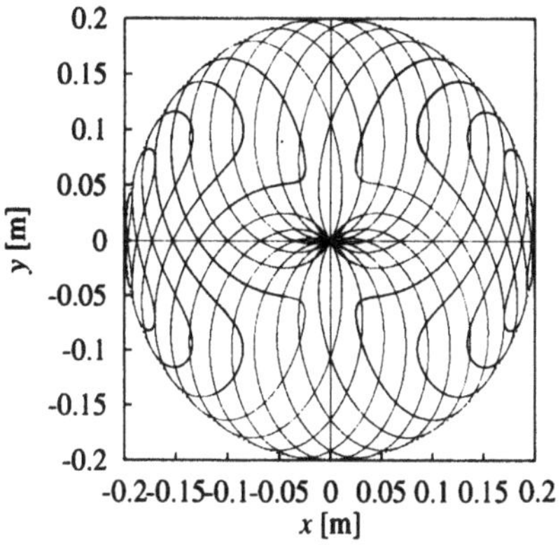

Figure 6 Reactionless paths of TREP

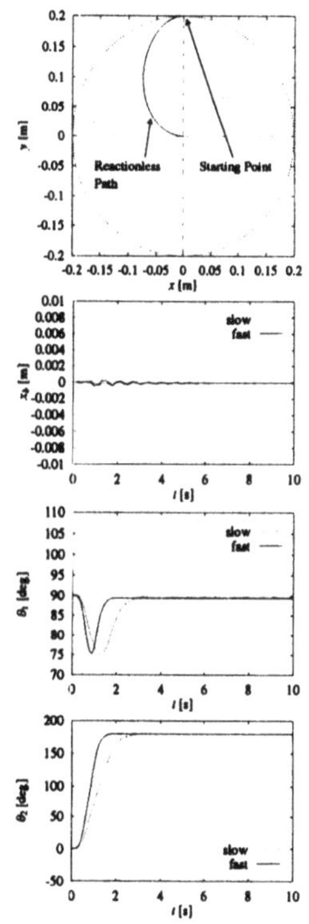

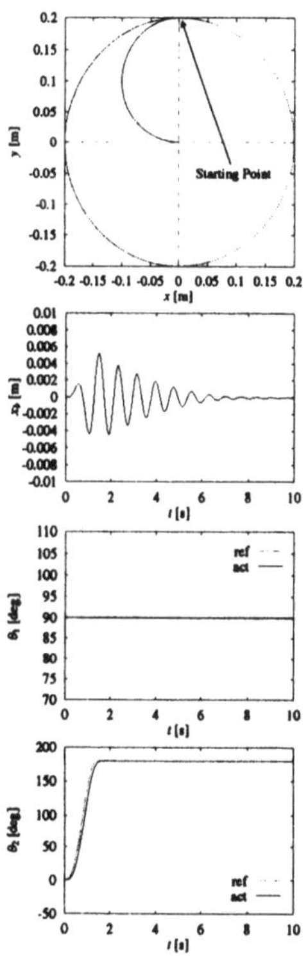

Figure 7 Reactionless operation

Figure 8 Point-to-point operation

(called fast and slow). The lower three graphs in the same figure show these two results. It is seen that almost no base vibration is excited in both cases, and in spite of the significant difference in the joint velocity.

For comparison, **Figure 8** shows a point-to-point motion path (in fact, just joint two rotation) with the same boundary conditions as in the previous motion. The base vibrates significantly.

5.2 Simulation of a 7DOF Flexible-Base Manipulator

We shall expand the application target to realistic 3 dimensional systems. One of the examples is a flexible-base manipulator, composed by a 7DOF rigid manipulator and a flexible structure in arbitrary shape. Such an example is found in manipulators used on a space station, e.g. the Space Station Remote Manipulator System (SSRMS) topped with Special Purpose Dextrous Manipulator (SPDM). The SSRMS, a macro arm, is a long-reach manipulator used for the relocation of the SPDM, a micro arm. The SSRMS plays the role of a supporting base for the micro arm when performing fine manipulation tasks. Note, however, that a space long-reach arm is usually designed as a light-weight arm. Thus, it comprises inherent flexibility which implies that vibrations will be excited due to the reaction from the micro arm. These vibrations are significant and degrade the operational accuracy and efficiency.

In order to avoid the excitation of vibrations, the end-point of the macro arm, i.e. the base of the micro arm should yield zero reactions in all 6DOF directions (3DOF translation and 3DOF orientation.) Such an operation will be possible if the arm has more than 6DOF and the reaction null-space control is applied.

We developed a dynamic simulation model of a 7DOF manipulator arm mounted on an L-shaped flexible structure (see **Figure 2**.) In the simulation, we use joint torque control (not velocity control) to realize the reactionless operation. The control torque is computed by substituting Equation (50) into the lower set of Equation (26), to obtain:

$$\boldsymbol{\tau} = \tilde{\boldsymbol{G}}_f \dot{\boldsymbol{x}}_b + \tilde{\boldsymbol{c}} + \boldsymbol{H}_m \boldsymbol{n} \qquad (55)$$

where

$$\tilde{\boldsymbol{G}}_f = -\boldsymbol{H}_m \boldsymbol{H}_{bm}^{+} \boldsymbol{G}_f,$$

$$\tilde{\boldsymbol{c}} = \boldsymbol{c}_m - \boldsymbol{H}_m \boldsymbol{H}_{bm}^{+} \boldsymbol{c}_b,$$

$$\boldsymbol{n} = (\boldsymbol{E} - \boldsymbol{H}_{bm}^{+} \boldsymbol{H}_{bm}) \boldsymbol{\xi}.$$

$\boldsymbol{G}_f$ denotes a constant feedback gain matrix and $\boldsymbol{n} \in R^n$ is an arbitrary vector from the reaction null space. This control law guarantees that system damping is achieved, resulting in zero vibrations or vibration suppression if any [16]. Here we assume zero gravity and no external forces at the manipulator end-point.

Figure 9 depicts one of the reactionless operations obtained by Equation (55). The upper graph shows joint operation of the arm. The lower shows resulting vibrations in the base structure, in this case no vibrations at all. During the operation, joint 7 makes several turns because it has relatively small moment of inertia. For comparison, **Figure 10** depicts a PD feedback operation to yield the same initial and final joint angles as **Figure 9,** except for joint 7. Since this operation doesn't take care of the reaction, the base vibrates significantly.

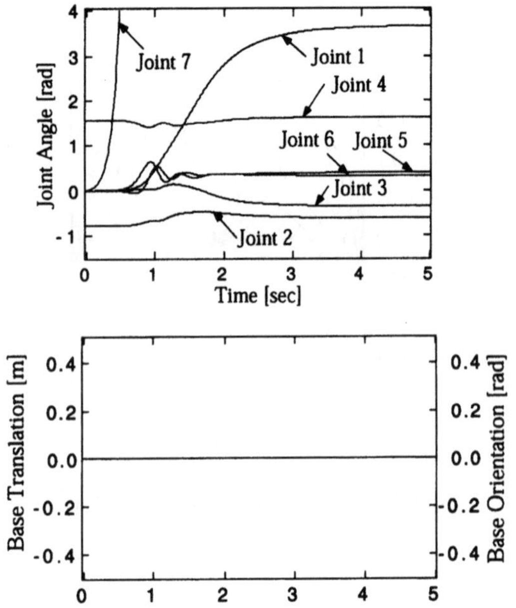

Figure 9 Reactionless operation for 7DOF flexible-base manipulator

5.3 Simulation of a 6DOF Floating-Base Manipulator

A typical example of a practical floating-base manipulator is the Experimental Test Satellite-VII (ETS-VII), developed and successfully launched by the National Space Development Agency of Japan, NASDA. The satellite carries a 2 meter-long 6DOF manipulator arm on a 2,500 kilogram satellite (see **Figure 1**.) 6DOF is not enough to have zero reaction in all directions at the base. However, for a satellite flying in orbit, translational deviation is less significant than attitude deviation. For the attitude deviation, if the satellite turns by 0.5 degrees, the communication link for manipulator operation will be disabled. We therefore pay attention to 3DOF attitude reaction of the base only, thus obtain a 3DOF reaction null-space with a conventional 6DOF manipulator.

We applied a control law similar to that in Equation (55), and obtained reactionless operation. **Figure 11** depicts the results. The upper graph shows joint operation of the arm. The lower shows the resulting base motion, in this case, almost zero attitude deviation while the translation is non-zero. For comparison, **Figure 12** depicts a conventional operation to yield the same initial and final joint angles as in **Figure 11.** In this case, the attitude deviation of the base becomes too significant to maintain the communication link.

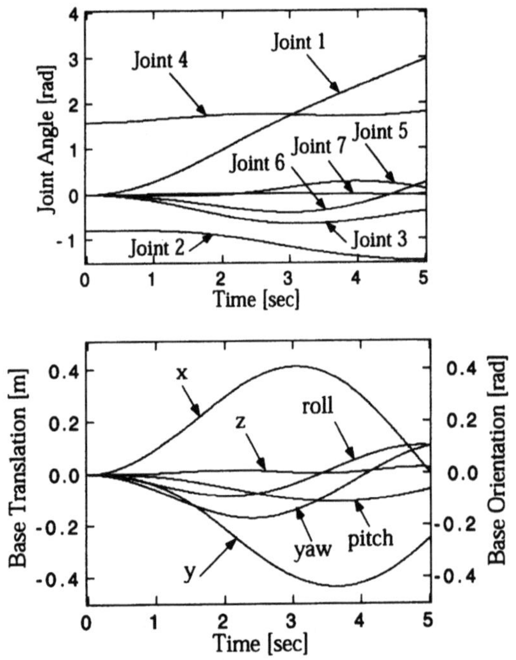

Figure 10 Conventional operation in 7DOF flexible-base manipulator

6 Conclusion

This paper discussed a general formulation both in kinematics and dynamics for under-actuated manipulator systems, represented by a free-joint manipulator, a free-floating manipulator, a flexible-base manipulator and a flexible-arm manipulator. We highlighted the fact that a concept originally developed for a particular system is applicable to the other systems as well, because the dynamic equations show common characteristics. Examples of such concepts are the generalized Jacobian matrix and the generalized inertia tensor which were originally developed for a free-floating manipulator. Another example is the reaction null-space developed for free-floating and flexible-base manipulators. We also emphasized

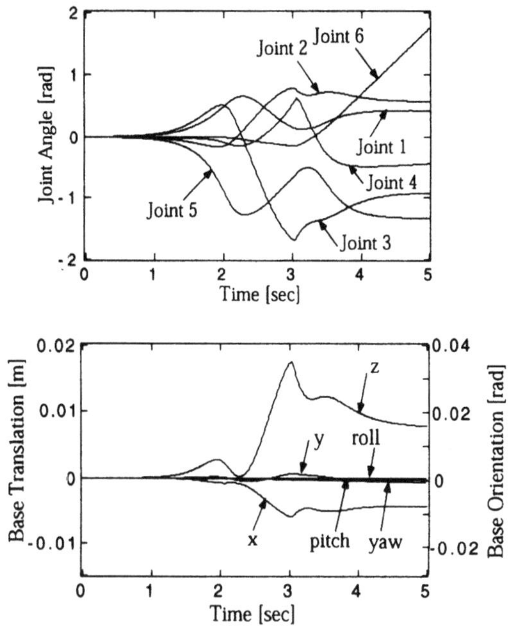

Figure 11 Reactionless operation for 6DOF floating-base manipulator

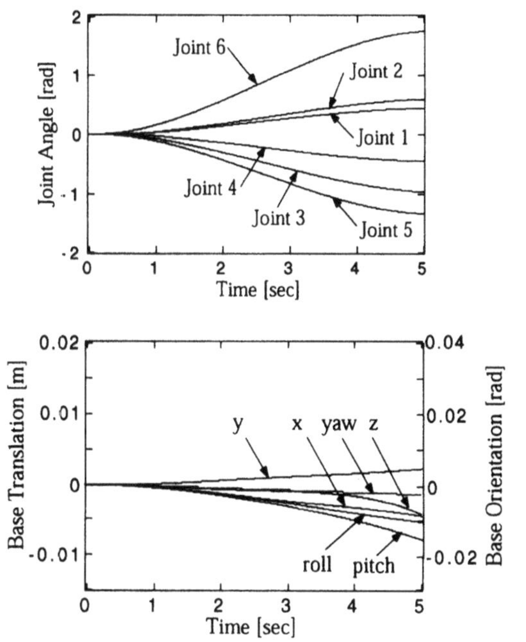

Figure 12 Conventional operation in 6DOF floating-base manipulator

the fact that differences among the four systems come from integrability conditions.

As one of the remarkable outcomes of this paper, the reactionless operation is emphasized by experiments and numerical simulations for flexible-base and floating-base manipulators. Such operations might be useful in space robotic missions, where the structural vibrations and the base attitude disturbance due to manipulator reaction will significantly degrade the quality and efficiency of the missions.

From the point of view of common dynamic characteristics, this paper suggested some new applications, including (a) force controls using Equation (7) for every manipulator, (b) inverse kinematic operations based on the generalized Jacobian (14) for free-joint, flexible-base and flexible-arm manipulators, and (c) tasks using the reaction null-space for free-joint and flexible-arm manipulators in case of $n > m$; these three problems are left for future study.

Acknowledgment

The authors acknowledge the contribution of Mr. Koichi Fujishima, a graduate student of Tohoku University, for his considerable effort in computer simulations, used in Section 5.

References

[1] A. Jain and G. Rodriguez, "An Analysis of the Kinematics and Dynamics of Underactuated Manipulators," *IEEE Trans. on Robotics and Automation*, vol. 9, no. 4, pp. 411–422, 1993.

[2] R. M. Murray, "Nonlinear control of mechanical systems: a Lagrangian perspective," in *Proc. 1995 IFAC Symp. Nonlinear Control Systems Design,* Lake Tahoe, CA, 1995.

[3] J. Ostrowski et al., "Nonholonomic mechanics and locomotion: The snakeboard example," in *Proc. 1994 IEEE Int. Conf. Robotics and Automation,* San Diego, CA, May 1994, pp. 2391–2397.

[4] J. F. Jansen et al., "Long-reach manipulation for waste storage tank remediation," DSC-Vol. 31, ASME, pp. 67–73, 1991.

[5] C. Mavroidis, S. Dubowsky, and V. Raju, "End-point control of long reach manipulator systems," in *Proc. IFToMM 9th World Congress*, Milan, Italy, 1995, pp. 1740–1744.

[6] K. Yoshida, D. N. Nenchev and M. Uchiyama, "Moving base robotics and reaction management control," in *Robotics Research: The Seventh International Symposium*, Ed. by G. Giralt and G. Hirzinger, Springer Verlag, pp. 100–109, 1996.

[7] Y. Nakamura and R. Mukherjee, "Nonholonomic Path Planning of Space Robot via a Bidirectional Approach," *IEEE Trans. Robotics and Automation,* vol. 7, no. 4, pp. 500–514, 1991.

[8] G. Oriolo and Y. Nakamura, "Free-Joint Manipulators: Motion Control under Second-order Nonholonomic Constraints," in *Proc. IEEE/RSJ Int. Workshop on Intelligent Robots and Systems '91,* pp.1248–1253, 1991.

[9] Y. Umetani and K. Yoshida, "Continuous Path Control of Space Manipulators Mounted on OMV," *Acta Astronautica,* vol.15, no.12, pp.981–986, 1987. (Presented at the 37th IAF Conf, Oct. 1986)

[10] Y. Umetani and K. Yoshida, "Resolved Motion Rate Control of Space Manipulators with Generalized Jacobian Matrix," *IEEE Trans. on Robotics and Automation,* vol.5, no.3, pp.303–314, 1989.

[11] *Space Robotics: Dynamics and Control,* ed. by Xu and Kanade, Kluwer Academic Publishers, 1993.

[12] K. Yoshida, "Impact dynamics representation and control with Extended Inversed Inertia Tensor for space manipulators," *Robotics Research: The Sixth International Symposium*, ed. by T.Kanade and R.Paul, pp. 453–463, 1994.

[13] K. Yoshida and D.N. Nenchev, "Space Robot Impact Analysis and Satellite-Base Impulse Minimization using Reaction Null Space," in *Proc. 1995 IEEE Int. Conf. Robotics and Automation*, Nagoya, Japan, May 21–27 1995, pp. 1271–1277.

[14] W.J. Book and S.H. Lee, "Vibration control of a large flexible manipulator by a small robotic arm," in *Proc. American Control Conf.*, Pittsburgh, PA, vol.2, pp. 1377–1380, 1989.

[15] I. Sharf, "Active Damping of a Large Flexible Manipulator with a Short-Reach Robot," in *Proc. American Control Conf.*, Seattle, WA, pp. 3329–3333, 1995.

[16] D.N. Nenchev, K. Yoshida, and M. Uchiyama, "Reaction Null-Space Based Control of Flexible Structure Mounted Manipulator Systems," in *Proc. IEEE 35th Conf. on Decision and Control,* pp. 4118–4123, 1996.

[17] T. Yoshikawa and K. Hosoda: "Modeling of Flexible Mnaipulator Using Virtual Rigid Link and Passive Joints," in *Proc. IROS'91*, 1991, pp. 967–972.

[18] M. Uchiyama and A. Konno, "Modeling, Controllability and Vibration Suppression of 3D Flexible Robots," in *Robotics Research: The Seventh International Symposium*, Ed. by G. Giralt and G. Hirzinger, Springer Verlag, pp. 90–99, 1996.

[19] M.A. Torres, "Modelling, path-planning and control of space manipulators: the coupling map concept," Ph.D. Thesis, MIT, 1993.

[20] Yangsheng Xu, "The Measure of Dynamic Coupling of Space Robot Systems," in *Proc. 1993 IEEE Int. Conf. Robotics and Automation*, 1993, pp. 615–620.

[21] M. Uchiyama, Z. H. Jiang: "Compensability of End-Effector Position Errors for Flexible Robot Manipulators," in *Proc. 1991 Amerian Control Conf.*, Boston, MA, June 1991, pp. 1873–1878.

[22] T. Yoshikawa, K. Hosoda, T. Doi, and H. Murakami: "Quasi-Static Trajectory Tracking Control of Flexible Mnaipulator by Macro-Micro System," in *Proc. 1993 IEEE Int. Conf. Robotics and Automation*, 1993, pp. 3-210–215.

[23] D.N. Nenchev, Y. Umetani, and K. Yoshida, "Analysis of a Redundant Free-Flying Spacecraft/Manipulator System," *IEEE Trans. on Robotics and Automation*, Vol. 8, No. 1, pp. 1–6, Febr. 1992.

[24] D.N. Nenchev, "A Controller for a Redundant Free Flying Space Robot with Spacecraft Attitude/Manipulator Motion Coordination," in *Proc.1993 IEEE/RSJ Int. Conf. Intelligent Robots and Systems (IROS'93)*, Yokohama, Japan, July 1993, pp. 2108–2114.

[25] D. N. Nenchev, K. Yoshida, P. Vichitkulsawat, A. Konno and M. Uchiyama, "Experiments on reaction null-space based decoupled control of a flexible structure mounted manipulator system," in Proc. 1997 IEEE Int. Conf. Robotics and Automation, Albuquerque, New Mexico, April 21-27, 1997, pp. 2528 – 2534.

Towards Precision Robotic Maneuvering, Survey, and Manipulation in Unstructured Undersea Environments

Louis Whitcomb*, Dana Yoerger†, Hanumant Singh, David Mindell‡§

Abstract

This paper reports recent advances in the precision control of underwater robotic vehicles for survey and manipulation missions. A new underwater vehicle navigation and control system employing a new commercially available 1,200 kHz doppler sonar is reported. Comparative experimental trials compare the performance of the new system to conventional 12 kHz and 300 kHz long baseline (LBL) acoustic navigation systems. The results demonstrate a hybrid system incorporating both doppler and LBL to provide superior tracking in comparison to doppler or LBL alone.

1 Introduction

Our goal is to develop new sensing and control systems for underwater vehicles with superior precision, reliability, and practical utility. While the analytical and experimental development of undersea robotic vehicle tracking controllers is rapidly developing, e.g. [14, 7, 5, 4, 12, 6], few experimental implementations have been reported other than for heading, altitude, depth, or attitude control. Conspicuously rare are experimental results for X-Y control of vehicles in the horizontal plane. This lacuna is a result of the comparative ease with which depth, altitude, heading, and attitude are instrumented in comparison to X-Y horizontal position. Precision vehicle position sensing is an often overlooked and essential element of precision control of underwater robotic vehicles. It is impossible, for example, to precisely control a vehicle to within 0.1 meter tracking error when its position sensor is precise only to 1.0 meter. This paper reports the design, implementation, and field-evaluation of a new navigation system for underwater vehicles. The new system utilized a bottom-lock doppler sonar system to provide order-of-magnitude improvements in the precision and update rates of vehicle position sensing and, in consequence, superior closed-loop vehicle positioning performance.

1.1 *Position Sensing for Underwater Vehicles*

At present, few techniques exist for reliable three-dimensional navigation of underwater vehicles. Table 1 summarizes the sensors most commonly used to measure a vehicle's six degree-of-freedom position. While depth, altitude, heading, and attitude are instrumented with high bandwidth internal sensors, X-Y position sensing is usually achieved by acoustically interrogating fixed seafloor-mounted transponder beacons [9]. Ultra-short baseline acoustic navigation systems are preferred for the task of docking a vehicle to a transponder-equipped docking station but are of limited usefulness for general long-range navigation [10]. Inertial navigation systems offer excellent strap-down navigation capabilities, exhibiting position errors that accumulate as a function of both time and distance travled. Their high cost has, however, generally precluded their widespread use in oceanographic instruments and vehicles. The U.S. spon-

*Whitcomb is with the Department of Mechanical Engineering, Johns Hopkins University, 123 Latrobe Hall, 3400 North Charles Street, Baltimore, Maryland, 21218 USA, email: llw@jhu.edu.

†Yoerger and Singh are with the Deep Submergence Laboratory, Department of Applied Ocean Physics and Engineering, Woods Hole Oceanographic Institution, Woods Hole, MA, 02543, USA, email: dyoerger@whoi.edu, hsingh@whoi.edu.

‡Mindell is with the MIT Program in Science, Technology, and Society, E51-194A, 77 Massachusetts Avenue, Cambridge, MA 02139 email: mindell@mit.edu.

§We gratefully acknowledge the support of the Office of Naval Research for the 1997 JASON field experiments under Grant #N00014-95-J-1237 and the National Science Foundation for operational support of the Deep Submergence Operations Group (DSOG) under Grant #OCE-9627160. This paper is Woods Hole Oceanographic Institution Contibution #9614.

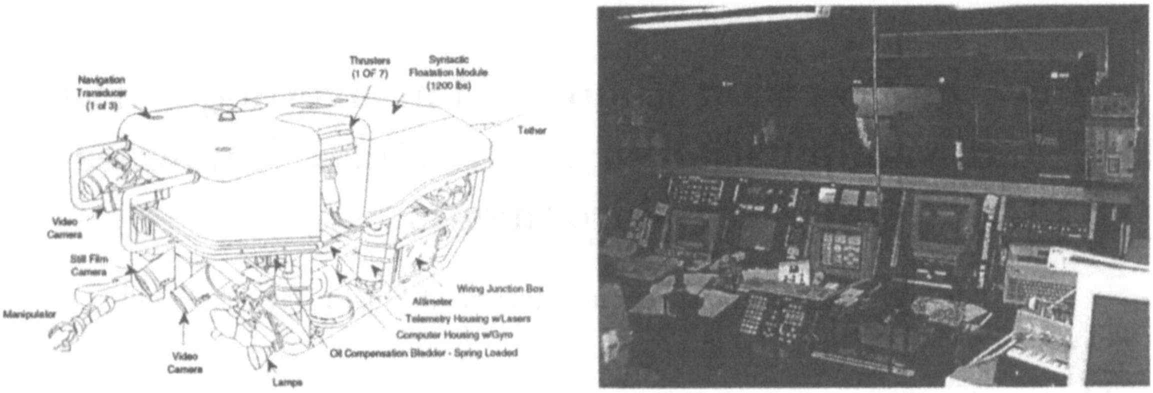

Figure 1: JASON, a 1200 Kg 6000 meter remotely operated underwater robot vehicle used in these experiments. Jason (left) is remotely operated from a control room (right) aboard the mother ship.

INSTRUMENT	VARIABLE	INTERNAL?	UPDATE RATE	RESOLUTION	RANGE
Acoustic Altimeter	Z - Altitude	yes	varies: 0.1-10Hz	0.01-1.0 m	varies
Pressure Sensor	Z - Depth	yes	medium: 1Hz	0.01-1.0 Meter	full-ocean
12 kHz LBL	XYZ - Position	NO	varies: 0.1-1.0 Hz	0.01-10 m	5-10 Km
300 kHz LBL	XYZ - Position	NO	varies: 1.0-5.0 Hz	+/-0.002 m typ	100 m
Mag Compass	Heading	yes	medium: 1-2Hz	$1 - 10^\circ$	360°
Gyro Compass	Heading	yes	fast: 1-10Hz	0.1°	360°
Inclinometer	Roll and Pitch	yes	fast: 1-10Hz	0.1° - 1°	$+/- 45^\circ$

Table 1: Commonly Used Underwater Vehicle Navigation Sensors

sored Global Positioning System (GPS) provides superior three-dimensional navigation capability for both surface and air vehicles, and is employed by all U.S. oceanographic research surface vessels. The GPS system's radio-frequency signals are blocked by seawater, however, thus GPS signals cannot be directly received by deeply submerged ocean vehicles.

Two problems with existing sensors severely limit the performance of fine maneuvering: *precision* and *update rate*. On-board depth, heading, and attitude sensors generally offer excellent precision and update rates. XY position, however, is generally instrumented acoustically and, over longer ranges, offers poor precision and low update rates. The standard method for full ocean depth XYZ acoustic navigation is 12 kHz long baseline (12 kHz LBL) acoustic navigation. 12 kHz LBL typically operates at up to 10 Km ranges with a range-dependent precision of +/- 0.1 to 10 Meters and update rates periods as long as 10 seconds (0.1 Hz) [9]. Although recent work suggests that the next generation of acoustic communication networks might provide position estimation [3, 10], no systems providing this capability are commercially available at present. At present, the best method for obtaining sub-centimeter precision acoustic XY subsea position sensing is to employ a high-frequency (300 kHz or greater) LBL system. Unfortunately, due to the rapid attenuation of higher frequency sound in water, high frequency LBL systems typically have a very limited maximum range. In addition to the standard long-range 12 kHz LBL system, in these experiments we employed a short-range 300 kHz LBL system called "Exact" (developed by the two of the authors) with a maximum range of about 100m. All absolute acoustic navigation methods, however, require careful placement of fixed transponders (i.e. fixed on the sea-floor, on the hull of a surface ship [9], or on sea-ice [2]) and are fundamentally limited by the speed of sound in water — about 1500 Meters/Second.

Our goal is to improve vehicle dynamic navigation precision and update rate by at least one order of magnitude over LBL and, in consequence, improve vehicle control. In the context of 1000Kg underwater robot vehicles, which typically exhibit limit cycles on the order of 0.1-1.0 meters, the goal is to provide position control with a precision of 0.01 meters. To achieve this requires vehicle navigation sensors precise to at least 0.005 meter, and an update rate of several Hz.

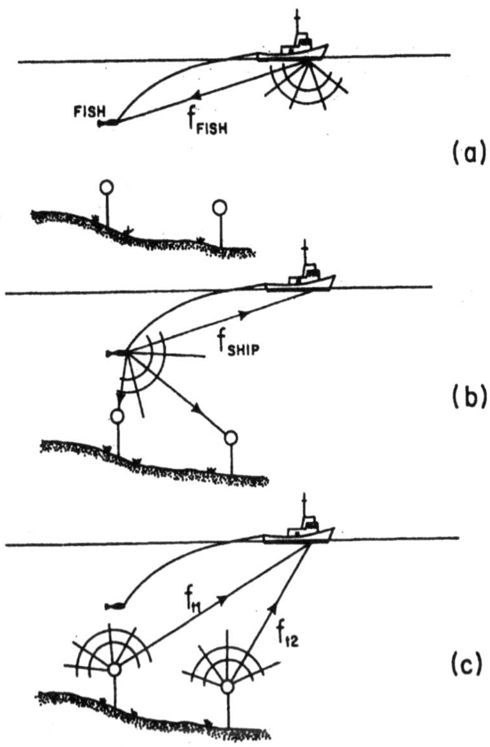

Figure 2: Long Baseline Navigation. This figure depicts typical LBL navigation cycle for determining an underwater vehicle's position.

1.2 *Review of Long Baseline Navigation*

Since its development over 30 years ago long baseline navigation (LBL) has become the *de-facto* standard technique for 3-dimensional acoustic navigation for full-ocean depth oceanographic instruments and vehicles [8].

LBL operates on the principle that the straight-line distance between two points in the ocean can be measured by the time-of-flight of an acoustic signal propagating between the two points. All LBL systems require an unobstructed line-of-sight between transmitting and receiving transducers and, as mentioned above, have an effective range that varies with frequency.

Figure 2 depicts a typical oceanographic deployment of a 8-12 kHz LBL system for navigating an underwater vehicle[1]. A typical LBL system is deployed and operated from the surface vessel as follows:

1. *Transponder Deployment:* Two or more acoustic transponders are dropped over the side of the surface ship at locations selected to optimize the acoustic range and geometry of planned subsea operations. Each transponder is a complete sub-surface mooring comprised of an anchor, a tether, and a buoyant battery-powered acoustic transponder. The tether's length determines the transponder's altitude above the sea-floor. Depending on range, local terrain, depth, and other factors, tether length might be chosen between 5 and 500 meters. The simplest transponders are designed to listen for acoustic interrogation "pings" on a specified frequency (e.g. 9 kHz), and to respond to each interrogation with a reply ping on a specified frequency (e.g. 10 kHz). It is common (but not universal) to set an entire network of transponders to listen on a single frequency, and to set each transponder to respond on a unique frequency.

2. *Sound-Velocity Profile:* An instrument is lowered from the surface ship to measure and tabulate the velocity of sound at various depths the water column. Sound velocity typically varies significantly with depth, and all subsequent computations use this sound velocity profile to compensate for the effects of variation in sound velocity.

3. *Transponder Survey:* The XYZ position of the sea-floor transponders is determined by maneuvering surface ship around each transponder location while simultaneously (*i*) acoustically interrogating the transponder and recording the round-trip acoustic travel time between the ship's transducer and the sea-floor transponder and (*ii*) recording the ship's GPS position, compass heading, and velocity. This data is processed to compute least-square estimate of the world-referenced XYZ position of each fixed sea-floor transponder. When using a full-precision P-Code GPS, the transponder's position can typically be estimated with a precision of just a few meters.

4. *Acoustic Navigation of Surface Ship Position:* First, the ship's acoustic signal processing computer transmits an interrogation ping via the ship's LBL transducer on a common

[1]A variety of LBL systems are commercially available. Vendors include Benthos Inc., 49 Edgerton Drive, North Falmouth, MA 02556 USA, phone: 508-563-1000, fax: 508-563-6444, http://www.benthos.com.

interrogation frequency, say 9.0 kHz. Second, each of the fixed sea-floor transponders replies with a ping on a unique frequency that is received by the ship's LBL transducer. The ship's computer measures the round-trip travel acoustic travel time between the ship's transducer and to two (or more) sea-floor transponders. Finally, the ship's computer computes the absolute ship position using (*i*) two or more measured round-trip travel times, (*ii*) the known depth of the ship's transducer, (*iii*) the surveyed XYZ position of the sea-floor transponders, and (*iv*) the measured sound-velocity profile.

5. *Acoustic Navigation of Underwater Vehicle Position:* Two general approaches are commonly employed for acoustic navigation of underwater vehicle position.

   The first general approach, often called "in-hull navigation", is used by an underwater vehicle to determine its own position without reference to a surface ship. The sequence is nearly identical to the surface ship navigation sequence described above, with the vehicle's actual time-varying depth (using a precision pressure-depth sensor) in place of the ship's constant transducer depth.

   A second general approach is used to determine the position of an underwater vehicle (or instrument) from the surface ship. This approach is depicted in Figure 2. First, the ship's acoustic signal processing computer transmits an interrogation ping via the ship's LBL transducer on special interrogation frequency, say 8.5 kHz (Figure 2.a). Second, the underwater vehicle's transponder responds to the ship's interrogation by generating a ping on a secondary interrogation frequency, say 9.0 kHz (Figure 2.b). Third, each of the fixed sea-floor transponders replies to the secondary interrogation by generating a ping on a unique frequency that is received by the ship's LBL transducer (Figure 2.c). The ship's computer measures (*a*) the direct round-trip travel acoustic travel time between the ship's transducer and the vehicle and (*b*) the indirect round-trip travel time from ship to vehicle to transponder to ship for two (or more) sea-floor transponders. Finally, the ship's computer computes the absolute ship position using (*i*) the measured round-trip travel times, (*ii*) the known depth of the ship's transducer, (*iii*) the surveyed XYZ position of the sea-floor transponders, and (*iv*) the measured sound-velocity profile. In the case of tethered underwater robot vehicles, the known depth of the vehicle is often used in the position computation.

6. *Transponder Recovery:* Most sea-floor acoustic transponders are equipped with an acoustically triggered device which releases the mooring tether in response to a coded acoustic release signal, thus allowing the transponder to float freely to the surface for recovery. In most oceanographic deployments the transponders are triggered, released, and recovered at the conclusion of operations.

The above description is typical for 8-12 kHz LBL systems in deep water where ranges may vary from about 1 to 10 Km. The details of deployments may vary when in shallow water, when operating over very short ranges, and when using high frequency LBL systems (100 kHz-1,000 kHz), but the essential steps of transponder placement, calibration, and operation remain invariant. As discussed previously, the precision and update rate of position fixes can vary over several orders of magnitude depending on the acoustic frequency, range, and acoustic path geometry. LBL navigation accuracy and precision can be improved to some extent by careful application of Kalman filtering techniques [1]. Figure 3 shows raw vehicle position fixes obtained simultaneously using a long-range 12 kHz LBL navigation system and a short-range 300 kHz LBL system.

2 Doppler-Based Navigation and Control

This section reports the design and experimental evaluation of a control system employing a new 6000 meter depth rated 1200 kHz bottom-lock doppler sonar[2]. to augment the standard vehicle navigation suite. The new doppler sonar precisely measures the UUV's velocity with respect to the fixed sea-floor. This promises to dramatically improve the vehicle navigation capabilities in two ways: First, use of the doppler velocity sensing in the vehicle control system will overcome the "weak link" of conventional velocity estimation

[2]The Workhorse 1,200 kHz doppler sonar was developed and is manufactured by RD Instruments Inc, 9855 Businesspark Ave., San Diego, CA 92131-1101 phone: 619-693-1178, web: http://www.rdinstruments.com.

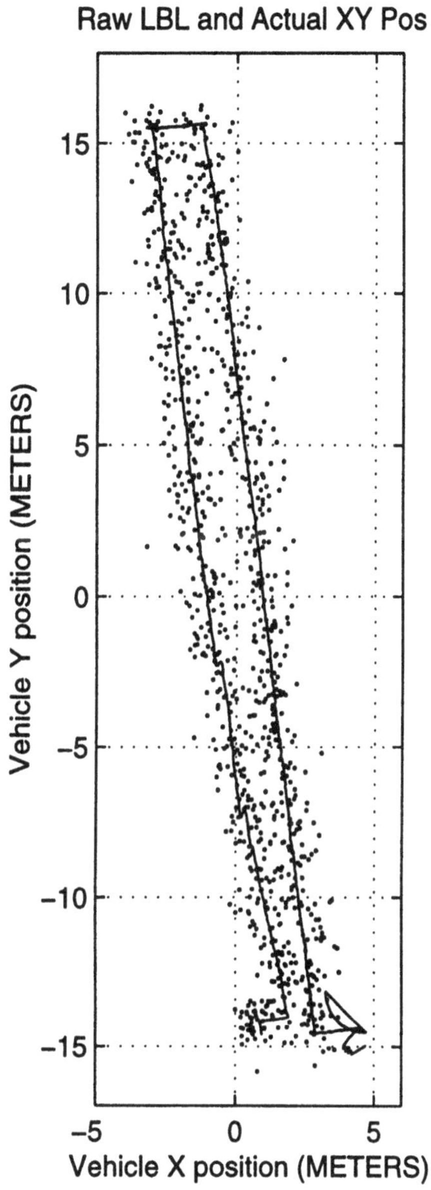

Figure 3: X-Y Plot of 12 kHz LBL Jason position fixes (dot cloud) fixes and 300 kHz LBL X-Y fixes (solid line). Data collected during a closed-loop sea-floor survey at approximately 850 meters depth, Jason dive 222, 24 June 1997.

techniques, and result in improved precision maneuvering. Second, by numerically integrating the vehicle velocity, the vehicle will for the first time be able to "dead reckon" in absence of external navigation transponders. This will enable missions in unstructured environments that were previously considered infeasible such as precision station-keeping and tracking; high-precision survey; improved terrain following; and combined vehicle-manipulator tasks such as sample gathering while precisely "hovering" at a site of interest.

2.1 System Design: A multi-mode vehicle navigation and control system

The new navigation system is configured by the pilot to operate in one of five modes detailed in Table 2. All of the control modes employ the same closed-loop control algorithms for vehicle heading and depth. The five control modes differ only in the type of control and sensing employed for the vehicle X-Y position.

Mode 1 employs manual X-Y positioning while Modes 2-5 employ closed-loop X-Y positioning. In all cases the vehicle heading position is instrumented by a heading gyroscope, and the vehicle depth is instrumented by a pressure depth sensor. The the X-Y control for the five modes are as follows:

2.1.1 *Mode 1: Manual X-Y*

X-Y is controlled manually, as the precision and update rate of 12 kHz LBL are insufficient to support closed-loop control. The pilot observes full real-time navigation data (including graphical bottom-track) and live video, and controls the vehicle X-Y thruster forces directly via joystick control. Mode 1 is the standard control mode employed in virtually all commercial remotely operated underwater vehicle (ROV) systems in which vehicle heading and depth are closed-loop controlled, while X-Y position is manually controlled.

2.1.2 *Mode 2: Closed Loop X-Y with 300 kHz LBL*

X-Y position is under PD control using a 300 kHz LBL transponder navigation system for X-Y state feedback. This acoustic navigation system provides sub-centimeter precision vehicle positions over 100 meter ranges with update rates of 1 to 5 Hz [13].

2.1.3 *Mode 3: Closed Loop X-Y with 1200 kHz Doppler*

X-Y position is under PD control using a 1200 kHz bottom-lock doppler sonar for X-Y state feedback. The vehicle referenced velocities are transformed to world coordinates using an on-board flux-gate heading compass and on-board attitude sensors, and then is integrated to obtain world-referenced vehicle position.

2.1.4 *Mode 4: Closed Loop X-Y with 1200 kHz Doppler and 12 kHz LBL*

X-Y position is under PD control using a combination of the 1200 kHz bottom-lock doppler sonar and 9 kHz LBL transponder navigation system for X-Y state feedback. To take advantage of the incremental precision of the doppler with the absolute (but noisy) precision of the LBL, we implemented a system utilizing complementary linear filters to combine low-passed LBL position fixes with high-passed doppler position fixes. The cutoff frequencies for both filters were set to 0.005 Hz. The Mode 4 system requires no additional fixed sea-floor transponders, in contrast to previously reported LBL+doppler systems which utilize fixed sea-floor mounted continuous-tone beacons [11].

2.1.5 *Mode 5: Mode 4: Closed Loop X-Y with 1200 kHz Doppler and 300 kHz LBL*

X-Y position is under PD control using a combination of the 1200 kHz bottom-lock doppler sonar and 300 kHz LBL transponder navigation system for X-Y state feedback. Here again, we utilized a system of complementary linear filters to combine low-passed LBL position fixes with high-passed doppler position fixes. The cutoff frequencies for both filters were set to 0.1 Hz.

2.2 Feedback Gains and Magic Parameters

In all of these experiments, the velocities used for feedback control are obtained by direct numerical differentiation of the corresponding position signal. All axes are controlled by standard Proportional-Derivative (PD) feedback laws. The feedback gains were tuned by normal pole-placement methods, based on estimated vehicle hydrodynamic parameters, to obtain approximately critically damped response. Identical feedback gains were used in all closed-loop control modes.

2.3 Experiments

This section reports experiments comparing the absolute precision of doppler-based, LBL, and hybrid navigation and control systems. Section 2.3.1 examines the absolute precision of the Mode 4 LBL+doppler navigation system in comparison to Mode 1 LBL-only navigation. Section 2.3.2 examines actual experimental closed loop tracking performance of the five control modes.

2.3.1 *Navigation Performance*

What is the absolute precision of Mode 1 (raw 12 kHz LBL) navigation? This is the de-facto standard technique for long-range 3-D underwater navigation of underwater vehicles. Mode 1 typically provides position fixes at 2-10 second intervals (too slow for closed-loop X-Y control) with precision that varies with network size, water depth, ambient noise, and a variety of other factors. This section reports an experimental evaluation of the absolute precision of both Mode 1 (LBL alone), and and Mode 4 (12 kHz LBL+Doppler) navigation systems. The experiments were conducted in June 1997 during a JASON field deployment at sea in approximately 850 meters depth. Mode 4 was first implemented and tested in these experiments. Our goal was to develop a new navigation system to provide update rate and precision suitable for closed-loop X-Y control, yet requiring no additional navigation sensors external to the vehicle itself. Mode 4 requires only a vehicle-mounted doppler sonar unit to augment the usual 12 kHz LBL navigation transponder system normally employed for Mode 1 deep-ocean navigation.

Figure 3 shows X-Y plot of 983 Jason XY position fixes obtained by Mode 1 12 kHz LBL navigation (dot cloud), and 8,845 highly precise actual Jason X-Y positions obtained by the 300 kHz LBL system (solid line). The geometry of the 12 kHz LBL transponders was nearly optimal in this deployment, yet LBL position errors of up to a meter were typical.

Figure 4 shows histogram plots of the X and Y navigation errors observed during these actual vehicle deployments using Mode 1 and Mode 4 navigation. The X and Y navigation errors under Mode 1 (LBL only) have a standard deviation of 0.50 meters and 0.46 meters, respectively. In con-

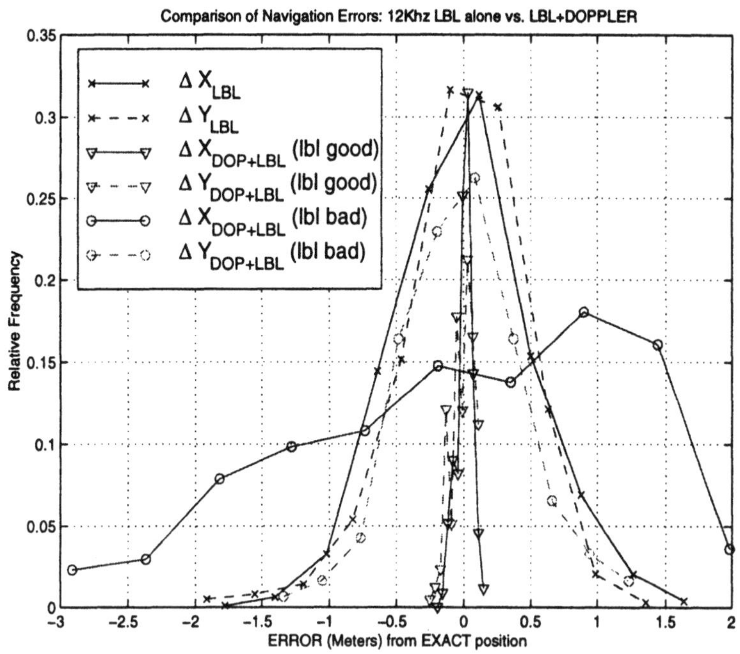

Figure 4: Histogram Plot showing X and Y position sensing errors of 12 kHz LBL showing (a) LBL alone (Jason dive 222), (b) LBL+Doppler with LBL working well (Jason dive 222), and (c) LBL+Doppler with LBL working poorly (Jason dive 219). The 12 kHz navigation errors were computed with respect to the 300 kHz LBL vehicle position.

trast, the X and Y navigation errors under Mode 4 (LBL+doppler) have a standard deviation of 0.06 meters and 0.08 meters, respectively, when the LBL system is receiving good fixes. Thus the Mode 4 (LBL+doppler) system is an order of magnitude more precise than Mode 1 (LBL only). Moreover, the Mode 4 system provides vehicle position fixes every 0.4 seconds, while the the Mode 1 system provides vehicle position fixes every 3.0 seconds — about an order of magnitude improvement.

It is common for a 12 kHz LBL navigation system to suffer from a variety of systematic errors that cause it to give imprecise readings. Typical LBL problems include acoustic multi-path, loss of direct acoustic path, and poor signal-to-noise due to machinery and electro-magnetic noise. As a result, it is typical for LBL systems to occasionally generate "bad fixes" for periods of time ranging from seconds to hours. These bad fixes are characterized by high, non-gaussian errors. The most difficult aspect of the errors is that they are not zero-mean. How do these "bad fixes" effect Mode 4 navigation precision? Figure 4 shows the Mode 4 X and Y navigation errors to be several meters when subject to bad LBL fixes.

We conclude Mode 4, a hybrid of doppler and 12 kHz LBL, provides order-of-magnitude improvement in vehicle navigation precision and update rate over Mode 1, yet requires the deployment of no additional transponders. Good 12 kHz LBL fixes are essential to Mode 4 precision; when LBL precision degrades, Mode 4 precision is proportionally diminished. Moreover, Mode 4 provides both the precision and update rate necessary for precision closed-loop X-Y vehicle control that is not possible with Mode 1.

2.3.2 *The Effect of Navigation Precision on Closed-Loop Positioning*

How do the various navigation modes effect underwater vehicle tracking performance? To answer this question we ran five experimental trials — one for each of the five modes described Section 2.1. In each trial, we commanded Jason to follow an X-Y trajectory in the shape of a 5 meter by 5 meter square at approximately 850 meters depth. In the Mode 1 trial the vehicle was under the manual X-Y pilot control. In the Mode 2, 3, 4, and 5 trials, the vehicle was under closed-loop X-Y control. The closed-loop trials all employed identical PD feedback control algorithms for X-Y motion; they differ only in their position sensing technique. In each case we recorded the ***actual*** vehicle position with the sub-centimeter precision

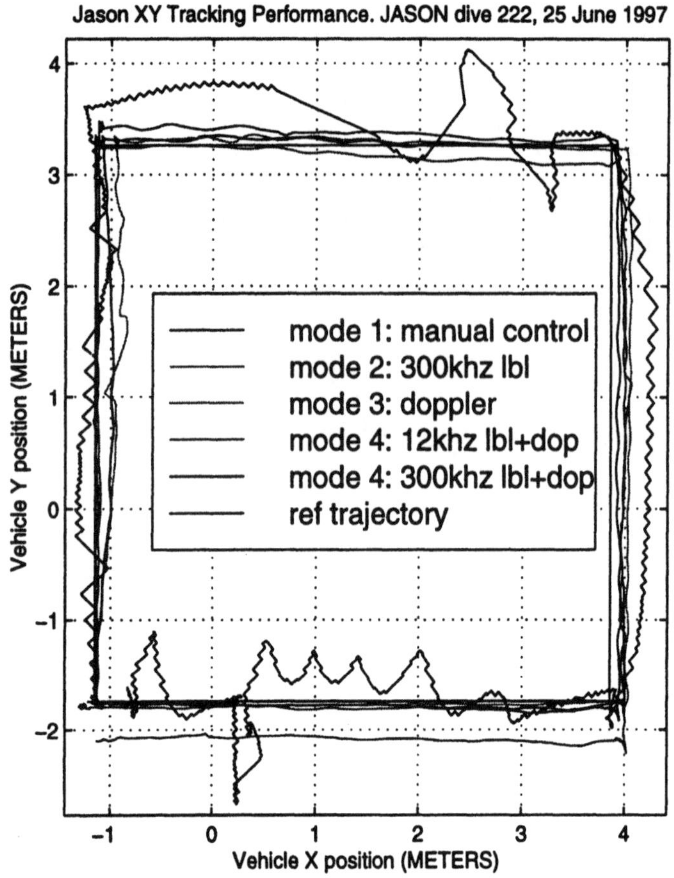

Figure 5: Jason X-Y Tracking Performance: Actual X-Y vehicle trajectories under Mode 1 (manual control) and Modes 2-5 (closed-loop control).

300 kHz LBL transponder navigation system.

Figure 5 shows the reference trajectory, a 5-meter square, and the *actual* Jason X-Y trajectories for each of the five trials. In manual X-Y control, Mode 1, the pilot could typically keep the vehicle within about 1 meter of the desired trackline. This manual tracking performance is typical of the Mode 1 tracking performance we have observed in hundreds of hours of Mode 1 deep-sea robot deployments. In closed-loop Mode 2, the vehicle remains within 0.1-0.2m of the desired trajectory. Here again, this closed-loop tracking performance is typical of the Mode 2 tracking performance we have observed in hundreds of hours of Mode 2 deployments.

Modes 3, 4, and 5 were implemented and tested for the first time on this deployment. Mode 3 (1200 kHz doppler alone) exhibits a tracking error roughly proportional to distance traveled — in this case we see errors up to 0.5 m, or about 2.5% of distance traveled. We observed two principal sources of error for pure-doppler navigation: First, the inherent 1% accuracy of the doppler velocity measurement is integrated directly into accumulated distance errors. To minimize this error it is essential to carefully calibrate the local sound velocity value used in the doppler velocity computation. Second, for longer tracklines (not shown) we observed that small errors in the doppler unit's on-board flux-gate magnetic compass will dramatically increase the accumulated XY position errors. To minimize this error, it is essential to have an absolutely stable earth-referenced heading sensor.

Mode 4 (12 kHz LBL and 1200 kHz doppler) and Mode 5 (300 kHz LBL and 1200 kHz doppler) provide the best tracking performance, with tracking errors within 0.1-0.2 m — commensurate to the performance of Mode 2. As indicated in the previous section, good 12 kHz LBL fixes are essential to Mode 4 precision; when LBL precision degrades and the error distribution becomes skewed, Mode 4 performance is diminished.

MODE	X-Y POSITION SENSING	CLOSED-LOOP XY?	TRACKING ERRORS	COMMENTS
Mode 1	12 kHz LBL	No	Worst	Industry standard. Only standard long-range 12 kHz sea-floor transponders required.
Mode 2	300 kHz LBL	Yes	Best	Requires deployment of additional short-range 300 kHz sea-floor transponders.
Mode 3	Doppler	Yes	Good	Tracking error increases as function of time and distance traveled due to integration errors. No additional sea-floor transponders required.
Mode 4	Doppler + 12 kHz LBL	Yes	Very Good	No additional sea-floor transponders required.
Mode 5	Doppler + 300 kHz LBL	Yes	Best	Requires deployment of additional short-range 300 kHz sea-floor transponders.

Table 2: Performance Summary: Five modes of underwater robot navigation and control.

3 Conclusion

The preliminary results are promising. We conclude Mode 4, a hybrid of doppler and 12 kHz LBL, provides order-of-magnitude improvements in vehicle navigation over Mode 1, yet requires the deployment of no additional transponders. Good 12 kHz LBL fixes are essential to Mode 4 precision; when LBL precision degrades, Mode 4 precision is diminished. Moreover, Mode 4 provides both the precision and update rate necessary for precision closed-loop XY vehicle control that is not possible with Mode 1. The principal error sources for any bottom-referenced doppler position estimation technique are (*i*) sound velocity calibration precision, (*ii*) heading reference precision. A companion paper (in preparation) describes high-precision sonar and optical surveys of sea-floor sites performed using closed-loop vehicle control with combined LBL/doppler navigation on the June 1997 Jason deployment.

We are presently pursuing several questions articulated in the present study: First, to what degree will an improved heading reference (e.g. north-seeking ring-laser gyroscope) improve the doppler XY position estimate? Second, how will variations in sea-floor composition and topography effect the X-Y position precision of doppler-based systems? This will be particularly important in the rough terrain typically found in geologically active seafloor sites. Third, bottom-referenced doppler fails at altitudes greater than about 30 meters (for 1200 kHz) to 100 meters (for 300 kHz). We expect that the techniques employed herein could be extended to use water-column referenced doppler for mid-water closed-loop navigation and control.

Acknowledgements

We gratefully acknowledge the invaluable support of WHOI Deep Submergence Operations Group and Dr. Robert D. Ballard; EM11(SS) Robert Stadel, U.S. Navy, who piloted the manual (Mode 1) Jason tracking trials; Captain Steve Laster, and the officers and crew of the S.S.V. Carolyn Chouest; and Officer-in-Charge LTCDR Charles A. Richard, and the officers and crew of the U.S. Navy Submarine NR-1.

References

[1] B. M. Bell, B. M. Howe, J. A. Mercer, and R. C. Spindel. Nonlinear kalman filtering of long-baseline, short-baseline, gps, and depth measurements. In *Conference Record of the Twenty-Fifth Asilomar Conference on Signals, Systems and Computers*, pages 131–136, Pacific Grove, CA, USA, November 1991.

[2] J. G. Bellingham, M. Deffenbaugh, J. Leonard, and J. Catipovic. Arctic under-ice survey operations. *Unmanned Systems*, 12(1):24–9, 1994.

[3] J. A. Catipovic and L. E. Freitag. High data rate acoustic telemetry for moving rovs in a fading multipath shallow water environment. In *Proceedings of the Symposium*

on Autonomous Underwater Vehicle Technology. AUV '90, pages 296–303, 1990.

[4] S. K. Choi and J. Yuh. Experimental study on a lerning control system with bound estimation for underwater robots. *Proc. IEEE Int. Conf. Robt. Aut.*, pages 2160–2165, April 1996.

[5] R. Cristi, F. A. Papoulis, and A. J. Healey. Adaptive sliding mode control of autonomous underwater vehicles in the dive plane. *IEEE Journal of Oceanic Engineering*, 15(3):152–160, June 1990.

[6] T. I. Fossen. *Guidance and Control of Ocean Vehicles.* John Wiley and Sons, New York, 1994.

[7] K. R. Goheen and E. R. Jeffereys. Multivariable self-tuning autopilots for autonomously and remotly operate underwater vehicles. *IEEE Journal of Oceanic Engineering*, 15(3):144–151, June 1990.

[8] M. M. Hunt, W. M. Marquet, D. A. Moller, K. R. Peal, W. K. Smith, and R. C. Spindell. An acoustic navigation system. Technical Report WHOI-74-6, Woods Hole Oceanographic Institution, Woods Hole, Massachusetts 02543 USA, December 1974.

[9] P. H. Milne. *Underwater Acoustic Positioning Systems.* Spon Ltd., New York, 1983.

[10] H. Singh, J. Catipovic, R. Eastwood, L. Freitag, H. Henricksen, F. Hover, D. Yoerger, J. Bellingham, and B. Moran. An integrated approach to multiple auv communications, navigation and docking. In *Proceedings of the OCEANS 96 MTS/IEEE Conference*, pages 59–64, Fort Lauderdale, FL, USA, September 1996.

[11] R. C. Spindel, R. P. Porer, W. M. Marquet, and J. L. Durham. A high-resolution pulse-doppler underwater acoustic navigation system. *IEEE Journal of Oceanic Engineering*, 1(1):6–13, September 1976.

[12] L. L. Whitcomb and D. R. Yoerger. Preliminary experiments in the model-based dynamic control of marine thrusters. In *Proc. IEEE Int. Conf. on Robotics and Automation*, 1996. (Invited paper).

[13] D. R. Yoerger and D. A. Mindell. Precise navigation and control of an rov at 2200 meters depth. In *Proceedings of Intervention/ROV 92*, San Diego, June 1992. MTS.

[14] D. R. Yoerger and J. E. Slotine. Adaptive sliding control of an experimental underwater vehicle. In *Proc. IEEE Int. Conf. Robt. Aut.*, Sacremento, CA, USA, April 1991.

Where does the Task Frame go?

H. Bruyninckx* J. De Schutter
Dept. of Mech. Eng., K.U.Leuven, Celestijnenlaan 300B, B-3001 Leuven (Belgium)
email: herman.bruyninckx@mech.kuleuven.ac.be

Abstract

This paper discusses the "Task Frame" (TF) as a central concept in (hybrid) robot force control and task specification. The title serves a double purpose: it refers to the desirable ability of a force controller to adapt on-line the motion constraint model on which the control is based, but also to the scientific evolution of the TF concept during the last two decades and its role in future developments.

1 Introduction

About two decades ago, Mason introduced a tool to model and specify "compliant motions," i.e., manipulations executed by a force-controlled robot in contact with its environment(Mason 1981). Mason's tool was called the "Compliance Frame;" "Task Frame" (TF) is a more neutral term, which this paper will use. The TF is attractive because it is an intuitive and manipulator-independent user interface to a robot controlled in the "Hybrid Control Paradigm" (HCP), (Raibert and Craig 1981). Within this HCP, the controller, the task specification module, and the force and motion sensing and interpretation modules, all use the fundamental assumption that the motion constraints imposed by the environment on the robot are *geometric* (or "infinitely stiff"). Theoretically speaking, this means that the constraints can be "integrated" in the configuration space of the robot (joint space, or Cartesian space). This means that some of the configuration variables could be eliminated. In HCP terminology, the robot has a number (say "n") of Cartesian degrees of freedom which are position controlled, while the other 6-n degrees of freedom are force controlled. Mason called both subspaces "orthogonal," an unfortunate terminology which proves hard to extinguish.

*Postdoctoral Fellow of the Fund for Scientific Research–Flanders (F.W.O.) in Belgium

The HCP is one of the two major force control paradigms, together with the Impedance Control Paradigm. The latter assumes ***dynamic*** interaction between robot and environment, instead of the geometric "interaction" model of the HCP. "Force-controlled" and "velocity-controlled" directions do not exist, since all degrees of freedom are controlled in both ways. Or rather, the dynamic relationships ("impedance") between forces and motions are controlled. This text does not aim at comparing both approaches (see e.g., (De Schutter, Bruyninckx, Zhu, and Spong 1997) for more details of the authors' viewpoints), but focusses on the role of the TF as the central modeling, specification, sensor processing and control concept in the HCP. It summarizes the authors' efforts to give the HCP theoretically more sound foundations, as well as to extend its capabilities. The major fundamental problems the HCP has been coping with are outlined in the following paragraphs; (partial) solutions to these problems are described in the rest of the paper.

First of all, no one has ever clearly defined the ***necessary and sufficient conditions*** to have a viable Task Frame motion constraint model and task specification, or has given an algorithm to produce them in a given contact situation. Hence, the current state of the art still requires human intervention and intuition to define and initialize the TF. This is just one of the many problems that hampers the design and implementation of "***intelligent***" force controlled robots (able to act (semi)autonomously in unstructured environments), as well as the introduction of commercial sensor-based robotic systems.

Second, the motion constraint imposed on the robot is not always known in advance, not even approximately. (For example, when tracing unknown surfaces during model building.) Hence, a controller in the HCP must be able to ***detect/adapt*** the geometric constraint model ***on line***.

Third, uncertainties in the geometric constraint

model imply that forces will occur in velocity-controlled directions, as well as motions in force-controlled directions. Hence, *robustness* of the HCP force controller is indispensable.

Fourth, Lipkin and Duffy (1988) and Duffy (1990) made clear that the TF's "orthogonality" concept of the HCP is *not invariant* with respect to changes in (i) a TF's position and/or orientation, (ii) the mathematical representation (choice of coordinates) and/or (iii) physical units. This non-invariance means that the overall behavior of a HCP force controller changes when one or more the above-mentioned changes are performed.

Finally, for *complex* contact situations (having more than one contact on the manipulated object at the same time), a TF compatible with Mason's original intuition is often impossible to find.

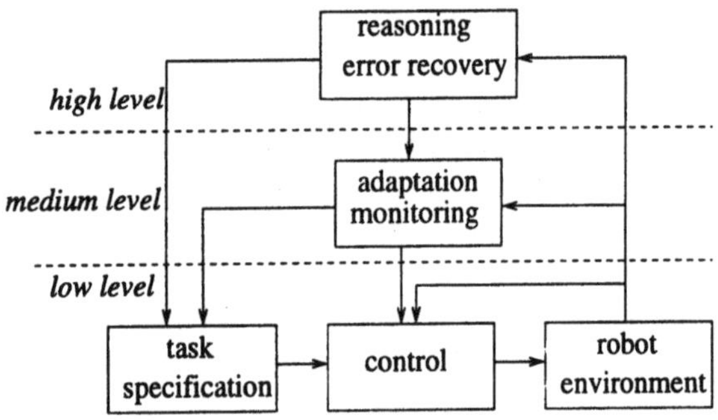

Figure 1: Three control levels for force-controlled compliant motion.

This paper summarizes the authors' research in the field of force-controlled compliant motion, based on the TF and HCP. This research evolved in a "bottom-up" fashion, within the framework sketched in Fig. 1:

1. The lowest level is the pure feedback-based set-point control, as treated in the majority of publications on force control.

2. The medium level uses the force and motion measurements for more than feedback control only, i.e., to adapt the motion constraint (TF) model and to check whether this model can still explain the current measurements ("*monitoring*"). The authors' contributions are mainly situated at this level.

3. Many current research efforts, by the authors as well as by other research groups, is to achieve the highest, "intelligent" level of force control. The major challenge is to reason automatically and reliably about contact situations, in order to recover from lower-level errors and to (re)plan corrective actions.

2 Modeling and specification

The classical goal of the TF is to give an intuitive yet complete *model* of (i) which forces can be generated in the contact between robot and environment, and (ii) which motion degrees of freedom are available (assuming that the constraint is infinitely stiff!). In screw theory terms, the TF gives a *basis* for the twist and wrench spaces associated to the motion constraint, (Lipkin and Duffy 1988). Mason implicitly suggested that this basis can always be chosen to be "minimal," in the sense that (i) forces and motions are either *zero* or *infinite* pitch screws, and (ii) they are represented by *three*-vectors along the orthogonal Task Frame axes. Such minimal bases can only represent a *subset* of all twist or wrench spaces, since twists and wrenches in general require *six* parameters. The user specifies the desired magnitudes of the forces and motions along each of the six TF *directions*; while the TF has only three axes, six "directions" can be defined since each axis of the TF is once used in *polar* form (rotation; moment) and once in *axial* form (translation; linear force).

2.1 Task Frame rules

In previous work, (Bruyninckx and De Schutter 1996), the authors have tried to formalize Mason's original intuitive ideas in the following four TF rules:

1. *Geometric compatibility.* The basis vectors in the force-controlled (velocity-controlled) directions are a basis of the wrench (twist) vector space corresponding to the current contact situation.

2. *Causal compatibility.* The TF motion specification should correspond to the goal the user wants to achieve during the task. And this goal is more than just satisfy the geometrical constraints. For example, transitions between two contact situations could be implemented by specifying TF set-points corresponding to the *desired* situation. Or, some of the geometrically available motion freedom should be used to satisfy secondary

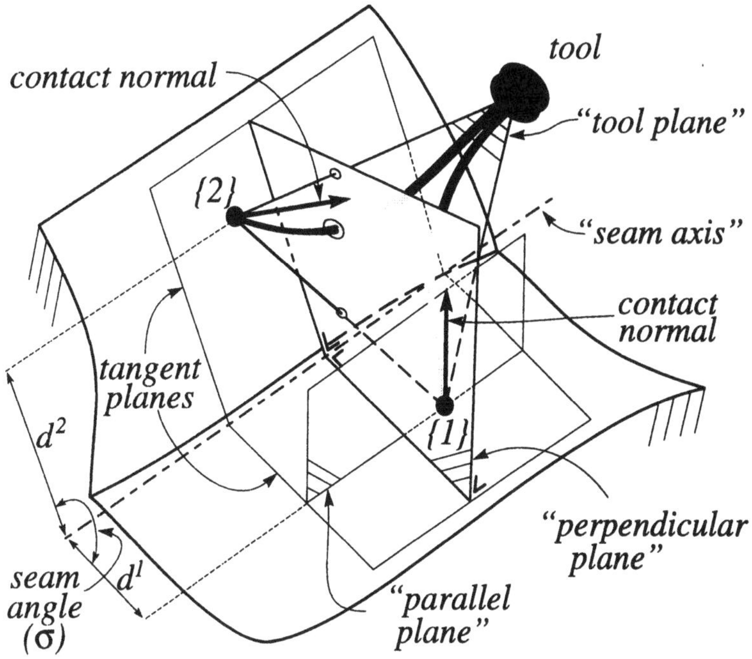

Figure 2: Two-point contact: geometric definitions.

task goals such as obstacle avoidance, actuator optimization, etc.

The major problem in achieving causal compatibility within a TF context is that a TF specification explains *how* the controller wants to reach a goal, but not *what* that goal is. Hence, the controller can only check the compatibility with the specification, not with the goal itself.

3. *Time-invariance.* While remaining geometrically and causally compatible, the TF moves as a virtual rigid body. This means that the constraint model (i.e., the choice of force and velocity controlled TF directions) as well as the motion specification (i.e., the choice of force and velocity set-points) remain unchanged during the task.

4. *Maximal decoupling.* (i) Kinetostatic decoupling: the twist and wrench spaces are spanned by screws with zero or infinite pitch, and can be described by three-vectors along mutually orthogonal axes; (ii) Control decoupling: the action of the force controller on one TF direction does not influence the control error in the other TF directions.

These rules cover the "classical" compliant motion tasks (e.g., peg-in-hole, turning a crank, opening a door, aligning a block with a plane, etc.), but not tasks in slightly more complex contact situations. This section discusses in what respect the above-mentioned TF rules have to be relaxed in the case of two contacts acting in parallel on the same manipulated object.

2.2 Two-point contact

The goal of the task is to maintain the two contacts and to move along the "seam" formed by the intersection of (the tangent planes to) the two contact surfaces, Fig. 2. (Note that only one single wrist force sensor is used.) The two-point contact is probably the simplest contact situation for which it is obvious that no classical TF approach to contact modeling and motion specification exists; it is also not possible to use two *independent* TFs, one at each contact. Section 2.2.1 gives a *geometric* (i.e., coordinate and reference frame independent) model of the two-point contact, taking into account the inevitable coupling between the two contacts in an intuitive yet unambiguous way. Section 2.2.2 gives corresponding motion specification approaches. Both sections try to follow the four TF rules of the previous Section as much as possible.

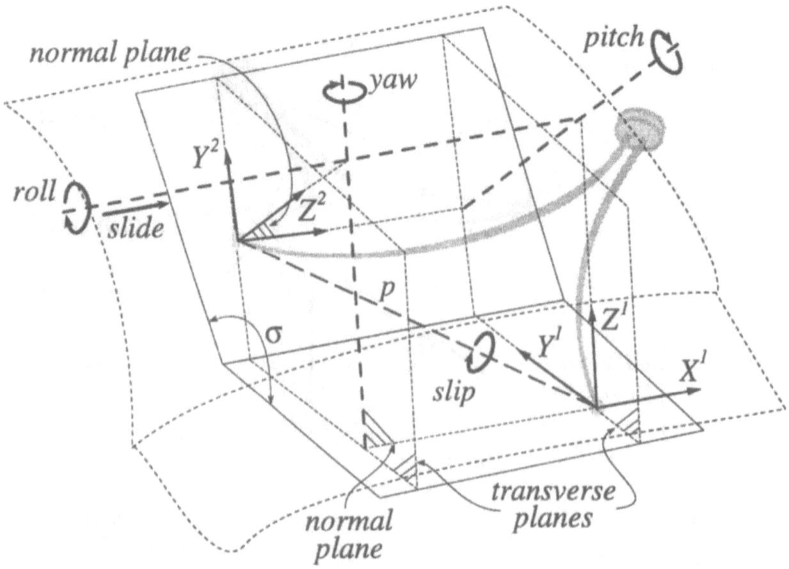

Figure 3: Elementary and coordinate-independent instantaneous motion degrees of freedom in the two-point contact: Roll, pitch, yaw, slip and slide.

2.2.1 *Modeling*

The ***seam axis*** is the intersection of the two tangent planes. A ***normal plane*** is each plane through a contact normal and parallel to the seam axis. A ***transverse plane*** is any plane perpendicular to the seam axis. The ***seam angle*** σ is the (free space) angle between the tangent planes. The contact points lie at distances d^1 and d^2, respectively, from the seam axis. All these parameters can be calculated if the tangent planes (or, equivalently, the normal vectors) in both contacts are known, as well as the vector $\boldsymbol{p}$ linking the two contact points, (Bruyninckx 1995). The ***tool plane*** is the plane through the two contact points and a third user-defined point $\boldsymbol{p}^t$ on the tool. The tool plane has two special coordinate-independent reference positions, one if it lies in a transverse plane, and a second if $d^1 = d^2$. The first situation is called the ***transverse tool*** position, the second one is called the ***symmetric tool*** position. (The latter is defined independently of the choice of the point $\boldsymbol{p}^t$.) Both situations can occur simultaneously. The tool is in a singular position if it is ***parallel*** to the seam, i.e., $d^1 = d^2 = 0$.

2.2.2 *Specification*

A basis of the four-dimensional twist space can be given by rotations about, and translations along, geometrically defined lines, Fig. 3:

1. ***Roll*** is rotation about the intersection of the normal planes through {1} and {2}.
2. ***Pitch*** is rotation about the intersection of the transverse plane in {1} and the normal plane in {2}.
3. ***Yaw*** is the rotation about the intersection of the normal plane in {1} and the transverse plane in {2}.
4. ***Slide*** is translation in the direction of the seam axis.

A rotation about the line between the two contact points does not move these points. This rotation is called ***slip***. It is a linear combination of roll, pitch, yaw and slide. The ***roll-pitch-yaw-slide*** and ***roll-pitch-yaw-slip*** bases are completely and unambiguously determined by the geometry of the environment, and hence independent of any choice of reference frames. But these bases are time-varying, i.e., the relative positions of the axes change during the motion (if the contact surfaces are curved). However, the ***procedure*** to obtain the bases is time-invariant.

The previous paragraphs describe bases allowing to specify unambiguously the instantaneous twist of the tool. However, a human user might like more intuitive ways of specifying this instantaneous twist, or even only the desired position. The following paragraphs describe two possible approaches, a ***local*** one and a ***global*** one.

The local specification approach follows the classical TF intuition: each individual contact gets its own TF, as if it were the only contact occurring on the manipulated tool. These local TFs obey the four TF rules, with the extra causality constraint that in total the user should not specify more than four independent motions, i.e., at least one of the five velocity-controlled "directions" in each local task frame should get a "*don't care*" specification. It is then the controller's job to translate both local specifications in one globally admissable instantaneous twist. This boils down to a ***projection*** onto the twist space of instantaneously available degrees of freedom, Sect. 3. Specification of the contact forces is straightforward: the wrench set-point is the resultant of the two specified ideal contact forces.

The global approach relies on a model of the remaining four motion degrees of freedom, for example ***roll-pitch-yaw-slide*** as described above. The advantage is that, *by construction*, any specified twist will be compatible with the modeled constraint. However, the resulting motion of each individual contact point might be less intuitive than in the local approach. If the user prefers to specify the desired *position* of the tool, instead of the desired instantaneous *velocity*, the *symmetric* and ***transverse tool*** positions are appropriate references for this purpose. One possible specification approach requires the user to define four geometrically defined distances that intuitively but unambiguously describe the desired tool position with respect the symmetric and transverse reference positions. For example, taking the TF in the first contact point as reference:

1. d_1: The distance of the first contact point to the seam axis.

2. p_z^t: The height of the (arbitrarily chosen) tool reference point.

3. p_x^2: The distance (parallel with the seam axis) between the two contact points.

4. The desired position along the seam.

Again, the controller is responsible for transforming these four numbers into a resultant motion that is instantaneously compatible with the contacts.

In summary, a "classical" single TF does not exist for the total motion constraint of the two-point contact (or any other complicated contact situation). The geometrical and causal compatibilities are maintained in both the global and the local approaches; time-invariance and (kinetostatic) decoupling are only maintained locally.

3 Invariant and decoupled hybrid control

This Section discusses some important aspects of the lowest control level in Fig. 1. The seminal papers on the HCP presented its concepts with simple examples only (point-plane contacts, kinematic joints, etc.), in which the choice of TF and task specification are straightforward. Most later papers blindly took this "orthogonality" concept for granted, until the above-mentioned publications by Lipkin and Duffy made clear that not "orthogonality" but "reciprocity" is the real physical concept underlying the HCP: the ideal contact forces produce no work against the allowed instantaneous velocities. Complex contact situations as in the previous section do not have twist and wrench space bases that consist of zero and/or infinite pitch screws only. Hence, no preferred TF reference frame exists, *and* tasks in these complex contact situations are inherently time-varying. This requires an *invariant* description of Hybrid Control, i.e., one that does not depend on a particular choice of reference frame, coordinate representation or physical units.

3.1 Invariant model and specification

Making a contact model and its corresponding motion specification invariant with respect to the above-mentioned changes is conceptually very simple: the coordinate expressions of the bases of the twist and wrench spaces are transformed from one reference frame to another with the well-known transformation formulas for twists and wrenches derived in robotic textbooks. However, these transformations have to be performed consistently. This violates the "basic" HCP custom of working with diagonal "***selection matrices***" with nothing but ones and zeros on the diagonal, see e.g. (Whitney 1987). The solution to this problem is straightforward: replace the selection matrices by what they intrinsically (and implicitly) stand for, i.e., ***projection matrices*** on the twist and wrench space, (De Schutter, Torfs,

Dutré, and Bruyninckx 1997). The transformation of these projection matrices uses the same twist and wrench transformations as mentioned above.

To solve the non-invariance problem under changes of physical units is equally simple: interpret the control gains between, on the one hand, force errors, position errors and/or velocity errors, and, on the other hand, the joint torques of the manipulator, according to their real physical meaning, i.e., as impedances or admittances. These also transform with the above-mentioned twist and wrench transformation formulas.

3.2 Decoupling

Classically, hybrid controllers aim at achieving closed-loop dynamics which are decoupled in the "orthogonal" TF directions. But this orthogonality concept is not invariant, so that it is not used in the context of the extensions discussed in this paper. However, hybrid controllers can achieve decoupled closed-loop dynamics between *any* set of independent basis vectors in the twist and wrench spaces. This is done by (i) choosing suitable basis vectors in those spaces, (ii) projecting the specified and the measured twists and wrenches onto the corresponding spaces, (iii) decomposing the resulting twist and wrench errors along the basis vectors, and (iv) applying decoupled feedback control along each twist or wrench basis vector. Note that:

1. The user not only influences the closed-loop dynamics through the choice of *control gains*, but also through the choice of suitable *basis vectors*. (For example, the classical peg-in-hole task gets different behavior when the decoupled directions are chosen at the top of the peg or at the robot end-point.) These two complementary choices should be made in a decoupled manner: *first* choose suitable bases for twist and wrench spaces, and only *then* design a controller on the *coordinates* in this bases. This has the advantage that the overall control dynamics do not change if it is necessary to express the twist and wrench spaces in other reference frames, or with other coordinate representations, provided that the bases are transformed consistently, as mentioned in the previous Section.

2. The projection operators on the twist and/or wrench spaces cannot be defined unambiguously, by lack of a natural metric on these spaces, see e.g., (Karger and Novak 1985), (Lončarić 1985), (Park 1995). The "weighting" that is implicitly introduced in the projection operation has the physical dimensions of *impedance*, e.g., stiffness or inertia, (De Schutter, Torfs, Dutré, and Bruyninckx 1997).

3.3 Invariance under changing contact impedance

In every contact situation, the interaction between the robot and its environment can be approximated by a linear impedance model; i.e., a mass-damper-spring system. If one knows this contact impedance, or the controller is able to identify it on line, it is worthwhile to include the inverse of this impedance into the force controller. This makes the overall closed-loop dynamics invariant for this (change in) contact impedance. This is one more interesting reason to disect the low-level set-point control into conceptually separated modules (impedance compensation, choice of basis, choice of control gains).

4 On-line adaptation

This Section discusses the authors' contributions to the medium control level in Fig. 1. Once the user has placed the TF (or its geometrically defined generalizations, such as in the two-point contact) in an appropriate position and orientation, and has specified desired motions and forces, the force controller can start executing the task. However, in general, the (basis screws of the) twist and wrench spaces do not remain at the same position and orientation with respect to either the executing robot or the environment. (The seminal papers in the HCP, as well as most new publications, give only examples where the TF *does* remain motionless with respect to one or both references.) Only few people started to develop on-line strategies to keep the TF's position and orientation compatible with the real, time-varying contact situation, e.g., (De Schutter and Leysen 1987) or (Wampler 1984). Moreover, these adaptation strategies have usually been developed for the simplest time-varying contact situations only (e.g., following an unknown planar contour with a point contact). Over the last ten years, the authors developed a fully general theory for this "TF tracking," and implemented many practical cases, (Bruyninckx, Demey, Dutré, and De Schutter 1995), (Bruyninckx and De Schutter 1996). This Section summarizes the main points of this evolution, as well as the newest results and the remaining research challenges.

4.1 Contact model

As discussed before in Section 2, the classical TF can only model a *subset* of all possible motion constraints. Therefore, the authors started to use the "*virtual contact manipulator*" (VCM) idea (see e.g. (West and Asada 1985)): each individual contact is modeled by a virtual kinematic chain between the manipulated object and the environment, giving the manipulated object (instantaneously) the same motion freedom as the contact. The VCM model is an interesting compromise between the intuitiveness and simplicity of the classical TF and the desire to be able to model *all* contact situations: it is easy to make a VCM for any possible contact, even between generally curved surfaces, Fig. 4, using the coordinate-independent directions and radii of minimum and maximum curvature.

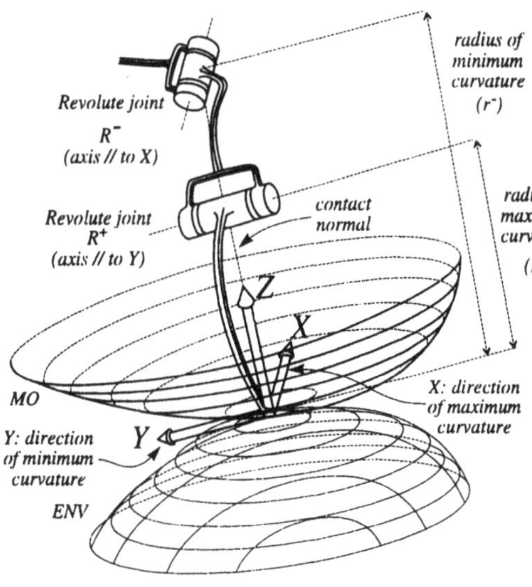

Figure 4: (Half of a) virtual contact manipulator for the contact between two arbitrarily curved surfaces.

The VCM fulfills the first two TF rules defined in Section 2.1; it remains intuitive, but gives up the decoupling and time-invariance properties in favor of generality and flexibility. A VCM as in Fig. 4 allows to model not only the instantaneous twist and wrench spaces ("first order" geometry) but also their nominal (local) *evolution* ("second order" geometry) as a result of the specified motion. The velocity and force kinematics of the parallel manipulator, formed by the virtual manipulators of all individual contacts, are found by procedures that are well known in kinematics of real manipulators; they allow to construct bases for the twist and wrench spaces of the total motion constraint. Note that the VCM for the frequently occurring *polyhedral* contact situations reduces to a classical TF.

4.2 Kalman Filter

When the contact situation evolves differently from what the model predicts, the measured and predicted motion and force will begin to deviate. These small incompatibilities can be measured, and can then be used to adapt the TF model on-line, with classical state-space "prediction–correction" estimators. The *Kalman Filter* is the work horse of *stochastic* on-line state-space estimation, due to its efficient linear structure. The basics of Kalman filtering can be found in textbooks such as, for example, (Bar-Shalom and Li 1993; Gelb 1978) or (Jazwinski 1970). (Removing the stochastic contents from the filter results in the deterministic *Luenberger observer*.) The basic algorithm is conceptually simple: (i) use a model of the system dynamics to predict its state at the next sample instant ("nominal evolution"); (ii) use a model of the relationships between sensor values and system state variables to predict the expected measurement values at the next sample instant (the "measurement" in this case is the value of the reciprocity relationship between measured force (velocity) and modelled velocity (force)); (iii) the difference between the new measurement and the predicted measurement is the new information (*innovation*) at this sample instant; (iv) calculate the new state variables as a *weighted* average of the predicted state and the innovation. The weights are determined by the noise behavior of the system dynamics and the sensors.

The authors applied the Kalman Filter to the estimation of the *geometric uncertainties* in the contact models, i.e., inaccuracies in the estimates of positions of contact points, directions of contact normals, or axes of kinematic joints This application is rather straighforward, except for one very important obstacle: the relationship between the state variables and the force/velocity reciprocity constraint is *non-linear*. The fact of being non-linear in itself is not the real problem, since the Kalman Filter theory is easily extended to work with *linearized* system and/or sensor mappings. The real problem lies in finding the linearizations! Bruyninckx, Demey, Dutré, and De Schutter (1995) and Dutré, Bruyninckx, and De Schutter (1997) give the theoretical solution to this problem for any possible complex contact situation. The major breakthrough was the use of the VCM as the general contact modeling tool:

- The state variables in the Kalman Filter are the uncertainties in the joint angles of the VCM.
- The measurement equations are simple functions of the VCM's Jacobian matrices.
- Linearization of these measurement equations boils down to calculating the derivatives of Jacobian matrices for serial and parallel manipulators.

Although maybe not apparent at first sight, the TF concept is not far away in this approach to on-line estimation and adaptation of contact uncertainties, since again each individual contact can be given its own VCM, i.e., its own simple TF extension. Once more, the intuitiveness of the user interface is maintained, while the tricky bookkeeping details are automated and hidden in the controller.

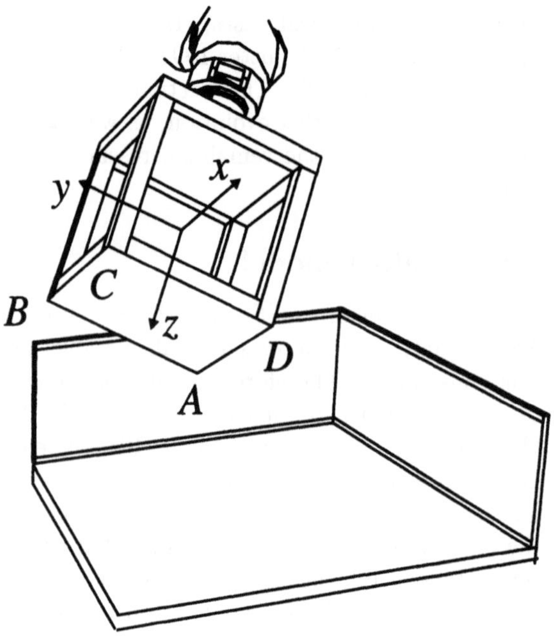

Figure 5: Sample task: putting the block in a corner while identifying all intermediate contact states. The uncertainties in the relative position and orientation of the block and the corner are such that, even if the block approaches the environment with a pure translation in the z-direction, it is not a priori known with which of the four vertices A, B, C or D it will come into contact first.

4.3 Experimental results

The theoretical developments summarized in the previous Sections have been applied to complex contact situations, involving multiple contacts between non-polyhedral objects, see e.g. (Dutré, Bruyninckx, and De Schutter 1996), (Dutré, Bruyninckx, Demey, and De Schutter 1997). The following paragraphs describe the results of the "block-into-corner" task, Fig. 5: the robot must manipulate a block until it is aligned with the three sides of a rectangular corner; the uncertainties in the relative position and orientation of the block and the corner are such that, even if the block approaches the environment with a pure translation in the z-direction, it is not a priori known with which of the four vertices A, B, C or D it will come into contact first. Once a contact has been established, and the *type* of the contact is known, the controller must identify the relative orientation of the block with respect to the environment. Figure 6 shows the two geometric uncertainty parameters θ_x and θ_y in the case of a vertex-face contact between the block and a surface of the environment. The results of the Kalman Filter based approach outlined in the previous paragraphs are depicted in Fig. 7. Deciding whether or not the type of the contact has been correctly chosen is the topic of the next Section.

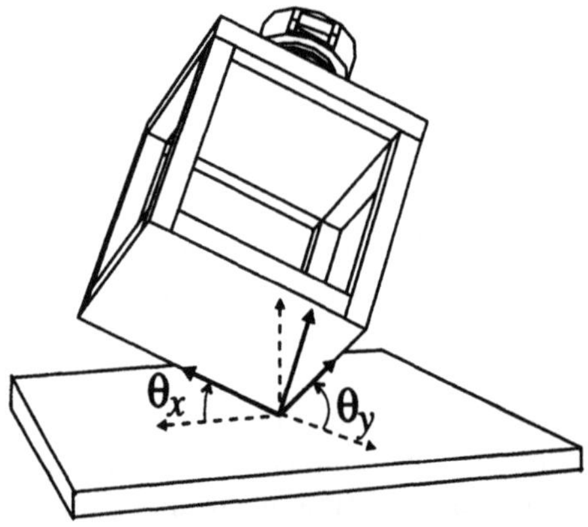

Figure 6: Vertex-face contact: uncertainty parameters.

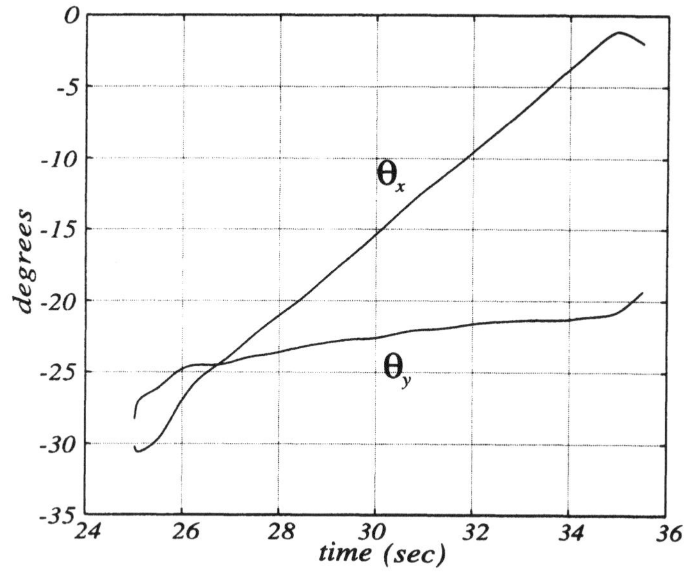

Figure 7: On-line estimates of parameters in Fig. 6. The contact model switch at instant 35 is easily detected by the Kalman Filter based statistical significance tests.

5 Contact model monitoring

Almost by definition, force control is used for tasks where the user has uncertain knowledge about (the geometry of) the environment, the manipulated object, and/or the relative position and orientation between both during the contact. It is in general not straightforward to detect if the current contact model is still valid, and, if not, what the new contact situation looks like.

The Kalman Filter parameter estimation introduced in the previous Section also offers a tool to tackle this model validation problem. The on-line filter updates of the system's stochastics reflect the evolution of the statistical "belief" in the parameter estimates. As soon as this belief drops below a statistical significance threshold, an alarm is raised to warn the controller that the contact model it is using *might* have become invalid. These statistical significance tests can be done in a variety of ways, depending on the level of a priori knowledge available about the contact situations. Note that these statistical tests can *never* guarantee 100% reliable decisions; not only because the methods are statistical in nature, but certainly also because the uncertainties in the system are never exactly described by the simple Gaussion probability distributions used in the Kalman Filters.

Figure 8 shows the outcome of such a statistical significance test applied to four possible contact hypotheses for the initial contact between block and environment, Fig. 5. Only the correct hypothesis has an acceptable significance level. Figure 9 collects the significance tests of all intermediate contact situations encountered during the complete task.

6 Conclusion

This paper discusses the authors' efforts to provide the Hybrid Control Paradigm with more solid theoretical foundations on the one hand, and more flexible and powerful medium-level control capabilities on the other hand. The treated topics include: (i) modeling and task specification for arbitrarily complex contact situations; (ii) an invariant HCP theory; (iii) stochastic on-line identification of the geometric uncertainties in the contact situation, which allows for tracking of time-varying contact situations; and (iv) the monitoring of a given contact model in order to assess its statistical acceptability based on a window of recent measurement samples.

The major remaining problems are: the persistent excitation of the uncertainties, reliable coping with high contact friction and compliance, and the robustness of detecting transitions in the contact situations. The Kalman Filter framework offers some partial solutions, but major breakthroughs on the "intelligent" control level of Fig. 1 are required to solve these problems satisfactorily.

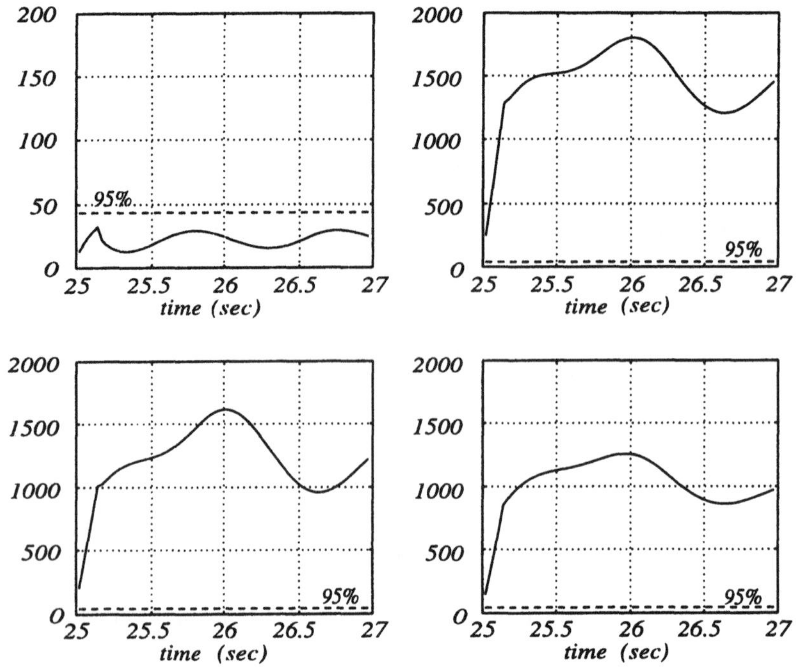

Figure 8: Statistical significance tests in the vertex-face contact of Fig. 6. Only one of the four possible contact model hypotheses has an acceptable statistical significance level.

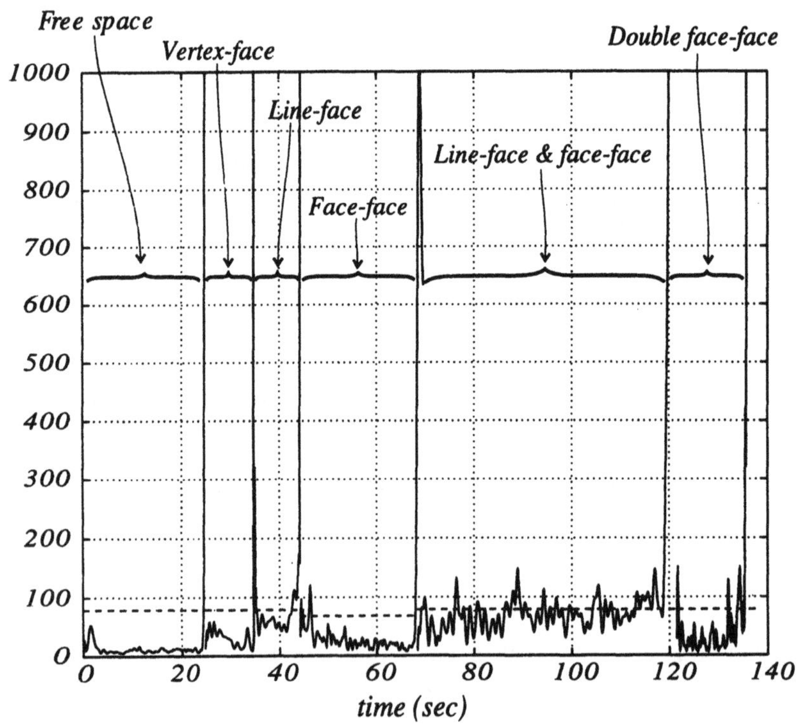

Figure 9: Sequence of statistical significance tests during complete task.

References

Bar-Shalom, Y. and X.-R. Li (1993). *Estimation and Tracking, Principles, Techniques, and Software*. Artech House.

Bruyninckx, H. (1995). *Kinematic Models for Robot Compliant Motion with Identification of Uncertainties*. Ph. D. thesis, Department of Mechanical Engineering, Katholieke Universiteit Leuven, Leuven, Belgium.

Bruyninckx, H. and J. De Schutter (1996). Specification of force-controlled actions in the "Task Frame Formalism": A survey. *IEEE Trans. Rob. Automation 12*(5), 581–589.

Bruyninckx, H., S. Demey, S. Dutré, and J. De Schutter (1995). Kinematic models for model based compliant motion in the presence of uncertainty. *Int. J. Robotics Research 14*(5), 465–482.

De Schutter, J., H. Bruyninckx, W.-H. Zhu, and M. W. Spong (1997). Force control: a bird's eye view. In B. Siciliano (Ed.), *IEEE CSS/RAS International Workshop on "Control Problems in Robotics and Automation: Future Directions"*. San Diego, CA: Springer Verlag.

De Schutter, J. and J. Leysen (1987). Tracking in compliant robot motion: Automatic generation of the task frame trajectory based on observation of the natural constraints. In R. Bolles (Ed.), *Proceedings of the 4th International Symposium of Robotics Research*, Santa Cruz, CA. MIT Press.

De Schutter, J., D. Torfs, S. Dutré, and H. Bruyninckx (1997). Invariant hybrid position/force control of a velocity controlled robot with compliant end effector using modal decoupling. *Int. J. Robotics Research 16*(3), 340–356.

Duffy, J. (1990). The fallacy of modern hybrid control theory that is based on "orthogonal complements" of twist and wrench spaces. *J. Robotic Systems 7*(2), 139–144.

Dutré, S., H. Bruyninckx, and J. De Schutter (1996). Contact identification and monitoring based on energy. In *IEEE Int. Conf. Robotics and Automation*, Minneapolis, MN, pp. 1333–1338.

Dutré, S., H. Bruyninckx, and J. De Schutter (1997). The analytical Jacobian and its derivative for a parallel manipulator. In *IEEE Int. Conf. Robotics and Automation*, Albuquerque, NM, pp. 2961–2966.

Dutré, S., H. Bruyninckx, S. Demey, and J. De Schutter (1997). Solving contact and grasp uncertainties. In *Int. Conf. Intel. Robots and Systems*, Grenoble, France, pp. 114–119.

Gelb, A. E. (1978). *Optimal Estimation* (3rd ed.). Cambridge, MA: MIT Press.

Jazwinski, A. H. (1970). *Stochastic processes and filtering theory*. New York, NY: Academic Press.

Karger, A. and J. Novak (1985). *Space kinematics and Lie groups*. New York, NY: Gordon and Breach.

Lipkin, H. and J. Duffy (1988). Hybrid twist and wrench control for a robotic manipulator. *Trans. ASME J. Mech. Transm. Automation Design 110*, 138–144.

Lončarić, J. (1985). *Geometrical Analysis of Compliant Mechanisms in Robotics*. Ph. D. thesis, Harvard University, Cambridge, MA.

Mason, M. T. (1981). Compliance and force control for computer controlled manipulators. *IEEE Trans. on Systems, Man, and Cybernetics SMC-11*(6), 418–432.

Park, F. C. (1995). Distance metrics on the rigid-body motions with applications to mechanism design. *Trans. ASME J. Mech. Design 117*, 48–54.

Raibert, M. and J. J. Craig (1981). Hybrid position/force control of manipulators. *Trans. ASME J. Dyn. Systems Meas. Control 102*, 126–133.

Wampler, C. W. (1984). Multiprocessor control of a telemanipulator with optical proximity sensors. *Int. J. Robotics Research 3*(1), 40–50.

West, H. and H. Asada (1985). A method for the design of hybrid position/force controllers for manipulation constrained by contact with the environment. In *IEEE Int. Conf. Robotics and Automation*, St. Louis, MS, pp. 251–259.

Whitney, D. E. (1987). Historical perspective and state of the art in robot force control. *Int. J. Robotics Research 6*(1), 3–14.

PART 3
EMERGENT MOTIONS
SESSION SUMMARY

Masaru Uchiyama
Department of Aeronautics and Space Engineering, Tohoku University
Aramaki-aza-Aoba, Aoba-ku, Sendai 980-77, Japan
e-mail: uchiyama@space.mech.tohoku.ac.jp

This session consists of two papers both of which deal with generation of robot motions in reactive behavior. Synthesis of physical motions of robots constitutes a major part of robotics to which enthusiastic research efforts have been directed; examples of the past research are trajectory planning for obstacle avoidance, kinematics and dynamics for trajectory control, sensor-guided motion generation, and so on. Rather simple feedback control strategy, of which stability analysis could be obtained easily, is assumed in those classical examples. Planning phase is separated from the control that is based on simple sensory feedback. The motion trajectory is planned with consideration of static environments. The approach is meaningful only for robots with "slow" intelligence, that is, robots to have sensory feedback of low frequency. To enhance reactive behavior with sensory feedback of higher frequency and to have faster robot behavior with "dynamic" intelligence, feedback strategy to generate motions should be more elaborated. The session title, "Emergent Motions," connotes motions of reactive behavior which are organized through sophisticated feedback strategies such as learning, abstraction, and graph theory that the two papers of the session address.

The first paper by Y. Nakamura, T. Yamazaki and N. Mizushima, entitled "Motion Synthesis, Learning and Abstraction through Parameterized Smooth Map from Sensors to Behaviors," discusses motion synthesis through mapping from sensors to behaviors. Integration of components such as sensors, actuators, and structures in recent robotic systems is becoming higher and higher. Theory of integration of reactive behaviors should be a challenge for efficient control of those sophisticated recent robotic systems. The authors of the paper take a unique approach in which they propose to use nonlinear functions of smooth mapping between sensory data to behaviors. The mapping gives a sophisticated structure to the feedback loop that provides parameters to be accessed and tuned by external agents. The parameters are weights for fusing primitive functions to emerge behavior. The structured mapping forms a brain-like system to organize behaviors. Learning and abstraction of physical skills are possible within the framework of the mapping. Experiments show that complicated behaviors such as multi-fingered grasping of different-shaped objects can be organized by this structured feedback control method.

The second paper by R. W. Ghrist and D. E. Koditschek, entitled "Safe Cooperative Robotic Patterns via Dynamics on Graphs," explores the possibility of using vector fields to design and implement reactive schedules for safe cooperative robot patterns on graphs. The physical system that they consider in the paper is AGV's (Automated Guided Vehicles) operating upon a predefined network of pathways. They use graphs on which reactive scheduling is implemented and thus a feedback controller is yielded. A geometrical model of the system, that is a graph in this paper, brings a controller with more complicated structure and with higher performance than a conventional feedback controller. Collision avoidance for safe motion is easily achieved via the vector fields. A detailed analysis of a pair of AGV's restricted to a Y-shaped graph, that is the simplest nontrivial situation, is presented in the paper. They obtained a gradient-style construction that brings all initial conditions of the two AGV's to any desired pair of goal points while guaranteeing safety.

Motion Synthesis, Learning and Abstraction through Parameterized Smooth Map from Sensors to Behaviors

Y. Nakamura T. Yamazaki N. Mizushima
Department of Mechano-Informatics
University of Tokyo
7-3-1 Hongo, Bunkyo-ku, Tokyo 113, JAPAN
email: nakamura@mech.t.u-tokyo.ac.jp

Abstract

The integration theory of reactive behaviors is to be discussed in this paper. A linear emerging model is adopted where the motion of a robot is represented as the weighted linear sum of reactive behaviors. The weights are defined as differentiable nonlinear functions of sensor signals and parameters. The functions can represent logical if-then rules as their extreme cases. The sensor space model is introduced to relate the sensors and the behaviors and to determine the parameters. We establish a learning method based on the sensor space model, where the parameters are systematically tuned through iteration of trials such that the sensor signals converge to the given teacher signals. A nonlinear dynamics in the sensor space model is also proposed to allow fluctuation for the future global search. The learning method is applied to the reactive grasp of a three-fingered robot hand. We integrate 48 kinds of sensor signals and 29 primitive behaviors. The experiments indicate that the emerging model allows us to use the semantics to initially program the nonlinear functions for the weights. The learning experiments successfully illustrate the usefulness of the proposed learning method.

1 Introduction

The synthesis of machine intelligence has been a central issue in robotics research. Many contributions have been made and accumulated by numerous researchers in image understanding, dexterous motion control, motion planning and optimization, and so on. A unique approach of Brooks (1) was based upon the principle that the essence of robotic intelligence comes out of the interaction with the real world through sensors, and proved that behaviors with biological complexity emerge not from the complexity of computation, but from the complexity of environments. More recent works (4; 3) agreed and showed that the networked behaviors and sensors with rather simple hardware and software generate animal-like instinctive and complex behaviors through the interaction with the environments.

The main issues of recent research in the behavioral robotics would be: (1) designing rational networks, (2) providing behaviors with objectivity, (3) structure for learning and self-organizing.

The most networks of the behavioral robotics adopted logical and selective structures. A behavior comes out of a set of if-then rules being conditioned with sensory information. The variety of behaviors is that of a series of discrete behaviors. This variety would be certainly the key to the machine intelligence, if we can *control* it by answering the above three research issues.

In this paper, we design the behavioral networks using smooth nonlinear functions, and propose to generate behaviors by smoothly blending. The smooth blending diversifies and enriches emerging behaviors. We use the radial base functions as nonlinear functions and parameterize the networks with their coefficients. The analytical and parameterized expression of the network structure enables the design of learning and self-organizing mechanism. Introduction of nonlinear chaotic dynamics in the parameter space is to be mentioned. We implemented the idea into the behavioral synthesis of grasping and experimentally discussed the usefulness.

Behavioral approaches to grasping include

Speeter (5) where manipulation by a Utah/MIT hand was implemented with more than 50 primitive behaviors. Matsui and Omata (6) designed selectively emerging mechanism adopting multi-agent architecture. On the other hand, Michelman and Allen (7) synthesized complex tasks of a Utah/MIT hand by combining two dimensional primitives of manipulation functions.

2 Behavioral Networks and Nonlinear Functions

2.1 Analytical expression of behaviors

A behavioral primitive B_j $(j = 1, \cdots, n)$ can be a functional unit of behavior, or a nonfunctional unit that has no linguistic meaning. Although the primitives are preferably to be represented in terms of acceleration, velocity, position and many other physical quantities, they could be easily integrated or blended if they all have an identical physical sense. We assume in this paper that B_j is represented only by $\dot{\boldsymbol{\theta}}_j$.

On the other hand, S_i $(i = 1, \cdots, m)$ represents a sensor. S_i does not necessarily correspond to a physical sensory device. It is one of sensory information obtained after processing raw sensory signals. For examples, a video image may offer many S_i's. Or one can make S_i using several physical sensors.

We call the normalized sensory information S_i the sensor intensity and represent it by α_i $(0 \leq \alpha_i \leq 1)$. The vector form of the sensor intensity is given by

$$\boldsymbol{\alpha} = (\alpha_1 \; \alpha_2 \; \cdots \; \alpha_m)^T \tag{1}$$

$\boldsymbol{\alpha}$ may include the internal states such as joint angles $\boldsymbol{\theta}$. $\boldsymbol{\alpha}$ may also include its history or integration.

Velocity $\dot{\boldsymbol{\theta}}_j$ of primitive behavior B_j can be a constant or a sensory feedback. Therefore, it is generally represented by

$$\dot{\boldsymbol{\theta}}_j = \dot{\boldsymbol{\theta}}_j(\boldsymbol{\alpha}) \tag{2}$$

where $\dot{\boldsymbol{\theta}}_j$ is differential with respect to $\boldsymbol{\alpha}$

Primitives $\dot{\boldsymbol{\theta}}_j$ $(j = 1, \cdots, n)$ are integrated and fused to emerge behavior $\dot{\boldsymbol{\theta}}$ in the following manner:

$$\dot{\boldsymbol{\theta}} = \sum_{j=1}^{n} \beta_j \dot{\boldsymbol{\theta}}_j \tag{3}$$

where β_j $(0 \leq \beta_j \leq 1)$ is a scalar named the behavior intensity. It is a function of $\boldsymbol{\alpha}$. Namely,

$$\beta_j = \beta_j(\boldsymbol{\alpha}) \tag{4}$$

β_j is assumed differentiable with respect to $\boldsymbol{\alpha}$. The vector form of β_j is represented by:

$$\boldsymbol{\beta} = (\beta_1 \; \beta_2 \; \cdots \; \beta_n)^T \tag{5}$$

It can be observed from Eq.(3) that $\dot{\boldsymbol{\theta}}_j$ does not appear when $\beta_j = 0$, while $\dot{\boldsymbol{\theta}}$ includes full of $\dot{\boldsymbol{\theta}}_j$ when $\beta_j = 1$. Even if a primitive $\dot{\boldsymbol{\theta}}_j$ is given constant, $\dot{\boldsymbol{\theta}}_j$ may change. $\boldsymbol{\beta}(\boldsymbol{\alpha})$ is a nonlinear function of $\boldsymbol{\alpha}$.

2.2 Sensor Intensity and Behavior Intensity

Behavior intensity β_j represents the blending rate of $\dot{\boldsymbol{\theta}}_j$. It has been common to use binary β_j in the selection of behaviors. Namely,

$$\beta_j = \begin{cases} 1 & \text{if } (\alpha_1 = \alpha_1^0 \text{ AND } \alpha_2 = \alpha_2^0) \\ & \quad \text{OR } (\alpha_3 = \alpha_3^0) \\ 0 & \quad \text{otherwise} \end{cases} \tag{6}$$

An advantage of representing β_j by if-then logic like Eq.(6) is the fact that β_j can be programmed based upon the semantic understanding of sensory information and primitive behaviors. In this paper, we seek to represent β_j by a differential function of $\boldsymbol{\alpha}$. We will later formulate learning and self-organizing as a problem of modifying and reforming the functions. It would be desirable if the initial forms of nonlinear functions can be programmed based on the semantic understanding.

First of all, α_{ij}^0 $(0 \leq \alpha_{ij}^0 \leq 1)$ is defined to be the sensor offset of B_j. Namely,

$$\bar{\boldsymbol{\alpha}}_j = \boldsymbol{\alpha} - \boldsymbol{\alpha}_j^0 \tag{7}$$

$$\boldsymbol{\alpha}_j^0 \triangleq (\alpha_{1j}^0 \; \alpha_{2j}^0 \; \cdots \; \alpha_{mj}^0)^T$$
$$\bar{\boldsymbol{\alpha}}_j \triangleq (\bar{\alpha}_{1j} \; \bar{\alpha}_{2j} \; \cdots \; \bar{\alpha}_{mj})^T$$

である.

$\bar{\alpha}_{ij}$ is said to be activable if it activates B_j when $\bar{\alpha}_{ij} \cong 0$. On the contrary, $\bar{\alpha}_{ij}$ is inhibitory if it activates B_j when $\bar{\alpha}_{ij} \cong 0$ is not satisfied. Like logical symbols, activable $\bar{\alpha}_{ij}$ is represented by A_{ij}, while inhibitory $\bar{\alpha}_{kj}$ is represented by $\bar{A}_{kj}$. We can describe semantic relationship between $\boldsymbol{\alpha}_j$ and primitive B_j in terms of the rules of binary logic. Such a relationship can be expressed by a logic equation. For example,

$$A_j = (A_{1j} * \bar{A}_{2j}) + A_{3j} + (A_{4j} * A_{5j}) * \bar{A}_{6j} \quad (8)$$

where $(*)$ and $(+)$ represent logical product (AND) and logical sum (OR) respectively.

We define the transformation from a logic equation to a function as follows:

$$\beta_j(\bar{\alpha}_j) = P(A_j) \quad (9)$$

where $P(\cdot)$ follows the expansion rules:

$$\begin{aligned} P(A * B) &= P(A) \cdot P(B) \\ P(A + B) &= 1 - (1 - P(A)) \cdot (1 - P(B)) \\ &= P(A) + P(B) - P(A) \cdot P(B) \\ P(\bar{A}) &= 1 - P(A) \\ P(E) &= 1 \\ P(O) &= 0 \end{aligned}$$

According to the above equations, $P(A_j)$ is transformed and represented as polynomials of $P(A_{ij})$ $(i = 1, \cdots m)$. Motivated by $A * A = A$, we now introduce the following reduction rule:

$$P(A) * P(A) = P(A) \quad (10)$$

The reduction rule reduces the order of polynomials. Each term becomes at most first-order of every $P(A_{ij})$.

We can easily prove that the transformation using the above expansion and reduction rules satisfy the axiom system of the binary Boolean algebra due to Huntington, if we assume $P(\cdot) \in X$, $X = [0, 1]$.

It implies that the transformation is closed under the basic theorems of the binary Boolean algebra, such as the associative law, the absorption law, the idempotency law, the identity element, and the de Morgan theorem. In other words, we can transform an arbitrary logic equation between $\bar{\alpha}_i$ and β_j like Eq.(8) into a polynomial equation of $P(A_{ij})$ with complete logical consistency.

Concretely, we choose $P(A_{ij})$ as follows:

$$P(A_{ij}) = \exp[-w_{ij}\bar{\alpha}_{ij}^2] \quad \in X \quad (11)$$

For examples, $P^j_{A\text{-}AND}$, $P^j_{A\text{-}OR}$, $P^j_{I\text{-}AND}$ and $P^j_{I\text{-}OR}$ are Activation (AND), Activation (OR), Inhibition (AND) and Inhibition (OR) and defined by

$$\begin{aligned} P^j_{A\text{-}AND} &\triangleq P\ (A_{1j} * A_{2j}) \\ &= \textstyle\prod_{i=1}^2 \exp[-w_{ij}\bar{\alpha}_{ij}^2] \\ P^j_{A\text{-}OR} &\triangleq P\ (A_{1j} + A_{2j}) \\ &= 1 - \textstyle\prod_{i=1}^2 (1 - \exp[-w_{ij}\bar{\alpha}_{ij}^2]) \\ P^j_{I\text{-}AND} &\triangleq P\ (\overline{A_{1j} * A_{2j}}) = 1 - P\ (A_{1j} * A_{2j}) \\ &= 1 - \textstyle\prod_{i=1}^2 \exp[-w_{ij}\bar{\alpha}_{ij}^2] \\ P^j_{I\text{-}OR} &\triangleq P\ (\overline{A_{1j} + A_{2j}}) = 1 - P\ (A_{1j} + A_{2j}) \\ &= \textstyle\prod_{i=1}^2 (1 - \exp[-w_{ij}\bar{\alpha}_{ij}^2]) \end{aligned}$$

The mappings of functions are shown in **Fig.1**.

Note that β_j in Eq.(9) is differentiable in terms of not only $\boldsymbol{\alpha}$, but also sensor offset $\boldsymbol{\alpha}_j^0$ and sensor weight w_{ij}. Namely,

$$\beta_j = \beta_j(\boldsymbol{\alpha}, \boldsymbol{\alpha}_j^0, \boldsymbol{W}_j) \quad (12)$$

where

$$\boldsymbol{W}_j = diag\{w_{ij}\} \in \mathcal{R}^{m \times m} \quad (13)$$

In this paper, we choose w_{ij} and $\boldsymbol{\alpha}_j^0$ as parameters for learning and self-organizing.

2.3 Sensory space model

w_{ij} in Eq.(11) yields the strength of influence from S_i to B_j. In order to describe the relationship between the sensors and the primitive behaviors, we use an Euclidean model of sensory space.

The sensory space is an Euclidean space with as many dimensions as the total number m of sensors S_i $(i = 1, \cdots, m)$. We assume that there exist point-masses as many as the total number n of primitives B_j $(j = 1, \cdots, n)$. Note that the "point masses" do not have physical dimension of mass. We introduced them to design the structure of mechanics.

A point mass is paired with a primitive behavior and called behavioral point-mass B_j. The position of the point-mass in the sensory space is represented by $\boldsymbol{b}_j$. Namely,

$$\boldsymbol{b}_j = (b_{1j}\ b_{2j}\ \cdots\ b_{mj})^T \quad (j = 1, \cdots, n) \quad (14)$$

Suppose sensor S_i activates when $b_{ij} > 0$, while it inhibits primitive B_j when $b_{ij} < 0$. In the scope of this paper, we also assume that b_{ij} does not change its sign. w_{ij} in Eq.(11) is given by

$$w_{ij} = b_{ij}^2 \quad (15)$$

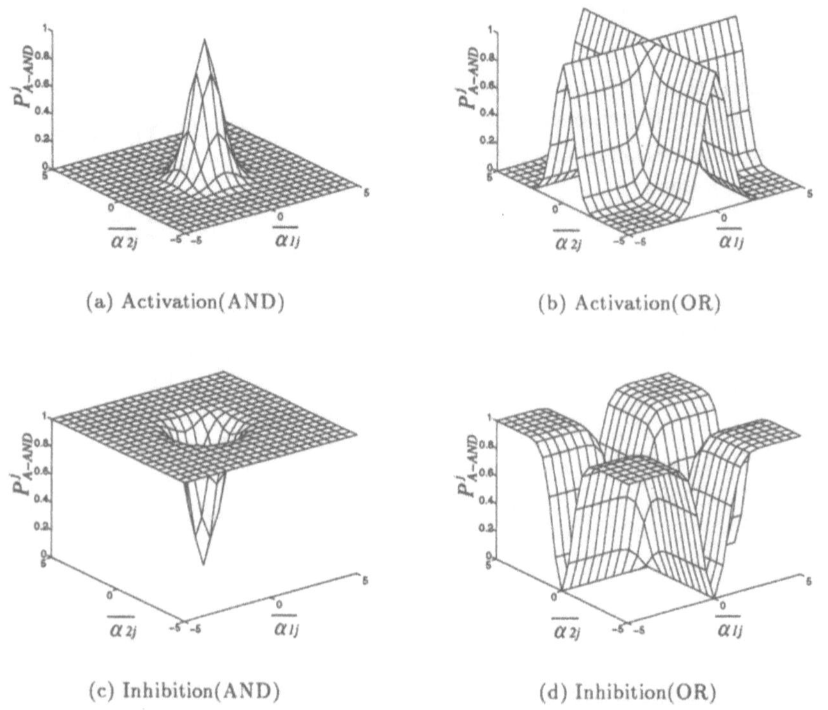

(a) Activation(AND) (b) Activation(OR)

(c) Inhibition(AND) (d) Inhibition(OR)

Figure 1: Nonlinear mapping for activation and inhibition

where

$$\begin{cases} i \in J^j_{A\text{-}AND} \text{ or } i \in J^j_{A\text{-}OR} & \text{if } b_{ij} > 0 \\ i \in J^j_{I\text{-}AND} \text{ or } i \in J^j_{I\text{-}OR} & \text{if } b_{ij} < 0 \\ i \in J^j_{A\text{-}AND} & \text{if } b_{ij} = 0 \end{cases}$$

If behavioral point-mass B_j is far from the origin along the $i-th$ coordinate, sensor S_i clearly affects primitive B_j. If B_j is near the origin, it becomes insensitive to sensor S_i. Likewise, by continuously changing the configuration of the behavioral point-masses, we can continuously change the mapping from sensor S_i $(i = 1, \cdots, m)$ to behavior $B_j (j = 1, \cdots, n)$.

3 Learning of Behavioral Networks

3.1 Ideal sensor intensity

In this section we discuss learning of behavioral networks. We assume that the sensory intensity $\alpha_d(t)$ $(0 \le t \le T)$ of the ideal behavior is known a priori, though we do not know what configuration of the behavioral point-masses generate the ideal mapping. We also assume the change of environments and their initial setting is precisely repeatable while learning. We determine the configuration of behavioral point-masses for the $k+1$-th learning from the sensor intensity α obtained after the k-th learning.

3.2 Learning

We prepare b from the configuration of the behavioral point-masses b_j $(j = 1, \cdots, n)$ as follows:

$$b = (b_1^T \; b_2^T \; \cdots \; b_n^T)^T \in \mathcal{R}^{m \cdot n} \quad (16)$$

The performance index of learning is defined as follows, where Q is a positive definite matrix:

$$I(b) = \int_0^T (\alpha(t) - \alpha_d(t))^T Q (\alpha(t) - \alpha_d(t)) \, dt \quad (17)$$

The index becomes zero only when the sensory signal completely agrees with the desired one. The change of the index ΔI is computed in response to a small perturbation of the behavioral point-masses from b to $b + \Delta b$

$$\Delta I(\boldsymbol{b}) = \frac{\partial I}{\partial \boldsymbol{b}} \Delta \boldsymbol{b} \tag{18}$$

Therefore, if we choose $\Delta \boldsymbol{b}$ as follows:

$$\Delta \boldsymbol{b} = -k \left(\frac{\partial I}{\partial \boldsymbol{b}} \right)^T , \tag{19}$$

then, we have the following result:

$$\Delta I(\boldsymbol{b}) = -k \frac{\partial I}{\partial \boldsymbol{b}} \left(\frac{\partial I}{\partial \boldsymbol{b}} \right)^T \leq 0 \tag{20}$$

Note that $\Delta I(\boldsymbol{b})$ is negative semi-definite, not negative definite. Therefore, it is anticipated that the learning will converge to a local minimum. Since we initially set the behavioral network based on the logical equations, the leaning here is considered to search for a local solution.

3.3 Computation of $\frac{\partial I}{\partial \boldsymbol{b}}$

The essential difficulty of learning is the computation of $\frac{\partial I}{\partial \boldsymbol{b}}$. As we discuss later more, the exact computation $\frac{\partial I}{\partial \boldsymbol{b}}$ requires the model of environments. We propose to compute $\frac{\partial I}{\partial \boldsymbol{b}}$ by approximation.

$\frac{\partial I}{\partial \boldsymbol{b}}$ is represented as follows:

$$\frac{\partial I}{\partial \boldsymbol{b}} = \int_0^T 2(\boldsymbol{\alpha} - \boldsymbol{\alpha}_d)^T \boldsymbol{Q} \frac{\partial \boldsymbol{\alpha}}{\partial \boldsymbol{b}} dt \tag{21}$$

Therefore,

we need to compute $\frac{\partial \boldsymbol{\alpha}}{\partial \boldsymbol{b}}$. $\frac{\partial \boldsymbol{\alpha}}{\partial \boldsymbol{b}}(t)$ $(0 \leq t \leq T)$ implies the rate of change of the sensor intensity in response to the configuration change of the behavioral point-mass.

Since $\boldsymbol{\alpha}$ changes it value by the motion of robot and the temporal change of environments it, $\boldsymbol{\alpha}$ is a function of these quantities. Accordingly,,

$$\boldsymbol{\alpha} = \boldsymbol{\alpha}(\ddot{\boldsymbol{\theta}},\ \dot{\boldsymbol{\theta}},\ \boldsymbol{\theta},\ t) \tag{22}$$

We now have

$$\frac{\partial \boldsymbol{\alpha}}{\partial \boldsymbol{b}} = \frac{\partial \boldsymbol{\alpha}}{\partial \ddot{\boldsymbol{\theta}}} \frac{\partial \ddot{\boldsymbol{\theta}}}{\partial \boldsymbol{b}} + \frac{\partial \boldsymbol{\alpha}}{\partial \dot{\boldsymbol{\theta}}} \frac{\partial \dot{\boldsymbol{\theta}}}{\partial \boldsymbol{b}} + \frac{\partial \boldsymbol{\alpha}}{\partial \boldsymbol{\theta}} \frac{\partial \boldsymbol{\theta}}{\partial \boldsymbol{b}} \tag{23}$$

- The 1st term

  Strictly speaking, $\boldsymbol{\alpha}$ is a function of $\ddot{\boldsymbol{\theta}}$ since the sensory information from the force sensor include the inertia force due to $\ddot{\boldsymbol{\theta}}$. However, we formulate the behavioral networks by Eq.(3) neglecting such dynamic quantities. The 1st term of Eq.(23) is neglected.

$$\frac{\partial \boldsymbol{\alpha}}{\partial \ddot{\boldsymbol{\theta}}} \cong 0 \tag{24}$$

- The 2nd term

$\frac{\partial \boldsymbol{\alpha}}{\partial \dot{\boldsymbol{\theta}}}$ implies the effect of the angular velocity of finger joints to all the sensors. Strict computation required the knowledge of shape and compliance. We cannot assume this, since we do not use the environmental model. We approximate as follows:

$\frac{\partial \dot{\boldsymbol{\theta}}}{\partial \boldsymbol{\alpha}}$ is a matrix whose row and column numbers are equal to those of the generalized coordinates and the total number of sensors. The column number is larger than that the row number. $\dot{\boldsymbol{\theta}}$ will widely change its value. Therefore, we can assume $\frac{\partial \dot{\boldsymbol{\theta}}}{\partial \boldsymbol{\alpha}}$ is row-full rank. Accordingly,

$$\frac{\partial \dot{\boldsymbol{\theta}}}{\partial \boldsymbol{\alpha}} \frac{\partial \boldsymbol{\alpha}}{\partial \dot{\boldsymbol{\theta}}} = \boldsymbol{E} , \tag{25}$$

From Eq.(25) we can approximate $\frac{\partial \boldsymbol{\alpha}}{\partial \dot{\boldsymbol{\theta}}}$ by the following least square solution:

$$\frac{\partial \boldsymbol{\alpha}}{\partial \dot{\boldsymbol{\theta}}} \cong \left(\frac{\partial \dot{\boldsymbol{\theta}}}{\partial \boldsymbol{\alpha}} \right)^{\#} \tag{26}$$

where $(*)^{\#}$ is the pseudoinverse. $\frac{\partial \dot{\boldsymbol{\theta}}}{\partial \boldsymbol{\alpha}}$ can be analytically computed from Eqs. (3) and (9) as follows:

$$\frac{\partial \dot{\boldsymbol{\theta}}}{\partial \boldsymbol{\alpha}} = \sum_{j=1}^{n} \frac{\partial \beta_j}{\partial \boldsymbol{\alpha}} \dot{\theta}_j + \beta_j \frac{\partial \dot{\theta}_j}{\partial \boldsymbol{\alpha}} \tag{27}$$

On the other hand, $\frac{\partial \dot{\boldsymbol{\theta}}}{\partial \boldsymbol{b}}$ is also analytically obtained.

$$\frac{\partial \dot{\boldsymbol{\theta}}}{\partial \boldsymbol{b}} = \sum_{j=1}^{n} \frac{\partial \beta_j}{\partial \boldsymbol{b}} \dot{\theta}_j \tag{28}$$

To summarize, the 2nd term is computed by Eq.(26), (27) and (28).

- The 3rd term
$\frac{\partial \boldsymbol{\alpha}}{\partial \boldsymbol{\theta}}$ represents the effect of the change of generalized coordinates on all the sensors. $\boldsymbol{\alpha}$ includes elements that is determined by robot itself. We reserve and consider those determined by robot itself in $\boldsymbol{\alpha}_\theta \in \mathcal{R}^m$ and set the other elements zero. Namely,

$$\frac{\partial \boldsymbol{\alpha}}{\partial \boldsymbol{\theta}} \cong \frac{\partial \boldsymbol{\alpha}_\theta}{\partial \boldsymbol{\theta}} \tag{29}$$

We can compute $\frac{\partial \dot{\boldsymbol{\theta}}}{\partial \boldsymbol{b}}$ as follows:

$$\begin{aligned} \frac{\partial \boldsymbol{\theta}}{\partial \boldsymbol{b}}(t) &= \frac{\partial}{\partial \boldsymbol{b}}\left[\boldsymbol{\theta}(0) + \int_0^t \dot{\boldsymbol{\theta}} dt\right] \\ &= \int_0^t \frac{\partial \dot{\boldsymbol{\theta}}}{\partial \boldsymbol{b}} dt \end{aligned} \tag{30}$$

where $\frac{\partial \dot{\boldsymbol{\theta}}}{\partial \boldsymbol{b}}$ is given by Eq.(28). To summarize, the 3rd term is computed using Eq.(29) and (30).

In this section, we developed the computation of $\frac{\partial I}{\partial \boldsymbol{b}}$ by adopting least-square solution and neglecting higher order terms. The differentiable analytical model such as Eqs.(3), (9), (15) enabled us these analytic computations.

The paradigm of learning here is different from that of Arimoto's (8) in the following two senses: (1) sensory information allows those that have dependency to unknown environments (2) we learn the parameters of parameterized behavioral networks, but not the input of robots.

4 Robot Hand Experimental Setup

4.1 Robot hand and sensors

Figure 2 shows the overview of the experimental setup.

We used a three-fingered robot hand (Yasukawa Co.) in **Fig.3**, which has a 3 axis force sensor at each finger-tip and a 6 axis force sensor in the wrist.

We adopted two vision systems. We obtained and used various features of the image from COGNEX4400 (COGNEX) every 120*ms*.

Tracking Vision (Fujitsu Co.), the second vision system, was used to detect whether the object entered the scope and to measure the velocity of the object. We used two CCD cameras (top and side views). The two images were combined into a single image by a quadrant picture unit and fed to the both vision systems.

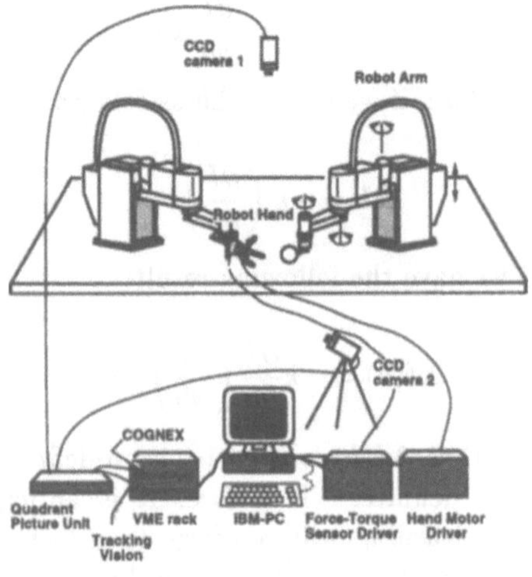

Figure 2: Experimental set-up

4.2 Sensory information

With internal sensors of the robot hand, we prepared 48 kinds of sensory information. They were normalized and used in the experiments. The list of sensory information follows and their reference numbers are shown in Table 1. **Figure 4** illustrates the physical meaning of the main information.

- area of object (S_u, S_s)
- ratio of object image's principal axes in top-view (γ)
- vertical distance from hand to object in side-view (H)
- horizontal distance from hand to object in top-view (O_d)
- object's velocity vector ($O_{vel_x}, O_{vel_y}, O_{vel_z}$)
- angle between finger tip and the major principal axis in top-view (ϕ_{fi})
- distance from object's center to finger tip in top-view (d_{fi})
- hiding ratio of current finger-tip location in top-view (ht_{fi})

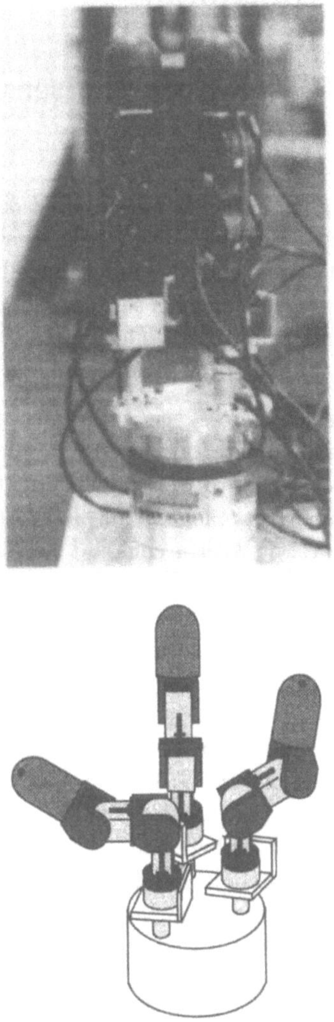

Figure 3: Three-fingered robot hand

- hiding ratio of point in top-view where finger-tip locates when 1st joint rotates +20° (hp_{fi})
- hiding ratio of point in top-view where finger-tip locates when 1st joint rotates −20° (hm_{fi})
- hiding ratio of point in top-view where finger-tip locates when 2nd and 3rd joints rotate 20° in opening direction (ho_{fi})
- finger joint angle ($\theta_{11}, \cdots, \theta_{33}$)
- finger joint velocity ($\dot{\theta}_{11}, \cdots, \dot{\theta}_{33}$)
- magnitude of finger force (F_{fi})
- task sensor (T)

The hiding ratio of a point is precisely the ratio of area occupied by the object in a square centered by the point. Although the task sensor is prepared to trigger the behavior after grasping, it is not used in the current experiments.

Table 1: Reference number of sensors

1	S_u	13	d_{f2}	25	ho_{f2}	37	θ_{12}
2	S_s	14	d_{f3}	26	ho_{f3}	38	$\dot{\theta}_{13}$
3	γ	15	ht_{f1}	27	θ_{11}	39	$\dot{\theta}_{21}$
4	H	16	ht_{f2}	28	θ_{12}	40	$\dot{\theta}_{22}$
5	O_d	17	ht_{f3}	29	θ_{13}	41	$\dot{\theta}_{23}$
6	O_{vel_x}	18	hp_{f1}	30	θ_{21}	42	$\dot{\theta}_{31}$
7	O_{vel_y}	19	hp_{f2}	31	θ_{22}	43	$\dot{\theta}_{32}$
8	O_{vel_z}	20	hp_{f3}	32	θ_{23}	44	$\dot{\theta}_{33}$
9	ϕ_{f1}	21	hm_{f1}	33	θ_{31}	45	F_{f1}
10	ϕ_{f2}	22	hm_{f2}	34	θ_{32}	46	F_{f2}
11	ϕ_{f3}	23	hm_{f3}	35	θ_{33}	47	F_{f3}
12	d_{f1}	24	ho_{f1}	36	$\dot{\theta}_{11}$	48	T

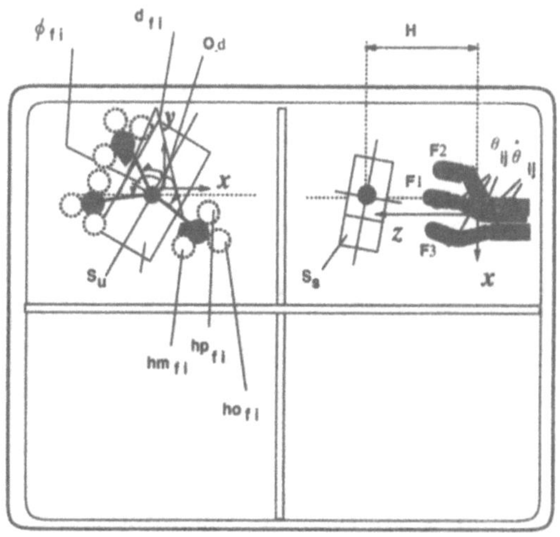

Figure 4: Sensor information

4.3 Behavioral primitives

We have prepared 29 kinds of behavioral primitives:

- OPEN1 (Behavior 1)
  Open three fingers when the object has large vertical distance to the hand (simultaneously).
- CLOSE1 (Behavior 2)
  Close three fingers when the object has short

- CLOSE1 (Behavior 2)
  Close three fingers when the object has short vertical distance to the hand (simultaneously).
- ADJUST (Behavior 3)
  The second joint determines its opening angle from the area of object and applies feedback control. The third joint stays at 0° (simultaneously).
- OPEN2 (Behavior 4,5,6)
  Open the finger that has short distance with the object, when the object has large horizontal distance from the hand origin (independently).
- CLOSE2 (Behavior 7)
  Close three fingers when the object has short horizontal distance with the hand origin (simultaneously).
- TWIST1 (Behavior 8,9,10)
- TWIST2 (Behavior 11,12,13)
- TWIST3 (Behavior 14,15,16)
  Keep the angle between the major axis and the hand at either 90° or 270° (independently). TWIST1, TWIST2 and TWIST3 have the sensor offset at $\phi = 0°$, $\phi = 180°$, $\phi = 360°$ respectively.
- AVOID1 (Behavior 17,18,19)
  If the object hides the finger-tip and if the point where the first joint rotates +20° is open, rotate the first joint (independently).
- AVOID2 (Behavior 20,21,22)
  If the object hides the finger-tip and if the point where the first joint rotates −20° is open, rotate the first joint (independently).
- AVOID3 (Behavior 23,24,25)
  If the object hides the finger-tip and if the point where the second joints and third rotates +20° is open, move there (independently).
- TOUCH (Behavior 26,27,28)
  Keep the joint angles, when the finger-tip touches with the object (independently).
- BTASK (Behavior 29)
  Task sensor, which is used to switch to the next task (currently unused).

Behaviors 1, 2, $4 \sim 7$, $17 \sim 25$ have constant $\dot{\theta}_j$. On the other hand, Behaviors 3, $8 \sim 16$, $26 \sim 28$ take feedback styles like $\dot{\theta}_j = \dot{\theta}_j(\alpha)$.

5 Experiments

5.1 Grasping via the sensor space model

Grasping experiments were carried out using the sensor space model. The initial locations of behavioral point-mass were determined by considering physical meanings of sensory information and behavioral primitives. The locations were then carefully tuned through experiments.

Figure 5 shows the experimental result of grasping a sphere of 70[mm] in diameter. A sphere was fixed to the endeffector of a SCARA robot and approached the hand on another SCARA robot with a constant velocity. Grasping was successfully done. The sensor intensity (α) and the behavioral intensity (β) are shown in the figure. Brightness of the figure (white implies 1.0 and black means 0.0) represents temporal change of the sensor and behavioral intensities. It is seen that the behaviors such as OPEN1(1), ADJUST(3), and CLOSE1(2) were activated in this order. TWIST was not activated since the sphere had a small ratio of principal axes. Although AVOID($17 \sim 25$) were active at the beginning since the hand was behind the object, their intensities were maintained zero afterward.

Grasping was also observed when the same sphere was hung by the SCARA robot with a string and approached the hand. Although the sphere rotated and vibrated, it was grasped successfully.

Then, we carried out the next experiment using a rectangular prism of 160[mm] × 70[mm] × 40[mm]. The prism was fixed to the endeffector of a SCARA robot and approached the hand with a constant velocity. Grasping was done as successfully as the case of fixed sphere. The sensor intensity and the behavioral intensity are shown in **Fig.6**. Since this object had the clear major principal axis, TWIST was always activated and rotated the first joints to maintain the fingers perpendicular to the major principal axis.

These experimental results imply that a tuned single configuration of the sensor space model can cover some variety of grasps. The extent of coverage would be determined by the set of primitive behaviors, the dimension of the sensor space model, and the complexity between them.

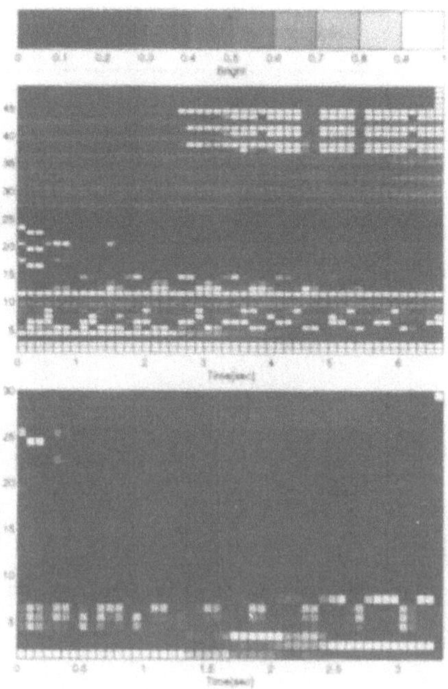

Figure 5: Reactive grasping of a sphere

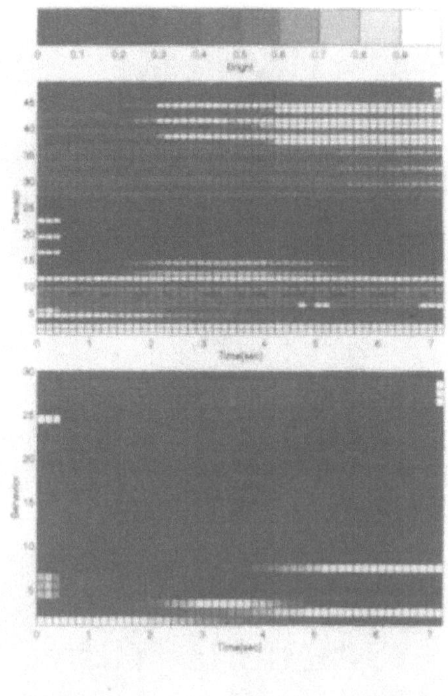

Figure 7: Desired sensor intensity (teacher) in grasping a sphere

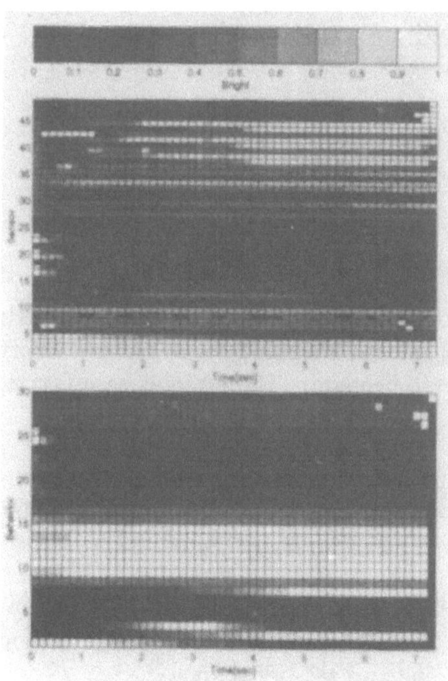

Figure 6: Reactive grasping of a rectangular parallelpiped

5.2 Self-organization of behavioral network through learning

The theory of section 3 was experimentally investigated. The ideal values of sensor intensity and the corresponding behavioral intensity were chosen as seen in **Fig.7**.

We changed and moved the behavioral point-mass configuration from the ideal configuration that results in **Fig.7**. The displacement of configuration was chosen to make (1) TWIST1, 2, 3 less sensitive to the ratio of principal axes (γ), (2) CLOSE1 more sensitive to the distance from hand to object (H), and (3) ADJUST less sensitive to the distance from hand to object (H). **Figure 8** shows the change of sensor intensity by the difference from the ideal one of **Fig.7**. The learning was iterated 20 times. The graphs of **Fig.8** are the results after the 1st, 2nd, 3rd, and 20th trials. The graph implies small error when it is gray, while error is large when the graph is white or black.

The convergence rate of the performance index can be found in **Fig.9**. **Figure 10** shows the joint motion of the first finger before and after the learning compared with the teacher motion. The real temporal motions after the 1st, 2nd, 3rd, 4th,

10th, and 20th trials are represented in **Fig.11**.

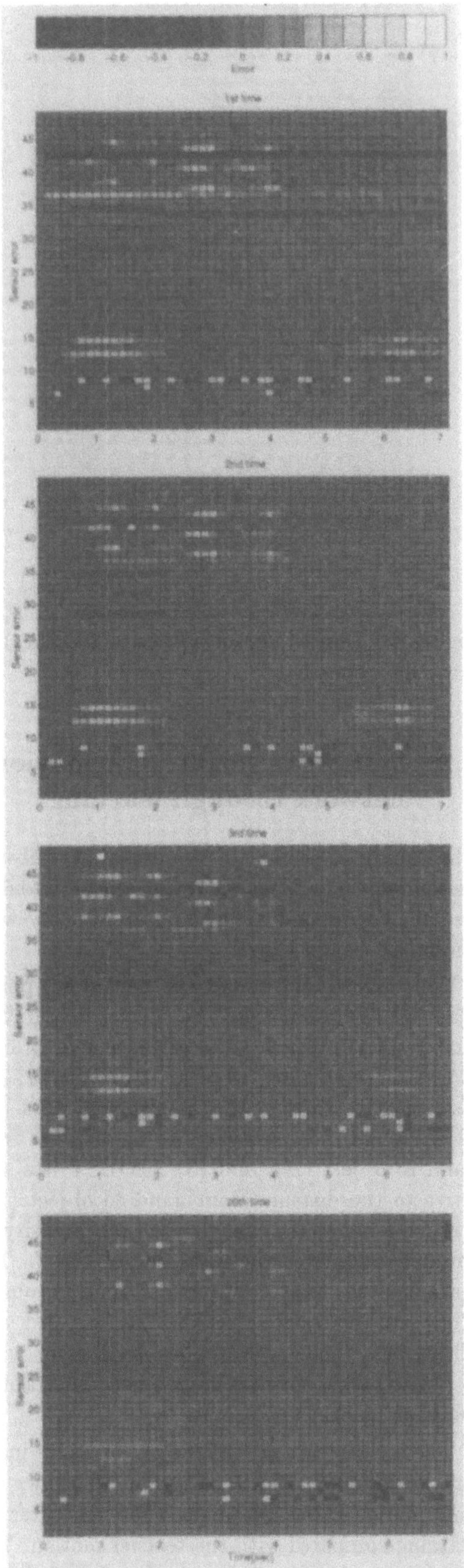

Figure 8: Change of sensor intensity (error from the teacher) in the process of learning

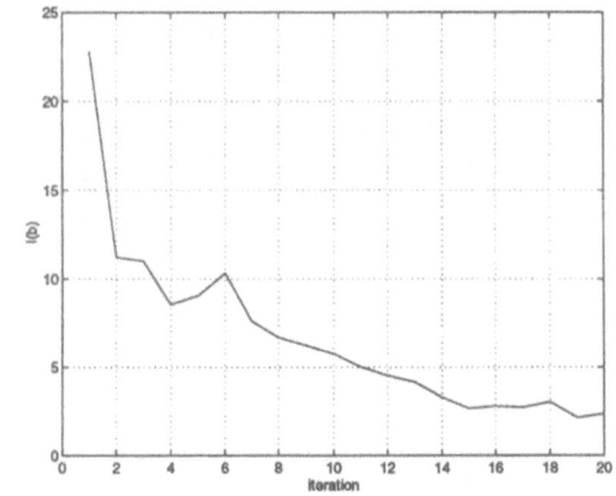

Figure 9: Convergence of the learning index of Eq.(3)

We can visually see the error of sensor intensity in **Fig.8**, which shows that $\boldsymbol{\theta}$ and $\dot{\boldsymbol{\theta}}$ improved significantly as learning proceeded. The error was clearly large before the 4th trial in **Fig.8**. **Figure 9** implies that learning successfully proceeded in spite of the computational approximation in learning algorithm. **Figures 10(a)** and **10(b)** also represent that the ideal joint motion is obtained with a good precision after 20 trials.

6 Conclusion

The conclusions of this paper is summarized in the following five items:

1. We proposed the behavioral networks with smooth mapping from sensory information to behaviors. The radial base function was adopted for the mapping function.

2. We showed that the mapping function can be programmed from the semantic understanding and its logical equations.

3. We proposed to parameterize the mapping function and to use the parameters for learning and self-organizing.

4. A learning algorithm was established using the parameterized behavioral networks and a simple gradient method. Unavailability of the environmental model requires approximate computation of the gradient.

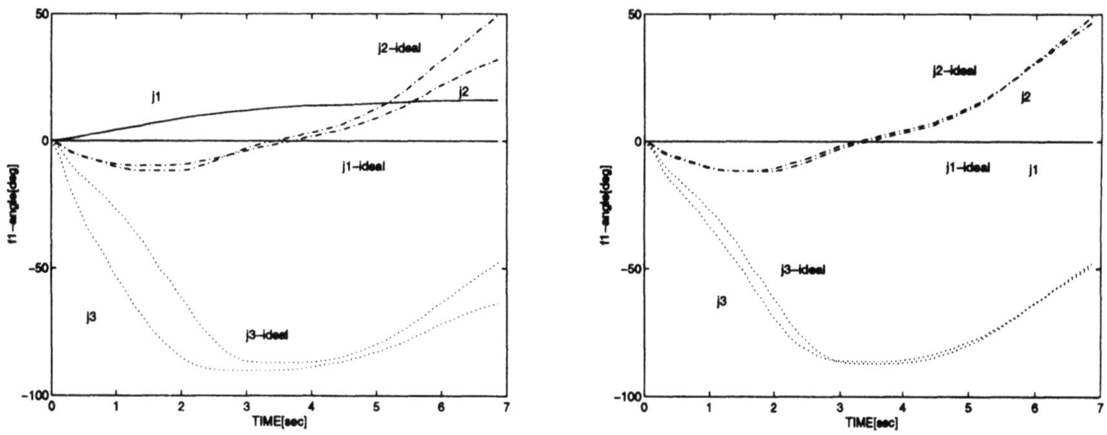

(a) Before learning (Error of the 1st trial from the teacher)

(b) After learning (Error of the 20th trial from the teacher)

Figure 10: Change of joint angles of Finger 1 (Error from the teacher)

5. Experiments of the proposed learning algorithm was done and showed that the learning was completed in spite of the approximation of computation.

This work was supported by the Ministry of Education, Culture, and Sports through an inter-university research program, "Research on Emergent Mechanisms for Machine Intelligence — A Tightly Coupled Perception-Motion Behavior Approach" (Principal Investigator H. Inoue, University of Tokyo). The authors would also like to acknowledge the supports in setting the experimental system by Yasukawa Co., COGNEX JAPAN Inc., and Komatsu Co.

References

[1] R. A. Brooks: *"A Robust Layered Control System for a Mobile Robot"*, IEEE Journal of Robotics and Automation, Vol.RA-2, No.1, pp. 14-23, 1986.

[2] J. J. Gibson: *"The Ecological Approach to Visual Perception"*, Boston: Houghton Mifflin, 1979.

[3] P. Maes: *"Situated Agents Can Have Goals"*, Designing Autonomous Agents, MIT press, pp. 49-70, 1990.

[4] A. Dubrawski, J. L. Crowley: *"Self-Supervised Neural System for Reactive Navigation"*, Proceedings of the IEEE International Conference on Robotics and Automation, pp. 2076-2081, 1994.

[5] T. H. Speeter: *"Primitive Based Control of the Utah/MIT Dextrous Hand"*, Proceedings of the IEEE International Conference on Robotics and Automation, pp. 866-877, 1991.

[6] T. Matsui, T. Omata: *"Multi-Agent Architecture for Multi-Fingered Hand Control"*, Proceedings of the 9th Annual Conference of Robotics Society of Japan, pp. 755-756, 1991.

[7] P. Michelman, P. Allen: *"Forming complex dextrous manipulations from task primitives"*, Proceedings of the IEEE International Conference on Robotics and Automation, Vol.4, pp. 3383-3388, 1994.

[8] S. Arimoto, S. Kawamura, F. Miyazaki: *"Bettering Operation of Robots by Learning"*, Journal of Robotic Systems, Vol.1-2, pp. 123-140, 1984.

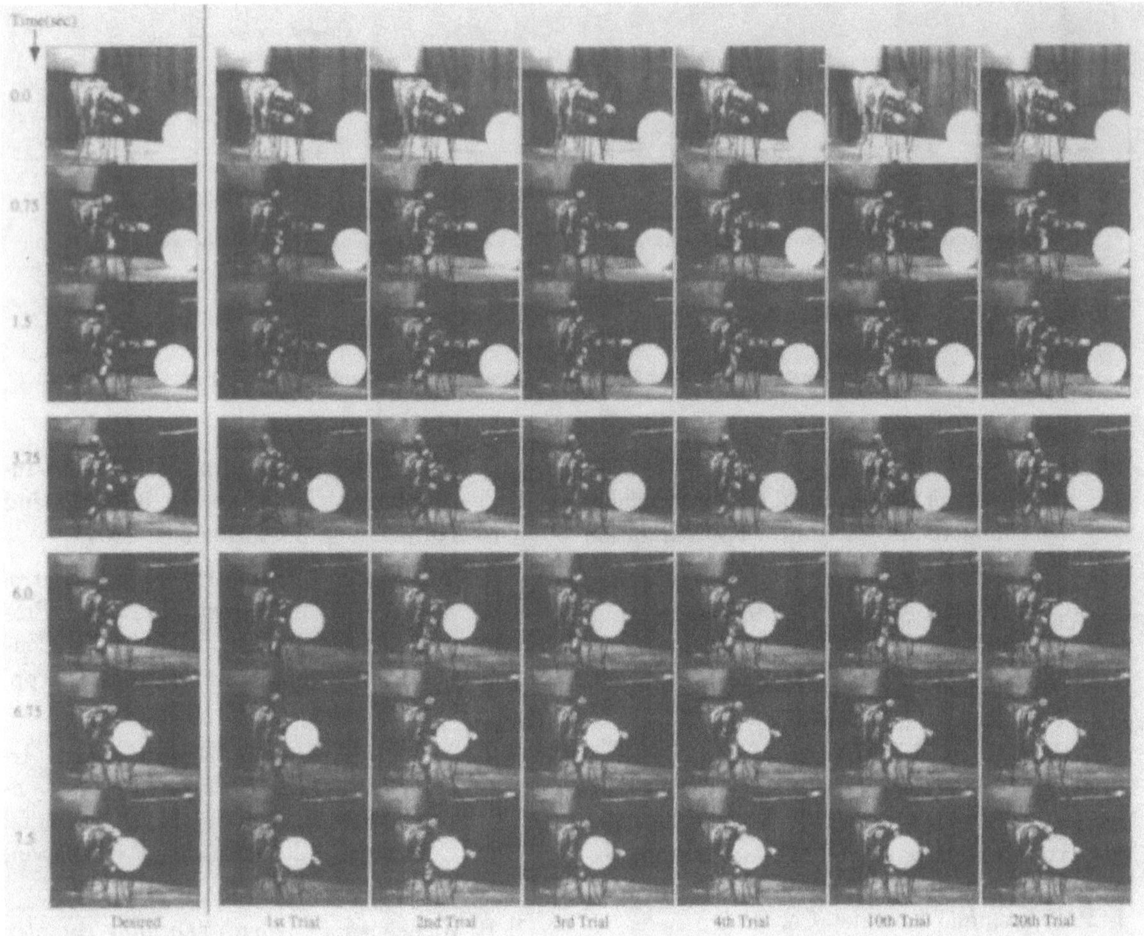

Figure 11: Change of grasping behavior in the process of learning

Safe Cooperative Robot Patterns via Dynamics on Graphs

Robert W. Ghrist*
Department of Mathematics
The University of Texas at Austin
Austin, TX 78712, USA

Daniel E. Koditschek†
Department of Electrical Engineering and Computer Science
The University of Michigan
Ann Arbor, MI 48109-2110, USA

Abstract

This paper explores the possibility of using vector fields to design and implement reactive schedules for safe cooperative robot patterns on graphs.

We consider Automated Guided Vehicles (AGV's) operating upon a predefined network of pathways, contrasting the simple cases of locally Euclidean configuration spaces with the more topologically intricate non-manifold cases. The focus of the present inquiry is the achievement of safe cooperative patterns by means of a succession of *edge point fields* combined with a *circulating field* to regularize collisions at non-manifold vertices.

1 Introduction

Recent literature suggests the growing awareness of a need for "reactive" scheduling wherein one desires not merely a single deployment of resources but a plan for successive redeployments against a changing environment [24]. But scheduling problems have been traditionally solved by appeal to a discrete representation of the domain at hand. Thus the need for "tracking" changing goals introduces a conceptual dilemma: there is no obvious topology by which proximity to the target of a given deployment can be measured. In contrast to problems entailing the management of information alone, problems in many robotics and automation settings involve the management of *work* — the exchange of energy in the presence of geometric constraints. In these settings, it may be desireable to postpone the imposition of a discrete representation long enough to gain the benefit of the natural topology that accompanies the original domain.

This paper explores the use of vector fields for reactive scheduling of safe cooperative robot patterns on graphs. The word "safe" means that obstacles — designated illegal portions of the configuration space — are avoided. The word "cooperative" connotes situations wherein physically distributed agents are collectively responsible for executing the schedule. The word "pattern" refers to tasks that cannot be encoded simply in terms of a point goal in the configuration space. The word "reactive" will be interpreted as requiring that the desired pattern reject perturbations: conditions close but slightly removed from those desired remain close and, indeed, converge toward the exactly desired pattern.

1.1 Motivation: Reactive Scheduling Suggests Feedback Controllers on Graphs

Graphs arise as a natural data structure by which to encode assembly plans, executive logic in machine tool cells, many other aspects of automation, and, indeed, many more diverse and abstract instances of plans and schedules. There is a well understood formal equivalence between graphs and finite automata [17], and the growing DES literature attests to the rich questions about controller synthesis that can arise in this context. The question further arises whether there might be merit to bringing scheduling problems into more intimate contact with the topology of con-

*Supported by National Science Foundation Postdoctoral Fellowship DMS-9508846.

†Supported in part by National Science Foundation Grant IRI-9510673.

tinuous spaces wherein such notions as feedback and asymptotic stability have proven so useful in conferring robustness against unforseen perturbations. This paper explores a domain of scheduling problems whose setting is so tightly bound to the continuous world that the resolution of this question is motivated by more than purely intellectual speculation.

Graphs, and the desirability of imposing (or analyzing already imposed) dynamics upon them arise in robotic contexts in several different settings:

- As a representation of some designated cover of state space. The nodes of the graph correspond to the cells (subsets of the state space) in the cover, and connecting arcs denote non-empty intersection. Dynamically induced movement along the graph corresponds to the construction of vector fields whose flow brings an entire cell into into its intersection with the adjacent cell to be visited.

  For example, in [9], the second author and colleagues define a *prepares* relation — a partial order that prunes this graph into a tree — via controllers with point attractors in a specified neighborhood whose domain of attraction includes the entire cell (which is also positive invariant). We will in fact adopt the framework of that paper as a point of departure in the present work.

- When the state space admits the structure of a simplicial complex. The nodes of the graph correspond to the (varying dimensional) cells of the complex, and the arcs denote adjacency.

  For example, in robotic manipulations requiring successively higher order contact, it has proven possible to actually compute a topologically valid representation of the resulting simplicial complex in some cases [8]. Brockett [7] has introduced the graph described above and explored some aspects of dynamics imposed upon it.

- When the workspace is itself organized as a graph whose one-dimensional "viaducts" connect workstation locations.

  For example, many factory materials handling systems are built using AGVs that must track a guidepath network embedded in the floor.

In all of these settings it makes sense to speak of patterns. In most of them are issues of safety, but typically passage to a graph presumes that the obstacles have been avoided by the manner in which the graph is embedded in the original configuration space. In some of these settings it makes sense to speak of multiple agents. In the last setting all of these considerations come together in a common problem, and the paper shall concentrate here in consequence.

1.2 Setting: AGV's on a Guidepath Network of Wires

An automated guided vehicle (AGV) is an unmanned powered cart "capable of following an external guidance signal to deliver a unit load from destination to destination" where, in most common applications, the guidepath signal is buried in the floor [10]. Thus, the AGV's workspace is a network of wires — a graph.[1] The motivation to choose AGV based materials handling systems over more conventional fixed conveyors rests not simply in their ease of reconfigurability but in the potential they offer for graceful response to perturbations in normal plant operation. In real production facilities, the flow of work in process fluctuates constantly in the face of unanticipated workstation downtime, variations in process rate, and, indeed, variations in materials transport and delivery rates [12]. Of course, realizing their potential robustness against these fluctuations in work flow remains an only partially fulfilled goal of contemporary AGV systems.

Choreographing the interacting routes of multiple AGVs in a non-conflicting manner presents a novel, complicated, and necessarily on-line planning problem. Nominal routes might be designed offline but they can never truly be traversed with the nominal timing, for all the reasons described above. Even under normal operating conditions, no single nominal schedule can suffice to coordinate the workflow as the production volume or product mix changes over time: new vehicles need to be added or deleted and the routing scheme adapted. In any case, abnormal conditions — unscheduled process down times; blocked work stations; failed vehicles — continually arise, demanding altered routes.

The traffic control schemes deployed in contemporary AGV systems are designed to simplify the real-time route planning and adaptation process by "blocking zone control" strategies. The workspace is partitioned into a small number of cells and, regardless of the details of their source and destination tasks, no two AGVs are ever allowed into the same cell at the same time [10]. Clearly, this simplification results in significant

[1] This is not necessarily the case for beacon guided vehicles. However, their obvious advantages in flexibility and reconfigurability notwithstanding, their greater cost, complexity and relative fragility conspire to relegate their use to a small minority of installations [10].

loss of a network's traffic capacity.

The contemporary robotic motion planning literature does not seem to offer much in the way of an alternative. Starting with pioneering work of Alami [1] there has been a small literature on multiple coordinated robots, but almost all papers seem to be concerned with offline versions of the problem. Latombe, in his excellent monograph [15] distinguishes between "centralized" and "decoupled" approaches to this problem. In the latter case, motion planning proceeds using multiple copies of the configuration space within which to situate a set of non-interacting robot vehicles. For a recent example of what Latombe terms a "coordinated" view of the decoupled case, Svestka and Overmars [25] introduce a "supergraph" on which multiple vehicles can be stepped through their individually specified paths, and vehicle-vehicle collisions prohibited by detaining one or another vehicle. For a recent example of what Latombe terms a "prioritized" view of the decoupled case, Lee et al. [16] compute k-shortest paths for each vehicle's source-destination pair, and work their way down the list of prefences based upon a vehicle's priority. It should be clear that in neither of these approaches has the recourse to blocking zone control been eliminated. Thus, while the decentralized approach side steps the inevitable curse of dimensionality, passing to the underlying configuration space seems to be required if the rigidity and inefficiency of blocking zone control strategies is to be eliminated.

The Industrial Engineering AGV literature seems chiefly concerned with higher level issues of layout and capacity [20] or dispatching and more general scheduling [2]. One interesting approach to layout seeks to avoid the subsequent traffic control problem entirely by clustering pickup and delivery stops in decoupled single vehicle loops [4]. In general, modeling the real-time factory floor is challenging enough that even the most recent treatments of these higher level layout and dispatching problems seem to rely on simulation rather than analysis for understanding the implications of one or another policy [3]. Here, again, is an indication that a dynamical systems point of view might shed additional light. For recent years have witnessed increasingly successful efforts to characterize such ensemble properties as the "mean transit time" induced by a flow. Thus, it seems to us entirely possible that a dynamics based network traffic control strategy might yield more readily to statistical analysis than present practice affords.

In this paper, we will consider a centralized approach that employs dynamical systems theory to focus on real-time responsiveness and efficiency as opposed to computational complexity or average throughput. No doubt, beyond a certain maximum number of vehicles, the necessity to compute in the high dimensional configuration space will limit the applicability of any algorithms that arise. However, this point of view seems not to have been carefully explored in the literature. Indeed, we will sketch some ideas about how an approach that starts from the coupled version of the problem may lend sufficient insight to move back and forth between the individuals' and the group's configuration spaces even in real time. For the sake of concreteness we will work in the so-called "pickup and delivery" (as opposed to the "stop and go" [4]) paradigm of assembly or fabrication, and we will not be concerned with warehousing style AGV applications.

In this context, a "pattern" amounts to a repetitive route through the graph (the particular sequence of workstations on the factory floor that the AGV must service). We desire a feedback policy that causes the robot to return to this pattern no matter what temporary obstructions or dislocations it experiences. We next desire to introduce two (or more) robots into the same graph, and seek a means of "juggling" them together that interleaves their patterns in an asymptotically stable and safe manner. That is, given a collection of patterns and a collection of corresponding feedback laws that achieve them for a single robot, we are interested in modifying the individual control strategies as modestly as possible so that a collection of robots can achieve those patterns simultaneously on the same graph with the guarantee of no collisions.

1.3 Contributions of the Paper

The paper is organized as follows. In §2, we review fundamental facts about the topology of graphs, without which, objects such as vector fields and gradients make little sense. We use this information to define the class of *edge point fields* — locally defined dynamics that realize single letter patterns. These act collectively as a toolbox from which to build a hybrid controller for achieving arbitrary patterns with a single AGV. This represents a slight generalization of the scheme the second author and colleagues have proposed in [9].

In §3, we turn to the problem of introducing multiple AGV's in the context of graphs which are manifolds. In this simpler case, it is often, but not always, possible to interleave controllers for single AGV's into a safe controller on the product configuration space.

The problem of dynamics and control on nonmanifold graphs is then considered in §4. We present a fairly detailed analysis of the configuration space for a pair of AGV's on a Y-shaped graph — the simplest nontrivial situation. Here, a clarification of the configuration space presentation leads easily to a vector field construction that brings all initial conditions of two robots on the graph to any desired pair of goal points while guaranteeing safety (i.e., no collisions along the way). The desire for a more decoupled controller — the hope of an "interleaving" of otherwise independent individual patterns — impels a revised approach to safe navigation leading to the construction of a vector field that enables the AGV's to "dance" about one other at a vertex.

The dynamical features of this *circulating field* are suggestive of future hybrid constructions that would allow multiple independent patterns to be safely interleaved. We comment on the form such juggling algorithms might take in §5.

2 Notation and Background

2.1 Graph Topology

A *graph*, Γ, consists of a finite collection of 0-dimensional vertices $\mathcal{V}:=\{v_i\}_1^N$, and 1-dimensional edges $\mathcal{E}:=\{e_j\}_1^M$ assembled as follows. Each edge is homeomorphic to the closed interval $[0,1]$ attached to $\mathcal{V}$ along its boundary points $\{0\}$ and $\{1\}$.[2] We place upon Γ the quotient topology given by the endpoint identifications [21]: Neighborhoods of a point in the interior of e_j are homeomorphic images of interval neighborhoods of the corresponding point in $[0,1]$, and neighborhoods of a vertex v_i consist of the union of homeomorphic images of half-open neighborhoods of the endpoints for all incident edges.

The configuration spaces we consider in §3-4 are self-products of graphs. The topology of $\Gamma \times \Gamma$ is easily understood in terms of the topology of Γ as follows [21]. Let $(x,y) \in \Gamma \times \Gamma$ denote an ordered pair in the product. Then any small neighborhood of (x,y) within $\Gamma\times\Gamma$ is the union of neighborhoods of the form $\mathcal{N}(u) \times \mathcal{N}(v)$, where $\mathcal{N}(\cdot)$ denotes neighborhood within Γ. In other words, the products of neighborhoods form a *basis* of neighborhoods in the product space.

Given a graph, Γ, outfitted with a finite number n of noncolliding AGV's constrained to move on Γ, the configuration space of safe motions is defined as

$$\mathcal{C}:=(\Gamma \times \ldots \times \Gamma) - \mathcal{N}(\Delta),$$

where $\Delta:=\{(x_i) \in \Gamma \times \ldots \times \Gamma : x_j = x_k \text{ for some } j \neq k\}$ denotes the pairwise diagonal and $\mathcal{N}(\cdot)$ denotes (small) neighborhood.

[2] We will assume away in the sequel the possibility of "homoclinic" edges whose boundary points are attached to the same vertex.

For general graphs, the topology of $\mathcal{C}$ can be extremely complicated, as measured by, say, the rank of the fundamental group (see [21] for definitions). Even in the case where the workspace, Γ, is contractible (and thus, the product of its n copies is contractible), removal of this collision diagonal often creates spaces with large fundamental group. For example, given a graph Γ_K with K edges all connected at a single point (forming an K-pronged "star"), we can show that the fundamental group of the configuration space $\Gamma_K \times \Gamma_K - \mathcal{N}(\Delta)$ is a free group on $K^2 - 3K + 1$ generators — i.e., the number of "independent" closed paths in this space (with respect to continuous deformation) grows quadratically with K.

We do not treat the general aspects of this problem comprehensively in this paper; rather, we restrict attention to several simple preparatory examples and one basic but nontrivial example which illustrates nicely the relevant features present in the more general situation.

In order to proceed, it is necessary to clarify what we mean by a vector field on a simplicial complex that fails to be a manifold. This is a nontrivial issue: for example, in the case of a graph, the tangent space to a vertex with incidence number greater than two is not well-defined. Clearly, graphs posessing such vertices are not manifolds. But every graph can be embedded in a Euclidean space with edges "pinched together" so that the tangent vectors at each vertex all lie within the same well-defined one-dimensional tangent space to the embedding [22]. The possible ambiguity in pairwise orientation between the tangent spaces to each edge endpoint is resolved by choice of some convention: here we will always assume that positive vectors are outward directed on the right boundary, 1, and inward directed on the left boundary, 0, of an edge. The pinched embedding yields an existence and (forward time) uniqueness guarantee for solutions to the differential equations defined in this manner [22].

For present purposes, we find it convenient to work with an intrinsic formulation (i.e., directly in the graph rather than via an embedding in a Euclidean space) of this property. To this end, denote by v a vertex with K incident edges $\{e_i\}_1^K$, and by $\{X_i\}_1^K$ a collection of nonsingular vector fields locally defined on a neighborhood of the endpoint of each e_i (homeomorphic to $[0,\epsilon)$). These meet the conditions for existence of solutions with respect to some "pinched" embedding of Γ if and only if (1) the magnitude of the endpoint vectors $\|X_i(0)\|$ (taken with respect to the attaching homeomorphisms) are all identical; and (2) the *signs* of the endpoint vectors $X_i(0)$ (either positive if pointing into $[0,\epsilon)$ or negative if pointing out) are not all the same. Heuristically, condition (1) implies that we may identify all the endpoints to the vertex while respecting the vector field, and condition (2) means that the negative directions flow into the vertex and the positive directions flow out. If there are only positive or only negative directions, we do not have a well-defined vector field at the vertex — i.e., the existence of solutions may not be guaranteed.

Since graphs need not be manifolds, results such as uniqueness of solutions are not guaranteed, and, in fact, are not true in general. The best one can hope for is to have a vector field generating a *semiflow* — that is, a continuous dynamical system with unique forward orbits. Pursuing the intrinsic formulation of such a "forward uniqueness property," observe that for a vector field on a graph, it is clear that a semiflow is defined if and only if there is a *unique* edge along which the vector field is positive. In other words, all orbits through the vertex emanate along a single edge. Note that running a semiflow in backwards time does not generate unique orbits, unless the semiflow is in fact a flow.

These notions of intrinsic vector fields generating semiflows on graphs extend naturally to the cross products of graphs we consider later. Again, the theme is to embed such spaces smoothly into a higher dimensional Euclidean space inducing well-defined tangent space. For the remainder of this work, we will define all vector fields intrinsically, without worrying about the specific embedding required.

2.2 Edge Point Fields

In the context of describing and executing *patterns* or periodic motions on a graph, one desires a set of building blocks for moving from one goal to the next. We thus introduce the class of *edge point fields* as a dynamical toolbox for a hybrid controller. Given a specified goal point $g \in e_j$ within an edge of Γ, an *edge point field* is a locally defined vector field X_g on Γ with the following properties:

Locally Defined: X_g is defined on a neighborhood $\mathcal{N}(e_j)$ of the goal-edge e_j within the graph topology. Furthermore, forward orbits under X_g are uniquely defined.

Point Attractor: every forward orbit of X_g asymptotically approaches the unique fixed point $g \in e_j$.[3]

Navigation-Like: X_g admits a C^0 Lyapunov function, $\Phi_g : \Gamma \to R$.

[3] When it is not clear from the context, we shall denote the goal point achieved by an edge point flow as $\mathbf{g}(X_g) = \{g\}$.

Lemma 1 *Given any edge $e_j \subset \Gamma$ which is contractible within Γ, there exists an edge point field X_g for any desired goal $g \in e_j$.*

Proof: Fix the desired goal $g \in \Gamma$. By hypothesis, e_j (and thus a neighborhood $\mathcal{N}(e_j)$) is contractible; hence, given any point $x \in \mathcal{N}(e_j)$, there exists up to reparametrization a *unique* one-to-one path from g to x in $\mathcal{N}(e_j)$. For, if there were two such paths, this would imply the existence of a one-to-one map of a circle into $\mathcal{N}(e_j)$, contradicting contractibility. Place any bounded metric d on Γ and define the function $\Phi : N(e_j) \to R$ via $\Phi(x)$ is equal to the d-length of the unique path from g to x. The properties of the metric then guarantee that Φ is a function whose gradient field is an edge point field for g, where we define the gradient to be the induced gradient on the interior of each edge under the homeomorphisms to $[0, 1]$ via the topology on Γ. At the vertex, the gradient is taken with respect to the unique edge along which the function descends. This is compatible with the given definitions for vector fields on graphs. □

The only occasion for which an edge e_j is not contractible in Γ is in the "homoclinic case" when both endpoints of e_j are attached to the same vertex, forming a loop. In such instances, one may avoid the problem by subdividing the edge to include more vertices, which is very natural in the setting of this paper, since vertices correspond to workstations along a path.

2.3 Discrete Regulation of Patterns

We adopt the standard framework of symbolic dynamics [17]. By an excursion on a graph is meant a (possibly infinite) sequence of edges from the graph, $E = e_{i_1} \dots e_{i_N} \dots \in \mathcal{E}^Z$, having the property that each pair of contiguous edges, e_{i_j} and $e_{i_{j+1}}$ share a vertex in common. The set of excursions forms a language, $\mathcal{L}$: the so-called *subshift* on the alphabet defined by the named edges (we assume each name is unique) [17]. The *shift operator*, σ, defines a discrete dynamical system on the set of excursions, mapping the set of infinite sequences into itself by decrementing the time index. An *M-block extension* of the original language arises in the obvious way from grouping together each successive block of M contiguous letters from an original sequence, and it is clear how σ induces a shift operator, σ^M on this derived set of sequences.

Given a legal block, $B = e_{i_1} \dots e_{i_M} \in \mathcal{L}$, we will say that an excursion realizes that pattern if its M-block extension eventually reaches the "goal" $BBBBB\dots$ under the iterates of σ^M. In other words, after some finite number of applications of σ, the excursion consists of repetitions of the block B (terminating possibly with the empty edge).

In a previous paper [9], the second author and colleagues introduced a very simple but effective discrete event controller for regulating patterns on graphs from all reachable initial edges by pruning the graph back to a tree (imposing an ordering). Of course, this simple idea has a much longer history. In robotics it was introduced in [18] as "pre-image backchaining;" pursued in [19] as a method for building verifiable hardened automation via the metaphor of a funneling; and in [11] as a means of prescribing sensor specifications from goals and action sets. In the discrete event systems literature an optimal version of this procedure has been introduced in [6] and a generalization recently has been proposed in [23].

Let $\mathcal{E}^0{:=}B \subset \mathcal{E}$ denote the edges of Γ that appear in the block of letters specifying the desired pattern. Denote by

$$\mathcal{E}^{n+1} \subset \mathcal{E} - \bigcup_{k \leq n} \mathcal{E}^k$$

those edges that share a vertex with an edge in $\mathcal{E}^n$ but are not in any of the previously defined subsets. This yields a finite partition of $\mathcal{E}$ into "levels," $\{\mathcal{E}^p\}_{p=0}^{P}$, such that for each edge, $e_i^p \in \mathcal{E}^p$, there can be found a legal successor edge, $e_j^{p-1} \in \mathcal{E}^{p-1}$, such that $e_i^p e_j^{p-1} \in \mathcal{L}$ is a legal block in the language. Note that we have implicitly assumed $\mathcal{E}^0$ is reachable from the entire graph — otherwise, there will be some "leftover" component of $\mathcal{E}$ forming the last cell in the partition starting within which it is not possible to achieve the pattern. Note as well that we impose some ordering of each cell $\mathcal{E}^p = \{e_i^p\}_{i=1}^{M_p}$: the edges of $\mathcal{E}^0 = B$ are ordered by their appearance in the block; the ordering of edges in higher level cells is arbitrary.

We may now define a "graph controller" law, $G{:}\mathcal{E} \to \mathcal{E}$ as follows. From the nature of the partition $\{\mathcal{E}^p\}$ above, it is clear that the least legal successor function,

$$L(e_i^p){:=}\left\{ \begin{array}{ll} i+1 \bmod M & :p=0 \\ \min\{j \leq M_p : e_i^p e_j^{p-1} \in \mathcal{L}\} & :p>0 \end{array} \right., \tag{1}$$

is well-defined. From this, we construct the graph controller:

$$G(e_i^p){:=}e_{L(p,i)}^{p-1}. \tag{2}$$

It follows almost directly from the definition of this function that its successive application to any edge leads eventually to a repetition of the desired pattern:

Proposition 2 *The iterates of G on $\mathcal{E}$ achieve the pattern B.*

2.4 Hybrid Edge Point Fields

A semiflow, $(X)^t$, on the graph induces excursions in $\mathcal{L}$ parametrized by an initial condition as follows. The first letter corresponds to the edge in which the initial condition is located (initial conditions at vertices are assigned to the incident edge along which the semiflow points). The next letter is added to the sequence by motion through a vertex from one edge to the next.

We will say of two edge point fields, X_1, X_2 on a graph, Γ, that X_1 *prepares* X_2, denoted $X_1 \succ X_2$, if the goal of the first is in the domain of attraction of the second,

$$\mathbf{g}(X_1) \subset \mathcal{N}(X_2).$$

Given any finite collection of edge point fields on Γ, we will choose some $0 < \alpha < 1$ and assume that their associated Lyapunov functions have been scaled in such a fashion that $X_1 \succ X_2$, implies

$$(\Phi_1)^{-1}[0,\alpha] \subset \mathcal{N}(X_2).$$

In other words, an α crossing of the trajectory $\Phi_1 \circ (X_1)^t$ signals arrival in $\mathcal{N}(X_2)$.

Suppose now that for every edge in some pattern block, $e_i^0 \in \mathcal{E}^0$, there has been designated a goal point, g_i^0, along with an edge point field X_i^0 taking that goal, $\mathbf{g}(X_i^0) = g_i^0$. Assume as well that the edge point field associated with each previous edge in the pattern prepares the flow associated with the next edge, in other words, using the successor function (1) we have,

$$\mathbf{g}\left(X_j^0\right) \subset \mathcal{N}\left(X_{L(j)}^0\right).$$

Now construct edge point fields on all the edges of Γ such that the tree representation of their $\succ$ relations is exactly the tree pruned from the original graph above — namely we have

$$\mathbf{g}\left(X_j^p\right) \subset \mathcal{N}\left(X_{L(j)}^{p-1}\right).$$

We are finally in a position to construct a hybrid semi-flow on Γ. This feedback controller will run the piece-wise smooth vector field, $\dot{x} = X$, as follows

$$X{:=}\left\{ \begin{array}{ll} X_j^p & :x \in e_j^p \text{ and } \Phi_j^p > \alpha \\ X_{L(j)}^{p-1} & :x \in e_{L(j)}^{p-1} \text{ or } \Phi_j^p \leq \alpha \end{array} \right.. \tag{3}$$

It is clear from the construction that progress from edge to edge of the state of this flow echoes the graph transition rule G, constructed above.

Proposition 3 *The edge transitions induced by the hybrid controller (3) are precisely the iterates of the graph map, G, (2) in the language, $\mathcal{L}$.*

3 Single Letter Patterns on Manifolds

In the voluminous literature concerned with artificial potential fields for robotic path planning, one often encounters the notion of a "locally repelling" field — a C^∞ (and typically piecewise analytic) gradient field capable of repelling in the neighborhood of an obstacle but otherwise (i.e., away from that neighborhood) completely goal-minded. In the context of multiple robots inhabiting the same workspace, this motivates the notion of "interleaving" controllers that promote the goals of each individual robot in isolation most of the time, reacting (in an appropriately repelling manner) to the presence of others only in configurations near the diagonal. We suspect that there may be fundamental obstructions to such strategies in configuration spaces with sufficiently large fundamental group.

In this section we explore the notion of "interleaving" in the simplest setting: where the goal of each robot consists of a single point and where the workspace is a manifold. Even here, it becomes apparent that there may exist intrinsic obstructions to matching up fully decoupled goal-seeking fields with repelling fields in the neighborhood of the diagonal.

3.1 The Line Segment

A segmented line, Λ, is any graph homeomorphic to $[-1,1]$ — for example, any connected graph with one edge and two vertices.

Suppose $g \in \Lambda$ is a point goal and let $h_g{:}\Lambda \to [-1,1]$ be a diffeomorphism that takes g to the origin. Then

$$\gamma_g(x){:=}(h_g(x))^2/2 \tag{4}$$

is a navigation function for g on Λ according to the definition in [14].

We now show how to build a safe cooperative version of this pattern with two robots. Define $\delta(x){:=}|x_1 - x_2|^2$. Since the diagonal,

$$\Delta{:=}\delta^{-1}[0] \subset \Lambda \times \Lambda \cong [-1,1] \times [-1,1],$$

disconnects the configuration space, it suffices to consider cooperation with two AGV's.[4] Consider, then, the case of a pair of AGV's within a connected component $\mathcal{C}_0$ of the configuration space,

$$x{:=}(x_1, x_2) \in \mathcal{C}_0.$$

[4]Note, however, looking ahead to a discussion to follow, that for m points on a line segment, the configuration space has $(m-1)!$ path components, given by the ordering of the points on the line.

Assume that there is some goal $g = (g_1, g_2) \in \mathcal{C}_0$. We have

$$\gamma_g(x){:=}\gamma_{g_1}(x_1) + \gamma_{g_2}(x_2),$$

as a safe navigation function for (g_1, g_2) on the square $[-1,1] \times [-1,1]$, but not safe, in general, on the diagonally severed $\mathcal{C}_0$.

It is established in [13] that

$$\Phi(x){:=}\gamma_g(x)/\delta(x)$$

is a safe navigation function for $g = (g_1, g_2) \in \mathcal{C}_0$. Thus, the gradient vector field associated with Φ produces the safe pattern defined by the point goal g.

One might have wished, instead, for a more "decoupled" controller than one arising from the vector field $\mathbf{grad}\,\Phi$. For example, variants on the scaled cross product vector field

$$\left(\mathbf{grad}\left(\gamma_{g_1}/\delta\right), \mathbf{grad}\left(\gamma_{g_2}/\delta\right)\right)$$

appear to yield safe navigation functions on $\mathcal{C}_0$ as well [5]. Notice that such a construction still presumes that centralized information regarding the state of each agent is available to all of them but relaxes their need to share information about their individual goals. Moreover, this formula might lend itself nicely to a C^∞ "blending" with the fully decoupled gradient field $\mathbf{grad}g_1 \times \mathbf{grad}g_2$ away from the boundary.

3.2 The Circle

Matters are much less satisfactory on the circle S^1 — a manifold which is not contractible. To begin with, every smooth edge point field will incur an unstable fixed point — only ***essential*** global attraction is possible. For example, identify S^1 with the unit circle in the complex plane $S^1 = \{e^{i\theta} : \theta \in [0, 2\pi)\}$. Then, given a goal θ_g, define the navigation function

$$\gamma_g(\theta){:=}1 - \cos(\theta - \theta_g). \tag{5}$$

Since every continuous function on a compact set takes both a maximum and a minumum on that set, this construction (and, indeed, any other) necessarily introduces a spurious unstable fixed point, in this case, at $\theta_u{:=}\theta_g - \pi$.

For understanding the configuration space of several AGV's on a circle, the following lemma (whose simple proof we omit) is key.

Lemma 4 *The configuration space $C_{C,n}$ of n points on a circle is homeomorphic to $S^1 \times C_{L,n-1}$, where $C_{L,m}$ denotes the configuration space of m points on the line segment.*

From the previous subsection, we know that $C_{L,m}$ is disconnected for $m > 1$. Since taking a cross product with S^1 does not change the number of connected components, we have that $S^1 \times S^1$ is not disconnected by the pairwise diagonal, but triple and higher cross products are.

We now explore the suggestion that the topology of the circle might preclude naive interleaving of individual navigation functions, in contrast to the situation for Λ. For example, consider the situation in which there are two AGV's with goals g_1 and g_2. By construction, we have a pair of navigation functions,

$$\gamma_1(\theta){:=}1-\cos(\theta-\theta_{g_1})\,;\;\gamma_2(\theta){:=}1-\cos(\theta-\theta_{g_2}).$$

One may interleave to form a navigation function

$$(\gamma_1(\theta_1)+\gamma_2(\theta_2))\,/\delta(\theta_1,\theta_2),$$

where δ is a nonegative function with $\delta^{-1}(0) = \Delta$, such as $1-\cos(\theta_1-\theta_2)$. Upon so doing, one can show that large regions of initial conditions do not reach the goal state. For example, in the case where $g_1 = 0$, $g_2 = \pi/2$, all initial conditions $\pi < \theta_1 < \theta_2 < 3\pi/2$ are trapped by the position of the sources and the diagonal repellor.

To construct safe navigation functions for attaining point goals of multiple AGV's on the circle, one may use the homeomorphism $h : C_{C,n} \to S^1 \times C_{L,n-1}$ of Lemma 4 to reduce to the cases already considered without the problems noted above. For example, in the case of two AGV's, the homeomorphism

$$h(\theta_1,\theta_2){:=}(\theta_1,\arg(\theta_2,\theta_1)),$$

takes $C_{C,2}$ to $S^1 \times (0, 2\pi)$. Here, we denote by $\arg(\theta_2, \theta_1)$ the unique number $z \in (0, 2\pi)$ such that $\theta_1 + z = \theta_2$ modulo 2π. Given a pair of goals $\theta_{g_1} \neq \theta_{g_2}$, we have the corresponding point $h(\theta_{g_1}, \theta_{g_2}) = (\theta_{g_1}, x_{g_2})$, where $x_{g_2}{:=}\arg(\theta_{g_2}, \theta_{g_1},)$. Construct the navigation function $\gamma_1(\theta)$ for θ_{g_1} as per (5). Then, construct the navigation function $\gamma_2(x)$ for x_{g_2} on the line segment $(0, 2\pi)$ using the obvious generalization of (4). Then, one obtains a safe navigation function Φ on $C_{C,2}$ by pulling back the sum via the homeomorphism:

$$\Phi(\theta_1,\theta_2){:=}(\gamma_1,\gamma_2)\cdot h(\theta_1,\theta_2).$$

This may be extended to the case of three (or more) AGV's on the circle by using the product navigation functions for multiple points on the line segment via a similar homeomorphism h.

Unfortunately, the decoupled field **grad** γ_1 + **grad** γ_2 on the "model space", $S^1 \times C_{L,n-1}$, is not at all decoupled when the pulled back to the physical setting in $C_{C,n}$. We shall explore a similar phenomenon in the far more complicated case of the Y-graph.

The skeptical pragmatic reader may find motivation to persevere through our account of this next complication by noting that even the apparently contrived problem of scheduling AGVs on a single loop discussed in this section may, in itself, hold some practical value. For example, the choice of optimal dispatching and scheduling rules for multiple AGVs on a single loop seems to be far from understood within the Industrial Engineering community [2].

4 The Y-Graph

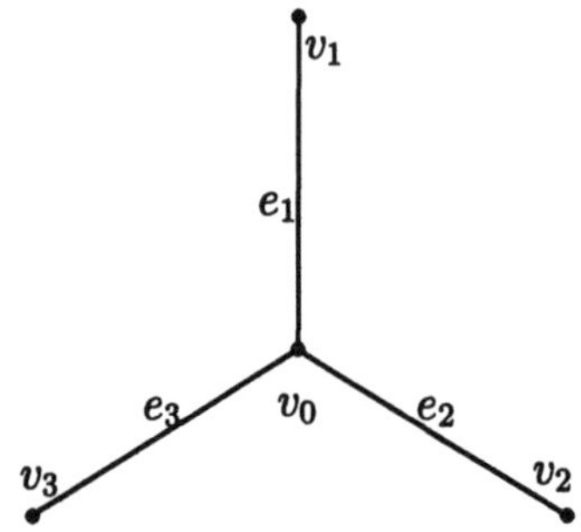

Figure 1: The Y-graph Υ.

In this section, we consider the simplest nontrivial example of a non-manifold configuration space: that associated to a *Y-graph*, Υ, having four vertices $\{v_i\}_0^3$ and three edges $\{e_i\}_1^3$, as illustrated in Figure 1. The topological features associated to such a system differ starkly from that of the previous examples. But this specical case has more general interest, since all graphs may be built up by gluing together K-pronged stars, of which the $K = 3$ model, Υ, is the simplest nontrivial example. In any case, this setting certainly suggests the richness of the problem of dynamics on general graphs.

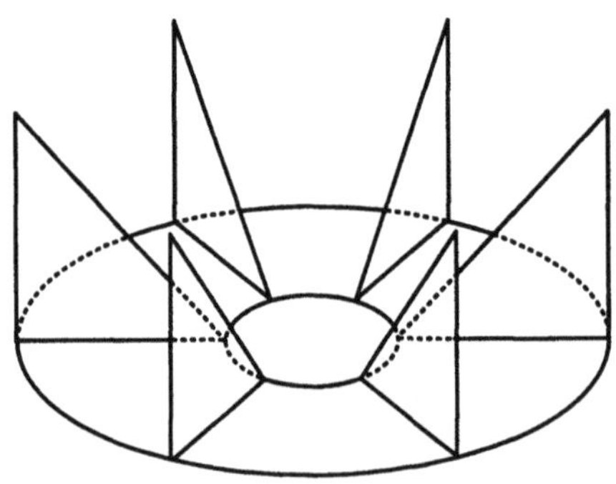

Figure 2: The configuration space C embedded in R^3.

Theorem 1 *The configuration space C associated to a pair of AGV's restricted to the Y-graph Υ is homeomorphic to an annulus with six 2-simplices attached as in Figure 2.*

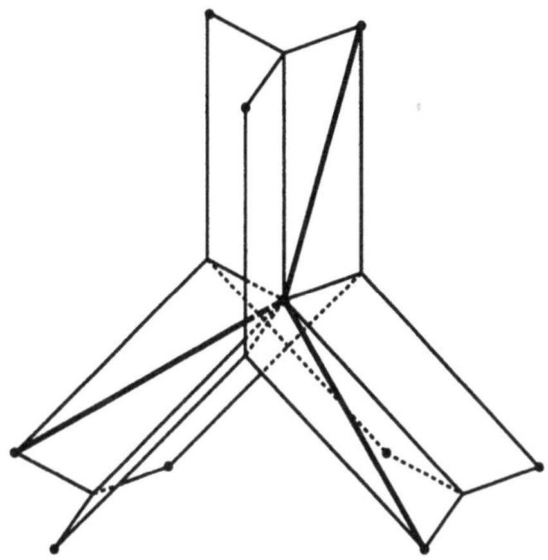

Figure 3: The space $\Upsilon \times \Upsilon - \Delta$.

Proof: The cross product $\Upsilon \times \Upsilon$ with the diagonal removed appears as in Figure 3: here, we have replaced every point of Υ with another copy of Υ and subsequently removed diagonal points, including the "center" of the graph. Note that this object as presented does not embed in R^3, but rather has an artificial self-intersection. To simplify the presentation, we deform the six "fins" created by the diagonal cuts as in Figure 4, yielding the simplicial complex of Figure 5. This can be easily deformed into the hexagonal "star" of Figure 6, where the six radial lines correspond to where the diagonally severed fins were retracted. The center point, all that remains of the punctured diagonal set, has been removed. Upon reattaching these fins, we may transform the configuration space by a homeomorphism which takes the hexagonal star to a smooth annulus, yielding the final form of Figure 2, which does embed in R^3. □

Corollary 5 *Given any pair of goals $g:=(g_1, g_2)$ where g_1 and g_2 live on different branches of Υ, there exists a navigation function (of class real-analytic off of the branch set) which sends all but a finite number of initial conditions to g under the gradient flow.*

Proof: On that portion of Figure 2 which is an annulus, the conditions for the theorems of Koditschek and Rimon [14] are met, since an annulus is a *sphereworld.* Hence, a navigation function on this subspace exists. Since the individual goals are not on the same branch of the graph, one may extend the navigation function to a function on the entire configuration space by defining

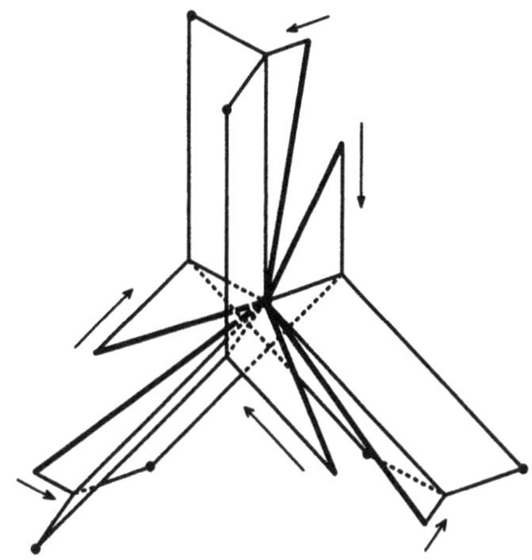

Figure 4: Retract the six fins of C.

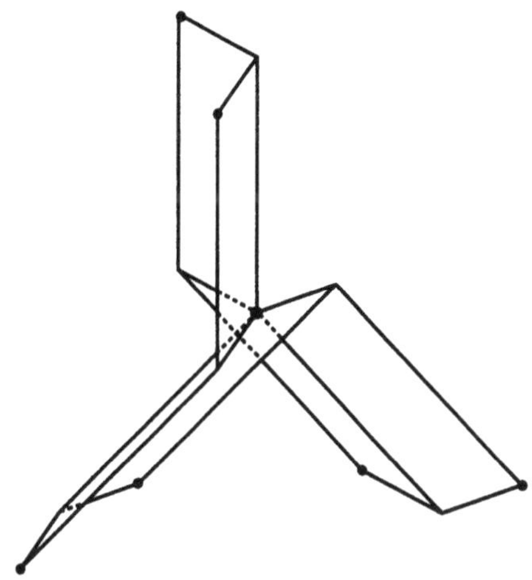

Figure 5: A simplicial complex with punctured center.

the action of the flow on the fins to monotonically "descend" away from the diagonal and onto the annulus, where the implicitly defined function takes over as per our definitions for gradients and vector fields on non-manifolds. Note that upon prescribing the flow on the fins to send orbits onto the annulus, we have defined a semiflow, and hence have a well-defined navigational procedure. □

This result is very satisfying in the sense that it guarantees a navigational function by applying existing theory to a situation which, from Figure 3 alone, would not appear to satisfy the conditions of being related to a sphereworld. However, it is not yet clear how such an application can be generalized to situations where the number of AGV's or the incidence number of the ambient graph increases. Hence, we consider an alternate

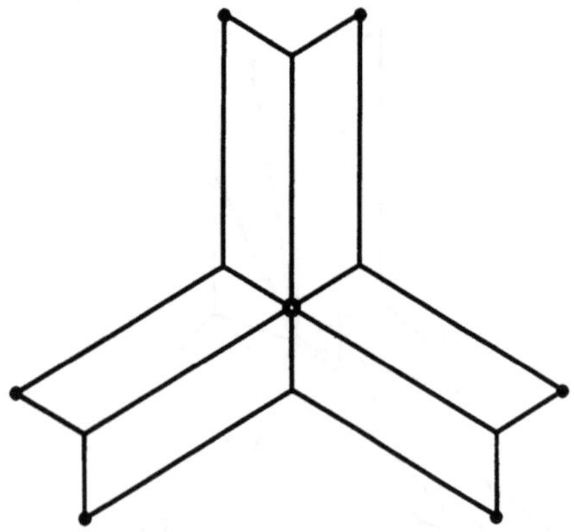

Figure 6: Flattening out yields a punctured hexagonal star.

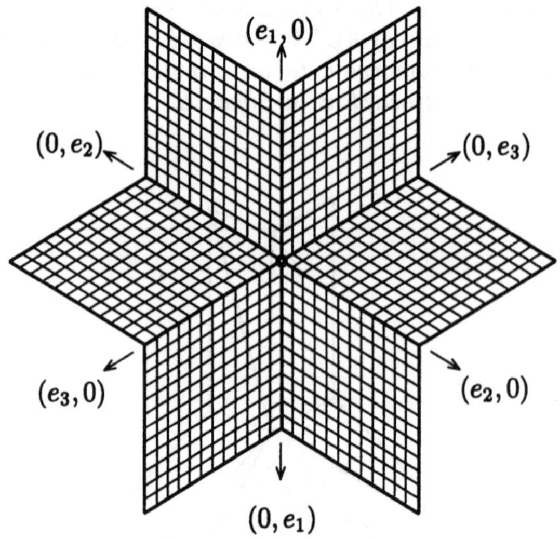

Figure 7: The coordinate system on the annular region of C.

solution to the problem of realizing compatible goals by means of a vector field on the configuration space. It is our belief that this method will readily adapt to complicated settings.

Theorem 1 and Figure 2 suggest a natural *circulating flow* on the configuration space C which has the effect of inducing a "dance" between the pair of AGV's until the appropriate configuration is reached.

Theorem 2 *There exists a piecewise-smooth vector field X on C which has the following properties:*

1. *X defines a nonsingular semiflow on C;*
2. *The diagonal Δ is repelling with respect to X; and*
3. *Every orbit of X approaches a unique attracting limit cycle on C which cycles through all possible ordered pairs of different edge-states.*

Proof: To construct the vector field, we introduce the following coordinate system fitted with respect to the topology of Υ. Let $\{e_i\}_1^3$ denote the three edges in Υ, parametrized so that $e_i \cong [0,1]$ with each $\{0\}$ identified at the center v_0 of Υ. Denote by $\hat{e}_i$ the unit tangent vector in each tangent space $T_x e_i$ pointing in the positive (outward) direction towards the endpoint v_i. Any point $x \in \Upsilon$ is thus given by a vector (x_1, x_2, x_3) in the $\{e_i\}$ frame, where $x_i \in [0,1]$ and at least two of these coordinates is zero. In other words, we are embedding Υ as the positive unit axis frame in R^3. Likewise, a point in C is given as a pair of distinct vectors (x, y), i.e., as a unit axis frame in $R^3 \times R^3 \cong R^6$ (see Figure 7). The reader should think of this as a collection of six unit coordinate planes, attached together pairwise along axes with the origin removed.

Any vector field on C may be uniquely represented as a pair of vectors in the $\{\hat{e}_i\}$ basis. Given a point $x \in \Upsilon$, denote by $\iota(x)$ the index of the nonzero coordinate of x (or by zero if $x = (0,0,0)$). Denote also by $|x|$ the value of the nonzero coordinate of x (or by zero if $x = (0,0,0)$). Thus, $x = |x|e_{\iota(x)}$. Any addition operation on the level of indices will always denote addition mod three.

The vector field we propose is the following: given $(x, y) \in C$,

1. If $\iota(x) = \iota(y)$ then

$$\left\{ \begin{array}{l} \dot{x} = -|y|\hat{e}_{\iota(x)} \\ \dot{y} = |y|(1-|y|)\hat{e}_{\iota(y)} \end{array} \right\} \quad 0 < |x| < |y|$$
$$\left\{ \begin{array}{l} \dot{x} = |x|(1-|x|)\hat{e}_{\iota(x)} \\ \dot{y} = -|x|\hat{e}_{\iota(y)} \end{array} \right\} \quad 0 < |y| < |x| \tag{6}$$

2. If $\iota(x) = \iota(y) + 1$ or $\iota(x) = 0$ then

$$\left\{ \begin{array}{l} \dot{x} = |y|\hat{e}_{(\iota(y)+1)} \\ \dot{y} = |y|(1-|y|)\hat{e}_{\iota(y)} \end{array} \right\} \quad 0 \le |x| < |y|$$
$$\left\{ \begin{array}{l} \dot{x} = |x|(1-|x|)\hat{e}_{\iota(x)} \\ \dot{y} = -|x|\hat{e}_{\iota(y)} \end{array} \right\} \quad 0 < |y| \le |x| \tag{7}$$

3. If $\iota(y) = \iota(x) + 1$ or $\iota(y) = 0$ then

$$\left\{ \begin{array}{l} \dot{x} = -|y|\hat{e}_{\iota(x)} \\ \dot{y} = |y|(1-|y|)\hat{e}_{\iota(y)} \end{array} \right\} \quad 0 < |x| < |y|$$
$$\left\{ \begin{array}{l} \dot{x} = |x|(1-|x|)\hat{e}_{\iota(x)} \\ \dot{y} = |x|\hat{e}_{(\iota(x)+1)} \end{array} \right\} \quad 0 \le |y| \le |x| \tag{8}$$

The vector field defines a nonsingular semiflow as follows: first, the vector field is by inspection nonsingular. Second, away from a neighborhood of the non-manifold points (where the fins attach to the annulus), the vector field is well-behaved and defines unique solution curves locally. Finally, at the non-manifold points, the vector field at any point is constructed so as to have two incoming directions and one outgoing direction; hence, the forward orbit of the vector field is uniquely determined.

This vector field admits a C^0 Lyapunov function $\Phi : C \to [0,1)$ of the form

$$\Phi(x,y) := \begin{cases} 1 - |(|x| - |y|)| & : \iota(x) = \iota(y) \\ 1 - \max\{|x|,|y|\} & : \iota(x) \neq \iota(y) \end{cases} . \tag{9}$$

A simple calculation shows that on the fins ($\iota(x) = \iota(y)$), one has $\dot{\Phi} < 0$, and furthermore that on the annulus ($\iota(x) \neq \iota(y)$), Φ changes as

$$\dot{\Phi}(x,y) = \Phi(\Phi - 1).$$

Hence, Φ strictly decreases off of the boundary of the annulus

$$L := \{(x,y) : |x| = 1 \text{ or } |y| = 1\} = \Phi^{-1}(0).$$

It follows from the computation of $\dot{\Phi}$ that the diagonal set Δ of $\Upsilon \times \Upsilon$ is repelling, and that the boundary cycle L is an attracting limit cycle. □

Figure 8: The circulating field on the annular region of C.

The action of the vector field is to descend off of the "fins" of C onto the annular region, and then to circulate about the annulus while pushing out to the boundary cycle L, as in Figure 8. The Lyapunov function Φ decreases as the "height" when $\iota(x) = \iota(y)$ (on the fins) and decreases as the "radius" when $\iota(x) \neq \iota(y)$ (on the annulus). From this, it can be seen that the vector field constructed is merely the image of the vector field

$$\dot{r} = r(1-r) \; : \; \dot{\theta} = 1$$

under the homeomorphism taking the closed unit disc in the plane to the hexagonal star of Figure 7 by rescaling linearly along rays emanating from the origin. Of course, we may tune the radius of the limit cycle easily within this context and push it forward to a modified vector field on C by changing every term of the form $(1 - |x|)$ and $(1 - |y|)$ to, respectively, $(R - |x|)$ and $(R - |y|)$, where $0 < R < 1$ is the radius of the desired limit cycle.

5 Control of Dynamics on Graphs

The circulating vector field of §4 can be incorporated into a controller for dynamics on the Y-graph and many generalizations thereof. In this paper, we restrict ourselves to a brief sketch of how this may be accomplished, leaving details for a more comprehensive work.

The principal features of the vector field of §4 are as follows:

1. It generates a well-defined nonsingular semiflow on Υ.
2. Orbits of the semiflow attract onto an orbit which cycles transitively through all ordered pairs of distinct edge combinations.

Consider the following simple hybrid controller on a general graph Γ with trivalent non-manifold points (i.e., there exists an atlas for Γ all of whose charts are homeomorphic copies of Υ). When the AGV's are geographically isolated (as measured by some natural metric d), use the edge point fields of §2 to send each AGV to its predetermined point goal, keeping track of the relative distances between AGV's. The distance between a pair of AGV's reaches a critical threshold when they are on a "collision course". Upon crossing the threshold, turn off the edge point fields, and turn on the circultaing field associated to the current chart on which the pair resides, using the chart homeomorphism to push forward the vector field of §4. The induced semiflow will by (1) and (2) above ensure that the AGVs' relative ordering on the troublesome edge are eventually permuted. When one of the AGV's has returned to its initial position with the prior obstruction removed, turn off the circulating field and resume the edge point field control.

In this way, the motion of the individual AGV's is to descend monotonically towards the goal except when a collision is imminent. Then, the circulating controller induces a "dance" between the pair which has the effect of moving the obstacle out of the way, allowing for further progress towards the goal. This prescribes an efficient, and courteous, traffic flow.

The future direction of this line of inquiry is clear: analogues of the circulating fields should be constructed for the more general cases of higher graph incidence and more AGV's. This presents a nontrivial topological problem, as noted earlier. Finally, it would be advantageous to implement a controller which is more modular — where incrementing the number of AGV's does not require a retooling of the entire controller. However, as noted in the case of dynamics on a circle, this is not always possible by straightforward interleaving of navigation functions.

Acknowledgements

We thank Professors Anthony Bloch, Philip Holmes and Stéphane Lafortune for a number of informative tutorial discussions bearing on the problems addressed in this paper.

References

[1] R. Alami, T. Simeon, and J.-P. Laumond. A geometrical approach to planning manipulation tasks. the case of discrete placements and grasps. In H. Miura and S. Arimoto, editors, *Robotics Research*, pages 453–463. MIT Press, 1990.

[2] J. J. Bartholdi and L. K. Platzman. Decentralized control of automated guided vehicles on a simple loop. *IIE Transactions*, 21(1):76–81, 1989.

[3] Y. A. Bozer and C. kuan Yen. Intelligent dispatching rules for trip-based material handling systems. *J. Manufacturing Systems*, 15(4):226–239, 1996.

[4] Y. A. Bozer and M. M. Srinivasan. Tandem configurations for automated guided vehicle systems and the analysis of single vehicle loops. *IIE Transactions*, 23(1):72–82, 1991.

[5] H. I. Bozma, C. S. Karagoz, and D. E. Koditschek. Assembly as a noncooperative game of its pieces: The case of endogenous disk assemblies. In *IEEE International Symposium on Assembly and Task Planning*, 1995.

[6] Y. Brave and M. Heymann. On optimal attraction of discrete-event processes. *Information Science*, 67(3):245–276, 1993.

[7] R. W. Brockett. Hybrid models for motion control systems. In H. L. Trentlman and J. C. Willems, editors, *Essays in Control: Perspectives in the Theory and Its Applications.* Birkhauser, Boston, MA, 1993.

[8] R. C. Brost. Computing metric and topological properties of configuration space obstacles. In *Proc. IEEE Conference on Robotics and Automation*, pages 170–176. IEEE, Scottsdale, AZ., 1989.

[9] R. R. Burridge, A. A. Rizzi, and D. E. Koditschek. Sequential composition of dynamically dexterous robot behaviors. *Int. J. Rob. Res.*, (to appear).

[10] G. A. Castleberry. *The AGV Handbook.* Braun-Brumfield, Ann Arbor, MI, 1991.

[11] M. Erdmann. Understanding action and sensing by designing actions-based sensors. *Int. J. Rob. Res.*, 14(5):483–509, 1995.

[12] S. B. Gershwin. *Manufacturing Systems Engineering.* Prentice Hall, Englewood Cliffs, NJ, 1994.

[13] D. E. Koditschek. An approach to autonomous robot assembly. *Robotica*, 12:137–155, 1994.

[14] D. E. Koditschek and E. Rimon. Robot navigation functions on manifolds with boundary. *Advances in Applied Mathematics*, 11:412–442, 1990.

[15] J.-C. Latombe. *Robot Motion Planning.* Kluwer Academic Press, Boston, MA, 1991.

[16] J.-H. Lee, B. H. Lee, M. H. Choi, J. D. Kim, K.-T. Joo, and H. Park. A real time traffic control scheme for a multiple agv system. In *IEEE Int. Conf. Rob. Aut.*, pages 1625–1630, Nagoya, Japan, 1995.

[17] D. Lind and B. Marcus. *Introduction to Symbolic Dynamics and Coding Theory.* Cambridge, 1995.

[18] T. Lozano-Perez, M. T. Mason, and R. H. Taylor. Automatic synthesis of fine-motion strategies for robots. *Int. J. Rob. Res.*, 3(1):3–23, 1984.

[19] M. T. Mason. The mechanics of manipulation. In *Proc. International Conference on Robotics and Automation*, pages 544–548, March 1985.

[20] W. L. Maxwell and J. A. Muckstadt. Design of automatic guided vehicle systems. *IIE Transactions*, 14(2):114–124, 1981.

[21] J. R. Munkres. *Topology, A First Course.* Prentice Hall, 1975.

[22] D. Ruelle. *Elements of Differentiable Dynamics and Bifurcation Theory.* Academic Press, NY, 1989.

[23] R. Sengupta and S. Lafortune. An optimal control theory for discrete event control systems. *SIAM J. Control and Optimization*, (to appear).

[24] S. F. Smith. Reactive scheduling systems. In D. E. Brown and W. T. Schering, editors, *Intelligent Scheduling Systems*, pages 155–192. Kluwer Academic Publishers, Boston, MA, 1995.

[25] P. Svestka and M. H. Overmars. Coordinated motion planning for multiple car-like robots using probabilistic roadmaps. In *IEEE Int. Conf. Rob. Aut.*, pages 1631–1636, 1995.

PART 4

MOTION PLANNING

SESSION SUMMARY

Georges Giralt
LAAS-CNRS
7, avenue du Colonel Roche, 31077 Toulouse.cedex 4 - France
giralt@laas.fr

The capacity for some sort of automated motion in our physical world is certainly a central feature present in all robotic systems, from the most simple application-tailored pick-and-place machine to the most versatile laboratory-born all-terrain mobile robot. This salient feature of any real-work robot expands to a key functionality for machine-intelligence based robots endowed with decisional autonomy and reasoning capacities including task planning.

Hence, robot motion planning has attracted in the recent years a large interest both from a theoretical standpoint and from applied, efficiently implemented, systems and this as well for manipulation tasks as for mobile robots.

This last category of machines encompasses the largest set of theoretical and implementation issues, including the front-line subjects related to motion planning in presence of uncertainty ([partially]-unknown environment, dynamic events, noise/errors...). The four papers, although based on methods and techniques which are of a general nature, are largely devoted to mobile robot topics, with a complete focus for papers number one (J.L. Lacombe) and number two (E. Rimon), and possibly to a lesser degree for paper number four (R. Chalita). Paper number three (V. Kumar) explicitly describes both manipulation and mobility as case studies.

Besides the algorithmic and computational intrinsical problems that are at the very core of motion planning and that still are source of many advanced research, the physical aspects involved in Robotics lead towards new and complex research issues.

In fact, the opening of novel real-world applications for mobile robots have set motion planning problems in a realistic context, with a key role for aspects related to imprecision and uncertainty, in Perception, Action and their interleaved role in Execution Control.

Any approach aiming at an effective implementation cannot oversee two very general questions:

1. **Perception**: beyond cost and reliability problems, all sensors that we can use are plagued by noise, errors and physical limitations. A relevant and spectacular example is actual range sensing of all sorts with specularity, beam-width/target-shape convolution, multiple reflections, ... We are even at a loss to guarantee the absence of collision for all **cases** and **circumstances** before actual contacts happen and there is the proper sensing system for this !

2. **Action**: all open-loop actions, or just servoed with proprioceptive sensors (articulation/wheel encoders...), will generate errors that will quickly built to be dramatic. Exteroceptive sensors are necessary for correct execution control with two possible modalities:

    - Perception-based execution control: here a classical sequential schema is implemented where the situation after an action step is evaluated by local perception and a decisional process commands the next step or, in case of impossibility to proceed with the current plan, calls for replanning.
    - Sensor-based execution control: here the planned actions are defined directly as feedback-loop modules that continuously servo the actions. Classical cases widely used are target/beacon/landmark tracking, wall following,...

The four papers presented in this session, although they do not cover, of course, the full scope and all the research topics of the domain, represent highly interesting contributions that provide an enlightening perspective for several research trends. Pure configuration space algorithmic questions are addressed by J.C. Latombe and co-authors in their paper. Similarly, E. Rimon and co-authors, although dealing with sensor-based motion planning, largely restrain their contribution to the algorithmic geometry aspects. In the following two papers, V. Kumar and R. Chatila explicitly address autonomous motion problems that entail execution control as an important feature.

Several interesting questions related to general subjects were discussed. We will only briefly refer in the following paragraph to issues and matters directly concerned with the central themes of computational complexity, uncertainty and execution control.

The first paper *"Motion Planning with Visibility Constraints: Building Autonomous Observers"* presented and discussed by J.C. Latombe concerns the concept of an autonomous observer as a basic tool to implement, in an indoors-like environment, tasks such as model acquisition/verification, surveillance (target finding), and target tracking. At the core of the paper, ***target finding*** which is presented in a detailed way, attracted questions and comments regarding the impact of less ideal perception models and uncertainty, e.g. environment model errors and robot position imprecision, on algorithms complexity. This – limited range perception, target bounded speed, imperfect observer localization – is part of the current work of the authors with no foreseen dramatic changes in complexity.

Limited range, but otherwise perfect, perception is chosen by E. Rimon at the outset of his work: *"Local and Global Planning in Sensor Based Navigation of Mobile Robots"*. Here, since the environment is a priori unknown at the start, planning and actual motion are continuously integrated with sensing and perception present at every step. Part of the discussion was concerned, beyond theoretical correctness, with the practical implementability of two critical concepts using physical sensors to measure ***critical points*** and ***minimum passage points*** which are so far detected by ***abstract sensors*** i.e. computed. This is a current research topic for the first author.

Departing completely from the approaches of algorithmic geometry used in these two papers, V. Kumar sets his work *"Motion planning in humans and robots"* in the frame of optimal control and variational calculus. The criteria to optimize is here inspired by human performance in object manipulation. The approach is extended to explicitly deal with environment uncertainty in the planning process by means of game theory. The problem to find an optimal feedback strategy can be deemed as difficult in almost all practical cases. An approximation is proposed as a tractable solution. The question of tractability and the computational scheme proposed was a matter of general discussion, particularly in relation to the on-line processus during motion execution with dynamic obstacle avoidance.

The critical issue, streaming from uncertain and time variant environments, of continuous dynamic adjustment and control of a mobile robot movement is addressed in the paper presented by R. Chatila *"Interleaving Motion Planning and Execution for Mobile Robots"*. The concepts of ***task potential*** and ***bubble bands*** are integrated in a comprehensive approach to deal with the problem of combining path planning in reactive sensor-based motion control including for non-holonomic vehicles. Thus, the full scope of the paper is devoted to the crucial problem of effective decisional autonomy in execution control, embedding, whenever necessary, partial replanning. Similarities, i.e. feedback control schemes, and differences, i.e. initial planning as a separate process, with V. Kumar's paper, were discussed as well as the computational problems related to the estimate of several module tuning parameters.

Motion Planning with Visibility Constraints: Building Autonomous Observers

H.H. González-Baños L. Guibas J.C. Latombe S.M. LaValle
D. Lin R. Motwani C. Tomasi
Department of Computer Science
Stanford University, Stanford, CA 94305, USA

Abstract

Autonomous Observers are mobile robots that cooperatively perform vision tasks. Their design raises new issues in motion planning, where visibility constraints and motion obstructions must be simultaneously taken into account. This paper presents the concept of an Autonomous Observer and its applications. It discusses three problems in motion planning with visibility constraints: model building, target finding, and target tracking.

1 Introduction

We are interested in mobile robots which autonomously perform vision tasks such as building 3-D models of unknown environments and finding/tracking unpredictable targets in cluttered environments. We call such robots *Autonomous Observers* (AOs for short). Multiple AOs may team up to accomplish tasks more quickly or to achieve goals that no AO could attain alone. E.g., it may not be possible for a single AO to reliably find a fast-moving target in a cluttered environment.

In the military domain, AOs may help assess the situation in a building by detecting potentially hostile targets and track their motions. In operating rooms, surgeons often operate by watching graphic displays of key tissues; AOs could automatically maintain visibility of the tissues in spite of obstructions caused by people and complex mechanical instruments. In robotics, researchers at one institution may want to conduct an experiment using hardware at another institution; AOs could be used to gather and transmit crucial real-time information over the Internet, allowing the remote researchers to effectively monitor their experiment. Other applications include remote monitoring of manufacturing operations in an assembly plant, search/rescue in a potentially hostile environment, and supervision of automated construction efforts in space.

AOs must execute motion strategies in which *visibility and motion obstructions are simultaneously taken into account.* In other words, they must not only avoid colliding with obstacles, an already well studied problem; they must also move so as to satisfy some visibility constraints. E.g., in the model building task, the AOs must eventually see the entire environment; in target finding, one AO must eventually see the target; and in target tracking, at least one AO must see the target at any one time. Concurrently, it is often desirable to minimize the number of AOs used. Note the relation with art-gallery problems [10], where the goal is to compute the locations of a minimal number of fixed guards that can collectively see all points in a given environment. In our case, AO mobility makes it possible to significantly reduce this number. Target tracking has an obvious connection to visual tracking of a moving object in an image sequence [6]. But, while the goal of the latter problem is to track the object *as long as it is visible in the images*, AOs must move to avoid potential visual obstruction by obstacles and keep the target in their field of view. Planning AO motions also relates to sensor placement [2] and active sensing [9].

In this paper we present our ongoing research on three specific planning problems related to the design of a team of AOs: i) *model building*, ii) *target finding*, and iii) *target tracking*. This sequence of problems corresponds to the following scenario: AOs are dropped into an unknown environment, of which they first have to build a model; then they have to find a target hiding among view-obstructing obstacles; finally, they must track this target's motions. However, the results obtained for each problem can be used independently. For lack of space, we only outline the main features of our approach for each problem and we show some experimental results. To give a better sense of the sort of algorithmic issues involved, we present one problem – target finding – in more detail.

2 Model Building

A basic task for an AO is to build a model of an environment using vision sensing. We wish this model to be usable for a variety of purposes, including future navigation (e.g., for target finding and tracking) and generation of realistic graphical renderings for virtual walkthrough operations. The model constructed by our AOs combines 2-D and 3-D geometry with texture maps.

A classical problem in automatic model building is known as the *next-best-view problem* [1, 9, 12]: Where to place the sensor next to maximize the amount of information that will be added to the partial model built so far? But existing techniques do not ideally suit AOs. One reason is inherent to the next-best-view problem itself: it is a *local* planning problem [7], so that a sequence of next-best views to build a complete model may yield too many sensing operations. In our case, each sensing operation is rather expensive: it requires acquiring 3-D and texture data, and merging this data with the current model. Hence, we wish to minimize the total number of operations. Moreover, a limitation of most next-best-view techniques is that they assume precise localization of the sensor. With our AOs, localization uncertainty must be taken into account. Finally, due to physical obstructions, a next-best-view technique may not suggest viewing positions that are accessible to the AOs.

These remarks led us to address model building in a different way. Assume for a moment that we are given a 2-D map describing the geometry of a horizontal cross-section of the environment at approximately the height of an AO's camera. An art-gallery algorithm computes the positions that the AOs must visit to eventually see the entire 2-D environment. We refine this approach as follows:

(1) The fact that the entire 2-D environment is visible from a set of positions does not in general entail that the entire 3-D environment is also visible. But this is almost true for many indoor environments. Some "holes" remain in the model built, but these are usually small and they can be filled by adding a few sensing operations at locations computed using a next-best-view technique.[1]

(2) Art-gallery algorithms use the simple "line-of-sight" visibility model: one point sees another if the line segment between them does not cross any obstacle. However, imperfections in vision sensors require that we use a more realistic definition taking distance and incidence into account. This yields new variants of the art-gallery problems.

(3) Art-gallery algorithms strive to minimize the number of positions to be visited by AOs. But 3-D sensing at these positions can yield partial models that have very small overlap between them. Uncertainty in AO localization requires that 3-D data from two partial models be aligned by partial shape matching before they are merged. Minimal overlap between the two models is needed.

(4) The 2-D model must be built in the first place. We do this by letting AOs navigate in the environment, each using a simple laser range sensor projecting a horizontal plane of light to obtain the environment's 2-D contour. Since this form of sensing is fast, the number of sensing operations is not critical, and a next-best-view technique is suitable to select on-line the successive viewing positions.

We are currently implementing this approach and experimenting with it. The planning component of our software computes the AO positions from a given polygonal map. Since the classical art-gallery problem is NP-complete, we have opted for a randomized algorithm that works as follows: First, it guesses many AO positions at random. Each position determines portions of the boundary of the 2-D map that an AO can see under the given distance and incidence constraints. The map's boundary is then partitioned into segments, each visible from the same subset of AO positions and labelled by this subset. If a segment has an empty label, more AO positions are picked. Finally, a greedy set-cover algorithm prunes the AO positions and selects a subset of them needed to see the entire boundary.[2] Fig. 1 shows an example of AO positions computed by

[1] Since a wheeled AO is not a free-flying device, some holes cannot be eliminated. This can be dealt with by limiting virtual walkthroughs to viewing positions that could be achieved by an AO.

[2] Set cover is NP-hard, but can be approximated within a log factor by a greedy algorithm.

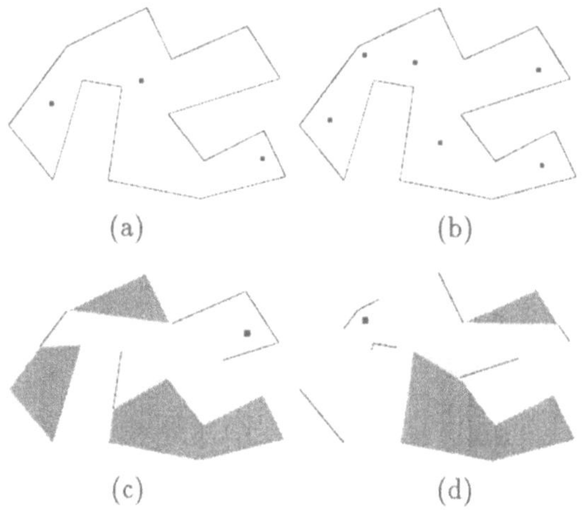

Figure 1: Computed AO positions in a 2-D map: (a) with no constraints; (b) with a minimum incidence of 60 degrees. The portions of walls seen from two AO positions are shown in solid lines in (c)-(d).

our software under incidence constraints only.

In addition to this planning component, our software also includes the construction of a partial 3-D/texture model at a given AO's position, and the fusion of multiple partial models. Fig. 2 shows the graphic rendering of a model constructed from two different positions. The scene is a pile of boxes with a book on the foreground and a whiteboard as a background.

Figure 2: 3-D/texture model

Figure 3: Geometry and number of AOs

3 Target Finding

Suppose that a model for the environment F is available. The target-finding problem is to plan a motion strategy for the AOs to eventually see a target. This strategy must be such that, as the AOs move, their visibility region (i.e., the region that they collectively see) deforms and sweeps F so that the target has eventually no remaining place where to hide. Non-geometric pursuit-evasion problems have been studied in graphs (e.g., [11]). A problem similar to ours is analyzed in [3].

In the following, we assume that the target is unpredictable, has unknown initial position, and can move arbitrarily fast. (Hence, time is irrelevant.) F is an arbitrary 2-D polygon with n edges and h holes, where each of the AOs and target is modelled as a point. The visibility model is the line-of-sight model N denotes the minimal number of AOs needed for the existence of a guaranteed strategy.

Number of AOs. Clearly, $N > 1$ if $h > 0$, since the target can always hide behind a hole to avoid being seen by a single AO. However, even when $h = 0$, small geometric differences may affect N; e.g., in Fig. 3 the rightmost environment requires two AOs, while the other two require a single AO. We have established the following worst-case bounds on N [4]:

- For simply-connected environments ($h = 0$), $N = O(\lg n)$; there exist environments such that $N = \Omega(\lg n)$.

- For multiply-connected environments ($h > 0$), $N = O(\sqrt{h} + \lg n)$; there exist environments such that $N = \Omega(\sqrt{h} + \lg n)$.

Note that the art-gallery problem in an n-sided polygon with h holes requires $\lfloor (n+h)/3 \rfloor$ static guards [10].

In [4] we also show that computing N for a given environment is NP-hard.

Single-AO Planner. The NP-hardness of computing N led us to investigate and develop a complete planner for the case of a single AO.

Let $V(q)$ be the AO's visibility region at position q in F. Each edge of $V(q)$ borders either an obstacle or free space. We call each edge bordering free space a *gap edge* and we associate a binary label with it: if the portion of free space

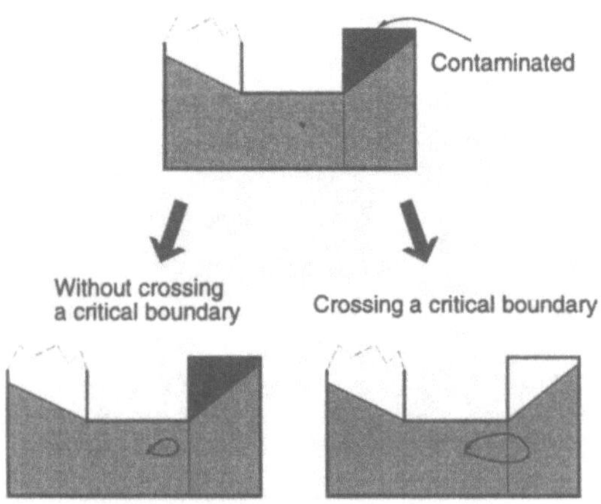

Figure 4: Critical event in information space

that borders the gap edge is *contaminated* (i.e., may contain the target), then the label is 1; otherwise, it is 0. Let $B(q)$ denote the circular sequence of the labels of the gap edges in $V(q)$. We call $(q, B(q))$ the AO's *information state*.

Let the AO move along a closed-loop path starting and ending at q. The information state at q may change only if gap edges appear or disappear during the AO's motion. To illustrate, consider Fig. 4 where the AO is approaching the end of a corridor. If the closed-loop path on the left is executed, the end of the corridor remains contaminated. If the path on the right is executed, a gap edge disappears and reappears, but with a different label (0 instead of 1).

We say that a subset S of F is *conservative* if no motion of the AO within S can cause a gap edge to appear or disappear. The critical places at which edge visibility changes form an arrangement of lines that decompose F into conservative cells. Such a decomposition has already been used in robot localization [5, 13]; it generates $O(n^3)$ cells for a simple polygon. To obtain conservative cells only, slightly fewer lines are needed than in a pure edge-visibility decomposition [4].

A directed *information state graph* G is built and searched using this cell decomposition. For each cell κ, a set of vertices are included in G for each possible labeling of the gap edges. An arc connects two vertices v_i and v_j of G if the two corresponding cells κ_i and κ_j are adjacent and if the information state of v_j is obtained from the information state in v_i when the AO crosses the boundary between κ_i and κ_j. E.g., in the case shown in the lower right of Fig. 4, assume that the gap edge on the left is initially labelled by 0 and the gap edge on the right is labelled by 1. Let the first bit denote the leftmost gap edge label. The corresponding arcs of the information state graph are $(\kappa_i, 01)$ to $(\kappa_j, 0)$, and $(\kappa_j, 0)$ to $(\kappa_i, 00)$. In other situations, if two gap edges are merged into one, the new gap edge receives the label 1 if any of the original gap edges is labelled by 1. The search terminates when an information state is reached in which all labels are 0. Our planner uses the Dijkstra's search algorithm with an edge cost that is the shortest distance between the centers of the two cells.

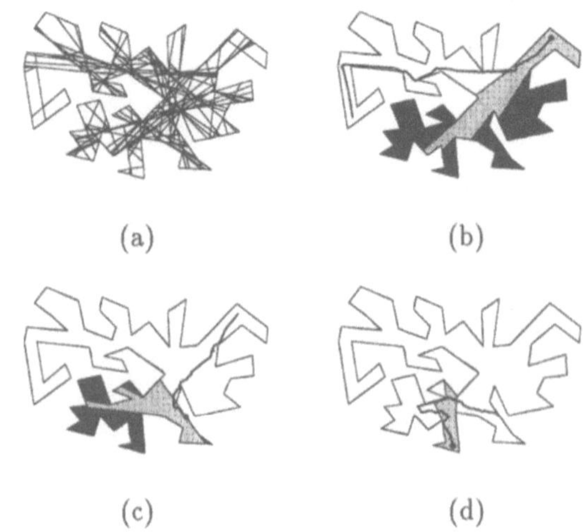

Figure 5: Target-finding strategy (example 1)

Experimental results. The planner is written in C++ and runs on an SGI Indigo2 Workstation with a 200 Mhz MIPS R4400 processor. Fig. 5 shows a path computed by this planner. The edge-visibility decomposition displayed in (a) contains 888 cells yielding an information state graph with 103,049 vertices. The path is the concatenation of the three bold lines in frames (b)–(d). The gray region is the visibility polygon at the AO's position attained in each frame (thick point); black regions are contaminated areas, while white regions are cleared areas (other than the visibility polygon). The total computation time is 171.63s. There are 130 conservative cells and the complete information state graph contains 1727 vertices. Fig. 6 shows another example which took 10.73s to compute. The environment yields 246 conservative cells and an information state graph with 18,830 vertices. Note that the AO must clear the "beak" at the top-left multiple times.

The planner is efficient in practice, but we have not established its precise complexity. Examples can be constructed that yield an exponential number of information states; but, whether

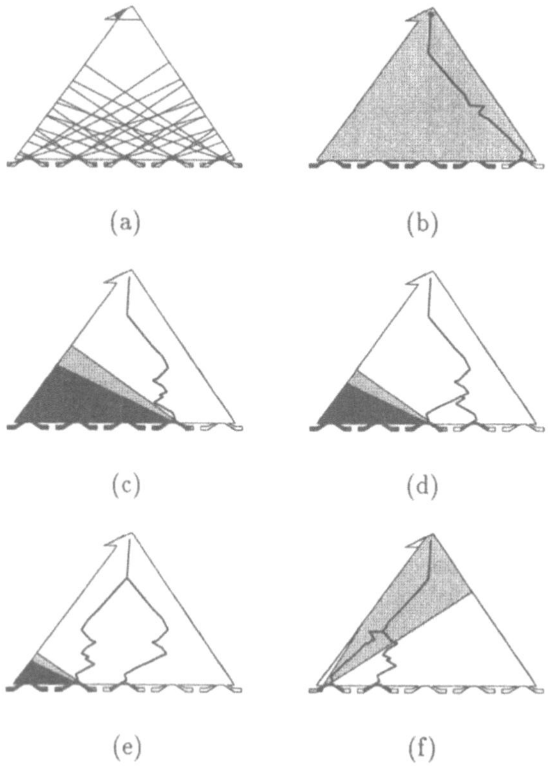

Figure 6: Target-finding strategy (example 2)

all these states may have to be considered to generate a path remains an open question. Fig. 6 shows, however, that there exist environments in which the same region is recontaminated $\Omega(n)$ times.

Multi-AO Planner. In theory, the techniques used in the single-AO planner also apply when $N > 1$, but they are likely to result in a very inefficient planner. Instead, we have developed a greedy multi-AO planner. This planner first plans for one AO, using the single-AO algorithm. If it succeeds, the problem is solved. Otherwise, the planner clears as many cells as possible. The visibility polygon of the AO at its final position partitions F into components. The planner recursively treats each contaminated component as a new environment. Let N' be the number of AOs needed to clear the component that requires the most AOs; $N = N' + 1$, since a subset of the N' AOs can also be used for clearing each of the other components in sequence.

This planner always returns a motion strategy. If this strategy uses one or two AOs, it is optimal; otherwise, a plan with fewer AOs may perhaps exist. Fig. 7 shows an example generated by the planner. In (a)-(b), the first AO clears as many cells as possible, leaving two contaminated components that are cleared by a second AO in (c)-(d).

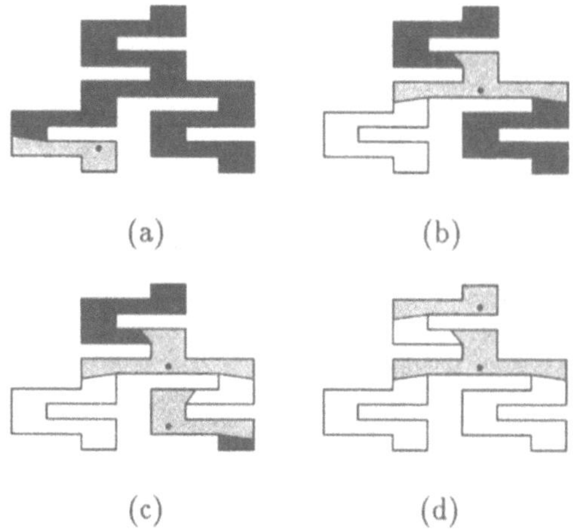

Figure 7: Target-finding strategy with two AOs

Current Work. We are studying two variations of the target-finding problem. In one we incorporate a more realistic visibility model in which the field of view of each AO is a cone with bounded depth. In the other we explore 3-D extensions that yield efficient algorithms. Other important variants include the cases where the target has bounded velocity and AO localization is imperfect; but such variants are more difficult to handle. For instance, when the target's velocity is bounded, time plays a critical role in AO strategies.

4 Target Tracking

Once a target has been found, the next step is to maintain visibility by appropriately moving the AOs, again taking visibility and motion obstructions into account. Unlike in target finding, time is now critical. The faster the planner and the more efficient the motion strategies, the better. This led us to develop two planning algorithms, depending on target predictability:

(1) For fully predictable targets, we have developed an off-line planner that computes an optimal solution for a given criterion [8]. Fig. 8 shows two examples computed by this planner. The target is displayed as a black disc and the AO as a white disc. In (a), the tracking trajectory minimizes the total distance traveled, while in (b) it minimizes the time during which the AO does not see the target under the additional constraint that the

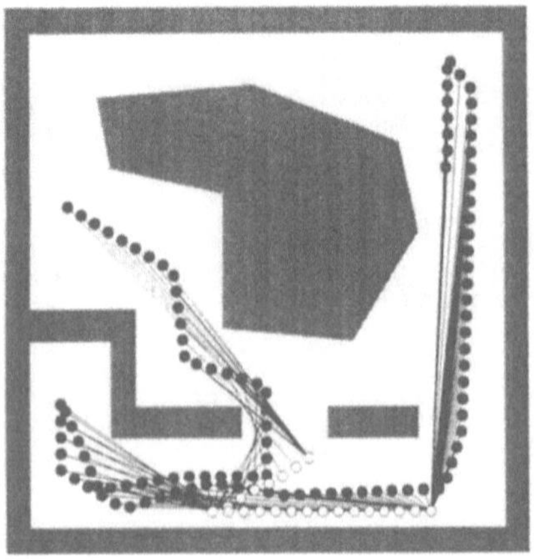

(a)

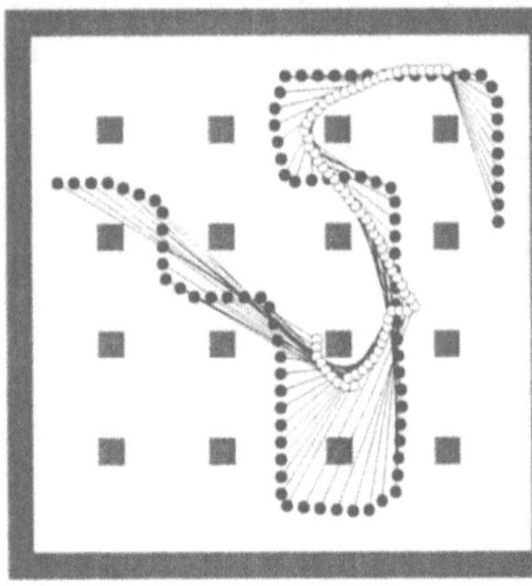

(b)

Figure 8: Optimal target tracking strategies

AO's speed is only half that of the target.

(2) For partially predictable targets (e.g., we may only know their maximum speed), we have designed an on-line planner. In one variant, this planner maximizes the probability that the target will remain in view at the next time step. In another variant, it maximizes the minimum time in which the target could escape the visibility region.

Our general approach for a single AO is the following. We represent each of the AO and target by a point, and we model its motions using discrete-time transition equations. Let each time step be of length δ. We denote the position of the AO (resp. the target) at time $k\delta$ by q_k (resp. t_k). The transition equation for the AO is $q_{k+1} = f(q_k, \phi_k)$, where ϕ_k is an action chosen from some given action space Φ. Constraints such as bounded velocity can be encoded

Figure 9: Experimental setup for target tracking

in f. Similarly. the equation for the target is $t_{k+1} = g(t_k, \theta_k)$, where θ_k is an action taken from some space Θ. When the target only partially pedictable, the AO knows Θ (and, possibly, a probabilistic distribution over Θ), but it does not know in advance the specific actions θ_k performed by the target.

At every time step the on-line planner computes a K-step motion strategy that optimizes a certain criterion over the next K time steps and the AO performs the first step of this strategy. In practice we take $K = 1$, as the cost of the computation increases dramatically with K. E.g., in the second variant of the on-line planner, the action ϕ_k is computed at each step in order to maximize the distance between t_k (as measured by the AO) and the boundary of the visibility region of the AO at its future position q_{k+1}. This choice, which corresponds to trying to minimize the time to escape, yields AO's motions that tolerate small errors in measuring t_k. It also easily extends to multiple AOs; we then maximize the distance between t_k and the boundary of the union of the visibility regions of the AOs at their future positions.

We have implemented these planners and performed numerous successful experiments both in simulation and with a real AO. Fig. 9 shows our experimental setup (the robot at the forefront is the AO; the other robot, with a distinctive "hat," is the target). Fig. 10 shows a run as it appears on the user's display (the gray disc is the target and the black disc the AO).

5 Conclusion

Motion planning with visibility constraints has received little attention so far, despite the fact that it has many potential applications. In this pa-

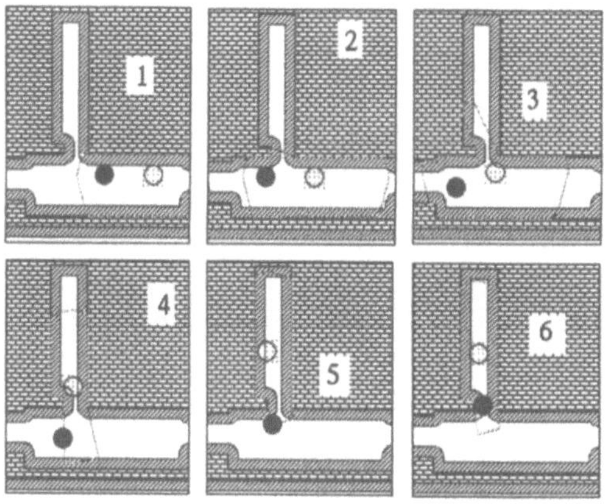

Figure 10: Target-tracking snapshots

per we presented three problems encountered in our AO project: model building, target finding, and target tracking. Our research combines theoretical investigation of planning problems with partly-idealized visibility models to produce guaranteed algorithms and pragmatic tradeoffs between algorithmic rigor and model realism to construct effective systems for realistic experimental setups.

Acknowledgments: This work is supported by ARO MURI grant DAAH04-96-1-007, ONR grant N00014-94-1-0721, and NSF grant IRI-9506064. We thank Prof. R. Bajcsy (U. of Pennsylvania) and Prof. J.L. Gordillo (ITESM, Monterrey, Mexico) and their students for having experimented with our AO over the Internet and for their useful suggestions.

References

[1] Banta, J.E., Y. Zhien, X.Z. Wang, G. Zhang, M.T. Smith, and M. Abidi (1995). A best-next-view algorithm for three-dimensional scene reconstruction using range images. In *Proc. SPIE*, Vol. 2588, 418–429.

[2] Briggs, A.J. and B.R. Donald (1997). Robust geometric algorithms for sensor planning. In *Algorithms for Robotic Motion and Manipulation (WAFR'96)*, 197–212. J.P Laumond and M. Overmars (eds.), A K Peters, Wellesley, MA.

[3] Crass, D., I. Suzuki, and M. Yamashita (1995). Searching for a mobile intruder in a corridor – the open edge variant of the polygon search problem. *Int. J. of Comp. Geom. and Appl.* *5*(4), 397–412.

[4] Guibas, L.J., J.C. Latombe, S.M. LaValle, D. Lin, and R. Motwani (1997). Visibility-based pursuit-evasion in a polygonal environment. In *Proc. 5th Workshop on Algorithms and Data Structures (WADS'97)*, Springer Verlag, 17–30.

[5] Guibas, L.J., R. Motwani, and P. Raghavan (1995). The robot localization problem. In *Algorithmics Foundations of Robotics (WAFR'94)*, 269–282. K. Goldberg et al. (eds.), A K Peters, Wellesley, MA.

[6] Hutchinson, S., G.D. Hager, and P. Corke (1996). A tutorial on visual servo control. *IEEE Tr. on Robotics and Automation 12*(5), 313–326.

[7] Kakusho, K., T. Kitahashi, K. Kondo, and J.C. Latombe (1995). Continuous purposive sensing and motion for 2D map building. In *Proc. IEEE Int. Conf. of Syst., Man and Cyb.*, 1472–1477.

[8] LaValle, S.M., H.H. Gonzalez-Bãnos, C. Becker, and J.C. Latombe (1997). Motion strategies for maintaining visibility of a moving target. In *Proc. IEEE Int. Conf. on Rob. and Automation.*

[9] Maver, J. and R. Bajcsy (1993). Occlusions as a guide for planning the next view. *IEEE Tr. on Pattern Analysis and Machine Intelligence 15*(5), 417–433.

[10] O'Rourke, J. (1997). Visibility. In *Handbook of Discrete and Computational Geometry*, 467–479. J.E. Goodman and J. O'Rourke (eds.), CRC Press, Boca Raton, FL.

[11] Parsons, T.D. (1976). Pursuit-evasion in a graph. In *Theory and Application of Graphs*, 426–441. Y. Alavi and D. Lick (eds.), Lecture Notes in Mathematics 642, Springer Verlag, Berlin.

[12] Pito, R. (1995). *A solution to the next best view problem for automated CAD model acquisition of free-form objects using range cameras.* Technical Report 95-23, GRASP Lab., U. of Pennsylvania.

[13] Talluri, R. and J.K. Aggarwal (1996). Mobile robot self-localization using model-image feature correspondence. *IEEE Tr. on Robotics and Automation 12*(1), 63–77.

Motion planning in humans and robots

Vijay Kumar Miloš Žefran J. Ostrowski
General Robotics and Active Sensory Perception (GRASP) Laboratory
University of Pennsylvania, 3401 Walnut St., Philadelphia, PA 19104-6228
email: kumar@central.cis.upenn.edu

Abstract

We present a general framework for generating trajectories and actuator forces that will take a robot system from an initial configuration to a goal configuration in the presence of obstacles observed with noisy sensors. Studies of human voluntary manipulation tasks suggest that human motions can be described as solutions of certain optimization problems. Motivated by these observations, we formulate the robot motion planning problem as a problem of finding the trajectory and the associated actuator inputs that optimize a performance criterion dictated by specific task requirements. We show that this approach can be extended to incorporate uncertainty by formulating the motion planning problem as a two-person, zero sum game with the optimal solution being a saddle-point strategy. We present several examples to illustrate our approach.

1 Introduction

In recent years, we have developed a computational framework for generating open-loop motion plans for manipulation tasks in a deterministic environment (Žefran and Kumar 1997; Žefran et al. 1996). The emphasis in these tasks is on the kinematic and dynamic interactions between the robot(s) and the environment. This paper summarizes the approach and an extension to uncertain environments.

Our approach is grounded in the framework of continuous mathematics. However, as shown later in this paper, this framework also allows the synthesis of behavioral patterns that are essentially discrete in nature. Such discrete behaviors can also be synthesized or analyzed in a discrete-event or hybrid systems framework. However, since our intention is to generate reference trajectories for closed loop control of systems governed by Newtonian dynamics, we use the methods of continuous mathematics.

The motivation for the proposed planning paradigm comes from our studies of human manipulation. Human motions appear to minimize a certain integral cost functional (Flash and Hogan 1985; Kawato 1990; Garvin et al. 1997; Desai et al. 1997). We use this idea to compute motion plans for robot systems governed by (nonlinear) dynamics, and moving amidst obstacles.

Optimal control and variational calculus have been extensively used for motion planning in the robotics literature (Vukobratović and Kirćanski 1982; Bobrow et al. 1985; Shin and McKay 1985; Buckley 1985; Nakamura and Hanafusa 1987; Singh and Leu 1989). Most of these works concentrate on a particular aspect of motion planning where they resolve the indeterminacy at a certain level (for example, kinematic redundancy), for a specific task (for example, point to point motion). Our goal is more general. In this paper, we formulate the motion planning problem in the setting of variational calculus and develop a set of tools that allow us to consider motion planning at all levels. Specifically, we strive for a method that has the following properties.

- It should be possible to incorporate the dynamics of the system and resolve redundancies at all levels. In other words, we must be able to compute the task space trajectory, the joint space trajectory, and the actuator forces for the system within the same framework.
- The method must allow equality and inequality constraints, such as kinematic closure equations, nonholonomic constraints, joint or actuator limits, and constraints due to obstacles, in the formulation.
- Since there are usually one or more natural measures of performance for the task, it is desirable to be able to find a motion with the best performance.

- The motion plans must be robust with respect to modeling errors. In other words, it should be possible to account for uncertainty in the system model.
- The method must be able to explicitly incorporate the robot's ability to use sensory information for error correction.

Depending on the level at which the system dynamics are modeled, we can obtain optimal kinematic trajectories or optimal actuator forces. In the next section, we address trajectory generation based on kinematic models. In Section 3, we discuss optimal motion plans for systems in which dynamic considerations are important. We show that the motion plans we develop can also encode discrete behaviors. Finally, in Section 4, we develop an extension to this method that allows us to handle uncertainty in the environment. In particular, we show that as more information about the world becomes available, the motion plans can be efficiently refined.

2 Trajectory planning

When studying motion planning for artificial systems it is beneficial to study the formation of motion in humans. We can model the human body as an articulated linkage of rigid bodies, similar to how we model robots. In humans, there are many instances of redundancies that make the mappings between the task, joint, and actuator spaces non-invertible, suggesting that humans possess mechanisms for resolving these redundancies. The principles that govern human motion planning can therefore provide insight into motion planning for robots.

The hypothesis that voluntary, planar reaching tasks performed under relaxed conditions can be described by optimality criteria has been explored by (Flash and Hogan 1985), (Kawato 1990) and others. Two natural questions that arise are (a) do such optimality criteria extend naturally to higher dimensions; and (b) can they explain tasks that involve constraints. In our studies of human manipulation, we have focused on tasks in $SE(2)$, the set of translations and rotations in a plane, and in which the left and right arm cooperatively grasp the object. In this case, the task space is no longer Euclidean. Further, the grasp introduces constraints due to the closed kinematic chain and the requirement of force closure. Our observations (Garvin et al. 1997) show that the kinematic properties of the measured trajectories exhibit a high degree of repeatability (within a subject and across subjects). The trajectories are approximately straight lines and have a smooth velocity profile. These trajectories depend only on the relative position and orientation of the initial and goal configurations and not on the global position and orientation. In other words, the trajectories are *left invariant*, or independent of the inertial reference frame.

In order to formalize planning of smooth motions that involve translations as well as rotations, and to study the invariance properties of different motion planning schemes, it is convenient to formulate kinematic motion planning in the framework of differential geometry and Lie groups. Using purely geometric ideas also makes the planning method independent of the description (parameterization) of the space[1]. Geometric analysis reveals that it is necessary to define the concept of distance (a Riemannian metric) and a method of differentiation (an affine connection) in space before the notion of smoothness can be defined (Žefran and Kumar 1997). Since there is a unique affine connection that is compatible with any Riemannian metric, the choice of a metric also leads to a natural definition for differentiation. Thus, once an appropriate metric is chosen, we can define a suitable measure of the smoothness of a trajectory and determine smooth trajectories between two points. Further, by properly choosing a metric, we can obtain trajectories with desired invariance properties. A metric that is particularly interesting for motion planning and produces left-invariant trajectories is the kinetic energy metric. It embodies the inertial properties of a rigid body for which we wish to plan the trajectories.

Given a Riemannian metric $<.,.>$, the problem of finding a smooth kinematic trajectory $\gamma(t)$ can be formulated as:

$$\min_{\gamma} \int_a^b < L(\gamma, \frac{d\gamma}{dt}), L(\gamma, \frac{d\gamma}{dt}) > dt. \qquad (1)$$

where L is a vector valued function that is a local measure of smoothness of the trajectory and usually depends on the affine connection corresponding to the chosen metric. For example, the general expression for the minimum-jerk cost functional (see (Flash and Hogan 1985) for the special

[1] This is particularly important in the context of rotations in three dimensions.

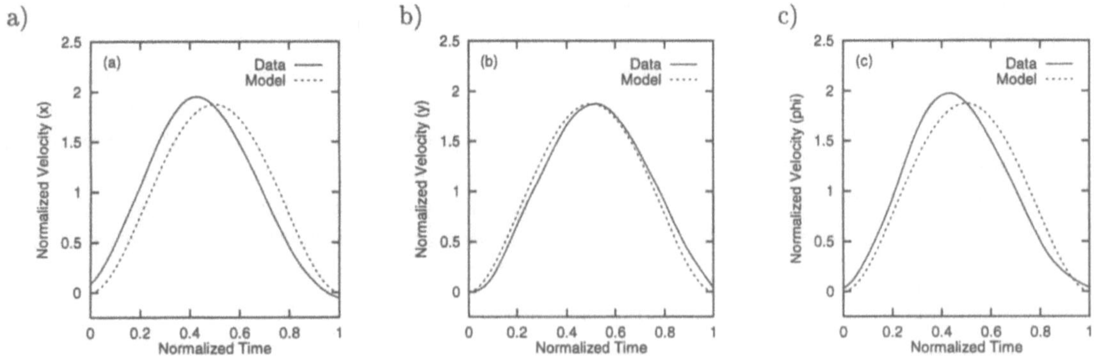

Figure 1: Measured velocity profiles for a planar two-handed manipulation by humans (solid) and those predicted by the minimum-jerk hypothesis (dotted) in the frontal plane (a), sagittal plane (b) and for rotation (c).

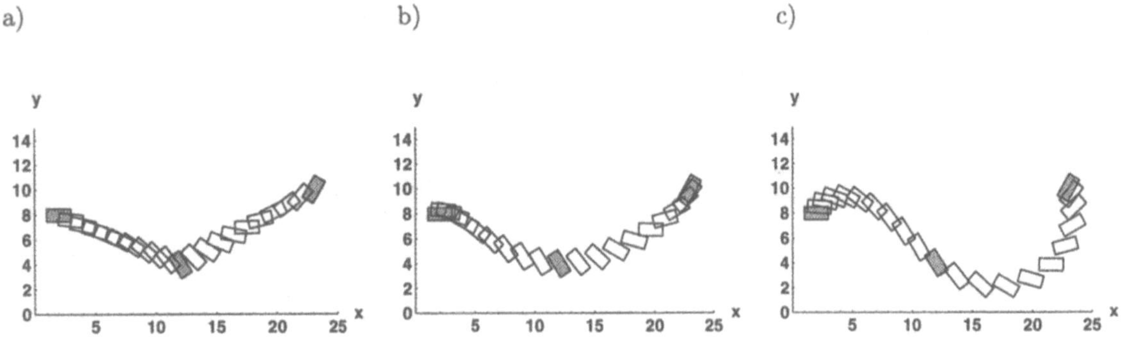

Figure 2: Examples of trajectories between specified configurations: (a) Piece-wise smooth geodesics; (b) Maximally smooth interpolant computed from a positive definite left-invariant metric; (c) Maximally smooth interpolant computed from an indefinite bi-invariant metric.

case of R^2) is:

$$J_{\text{jerk}} = \int_a^b < \nabla_V \nabla_V V, \nabla_V \nabla_V V > dt. \quad (2)$$

In the equation, $V = \frac{d\gamma}{dt}$ is the velocity vector field and ∇ is the affine connection obtained from the chosen Riemannian metric. The resulting trajectories are given by the Euler-Lagrange equations (Camarinha et al. 1995; Žefran and Kumar 1996):

$$\nabla_V^5 V + R(V, \nabla_V^3 V)V - R(\nabla_V V, \nabla_V^2 V)V = 0, \quad (3)$$

where R denotes the metric-dependent tensor describing the curvature properties of the space. These equations and the initial and final constraints lead to a boundary value problem that can be solved using the finite-difference method (Žefran and Kumar 1997).

The measured trajectories (averaged across 50 trials and 10 human subjects) and the computed trajectories that minimize the jerk cost functional (obtained by solving (1-2)) are shown in Figure 1 for a typical motion. This suggests that the observed motions are well predicted by the minimum-jerk hypothesis proposed by (Flash and Hogan 1985). More details are provided in (Garvin et al. 1997).

The cost functional of the form (1) is not completely general, but it allows us to obtain generalized spline motions (Noakes et al. 1989). For example, a maximally smooth trajectory that is C^1 continuous (*i.e.*, it can satisfy arbitrary boundary conditions on the velocities) is a generalization of a cubic spline and can be obtained by minimizing the integral of the acceleration along the trajectory:

$$J_{\text{acc}} = \int_a^b < \nabla_V V, \nabla_V V > dt. \quad (4)$$

Figure 2 shows generalized cubic splines that satisfy boundary conditions on positions and velocities and pass through a given intermediate configuration for two different choices of the metric for the space[2]. A left-invariant positive definite product metric is used in Figures 2(a) and 2(b). According to this metric, the length of a twist or a generalized vector is computed by adding the lengths (using the standard Euclidean norm) of the angular velocity vector multiplied by a suitable scaling factor and the linear velocity vector in the body-fixed frame. The resulting geodesics are shown in 2(a). The geodesics of this product metric are independent of the choice of the scaling factor (Žefran and Kumar 1996). The trajectory minimizing J_{acc} in (4) while passing through the intermediate configuration is shown in 2(b). The trajectory in 2(c) is similar to 2(b), but it is computed from the bi-invariant, indefinite metric that can be derived from the Klein form (see (Murray et al. 1994) for for the definition).

While kinematic motion plans may not be adequate for some applications, they have the advantage that they can be easily computed. In fact, some important problems have explicit, closed-form solutions (Žefran and Kumar 1997). Also, in many cases, when a detailed dynamic model of a robot system is not available, it may be desirable to simply determine the smoothest trajectory that satisfies the required boundary conditions. In such a situation, left-invariance (invariance with respect to the choice of the inertial frame) is desirable. Further, when there is a 1-1 map between trajectories in the task space and the actuator forces that generate them, the kinematic motion plan uniquely determines the motion.

3 Dynamic constraints

Walking, grasping, and cooperative manipulation are tasks in which multiple articulated linkages are strongly dynamically coupled and the coordination of the interactions between them becomes critical for motion planning. A further characteristic of these tasks is that the dynamic equations may change as the system moves. For example, in a grasping task, the dynamic equations change when a contact between a finger and the object is broken or when a new contact is established.

We formulate the motion planning as a variational problem. Given a performance criterion in the form of an integral cost functional:

$$J = \Psi(x(t_1), t_1) + \int_{t_0}^{t_1} L\left(x(t), u(t), t\right)\, dt, \qquad (5)$$

[2]More complicated, three-dimensional examples are presented in (Žefran and Kumar 1997).

and given the dynamic equations of the system,

$$\dot{x} = f(x, u, t), \qquad (6)$$

with geometric, kinematic and dynamic constraints,

$$g(x, u, t) \leq 0, \qquad (7)$$

the problem becomes that of finding a trajectory in the time interval $[t_0, t_1]$, that satisfies the dynamic equations (6), minimizes the cost functional (5) and satisfies the equality and inequality constraints (7).

We have developed an efficient numerical technique for solving such problems based on finite-element techniques and finite-dimensional optimization. We approximate the unknown (state vector) function with a set of basis functions:

$$x_i(t) \approx \sum_{j=0}^{N} p_i^j \phi_j(t), \qquad (8)$$

where $\phi_j(t)$ is a complete set of basis functions. The approximation of the function is required to be exact on the chosen set of grid points:

$$x_i(a_j) = \sum_{j=0}^{N} p_i^j \phi_j(a_j). \qquad (9)$$

For simplicity, we assume that $a_j - a_{j-1} = h$ for all $j = 1, \ldots, N$. It is convenient to choose the triangular shape functions as the basis functions:

$$\phi_k(t) = \begin{cases} \frac{t-a_{k-1}}{h} & \text{if } a_{k-1} < t \leq a_k, \\ \frac{a_{k+1}-t}{h} & \text{if } a_k < t \leq a_{k+1}, \\ 0 & \text{otherwise.} \end{cases} \qquad (10)$$

In this way, the function x is approximated by a piecewise linear function and $p_i^j = x_i(a_j)$. With some work, one can develop a variable step size (h) approximation that allows a finer mesh at points where x changes rapidly. The discretization in (8) leads to a finite-dimensional nonlinear programming problem that can be solved using any of the methods available in the literature. In this work we use the method of augmented Lagrangian (Bertsekas 1982) to convert the constrained problem into an unconstrained one and Newton's minimization method to solve the ensuing unconstrained minimization problem. We refer the reader to (Žefran 1996) for details.

This framework has been applied to several practical examples such as two-arm coordinated manipulation with frictional constraints, multi-fingered grasping and coordination of multiple mobile manipulators (Žefran 1996; Žefran et al. 1996). The approach allows us to solve in reasonable time complex motion planning problems that have not been attempted before[3].

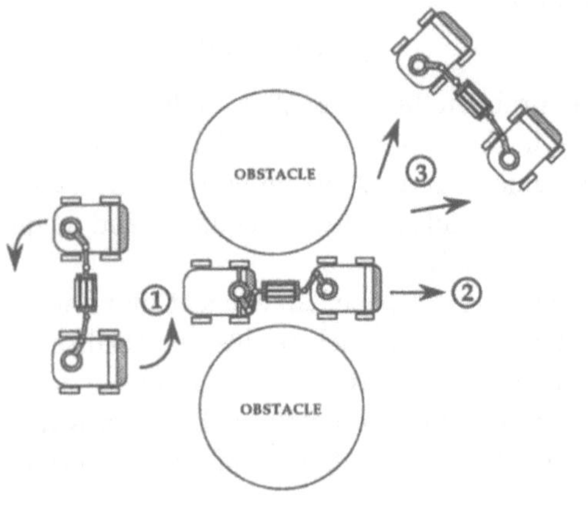

Figure 3: Two mobile manipulators negotiating a narrow corridor. The manipulators must change their formation before entering the corridor.

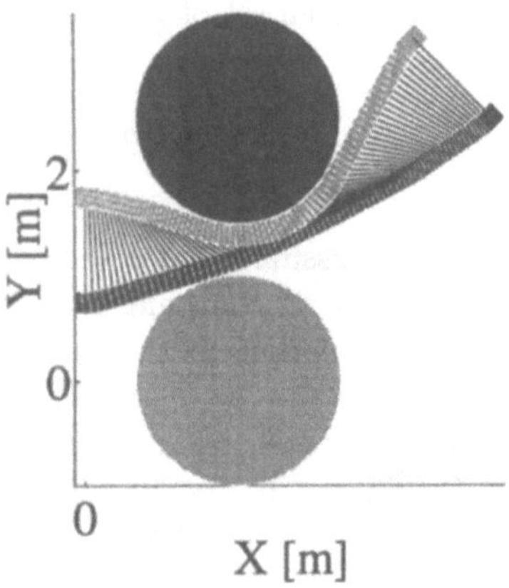

Figure 4: The motion plan for the two cooperating mobile manipulators. Snapshots at equal time intervals are shown in the figure.

Figure 3 shows an example [4] of the motion plan for two mobile manipulators negotiating the free space between two obstacles. The two manipulators are three degree-of-freedom planar manipulators holding an object in a friction-assisted grasp (Desai and Kumar 1997). They are required to keep the object in a stable grasp which means that they cannot drift too far apart. Further, to avoid collisions, they cannot come too close together. Our intuition would suggest that the manipulators change formation before negotiating the corridor and change back to the original formation. In human, planar grasping and manipulation tasks with two arms, the observed trajectories and the distribution of forces suggest a maximally smooth variation of the joint torques (Desai et al. 1997). Letting the integrand in Equation (5) be the norm of the vector of the rate of change of torques, and imposing constraints due to the dynamics, the obstacles, and the nonholonomic nature of the mobile platforms, yields the trajectory shown on the right in Figure 4. The shaded rectangular boxes show the centers of the mobile platforms and their orientations. The configuration of the manipulators is not shown in the figure. It is interesting to note that the system shows some apparently discrete behaviors. The two mobile manipulators reconfigure from a "march-abreast" formation to a "follow-the-leader" formation, then follow a straight-line to the closest obstacle, pursue a trajectory that hugs that obstacle until they are beyond the constriction, before reconfiguring back to the "march-abreast" formation and taking the unconstrained straight-line path to the goal. These behaviors can be represented in a discrete-event framework, but they are naturally generated with a continuous method by requiring the motion and actuator forces to be smooth.

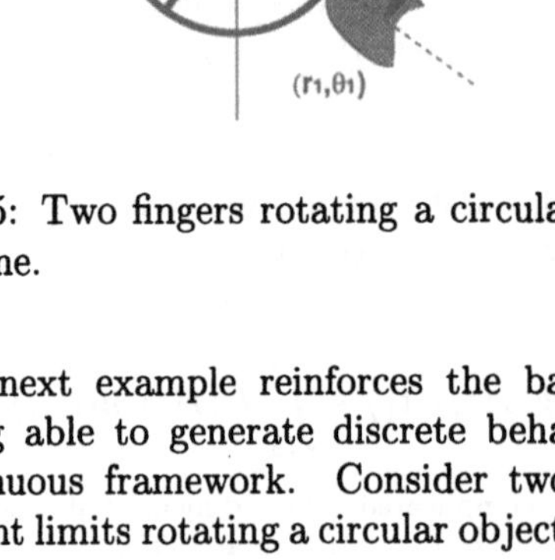

Figure 5: Two fingers rotating a circular object in a plane.

The next example reinforces the basic idea of being able to generate discrete behaviors in a continuous framework. Consider two fingers with joint limits rotating a circular object pivoted

[3]The typical time for a system with a 20-dimensional state space on a 200 MHz SGI (for example, the system shown in Figure 3) is two minutes.

[4]In more complicated examples, we assume that an initial path that satisfies the geometric constraints is available. There is extensive literature that addresses the generation of such a path (Latombe 1991). Our method is not specifically designed to solve such problems as the piano-movers problem.

about a fixed axis in a horizontal plane (Figure 5). The workspaces of the fingers are cones of angles $2\,\alpha$ centered along the x axis, located at the origin diametrically across each other ($-\alpha \leq \theta_1 \leq \alpha$, $\pi - \alpha \leq \theta_2 \leq \pi + \alpha$). The task is to rotate the object through an angle $\Delta\varphi$, but during any finite time interval only one finger is allowed to be on the object. If there are limits on finger movement ($2\alpha < \Delta\varphi$), it is necessary to regrasp to complete the task. To guarantee the continuity of the positions and velocities, the cost functional is chosen to be the L^2 norm of the actuator torques.

In this example, the dynamic equations change every time a finger comes in contact with the object or breaks the contact with the object (Žefran et al. 1996). We assume that the sequence of moves of the fingers is known. In other words, the sequence of discrete states is assumed to be known, but we use our computational scheme to solve for the history of the continuous state through the switches between the discrete state. Specifically, we obtain the optimal times for establishing or breaking the contact between a finger and the object, the locations of the fingers at these events, as well as the continuous trajectories and actuator forces for the system.

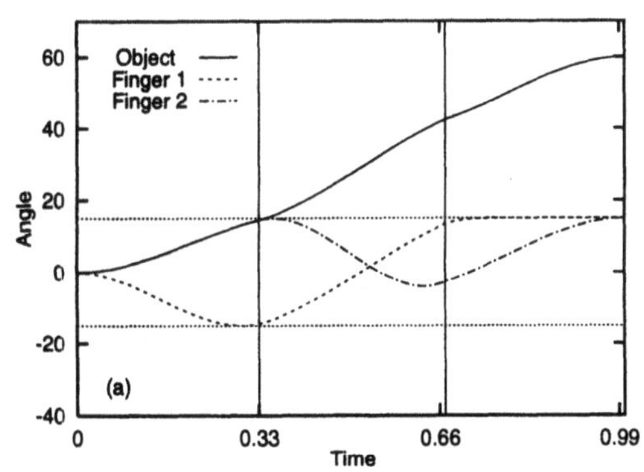

Figure 6: The angles φ (solid), θ_1 (dashed) and $\theta_2 - \pi$ (dot-dash) for the workspace $\alpha = 15°$ (shown dotted).

Figure 6 shows the results for the workspace $\alpha = 15°$, $\Delta\varphi = 60°$. During the first third of the maneuver, finger 2 rotates the object through 15° before it reaches a workspace limit. Meanwhile, finger 1 positions itself at $-15°$ and subsequently rotates the object through almost 30° in the second stage of the task. While finger 1 is rotating the object, finger 2 moves back towards the middle of its allowable workspace so that it can complete the rotation of the object in the third stage. During this stage, finger 1 stays

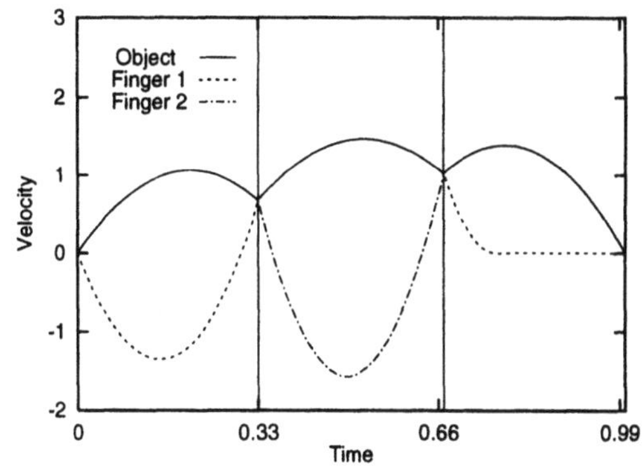

Figure 7: The velocities $\dot{\varphi}$ (solid), $\dot{\theta}_1$ (dashed) and $\dot{\theta}_2$ (dot-dash) for the example in Figure 6.

at the upper edge of its workspace. The results are consistent with our intuition of what the best solution would be - both fingers move through the minimal distance necessary, but in a smooth fashion, to complete the task. As seen in Figure 7, the velocities of the contact points on the object and on the finger at the time of establishing or breaking the contact are equal so no impact occurs. Hence, the continuous state is continuous across the switches between discrete states. We also note that the positions of the fingers on the object at the switch are computed as part of the motion plan.

4 Planning with uncertainty

Game theory provides the natural framework for solving motion planning problems with uncertainty. The motion planning problem can be formulated as a two-person zero sum game (Basar and Olsder 1982) in which the robot is a player and the environment consisting of the obstacles and the other robots is the adversary. The goal is to find a control law that yields the best performance in the face of the worst possible uncertainty (a saddle-point strategy). The motion planning problem can be solved with open loop control laws (as we have done thus far) or with closed loop control laws (feedback policies). Rather than develop the notation and the theory that is required to cast the motion planning problem in the framework of game theory (see (Lygeros et al. 1995) for the formulation for a similar problem in the continuous setting, and (LaValle and Sharma 1995) for the formulation in a discrete setting), we present representative examples and discuss optimal open loop and closed loop control laws.

In order to explore how the framework of the

previous section may be extended, consider the situation where we have a deterministic model of the robot and exact state information, but there is uncertainty in the environment. Except in very structured environments, it may be inappropriate to assume a prior distribution for the noise or uncertainty in the environment. It is more realistic to assume a sensor system and estimation algorithms that return, along with the measurement of each parameter in the model, a confidence interval for each parameter in which the true value lies. Examples of such confidence set-based estimation algorithms are discussed in (Kamberova et al. 1997).

4.1 Open loop motion plans

The approach in the previous sections for generating open-loop trajectories for deterministic systems can be extended in an obvious fashion to problems with uncertainty. The uncertainty in the environment is incorporated through conservative bounds on the feasible regions in the state space. This effectively amounts to making the obstacles bigger, reflecting the uncertainty in the position and the geometry of the obstacles[5]. With the additional sensory information that becomes available during the execution of the plan, the bounds on the feasible regions can be made less conservative and the open-loop plan can be refined accordingly. This method is attractive because our numerical method lends itself to efficient remeshing and refining and the computational cost of refining an open loop plan is an order of magnitude less than the cost of generating an initial open loop plan, when the changes in the model remain small. Thus, open loop plans may be recursively refined.

We illustrate this approach with the help of a simple example in which two nonholonomic robot vehicles navigate an environment with two obstacles. Each robot is a wheeled platform driven by two coaxial, powered wheels and has four passive castors for support. A schematic is shown in Figure 8. The robot's position, $(x, y, \theta) \in SE(2)$, is measured via a frame located at the center of the wheel base. The robot's motion is driven by the movements of its two wheels, whose angles, (ϕ_1, ϕ_2), are measured relative to vertical. If the wheels do not slip with respect to the ground, we can describe the system in terms of three first-

[5]The resulting motion plan can be shown to be a saddle-point solution in the set of all open-loop strategies.

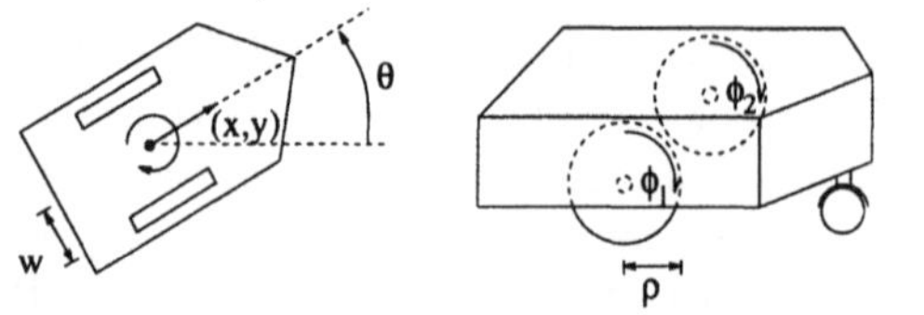

Figure 8: Two wheeled planar mobile robot.

order ODE's in (x, y, θ):

$$\begin{bmatrix} \dot{x} \\ \dot{y} \\ \dot{\theta} \end{bmatrix} = \begin{bmatrix} \cos\theta & -\sin\theta & 0 \\ \sin\theta & \cos\theta & 0 \\ 0 & 0 & 1 \end{bmatrix} \begin{bmatrix} \frac{\rho}{2}(\dot{\phi}_1 + \dot{\phi}_2) \\ 0 \\ \frac{\rho}{2w}(\dot{\phi}_1 - \dot{\phi}_2) \end{bmatrix}$$

or,

$$\begin{bmatrix} \dot{x} \\ \dot{y} \\ \dot{\theta} \end{bmatrix} = \frac{\rho}{2} \begin{bmatrix} \cos\theta \\ \sin\theta \\ \frac{1}{w} \end{bmatrix} u_1 + \frac{\rho}{2} \begin{bmatrix} \cos\theta \\ \sin\theta \\ -\frac{1}{w} \end{bmatrix} u_2 \quad (11)$$

where the two inputs for each robot are $u_1 = \dot{\phi}_1$, $u_2 = \dot{\phi}_2$.

In the example, the two vehicles, Robot 1 and Robot 2, are to interchange their positions while moving through a narrow corridor formed by two obstacles as shown in Figure 9. Each robot replans (refines) the initial open loop trajectory at the points shown by the markers. The shading of each robot in this diagram represents the time elapsed, moving from an initial dark shading (at $t = 0$) to a light shading (at $t = 1$). Neither robot knows the other's task or planned route. Each robot determines its open loop control based only on its estimate of the current position and orientation of the other robot and the obstacle. Thus, the robots change their motion plans only when they are about to collide. While the refinement of the plans is locally optimal, it is clearly not globally optimal and the resulting trajectories are more expensive than needed. In this simulation, Robot 1 is given a priority over Robot 2, and so follows a path that is closer to being optimal.

This basic procedure can be used for the motion planning and control of many robots. In Figure 10, four robots designated 1 - 4 are moved from a random initial configuration into a desired in-line formation. Again, the robots are prioritized so that the highest priority agent (1 in this example) calculates its optimal path first. Thus, each robot chooses an optimal solution based on its current set-valued estimates of the positions of the other robots or obstacles. The optimal solution is a saddle-point strategy in the set of all open loop motion plans. In other words, it is the best open loop trajectory under the worst possible locations of the other robots and the obstacles

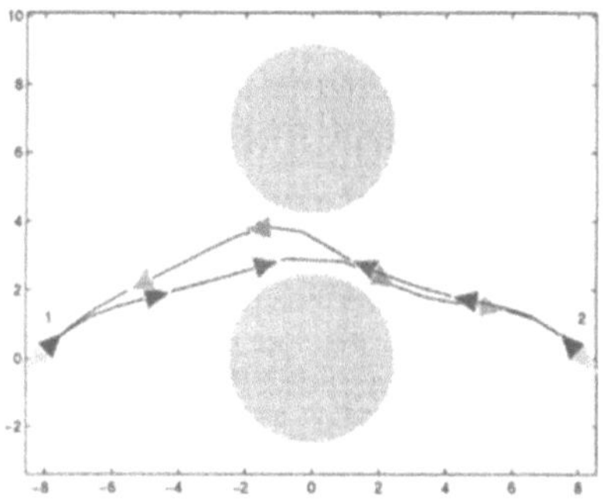

Figure 9: Successive refinement of the plans for two autonomous robots A and B.

in the environment. In this case, the adversary's optimal strategy, or the worst possible choice of positions for the obtacles (and the other robots), is not deterministic. Note however that the open loop plans are only optimal with respect to the *ad hoc* prioritization scheme imposed on the multirobot system.

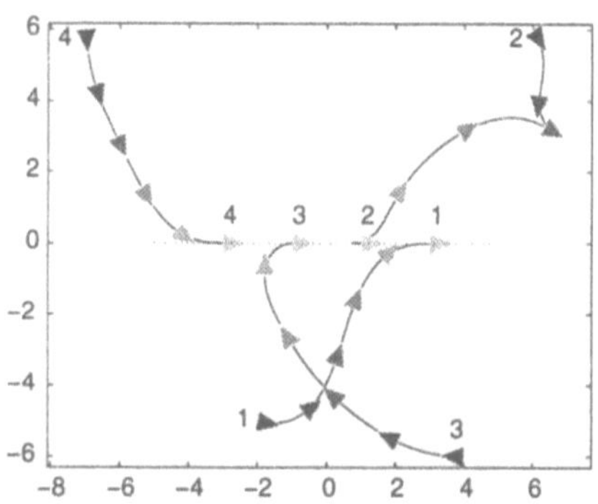

Figure 10: Motion planning for moving into formation based on the successive refinement approach.

4.2 Closed loop motion plans

While it is possible to incorporate modeling uncertainties using such approaches, they invariably lead to sub-optimal paths. Further, these paths are designed to stay away from areas that have even a very small probability of being occupied by an obstacle. There is clearly a trade-off between the conservative strategy of skirting the uncertain boundaries of the obstacle and the more aggressive strategy that incorporates better sensory information about the obstacle as it gets closer to it. Such an aggressive strategy requires feedback control, suggesting that the motion planning should be reformulated as a search for the optimal feedback control law. In this formulation, it is necessary to concurrently consider the dynamics of the system and the problem of estimating the geometry of the environment.

A simplified but realistic problem that is mathematically tractable is discussed below. We assume a single robot and an obstacle (or obstacles) that can be observed with a sensor. The sensor estimates the position of the obstacle with some uncertainty bounds depending on the distance between the robot and the obstacle. We consider a simple model for the robot dynamics:

$$\dot{x} = u, \tag{12}$$

where the vector x is the position of the robot and u is the vector of inputs. The obstacles (including other robots) define a set of points in R^2 parameterized by a vector $y \in R^m$. The initial model of the environment is denoted by y_0. $d(x,y)$ is a distance function whose value is the Euclidean distance between the nearest pair of points, one on an obstacle and the other on the robot. We use $\hat{y} \in R^m$ to denote the estimated obstacle(s). The basic idea is that $\hat{y} \to y$ as $d(x,y) \to 0$. An example is provided by a simple sensor model:

$$\hat{y}_i = y_i + (y_{0,i} - y_i)e^{-\beta d(x,y)^{-1}}, \tag{13}$$

where the exponential law is scaled by the parameter β so that the effect is approximately linear in an appropriate neighborhood of the obstacle, and levels off to the initial (worst-case) value further away.

For this problem, we are interested in obtaining a (static) feedback control law[6] $u^* = u(x,y)$ that will minimize the traveled distance as well as ensure that the robot avoids the obstacle and reaches the goal. We can allow the robot to come arbitrarily close to the obstacle, but we want to prevent collisions. Thus the allowable feedback policies are restricted to ones for which $d(x(t), \hat{y}(t)) \geq 0$, through the time interval $[0, T]$.

In general, it is difficult to find even a sufficing feedback strategy $u(x,y)$ for the above problem (Rimon and Koditschek 1992). One way to simplify the computation is to parameterize the control law and find the optimal values for the parameters. For example, we can try to find the optimal linear feedback control law:

$$u(x,y) = A(x^d - x) + B(x - y). \tag{14}$$

[6]Strictly speaking, u is a function of x and $\hat{y}$.

The task then becomes to find the optimal values for the matrices A and B. For even very simple problems there may not exist any feasible, linear feedback law. The set of all piecewise linear feedback laws affords a richer set of control policies, also one that happens to be complete. In order to find the optimal piece-wise linear feedback policy, we can divide the path into discrete intervals in each of which the feedback parameters A and B are held constant. The task now is to determine the values of A and B in each time interval.

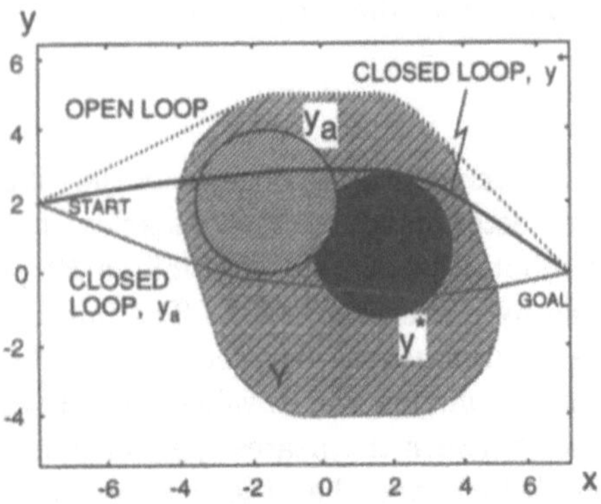

Figure 11: A comparison of the worst-case path with the optimal feedback law with that generated by the open loop control law for a circular obstacle in the the set Y. The trajectory for the worst case obstacle location y^* (shaded black) is shown in solid black and the trajectory for the actual obstacle location y^a is shown in gray. The optimal open loop policy is shown dotted.

In Figure 11 we show the motion plan when an obstacle is known to belong to a compact subset Y in R^m, but the exact location is unknown. The set Y is shown shaded, the worst obstacle location y^* is shown in black, and the actual object location $y_a \in R^m$ is the hatched circle. The figure shows three trajectories. The longest path (shown dotted) is the most conservative one that is optimal in the set of all open-loop control laws. The intermediate path (shown solid black) is the worst-case path that could possibly arise with the optimal feedback law. In other words, this is the path corresponding to the min-max solution. It is given by $u^*(x, y)$ for the worst-case object location y^*. Finally, the shortest path (shown gray) is the path followed for the obstacle location y_a, under the optimal feedback law, u^*.

This approach can be used to solve more complicated min-max (or inf-sup) motion planning problems. However, while the simplified model (12-13) may guarantee the existence of a saddle-point (Basar and Olsder 1982), this is not the case in more complicated situations. Even if there are saddle-point solutions, there may be many such solutions. Finally, the computational cost of generating a min-max solution is an order of magnitude higher than solving the open loop problem (essentially a single person game).

5 Conclusion

We have summarized our previous and ongoing work on motion planning[7]. This work has been motivated by studies of human manipulation that suggest existence of a repeatable optimal strategy for reaching and grasping. We have established a continuous framework that allows us to apply a similar strategy to robot motion planning and discussed how it can be extended to tasks in uncertain environments.

References

Basar, T. and G. J. Olsder (1982). *Dynamic noncooperative game theory.* London: Academic Press.

Bertsekas, D. P. (1982). *Constrained optimization and Lagrange multiplier methods.* New York: Academic Press.

Bobrow, J., S. Dubowsky, and J. Gibson (1985). Time-optimal control of robotic manipulators along specified paths. *Int. J. Robotic Research 4*(3), 3–17.

Buckley, C. E. (1985). *The application of continuum methods to path planning.* Ph. D. thesis, Stanford University, Stanford, CA.

Camarinha, M., F. S. Leite, and P. Crouch (1995). Splines of class c^k on non-Euclidean spaces. *IMA J. Math. Control Inform. 12*(4), 399–410.

Desai, J. P. and V. Kumar (1997). Nonholonomic motion planning for multiple mobile manipulators. In *Proceedings of 1997 International Conference on Robotics and Automation,* Albuquerque, NM.

Desai, J. P., M. Žefran, and V. Kumar (1997, September). Two-arm manipulation tasks

[7]We acknowledge the discussions with Max Mintz on game theory and sensor models. We also thank Jaydev Desai and Hong Zhang for providing the examples used in this paper.

with friction assisted grasping. In *IROS'97*, Grenoble, France.

Flash, T. and N. Hogan (1985). The coordination of arm movements: An experimentally confirmed mathematical model. *The Journal of Neuroscience 5*(7), 1688–1703.

Garvin, G. J., M. Žefran, E. A. Henis, and V. Kumar (1997). Two-arm trajectory planning in a manipulation task. *Biological Cybernetics 76*, 65–71.

Kamberova, G., R. Mandelbaum, and M. Mintz (1997). Statistical decision theory for mobile robots. Technical report, GRASP Laboratory, University of Pennsylvania, Philadelphia, PA.

Kawato, M. (1990). Optimization and learning in neural networks for formation and control of coordinated movement. Technical Report TR-A-0086, ATR Auditory and Visual Perception Research Laboratories, Kyoto, Japan.

Latombe, J.-C. (1991). *Robot motion planning.* Boston: Kluwer Academic Publishers.

LaValle, S. M. and R. Sharma (1995). A framework for motion planning in stochastic environments: applications and computational issues. In *Proceedings of 1995 International Conference on Robotics and Automation*, Volume 3, San Diego, CA, pp. 3063–3068.

Lygeros, J., D. N. Godbole, and S. Sastry (1995). A game-theoretic approach to hybrid system design. In R. Alur, T. A. Henzinger, and E. D. Sontag (Eds.), *Hybrid Systems III. Verification and Control*, pp. 1–12. Berlin, Germany: Springer-Verlag.

Murray, R. M., Z. Li, and S. S. Sastry (1994). *A Mathematical Introduction to Robotic Manipulation.* CRC Press.

Nakamura, Y. and H. Hanafusa (1987). Optimal redundancy control of robot manipulators. *International Journal of Robotics Research 6.*

Noakes, L., G. Heinzinger, and B. Paden (1989). Cubic splines on curved spaces. *IMA J. of Math. Control & Information 6*, 465–473.

Rimon, E. and D. E. Koditschek (1992). Exact robot navigation using artificial potential functions. *IEEE Transactions on Robotics and Automation 8*(5), 501–518.

Shin, K. G. and N. D. McKay (1985). Minimum-time control of robotic manipulators with geometric constraints. *IEEE Transactions on Automatic Control AC-30*(6), 531–541.

Singh, S. and M. C. Leu (1989). Optimal trajectory generation for robotic manipulators using dynamic programming. *ASME Journal of Dynamic Systems, Measurement, and Control 109.*

Vukobratović, M. and M. Kirćanski (1982). A method for optimal synthesis of manipulation robot trajectories. *ASME Journal of Dynamic Systems, Measurement, and Control 104.*

Žefran, M. (1996). *Continuous methods for motion planning.* Ph. D. thesis, U. of Pennsylvania, Philadelphia, PA.

Žefran, M., J. Desai, and V. Kumar (1996). Continuous motion plans for robotic systems with changing dynamic behavior. Proceedings of 2nd Int. Workshop on Algorithmic Foundations of Robotics.

Žefran, M. and V. Kumar (1996, April). Planning of smooth motions on se(3). In *Proceedings of 1996 International Conference on Robotics and Automation*, Minneapolis, MN, pp. 121–126.

Žefran, M. and V. Kumar (1997). Rigid body motion interpolation. *Computer Aided Design.* Accepted for publication.

Local and Global Planning in Sensor Based Navigation of Mobile Robots

Elon Rimon, Dept. of ME
Ishay Kamon, Dept. of CS
Technion, Israel Institute of Technology
John F. Canny, Dept. of CS, U.C. Berkeley

Abstract *Recent research in mobile robots is concerned with sensor based navigation. In this problem the robot starts with no apriori information about the environment, and must use its sensors to perceive the environment and plan a collision-free path to a target configuration. Algorithms for this problem are classified into* local *and* global *planners. Local sensor-based planners strive to be reactive, while still guaranteeing convergence to the target. We describe a local planner called* TangentBug, *which uses local range data to navigate two degrees-of-freedom mobile robots. In contrast, global sensor-based planners incrementally construct a global model of the environment, and use this model for path planning. We describe a global planner called* IRoadmap, *which constructs a roadmap based on range data encoded as a repulsive potential field. The algorithm uses two novel abstract sensors: a* critical point detector *and a* minimum passage detector, *and we show by examples how to implement these detectors for the navigation of three degrees-of-freedom mobile robots.*

1 Introduction

Autonomous sensor-based navigation of mobile robots has received considerable attention in recent years, e.g. [1, 3, 21]. Work in this area is motivated by applications such as cargo delivery in office environment, where the robot cannot base its planning on complete apriori knowledge of the environment. Rather, the robot must use its sensors to perceive the environment and plan its path on-line. The two main sensor-based motion planning approaches use either local or global planning. We describe the two approaches and present an algorithm of each category.

In the local sensor-based planning approach, the mobile robot strives to use the local sensory information in a reactive fashion. In every control cycle the robot uses its sensors to locate nearby obstacles, and to plan its next action based on this local information. However, to guarantee global convergence, local planners augment the local planning with a globally convergent criterion [6, 15, 17, 18, 24].

We present a local sensor-based planner for two degrees-of-freedom mobile robots called *TangentBug*. The algorithm uses local range-data to navigate a mobile robot in two reactive modes of motion: motion towards the target and motion along obstacle boundary. The robot initially moves towards the target. When it encounters an obstacle, it switches to motion along the obstacle boundary. The robot follows the obstacle boundary until a *leaving condition*, which monitors a globally convergent criterion, holds. The leaving condition ensures that the distance to the target decreases at successive leave points, and thus guarantees convergence to the target.

In the other sensor-based planning approach, the mobile robot incrementally builds a global model of the environment and uses it for path planning [4, 8, 25, 26]. We present a global sensor-based planner called *IRoadmap*. The algorithm encodes range data as a repulsive potential field, and uses it to incrementally construct a network of curves called a *roadmap*. The roadmap represents the robot free configuration space, and the search for the goal is reduced to a search along the roadmap. The roadmap consists of a collection of curves related to the "ridges" of the repulsive potential, and termed *ridge curves*. The ridge curves are interconnected by *linking curves*, through two types of special points. The first kind are critical points where the connectivity of the free configuration space changes. The second kind are minimum clearance configurations, that always lie between adjacent ridge curves. Both types of points are detected by specialized sensors, the *critical point detector* and the *minimum passage detector*, respectively.

Each of the two sensor-based approaches has some advantages and limitations. Local planners are simpler to implement, since they typically use navigation vector-fields which directly map sensor readings to actions. Moreover, local planners strive to minimize the reliance on global positioning information, which is difficult to obtain with practical sensors [7, 13]. However, local planners may generate wasteful backtracking and unnecessary long paths. In contrast, global planners use a global data structure to minimize backtracking and have a better chance at producing efficient paths in congested environments. But the construction and maintenance of a global model based on sensory information imposes a heavy computational burden on the robot. Moreover, the reliance on a global model for the navigation requires frequent localization of the robot relative to the model—a process which is difficult to attain with today's sensors. Thus the two approaches are useful in practical applications, and we present an algorithm of each category.

We begin with the *TangentBug* algorithm. After presenting the algorithm, we discuss an example and bounds on the algorithm's path-length. Then we discuss simulation results, which show that *TangentBug* consistently performs better than the classical local planners. The simulations also show that *TangentBug* produces paths that in simple environments approach the globally optimal path, as the range-sensor's detection range increases. Next we present the *IRoadmap* algorithm. First we review Morse theory and bifurcation theory, as these theories form the foundation for the algorithm. Next we describe the naive version of *IRoadmap*, where only the critical point detector is used. Then we show that the minimum passage detector must be added. A proof that the algorithm based on the two detectors always finds a path to the goal is then sketched. We conclude with a discussion of the two algorithms.

2 The TangentBug Algorithm

Classical local planners, such as the *Bug2* algorithm [17], were designed for using contact sensors, not range sensor. The only exception is the *VisBug* algorithm [16], but even *VisBug* uses the range data only to find shortcuts relative to the path generated by *Bug2*, which relies on contact sensors. The new algorithm *TangentBug* is specifically designed for using range data. The algorithm uses a *local tangent graph* to produce locally optimal paths, and we first define this data structure.

2.1 The Local Tangent Graph

The tangent graph, denoted TG, was introduced in [23] and subsequently extended by [14]. In a polygonal environment, TG is a subgraph of the visibility graph whose vertices are the convex vertices of the obstacles, together with the start and target points, denoted S and T. The edges of the tangent graph are straight lines that connect the vertices of TG, such that the edges lie in the free space and are tangent to the obstacles at their endpoints. Like the visibility graph, the tangent graph contains the shortest path in the environment. But the tangent graph is the *minimal* such graph, as the removal of any of its edges would destroy its optimality. The tangent graph typically has a significantly smaller number of vertices and edges than the visibility graph, thus searching for the shortest path is more efficient on the tangent graph.

We assume a range sensor which provides readings of the distance to the obstacles within the *visible set*. This is the star-shaped set centered at the current robot location x, whose maximal radius is R, where $0 \leq R \leq \infty$ is the sensor maximal detection range. The *local tangent graph*, or LTG, is a tangent graph that includes only the portion of the obstacles which lie in the visible set. The local range data is first divided into distinct *sensed obstacles* at discontinuity points of the range readings, then each sensed obstacle is modeled as a thin wall in the environment. The nodes of the LTG are the current robot location x, the endpoints of the sensed obstacles, and optionally an additional node in the direction of the target called T_{node}. If the line segment from x to T is not blocked by any obstacle, T_{node} is at the furthest point on this line within the visible set. As shown in Fig. 1(a), the edges of the LTG connect the robot location x with all the other nodes of the LTG. Last, when the robot touches an obstacle edge, the portion of the edge within the visible set is considered a sensed obstacle, as shown in Fig. 1(b).

2.2 Algorithm Description

The algorithm navigates a point robot in a planar unknown environment populated by stationary

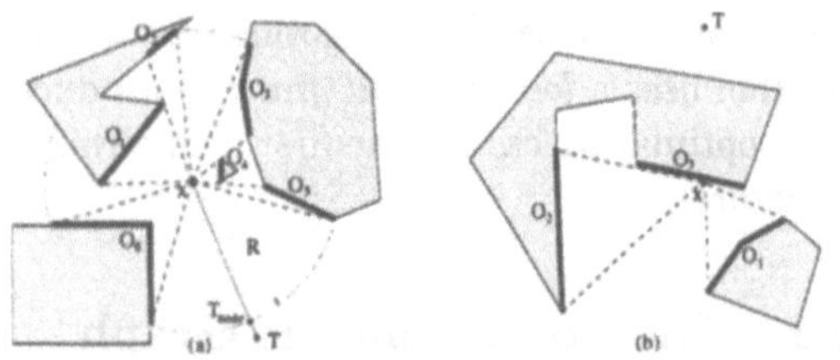

Figure 1: (a) Using a detection range R, the LTG edges connect the robot location x with the endpoints of the sensed obstacles and T_{node} (bi-tangent edges are not shown). (b) The LTG when the robot touches an obstacle edge.

polygonal obstacles. The sensory input consists of the robot current position x, and the distance from x to the obstacles within a detection range R. *TangentBug* uses two basic behaviors: motion towards the target and motion along obstacle boundary. In every step the robot constructs the LTG based on the current range readings. Then the robot uses a subgraph of the LTG described below to determine a locally optimal direction as follows. During motion towards the target, the algorithm adds a ***virtual edge*** from each node V of the LTG to T, and assigns to the edge the length of the shortest path from V to T based on the currently visible obstacles. The shortest path from the robot location x to T is computed on the resulting augmented LTG graph, and the robot moves is the ***locally optimal direction*** along the shortest path.

The robot keeps moving towards the target until it is trapped in a local minimum of the function measuring the distance of the robot from the target, denoted $d(x,T)$. This situation occurs precisely when an obstacle lies between the robot and the target. The robot then switches to boundary following behavior. It chooses a boundary following direction and moves along the boundary while using the LTG to make local shortcuts. But the robot may not leave the boundary before the following ***leaving condition*** is met. While the robot is following the boundary, it records the minimal distance to the target, $d_{min}(T)$, observed so far along the obstacle boundary. The robot leaves the obstacle boundary and resumes its motion towards the target when there is a node V_{leave} in the current LTG which satisfies $d(V_{leave},T) < d_{min}(T)$. After leaving the obstacle, the robot performs a transition phase before it resumes its motion towards the target. In the transition phase, the robot moves directly towards V_{leave} until it reaches a point Z which satisfies $d(Z,T) < d_{min}(T)$. At this point the robot resumes its motion towards the target. A summary of the algorithm now follows.

1. Move towards T along the *locally optimal* direction on the current LTG, until one of the following events occurs:
   - The target is **reached**. Stop.
   - A local minimum is detected. Go to step 2.
2. Choose a boundary following direction. Move along the boundary using the LTG while recording $d_{min}(T)$, until one of the following events occurs:
   - The target is **reached**. Stop.
   - The leaving condition holds: $\exists V_{leave} \in LTG$ s.t. $d(V_{leave},T) < d_{min}(T)$. Go to step 1.
   - The robot completed a loop around the obstacle. The target is **unreachable**. Stop.
3. Perform the transition phase: Move directly towards V_{leave} until reaching a point Z which satisfies $d(Z,T) < d_{min}(T)$. Go to step 1.

During motion towards the target, the robot moves in the direction of the shortest path to the target based on the following subgraph of the current LTG. An LTG edge which emanates from the current robot location is *admissible* if the motion of the robot along the edge decreases the robot's distance to the target. The subgraph, denoted G, consists of the admissible edges and the LTG nodes associated with these edges. Thus the nodes of G, denoted V_G, are given by $V_G = \{x\} \cup \{V \in LTG : (V - x) \cdot (T - x) > 0\}$, where x is the current robot location. By limiting the computation of the shortest path to the subgraph G, we are able to incorporate local shortest-path considerations with global convergence considerations.

For clarity, we note that the motion towards the target in step 1 of the algorithm includes both motion between obstacles and sliding along obstacle boundaries (Fig. 3). This motion mode persists as long as $d(x,T)$ decreases. The robot is trapped in the basin of a local minimum of $d(x,T)$ when the subgraph G becomes empty. At this point the robot switches to step 2 of the algorithm, where it follows an obstacle boundary. The full details of the algorithm appear in Ref. [11].

2.3 An Example

Figure 2 illustrates the paths generated by *TangentBug* and the classical *VisBug* algorithm [16]. The path planned by *VisBug* using a con-

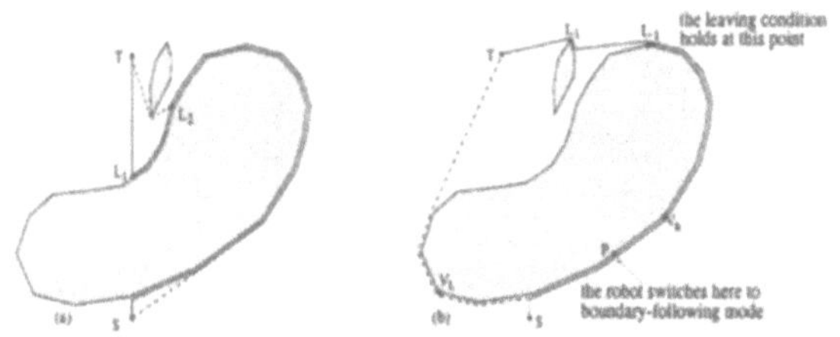

Figure 2: An example comparing *TangentBug* with *VisBug*.(a) The path generated by *VisBug* using a contact sensor (solid line), and a sensor with unlimited detection range (dashed line). (b) The path generated by *TangentBug* using the same sensors.

tact sensor is shown with solid line in Fig. 2(a) (it is identical to *Bug*2's path). Using unlimited sensor range, *VisBug* plans local shortcuts relative to *Bug*2's path, as shown with dashed line in Fig. 2(a). Using *VisBug*, the robot leaves the obstacle boundary at the point L_2, as soon as the straight line $[L_1, T]$ is visible to the robot. The path planned by *TangentBug* using a contact sensor is shown with solid line in Fig. 2(b). The robot switches from motion towards the target to boundary following at the point P, where it detects a local minimum of $d(x, T)$. The robot leaves the first obstacle at the point L_1 where $T_{node} \in$ LTG and $d(T_{node}, T) < d_{min}(T)$. When unlimited sensor range is used, as depicted with dashed line in Fig. 2(b), the resulting path is completely different. The LTG at S consists of the two endpoints of the blocking obstacle, V_L and V_R, and the locally optimal direction is towards V_L. In this case the robot continuously uses the motion-towards-the-target behavior until it reaches the target. A more detailed comparison of *TangentBug* with *VisBug* is carried out in the simulations section.

2.4 Algorithm Analysis

A detailed analysis of *TangentBug*, for the cases of a contact sensor ($R = 0$) and a sensor with unlimited detection range ($R = \infty$), appears in Ref. [11]. We show that *TangentBug* terminates after a finite-length path, and finds the target whenever it is reachable from the start point. Of special interest are the following upper bounds on the path-length of *TangentBug*. First we consider the case of a contact sensor. In the following, D_T denotes the disc with center at T and radius $\|S-T\|$.

Proposition 2.1 *Using a contact sensor, an upper bound L_{max} on the path length of TangentBug is:*

$$L_{max} = \|S-T\| + \sum_{i \in \mathcal{I}} \Pi_i \times \sharp Minima_i,$$

where $\mathcal{I}$ is the index set of the obstacles which intersect the disc D_T, Π_i is the perimeter of the ith obstacle, and $\sharp Minima_i$ is the number of local minima of $d(x, T)$ in D_T along the ith obstacle boundary.

In the case of a range-sensor with unlimited detection range, we make the following two assumptions. Let $\hat{v}$ denote a unit magnitude vector. The nodes V of the subgraph G satisfy $(\widehat{V-x})\cdot(\widehat{T-x}) > 0$, where x is the current robot location. We assume that the subgraph G is further restricted to the nodes V which satisfy $(\widehat{V-x})\cdot(\widehat{T-x}) > \gamma$, where γ is a small positive parameter. The second assumption is concerned with the leaving condition. The robot is allowed to seek candidate nodes which satisfy the leaving condition only within a range of ρ, where ρ is a positive and finite parameter.

Proposition 2.2 *Using a range sensor with unlimited detection range, and under the above assumptions, an upper bound L_{max} on the path length of TangenBug is:*

$$L_{max} = \left(1+\frac{1}{\gamma}\right)\|S-T\| + \sum_{i \in \mathcal{I}} \sharp Minima_i \times (\Pi_i + \rho),$$

where $\mathcal{I}$ is the index set of the obstacles which intersect the disc D_T, Π_i is the perimeter of the ith obstacle in $\mathcal{I}$, and $\sharp Minima_i$ is the number of local minima of $d(x, T)$ in D_T along the ith obstacle boundary.

Note that the bounds are only worst-case bounds on the performance of *TangentBug*. The average performance of *TangentBug* is characterized in the next subsection.

2.5 Simulation Results

The simulations compare *TangentBug* with the classical *VisBug* in two classes of simulated environments. The environments **world1** consist of convex non-intersecting obstacles (Fig. 3). The environments **world2** consist of concave obstacles with "office-like" shape (Fig. 4). All environments are of size 800×700 units. The results in Table 1 present the average path length relative to the globally shortest path. The paths produced

	world1		world2	
R	*VisBug*	*TangentBug*	*VisBug*	*TangentBug*
0	1.56	1.09	9.86	7.10
50	1.36	1.04	8.78	5.52
100	1.31	1.03	7.28	3.64
200	1.29	1.03	5.22	1.48
∞	1.28	1.03	4.14	1.38

Table 1: *TangentBug* compared to *VisBug*.

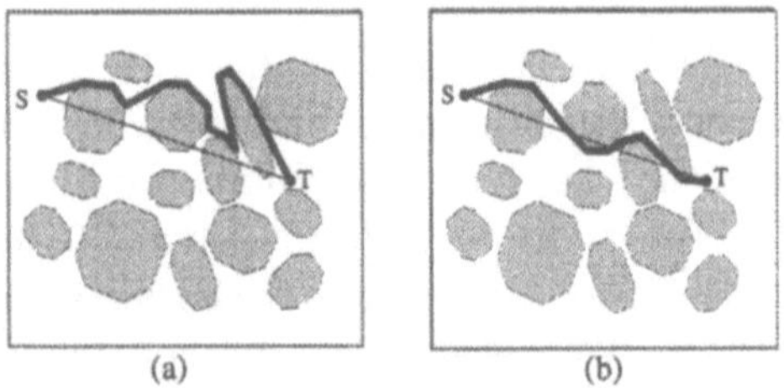

Figure 3: Using unlimited sensor range among convex obstacles. (a) *VisBug* (path length 1.52). (b) *TangentBug* (path length 1.00).

by *TangentBug* are shorter than the paths produced by *VisBug* in all tested scenarios. Increasing the sensor detection range, R, improves the performance of both algorithms. But *TangentBug* produces shorter paths for all values of R.

Furthermore, in world1 the paths produced by *TangentBug* using unlimited sensor range resemble the ***globally optimal*** paths. In these simple environments the algorithm uses mostly the motion-towards-the-target mode, implying that the range data is continuously used for choosing the locally optimal direction. The minor improvement in the path length as R increases suggests that ***in simple environments small sensor range is sufficient***, since in such environments the local data usually leads to the globally correct decisions.

In the class of more complex environments, world2, the two algorithms achieve significant local shortcuts by scanning the boundaries of concave obstacles instead of actually following them.

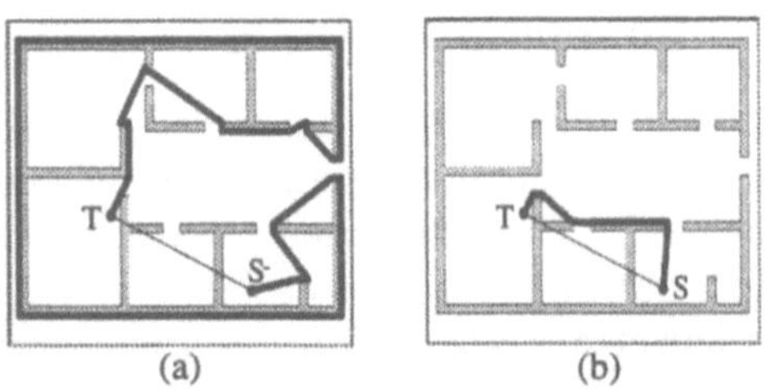

Figure 4: Using unlimited sensor range in an office-like environment. (a) *VisBug* (path length 7.92). (b) *TangentBug* (path length 1.03).

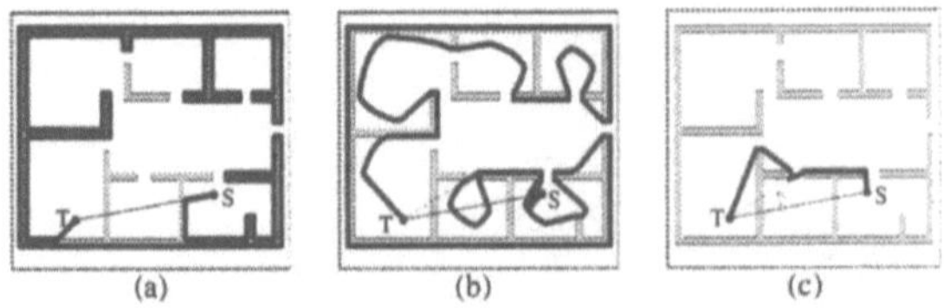

Figure 5: Effect of increasing R in an office-like environment. *TangentBug* with a detection range of (a) zero (path length 12.0), (b) 100 (path length 10.92), (c) 200 units (path length 1.08).

However, *TangentBug* uses the range data more effectively than *VisBug*. For example, Fig. 5 shows the paths generated by *TangentBug* in an office-like environment for increasing values of R. One important reason for the effectiveness of *TangentBug* in world2 is that office-like environments contain some built-in regularity, which makes the local information more relevant to the globally correct decisions.

3 The IRoadmap Algorithm

The *IRoadmap* algorithm is a global planner which uses range data to incrementally construct a roadmap for the robot free configuration space. The algorithm has its roots in the work of Canny and Lin [2], who first proposed the idea of constructing a roadmap using potential fields. However, their algorithm relies on range data collected ***off-line*** for the robot environment. In contrast, *IRoadmap* is a purely ***on-line*** algorithm. The principal formulation of *IRoadmap* applies to any robot [22], but here we describe the algorithm in the context of mobile robot navigation.

3.1 Preliminaries

This subsection reviews relevant results from Morse theory and bifurcation theory. The mobile robot configuration space, called ***c-space***, is parametrized by the coordinates $q = (d_x, d_y, \theta) \in \mathbb{R}^3$, where θ represents the robot orientation and is periodic in 2π. Given a physical obstacle, its ***c-obstacle*** is the set of configurations at which the robot intersects the obstacle. The free configuration space, called the ***freespace*** $\mathcal{F}$, is the complement of the c-obstacles' interiors.

3.1.1 *Morse Theory on Stratified Sets*

The freespace $\mathcal{F} \subset \mathbb{R}^3$ is typically a stratified set—a union of smooth manifolds[1] called ***strata.*** The connected components of the interior of $\mathcal{F}$ form 3-D strata. The boundary of $\mathcal{F}$ is a union of 2-D strata, formed by portions of the individual c-obstacles boundaries and their respective intersection along lower dimensional strata. This is illustrated in Fig. 6, which shows the c-space of a planar body and three obstacles.

Let f be a smooth real-valued function on a smooth manifold $\mathcal{M} \subset \mathbb{R}^3$. A point $x \in \mathcal{M}$ is a *critical point* of f if its derivative at x, $Df(x)$, vanishes there. A *critical value* of f is the image $c = f(x) \in \mathbb{R}$, of a critical point x. f is a *Morse function* if all its critical points are *non-degenerate* i.e., its second derivative matrix $D^2 f(x)$ is non-singular at the critical points. When f is defined on a stratified set $\mathcal{S}$, its critical points are the *union* of the critical points obtained by restricting f to the individual strata. f is a Morse function on $\mathcal{S}$ if it is, first, Morse in the classical sense on the stratum containing the critical point x. And, second, if $\nabla f(x) = Df(x)$ is *not* normal to any of the other strata meeting at x.

Stratified Morse theory is concerned with such Morse functions [9]. The theory guarantees that *as c varies within the open interval between two adjacent critical values of f, the level-sets $\mathcal{S}_c = \{x \in \mathcal{S} : f(x) = c\}$ are topologically equivalent* (homeomorphic) *to each other.* In particular, the path-connectivity of the level-sets $\mathcal{S}_c$ is preserved between critical values. In our algorithm f is the linear function $\sigma(d_x, d_y, \theta) = \theta$. It is called the *sweep function,* since the algorithm uses this function to sweep $\mathcal{F}$ with hyperplanes along the θ axis.

Any connectivity change of the slices $\mathcal{F}|_c = \{q \in \mathcal{F} : \sigma(q) = c\}$ must occur locally, in a neighborhood of a critical point of σ in $\mathcal{F}$. But only some of the critical points correspond to change in the path-connectivity of the slices $\mathcal{F}|_c$. Of those, only the following two types are of interest to *IRoadmap.* Critical points at which locally distinct connected components of the freespace slices meet, called *join points,* and critical points where two connected components split, called *split points.* Their union is called the *interesting* critical points. In Fig. 6, c-space is swept along the orientation axis θ. A *non*-interesting critical point, such as the local minimum in the figure,

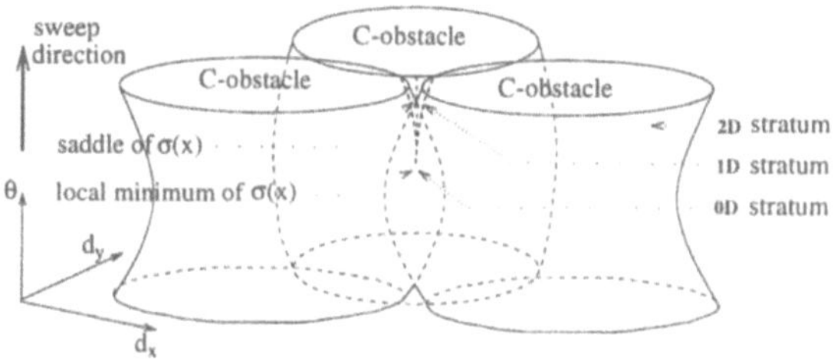

Figure 6: For a plane-sweep along θ, only the saddle is an interesting critical point.

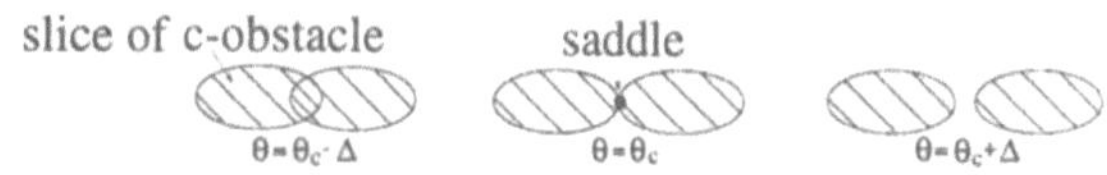

Figure 7: C-space slices centered at the saddle.

may involve the appearance (or disappearance) of a new connected component of the freespace slices. The saddle shown in the figure is an interesting critical point, since two connected components of the freespace slices meet there and become a single components in the slices above it. For clarity, the slices centered at the saddle are shown in Fig. 7.

3.1.2 *Range Sensor Based Potentials*

In addition to the sweep function σ, the algorithm uses a range-sensor based *distance function,* $\delta : \mathcal{F} \to \mathbb{R}$, that serves as a repulsive potential. Let $\mathcal{A}(q)$ be the set occupied by the robot when it is at a configuration q. Then

$$\delta(q) = \min_{a \in \mathcal{A}(q), b \in \mathcal{B}} \{\|a - b\|\}, \qquad (1)$$

where $\mathcal{B}$ is the union of the physical obstacles. An extension of [5, Proposition 2.4.1] yields that $\delta(q)$ is differentiable almost everywhere. We, however, shall treat $\delta(q)$ as a smooth function.

We will need to evaluate δ on the individual level-sets of σ. Since $\sigma(q) = \theta$, the restriction of δ to the slice $\mathcal{F}|_{\theta=c}$ is simply $\delta_c(d) \triangleq \delta(d, c)$, where $d = (d_x, d_y)$ and $\theta = c$. The algorithm constructs curves, called *ridge curves,* by tracing the *local maxima* of $\delta_c(d)$ as c varies in $\mathbb{R}$. (The points d are restricted to the slice $\mathcal{F}|_c$, while $\delta_c(d)$ is measured with respect to the entire obstacle set $\mathcal{B}$.) Fig. 9 illustrates the ridge-curves in a planar c-space swept along the horizontal axis.

3.1.3 *Bifurcation Theory*

If a ridge-curve has endpoints, they lie on the boundary of $\mathcal{F}$, or they are bifurcation points

[1] A manifold $\mathcal{M} \subset \mathbb{R}^k$ of dimension d is a hypersurface that locally looks like $\mathbb{R}^d$, for a fixed d, $0 \le d \le k$.

that lie in the interior of $\mathcal{F}$. *Bifurcation theory* characterizes the behavior of the equilibria of one-parameter families of smooth gradient vector-fields $\nabla\delta_c(d)$, as c varies in $I\!R$ [20]. It tells us the following facts. First, the loci of the equilibria of a generic one-parameter family $\nabla\delta_c(d)$ form a collection of smooth one-dimensional curves. This is illustrated in Fig. 9, where the solid curves are the local maxima and the dashed curves are the local minima of $\delta_c(d)$. Second, the equilibria curves meet each other at isolated points, called ***bifurcation points***. Third, a generic $\nabla\delta_c(d)$ has only one type of bifurcation, a *fold* or ***saddle-node*** bifurcation, where two equilibria curves meet each other smoothly at some parameter value $c = c_0$, and cease to exist for parameter values beyond c_0. Indeed, all the bifurcation points in Fig. 9 are of this type.

3.2 The Naive Algorithm

A ***roadmap*** for the robot freespace is a graph $\mathcal{R} \subset \mathcal{F}$ satisfying three properties: 1) *Connectivity*: every connected component of $\mathcal{F}$ contains exactly one component of $\mathcal{R}$; 2) *Accessibility*: $\mathcal{R}$ can be effectively reached from any start configuration; 3) *Departibility*: every target configuration can be effectively reached from $\mathcal{R}$. Traditionally, when a roadmap is built off-line with complete information about the environment, there is no distinction between accessibility and departibility. In our case accessibility is achieved by uphill climb along the direction of increasing distance from the obstacles, until a ridge-curve is reached. Departibility is achieved by invoking the algorithm on a sequence of c-space slices of decreasing dimension.

The naive algorithm uses the ***sweep function*** σ, the ***distance function*** δ, and a critical point detector described below. It constructs a roadmap which consists of two types of curves. The first are the ***ridge curves***, traced by the local maxima of $\delta_c(d)$ as c varies in $I\!R$. The second are the ***linking curves.*** Each linking-curve passes through an interesting critical point of σ in $\mathcal{F}$, and connects two ridge-curves. Let Σ be the set of interesting critical points of σ in $\mathcal{F}$. Since $\nabla\sigma(q)$ never vanishes in the interior of $\mathcal{F}$, these points occur on the boundary of $\mathcal{F}$. By definition, an interesting critical point q^* of Σ is a common boundary point of two locally distinct connected components of $\mathcal{F}|_{q^*}$, where $\mathcal{F}|_{q^*}$ denotes the slice $\mathcal{F}|_{\sigma(q^*)}$. Assuming that $\mathcal{F}$ is bounded, $\delta_c(d)$ must attain a maximum on each connected component of $\mathcal{F}|_{q^*}$.

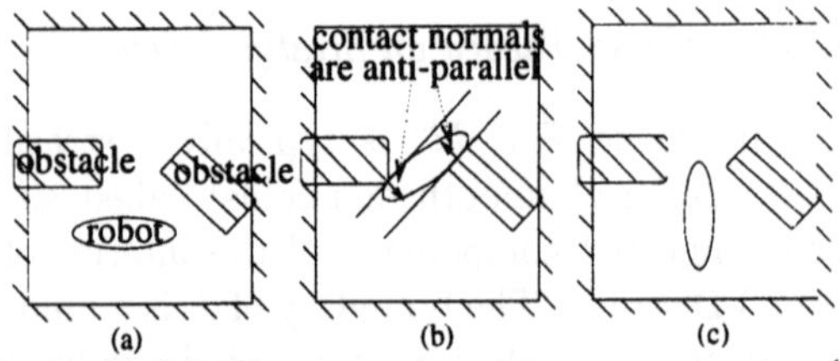

Figure 8: (a) Freespace has two components. (b) Critical configuration. (c) Freespace is connected.

Hence q^* is also a common boundary point of two locally distinct basins of attraction of $\nabla\delta_c(d)$. The linking-curve is a union of two curves, each starting at q^* and moving uphill in one of the two basins, until a ridge-curve is reached.

The robot must detect points q^* of Σ as it traverses a ridge-curve. This is achieved by means of the ***critical-point detector***. It is required of this "sensor" that: *from a ridge-curve, it must detect all the interesting critical points associated with the ridge-curve's basin of attraction.* The latter term needs explanation. If $y(c)$ is a ridge-curve, its ***basin of attraction*** is the union of the individual basins of the points $y(c)$ with respect to the flow of $\nabla\delta_c(d)$, as c varies in the domain of $y(c)$.

Implementation of the critical-point detector: For mobile robots, the points of Σ correspond to passages between obstacles in the environment. This is exemplified in Fig. 8 for an ellipse robot. It can be shown that for a sweep along the θ axis, the interesting critical points are precisely the configurations where the ellipse maintains contact with two disjoint obstacles, such that the contact normals are anti-parallel. Fig. 8(b) shows the ellipse at a critical configuration. The fixed-orientation slices of the freespace are disconnected just below the critical orientation, Fig. 8(a), and comprise a single connected component just above it, Fig. 8(c).

Given a detected critical point $q^* \in \Sigma$, the robot must depart from the ridge-curve and reach q^*, from which it continues with uphill motion to the other ridge-curve. To reach q^*, the robot first moves to a point q on the ridge-curve that lies in the slice $\mathcal{F}|_{q^*}$. A curve from q to q^* is then constructed within $\mathcal{F}|_{q^*}$ by a recursive call with the start point $\hat{q}_S = q$, target point $\hat{q}_T = q^*$, and $\hat{\mathcal{F}} = \mathcal{F}|_{q^*}$. Note that q^* is guaranteed to be reachable from q within $\mathcal{F}|_{q^*}$, since by construction q^* belongs to the ridge-curve's basin of attraction. The recursion ends when the dimension of the slice becomes unity, since the roadmap of a one-dimensional slice is identical to the slice itself.

The target configuration, denoted $\boldsymbol{q}_T$, is found by identifying departure points at points where the roadmap crosses the slice through the target, $\mathcal{F}|_{\boldsymbol{q}_T}$. Every such point q serves as an intermediate starting point, from which a search after a path to $\boldsymbol{q}_T$ is executed within $\mathcal{F}|_{\boldsymbol{q}_T}$ by a recursive call with $\hat{\boldsymbol{q}}_S = q$, $\hat{\boldsymbol{q}}_T = \boldsymbol{q}_T$, and $\hat{\mathcal{F}} = \mathcal{F}|_{\boldsymbol{q}_T}$. While $\boldsymbol{q}_T$ is *not* automatically reachable from q within $\mathcal{F}|_{\boldsymbol{q}_T}$, it is shown in Ref. [22] that the algorithm always finds one such q from which $\boldsymbol{q}_T$ is reachable. Here is a schematic description of the *naive* algorithm, starting with the required sensors.

Sensors: 1) A sensor that measures the robot configuration q; 2) A range-sensor that measures the distance function $\delta(q)$; 3) A *critical-point detector*, that detects from a ridge-curve the interesting critical points associated with the ridge-curve's basin of attraction.
Input: A target configuration $\boldsymbol{q}_T$.
Output: A path from $\boldsymbol{q}_S$ to $\boldsymbol{q}_T$ if $\boldsymbol{q}_T$ is reachable. Otherwise, a roadmap for the connected component of $\boldsymbol{q}_S$.
Data structure: $\mathcal{R}$—the explored roadmap.
IRoadmap Algorithm:
1. Connect $\boldsymbol{q}_S$ to a ridge-curve by following $\nabla\delta_c(d)$, where $c = \sigma(\boldsymbol{q}_S)$. Put $\boldsymbol{q}_S$, the curve, and its ridge-curve endpoint in $\mathcal{R}$.
2. Repeat: Move to an explored node of $\mathcal{R}$ that has an unexplored curve emanating from it. Explore the curve until its endpoint q is found according to one of the following events.
• An interesting critical point q^* is detected: Move on the ridge-curve to a point q in $\mathcal{F}|_{q^*}$. Make q a node of $\mathcal{R}$. Decide whether to make a recursive call with $\hat{\boldsymbol{q}}_S = q$, $\hat{\boldsymbol{q}}_T = q^*$, $\hat{\mathcal{F}} = \mathcal{F}|_{q^*}$; or to continue exploring the ridge-curve first. If the first option is taken, after q^* is reached, complete the linking-curve by following $\nabla\delta_c(d)$ to the other ridge-curve. Add the linking-curve and its other endpoint to $\mathcal{R}$.
• The target slice $\mathcal{F}|_{\boldsymbol{q}_T}$ is encountered at q: Make q a node of $\mathcal{R}$. Decide whether to make a recursive call with $\hat{\boldsymbol{q}}_S = q$, $\hat{\boldsymbol{q}}_T = \boldsymbol{q}_T$, and $\hat{\mathcal{F}} = \mathcal{F}|_{\boldsymbol{q}_T}$; or to continue exploring the ridge-curve first. If a recursive call is made, stop if $\boldsymbol{q}_T$ is **found**.
• A ridge-curve endpoint encountered: Make q a node of $\mathcal{R}$ (no other curves emanate from q).
• An explored node is encountered again: a full circle along $\mathcal{R}$ has been completed. Add the explored curve to $\mathcal{R}$.
Until the explored roadmap contains no nodes with unexplored curves. In that case the target is **unreachable**.

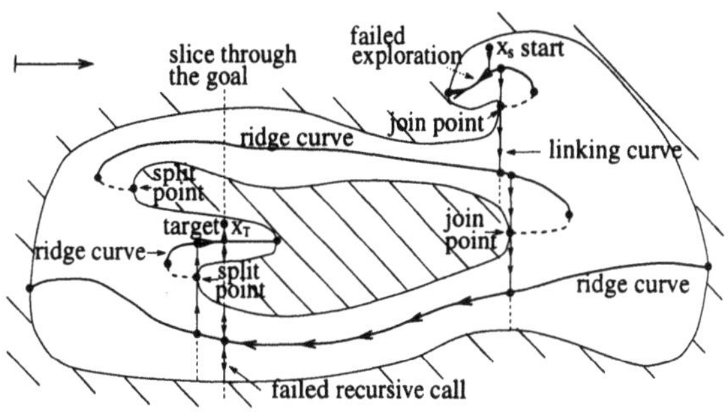

Figure 9: *IRoadmap* in a 2-D configuration space.

3.3 An Example

Figure 9 shows a planar configuration space which is swept along the horizontal axis. The ridge-curves are shown as solid lines. The robot reaches the ridge-curve basin which contains the target through a series of recursive calls triggered by detection of interesting critical points. The ridge-curve of the goal's basin crosses the slice $\mathcal{F}|_{\boldsymbol{q}_T}$ at a point that triggers a final recursive call, which finds the goal.

3.4 The Complete Algorithm

It can be verified that the naive algorithm always finds the target in two-dimensional configuration spaces. However, we now show that in three-dimensional configuration spaces the naive algorithm must be augmented with a *minimum passage detector.*

3.4.1 *The Naive Algorithm Can Fail*

The naive algorithm employs two means for departing from a ridge-curve: recursive calls in $\mathcal{F}|_{q^*}$ and recursive calls in $\mathcal{F}|_{\boldsymbol{q}_T}$. It might happen that all the interesting critical points lie beyond the closure of the current ridge-curve basin. It might also happen that the same ridge-curve doesn't cross the slice $\mathcal{F}|_{\boldsymbol{q}_T}$. In Fig. 10, the freespace is the interior of two vertical pipes joined by a horizontal pipe. There is an inner c-obstacle in the shape of a plate "hiding" the entrance to the connecting pipe. For the z sweep direction, there are three ridge-curves: R_1, R_2 (associated with the plate), and R_3. Paths from the basin of R_1 to the basin of R_3 must pass through the basin of R_2. However, the interesting critical points p_1 and p_2 lie between the basins of R_2 and R_3 i.e., outside the closure of the basin of R_1. The algorithm, after exploring the entirety of R_1, erroneously concludes that the goal is not reachable.

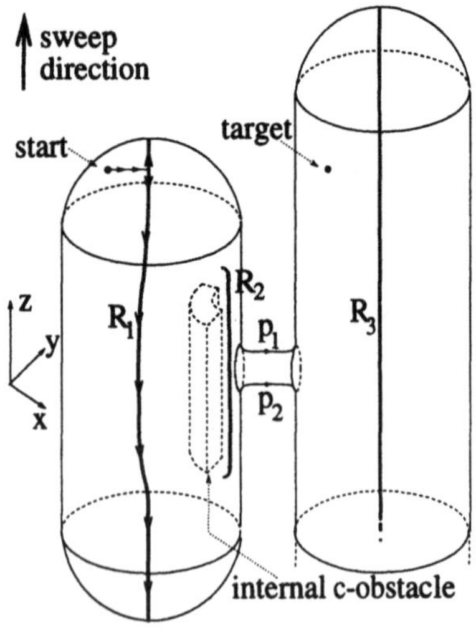

Figure 10: There are no interesting critical points between the basins of R_1 and R_2.

This false conclusion persists even if we make local changes in the sweep direction.

3.4.2 *The Minimum Passage Detector*

The minimum passage detector we now describe enables the robot to move between adjacent basins whose common boundary is not marked by interesting critical points. Let $\delta(q)$ be a smooth repulsive potential. Bifurcation theory (Section 3.1) guarantees that the equilibria of a generic $\nabla\delta_c(d)$ form smooth curves. Let the *saddle curves* be the ones traced by saddles of $\delta_c(d)$, as c varies in $\mathbb{R}$. (Each saddle is a local minimum along one direction and a local maximum along another direction in $\mathcal{F}|_c$.) It is shown below that between any two adjacent ridge-curve basins lies at least one interesting saddle curve.

It is required of the minimum passage detector that: *from a ridge-curve, it must detect the local minima of $\delta(q)$ on the saddle curves associated with the ridge-curve's basin of attraction.* (Points q are restricted to the saddle curve, while $\delta(q)$ is the distance between the robot $\mathcal{A}(q)$ and the entire obstacle set $\mathcal{B}$.) The local minima of $\delta(q)$ along the saddle curves are termed *minimum passage points* (Fig. 12).

Implementation of the minimum passage detector: Consider as an example the ellipse robot shown in Fig. 11, where the ellipse major axis is M. The repulsive potential is the function $\delta(q)$ defined in (1). For a sweep along θ, the minimum passage points are characterized as follows. Consider a pair of disjoint obstacles, $\mathcal{O}_1$ and $\mathcal{O}_2$ shown in the figure. Let $\mathsf{gap}(\mathcal{O}_1, \mathcal{O}_2)$ be the minimal distance between them, and let u be the direction of the line along which $\mathsf{gap}(\mathcal{O}_1, \mathcal{O}_2)$

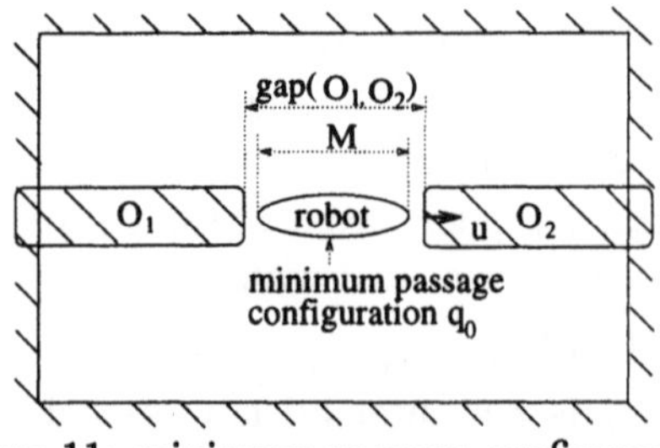

Figure 11: minimum passage configuration.

is attained. $\mathsf{gap}(\mathcal{O}_1, \mathcal{O}_2)$ is larger than M, hence the c-obstacles corresponding to $\mathcal{O}_1$ and $\mathcal{O}_2$ are disjoint. The ellipse orientation that maximizes its width along u is the one where its major axis is collinear with u. Let c_0 be the corresponding ellipse orientation. Further, let $q_0 = (d_0, c_0)$ be the ellipse configuration where its center is at the center of the gap.

Intuitively, d_0 is a local minimum of $\delta_{c_0}(d)$ for translations perpendicular to u, and is a local maximum for translations parallel to u. Hence d_0 is a saddle point. Moreover, q_0 is a local minimum of $\delta(q)$ along the saddle curve containing q_0. To see this, consider translations of the ellipse with its orientation fixed to c close to c_0. Consider, in particular, translations perpendicular to u, such that the ellipse crosses the gap from one side to the other, while maximizing its distance from both obstacles. It can be intuitively observed that the minimum value of δ along this motion is larger than $\delta(q_0)$. Hence q_0 is a local minimum of $\delta(q)$ on the saddle curve. Research under progress will make this statement precise.

The complete algorithm is identical to the one described in Section 3.2. Only that the sensor list is augmented by the minimum passage detector, and step 2(a) is augmented with a recursive call for construction of linking-curves through minimum passage points. Last, we mention the following connection between the interesting critical points and the minimum passage points. It is shown in Ref. [22] that the interesting critical points are exactly the minimum passage points that occur on the boundary of $\mathcal{F}$ (Fig. 12(b)). This result implies that *the minimum passage detector generalizes the critical point detector.*

3.5 Correctness of IRoadmap

A detailed analysis of *IRoadmap* appears in Ref. [22]. In this analysis, we assume the generic situation where the freespace $\mathcal{F}$ has no special symmetries, so that the one-parameter family $\nabla\delta_c(d)$ is generic (Section 3.1). A curve traced by an equi-

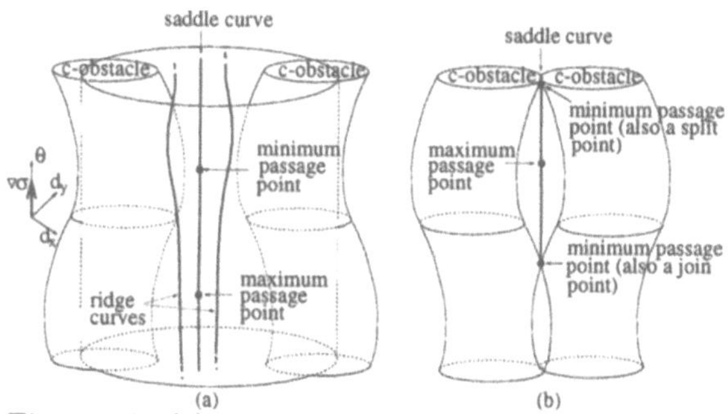

Figure 12: (a) minimum passage point in the interior of $\mathcal{F}$, (b) and on the boundary of $\mathcal{F}$.

librium of $\nabla\delta_c(d)$ of a specific type, as c varies in $I\!R$, is termed ***equilibria curve***. For instance, ridge-curves are equilibria curves of local maxima. The assumption that $\nabla\delta_c(d)$ is generic implies that there are finitely many equilibria curves in $\mathcal{F}$. We use this fact to show that *IRoadmap* terminates in every bounded workspace.

To show that *IRoadmap* always finds the target, we use the following mountain pass theorem, adapted for our purposes from [19]. Let $\mathcal{X} \subset I\!R^2$ be a stratified set with non-empty interior, which is a closed set in $I\!R^2$. Let f be a Morse function on $\mathcal{X}$, such that f is zero on the boundary of $\mathcal{X}$ and is strictly negative in its interior. Let $\mathcal{B}_1, \mathcal{B}_2 \subset \mathcal{X}$ be two distinct basins of attraction of the flow of $-\nabla f$, each attracted to a local minimum of f. Let $\overline{\mathcal{B}}_i$ denote the closure of $\mathcal{B}_i$, for $i = 1, 2$. The set $\mathcal{V} = \overline{\mathcal{B}}_1 \bigcap \overline{\mathcal{B}}_2$, if non-empty, represents a "mountain range"—a set separating $\mathcal{B}_1$ from $\mathcal{B}_2$ on which f attains higher values.

Theorem 1 (Mountain pass theorem) *If $\mathcal{V}$ is non-empty, f has a* **saddle point** *in $\mathcal{V}$.*

In our case the stratified set $\mathcal{X}$ is the slice $\mathcal{F}|_c$ with coordinates $d = (d_x, d_y)$, and the function f is $f(d) = -\delta_c(d)$. The basins $\mathcal{B}_1$ and $\mathcal{B}_2$ are two "mountains" separated by "valleys". The saddle occurs along the path connecting the two mountain tops that minimizes the amount of descent. The following theorem establishes that *IRoadmap* always finds the target.

Theorem 2 ([22]) *Let the robot be a planar rigid body with three degrees of freedom (d_x, d_y, θ). If q_T is reachable from q_S, IRoadmap finds at least one departure point in $\mathcal{F}|_{q_T}$ from which q_T is* **reachable**. *Through recursive invocation of the algorithm from departure points in $\mathcal{F}|_{q_T}$,* **a path to q_T is found.**

Sketch of proof: First we show that every point in $\mathcal{F}$ lies in the closure of some ridge-curve basin. Then we invoke the mountain pass theorem, to show that a saddle curve lies between every two adjacent ridge-curve basins. The critical-point detector and the minimum-passage detector locate at least one point on each saddle curve. Hence linking-curves are constructed to every ridge-curve reachable from q_S. □

4 Conclusion

We presented two range-sensor based navigation algorithms for mobile robots. *TangentBug* is a local planner that strives to act reactively while guaranteeing convergence to the target. *IRoadmap* is a global planner that incrementally constructs a global roadmap for the freespace while searching for the target. We have implemented both algorithms on a cylindrical Nomad200 mobile robot, using a sonar ring as a range sensor.

Let us discuss our current work on sensor-based mobile robot navigation. First we consider local reactive-like navigation. *TangentBug* and other sensor-based algorithms of its type are suitable only for two-dimensional configuration spaces, for the following reason. These algorithms determine target unreachability when the robot completes a loop around an obstacle without finding a suitable leave point. However, in three-dimensions the boundary of a c-space obstacle is a surface, not a curve. The robot therefore must inspect the entire surface of an obstacle before determining target unreachability. As Kutulakos and coworkers have noted [12], the process of surface exploration requires maintenance of a model of the surface. Unfortunately, the use of such a model in the navigation process cannot be considered as being reactive.

In Ref. [10], we propose a new local planner for a point robot navigating in a three-dimensional polyhedral environment. The robot is equipped with an omni-directional range or vision sensor, which measures distance to all the currently visible obstacles. The planner, called *3DBug*, estimates the locally shortest path based on the currently visible obstacles, and moves in a reactive fashion along the locally shortest path. Similarly to *TangentBug*, when the robot detects that it is trapped in the basin of a local minimum of the distance function $d(x, T)$, it switches to obstacle surface transversal. However, this time the robot incrementally constructs a model of the obstacle surface while attempting to find suitable leave points. The robot determines target unreachability when it has scanned the entire surface of the obstacle without finding a suitable leave point. Simulation results show very encouraging behavior of the algorithm, and these results appear in Ref. [10].

Current research also involves the implementation of the global planner *IRoadmap* on three degree-of-freedom mobile robots, equipped with sonar and laser range sensors. The following three topics are among the issues currently under investigation. First, how to physically measure the interesting critical points using the θ-axis a s weep direction. Second, how to characterize the minimum passage points in terms of readily measurable features in the robot's environment. Third, how to guarantee that while traveling along a ridge-curve, the robot sensors scan the entire basin of attraction of the ridge-curve, as required by the definition of the critical point and minimum passage point detectors. In particular, we seek a suitable representation of the physically visible set as a region in the robot's (d_x, d_y, θ) configuration space. These and other topics must be resolved before *IRoadmap* can become a fully implementable global sensor-based planner.

References

[1] R. C. Arkin. Motor schema based mobile robots navigation. *International journal of robotic research*, 8(4):92–112, 1989.

[2] J. F. Canny and M. C. Lin. An opportunistic global path planner. *Algorithmica*, 10:102–120, 1993.

[3] R. Chatila. Deliberation and reactivity in autonomous mobile robots. *Robotics and Autonomous Systems*, 16:197–211, 1995.

[4] H. Choset and J. W. Burdick. Sensor based planning, part ii: Incremental construction of the generalized voronoi graph. In *IEEE Conference on Robotics and Automation*, pages 1643–1649, 1995.

[5] F. H. Clarke. *Optimization and Nonsmooth Analysis*. SIAM Publication, 1990.

[6] J. Cox and C. K. Yap. On-line motion planning: moving a planar arm by probing an unknown environment. Technical report, Courant Institute of Mathematical sciencs, New York University, July 1988.

[7] James L. Crowley and Yves Demazeau. Principles and techniques for sensor data fusion. *Signal Processing*, 32:5–27, 1993.

[8] G. Foux, M. Heymann, and A. Bruckstein. Two-dimensional robot navigation among unknown stationary polygonal obstacles. *IEEE Transactions on Robotics and Automation*, 9(1):96–102, 1993.

[9] Goresky and Macpherson. *Stratified Morse Theory*. Springer-Verlag, New York, 1980.

[10] I. Kamon, E. Rimon, and E. Rivlin. Range-sensor based navigation in three dimensions. CIS 9712, Center of Intelligent Systems, Dept. of Computer Science, Technion, Israel, 1997.

[11] I. Kamon, E. Rimon, and E. Rivlin. Tangent-bug: A range-sensor based navigation algorithm. *Int. Journal of Robotics Research*, (to appear).

[12] K. N. Kutulakos, V. J. Lumelsky, and C. R. Dyer. Vision guided exploration: a step toward general motion planning in three dimensions. In *IEEE Conference on Robotics and Automation*, pages 289–296, 1993.

[13] J. J. Leonard and H. F. Durrant-Whyte. *Directed sonar sensing for mobile robots navigation.* Kluwer academic publishers, Boston, London, Dordrecht, 1992.

[14] Y. H. Liu and S. Arimoto. Path planning using a tangent graph for mobile robots among polygonal and curved obstacles. *International Journal of Robotic Research*, 11(4):376–382, 1992.

[15] V. J. Lumelsky. A comparative study on the path length performance of maze-searching and robot motion planning algorithms. *IEEE Transactions on on Robotics and Automation*, 7(1):57–66, 1991.

[16] V. J. Lumelsky and T. Skewis. Incorporating range sensing in the robot navigation function. *IEEE Transactions on Systems, Man, and Cybernetics*, 20(5):1058–1068, 1990.

[17] V. J. Lumelsky and A. A. Stepanov. Path-planning strategies for a point mobile automaton moving amidst obstacles of arbitrary shape. *Algoritmica*, 2:403–430, 1987.

[18] H. Noborio and T. Yoshioka. An on-line and deadlock-free path-planning algorithm based on world topology. In *IEEE/RSJ Conference on Intelligent Robots and Systems, IROS*, pages 1425–1430, 1993.

[19] R. S. Palais and C.-L. Terng. *Critical Point Theory and Submanifold Geometry*, volume 35 of *Lecture Notes in mathematics*. Springer-Verlag, 1988.

[20] T. Poston and I. Stewart. *Catastrophe Theory and its Applications*. Pitman, England, 1978.

[21] N. S. V. Rao and S. S. Iyengar. Autonomous robot navigation in unknown terrains: visibility graph based methods. *IEEE transactions on Systems, Man and Cybernetics*, 20(6):1443–1449, 1990.

[22] E. Rimon. Construction of c-space roadmaps from local sensory data. what should the sensors look for? *Algorithmica*, 17(4):357–379, 1997.

[23] H. Rohnert. Shortest paths in the plane with convex polygonal obstacles. *Information Processing Letters*, 23:71–76, 1986.

[24] A. Sankaranarayanan and M. Vidyasagar. Path planning for moving a point object amidst unknown obstacles in a plane: the universal lower bound on worst case path lengths and a classification of algorithms. In *IEEE Conference on Robotics and Automation*, pages 1734–1941, 1991.

[25] A. Stentz. Optimal and efficient path planning for partially known environments. In *IEEE Conference on Robotics and Automation*, pages 3310–3317, 1994.

[26] C. J. Taylor and D. J. Kriegman. Vision based motion planning and exploration algorithms for mobile robots. In *Workshop on Algorithmic Foundations of Robotics*, pages 69–83. A K Peters, 1995.

Interleaving Motion Planning And Execution For Mobile Robots

Raja Chatila and Maher Khatib
LAAS-CNRS
7, Ave. du Colonel Roche 31077 Toulouse Cedex 4 - France
email: {raja,maher}@laas.fr

Abstract

Motion planning and control for a mobile robot are often considered as two different and sequential processes. However, in a real environment where the robot is subject to perturbations induced by other moving agents or incomplete and uncertain models, it appears necessary to bridge the gap between planning and execution. The presented approach consists in considering the planned path as an indication by an execution system that will have some freedom in the actual motion control. The questions discussed in this paper are then: What is a planned path? How does the execution system modify it on-line? How are the kinematic features of the robot taken into account? The approach relies on an extensive use of potential fields for defining sensor-based motions, as well as for dynamic path modification. Experimental results show the effectiveness of the algorithms.

1 Introduction

Motion planning and control for a mobile robot are often considered as two different and sequential processes. However, in a real environment where the robot is subject to perturbations induced by other moving agents or incomplete and uncertain models, it appears necessary to bridge the gap between planning and execution. Otherwise, the planner will produce non executable paths, and the robot will often be trying to circumnavigate obstacles and to execute "stubbornly" the planned path, which may produce very inefficient motions.

When the problem is considered from the standpoint of motion planning, the planners need complete and precise knowledge on the geometry of obstacles, which is often not available, and when it is, it will necessarily be an idealization of the real world. The resulting path is often not executable as such. On the other hand, approaches using only sensor based-motions which do not rely on a planned path will produce a non efficient behavior (i.e. far from optimal paths) and being not complete, they do not guarantee the execution of the task.

Considering planning and execution sequentially is not a solution to the problem as the system will be bound to revise its plans more or less often according to the evolution of the environment and the output of the execution process. There will be several time consuming plan/execute cycles.

In general, the robot should be able to achieve motion both in rather open spaces, or in the close vicinity of obstacles. In both cases, there are perturbations induced by other moving agents, non modeled obstacles, and uncertainties.

In order to address the general problem of collision-free motion, we consider motion planning and sensor-based execution to be interleaved. Our approach is twofold:

• Planning may be achieved in terms of a sequence of primitive sensor-based motions, and the presence of obstacles, some used as landmarks, is used to servoe the motion by sensory feedback; this is classical in several systems for motion planning with uncertainty [12, 21, 14, 1, 2]. We have devised for such systems several sensor-based motion primitives making extensively use of potential fields [11]. We introduce the notion of "Task Potential" (section 2) which is a convex potential function defined such that to enable the robot to achieve a sensor-based motion under some conditions represented by events detected during the control loop. The execution of the task is a gradient descent of a global potential function composed linearly from several convex potential functions with coefficients that are scalar positive and monotonous functions of time. Several task potentials can be defined such as "wall following", "move until contact", "rotate until parallel to wall", etc.

• Planning may also produce a trajectory composed of a sequence of geometric primitives such as segments and arcs of circle [19], that can be smoothed for execution by clothoids for example, or in terms of parametric analytic functions such as splines or Bezier curves [7, 6, 4]. In this case, the system should be able to cope with transient obstacles, or with local modifications of the trajectory due to uncertainties.

We rely then on sensor-based dynamic path modification (section 3), using an "elastic band", introduced in [17, 18, 16], which consists in maintaining a sequence of connected local subsets of free-space called bubbles (basically the largest free-space subset centered on the robot). The resulting global path is deformable and is the concatenation of paths joining bubble centers. We have further developed this idea to car-like nonholonomic robots. The metric structure of feasible paths associated to the robot defines the bubbles and their interactions. The bubble shape is given by the Reed and Shepp metric. The dynamic modification of the path, i.e., band optimization and adaptation to cope with new obstacles, is based on convex potential functions ensuring path stability. But the resulting trajectories are yet to be smoothed for execution to remove any path curvature discontinuity. We then use Bezier curves (section 4) to smooth the curvature at junctions. Bezier curves are guaranteed to fully lie within a convex envelop included in the bubbles.

One important feature in this approach is that when the path is subject to a perturbation, even if this is yet away from the current robot position, this propagates to the whole path which is permanently dynamically modified optimally to cope with the perturbation. There is no "pure" local avoidance, and no replanning is necessary, except when the connectivity of the path is suppressed.

2 Sensor-based actions

2.1 The "Task Potential"

To bring the robot to a given configuration, we classically apply an attractive potential which depends on the distance function to this configuration and which have a single minimum. The force, gradient of the potential, will make the robot move towards the goal. Note that the only useful state feedback is the distance to the goal. Using an odometric feedback or external perception, the robots position is regularly updated and the distance to the goal computed. Displacement may be generated by a command using exteroceptive or proprioceptive feedback.

The potential function is defined on the operational space as function of the current position of the robot. In the context of sensor-based commands, the environment is not (totally) known and it is not possible to define the potential function on the whole space (or subspace). If the perceptual data may determine the position of the robot at every moment in some global frame, we can define a global potential which evolves over time in order to realize the required action. The realization of the task is insured if the evolving potential minimum converges towards a unique global minimum. The attractive potential linked to a mobile goal is a typical example of such a potential. On the other hand, if the perceptual data do not permit the expression of the robot position in a global referential, the potential must depend directly on the data and the generation of the displacement must be done incrementally according to the evolution of the environment. This is the case of the repulsive potential associated with the distance to the obstacle.

The principle of the task potential is to associate with each sensor-based action (or task), such as "move to contact" or "follow wall", a time dependent potential. The evolution of the potential function being dependent on the environment, it is controlled by control functions fixing the global behavior of the robot according to the task it should realize.

The task control may be expressed by the decent of the gradient of a potential function $\mathcal{P}(q,t)$, $t \in [0,T]$, composed linearly of several ***convex*** potential functions whose coefficients are scalar functions of time, positive and monotonous.

Let $\mathcal{F}_i(t)$, $t \in [0,T], i \in \{0,\ldots,n\}$ be scalar functions of time, positive and monotonous. Let q be the robot configuration vector. Let $\mathcal{P}_i(q)$ be positive differentiable convex potentials. The potential $\mathcal{P}(q,t)$ produced by a linear application of the form:

$$\mathcal{P}(q,t) = \sum_i \mathcal{F}_i(t)\mathcal{P}_i(q), \quad \forall t \in [0,T] \qquad (1)$$

is a convex potential whose unique solution, q_m, verifies:

$\frac{\partial \mathcal{P}}{\partial q}|_{q_m,t=T} = 0$

Indeed, according to convex analysis, a linear composition of convex functions is a convex function [5]. Thus $\mathcal{P}$ has for each $t \in [0,T]$ a unique minimum. The task to realize corresponds to the global minimum at $t = T$.

If the linear application is a pure convex composition, we can write:

$$\sum \mathcal{F}_i(t) = 1, \quad \forall t \in [0,T]$$
$$\mathcal{F}_i(t) \in [0,1], \quad \forall t \in [0,T]$$

Different task potentials can be constructed from a basic, simple and generic potential functions. We give here the example of wall following.

2.2 Example: Wall following

The aim of this action is to make the robot advance parallel to a wall at a fixed distance from it while avoiding obstacles. Obstacle avoidance may temporally violate the distance condition.

2.2.1 *Action Definition*

Let $R_g(Oxy)$ be a global frame in the operational space related to the configuration space $\mathcal{A} \subset \Re^3$, $X = (xy\theta)^T$ be the robot configuration vector and $\mathcal{O}$ the set of stationary objects in $\mathcal{A}$.

Let $R_w(Oxy)$ be a frame associated with a segment $S\{O, x_p\}$ such as shown in figure 1. Let d be the distance from the center of the robot to the segment, $y = d$. Consider the frame R_r centered on the robot such as the x-axis which corresponds to the robot direction is perpendicular to the wheel axis. Note $\vec{\imath}$ the unitary vector in this direction. Let S_o be a segment delimiting the obstacle represented in the frame R_r and let V_s be its direction vector. Define the angle α as:

$$\alpha = Arcsin\left(\frac{\mathcal{K}(V_s \times \vec{\imath})}{|V_s|}\right) \qquad (2)$$

where $\mathcal{K} : \mathcal{R}^3 \rightarrow \mathcal{R}$ is the projection of a point $(x,y,z)^T$ on its third component z. α has thus its values in the interval $]-\frac{\pi}{2}, \frac{\pi}{2}[$. It is null when the robot is parallel to the segment.

The parameter α will be used to modify the obstacle force intensity in function of the orientation of the robot. Thus it represents the "danger" due to the obstacle orientation relative to the robot.

If d_w is the required distance from the wall, and if α represents the segment orientation in the robot frame, a **Follow Wall** action is represented by the curve $S(t) = [x(t)y(t)\theta(t)]^T$, $t \in [0,T]$ such that:

$$S(0) = X_0 \qquad (1)$$
$$\{S[0,T]\} \cap \mathcal{O} = \emptyset \qquad (2)$$
$$\exists t_m \forall t > t_m,\ y(t) = d_w\ \theta(t) = 0 \qquad (3)$$
$$\dot{x}(t) > 0,\ \forall t < T \qquad (4)$$
$$S(T) = X_T = \begin{pmatrix} x_p \\ d_w \\ 0 \end{pmatrix},\ \dot{S}(T) = 0 \qquad (5)$$

The first condition represents the initial state of the robot in the frame R_w. The non-collision is guaranteed by the second condition all along the curve $S(t)$. Condition (3) characterizes the robot configuration relative to the wall to follow after an interval of time fixed by t_m (figure 1). The movement direction and the requirement of a longitudinal velocity component are expressed by condition (4). The final state of the action is provided by the last condition.

2.2.2 *The Task Potential*

An attractive force F_x in the direction of the x-axis, function of the distance $x_p - x(t)$, x_p being the (possibly time varying) abscissa of the edge of segment S, is derived from a potential defined on the x-axis with a minimum at $x(t) = x_p$. The motion along the y-axis at d_w is due to another potential defined on this axis function of the distance $y(t) - d_w$. Robot orientation is controlled by a rotational potential $\mathcal{P}_\alpha$, function of the parameter α which represents the wall orientation. The induced torque τ_α will also contribute to optimize the distance to the wall.

Figure 1 shows the potentials $\mathcal{P}_x$ and $\mathcal{P}_y$ with their minima. Note that the potential $\mathcal{P}_y$ tends to infinity on the wall surface to guarantee non-collision.

The different potentials that realize this task are:

$$\mathcal{P}_x = \frac{k_x}{2}(x - x_p)^2$$
$$\mathcal{P}_y = \begin{cases} \frac{k_y}{2}(y - d_w)^2 + \frac{\eta}{2}\left(\frac{1}{y} - \frac{1}{d_w}\right)^2 & \text{if } y < d_w \\ \frac{k_y}{2}(y - d_w)^2 & \text{otherwise} \end{cases}$$
$$\mathcal{P}_\alpha = \frac{k_\alpha}{2}\alpha^2$$

where k_x, k_y, k_α, η are positive gain constants. The **Task Potential** is thus:

$$\mathcal{P} = \mathcal{P}_x + \mathcal{P}_y + \mathcal{P}_\alpha$$

figure 2 shows the distribution of this potential in function of the variation of y and θ. Note that

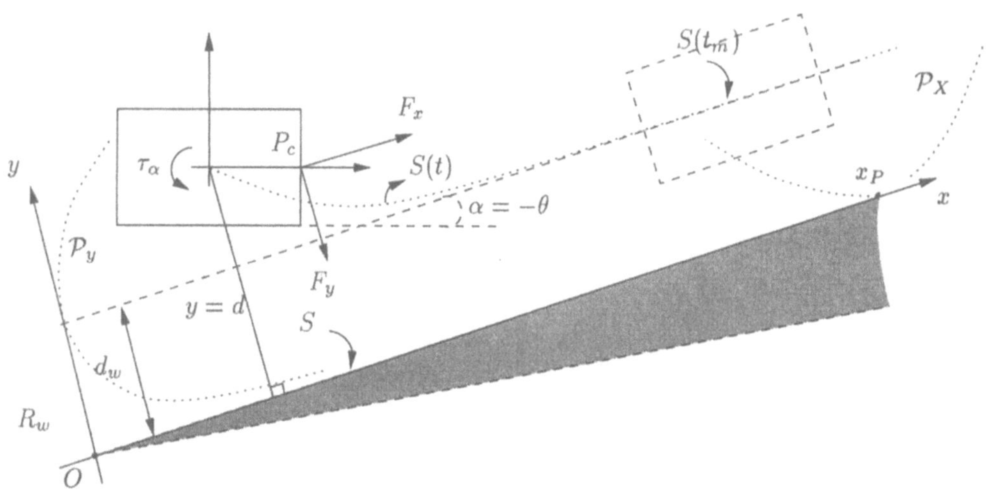

Figure 1: *Parameters defining the* Follow Wall *action.*

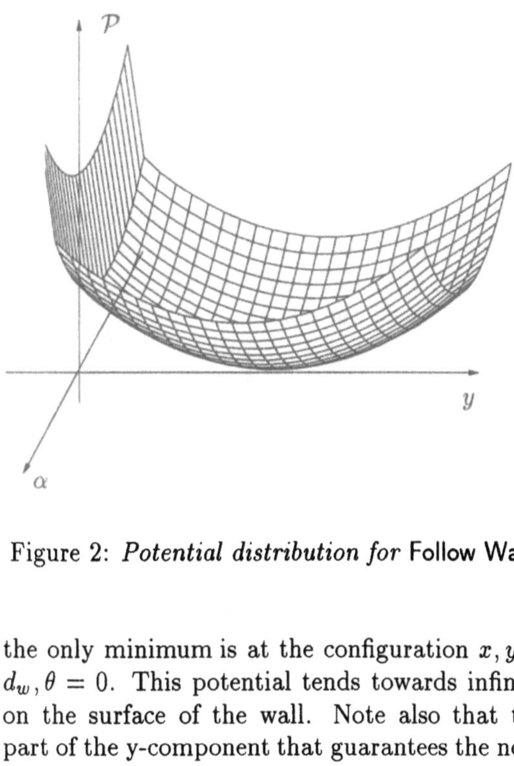

Figure 2: *Potential distribution for* Follow Wall.

the only minimum is at the configuration $x, y = d_w, \theta = 0$. This potential tends towards infinity on the surface of the wall. Note also that the part of the y-component that guarantees the non-collision is null when $y > d_w$.

The derived force is:

$$\mathcal{F} = \begin{pmatrix} -k_x(x - x_p) \\ -k_y(y - d_w) + \eta\left(\frac{1}{y} - \frac{1}{d_w}\right)\frac{1}{y^2} \\ -k_\alpha \alpha \end{pmatrix}$$

The asymptotic stability of this force is insured by the addition of a dissipation:

$$\mathcal{F}^* = \mathcal{F} - \bar{K}\dot{X}$$

where $\bar{K}$ is a diagonal matrix of the form:

$$\bar{K} = \begin{pmatrix} k_{\dot{x}} & 0 & 0 \\ 0 & k_{\dot{y}} & 0 \\ 0 & 0 & k_{\dot{\theta}} \end{pmatrix}$$

$k_{\dot{x}}, k_{\dot{y}}, k_{\dot{\theta}}$ are positive dumping gain constants. In fact these constants depend upon the proportional gain constants of the different components of the force $\mathcal{F}$. The stability of the system is demonstrated in [8].

2.2.3 *Collision Avoidance*

One of the fundamental interests of the Task Potential formulation is the possibility of taking obstacles into account permanently. The additivity of convex potentials enables to associate a repulsive potential function of a minimum distance from the robot and to add this potential to the Task Potential. The global convergence is no longer insured (because of possible local minima) but the task becomes more robust.

In order to include collision avoidance in this action, we define a distance $d_e : \mathcal{A} \times \mathcal{O}_{\bar{\Omega}} \to \Re^+$ that returns the shortest distance between the robot and the objects $\bar{\Omega}$ in a given angular sector Ω in front of the robot. This distance is associated to a repulsive potential that insures non-collision.

2.2.4 *Experimentation*

Figure 3 show an example of wall following in presence of obstacles. The first obstacle near the wall is practically considered as a part of it. To avoid the second obstacle the robot violates the follow distance. A local minimum may occur with an obstacle whose position makes avoidance impossible.

Figure 3: *Collision avoidance during* Follow Wall.

Figure 4 shows an experiment realized with a sensor-based motion planner considering uncertainty [9]. We can note the important ratio of the distance covered by the sensor-based actions to the total distance. **Note1** represents the robot absolute position error w.r.t to its real position in the model.

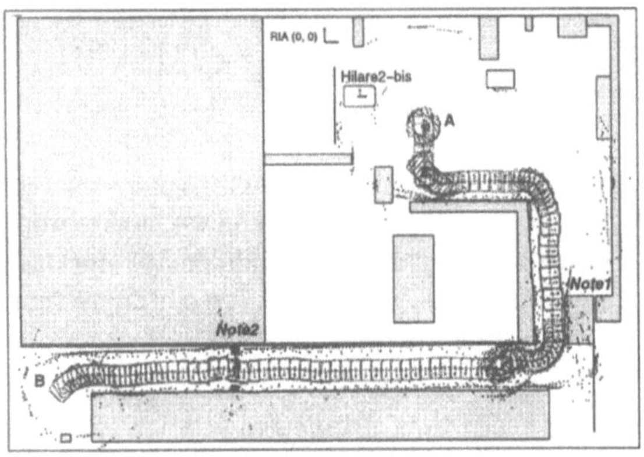

Figure 4: *Motion planning using sensor-based actions.*

The first important action in this experiment is the following of three consecutive walls. Wall extremity detection and switching walls are done dynamically thanks to the continuous composition of potentials. During the second following action the robot crosses a door which is not integrated in the model by considering its edges as unexpected obstacles to be avoided (Note2).

3 Bubble Bands

As we discussed previously, it is important for the robot system to plan paths and yet to be able to modify them during execution. In this section, we present an approach to this issue based on the bubble band concept that we extend to car-like non-holonomic robots.

A bubble, as introduced in [16], is the maximal free local subset of the free-space around a given configuration of the robot. It is generated from distance information about the robot model and the objects in the environment. Given a metric space (a configuration space and a distance `Dis_CC`) based on elementary feasible paths, and a distance to obstacles function (`Dis_CO`) [15], a bubble $B(\mathbf{p},\mathbf{r})$ is characterized by its center $\mathbf{p}$ and its radius, i. e., the distance to the closest obstacle.

Given $\{\mathbf{p_i}\}_{\mathbf{i=1...n}}$ configurations in free-space such that the bubbles defined around two consecutive configurations, $B(\mathbf{p_i})$ and $B(\mathbf{p_{i+1}})$ overlap, it is possible to construct a collision-free path from $\mathbf{p_1}$ to $\mathbf{p_n}$. This path is an elastic band paved with bubbles, or a bubble band. This band will be deformed according to two kinds of forces: internal and external. The band motion at each configuration in the path is determined by calculating a force composed of these two parts. The internal forces act to remove slack in the band (energy minimization), the external forces move the band away from obstacles (thus maintaining the path free from collision).

To maintain the band in free-space we should insure that adjacent bubbles overlap. Before two adjacent bubbles become disconnected because of the action of the forces, a new bubble is created between them. Another modification is to remove redundant bubbles. If two non-adjacent bubbles overlap, all the intermediate bubbles are suppressed from the band.

3.1 Nonholonomic bubbles

We consider the metric space associated with a car-like nonholonomic mobile robot with a non-null minimal steering radius R. A configuration of the robot is defined in $\mathcal{R}^2 \times \mathcal{S}^1$. The shortest paths between two configurations is composed of sequences of straight line segments and arcs of a circle, with at most two cusp points [19]. Let d be the metric associated with this space [13].

A *bubble* $B(\mathbf{p},\mathbf{r})$ is the set of configurations at a distance smaller than r from $\mathbf{p}$:

$$B(\mathbf{p},\mathbf{r}) = \{\mathbf{q} \in \mathcal{R}^2 \times \mathcal{S}^1 \mid \mathbf{d}(\mathbf{p},\mathbf{q}) < \mathbf{r}\}.$$

Two bubbles $B(\mathbf{p},\mathbf{r})$ and $B(\mathbf{q},\mathbf{s})$ are said to be *connected* if $B(\mathbf{p},\mathbf{r}) \cap \mathbf{B}(\mathbf{q},\mathbf{s}) \neq \emptyset$ i.e., if and only if $d(\mathbf{p},\mathbf{q}) < \mathbf{r}+\mathbf{s}$.

As a consequence, a Reeds and Shepp path, $\gamma[\mathbf{p},\mathbf{q}]$, joining the two centers of connected bubbles is wholly included in the union of these bubbles.

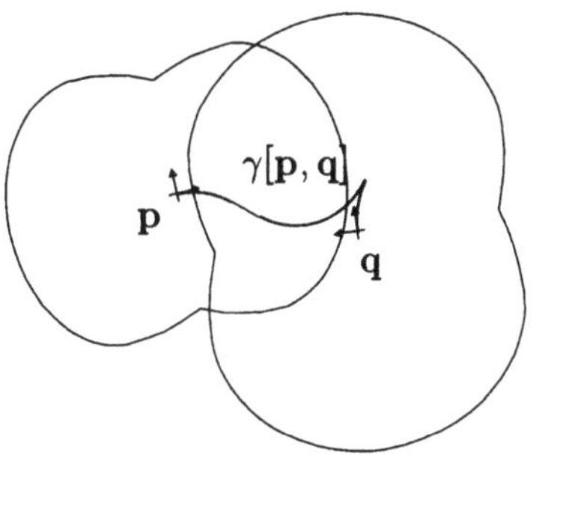

Figure 5: *Path between two consecutive centers*

This property enables to prove the existence of a feasible path from the initial to the final configuration by finding a sequence of connected bubbles joining them (fig. 5).

A bubble may be of two types, *movable* or *immovable*. A movable bubble is a bubble whose center can be modified according to the forces applied to it. An immovable bubble keeps the same center throughout the execution (for example the start configuration or the goal when it is fixed, or mendtory waypoints).

A bubble is said to be smaller than another if they have the same center and if the first is wholly included in the second.

Based on [20, 22], the distance between a robot configuration and a point or segment obstacle can be computed in the metric. This allows to define the biggest collision-free bubble around a configuration as $B(\mathbf{p}) = \mathbf{B}(\mathbf{p}, \mathbf{d}(\mathbf{p}, .\mathcal{O}))$, where $\mathcal{O}$ is the set of segment obstacles.

3.2 Forces

The forces applied to the bubbles are of two 1 types. The external forces due to the repulsion of the obstacles and the internal forces which work to keep the bubbles connected, as long as possible, and optimize the form of the band.

3.2.1 *Internal Forces*

Let $B(\mathbf{p_{i-1}}, \mathbf{r_{i-1}})$, $B(\mathbf{p_i}, \mathbf{r_i})$ and $B(\mathbf{p_{i+1}}, \mathbf{r_{i+1}})$ be three consecutive bubbles. Adjacent bubbles are connected:

$$d^+(\mathbf{p_i}) = \mathbf{d}(\mathbf{p_i}, \mathbf{p_{i+1}}) \leq \mathbf{r_i} + \mathbf{r_{i+1}} - \epsilon_c. \quad (3)$$

for some small $\epsilon_c \in \mathcal{R}^{+*}$.

To optimize bubble number, adjacent bubbles should not overlap too much. Thus

$$d^+(\mathbf{p_i}) = \mathbf{d}(\mathbf{p_i}, \mathbf{p_{i+1}}) \geq \mathbf{r_i} + \mathbf{r_{i+1}} - \epsilon_o, \quad (4)$$

for some small $\epsilon_o \in \mathcal{R}^{+*}$ greater than ϵ_c.

The band should tend to contract to optimize the global path. Thus the center of a middle bubble should try to place itself on a shortest path joining its two neighbors, on the radius-weighed barycenter of their centers. If $\gamma_p[\mathbf{p_{i-1}}, \mathbf{p_{i+1}}] : [0,1] \longrightarrow \mathcal{R}^2 \times \mathcal{S}^1$ is a parameterization of $\gamma[\mathbf{p_{i-1}}, \mathbf{p_{i+1}}]$, the path joining $\mathbf{p_{i-1}}$ and $\mathbf{p_{i+1}}$, and if $\tilde{r}_i = \frac{r_{i-1}}{r_{i-1}+r_{i+1}}$, then

$$d^\gamma(\mathbf{p_i}) = \mathbf{d}(\mathbf{p_i}, \gamma_\mathbf{p}[\mathbf{p_{i-1}}, \mathbf{p_{i+1}}](\tilde{\mathbf{r}}_\mathbf{i})) \leq \epsilon. \quad (5)$$

where $\epsilon \in \mathcal{R}^+$ is arbitrarily small.

Using the constraints 3 to 5 and noting $d^-(\mathbf{p_i}) = \mathbf{d}^+(\mathbf{p_{i-1}})$, the following potential fields are built to produce the internal forces:

$$\begin{aligned} \mathcal{P}_f(\mathbf{p_i}) &= \tfrac{\mathbf{K_f}}{\mathbf{2}} \left(d^+(\mathbf{p_i}) - (\mathbf{r_i} + \mathbf{r_{i+1}}) + \epsilon_c\right) \\ &\quad \times \left(d^+(\mathbf{p_i}) - (\mathbf{r_i} + \mathbf{r_{i+1}}) + \epsilon_o\right) \\ \mathcal{P}_b(\mathbf{p_i}) &= \tfrac{\mathbf{K_b}}{\mathbf{2}} \left(d^-(\mathbf{p_i}) - (\mathbf{r_{i-1}} + \mathbf{r_i}) + \epsilon_c\right) \\ &\quad \times \left(d^-(\mathbf{p_i}) - (\mathbf{r_{i-1}} + \mathbf{r_i}) + \epsilon_o\right) \\ \mathcal{P}_c(\mathbf{p_i}) &= \tfrac{\mathbf{K_c}}{\mathbf{2}} \left(d^\gamma(\mathbf{p_i})\right)^2. \end{aligned} \quad (6)$$

where $\mathcal{P}_f$ is the forward connection potential, with the next bubble, $\mathcal{P}_b$ is the backward connection potential and $\mathcal{P}_c$ is the contraction potential.

3.2.2 *External Forces*

These forces can be simply derived from repulsive potential fields.

The application of the ∇ operator to derive the force expressions from the potentials brings us to calculate $\frac{\partial d}{\partial \mathbf{p}}$ where $\mathbf{p}$ is the configuration that varies. In fact this last expression gives the direction of the force. Unlike the Euclidean distance, the car-like compatible distances, defined previously, have no normalized differentials. They contribute thus to the magnitude of the force. As well, they do not vary smoothly with $\mathbf{p}$, with consequent effects on stability (see section 3.4).

3.3 Creation and Modification of a Bubble Band

As the band expands, bubbles should be created to guarantee the connectivity. Bubbles are also created when an obstacle approaches the band obliging the bubbles to shrink. Inversely, when the band is contracted or when obstacles move away, bubbles may overlap too much, so that some of them should be eliminated.

To construct the original band, we start with a path which may be given by a motion planner. This path is firstly sampled into a sequence of configurations. Then the first bubble is defined to be the greatest bubble centered at the first configuration in the sequence. The sequence is then scanned, by dichotomy, to find the last configuration which still belongs to the last constructed bubble. This configuration is thus taken to be the center of the following bubble which also has the maximum size. The previous step is repeated till the goal (the last configuration) is found to belong to a bubble. This insures the connectivity of the bubble band, as well as its total presence in free-space while making it redundant, obliging the center of each bubble, but the first, to belong to its previous neighbor. This redundancy will be reduced after the first application of the elimination algorithm and of the contraction forces.

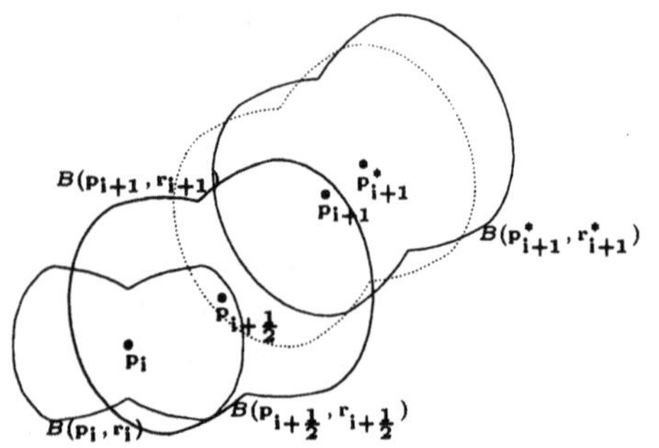

Figure 6: *Adding a bubble to avoid disconnection.*

If three consecutive bubbles, $B(\mathbf{p_{i-1}}, \mathbf{r_{i-1}})$, $B(\mathbf{p_i}, \mathbf{r_i})$ and $B(\mathbf{p_{i+1}}, \mathbf{r_{i+1}})$ have the property

$$d(\mathbf{p_{i-1}}, \mathbf{p_{i+1}}) \leq \mathbf{r_{i-1}} + \mathbf{r_{i+1}} - \epsilon_o,$$

where $\epsilon_o > 0$, then there is redundancy and the middle bubble $B(\mathbf{p_i}, \mathbf{r_i})$ is eliminated.

On the other hand, for each pair of adjacent bubbles $B(\mathbf{p_i}, \mathbf{r_i})$ and $B(\mathbf{p_{i+1}}, \mathbf{r_{i+1}})$, if the following inequality is satisfied

$$d(\mathbf{p_i}, \mathbf{p_{i+1}}) \geq \mathbf{r_i} + \mathbf{r_{i+1}} - \epsilon_c,$$

where ϵ_c is a positive constant, then an intermediate bubble should be created between them (fig 6). The center of this new bubble should be on the path, $\gamma_p[\mathbf{p_i}, \mathbf{p_{i+1}}]$, relating the two neighboring centers such that:

$$\mathbf{p_{i+\frac{1}{2}}} = \gamma_p[\mathbf{p_i}, \mathbf{p_{i+1}}] \left(\frac{\mathbf{r_i}}{\mathbf{r_i} + \mathbf{r_{i+1}}} \right).$$

To prevent the creation-elimination loop from endlessly creating and eliminating the same bubble, the condition $\epsilon_o > \epsilon_c$ should be satisfied.

3.4 Stability

The evolution of the bubbles on the band is governed by conservative forces derived from convex potential fields. For contraction forces, the potential fields are functions of the metric distance which is a true distance and conserves thus the convexity property. These potential fields have thus a unique minimum. For obstacle repulsion forces, the potential fields are functions of the configuration to point distance which is not a true distance. In fact, this distance presents a problem for lengths inferior to $\sqrt{2}R$, as discussed in section 3.2.2. This is due to the discontinuity of the differential of this distance and to the fact that the shortest path, from a configuration to *a point*, does not have the property: *All sub-paths of shortest paths are shortest paths.* In fact, sub-paths of length inferior to $\sqrt{2}R$ may not be shortest paths. To guarantee the convergence of the bubble band we impose thus a minimum radius, $R_{\min} = \sqrt{2}R$, on the bubbles.

4 Smoothing and Execution

The path to execute is the concatenation of the Reeds and Shepp paths $\{\gamma[\mathbf{p_i}, \mathbf{p_{i+1}}], \mathbf{i} = \mathbf{1} \ldots \mathbf{n} - \mathbf{1}\}$ between the centers of n bubbles. As already mentioned, the path curvature is discontinuous forcing the robot to numerous stops. To smooth these paths we use Bezier polynomials. The advantage of Bezier polynomials is that they can be wholly defined in a convex envelop as well as their derivatives. We can thus guarantee that the executed path lies within the union of the bubbles. The velocity and acceleration on these curves are bounded. To insure that the curvature lies in the neighborhood of R, the steering radius, we use a set of control points.

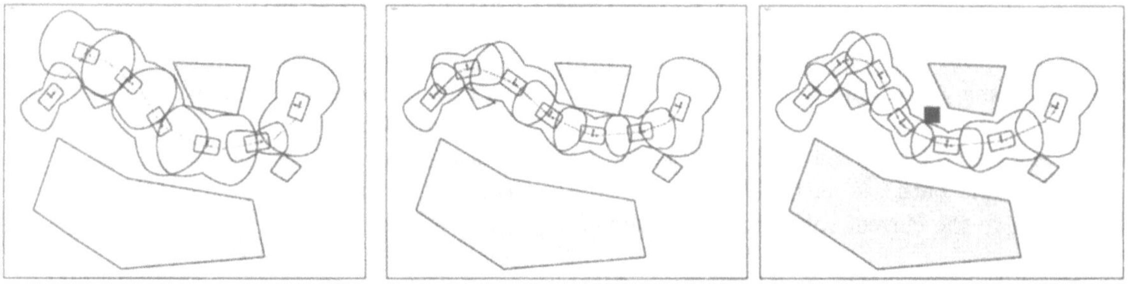

Figure 7: *Bubble band creation, evolution and deformation to avoid a mobile obstacle.*

4.1 Approximation by Bezier Curves

A Bezier curve of *order* $n+1$ and *control points* $\mathbf{p} = \{\mathbf{p_0}, \dots, \mathbf{p_n}\}$ is defined by

$$\mathcal{BP}[\mathbf{p}](\mathbf{t}) = \sum_{\mathbf{i=0}}^{\mathbf{n}} \mathbf{B_i^n(t)p_i},$$

where $B_i^n(t) = \binom{n}{i}(1-t)^{n-i}t^i, \qquad i = 0 \dots n.$

We use the following properties of Bezier curves [3].

- A Bezier curve lies in the convex envelop of its control points $\mathcal{BP}[\mathbf{p}] \subset \mathbf{co}(\mathbf{p})$.
- The derivative of a Bezier curve is a Bezier curve whose control points are finite differences of the original control points.

According to the first property, choosing the control points in a bubble or, a region, guarantees that all the curve is in that bubble, or region.

The second property combined with the first one insures that the derivatives $\frac{\partial^k}{\partial t^k}\mathcal{BP}[\mathbf{p}]$ are contained in the convex envelops $co(\Delta^k \mathbf{p})$ and thus that the velocity and the acceleration are bounded.

To join smoothly Reeds and Shepp curves, we approximate the arcs of circle by Bezier curves whose curvature is null at the two extremities. To satisfy this condition and fix the starting and finish positions and velocities, at least six control points are necessary $(\mathbf{p_i}, \mathbf{i} = \mathbf{0} \dots \mathbf{5})$. For each extremity, the first point is on the extremity, the second in the direction of the velocity and the third collinear to the preceding two, to insure null curvature. The initial and end velocities can be controlled by an appropriate parameterization, and four distances have to be determined. Due to symmetry, these four degrees of freedom are reduced to two: $l = |\overline{\mathbf{p_0 p_1}}| = |\overline{\mathbf{p_4 p_5}}|$ and $k = |\overline{\mathbf{p_1 p_2}}| = |\overline{\mathbf{p_3 p_4}}|$.

The problem is thus reduced to find

$$\min_{l>0,k>0} \max_{t\in[0,1]} \frac{\|\mathcal{BP}'[\mathbf{p}](\mathbf{t}) \times \mathcal{BP}''[\mathbf{p}](\mathbf{t})\|}{\|\mathcal{BP}'[\mathbf{p}](\mathbf{t})\|^3}.$$

This optimization problem has not been solved yet. For the time being we use empirical values of l and k that bound the curvature to $\frac{6}{5}\rho$ where ρ is the curvature of the arc.

The parameters l and k determine the shape of a Bezier curve. This curve can then be followed in different ways by changing its parameterization. In fact, if $\varphi : [t_0, t_1] \longrightarrow [0,1]$ is a strictly increasing onto function, except perhaps on t_0 and/or t_1 then $t \longrightarrow \mathcal{BP}[\mathbf{p}](\varphi(\mathbf{t}))$ is a new parameterization of $\mathcal{BP}[\mathbf{p}]$. If we have a predetermined starting velocity v_s, this parameterization should verify

$$\mathcal{BP}'[\mathbf{p}](\varphi(\mathbf{t_0}))\frac{\partial}{\partial \mathbf{t}}\varphi(\mathbf{t_0}) = \mathbf{v_s}.$$

Similarly, if the finishing velocity is known,

$$\mathcal{BP}'[\mathbf{p}](\varphi(\mathbf{t_1}))\frac{\partial}{\partial \mathbf{t}}\varphi(\mathbf{t_1}) = \mathbf{v_{finish}}.$$

The reparameterization can also be used to modify the maximum velocity. In fact, due to the convex envelop property, to guarantee that the velocity does not exceed v_{max}, φ should verify

$$\left|\frac{\partial}{\partial t}\varphi(t)\right| \le \frac{v_{max}}{5 \max_{i=0\dots4} \|\Delta \mathbf{q_i}\|}, \qquad \forall t \in [t_0, t_1].$$

A similar reasoning can also be applied to bound the acceleration, the angular velocity and the angular acceleration [15].

4.2 Execution controller (kinan)

We discuss in this section the different aspects related to the execution of the bubble band. In fact, the bubble band is a dynamic system, evolving permanently in function of environment changes. To execute the varying path, we have to satisfy the following constraints:

- The current executed configurations on the dynamic path are, either dynamically joinable to the current robot state or immobile.
- Sensor-based path modification is coherent (correctly synchronized) with the current robot state in function of environment changes.
- Collision-free robot stopping is guaranteed, in case of path modification failure (obstacle avoidance failure).

Beforehand we suppose the existence of a parameterization unit able to compute the robot vector state profile on a given part of the path in function of time [10]. On the other hand, a control module is available to execute the parameterized part of the path. The role of the execution controller (kinan) is now to link the three modules (bubble band, parameterizer, and control module), while satisfying the above-mentioned constraints.

The key idea of kinan is based on the **Freeze_P** operator of the bubble band module. Within this operator, the bubble band is able to freeze a part, on the global path, defined by its length. Actually, the initial configuration of the frozen part is automatically considered as the center of the last immobile bubble. Thus the global path is decomposed into two parts, the first is immobile and the second remains flexible. We can note that this operation guarantees path continuity at the junction configuration. The parameterizer is able to compute the robot state vector profile between any two states. We will discuss now how to compute the length of the part to be freezed to satisfy the robot kinematic constraints, cycle time compatibility between modules, and how to deal with the avoidance failure problem.

4.2.1 *Partition*

Suppose that T_b, T_p, T_c are respectively the cycle time of the bubble band, parameterization, and control modules. T_b is the time necessary to update bubble band which includes distance computation, force computation, bubbles movement and connection check. T_p is the parameterization time of a part of the path and T_c is the robot control level period. In fact, it is easy to remark that T_b is considerably larger than T_p and T_c. Moretheless, if we denote by T_u the necessary communication time to handle all data flow between modules, updating thus a global cycle time of kinan system, we can deduce the first following temporal constraint:

$$T_{part} > T_b + T_p + T_c + T_u$$

where T_{part} is the needed time to execute the frozen part. This constraint expresses the minimum frozen part length, ℓ_f, for a given nominal execution speed. On the other hand, given the maximum acceleration, ℓ_f should be long enough for the robot to be able to stop in case of a normal stop (cusp point, end of the path, ...) or an emergency. In fact, the robot should always move at the nominal desired speed that its path form allows. Then the remaining length on the path should be greater than the minimal length in which the robot can stop completely if it started this part with the nominal speed. If this is not verified the remaining part of the path has to be frozen also.

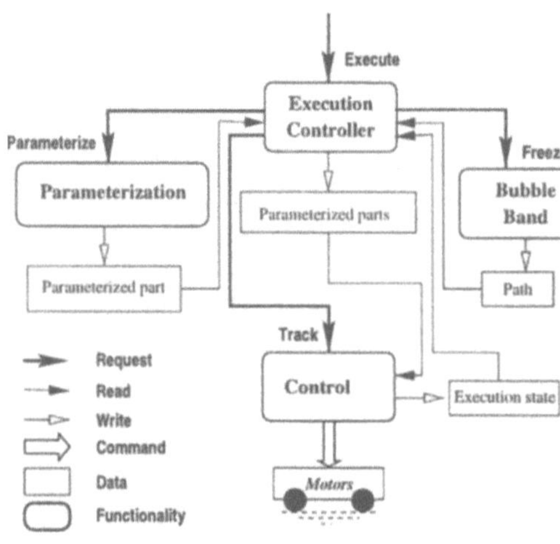

Figure 8: *Architecture of* Kinan *system*

The path is always available in a memory zone, called poster, updated by the bubble band module. When a part is said to be frozen, its length satisfies the previous constraints, and a **Freeze_P** request is sent to the bubble band module (figure 8) which then updates the flexible remaining part of the path in its own poster.

4.2.1.1 Parameterization For a frozen part to be parameterized, its final speed should be determined. To do this, a forward look-up should test for a stop point, such as a cusp point or the end of the path. In both cases the final speed is null. Otherwise it is the maximum speed variation that the parameterization algorithm can afford to reach the desired speed.

4.2.1.2 Control module In order to simplify communication handling operations we will suppose that the control module control has the following behavior: the execution of the path is activated by a Track request. The module waits till one or more new parameterized parts are written in the execution controller poster (kinan poster).

The execution of the first parameterized part is started only if the last one ends with a null speed. While executing a part, control updates robot state, part identifier, and especially the current execution time in its own poster. The execution continues as long as kinan poster is updated with new parameterized parts.

4.2.1.3 kinan execution algorithm Having defined the different requests and operators, we present here the general algorithm that controls the execution.

Initialization
  $s_i = 0$
  $v_i = 0$
-- *Execution time, initial speed and* kinan
-- *poster are initialized.*
  Track
-- *A* Track *request is sent to* control.
-- *The current execution time* s_c *is reset.*
Start
  While ($Path \neq \emptyset$)
    Part$_1$ = Freeze_P -- *Freeze a part*
      -- *for the nominal execution.*
    Get v_f, s_f -- v_f *is the final speed and*
      -- s_f *is the expected execution time.*
    Parameterize (Part$_1$, v_i, v_n)
      -- v_n *is the nominal speed.*
    If $v_f \neq 0$ **Then**
      Part$_2$ = Freeze_P
      -- *Freeze a part to stop.*
      Parameterize (Part$_1$, v_n, 0)
      -- *Parameterize this part to stop.*
    End If
    Do
      Read s_c
    While ($s_i - s_c > \epsilon$)
    Write Part$_1$ and/or Part$_2$
    $v_i = v_f$
    $s_i = s_f$
  End While
End

Discussion This algorithm presents a drawback which is mostly due to parameterization constraints. The speed limits on the ends of a part of a path are sometimes bounded by the parameterization module in function of the geometry of the part. Thus the initial speeds given to the parameterizer may be infeasible in some cases. To avoid this Path$_2$ can be parameterized before Path$_1$ with the nominal speed at the beginning and a stop at the end. The actual initial speed computed by the parameterizer is then taken to be the final speed of Path$_1$.

5 Experimentation

The car-like bubble band and the execution method were implemented on the robot Hilare2 . The following experiment illustrates the behavior of the bubble band.

The first pavement of an arbitrary trajectory is usually redundant. This is immediately corrected by the application of internal forces which tend also to optimize the initial path.

Figure 9 shows the result of the application of internal and external forces on the band. The resultant path is not only geometrically feasible but it is also simpler than the initial one. In fact the bubble band tends to minimize a cost which is a complex combination of obstacle repulsion and path length. Thus the resultant path is usually simpler (at least shorter) and safer than the initial one.

An expected obstacle (the transparent square in the middle of figure 10) modifies the band locally and this disturbance is propagated along the band, with an influence decaying with distance.

Figures 10 and 11 show respectively the executed path and the execution trace. After passing the mobile obstacle, a cusp point enables the robot to re-take the direction of the goal.

6 Conclusion

We have discussed two approaches using potential functions for combining path planning and reactive motion control for non-holonomic car-like mobile robots. According to the environment configuration, either a sequence of sensor-based motions or a modifiable path can be produced by

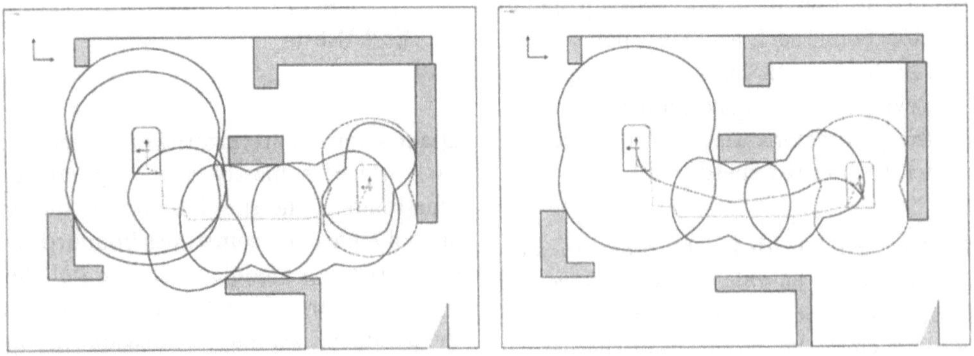

Figure 9: *Initial holonomic path, application of internal and external forces.*

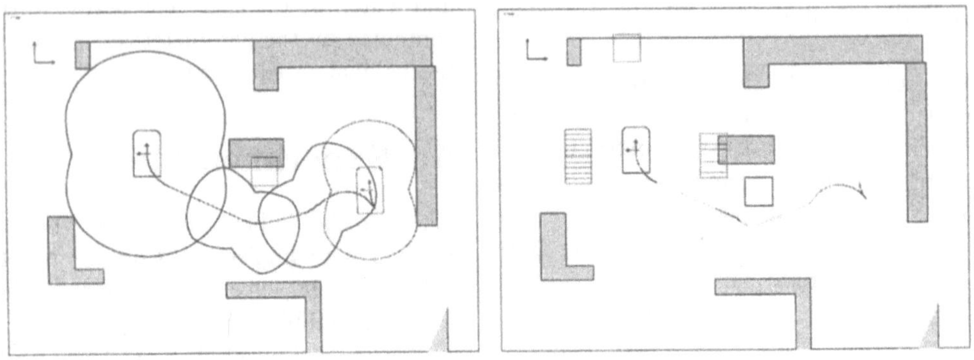

Figure 10: *Path modification and execution.*

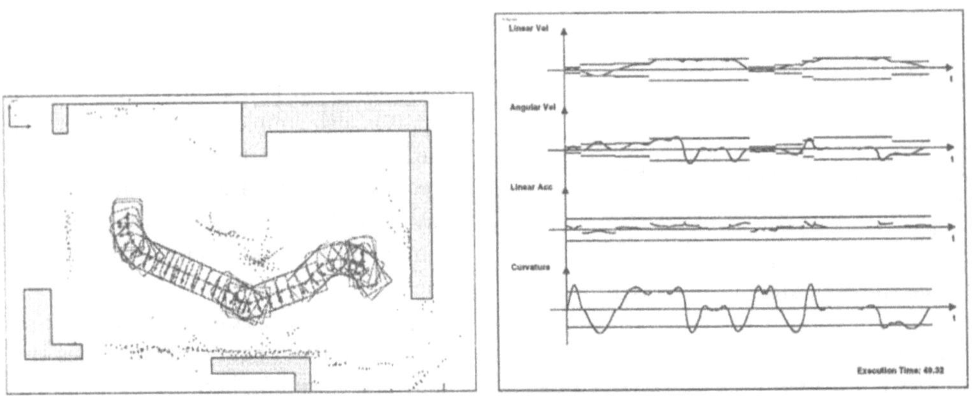

Figure 11: *Execution trace and speed profile.*

a planner and executed by the robot with efficient real time compliance with environment changes.

The theoretical results, implemented in the integrated system kinan, enabled the coexistence of activities with different cycle periods. The experiments confirm the necessity as well as the success of this execution system.

This, brings us to the conclusion that the choice of the navigation method to use must depend on the nature of the desired task and the constraints relative to its application context. This is the reason why the decisional level in the control architecture of an autonomous mobile robot should have available a set of execution techniques and controls compatible with its needs.

Acknowledgments The authors thank H. Jaouni, J-P. Laumond and M. Vendittelli for contributions and developments of some aspects of non-holonomic bubble bands.

References

[1] B. Bouilly and T. Siméon. A sensor-based motion planner for mobile robot navigation with uncertainty. In L. Dorst, M.V. Lambalgen, and F. Voorbraak, editors, *Reasoning with uncertainty in robotics*, pages 235–247. Springer, 1995.

[2] F. De la Rosa, C. Laugier, and J. Najera. Robust path planning in the plane. *IEEE Transactions on Robotics and Automation*, 12(2):347–352, 1996.

[3] J. Fiorot and P. Jeannin. *Courbes et surfaces rationelles*. Masson, Paris, 1989.

[4] S. Fleury, P. Souères, J.-P. Laumond, and R. Chatila. Primitives for smoothing mobile robot trajectories. *IEEE Transactions on Robotics and Automation*, 11(3):p.441–448, June 1995.

[5] R. V. Gamkerlidze. *Analysis II - Convex Analysis and Approximation Theory*. Springer-Verlag, 1980.

[6] Y. Kanayama and B. Hartman. Smooth local planning for autonomous vehicles. In *Proc. IEEE International Conference on Robotics and Automation, Scottsdale, (USA).*, 1989.

[7] K. Kant and S. W. Zucker. Planning smooth collision-free trajectories: path, velocity, and splines in free-space. *The International Journal of Robotics Research*, 2(3):117–126, 1987.

[8] M. Khatib. *Sensor-based motion control for mobile robots*. PhD thesis, LAAS-CNRS, Toulouse, France, December 1996. - Ref.: 96510.

[9] M. Khatib, B. Bouilly, T. Siméon, and R. chatila. Indoor navigation with uncertainty using sensor-based motions. In *IEEE, International Conference on Robotics and Automation, Albuquerque (USA)*, pages 2920–2925, April 1997.

[10] M. Khatib, H. Jaouni, R. Chatila, and J.P. Laumond. Dynamic path modification for car-like non-holonomic mobile robots. In *IEEE, International Conference on Robotics and Automation, Albuquerque (USA)*, pages 2920–2925, April 1997.

[11] O. Khatib. Real-time obstacle avoidance for manipulators and mobile robots. *The International Journal of Robotics Research*, 5(1):90–98, 1986.

[12] J.C. Latombe. Robot motion planning with uncertainty in control and sensing. *Artificial Intelligence*, 52(1):1–47, 1991.

[13] J.-P. Laumond and P. Souères. Metric induced by the shortest paths for a car-like mobile robot. In *IEEE International Workshop On Intelligent Robots and Systems, Yokohoma, (Japan)*, 1993.

[14] A. Lazanas and J.C. Latombe. Motion planning with uncertainty: a landmark approach. *Artificial Intelligence*, 76:287–317, 1995.

[15] M. Khatib and H. Jaouni. Kinematics integrated in non-holonomic autonomous navigation. Technical Report 96346, LAAS-CNRS, September 1996.

[16] S. Quinlan. *Real-Time Collision-Free Path Modification*. PhD thesis, Stanford University, CS Departement, January 1995.

[17] S. Quinlan and O. Khatib. Towards real-time execution of motion tasks. In R. Chatila and G. Hirzinger, editors, *Experimental Robotics 2*. Springer-Verlag, 1992.

[18] S. Quinlan and O. Khatib. Elastic bands: connecting path planning and control. In *IEEE International Conference on Robotics and Automation, Atlanta, (USA)*, 1993.

[19] J.A. Reeds and L.A. Shepp. Optimal paths for a car that goes both forwards and backwards. *Pacific Journal Mathematics*, 145(2):367–393, 1990.

[20] P. Souères, J.-Y. Fourquet, and J.-P. Laumond. Set of reachable positions for a car. *IEEE Transactions on Automatic Control*, 39(8), August 1994.

[21] H. Takeda, C. Facchinetti, and J.C. Latombe. Planning the motions of a mobile robot in a sensory uncertainty field. *IEEE Trans. on Pattern Analysis and Machine Intelligence*, 16(10):1002–1017, 1994.

[22] M. Vendittelli and J.-P. Laumond. Visible position for car-like robot amidst obstacles. In *Workshop on Algorithmic Foundations of Robotics, WAFR'96*, July 1996.

PART 5
MANUFACTURING
SESSION SUMMARY

Bernard Roth
Dept. of Mechanical Engineering, Stanford University, Stanford, CA 94305 (U.S.A)
e-mail: roth@robotics.stanford.edu

1 Presentation by Hollis

This session dealt with advanced concepts for using robotics in manufacturing. The first paper was presented by Ralph Hollis. In his presentation he outlined the major concepts of the Agile Assembly Architecture approach currently being developed by his group at CMU. They are developing hardware and software technologies and strategies for automatic assembly of precision products. Their concepts are reminiscent of Victor Scheinman's "Robot World" but on a much larger scale. Hollis describes his basic workstation as consisting of four degrees of freedom divided into two two-degree-of-freedom components. One of these is movable x-y courier, and the other is an overhead insertion, manipulator capable of rotation and translation about a fixed vertical axis.

Hollis' objective is to within one week have his minifactory be able to be reconfigured to start assembling a new product. He hopes to accomplish this by powerful software and mechanical modularity and versatility. His system would use overhead processing, and he hopes to achieve micron-level precision for parts placement during assembly operations.

The following summarizes the discussion that followed Hollis' presentation: Bolles asked how Hollis expects his system to deal with failures, such as when parts accidentally slip from a fixture. Hollis replied that they hope to have the system heavily enough instrumented to handle many situations automatically. As a last recourse, if the system detects an error it cannot correct, it would "raise a flag."

Ikeuchi wondered why Hollis chose the approach of having only two degrees-of-freedom in his manipulator. He also wondered why Hollis thought he would have more success than Robot World has had. Hollis replied that while of course four degrees-of-freedom gives more capability than two, he felt that his choice results in a device with much more accuracy. In regard to Robot World, Hollis feels that one of its largest drawbacks is that its sandwich type construction limits the size of the tools, and, for example, important operations such as driving screws cannot be accomplished with existing tools. He believes his system will have the added ability because of its overhead manipulation configuration.

Carlisle asked about the limitations inherent in using tethered carriers. Hollis replied that although he would of course like to get rid of the tethers, for the while they have no plans to do without them. Hollis pointed out that limitations due to the system geometry are even more limiting than the tethers to the range of carrier travel at this time.

Trevelyan asked if Hollis had considered a means of conveying parts between work stations and then moving them onto a local carrier. Hollis replied that that is essentially what is done now, and that Hollis feels the current method has problems with maintaining alignment and accuracy which is where he hopes his system will prove superior.

2 Presentation by Carlisle

The second paper in this session was presented by Brain Carlisle. His main theme was the same as Hollis' - the need to shorten the lead time for installing robotics on the factory floor. Carlisle's goals seem more modest than Hollis', since he is talking in terms of 2-3 months in contrast to Hollis' one week. However, Carlisle is talking in terms of a broader scope, with more varied operations, and he is looking at a more generic situation. If we look just at the part of Carlisle's presentation where he is concerned with the design of the automation system and its construction and installation then he too is talking in terms of weeks, not months.

Carlisle's overall goal is to shorten the time it takes to sell and install a robot system by a factor ten. In his talk and paper he outlined Adept's approach to shortening this time cycle by using what Carlisle calls Rapid Deployment Automation. He also pointed out that safety is perhaps the most critical technical need regarding industrial acceptance of robotics.

The following is a summary of the discussion which followed Carlisle's presentation: Shirai agreed with Carlisle's observation that practical factory floor 3-D machine vision was not yet available. He wondered about what applications Carlisle foresaw other than bin-picking. Carlisle felt that it would be useful to have 3-D vision for aligning parts, for recognizing tipped parts, and beyond that for bin-picking types of operations.

Hollerbach agreed with Carlisle's point that there is still much work to be done in modeling of systems.

Koditschek wondered about whether any work had been done on modeling in a way that would give information as to the current state of a given assembly. Carlisle replied that they mainly simply model a small portion of an entire assembly, for example the assembly of two parts, and do not model the entire process. They do however concern themselves with the error budget in each small set of steps.

Inoue agreed that safety is very important for future applications, especially for home robots. He noted that in industry the problem is solved by excluding the human from the robot's environment. Carlisle noted that operators do get in the robot's space during teaching operations, and the current practice of the use of human exclusion zones in industry is costly in terms of space efficiency. He felt that a much better approach to safety might be a "watchdog" system where a third system monitors both the human and the robot and takes action when the situation starts to look dangerous -such as a mother watching a child at play.

Chatila pointed out that although one wants the human operator to be in command, there are times when the human makes a bad decision, and then ideally the robotic device should act autonomously. Carlisle commented that the watchdog approach would be a desirable way to handle such difficulties. At this point the discussion was cutoff due to the press of time.

3 Presentation by Mosemann

The third paper in this session was presented by Heiko Mosemann. This paper dealt with a mathematical model of the contact between two bodies during the assembly process. The specific concern here is to determine a set of potentially stable assembly orientations, and to use this information to determine the best configurations to use in a high level autonomous assembly planning system. The model includes friction at the contacts and allows for the computation of the contact forces.

The following discussion followed this presentation: Latomb felt that because of the computational complexity, the claim that the problem is "solved" is premature. He pointed out that the authors have used an exponential algorithm which requires a linear programming problem to be solved for every subassembly. Mosemann responded that, of course they would welcome a more efficient method, but this is the best they have been able to do with such a complex problem.

Kaneko raised the question as to the sensitivity of the choice of contact point for the model, and wondered how the authors determined which contact point to use in modeling a given assembly. Mosemann replied that the use of friction in the model decreases the sensitivity of the choice of contact point. He affirmed that the choice is in general a difficult question, and that in their work they relied on the choice being made automatically by the geometric modeling program.

Rimon and Burdick noted that the stability criterion used by the authors is a necessary, but not a sufficient condition, and that this might effect the validity of the model when there are surface or edge contacts rather than just point contacts. Burdick asserted that this is really a quadratic programming problem. Mosemann replied that he agrees that his model can not guarantee the stability of every assembly state it choses as being stable.

4 Presentation by Hirzinger

The fourth and final presentation of this session was Gerhard Hirzinger's overview of the work in his laboratory. His talk covered the following major topics: 1. His laboratory has cooperated with the robot manufacturer KUKA in a very successful implementation of real-time inverse dynamics for KUKA's manipulator. They have also developed a space mouse for robot teaching, and integrated sensors into the KUKA robot arms. Overall he feels industry is now more willing to incorporate advanced technology onto their robotics.

2. His laboratory has developed a four-fingered hand based on small high performance motors. This hand has three actuators per finger, and 112 sensors.

3. In regard to telerobotics they have developed a hierarchical system with four layers which hides the details at lower levels from the operators. They have a real time 3-D vision system running on a Data Cube. In addition they have developed teaching systems for space exploration which take into account the intervening system dynamics, and have data gloves for force detection.

4. They have worked on developing robotic devices for repair of satellites in flight.

5. Finally, they are working on minimally invasive surgery devices. One such device is a remote endoscope with a servo controlled video camera.

At the end of the presentation, Nakamura asked if Hirzinger was willing to share the very nice equipment his laboratory has developed with the research community. Hirzinger replied that he thinks it is very important to do so, and that he is planning to make his force sensors available.

5 Summary

The first two presentations dealt with the important issue of making factory robotics more practical and less time consuming and expensive to develop and implement. Although, both Hollis and Carlisle have their own agendas and pet solutions, it is clear that they agree on the issues and the main areas of concern.

The third paper will be of interest to those working on the question of generating autonomous assembly plans. Clearly this is a very complex and detailed modeling problem.

The Hirzinger summary, points out once again just how vibrant and varied the robotics activities are at DLR, and although they are not mainly at first related to manufacturing, the cross fertilization from space to manufacturing is apparent.

The written papers of the presenters, and their co-authors, follow.

Opportunities for Increased Intelligence and Autonomy in Robotic Systems for Manufacturing

Alfred A. Rizzi and Ralph L. Hollis

in collaboration with

Ben Brown, Zack Butler, Greg Fries, Jay Gowdy, Wing-Choi Ma, Patrick Muir, Arthur Quaid and Xing-Xing Yu

The Robotics Institute
Carnegie Mellon University
5000 Forbes Avenue
Pittsburgh, PA 15213 USA

URL: http//www.cs.cmu.edu/~msl

1 Introduction

Since the introduction of programmable industrial robots at the beginning of the 1970s, an industry based on these machines grew steadily to a point in the mid-1980s where it was poised for an explosion of huge proportions. This explosion failed to happen as a "robotics backlash" took hold, with the strong perception held by many that these machines did not, and would not, live up to the high expectations held by their users.

While the interest in robotics by industry waned, it continued to grow tremendously within the academic community and many difficult and fundamental problem areas began to be investigated. The emphasis has been on key components, (*e.g.* kinematics, dynamics, control, 3D vision, planning, etc.) but not complete systems, and integration of these results into industrial practice has been slow.

One result of this history, in the opinion of the authors, is that academic robotics researchers—perhaps faced with the difficulties of technology transfer to industry—have "turned their backs" on industrial robotics in favor of working in other, more exploratory, areas such as mobile, field, medical, space, and service robotics. Meanwhile, worldwide sales of industrial robots has steadily grown during the past decade. To enable further growth, there must be significant changes in the way robotic systems are deployed in manufacturing environments. It would appear that there are now significant opportunities for applying increased *intelligence* and *autonomy* to industrial robot systems. This observation derives from several factors, including:

(i) Increasing demand for ever smaller and more complex products whose lifetimes are ever shortening.

(ii) The need to remove humans from the immediate vicinity of the manufacturing process because of scale and cleanliness requirements.

(iii) The recent widespread and ubiquitous availability of significant computing power at reasonable costs.

(iv) The Internet explosion.

1.1 Practical Difficulties with Automated Manufacturing Systems

The fundamental problem with "modern" robotic manufacturing systems, is that the individual components (robots, part-feeders, conveyor systems, etc.) are generally designed as stand-alone devices. As a result, little or no explicit effort is dedicated to enabling the integration of these factory components into a complete manufacturing system. Similarly, information about the design of most complete systems is scarce, as there are few incentives for system integrators to document their work, and it is difficult to extract durable truths from case studies. There remain prohibitively high economic and technical costs associated with the factory integration process that in turn severely limit the utilization of robotic elements in many practical applications.

Meanwhile, work on programming robotic assembly systems has progressed at both the task-level (*e.g.* "place part A on part B") and at the manipulator level (*e.g.* "move joint 3 10.15 inches; close gripper; etc."). To date, task-level systems have not moved out of the laboratory, while the manipulator-level systems, despite their enormous programming complexities, have become widely accepted.

1.2 Related Efforts

A number of academic and industrial groups have attempted to provide partial solutions to these fundamental problems over the past decade.

A leading-edge benchmark, which attempts to address some of these issues, is Sony's SMART flexible assembly line. It makes use of SCARA robots equipped with indexing multiple grippers and modular product and part transport systems to simplify the mechanical problems of factory reconfiguration. Unfortunately the individual modules are physically large, and the problems of programming and tuning a complete factory system are still daunting.

A key study was the DARPA microfactory demonstration [4], developed as part of the Defense Department's Intelligent Task Automation program. This work emphasized operation in unstructured environments, recognition and grasping of overlapping parts, semiautomatic planning, and geometric reasoning. The system which resulted used parts kitting, and sensor-moderated motion to assemble a precision microswitch. While meeting many objectives, the system required 18 minutes to complete the microswitch assembly task.

A recent significant trend is the notion of programming and operating robots over the Internet. For example, a concept of virtual laboratories was recently demonstrated [5], showing that a robot in one laboratory can be programmed and controlled from another laboratory thousands of miles away. In another case, exploration and tele-gardening [6] was demonstrated. Both of these studies show that it is now possible to allow meaningful remote (Internet based) interaction between a robot and a programmer or operator. However, neither demonstrates the level of expressiveness required to undertake a significant practical manufacturing task.

Sandia National Laboratory has developed its "Agile Manufacturing Prototyping System," (AMPS) comprised of robotic cells supplied by various vendors[1]. Simultaneously, industrial robot producers have begun to service the demand for increased flexibility and precision. Adept has developed a concept of "rapid deployment automation" [3] that embraces key elements of modularity and off-line programming. Megamation and Yaskawa have produced systems of small, modular, easily programmed robots capable of moderately precise assembly.

Our vision of an agile manufacturing system is one that provides for a large pallet of modular robotic processing and product transport systems from a wide variety of vendors; with each module presenting a standard mechanical, computational, and algorithmic interface enabling their simple and rapid integration (both physical, and programatic) into a complete factory system. In contrast to much "academic" research on agile manufacturing systems, we are not striving to provide a "universal assembly machine," but rather wish to adhere to the industrially accepted model of flow-through (assembly line) processing while providing for the rapid deployment and reconfiguration of such systems. We forsee this being achieved through the use of compact, mechanically simple elements whose customizable combined behaviors provides the specific complex capabilities required for a specific application. Furthermore, we do not foresee these modules being used to form a "lights-out" factory, but rather we expect them to act smoothly in concert with humans, serving as highly capable and intelligent tools for their operators.

[1] See http://www.sandia.gov/AMPSfact.html

2 The Agile Assembly Architecture

As part of a multi-million dollar four-year project funded under the NSF Multi-Disciplinary Challenges component of the High Performance Computing and Communication program, we are developing new hardware and software technologies and strategies for automated assembly of precision high-value products such as magnetic storage devices, palm-top and wearable computers, and other high-density equipment [7]. We envision a design cycle focused on the development of virtual factories which could combine resources for producing products from geographically distant locations. Our approach draws extensively on high-speed wide-area communication and intensive distributed computation. The Agile Assembly Architecture (AAA) supports the creation of miniature manufacturing systems (minifactories), built from small modular robotic components, which will occupy drastically less floor space than today's automated assembly lines. Our goals are to reduce assembly system changeover times, facilitate geographically distributed design and deployment of assembly system, and to increase product quality levels.

We are developing AAA as a distributed system of tightly integrated mechanical and computational robotic modules endowed not only with information about their own capabilities but also with the ability to appreciate their role in the factory as a whole and negotiate with their peers in order to participate in flexible factory-level cooperation [12]. A unified interface tool will allow a user to select and order these mechanisms over the Internet and to assemble, program, and monitor them in both a simulated factory environment and the real factory environment.

AAA relies on factory-wide standard procedures and protocols, and well-structured autonomy to simplify the process of designing and programming high-precision distributed assembly systems. The architecture makes use of modules' self knowledge and ability to explore their environments to make the transition between simulation and reality as painless and seamless as possible.

Our sample instantiation of these ideas is a modular tabletop factory that we refer to as *minifactory*. The key technical ideas include the use of distributed low-DOF robotic modules or agents[2], and integrated product transport and manipulation subsystems. The entire system is composed of compact elements with standardized mechanical and electrical interconnects allowing for the rapid setup and adjustment of a factory system.

Figure 1: Cartoon characterization of a state of the art modular manufacturing system.

Figure 1 depicts a cartoon chacterization of what we would consider a state of the art modular manufacturing system. Whereas this approach is widely regarded as a significant advance beyond older non-modular systems, there are significant characteristics which limit its agility. In the figure, the large and fairly complex factory modules (represented by buildings) all tend to be produced by a single vendor, and are separated from one another by fairly rigid interfaces (the high fences) greatly limiting flexibility. Furthermore, each module requires a semi-custom interface (walkway) to the essentially inflexible product transport system (railway). Conveyors belts and similar mechanisms capable of only one-way flow are the norm in such systems. This makes for a system which can be very efficient (haul a lot of freight) in production, but which is difficult to change.

Modules tend to be of fixed size, independent of the functionality provided, leading to inefficient use of floor space. If changing market conditions require modifications to the manufacturing system the factory designer is faced with the choice of either modifying the internal functions of modules or adding/subtracting modules. In the first

[2]By agent we explicitly mean a mechanically, computational, and algorithmically modular manufacturing entity, *e.g.* robot, capable of both communication and physical interaction with its peers.

case, altering the module severely interferes with its modularity and can be complicated by the module's inherent complexity. In the second case, inserting or deleting modules requires either making space by physically moving all upstream or downstream modules, or closing up gaps created when modules are removed.

Finally, it is extremely difficult for more than one module vendor to participate in this scheme. Because of module complexity, the cost of entry is high. There is a "winner take all" force at work here where there is every incentive for a given vendor to offer only a "complete" line of modules (even if some may not employ the best technology available), and really little incentive for a company which has only the capability to field a few modules to participate in a modular system controlled by another company. It can be argued that this situation leads to inefficiencies, less than optimum performance, and poor agility with respect to the marketplace.

Figure 2: Cartoon characterization of an AAA based manufacturing system.

In contrast to this highly structured model of factory automation, Figure 2 depicts a similar rendition of a manufacturing system based on the ideas of the AAA. Here, the factory processing elements (large regularly placed and isolated tenements) have been replaced by a collection of modules (small cottages) from a variety of vendors, which are placed as needed in the manufacturing system. The high fences between these modules have been removed since each module is designed with the explicit intent of interacting with their neighbors. Finally the product transport system (centralized railroad) has been replaced by a large collection of moving modules (bicycle couriers). These capable and independent agents, who explicitly coordinate their actions with both the factory processing elements (home owners) and each other both form a highly flexibly product transport system and can participate in local manipulation and processing tasks. This approach is less efficient than the state of the art systems in production, but on the other hand, it is much easier to set up and change.

Adding or deleting functionality is relatively straightforward in this approach. New modules can be added to the system almost anywhere (picture additional cottages placed in front, back, or side yards with perhaps some small adjustments to the positions of the surrounding cottages), and taking modules away will have no effect on the existing ones. We propose that a module's internal functionality never be subject to modification by a factory designer, but rather only by the module's vendor. Adhering to this stricture ensures that modularity, and hence the system's inherent agility, will not be broken.

Because each factory module is fairly simple and limited in functionality, the cost of entry for module suppliers is kept low. In this way, each modular component can employ the latest and best technology from that particular company. We argue that this sort of approach can respond more quickly to changing market opportunities than can today's state of the art modular systems.

2.1 Underlying Challenges

A fundamental component of our long term goal is to elevate the design of automated assembly systems from the detailed technical problems associated with designing and integrating independent mechanisms to the more salient problem of designing the factory as a whole. We see this as a complimentary effort to that provided by the industrial engineering and operations research communities, but one that can provide a natural mechanism for the widespread application of new factory design methodologies. Ideally the collections of machines that follow the framework presented here will form a natural template onto which the more abstract ideas of factory design and optimization can be applied.

From our perspective there are several key barriers that currently stand between current best practice and a more agile and open manufacturing infrastructure. Fundamentally these barriers all relate to the need for standard mechanisms to support interaction between agents, designers attempting to integrate the agents into a system, programmers developing control software for a factory involving the agents, and operators whose task is to monitor performance and provide support when an agent is unable to cope with its environment.

As mentioned in Section 1 the current practice in the robotics and automation community is to focus on the engineering of individual robots and mechanisms, with little or no consideration for how they are later integrated into a complete factory system. Only by designing robotic modules that are explicitly prepared to participate in a larger factory system can we begin to provide the types of tools necessary to move towards fundamentally more useful systems of machines. This represents what we see to be an important goal of the robotics field: the construction of mechanisms capable of both physical and "social" interaction. Physical manipulation has been the province of both the academic and industrial robotics communities from their inception. While "social" interaction has been an academic goal [9] which has produced several novel and interesting systems [1, 8]. There have been, however, few practical applications for these systems.

We believe our efforts to design rapidly reconfigurable and "user friendly" factories represents a modest step towards achieving this goal. The scope of "social" interaction has been explicitly limited to two well defined domains: inter-agent interaction for factory coordination, and interaction between agents and the factory personnel. Given even this limited scope, the details underlying the definition of these agent interfaces are not obvious and comprises a significant portion of the Agile Assembly Architecture.

2.1.1 *Factory Interaction*

The definition of a suitable "machine language," for inter-agent communication is a central issue in enabling the type of interaction under discussion between both multiple robotic agents, and humans and agents. For inter-agent communication the basic requirements include:

Extensibility: Whatever the actual format of messages, the underlying media must efficiently allow for the introduction of not only new message formats, but also the negotiation of completely separate communication modalities. In principle, these allow the natural growth and development of new methods for inter-agent coordination, and with responsible classification of which protocols are to be considered "required" and which are "optional" for an agent it is reasonable to expect long term compatibility through use of the required basic protocols.

Real-Time Coordination: Sufficient communications capability (enough bandwidth with sufficiently low latency) is essential to allow the tight coupling and integration of agents which are incapable of performing complete manufacturing tasks in isolation. In the sample system described in Section 3 it is clear that there are significant advantages both in terms of flexibility and simplicity inherent in supporting such distributed mechanisms. This issue is mitigated by the fact that in general an individual member of a factory is likely to only interact with a well defined "neighborhood" of peers—*e.g.* an insertion mechanism need only perform precise coordination with a part feeder (providing the part to insert) and with a product transport mechanism (presenting the subassembly to be operated on)—greatly limiting the scope of high performance communication by nature of its locality.

Factory Communication: Conversely, there is a need to provide a standard means for factory wide control and monitoring, and hence the need for a standard interface to join every robotic module in a factory together with each other and an arbitrary number of control and monitoring workstations. The intent is to provide a common medium and basic interchange format for the most rudimentary forms of factory control while simultaneously providing a means by which modules can negotiate for the use of more application specific interchange formats. By requiring every element of a factory system to "understand" this basic level of interaction, we strive to ensure that each and every machine is capable of participating in the factory at a minimum level.

2.1.2 *Integrated Design, Simulation, and Evaluation Tools*

Not only is it necessary to require a facility for interaction between machines, it is equally important to support interaction between humans acting as factory designers and the agents. In a traditional design process there are three major classes of interaction between the designer and a component under consideration.

(i) **Preliminary selection:** Initial evaluation of a manufacturing component for its suitability to a problem.

(ii) **Detailed evaluation:** Iterative validation and discard of candidate solutions and components based on analysis, simulation, and mock-up of proposed designs.

(iii) **Integration and refinement:** Detailed analysis, design, construction, and test of a working system.

While the distinctions are somewhat arbitrary, they highlight fundamentally different forms of inquiry performed on candidate components by a designer, and provide a model under which we can explore the interactions necessary to reduce the designers' uncertainty about the factory system they are working on.

Given the widespread acceptance and rapid development of high-performance computation and communication systems, particularly as embodied in the Internet, we foresee the integration of such capabilities with robotic agents as enabling a new kind of relationship between the designer and the component. In stage (i) of the design process where traditionally decisions would be made based primarily on vendor-provided catalogs, it now becomes possible for the designers (or some agent acting on their behalf) to directly interrogate an actual mechanism (probably at a vendors facility) for relevant properties. There are a myriad of options for exactly what remote entity answers such queries depending on the nature of the component under scrutiny. In the case of "brainless" components this would be similar to a catalog search, but for full-fledged factory agents the designer could interact directly with the specific machine under consideration, potentially providing a significantly more accurate representation of the actual mechanism and its capabilities.

The implications of this model on phase (ii) of the design process are more significant. It now becomes possible for the component under evaluation to provide a number of different "renderings" of its physical and behavioral models for use by a designer. With the careful integration of tools for either retrieving down-loadable representations from factory components, or remotely involving the component in a distributed simulation or analysis process it becomes possible for the item under review to provide a model with an appropriate level of fidelity to its actual performance. It is easy to imagine a broad range of models ranging from trivial kinematic representations to highly detailed physics-based distributed simulations or even remote experimental environments being made available to a designer through a single and consistent set of design and simulation tools capable of allowing the construction and interrogation of a highly accurate "virtual" factory identical to the design under evaluation.

Finally, in phase (iii) as a design is refined and physical experiments are undertaken, it is through these same tools that the designer will continue to interact with the evolving factory design. The ultimate goal being the truly seamless transition from factory simulation to operation, but with enough expressiveness and flexibility in the underlying components and representations so as not to unduly constrain the behavior and performance of the final system.

2.1.3 *Programming Interaction*

The key goal in simplifying human/machine interaction is one of providing a simple and natural language for specifying complete machine behavior. Further complicating the problem is that given a large collection of disparate agents, it will be necessary to distribute the "factory program" among the various agents. However, in contrast to current practice, it will be necessary to provide tools and highly expressive, yet convenient, languages that aid a factory programmer in developing and debugging the agent level programs which instantiate a specific solution to a manufacturing problem. Most of our effort in this domain is being directed at understanding and developing appropriate representations for machine behavior in a factory setting—hoping to take advantage of the rapidly developing field of human computer interaction to provide flexible and expressive interfaces between our representations and the factory programmers. We feel this aspect of the larger problem is most suitable for immediate investigation, as it is the natural avenue through which to explore the advantages of increased autonomy on the part of the individual robotic agents, while still being closely related to the more well understood problems of assembly planning and factory optimization.

2.1.4 *Operator Interaction*

Finally, as we do not see the near-term future of automated manufacturing to be "lights-out," it is important to consider the role played by factory operators and their interaction with the agents that make up the factory. Predominantly we see operators serving as aids to the factory, acting to help agents recover from and avoid situations that they are unable to manage in an automatic manner. This includes such mundane tasks as managing factory supplies by refilling part supplies and removing finished products, possibly for additional processing by a more traditional factory system. Furthermore, we foresee operators being called to the aid of agents that recognize factory difficulties that they are unable to recover from. This form of interaction should include the ability of an agent to notify an operator of the difficulty, allowing the operator to remotely (from across the room or facility) interact with the agent in question and its peers via a set of agent "front-panels" or "dashboards" — remotely rendered presentations of the agent's status — to diagnose the problem and choose a corrective course of action.

3 Instantiation

We are currently designing and building a working example of a modular tabletop factory or *minifactory* conforming to our notion of the agile assembly architecture. This minifactory incorporates planar robot couriers that travel on connected tabletop platen surfaces. These robots are derived from planar linear motors that float on air bearings and translate along the platens in two directions with micron-level precision. Other devices, including 2-DOF "overhead manipulators," are mounted on modular bridges above the platens (see Figure 3). The couriers are responsible for both carrying the product subassemblies from one overhead device to another, and cooperating with the overhead devices to execute assembly operations. Limiting the robots to 2 DOF has advantages in terms of modularity, reliability, and performance [11], but allows the minifactory to perform 4-DOF assembly tasks through the use of robot cooperation. Each low-DOF device will have integrated high-performance computing resources, and serve as an agent in the AAA context. This eliminates the need for central resources that would degrade the modularity and scalability of the system.

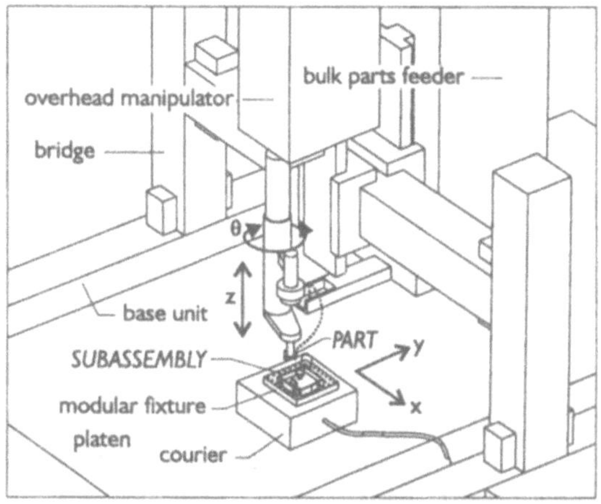

Figure 3: Basic components of a minifactory.

We are currently developing the key electromechanical elements, including couriers, manipulators, precision parts feeders, and other modular components. We are implementing a distributed realtime computer architecture, modeling and simulation software, high-level network communication protocol, and graphical programming tools to support the long-term vision. We believe this type of system can only be developed through careful integration of hardware and software tools in a manner heretofore unseen in the robotics and automation community.

3.1 What is a Minifactory?

In addition to the limitations on the forms of interaction between factory elements implied by Section 2, we have deliberately chosen to restrict the scope of mechanical capabilities we wish a minifactory to perform to afford both analytic tractability and design practicality. Toward this end we have limited the class of tasks to assembly and processing operations requiring four or fewer degrees of freedom. Specifically we want to construct systems capable of:

- Four-degree-of-freedom vertical insertion.
- Easily integrating overhead processing (*e.g.* laser processing or material/glue deposition).

- Micron-level part placement accuracy.
- Factory design and programming in less than a week.

To provide this functionality, a minifactory consists of a potentially large collection of mechanically, computationally, and algorithmically distributed modules. Each element in this collection is responsible for providing a minimum level of cooperation and communication in order to participate in the most basic minifactory operations.

The most obvious departure from traditional automation systems and one of the most obvious embodiments of our philosophy of factory level integration can be seen in our choice to integrate product transfer and local manipulation. As such, we have eschewed the traditional use of SCARA manipulators coupled with part conveyor systems and local fixtures. Alternatively, as depicted in Figure 3, we have chosen to make use of two-DOF manipulators and two-DOF planar couriers moving over a high-precision platen surface. The couriers are thus responsible both for product transport within the factory and for transiently forming cooperative four-DOF manipulators when they present sub-assemblies to a stationary manipulator.

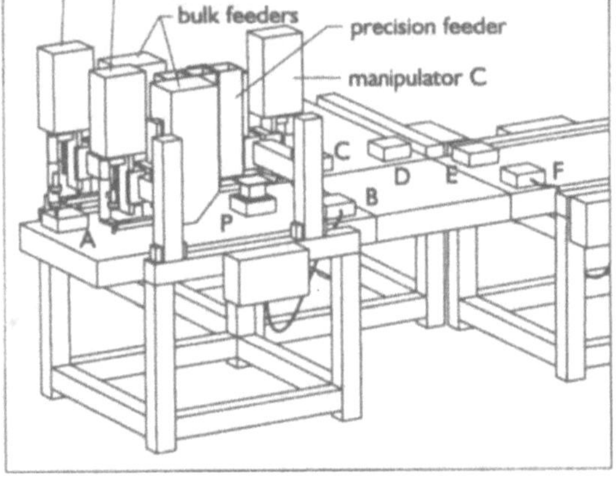

Figure 4: View of a "typical" section of a minifactory assembly system, including a "tee" junction.

Perhaps the best way to appreciate the implications of this approach to the design of factory level assembly systems is to consider a somewhat contrived but illustrative example. Figure 4 depicts a view of a small section of a fictitious minifactory. The system pictured includes six couriers and three manipulators with two bulk-random parts feeders and one precision feeder. Couriers begin on the left of this system, present their subassemblies to the first two manipulators where two new components are added; the resultant sub-assemblies then travel to the right where the final manipulator is responsible for both placing a precision component and transferring the final assembly to one of four couriers.

3.2 Run-Time Coordination and Communication

Any element of a minifactory, be it a courier, a manipulator, or some custom designed module, must provide a minimal level of capability in order to participate in the minifactory "society." Currently we foresee there being three general classes of capability every agent must reliably provide: *basic trustworthiness*, *self initialization*, and *inter-agent coordination*, the latter which includes facilities for resource negotiation.

3.2.1 *Basic Trustworthiness*

For an agent to be a successful member of a factory community its peers must be able to trust it to reliably represent itself. Practically, we see this manifesting itself in the form of three fundamental capabilities.

- All agents must advertise their basic capabilities and the protocols they understand to their peers.
- Every agent must be capable of reporting its current status and its understanding of its environment.
- Each agent must implement reliable and safe failure detection and recovery schemes.

The first two of these requirements are essential to address the issues of Section 2.1.1 and support the graceful coordination between minifactory components, their peers, and factory monitoring tools. Furthermore, the ability to advertise capabilities addresses the need for a predefined extensible protocol suitable for the exchange of such information between agents. The next two capabilities may well be the most important, and quite possibly the most difficult to precisely define and implement. The assertions demand that agents be capable of constantly monitoring the state of the factory available to them. Furthermore, when an agent detects conditions outside the norm it must be capable of independently correcting the aberration, negotiating with its peers to recover from the fault, or broadcasting its inability to proceed thus bringing the factory to an orderly stop. Although it is potentially difficult to guarantee this level of capability in an arbitrary system we feel that through judicious use of a combination of traditional AI reasoning [10] and reactive behaviors [2] that it can be achieved in the highly-constrained domain of minifactory.

3.2.2 *Factory Calibration/Initialization*

Integral to the rapid deployment of an assembly system is the need for precise and automatic calibration and initialization whenever a factory is "turned on." There are three interrelated tasks that must be collectively undertaken by the minifactory components to successfully initialize a factory system. This process will begin with agents identifying their peers through the use of messages broadcast to the factory at large. Following this, couriers must explore their environs to discover both the exact geometry of the platen surfaces, as well as the positions of any stationary agents within their range of motion. Finally, through a careful exchange of this information between agents, a complete map of the minifactory can be constructed both in the agents and in a monitoring interface tool.

3.2.3 *Robotic Agent Coordination*

Since individual elements of the minifactory are rarely capable of performing "useful" tasks alone, it is essential to include standard mechanisms for orchestrating their coordination. Fortunately, the locality of action performed by individual agents provides a natural locality of communication and coordination. For example, manipulators A and B in Figure 4 both only interact with couriers A and B, while couriers C through F interact with everything other than manipulators A and B. To help alleviate the problems associated with manually coordinating the motions of all of these machines, we have chosen to make use of a *geometry reservation system*. Under this system, factory elements which are potential competitors for a specific predefined segment of space are grouped and required to negotiate for the use of that shared resource. In principle, an individual machine may well be a member of several different "groups of agents" sharing a myriad of resources associated not only with physical resources but potentially with more abstract factory goals. It is the neighborhood groupings of machines that form the basic fabric for cooperation between the elements of the factory system.

The most fundamental form of this cooperation will happen whenever a courier and manipulator transiently form a four-degree-of-freedom system to perform a part placement task. The most basic mode for such cooperation will take the form of a virtual linkage between two machines where one agent is effectively slaved to the state of the other, allowing for simple coordinated movement. Other modes of cooperation will include coordinated behavior changes and cooperative sensor-based action. Behavior changes will be used to encode the sequence of operations necessary for a high-precision force-controlled insertion task (*e.g.* manipulator exerts low vertical force while the courier "finds" the hole, followed by the courier becoming compliant while the manipulator exerts higher forces to perform the insertion).

3.3 User-Level Design, Programming, and Monitoring Tools

In the absence of a centralized controller, a minifactory will have a centralized *user interface tool* capable of supporting the design, simulation, and run-time monitoring and control of a minifactory. Each element of a minifactory—whether it be a courier, manipulator, or other custom robot—is an independent computational entity. The overall behavior of the minifactory results from the interaction between these elements and their environments. The central challenge for the minifactory simulation and programming environment is to provide the services of Section 2.1.2, facilitating the development of well-debugged distributed programs, while simultaneously easing the difficult transition from the simulated world of bytes and pixels to the real world of actuators and sensors.

3.3.1 *Design and Programming*

The goal of the minifactory programming environment is to simplify the difficult problem of selecting and integrating the components of a factory while generating the distributed programs for every agent in the system. The desired outcome is that a minifactory system will be able to be designed and programmed by an expert in the domain of the assembly problem at hand without requiring expertise in "minifactory programming."

Within this framework we foresee the use of constraints, such as local frames of reference, and abstractions, such as coordinating the gross motion of couriers through the use of distributed resource management rather than considering it as a global allocation problem. Such uses of constraints guides the user to construct robust systems while abstractions hide details that the user cannot afford to be concerned with if correct programs are to be rapidly built.

This approach to geometry management and gross motion planning provides several advantages to the user:

- **Abstraction:** The user does not have to specify that a specific courier must explicitly contact some other courier and/or manipulator for permission to move into a manipulator's workspace; all of these negotiations are hidden through the use of a reservation area.
- **Modularity:** Rather than a manipulator knowing it has to interact with a particular courier, it just has to know that it interacts with whatever courier has reserved its workspace.
- **Robustness:** Since the reservation areas are referenced to physical components of the factory, if these components move slightly, the various agent programs will continue to function properly.

3.3.2 *Simulation*

Minifactory simulations allow a user to explore the application of minifactory technology to a particular assembly problem, and to do much of the development and debugging of the factory programs off line in a virtual environment. A key to a minifactory's rapid and successful deployment is the nearly seamless transition of a factory program from this virtual environment to the actual machines. The two facets of our architecture that make this transition possible are ***fidelity*** of the simulation and ***robustness*** of the underlying agents. Fidelity demands that the simulated factory will behave reasonably closely to a physical system under similar conditions. Robustness of the agents acknowledges the inability to configure an actual factory in exactly the same manner as it was simulated, but that the differences can be detected and accounted for.

The issue of fidelity is addressed through an agent's self knowledge and self representation. Each minifactory module provides a representation of its own geometry, behavior, and integration constraints. Thus, the simulator will not use a catalog to look up the characteristics of a typical manipulator, but rather will query an actual manipulator via an Internet connection for its own self representation. This reliable representation of an agent's characteristics eliminates many inaccuracies that would otherwise occur. Robustness is provided through the inclusion of additional sensing resources that enable the individual agents to self-calibrate and explore their environment (as described in Section 3.2.2).

Additionally we foresee simplifying the transition from simulation to reality by allowing mixed operation of the simulation system in conjunction with running hardware. In full simulation mode, most of the agent models and programs will be internal to the simulation environment itself, each of them having been constructed from the description provided by the agents themselves. In practice there is no reason beyond efficiency why the implementation of these agent models could not be performed by the remote agents themselves rather than internal to the simulator. Thus, simply by mixing internal agent models and external agents, a real agent could be put through its paces in isolation, with all its actuators working and sensors gathering data, but within the context of a greater simulated system.

4 Conclusion

The AAA project began in November, 1995. So far, we have concentrated most of our efforts on developing the engineering technologies needed to build the prototype minifactory. We are also well into the tasks of developing comprehensive environments for modeling, simulation, and programming. A prototype 3D interactive user interface package has been implemented which allows the user to look at a detailed running minifactory simulation with zoom, pan, and other controls. This factory simulation can be downloaded from our web site and viewed by anyone. Many of the AAA design, operation, and agent coordination issues have been formulated [12].

Looking to the future, we foresee AAA and minifactory serving both as a research testbed and as an exemplar of what we see as one potential path for the future of automated assembly systems. Whereas our focus has been on rapidly reconfigurable assembly systems for precision assembly, this approach may also have utility for agile parts fabrication, chemical synthesis, pharmaceutical manufacturing, and other such applications.

We are encouraged by the positive response received to date from our colleagues in both the research and the industrial communities. In particular, it would seem the notions of increasing both *intelligence* and *autonomy* through the use of well-crafted modular building building blocks to realize rapid deployment is very attractive—provided that it can be made to work in the real world, and that it can provide realizable benefits in the marketplace.

It is our hope that solving many of the complex problems faced by modern manufacturers will be attractive enough to receive increased attention from robotics researchers, thereby re-establishing more active collaboration between these communities.

Acknowledgments

This work is supported in part by NSF grants DMI-9527190, DMI-9523156, and CDA-9503992. The authors would like to thank Arthur Quaid for providing figures 3 and 4.

References

[1] R.A. Brooks and L.A. Stein. Building brains for bodies. *Autonomous Robots*, 1(1), 1994.

[2] R. R. Burridge, A. A. Rizzi, and D. E. Koditschek. Sequential composition of dynamically dexterous robot behaviors. submitted to the International Journal of Robotics Reaserch, 1996.

[3] John J. Craig. Simulation-based robot cell design in AdeptRapid. In *Proceedings IEEE International Conference on Robotics and Automation*, pages 3214–3219, 1997.

[4] W. Foslien and V. Nibbe. A robotic workcell for small-batch assembly. *Robotics Today*, pages 1–5, 2nd quarter 1990.

[5] M. W. Gertz, D. B. Stewart, B. J. Nelson, and P. K. Khosla. Using hypermedia and reconfigurable software assembly to support virtual laboratories and factories. In *Proc. 5th Int'l Symp. on Robotics and Manufacturing*, Maui, Hawaii, August 15-17 1994.

[6] K. Goldberg, M. Mascha, S. Gentner, N. Rothenberg, C. Sutter, and J. Wiegley. Desktop teleoperation via the world wide web. In *Proceedings IEEE International Conference on Robotics and Automation*, 1995.

[7] R. L. Hollis and A. Quaid. An architecture for agile assembly. In *Proc. Am. Soc. of Precision Engineering, 10th Annual Mtg.*, Austin, TX, October 15-19 1995.

[8] D.B. Lenat. Cyc: a large-scale investment in knowledge infrastructure. *Communications of the ACM*, 38(11), Nov. 1995.

[9] Marvin Minsky. *The Society of Mind.* Simon and Schuster, 1986.

[10] D. J. Musliner, E. H. Durfee, and K. G. Shin. World modeling for the dynamic construction of real-time control plans. *Artificial Intelligence*, 74(1):83–127, March 1995.

[11] A. Quaid and R. L. Hollis. Cooperative 2-DOF robots for precision assembly. In *Proc. IEEE Int'l Conf. on Robotics and Automation*, Minneapolis, May 1996.

[12] A. A. Rizzi, J. Gowdy, and R. L. Hollis. Agile assembly architecture: An agent-based approach to modular precision assembly systems. In *IEEE Int'l. Conf. on Robotics and Automation*, Albuquerque, April 1997.

Rapid Deployment Automation: Technical Challenges

B. Carlisle
Adept Technology, Inc.
150 Rose Orchard Way, San Jose California 95134
email: brian.carlisle@adept.com

Abstract

Product life cycles have shortened to 1-2 years in many industries, causing increased interest in flexible automation. However it still often takes 1-2 years to sell and install a robot system. We would like to shorten this time by a factor of ten to 2-3 months. In order to achieve this we must be able to design automation systems in a matter of days, build and install them in a matter of weeks, and train them in a few hours. Adept is pursuing a technical approach to this challenge based on geometric modeling and reasoning, sensor- based part feeding and assembly, and smart application packages. Significant technical challenges remain, including acquiring models or parts quickly and automatically, 3-D machine vision, assembly strategies which reason about geometric uncertainties, error representation and recovery, and simple teaching techniques. Adept's current approach is described and technical challenges are discussed.

1 Introduction

To date, applications for robots have been primarily industrial. Many potential new applications are emerging, however they face many of the same types of barriers that faced industrial applications. This paper will examine one industrial application, assembly, and discuss factors which have limited its rate of acceptance. Readers may wish to examine their own applications with these factors in mind.

Market demand for flexible assembly automation is driven by several factors. These include rapidly shrinking product life cycles, miniaturization, increased demands for quality and cleanliness, and a constantly rising cost of labor. Flexible automation using industrial robots is replacing both conventional automation (which used cams and air cylinders) and manual labor.

It is interesting to note that robotic assembly was accepted more rapidly in Japan than in the U.S. or Europe (International Federation of Robotics, 1996). In Japan assembly robots account for over 40% of the robot market whereas in the U.S. and Europe assembly is 10-15% of the total robot market. Japan initially used assembly robots largely in high-volume production applications, essentially to replace hard automation. The U.S. and Europe looked for assembly robots to replace expensive labor. It was easier to replace hard automation than manual labor resulting in broader proliferation of assembly robots in Japan.

Today, with the continuing demands for shorter product life cycles and more product variety, Japan is beginning to require increasing levels of flexibility from assembly systems. At least one manufacturer (Sony) which has invested extensively in flexible automation, is discussing whether manual labor might be the better solution for their flexible assembly needs.

Adept has identified a number of barriers which have limited the widespread adoption of robotic assembly and has developed a plan to address these barriers.

2 Time and Risk

An informal survey conducted at General Motors several years ago found that the typical implementation time for a "hard" automation assembly system was three years. While this was acceptable when automobiles had product lives of 7-10 years, it is not acceptable as the auto industry compresses its product lives to three years.

Three years ago Adept surveyed a number of its customers and determined that the average time to sell and install a "flexible" automation assembly system was 1.5 - 2 years. This was not dramatically better than "hard" automation.

This implementation cycle could be divided into two parts: the sales cycle and the system build cycle. The sales cycle is defined as the time from first contacting a prospective customer to the time the purchase order is received. The system build cycle starts with the receipt of a purchase order and ends when the customer accepts the system running production on the factory floor. The sales cycle was taking 9-12 months and the build cycle an additional 9-12 months. This contrasts with installing a manual assembly line in 2-3 months.

The sales cycle includes time for the following activities: determining the feasibility of the application, developing one or more detailed automation system proposals for the application, determining the cost and throughput of the proposed system (justification), and convincing the customer that the system will work (addressing customer perceived risk).

The build cycle includes time for detailed system design and layout, ordering parts, building and installing system components, writing software for the system, and running and debugging the system until it meets acceptable levels of performance.

3 Rapid Deployment Automation

Adept asked the question, "What is required to reduce the sales and build cycle by a factor of ten, from several years to several months?" This magnitude of reduction would make buying and installing a robot assembly line as fast as installing a manual assembly line and consequently would reduce the risk associated with buying automation and would increase the ability of automation to meet the rapidly changing needs of the marketplace.

A technical strategy was developed to address this goal, resulting in a four- layer product hierarchy. We call the top layer in this hierarchy "The Design Layer". It includes 3-D simulation products for the design and off-line programming of robot systems. For example a simulation product called "RAPID" allows virtual workcells to be composed from 3-D CAD models of robots, part feeders, conveyors, etc. and for CAD models of customers parts to be imported into the virtual workcell. The workcell can then be animated, and can provide a 3-D visualization of the proposed system working with the customers parts, accurate cycle times, a list of workcell components required, a physical layout of the system, a wire list to hook up the system and the software to download to the actual system. The design layer is intended to compress the proposal and sales cycle by allowing 3-D animated system concepts to be developed and verified in a matter of hours or days. The efficiency of this tool allows the customer to easily interact with the design process and results in a video tape which both documents the concept and becomes an internal sales and education tool for the customer.

The second layer in the Rapid Deployment Automation (RDA) hierarchy is called "The Process Knowledge Layer". This layer is composed of an application development environment and a series of application packages which contain process knowledge and allow programming at the "task" level. This environment is called "AIM". Task level application instructions separate "WHAT" is to be done from "HOW" to do it from "DATA" necessary to perform the task. This allows the application developer to embed process knowledge ("HOW") in task-level instructions. The "DATA" necessary for the particular instance can be trained or downloaded from a model, i.e. the RAPID simulation system. The process knowledge layer allows application programs to be quickly developed by less skilled programmers, and eventually to be semi-automatically generated from the system design process.

The third layer in the RDA hierarchy is called "The Real-Time Control Layer". This layer includes the robot programming language, sensors such as machine vision and force, and motion control hardware. We now view this layer as a building block rather than as an end in itself.

The fourth layer is called "The Mechanism Layer". It includes the mechanical devices necessary to construct an assembly system. Adept has recently expanded its product line to include part feeders which use machine vision to enable them to feed a wide variety of parts with no mechanical adjustments. As custom part feeders such as bowl feeders or pallets were dedicated to a specific part, they limited the flexibility of the system. To change from one product to another, any part-specific feeders must be changed. This is expensive and time -consuming. Flexible part feeders may be programmed in a few minutes and as a result shorten both system development time and system changeover time.

The set of products developed under the RDA strategy has allowed Adept to shorten the sales and implementation cycle by roughly a factor of two to date by addressing the issues of system design at the start of a project, reducing perceived customer risk, and providing "smart" robots with built-in application knowledge. This has been an important element in growing the market for robotic assembly systems. We believe a similar approach may benefit other emerging robot application areas, where it often appears that "the right robot" is viewed as the only solution necessary to enable new markets such as surgery, agriculture, or service.

Developing the product hierarchy described above has pointed out a number of technical challenges which continue to impede progress in robotic assembly automation. Many of these technical challenges are of course generic to intelligent machine applications.

4 Geometric Modeling and Reasoning

It is increasingly apparent that geometric modeling and reasoning about these models will be critical not only to enable design systems such as RAPID described above, but to enable robots to make better use of sensors to respond to deviations from a nominal plan or to work in less structured environments than presently possible.

As a first step, it is desirable to create 3-D surface models of objects very quickly, without the laborious methods offered by computer aided design. Many control applications do not require highly accurate geometric models; we can avoid a tree without knowing its exact dimensions. Tomasi (1993) has worked on creating surface models from video motion; incorporating some type of approach similar to this would be very useful as we seek to quickly develop approximate models for control systems.

We need to be able to attribute behaviors to geometric models. Today certain systems such as RAPID allow us to attach time-based motion to models, however we have no tools to model non-rigid objects, chemical or process changes, attributes such as "fragile" or "dangerous", or even constraints related to the safety of humans (the fork must never contact the eyeball). Goad (1997) is beginning some work in this area.

5 3-D Machine Vision

Adept is still not aware of a machine vision system that can recognize a 3-D object in an arbitrary pose with a high degree of reliability (¿99%). This may simply be a matter of time, however it is disturbing that at least in the United States, a good deal of machine vision research funding over the last ten years was directed at navigation, rather than recognition. Also, many vision researchers seem to be pleased with algorithms that work 60-70% of the time. This level of performance is not sufficient for commercial applications, or even for commercial firms to become very interested in refining the algorithms. It is possible that further progress in this area depends on progress in geometric modeling and reasoning, as we seem to have made progress in developing range maps of objects, but not corresponding progress in recognizing these objects.

6 Planning for Geometric Uncertainty

Assembly is perhaps the most demanding of current robot appplications with respect to dealing with geometric uncertainty. The position tolerance stack-up involved in grasping a part and inserting into a second part held by a fixture almost always exceeds the clearances between the two mating parts. This is currently addressed by a variety of ad-hoc methods, including chamfers, carefully calibrated vision systems to measure offsets, force-based "hoparound" insertion strategies, etc. We have no generic assembly strategy planning tools to determine a motion and sensing strategy based on the geometry of the mating parts and the potential tolerance stack-up in the workcell. In robot applications less structured than factory assembly systems this problem will be exacerbated. This may be an interesting area in which to examine the tradeoffs between deterministic efforts at reasoning about a large number of variables and less deterministic efforts at "learning" successful strategies.

7 Error Representation and Recovery

As we strive to build increasingly robust systems which are insensitive to noise and uncertainty, the issue of how we think about errors arises. This is a system programming issue and does not appear to be commonly addressed in the literature, or viewed as a research area. One system programming approach (Costes-Maniere, 1997) requires a programmer to anticipate all possible errors and develop a set of pre-conditions and post-conditions for any program segment to execute. If the pre-conditions are not met, the program does not begin, if the post-conditions are not met, the programmer tries to anticipate all possible contingencies and develop a branching strategy to recover from any contingency. This approach, while a practical way to address the issue with today's technology, does not address the more fundamental questions of "How do we define errors?" Should we try to anticipate all errors in advance and develop plans to address them or should we ignore errors in global planning and use sensors, models, and geometric reasoning to describe them only as they occur and to do local planning in real time to recover?

8 Simple Teaching Techniques

As the technology for intelligent machines becomes more sophisticated, we cannot depend on legions of PhDs to program them. The amount of software involved in typical robot applications has increased by many orders of magnitude over the last fifteen years. In 1975 a typical robot program consisted of a few tens of lines of code, primarily simple motion commands and input/output signal sequencing. Todays robot programs can contain hundreds of thousands of lines of code, and the technologies advocated above will only increase this number. While we develop these sophisticated technologies, we have to remember that people struggle to program their VCRs. If we are to set up and run a robot application in a matter of minutes, we must spend as much effort on the user interface as on the technical depth. While this topic may be viewed as application specific and therefore not worthy of generic research, the general topic of conveying human intent to machines and vice versa is just emerging and incorporates broad and deep areas such as natural language recognition, haptic interfaces, and interactive man-machine planning. It would be nice to see collaborative research attempt to integrate work in these areas into simple and effective teaching of robot applications.

9 Safety

Safety is perhaps the single most critical technology needed for robots to be widely accepted. It is time to recall Issaac Assimov's three laws of robotics:

"A robot may not harm a human being or through inaction allow a human being to come to harm."

"A robot must obey the orders given it by human beings except where such orders would conflict with the first law."

"A robot must not let itself come to harm unless that is in conflict with the first two laws."

It appears that ensuring that robots never harm humans is not regarded as a topic worthy of PhD research. At the recent International Conference on Advanced Robotics (Monterey, CA, 1997) only one professor in the room had a student working on this topic. In industry we build cages around robots to keep people away from them. How will we ever develop new applications which require people and robots to be in close proximity without addressing safety?

Safety has two elements: hardware and software. What few efforts exist in safety tend to focus on hardware reliability and redundancy. Software will soon be the bigger issue if it is not already. What are the technolgies needed to meet Assimov's laws?

10 Conclusion

Activities in developing robot technology which can be implemented quickly and easily for applications in assembly point to generic technical challenges which will benefit many other robot applications. The author has limited his comments to areas where he feels insufficient progress is being made. While these are very broad research areas, they represent an opportunity for the robot community to collaborate with other computer science researchers as we attempt to develop safe and friendly intelligent machines.

References

[1] The International Federation of Robotics: World Industrial Robots 1996: Statistics 1983-1996 and Forecasts to 1999, United Nations Economic Commission for Europe, 1996.

[2] Tomasi, Carlo: International Symposium on Robotics Research, Pittsburgh, Pennsylvania, 1993.

[3] Goad, Chris: Infloresence Inc., Portland, Oregon.

[4] Costes-Maniere, E.: International Conference on Advanced Robotics, Monterey, California, 1997.

Stability of Assemblies as a Criterion for Cost Evaluation in Robot Assembly

H. Mosemann, F. Röhrdanz, and F. Wahl
Institute for Robotics and Process Control
Technical University of Braunschweig
Hamburger Str. 267, 38114 Braunschweig (Germany)
email: H.Mosemann@tu-bs.de

Abstract

In this paper we discuss assembly stability as a criterion for cost evaluation in robot assembly. We propose an algorithm for the calculation of the set of potentially stable orientations of arbitrary configurations of rigid bodies considering static friction under uniform gravity. The algorithm determines the magnitudes of the contact forces leading to potential assembly stability using linear programming techniques. A new evaluation function based on the set of potentially stable assembly orientations is proposed and integrated into the assembly cost evaluation of a high level assembly planning system. The proposed stability analysis is an indispensible prerequisite for the execution of robot assembly operations generated by a task-level programming system.

1 Introduction

The goal in robot assembly is to automatically build a mechanical product consisting of several parts. A high level assembly planning system generates sequence plans specifying the order in which parts are to be assembled to form the desired product and computes the trajectories to bring the parts together. In addition to such sequence plans, task plans for actually performing the assembly operations must be generated. The ***high level assembly sequence plan*** is the basis for such ***lower level plans*** generated by task planning systems, and taken together they ensure the efficient and flexible assembly of a mechanical product by robots. Early assembly reasoning systems were strongly based on user interactions [4] [5]. One basic idea in most of the work on assembly planning is the ***assembly by disassembly strategy.*** Much research work has been done to reduce the necessary user interaction [6], [9], [11], [14], [22], [31], [38], [40]. But even today it is very difficult for an assembly planning system to automatically generate assembly sequences in general. Kavraki et. al. [12] proved, that the problem of automatically generating assembly sequences is NP-complete in the two-dimensional case. As a matter of fact, past and present work in this research area make simplifications in order to cope with complexity. Other researchers focus their attention onto the physical reasoning needed to accomplish single steps in an assembly plan. Some authors studied motion planning to assemble subassemblies [17], [37]. Grasp [26], [27], [33] and fixture planning [39] are also important for a high level assembly planning system.

We embedded the stability analysis of assemblies in our **high l**evel **a**ssembly **p**lanning system *HighLAP* [29]. *HighLAP* has a modular structure and an open architecture. It uses an assembly hierarchy, an extended cycle finder and physical reasoning to simplify the search for an optimal plan. The system covers all modern aspects of high level assembly planning and introduces new approaches for physical evaluation of sequence plans like the stability analysis. We integrated our high level assembly planning system into the commercial robotic simulation system *IGRIP* (Interactive Graphics Robot Instruction Program [35]) for flexible robot-based manufacturing. Figure 1 depicts the system architecture of *HighLAP* and the integration into *IGRIP* . *HighLAP* takes CAD descriptions of assembly components and high level assembly specifications using symbolic spatial relationships as input. The generated assembly sequences are stored in AND/OR-Graphs which constitute a compact representation of all feasible assembly sequences.

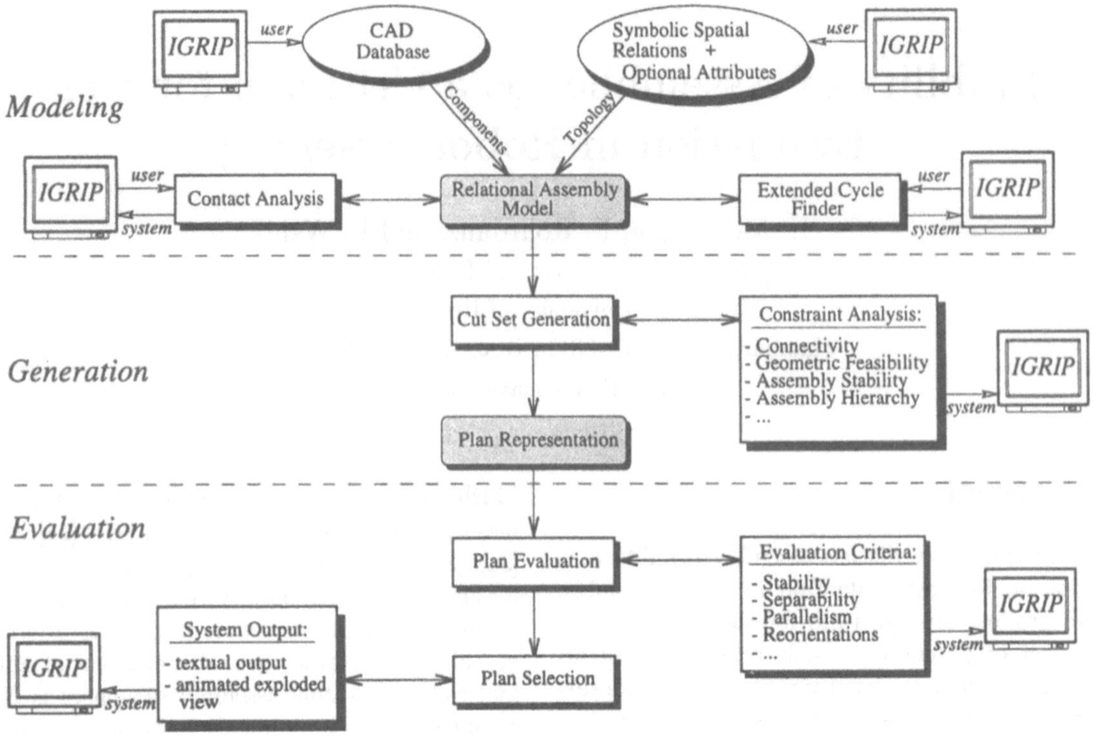

Figure 1: System architecture and user interface based on the commercial simulation system *IGRIP*.

Evaluation of assembly sequences is important for high productivity, quality control and high flexibility. For the generation of assembly sequences, many constraints have to be taken into account, for example separability of two parts/subassemblies, manipulability of the generated assembly, necessity of reorientations, and stability of the created (sub)assemblies. In this paper we discuss and solve the problem of determining stability of assemblies consisting of arbitrary configurations of rigid bodies considering their frictional properties. The following assumptions will apply for the reminder of this paper:

- The bodies are rigid and bounded by surface primitives like planes, cylindrical and spherical surfaces.
- The bodies are initially at rest and we take into account static friction only.
- One or more bodies are fixed by a supporting surface or a gripper.
- The normal directions at each contact are well-defined (degenerated contacts are approximated).
- Potential stability is calculated under gravitational forces.

For assembly sequence planning it is important to compute stable assembly orientations, as any reorientation of an assembly or the additional usage of fixtures reduce the flexibility and efficiency of the corresponding assembly process and thus induce additional costs. After generation and selection of the best assembly sequence plan, the objects in the workspace have to be assembled by a robot in real world. Determination of the set of Potentially Stable Assembly Orientations (PSAO) is an indispensable prerequisite for plan execution. We introduce a new algorithm to solve the problem of calculating the PSAO. The algorithm calculates the magnitudes of the contact forces leading to potential stability [25] and analyses these forces. Therefore, we are able to drop, for example, orientations with forces exceeding a given limit or violating certain stability criteria. Furthermore, we discuss how to use the PSAO for assembly cost evaluation. An evaluation function is derived and integrated into a high level assembly planning system [30].

2 Previous Work

Blum et al. [2] proposed and implemented a stability test for configurations of blocks. In their approach the unknown interaction forces between the blocks arising at contact points are determined. Equations for the equilibrium of the system are formulated in terms of unknown interaction forces and known gravitational force. A configuration of blocks is said to be stable if there exists a set of forces satisfying the equilibrium equations. This set can be found with linear programming techniques. When using a Coulomb friction model, this approach is not limited to frictionless assemblies. Nevertheless, this method neither determines which objects are unstable nor does it determine the full set of stable orientations of an assembly.

Palmer [25] investigated the computational complexity of the assembly stability problem in two-dimensional space, i. e. , rigid polygons in the plane, and gave several stability definitions. An assembly is said to be *potentially stable* if there exists a set of interaction forces resulting in a static equilibrium. *Guaranteed instability* (no stable orientation can be found) is the complement of potential stability. If there exists no infinitesimal small motion for which the assembly is unstable, then the assembly is denoted to be *infinitesimally stable.* Palmer showed, that frictional potential stability is **P** complex, while the frictional infinitesimal stability is **NP-hard**. Following the notation of Palmer we compute the potential stability of assemblies.

Boneschanscher et al. [3] presented a method to determine the stability of an assembly under the gravitational force and the insertion force of the next object to be assembled. The assembly in question is placed on a table and the insertion force is given. Friction is considered to compute the stability of an object with respect to the table. A system of inequalities describes conditions of forces between the objects and the table. A contact graph gives information about the relation at each contact point in the assembly. With merge and compound operations, the contact graph is reduced for each object in order to obtain a single relation between the object and the table. Using linear programming, this single relation is tested for stability. However, as Boneschanscher et al. pointed out, a significant shortcoming is the fact, that their technique to determine assembly stability does not work, if there are loops in the contact graph.

Sukhan Lee [13] described a heuristic approach to (sub)assembly stability. He uses the definitions of *floating clusters* and *disconnected clusters* to define the internal freedom of motion of (sub)assemblies. The internal freedom of motion is defined as the collections of directions to which the (sub)assembly can be broken apart. An (sub)assembly is stable if the internal freedom of motion is null or has at most a single translational freedom of motion. Sukhan Lee integrated this stability test in an assembly planning system but the test is very restrictive and classifies many (sub)assemblies as unstable which are stable in reality.

In [19] Mattikalli et al. presented a method to determine frictionless stability of assemblies under uniform gravity by considering changes in potential energy. One or more of the assembly's components are fixed by a gripper or a table. An assembly is said to be stable if all objects remain at rest under the influence of gravity. An expression is formulated for the change in potential energy of a system of bodies subject to a given virtual displacement. Inequalities for virtual translations and rotations are formulated for each contact point. If there exists no valid displacement which decreases the potential energy, the system is stable. Linear programming is used to check whether such displacements exist. If there exists a displacement which decreases the potential energy the impending motions of the objects are calculated. The problem to find a frictionless stable orientation is formulated as a constrained maximin problem. In [19] a solution to this maximin problem, found by solving a single linear program is shown. Another advantage of this method is the possibility to find the most frictionless stable orientation for unstable assemblies.

In [21] De Meter gives a technique for determining whether carrier clamps and locators may resist arbitrary external forces throughout assembly transport. Furthermore, he presents a model for calculating the minimum clamp actuator intensities necessary to restrain parts throughout transport an assembly. Wolter and Trinkle [39] presented a stability test and a new approach to automatically generate the positions of a small set of fixture elements that will stabilize an assembly. The linear functions of the contact forces and the components of the velocities of the parts are solved with linear programming techniques. Schimmels and Peshkin [32] discussed how to guide a workpiece into a fixture using only the contact forces (force-assembly) considering friction. They obtain a single accommodation control law by an optimization procedure. A set of linear sufficient conditions is used to generate accommodation basis matrices. In [36] Trinkle et al. discuss the problem of predicting the possible acceleration of a set of rigid, three-dimensional bodies in contact in the presence of Coulomb friction. They present two novel complementary formulations for the contact problem under Coulomb's Friction Law and an analogous law in which Coulomb's quadratic friction cone is approximated by a pyramid.

In [20] Mattikalli et al. discuss the problem of finding all potentially stable orientations of assemblies including Coulomb friction. This work bases on [18]. Assemblies consist of contacting rigid bodies. One or more of the bodies are fixed by a gripper or a table. A uniform gravity field is acting on the assembly. Considering the forces arising at the contact points of the bodies, equations for the net force and net torque for each body are formulated. The assembly is potentially stable if there exist forces such that the net force and net torque is zero. The existence of such forces is determined by linear programming. To compute the set of all stable orientations of an assembly the direction of the gravity vector $\vec{g}$ is changed. Searching among all possible directions of $\vec{g}$ the set of stable orientations is determined by sensitivity analysis. But the important aspect of degeneracy of the corresponding linear program is not considered.

The cited works except [13] do not discuss the integration of a stability analysis in an assembly planning system.

3 Modeling Friction

Implementation of friction models which reflect the precise physical behavior is quite difficult if not impossible. We adopted a deterministic static friction model commonly known as *Coulomb friction* (see e. g. [7],[16]). In Figure 2 the contact frame of the ith contact point of two touching objects is depicted. The vectors $\vec{n}_i$, $\vec{t}_{x_i}$, and $\vec{t}_{y_i}$ denote mutually perpendicular unit vectors. The vector $\vec{n}_i$ denotes the surface normal of the contact plane while $\vec{t}_{x_i}$ and $\vec{t}_{y_i}$ span the plane tangent to the contact surface. The Coulomb model of friction is an accepted empirical relationship between the normal force magnitude f_n and the tangential force magnitude f_t.

The proportionality constant μ is known as the coefficient of static friction. Considering Coulomb's law of friction, the following inequality must be satisfied by the tangential force magnitudes $f_{t_{x_i}}, f_{t_{y_i}}$ at each contact point i of the assembly[1]:

$$(f_{t_{x_i}}^2 + f_{t_{y_i}}^2)^{\frac{1}{2}} \leq \mu f_{n_i} \tag{1}$$

Since we are using linear programming to solve the stability problems mentioned in section 1, we approximate inequality (1) by an adaptively chosen number l ($l = 2^n, n \geq 2$) of linear inequalities. This approximation is adopted from [36]. The vertices $\vec{v}_i$ of the approximated unit friction cone orthogonally projected onto the contact surface are:

$$\vec{v}_i = \begin{pmatrix} \cos a_i \\ \sin a_i \end{pmatrix}, \quad a_i = \frac{2\pi i}{l}, \quad i \in [0, l-1] \tag{2}$$

If the slopes s_i of the lines passing through the first $l/2$ successive points $\vec{v}_i$ and $\vec{v}_{i+1}$ are defined by[2]

$$s_i = -\cot(a_i + \frac{\pi}{l}), \quad i \in [0, \frac{l}{2} - 1] \tag{3}$$

and the crossings c_i of the t_{y_i}-axis are determined by

$$c_i = \vec{v}_i \circ \begin{pmatrix} -s_i \\ 1 \end{pmatrix}, \quad i \in [0, \frac{l}{2} - 1] \tag{4}$$

the following $l \times 3$ matrix $\mathbf{C}$ defines the approximated friction cone:

$$\mathbf{C} = \begin{pmatrix} s_0 & 1 & c_0 \\ s_1 & 1 & c_1 \\ \vdots & \vdots & \vdots \\ s_{l/2-1} & 1 & c_{l/2-1} \\ s_0 & -1 & -c_0 \\ s_1 & -1 & -c_1 \\ \vdots & \vdots & \vdots \\ s_{l/2-1} & -1 & -c_{l/2-1} \end{pmatrix} \tag{5}$$

Thus, considering friction, l linear constraints are induced by the approximated friction cone at each contact point i for the tangential force components $f_{t_{x_i}}, f_{t_{y_i}}$ which are dependent on the normal force magnitude f_{n_i} and the friction coefficient μ:

$$\mathbf{C}\vec{\mu}_i \leq \vec{0}, \quad \vec{\mu}_i = \begin{pmatrix} f_{t_{x_i}} \\ f_{t_{y_i}} \\ \mu f_{n_i} \end{pmatrix}, \quad i \in [1, k] \tag{6}$$

We are not limited to a fixed number of facets l for the approximation of the friction cone. Our planning system changes l automatically with respect to the number of contact points k of the assembly to resolve the trade-off between complexity and accuracy of the approximation. We used $l = 8$ in our simulation experiments.

[1] For simplicity of notation we assume an identical coefficient of static friction at each contact.

[2] Note that $a_i + \frac{\pi}{l} \neq 0 \wedge a_i + \frac{\pi}{l} \neq \pi$ for $i \in [0, \frac{l}{2} - 1]$.

3.1 Critical Reflections on Friction Coefficients

The friction coefficient μ is physically not well-defined; rather it depends on the local material properties of each microscopic area. Thus, it is not possible to describe μ exactly for every situation, because the nature of the surface of two contacting objects is not clearly determined. I. e. , to some extend μ is of stochastic nature. As we need a reliable quantity μ for the integration of the stability module in an assembly planning system [29] reflecting real world behaviours of real objects, we conducted many experiments to determine μ for different objects made from different materials with differently machined surfaces. As mentioned above, we approximated the nonlinear inequality (1) by the linear inequalities (6) by means of a finite number of facets of the friction cone. How does this affect our results? It is obvious, that the choice of μ results in a much greater effect on the outcome of the frictional stability analysis than the non isotropic friction behavior induced by the linear approximation of the friction cone. Our approximation will not degrade the overall performance decisively at this point. An exact model of the friction cone, using for example nonlinear programming techniques, will not give rise to more reliable results because of the significant variation of μ in real world. Besides this, the complexity of the stability computation increases; so we are confident that our model of friction is a good compromise between complexity and reliability.

4 Stability of Assemblies

In order to analyze the stability of an assembly, we consider the contact forces arising at the contact points of an assembly. Since we are dealing with static friction, a contact force is the sum of a normal force, which prevents interpenetration of the bodies, and a frictional force, acting in the contact plane [28]. We assume that n objects contact at finite many points, which are indexed from 1 to m. Let $\vec{n}_i$, $\vec{t}_{x_i}$, and $\vec{t}_{y_i}$ denote the unit vectors of the ith contact frame depicted in Figure 2. We consider a contact force $f_{n_i}\vec{n}_i$ of the ith contact point that acts normal on body B, contact forces $f_{t_{x_i}}\vec{t}_{x_i}$, $f_{t_{y_i}}\vec{t}_{y_i}$ of the ith contact point acting on body B tangential in the contact plane and the corresponding reactive forces $-f_{n_i}\vec{n}_i$, $-f_{t_{x_i}}\vec{t}_{x_i}$, $-f_{t_{y_i}}\vec{t}_{y_i}$ of the ith contact acting on body A. f_{n_i}, $f_{t_{x_i}}$, and $f_{t_{y_i}}$ denote the

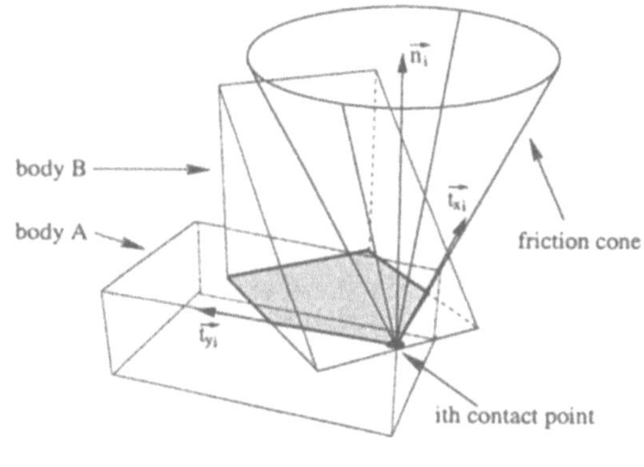

Figure 2: Contact frame of the ith contact point of two objects with its corresponding friction cone.

unknown scalar magnitudes of the forces. The scalars f_{n_i} must be non-negative since normal forces must be repulsive; that is $f_{n_i} \geq 0$. The net force $\vec{F}_j \in \mathbb{R}^3$ acting on the jth body of the assembly can be written as:

$$\vec{F}_j = \sum_{i=1}^{m} s_{ji}(f_{n_i}\vec{n}_i + f_{t_{x_i}}\vec{t}_{x_i} + f_{t_{y_i}}\vec{t}_{y_i}) + \mathcal{M}_j\vec{g} \quad (7)$$

The scalars s_{ji} are zero, if the jth body is not involved in the ith contact; s_{ji} is 1, if the contact force exerted on the jth body from the ith contact point is $f_{n_i}\vec{n}_i + f_{t_{x_i}}\vec{t}_{x_i} + f_{t_{y_i}}\vec{t}_{y_i}$; otherwise, the contact force acting on the jth body is $-(f_{n_i}\vec{n}_i + f_{t_{x_i}}\vec{t}_{x_i} + f_{t_{y_i}}\vec{t}_{y_i})$ and s_{ji} is -1 since normal forces must be repulsive. The mass of body j is denoted as $\mathcal{M}_j$ and $\vec{g} \in \mathbb{R}^3$ describes the gravitational force of a uniform gravity field acting on the jth body. The net torque $\vec{\tau}_j \in \mathbb{R}^3$ acting on the jth body of the assembly can be written as:

$$\vec{\tau}_j = \sum_{i=1}^{m} s_{ji}(\vec{d}_i - \vec{c}_j) \times (f_{n_i}\vec{n}_i + f_{t_{x_i}}\vec{t}_{x_i} + f_{t_{y_i}}\vec{t}_{y_i}) \quad (8)$$

The vector $\vec{d}_i$ denotes the location of the ith contact point, and $\vec{c}_j$ is the location of the center of mass of the jth body. The scalars s_{ji} are the same as in equation (7). The $\vec{\tau}_j$ are independent of $\vec{g}$ since a uniform gravity field does not exert a torque. If we define the vectors $\vec{r}$, $\vec{e} \in \mathbb{R}^{6n}$ and $\vec{f} \in \mathbb{R}^{3m}$ as the collections

$$\vec{r} = \begin{pmatrix} \vec{F}_1 \\ \vec{\tau}_1 \\ \vdots \\ \vec{F}_n \\ \vec{\tau}_n \end{pmatrix} \quad \vec{e} = \begin{pmatrix} \mathcal{M}_1\vec{g} \\ \vec{0} \\ \vdots \\ \mathcal{M}_n\vec{g} \\ \vec{0} \end{pmatrix} \quad \vec{f} = \begin{pmatrix} f_{n_1} \\ f_{t_{x_1}} \\ f_{t_{y_1}} \\ \vdots \\ f_{n_m} \\ f_{t_{x_m}} \\ f_{t_{y_m}} \end{pmatrix} \quad (9)$$

we can write $\vec{r} = \mathbf{A}\vec{f}+\vec{e}$ where $\mathbf{A}$ is a $6n\times 3m$ matrix whose coefficients are given by equations (7) and (8). An assembly is said to be potentially stable if there exist contact forces such that the net force and net torque on every body is zero. External forces arise from task forces and gravitation, and internal forces from the mutual contacts of the objects. If we define the vector $\vec{f_N} \in \mathbb{R}^m$ as the collection $\vec{f_N} = (f_{n_1}, f_{n_2}, \dots, f_{n_m})^T$, a given gravity vector $\vec{g}$ induces potential stability if there exists $\vec{f}$ such that[3]

$$\begin{aligned} \vec{r} &= \mathbf{A}\vec{f}+\vec{e}=\vec{0} \text{ and } \vec{f_N} \geq \vec{0} \\ &\text{and } \mathbf{C}\vec{\mu_i} \leq \vec{0},\ i \in [1,m] \end{aligned} \quad (10)$$

Whether such a vector $\vec{f}$ exists can be decided with *linear programming* [24]. We introduce the components of the gravity vector as variables in the linear program and search among all possible directions of the vector $\vec{g}$. For that purpose we apply the constraint $\|\vec{g}\|_\infty = 1$ (for a vector $\vec{g} = (g_0, g_1, g_2)^T \in \mathbb{R}^3$, $\|\vec{g}\|_\infty = max_i \mid g_i \mid$). This forms a unit cube around the origin. The following linear program describes the search for a potential stable orientation in the jth facet, $j \in [0,5]$, of the unit cube:

Minimize:

$$z = \sum_{i=1}^{m}(f_{n_i}) \quad (11)$$

subject to:

$$\begin{aligned} \mathbf{A}\vec{f}+\vec{e} &= \vec{0} \\ \vec{f_N} &\geq \vec{0} \\ \mathbf{C}(f_{t_{x_i}}, f_{t_{y_i}}, \mu) &\leq \vec{0}, \quad i \in [1,m] \\ g_{(j+1)mod3} &\leq 1, \quad g_{(j+1)mod3} \geq -1 \\ g_{(j+2)mod3} &\leq 1, \quad g_{(j+2)mod3} \geq -1 \\ g_{jmod3} &= (-1)^j \end{aligned}$$

Up to this point we follow similar lines like Mattikalli et al. [20] which in turn bases on prior work of Blum et al. [2]. However, we suggest a new algorithm to calculate the PSAO. In contrast to former approaches,this algorithm handles all degenerate cases, allows a decomposition to handle very large problems and calculates the magnitudes of the forces leading to potential stability. Simply using potential stability as the criterion for computing the PSAO results in solutions which are not guaranteed to be stable in reality. Potential stability is only a necessary but not a sufficient condition for assembly stability: The consideration of static friction leads to the problem of *static indeterminacy* where the distribution of normal forces is indeterminate [36]. In this paper we discuss potential stability. If we define an assembly to be stable under any legal distribution of forces at the contact points we have to perform an exhaustive search method described in [1]. Figure 3(a) shows a peg in hole assembly[4] and Figure 3(b) illustrates the PSAO computed by [20] projected onto the unit sphere. According to Figure 3(b) the assembly is stable in all orientations which obviously does not comply with reality. In this example an arbitrary contact force acting over one of the sides of the upper block can be balanced by an equal and opposite force on the parallel opposite contact face. This results in the incorrect PSAO. In contrast to [20] we calculate all basic solutions to the stability problems (11) using an algorithm based on the Double Description Method (DDM) [23]. The main advantage of this algorithm is the ability to calculate the magnitudes of the forces leading to stability. Therefore, we can eliminate most of the incorrect orientations depicted in Figure 3(b) performing a postprocessing. This leads to the reduced PSAO visualized on the unit sphere as depicted in Figure 3(c).

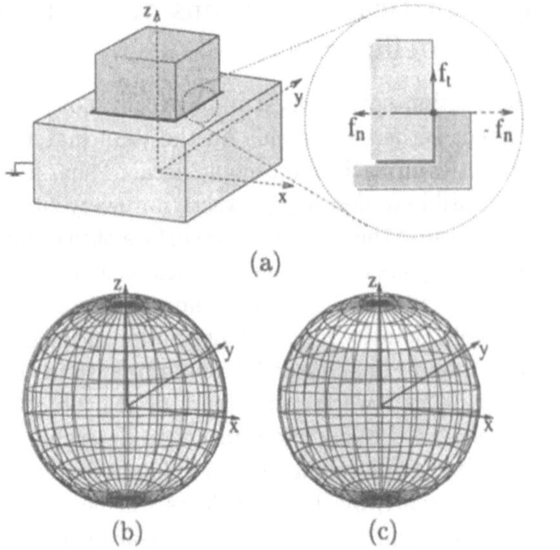

Figure 3: (a) Example assembly and (b) corresponding incorrect PSAO computed according to [20]; (c) the reduced PSAO computed with our approach (friction coefficient $\mu = 0.5$).

[3]The matrix $\mathbf{C}$ is defined for the friction cone approximation by an inscribed pyramid and μ is the coefficient of static friction [28].

[4]The lower block is fixed on a work table as indicated by the ground symbol.

To determine the PSAO we consider each of the solution sets P_j, $j \in [0,\dots,5]$, of problem (11). We calculate the vertices of the solution sets P_j by the DDM. The main advantages of this algorithm are:

- The algorithm handles all degenerate cases
- The algorithm allows a decomposition strategy to handle very large problems
- The magnitudes of the forces leading to potential stability can be determined

To understand the DDM we have to define a Double Description Pair (DDP):

Definition 4.1 (Double Description Pair) *A pair $(\mathbf{A},\mathbf{R})$ of real matrices A and R is said to be a DDP if the following relationship holds:*

$$\mathbf{A}\vec{x} \geq \vec{0} \quad \textit{iff} \quad \vec{x} = \mathbf{R}\vec{\lambda} \quad \textit{for some } \vec{\lambda} \geq \vec{0} \qquad (12)$$

For a DDP $(\mathbf{A},\mathbf{R})$ it is necessary, that the column size d of $\mathbf{A}$ is equal to the row size of $\mathbf{R}$. The *representation matrix* $\mathbf{A}$ describes the set $P(\mathbf{A})$:

$$P(\mathbf{A}) = \{x \in \mathbb{R}^d : \mathbf{A}x \geq 0\} \qquad (13)$$

and the set $P(\mathbf{A})$ is simultaneously described by the *generating matrix* $\mathbf{R}$:

$$P(\mathbf{A}) = \{x \in \mathbb{R}^d : x = \mathbf{R}\lambda \quad \text{for some } \lambda \geq 0\} \qquad (14)$$

A subset P of $\mathbb{R}^d$ is called *polyhedral cone* if $P = P(\mathbf{A})$ for some matrix $\mathbf{A}$. Each column vector of $\mathbf{R}$ lies in the cone P and every vector in P is a non-negative combination of some columns of $\mathbf{R}$.

Theorem 4.1 (Minkowski's Theorem) *For any $m \times d$ real matrix $\mathbf{A}$, there exists some $d \times m$ real matrix $\mathbf{R}$ such that $(\mathbf{A},\mathbf{R})$ is a DDP. The cone $P(\mathbf{A})$ is generated by $\mathbf{R}$.*

Theorem 4.2 (Weyl's Theorem) *For any $d \times m$ real matrix $\mathbf{R}$, there exists some $m \times d$ real matrix $\mathbf{A}$ such that $(\mathbf{A},\mathbf{R})$ is a DDP. The set generated by $\mathbf{R}$ is the polyhedral cone $P(\mathbf{A})$.*

To solve the problem of determining the set of stable orientations of assemblies we have to transform the linear inequalities of the six problems (11) into the form $P_j(\mathbf{A}^j) = \{x : \mathbf{A}^j x \geq 0\}$ for $j = 0,\dots,5$. The DDM constructs a $d \times n$ matrix $\mathbf{R}^j$ such that $(\mathbf{A}^j,\mathbf{R}^j)$ is a DDP. With the matrices $\mathbf{R}^j$ we get the solution set of our stability problem. In the following we describe the algorithm to construct the matrices $\mathbf{R}^j$ for one fixed j and for simplicity we drop the subscript. Let K be a subset of the row indices of $\mathbf{A}$. The matrix $\mathbf{A}_K$ denotes the submatrix of $\mathbf{A}$ consisting of rows indexed by K. Once we have found a matrix $\mathbf{R}$ for $P(\mathbf{A}_K)$ and $\mathbf{A} = \mathbf{A}_K$ we are done. In the other case we select a row index i not in K and try to construct a DDP $(\mathbf{A}_{K+i},\mathbf{R}')$ using the information of $(\mathbf{A}_K,\mathbf{R})$. The introduced inequality $\mathbf{A}_i x \geq 0$ partitions $\mathbb{R}^d$ into three parts:

$$\begin{aligned} H^+{}_i &= \{x \in \mathbb{R}^d : \mathbf{A}_i x > 0\} \\ H^0{}_i &= \{x \in \mathbb{R}^d : \mathbf{A}_i x = 0\} \\ H^-{}_i &= \{x \in \mathbb{R}^d : \mathbf{A}_i x < 0\} \end{aligned} \qquad (15)$$

We define J as the column indices of $\mathbf{R}$. Then the induced rays r_j with $j \in J$ are partitioned into three parts:

$$\begin{aligned} J^+{}_i &= \{j \in J : r_j \in H^+{}_i\} \text{ positive rays} \\ J^0{}_i &= \{j \in J : r_j \in H^0{}_i\} \\ J^-{}_i &= \{j \in J : r_j \in H^-{}_i\} \text{ negative rays} \end{aligned} \qquad (16)$$

By generating new $\|J^+\| \times \|J^-\|$ rays lying on the ith hyperplane[5] $H^0{}_i$ we construct a matrix $\mathbf{R}'$ from $\mathbf{R}$. The next lemma is the key procedure for the DDM:

Lemma 4.1 (Main Lemma for the DDM) *Let $(\mathbf{A}_K,\mathbf{R})$ be a DDP and let i be a row index of $\mathbf{A}$ not in K. Then the pair $(\mathbf{A}_{K+i},\mathbf{R}')$ is a DDP, where $\mathbf{R}'$ is the $d \times \|J'\|$ matrix with column vectors r_j ($j \in J'$) defined by*

$$\begin{aligned} J' &= J^+ \cup J^0 \cup (J^+ \times J^-) \\ r_{jj'} &= (\mathbf{A}_i r_j) r_{j'} - (\mathbf{A}_i r_{j'}) r_j \\ &\qquad \textit{for each}(j,j') \in J^+ \times J^- \end{aligned} \qquad (17)$$

To determine the PSAO we transform the linear inequalities of the six problems (11) into the form $P_j(\mathbf{A}^j) = \{\vec{x} : \mathbf{A}^j\vec{x} \geq \vec{0}\}$ for $j = 0,\dots,5$. The DDM constructs a matrix $\mathbf{R}^j$ such that $(\mathbf{A}^j,\mathbf{R}^j)$ is a DDP. With the matrices $\mathbf{R}^j$ we get the solution set of our stability problem. We give the DDM algorithm [8] in procedural form:

[5]To this end, we take an appropriate positive combination of each positive ray r_j and each negative ray r'_j discarding all negative rays.

DDM($\mathbf{A}$) {
  start with any initial ($\mathbf{A}_K, \mathbf{R}$);
  while $K \neq \{1, \ldots, l\}$ {
    select any i from $\{1, \ldots, l\} \setminus K$;
    construct double description pair
    ($\mathbf{A}_{K+i}, \mathbf{R}'$)
    from ($\mathbf{A}_K, \mathbf{R}$) using the DDM Lemma;
    $\mathbf{R} = \mathbf{R}'$;
    $K = K + i$;
  }
  solution is $\mathbf{R}$;
}

With the DDM algorithm we solve the PSAO stability problem with the following algorithm:

PSAO(Assembly) {
  for ($j = 0; j \leq 5; j++$){
    calculate P_j /* using (11) */;
    transform P_j to $P_j(\mathbf{A}^j) = \{\vec{x} : \mathbf{A}^j \vec{x} \geq 0\}$;
    DDM($\mathbf{A}^j$) /* calculates solution $\mathbf{R}^j$ */;
    analyse forces given by $\mathbf{R}^j$;
  }
}

With the matrices $\mathbf{R}^j$ we get all the magnitudes of the normal and tangential forces leading to a static equilibrium of the given assembly and the components of the gravity vector; in other words, $\mathbf{R}^j$ represents all stable orientations of the given assembly. Now we can analyze the distribution of the forces given by $\mathbf{R}^j$ and reject for example potential stable orientations with very large magnitudes of normal forces which are balanced by an equal and opposite force on parallel contact faces as indicated in Figure 3(a).

5 PSAO for Assembly Cost Evaluation

Due to the large number of feasible assembly sequences often represented by an AND/OR-graph [10] it is desirable to select a few best sequences or the best assembly sequence. This however, has been hampered by the difficulty of selecting proper performance criteria (see e. g. [15], [34]) and relating them directly to assembly cost. We propose a new evaluation function based on the PSAO for assembly cost evaluation. This function assigns weights to the nodes of the AND/OR-graph to select an assembly plan minimizing the costs from the initial nodes up to the goal node taking into account the PSAO.

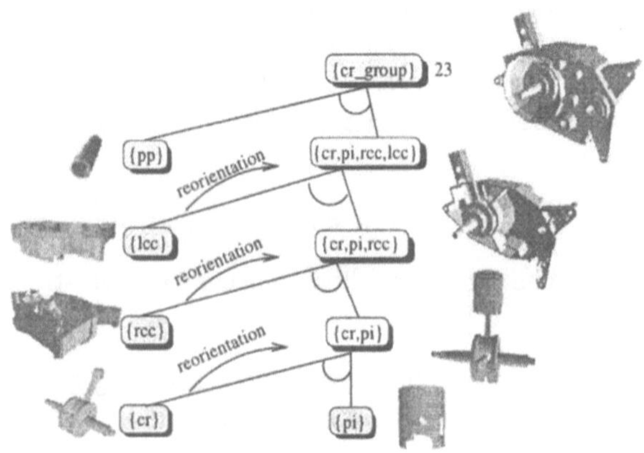

Figure 4: Alternative sequences for the {cr_group} subassembly with less total stable nodes.

For each AND/OR-graph node our assembly planning system rates the stability of the corresponding assembly using a stability evaluation function. The stability function bases on the PSAO of an assembly A denoted as $psao(A) \subseteq \mathbb{R}^3$. The system considers each of the solution sets P_j, $j \in [0,5]$, to problem (11) to determine $psao(A)$. Having found all vertices of the solution sets, the g_x, g_y, g_z components of the gravity vector are transformed to the unit sphere U to mark the stable region on the sphere, denoted as $\Pi(stable\_range(A), U)$ (see [28] for more details). We use the following stability function σ for any ***valid assembly*** $A \in \mathcal{P}(C)$ represented by an AND/OR-graph node[6]:

$$\begin{aligned} \sigma : \mathcal{P}(C) &\longrightarrow [0,1] \\ A &\longmapsto \sigma(A) = 1 - \frac{Area[\Pi(psao(A), U)]}{4\pi} \end{aligned} \tag{18}$$

Table 1 summarizes the stability costs assigned to the nodes of the AND/OR-graph of the assembly depicted in Figure 5. The gray shaded nodes in the AND/OR-graph represent subassemblies which are stable under any orientation. Therefore, the evaluation function σ assigns no costs to these nodes.

[6] $\mathcal{P}(C = \{c_1, c_2, \ldots, c_n\})$ denotes the set of all subsets of C representing the set of the assembly components.

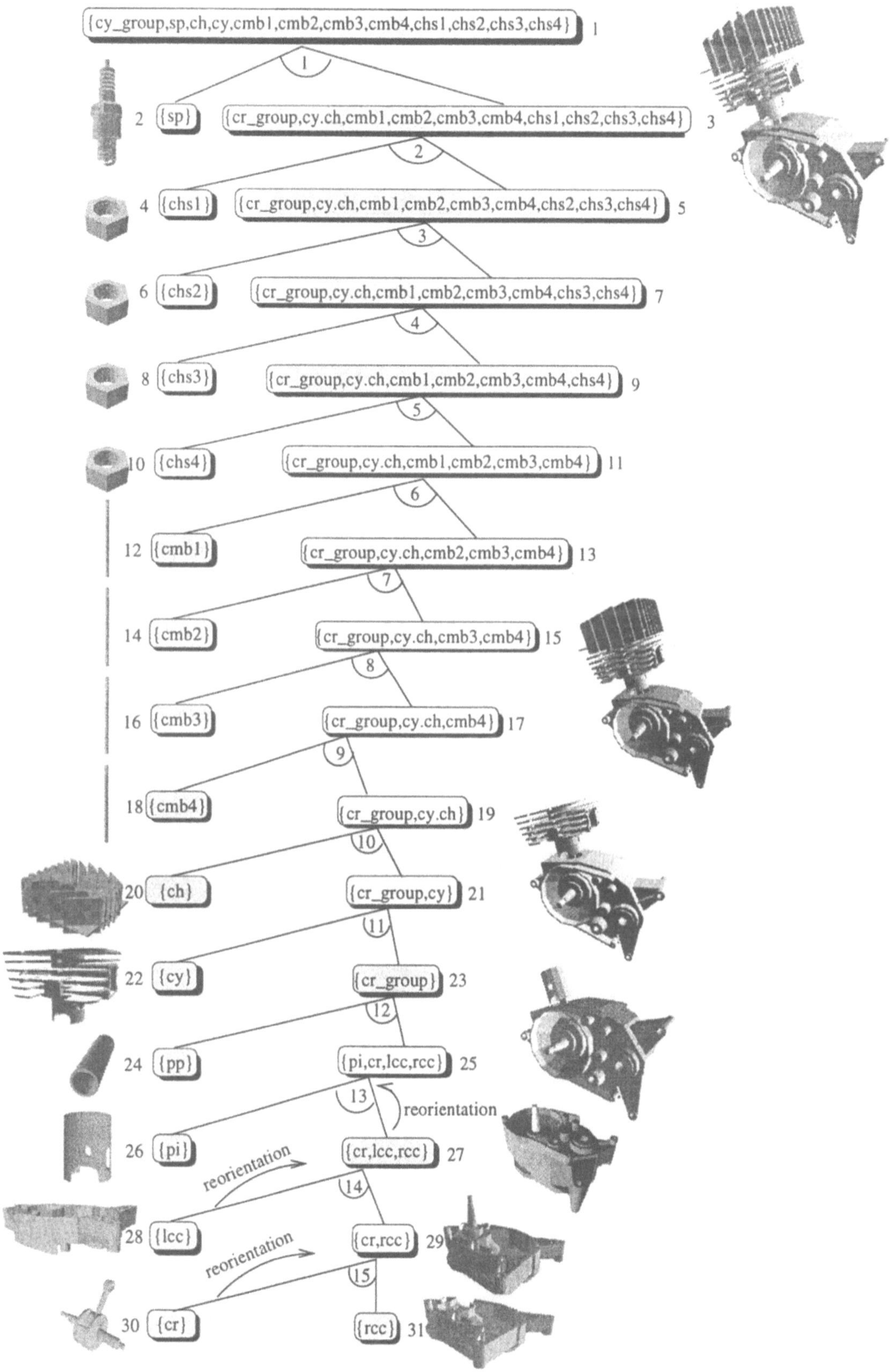

Figure 5: AND/OR-graph of the best assembly sequence for the root node assembly.

Table 1: Stability costs assigned to the nodes of the AND/OR-graph depicted in Figure 5; each node is identified by its corresponding number.

node number	1	2	3	4	5	6	7	8	9	10	11	12	13	14	15	16
#components	16	1	15	1	14	1	13	1	12	1	11	1	10	1	9	1
stability costs σ	0.0	0.0	0.0	0.0	0.0	0.0	0.0	0.0	0.0	0.0	0.48	0.0	0.48	0.0	0.48	0.0
node number	17	18	19	20	21	22	23	24	25	26	27	28	29	30	31	
#components	8	1	7	1	6	1	5	1	4	1	3	1	2	1	1	
stability costs σ	0.48	0.0	0.96	0.0	0.48	0.0	0.0	0.0	0.48	0.0	0.0	0.0	0.48	0.0	0.0	

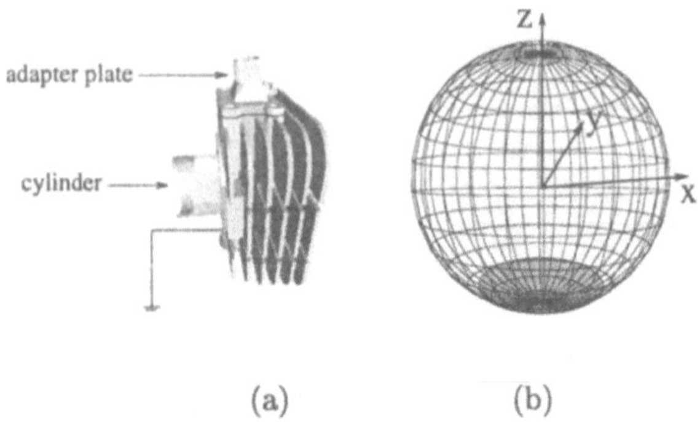

Figure 6: PSAO (*b*) for the assembly (*a*).

6 Results

Figure 6(b) shows the PSAO of the two-part assembly depicted in Figure 6(a). This set was calculated with the methods described in section 4 and is mapped onto the unit sphere. The cylinder is fixed by a gripper as indicated by the ground symbol. To illustrate the efficiency of our evaluation function for the selection of best assembly sequences we give an assembly planning example of a Yamaha RD80 motorcycle engine. Figure 5 illustrates the best (dis)assembly sequence for a subassembly of the motor represented by the root node. Furthermore, for the selection of the best assembly sequences our assembly planning system [29] considers other evaluation criteria like necessity of reorientations, separability and parallelism of subassemblies. Among these, (sub)assembly stability is a very important criteria for the evaluation. Figure 4 shows an alternative sequence for the *cr_group* subassembly of the motor. Here, the evaluation function σ assigns costs higher than 0.0 to three nodes instead of two nodes like in Figure 5. Therefore, our system selects the sequence with more total stable nodes in the AND/OR-graph thus reducing the stability costs.

7 Conclusions and Future Work

We discussed the problem of stability of assemblies consisting of arbitrary configurations of rigid bodies. We gave methods to determine whether an assembly in a given orientation is frictional stable andshowed how to compute a frictional stable orientation. Furthermore, an algorithm to compute the complete set of all frictional stable orientations of assemblies based on the DDM has been presented. We proposed a function based on the PSAO for the evaluation of assembly sequences. We integrated the stability analysis in our high level assembly planning system as an evaluation tool for assembly sequences. Simulation experiments have been presented and the results illustrate the applicability of the algorithms for plan evaluation within an assembly planning system. Our experiments have shown that, it is practicable to compute the frictional stable set of our test assemblies with the presented algorithm. The performance of this algorithm depends on both, the number of contact points and the topology of the assembly, i. e. the position and orientation of the contact surfaces' normals of the involved objects. The frictional stable set of an assembly, with for example 100 contact points, can be calculated in well under one second of CPU time on a SPARCstation 5 computer. For degenerate cases, a calculation time of up to a few minutes may be required after the decomposition of the problem. In future work the contact analysis will be extended to study frictional stability of assemblies in the presence of external forces other than gravitational forces. Moreover, we plan the development of a fixture algorithm taking into account the PSAO. Components with maximal emerging tangential forces will be fixed.

Acknowledgements

The authors would like to thank Marco Lübbecke of the Department of Mathematical Optimization at the Technical University of Braunschweig for many helpful discussions about the Double Description Method.

References

[1] D. Baraff. Issues in Computing Contact Forces for Non-Penetrating Rigid Bodies. *Algorithmica*, 10:292–352, 1993.

[2] M. Blum, A. Griffith, and B. Neumann. A stability test for configurations of blocks. Artificial Intelligence Memo No. 188, Massachusetts Institute of Technology, February 1970.

[3] Nico Boneschanscher, Hans van der Drift, Stephen J. Buckley, and Russell H. Taylor. Subassembly stability. *Proc. National Conf. on Artificial Intelligence*, pages 780–785, August 1988.

[4] A. Bourjault. Methodology of assembly automation: A new approach. pages 37 - 45. International Society for Productivity Enhancement, Springer-Verlag, July 1987.

[5] T. L. De Fazio and D. E. Whitney. Simplified Generation of All Mechanical Assembly Sequences. *IEEE Journal of Robotics and Automation*, RA-3(6):640 - 658, December 1987.

[6] L. S. Homem de Mello and S. Lee, editors. *Computer-Aided Mechanical Assembly Planning*. Kluwer Academic Publishers, 1991.

[7] M. Erdmann. On a Representation of Friction in Configuration Space. *International Journal of Robotics Research*, 13(3):240–270, 1994.

[8] K. Fukuda and A. Prodon. Double description method revisited. Technical report, Institute for Operation Research, ETHZ, Zurich, Switzerland, 1995.

[9] R. Hoffman. *Computer-Aided Mechanical Assembly Planning*, chapter A common sense approach to assembly sequence planning. Kluwer Academic Publishers, 1991.

[10] L. S. Homem de Mello and A. C. Sanderson. AND/OR-Graph Representation of Assembly Plans. *IEEE Trans. Robotics and Automation*, 6(2):188 - 199, April 1990.

[11] S. G. Kaufman, R. H. Wilson, R. E. Jones, and T. L. Calton. The archimedes 2 mechanical assembly planning system. In *IEEE International Conference on Robotics and Automation*, pages 3361–3368, 1996.

[12] L. Kavraki and M. Kolountzakis. Partitioning a planar assembly into two connected parts is np-complete. *Information Processing Letters*, 55:159–165, 1995.

[13] S. Lee. *Computer-Aided Mechanical Assembly Planning*, chapter Backward assembly planning with DFA analysis. Kluwer Academic Publishers, 1991.

[14] S. Lee and Y. G. Shin. Assembly Planning based on Geometric Reasoning. *Computers and Graphics*, 14(2):237 - 250, 1990.

[15] S. Lee and Chunsik Yi. Assemblability Evaluation Based on Tolerance Propagation. *Proc. IEEE Int. Conf. on Robotics and Automation*, 2:1593 - 1598, May 1995.

[16] P. Lötstedt. Coulomb friction in two-dimensional rigid body systems. *Zeitschrift für Angewandte Mathematik und Mechanik*, 61:605–615, 1981.

[17] T. Lozano-Pérez. Spatial Planning: A Configuration Space Approach. *IEEE Trans. Computers*, C-32(2):26–38, February 1983.

[18] R. Mattikalli, D. Baraff, and P. Khosla. Finding all gravitationally stable orientations of assemblies. In *Proc. IEEE Int. Conf. on Robotics and Automation*, pages 251–257, March 1994.

[19] R. Mattikalli, D. Baraff, P. Khosla, and B. Repetto. Gravitational stability of frictionless assemblies. *IEEE Trans. Robotics and Automation*, 11(3):374–388, 1995.

[20] R. Mattikalli, D. Baraff, P. Khosla, and B. Repetto. Finding All Stable Orientations of Assemblies with Friction. *IEEE Trans. Robotics and Automation*, 12(2):290–301, 1996.

[21] E. C. De Meter. Restraint analysis of assembly work carriers. *Robotics and Computer Integrated Manufacturing*, 1993.

[22] J. M. Miller and R. L. Hoffman. Automatic assembly planning with fasteners. In *IEEE International Conference on Robotics and Automation*, volume 1, pages 69–74, 1989.

[23] T.S. Motzkin, H. Raiffa, GL. Thompson, and R.M. Thrall. *Contributions to the theory of games, Vol. 2*, chapter The Double Description Method. Princeton University Press, 1953.

[24] K. Murty. *Linear Programming*. Wiley, 1983.

[25] R. S. Palmer. *Computational Complexity of Motion and Stability of Polygons*. PhD thesis, Cornell University, January 1987.

[26] J. Pertin-Troccaz. *Grasping: A State of the Art*, chapter Programming, Planning, and Learning, pages 71–98. The Robotics Review. MIT Press, 1989.

[27] F. Röhrdanz, R. Gutsche, and F. M. Wahl. Assembly Planning and Geometric Reasoning for Grasping. In I. Plander, editor, *Sixth Int. Conf. on Artificial Intelligence and Information-Control Systems of Robots*, pages 93–106. World Scientific, September 1994.

[28] F. Röhrdanz, H. Mosemann, and F. M. Wahl. Stability Analysis of Assemblies Considering Friction. Technical Report 5-1996-1, Institute for Robotics and Process Control, Braunschweig, Germany, May 1996.

[29] F. Röhrdanz, H. Mosemann, and F. M. Wahl. Generating und Evaluating Stable Assembly Sequences. *Journal of Advanced Robotics*, 11(2):97–126, 1997.

[30] F. Röhrdanz, H. Mosemann, and F. M. Wahl. HighLAP: A High Level System for Generating, Representing, and Evaluating Assembly Sequences. *International Journal on Artificial Intelligence Tools*, 6(2):149–163, 1997.

[31] B. Romney, C. Godard, M. Goldwasser, and G. Ramkumar. An efficient system for geometric assembly sequence generation and evaluation. In *ASME International Computers in Engineering Conference*, pages 699–712, 1995.

[32] J. M. Schimmels and M. A. Peshkin. Force-assembly with friction. *IEEE Transactions on Robotics and Automation*, 1994.

[33] K. B. Shimoga. Robot Grasp Synthesis Algorithms: A Survey. *International Journal of Robotics Research*, 15(3):230–266, June 1996.

[34] C. K. Shin, D. S. Hong, and H. S. Cho. Disassemblability Analysis for Generating Robotic Assembly Sequences. *Proc. IEEE Int. Conf. on Robotics and Automation*, 2:1284 - 1289, May 1995.

[35] *User Manual and Tutorials, IGRIP Version 3.0*. Deneb Robotics, 3285 Lapeer Road West, P.O. Box 214687, Auburn Hills, MI 48321-4687, 1994.

[36] J. Trinkle, J. S. Pang, S. Sudarsky, and G. Lo. On dynamic multi-rigid-body contact problems with coulomb friction. Technical report, Department of Computer Science, Texas A&M University, 1995.

[37] J. M. Valade. Geometric reasoning and automatic synthesis of assembly trajectory. In *International Conference on Advanced Robotics*, pages 43–50, 1985.

[38] J. Wolter. On the Automatic Generation of Assembly Plans. In *Proc. IEEE Int. Conf. on Robotics and Automation*, pages 62–68, 1989.

[39] J. D. Wolter and J. C. Trinkle. Automatic selection of fixture points for frictionless assemblies. In *IEEE International Conference on Robotics and Automation*, 1994.

[40] T. C. Woo and D. Dutta. Automatic disassembly and total ordering in three dimensions. *Transactions of the ASME*, 113(2):207–213, 1991.

Towards a new Robot Generation

G. Hirzinger, B. Brunner, S. Knoch, R. Koeppe, M. Schedl
Deutsches Zentrum für Luft und Raumfahrt e.V. (DLR)
Oberpfaffenhofen, D-82234 Wessling
email: Gerd.Hirzinger@dlr.de

Abstract: Key items in the development of a new smart robot generation are explained by hand of DLR's recent activities in robotics research. These items are the design of multisensory gripper and articulated hands systems, ultra-light-weight links and joint drive systems with integrated joint torque control, learning and self-improvement of the dynamical behaviour, modelling the environment using sensorfusion, and new sensor-based off-line programming techniques based on teaching by showing in a virtual environment.

1 Introduction

In the past there has been kind of a very general disappointment about the fairly slow progress in robotics compared to human performance - despite of many years of robotics research involving a large number of scientists and engineers. Robots today in nearly all applications are still purely position controlled devices with may be some static sensing, but still far away from the human arm's performance with its amazingly low *own weight against load* ratio and its online sensory feedback capabilities involving mainly vision and tactile information, actuated by force-torque-controlled muscles.

Space robotics (and service robotics in general) might become a major driver for a new robot generation. The experience we made with ROTEX, the first remotely controlled robot in space, has strongly underlined this. As has been outlined in different papers [1], ROTEX flew with spacelab mission D2 in April 93, performed several prototype tasks (assembly, catching a floating object etc.) in a variety of operational „telerobotic" modes, e.g. on-line teleoperation on board and from ground (via operators and pure machine intelligence) as well as off-line programming on ground. Key technologies for the big success of this experiment have been

- the multisensory gripper technology, which worked perfectly during the mission (redundant force-torque sensing, 9 laser range finders, tactile sensing, stereo TV).
- local (*shared autonomy*) sensory feedback, refining gross commands autonomously
- powerful delay-compensating 3D-stereo-graphic simulation (predictive simulation), which included the robot's sensory behaviour.

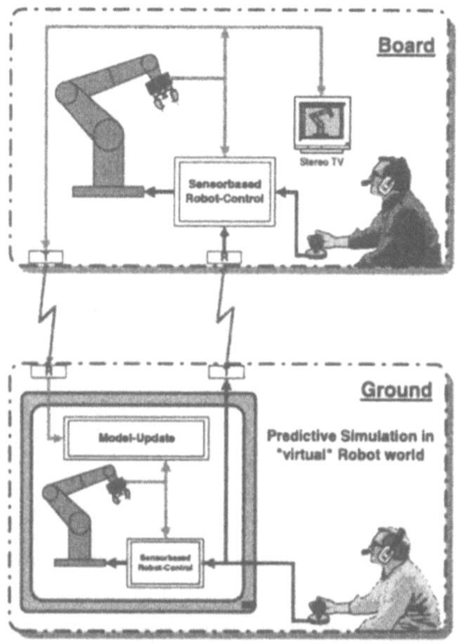

Fig. 1 ROTEX telerobotic control

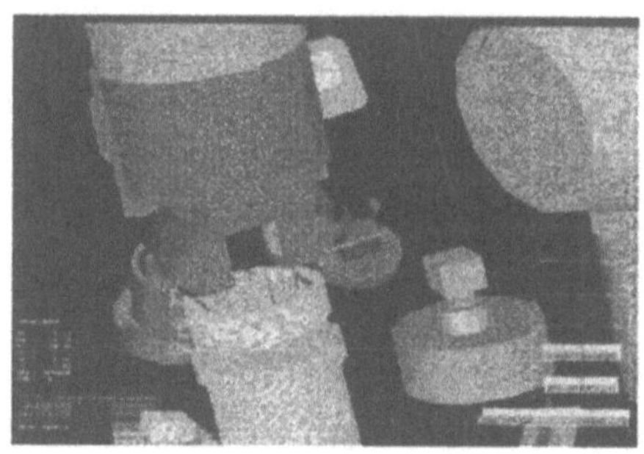

Fig. 2 Predictive Simulation of sensory perception in the ground station

With this background we are now developing a new light-weight robot generation with manipulative skills and multisensory perception (for space and terrestrial service applications) from bottom up in a unified and integrated, mechatronics approach. Its main features are

- Ultra light carbon fibre grid structures for the links, with structurally optimized integration of torque-controlled actuators, and with all electronics integrated into the arm, aiming at a 1:1 ratio of weight/maximum load. For us new joint actuators with a higher torque to weight ratio are still one of the key research goals.

- Multisensory grippers and articulated hands that are not just based on an assembly of available sensory and actuator components, but are designed from scratch following a unified multisensory hard- and software design philosophy.
- Learning and self-improvement for internal and external sensor control loops.
- Model-based, real-time 3D vision and environment modelling (sensor fusion).
- Implicit, task-level-programming, a sensor-based off-line programming technique (strongly related to learning by showing sensor-based elemental moves).

Let us describe these features in more detail in the subsequent sections.

2 Our mechatronics approach to robot arm and hand design

2.1 General remarks

Our approach in designing *multisensory light-weight robots* and *is an integrated, mechatronics one*. The new sensor and actuator generation developed in the last years does not only show up a high degree of electronic and processor integration, but also a fully modular hard - and software structure. Analogous signal conditioning, power supply and digital preprocessing are typical subsystem modules of this kind. The 20 kHz power supply line connecting all sensor and actuator systems in a galvanically decoupled way, and high-speed (optical) serial data bus systems (SERCOS or CAN) are typical features of our multisensory and multi-actuator concept.

2.2 DLR's first light weight robot

The fully modular, highly integrated hardware design

Ultra-light-weight mechanisms are indispensable for future space as well as terrestrial service robots [3]. Our light-weight robot (Fig. 3, Fig. 4) was designed from scratch in a modular way, each module consisting of a joint, the (carbon fiber grid) link-structure-element and the embedded electronics [4]. Each joint electronic module is of the same type and consists of a power inverter, the joint controller module and the joint-torque-sensor.

The design-philosophy of this light-weight-robot [4] was to achieve a type of manipulator similar to the kinematic redundancy of human arm , i.e. with seven degrees of freedom, a load to weight ratio of at least 1:2 (industrial robots ≈ 1:20), a total system-weight of about 15 kg, , no bulky wiring on the robot (and no electronics cabinet as it comes with every industrial robot), and a high dynamic performance. As all modern robot control approaches are based on commanding joint torques, we have developed an inductive torque-measurement system that may be seen as an integral part of our double-planetary gearing system (Fig. 5). Its core elements are a six-spokes wheel forming a rotational spring and a differential position measurement sensor (13 bit resolution now with 1 kHz bandwidth!).

Since each joint has its own control integrated, only a few power and information lines (using a fiber-optical ring) have to be distributed between the joints and the robot-controller.

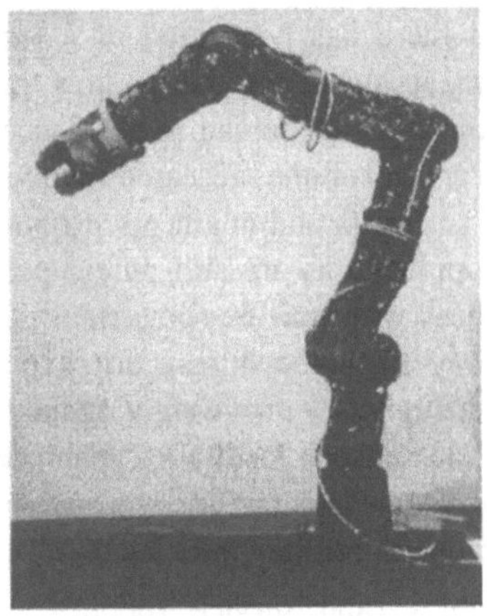

Fig. 3 The 7 axis version of DLR's light-weight-robot with integrated electronics

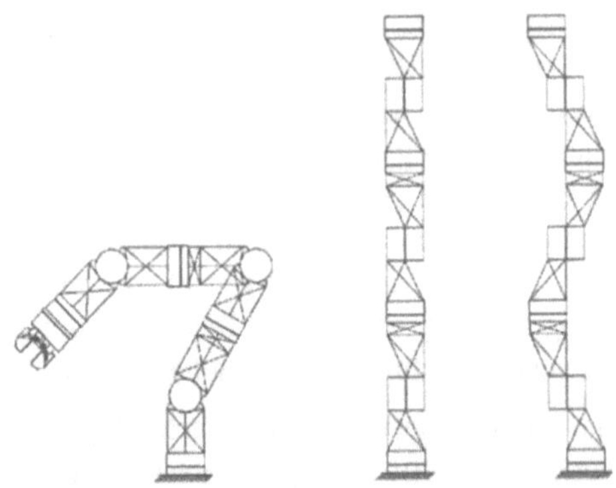

Fig. 4 Kinematic structure and two different assemblies. The right type is optimally foldable for stowing in space applications

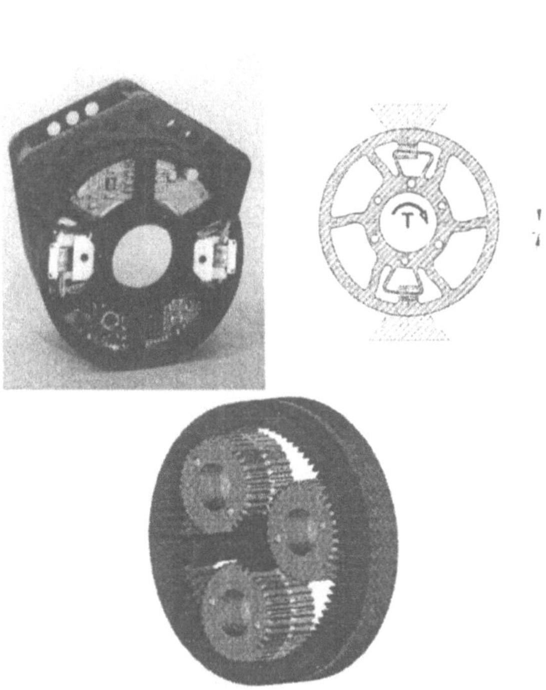

Fig. 5 Our 13 bit inductive joint torque sensor (upper), an integral part of the gearing (lower)

Dynamic feedback control

From the control point of view [18], [19] the DLR light weight robot belongs to the category of flexible joint robots due to the structure of the gear box and the integrated torque sensor [8]. The dynamic model can be established by applying Lagrange's Equation. For the decoupling of the manipulator dynamics this model is transformed into a new coordinate system in which the joint torque is treated as a state variable instead of the motor position. This leads to the so-called singular perturbation formulation of the robot dynamics. As a result, the fast motion corresponds to the joint-torque loop and the slow motion corresponds to the dynamic path concerned with the link position. On the higher levels, particularly interesting control results so far have been achieved with a hybrid learning approach, it is based on a full inverse dynamic model providing torque control; but as any model never will be perfect, the remaining uncertainties are learnt via back-propagation neural nets (Fig. 6). A first impressive demonstration of this type was learning zero-torque control, i.e. pure gravity compensation so that the arm was just able to sustain itself against gravity, but reacted softly to any external force at any link without additional force sensing [3].

Robot control architecture

Our robot control system is based on VME bus boards, the 4MBaud optical high-speed bus SERCOS and the industrial standard Real-time Multitasking Operating System VxWorks with its powerful development framework. The basic software structure idea is a data flow driven intertask communication approach. That means the necessary software modules (tasks), e.g. path interpolator, coordinate transformation, and drive controller are synchronized via the message passing mechanism supported by VxWorks.

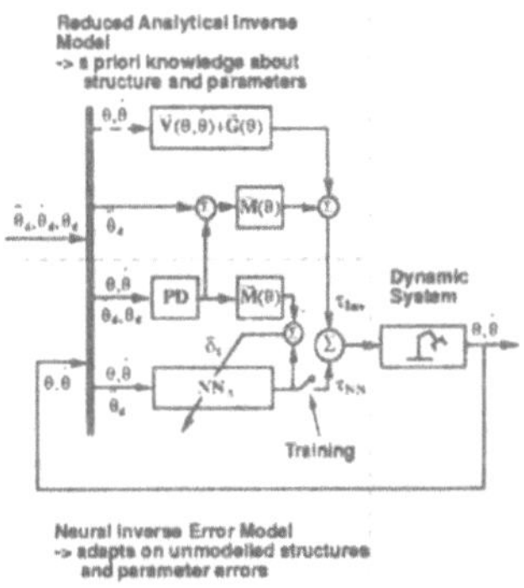

Fig. 6 „Hybrid" learning control scheme applied to gravity compensation of the light weight robot

The architectural design is based on a model of four layers connected by a global database. Data representation and access are organized by „object oriented programming techniques".

Singularity treatment, i.e. when the Jacobian looses full rank, has always been a problem in robotics. Singular robust Jacobian by damped least squares tends to lead to insufficient behavior within singular regions. We have developed algorithms that make efficient use of redundant motions in singular configurations [7] even with non-redundant robots (using symbolic expressions for the inverse Jacobian) and as an alternative, an extension of the so-called transpose Jacobian approach which we call *enhanced* transpose Jacobian (Fig. 7).

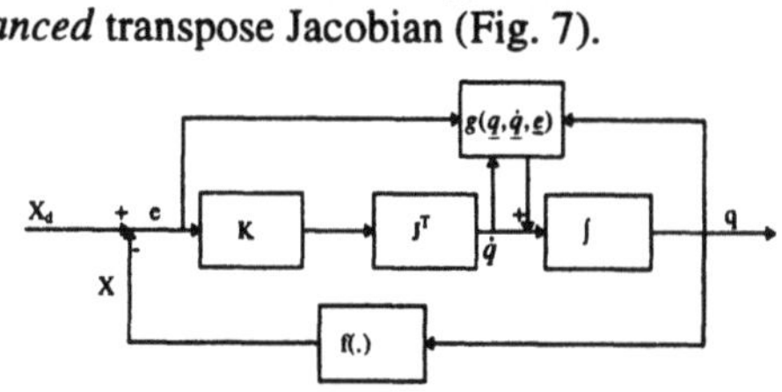

Fig. 7 Block scheme of the inverse kinematics algorithm with the *enhanced* Jacobian transpose; function g is chosen via Liapunov theory to optimize diverse criteria (e.g. joint limit avoidance).

2.3 Sensors and sensorbased 3D-interfaces

A number of essential improvements have been achieved since ROTEX. The patented 6 dof optoelectronic measurement system], used in the compliant force torque sensor (Fig. 13) as well as in the teleoperational control balls of ROTEX, has already been optimized a few years ago for use in the low-cost 6 dof controller Space Mouse LOGITECH's MAGELLAN (Fig.8 or Fig. 9).

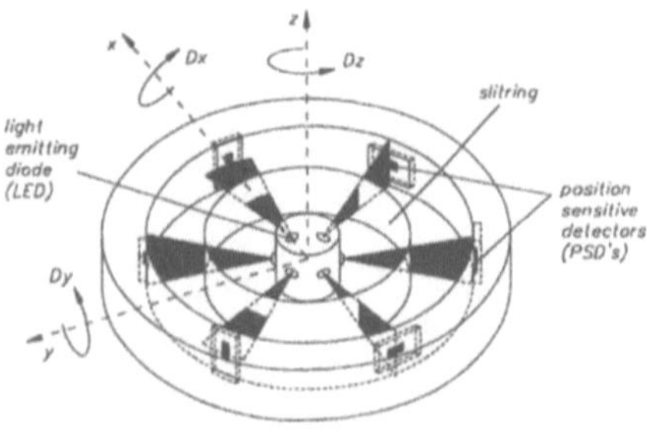

Fig.8 The patented opto-electronic measuring system

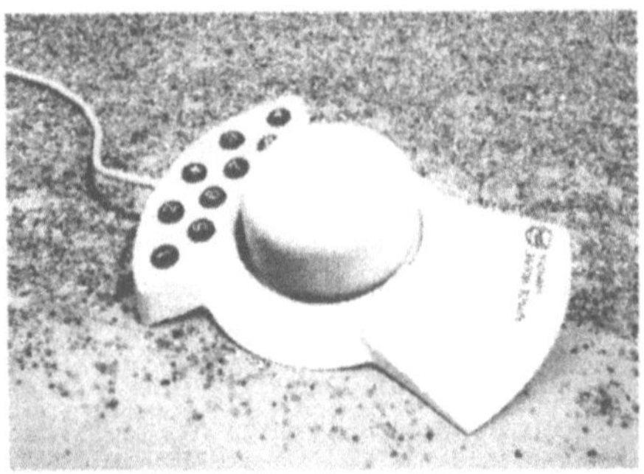

Fig. 9 The Space Mouse

It has become Europe`s most popular 3D computer input device. More than twenty thousand installations in drawing offices (especially of the car manufacturers) make 3D-CAD-constructions faster and more creative (Fig. 10). Moreover, this technology moves back to its origins; companies in medical industry like ZEISS are guiding their surgical microscopes (Fig. 11) by integrating our device into the handles, and the first robot manufacturers have integrated Space Mouse into their programming panels, like STÄUBLI and KUKA (the leading German manufacturer), thus, realizing our very early vision on how to move and program robots in 6 dof intuitively (Fig. 12).

We are meanwhile using the optimization experience of the SPACE MOUSE for the design of a low-cost compliant force-torque wrist sensor, which might lead to kind of a breakthrough with respect to the use of force-torque sensing with industrial robots (e.g. realizing „soft servos"). This new sensor will be equipped with a pneumatic locking mechanism and a so-called CAN bus interface as widely used by robot industry.

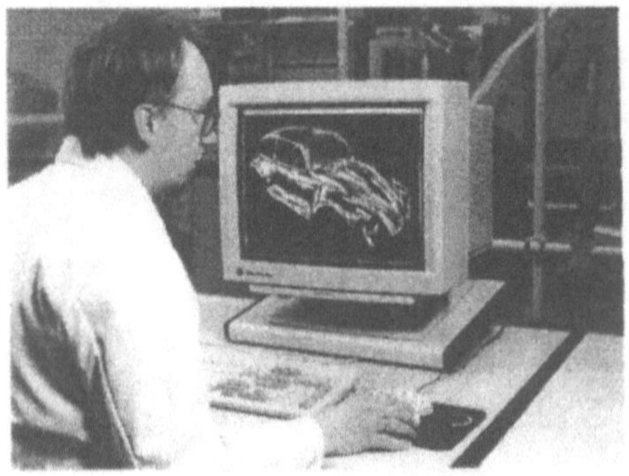

Fig. 10 Space Mouse is going to become a standard interface in 3D-graphics

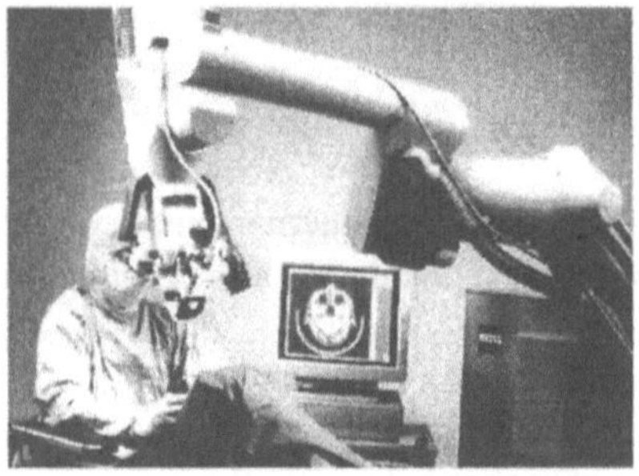

Fig. 11 Guiding surgical microscopes via Space Mouse integrated in the handle

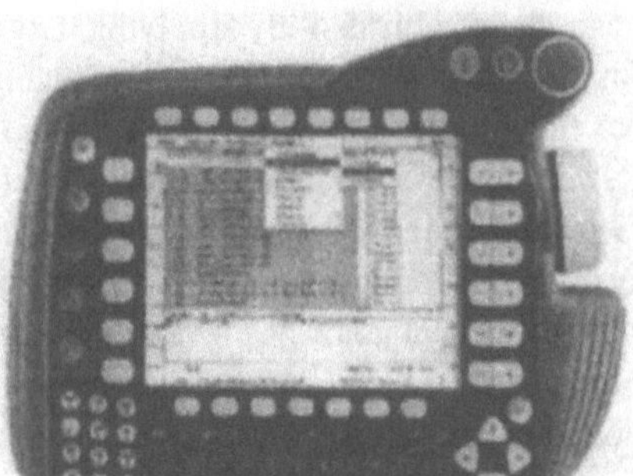

Fig. 12 The Space Mouse is used now by different robot manufacturers e.g. Germany`s leading manufacturer KUKA in their control panels

(a) (b)

Fig. 13 The compliant force-torque sensor (a) and the stiff force-torque-sensor (b)

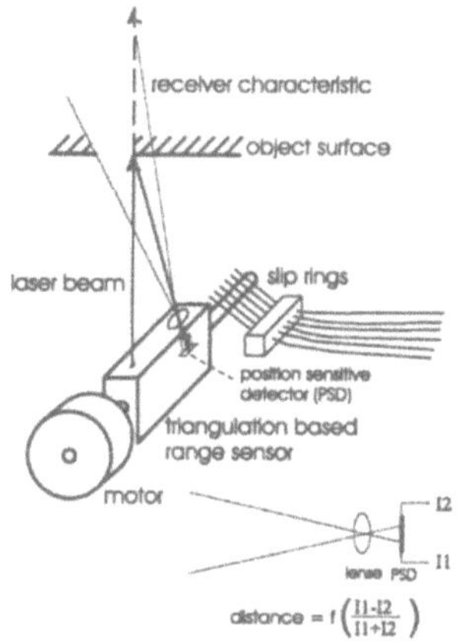

Fig. 14 Our rotating laser scanner

The stiff strain gauge based force-torque sensors which are perfectly temperature stable show up cycle times of 1 msec (Fig. 13). The medium range (3-40 cm) laser distance sensor based on triangulation as integrated in the ROTEX gripper has been augmented by a tiny motor drive, thus, generating a rotating (or oscillating) miniaturized 2D-Scanner without any mirrors (Fig. 15). Signal and power transfer presently is realized using slip rings which will be replaced by optical and inductive contactless techniques in the next step.

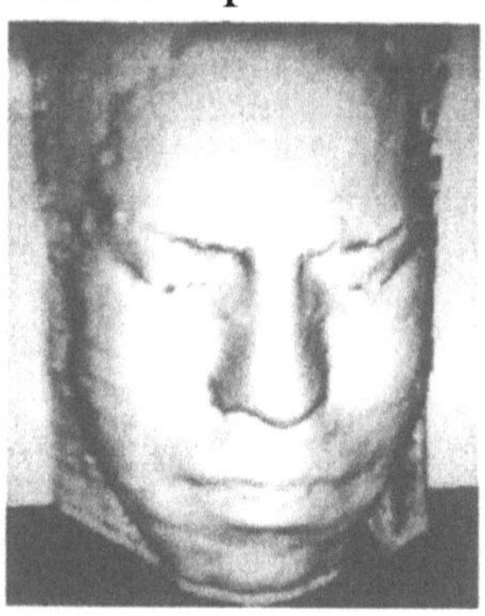

Fig. 15 Scanning a human face with a laser range finder

2.4 A four-fingered articulated hand

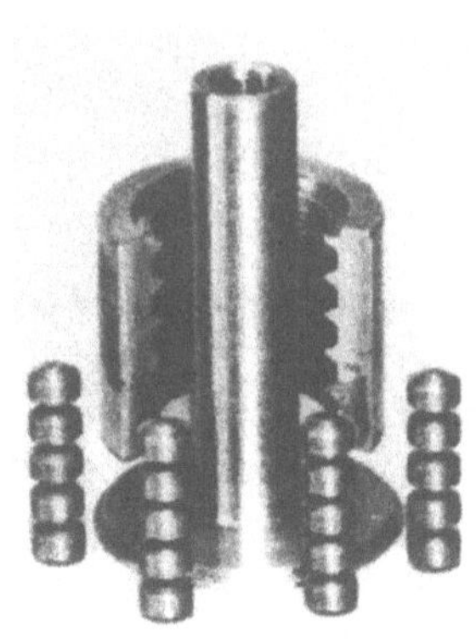

(a)

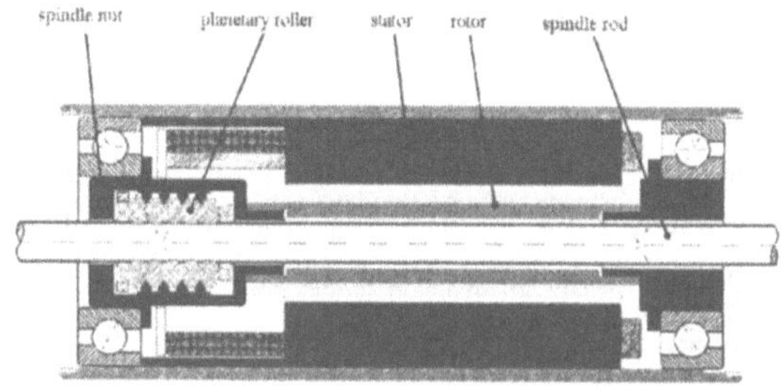

(b)

Fig. 16 DLR's planetary roller screw (a) integrated into tiny motors yields the artificial muscle (b)

Impressive dexterous robot hands have been built in the past, e.g. the MIT / UTAH hand, the JPL/Stanford hand, the Belgrade hand etc.. However, all of them suffer from one main drawback: if the number of active degrees of freedom exceeds a fairly small number (e.g. 6 dof), there was no chance so far to integrate the drives in the hands wrist or palm: either a number of cables or tubes leads to a separate pneumatic or hydraulic actuator box or a mass of bulky motors is somehow mounted at the robot arm, so that the practical use of articulated multifinger hands has often been called into question. Thus, it was our declared goal to build a **multisensory 4 finger hand** with in total twelve degrees of freedom (3 active dof in each finger), **where all actuators (uniformly based on the artificial muscle) are integrated in the hand's palm or in the fingers directly** (Fig. 17). Miniaturizing the artificial muscle down to a nut diameter of 10 mm in combination with a specially designed brushless DC-motor with hollow shaft was a first important step. Force transmission in the fingers is realized by special tendons (highly molecular polyethylene), which are optimal in terms of low weight and backlash despite of fairly linear behavior.

Each finger shows up a 2 dof base joint realized by two artificial muscles (AM) and a third actuator of this type integrated into the bottom finger link (phalanx proximal), thus, actuating the second link (phalanx medial) actively and, by elaborate coupling via a spring, the third link (phalanx distal) passively. The anthropomorphic fingertips are of crucial importance for grasping and manipulation, thus, they are modular and easily exchangeable with specially adapted versions. Following our mechatronic design principles, literally every millimeter in the fingers is occupied by sensing, actuation and electronic preprocessing technology. **Every finger unit with its 3 active degrees of freedom integrates 28 sensors(!).**

With 112 sensors, around 1000 mechanical and around 1500 electrical components the new hand is one of the most complex robot hands ever built. The fingers are position-force-controlled (impedance control) they are gravity compensated and they are prevented from colliding by appropriate collision avoidance algorithms. For more details see [2].

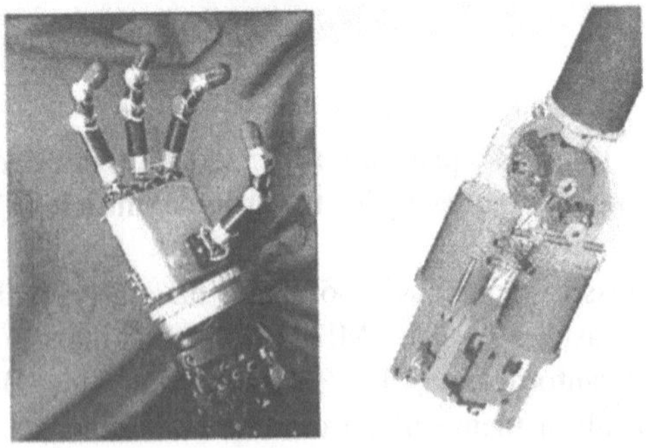

Fig. 17 Our 4 finger hand with its 12 actuators and 112 sensors integrates 1000 mechanical and 1500 electronic components

Sensor type	nr.	data
laser diode	1	wavelength 670 nm
torque sensors	5	range 0-1,8 Nm (9 Bit)
joint position	4	range 110° (9 Bit)
temperature	5	0 - 100°C, res. 0,1°C
rotor position	3	3072 pulses/rotation
light barriers	6	infrared
tactile sensors	4	0,5 - 10 N, res. 35mN

Table 1: A finger integrates 28 sensors

3 High-level robot programming and information processing

The robots we are developing are supposed to have multisensory (especially visual) capabilities and, thus, should not only be able to on-line react on sensory information, but also to learn about their environment, to update world models as well as improve their dynamic behavior. In the sequel we outline our approaches to telerobotics, visually guided behavior and learning.

3.1 Advances in telerobotics: Task-directed sensor-based tele-programming

After ROTEX we have focused our work in telerobotics on the design of a high-level task-directed robot programming system, which may be characterized as **learning by showing in a virtual environment** [5] and which is applicable to the programming of terrestrial robots as well. The goal was to develop a unified concept for

- a flexible, highly interactive, **on-line programmable teleoperation station** as well as
- an **off-line programming tool**, which includes all the sensor-based control features as tested already in ROTEX, but in addition provides the possibility to program a robot system on an **implicit, task-oriented level.**

A non-specialist user - e.g. a payload expert - should be able to remotely control the robot system in case of internal servicing in a space station (i.e. in a well-defined environment). However, for external servicing (e.g. the repair of a defect satellite) high interactivity between man and machine is requested.

To fulfill the requirements of both application fields, we have developed a 2in2-layer-model, which represents the programming hierarchy from the executive to the planning level.

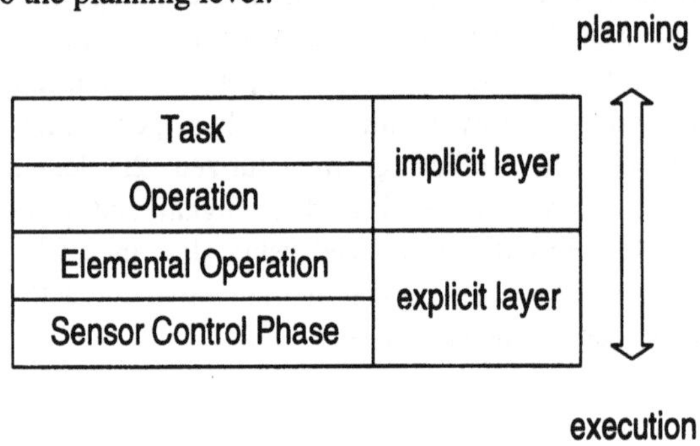

Fig. 18 2in2-layer-model

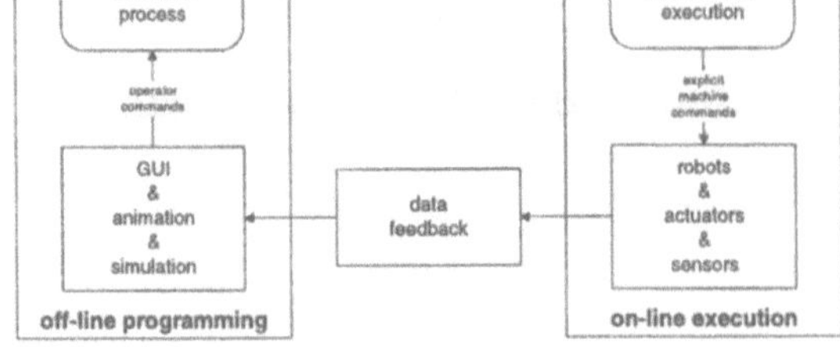

Fig. 19 Task-directed sensor-based programming

Sensor controlled phases

On the lowest programming and execution level our **tele-sensor-programming** (TSP) concept [15] consists of so-called *SensorPhases*, as partially verified in the local feedback loops of ROTEX. They guarantee the local autonomy at the remote machine's side. TSP involves ***teaching by showing*** the reference situation, i.e. by storing the nominal sensory patterns in a virtual environment and generating reactions on deviations. Each SensorPhase is described by

- a controller function, which maps the deviations in the sensor space into appropriate control commands.
- a **state recognition component**, which detects the end conditions and decides with respect to the success or failure of a SensorPhase execution
- the **constraint frame information**, which supports the controller function with task frame information to interpret the sensor data correctly (shared control)
- a **sensor fusion** algorithm, if sensor values of different types have to be transformed into a common reference system (e.g. vision and distance sensors) [6].

Elemental operations

The explicit programming layer is completed by the Elemental Operation (*ElemOp*) level. It *integrates the sensor control facilities with position and end-effector control.* According to the constraint frame concept, the non-sensor-controlled degrees of freedom (dof) of the cartesian space are position controlled

- in case of *teleoperation* directly with a telecommand device like the SpaceMouse
- in case of *off-line programming* by deriving the position commands from the selected task. Each object, which can be handled, includes a relative approach position, determined off-line by moving the end-effector in the simulation and storing the geometrical relationship between the object's reference frame and the tool center point, including the relevant sensory patterns.

A model-based ***on-line collision detection*** algorithm supervises all the robot activities, it is based on a discrete workspace representation and a distance map expansion. For global transfer motions a ***path planning*** algorithm avoids collisions and singularities.

Operations

Whereas the SensorPhase and ElemOp levels require the robot expert, the implicit, task-directed level provides a powerful man-machine-interface for the non-specialist user. We divide the implicit layer into the Operation and the Task level.

An Operation is characterized by a sequence of ElemOps, which hides the robot-dependent actions. Only for the specification of an Operation the robot expert is necessary, because he is able to build the ElemOp sequence. For the user of an Operation the manipulator is ***fully transparent***, i.e. not visible.

Tasks

Whereas the Operation level represents the subtask layer, specifying complete robot tasks must be possible in a task-directed programming system. A ***Task*** is described by a consistent sequence of Operations. To generate a Task, we use the VR-environment as described above. All the Operations, activated by selecting the desired objects or places, are recorded with the respective object or place description.

Our task-directed programming system with its VR-environment provides a man-machine-interface at a very high level i.e. without any detailed system knowledge, especially w.r.t. the implicit layer. To edit all four levels as well as to apply the SensorPhase and ElemOp level for teleoperation, a sophisticated graphical user interface based on the OSF/Motif standard has been developed (Fig. 23 bottom, screen down on the left). This GUI makes it possible to switch between the different execution levels in an easy way. Fig. 23 shows different views of the simulated environment (far, near, camera view), the Motif-GUI, and the real video feedback, superimposed with a wireframe world model for vision-based world model update („augmented reality“, top screen up on the right).

Fig. 20 VR-environment with the ROTEX-workcell and the Universal Handling Box, to handle drawers and doors and peg-in-hole-tasks

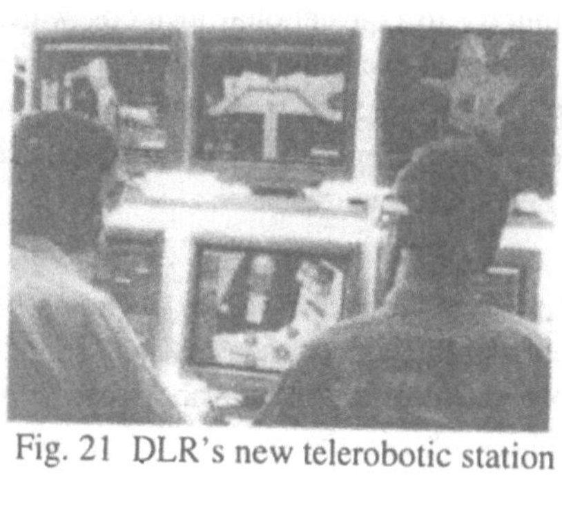

Fig. 21 DLR's new telerobotic station

3.2 Robotics goes WWW - Application of Telerobotic Concepts to the Internet using VRML and JAVA

New chances towards standardization in telerobotics arise with VRML 2.0, the newly defined Virtual Reality Modeling Language. In addition to the description of geometry and appearance it defines means of animations and interactions. Portable Java scripts allow programming of nearly any desired behavior of objects. Even network connections of objects to remote computers are possible. Our new telerobotic station is fully remotely programmable via VRML now.

A commercial application which gains increasing interest is **teleservicing** of industrial robots. Teleservicing through computer networks could dramatically reduce costs and simultaneously improve customer support in this field. Teleoperation in this case mainly requires remote execution of off-line generated robot programs. VRML will be used as an online visualization tool for the entire state of the remote real robot even during fast motions. Using VRML compared to a video feedback channel has the advantage of consuming less bandwith (about 5kByte/s compared to 384 kByte/s) and the possibility to show the scenery from arbitrary points of view.

3.3 Robot vision

The focus of our work is on real-time image postprocessing for finding the *pose* of objects in order to enable the robot to grasp them, or to get their *shape* to perform object recognition and world model update. We can only give a few examples here.

Multisensory robot servoing - sensorfusion

The purpose of *robot servoing* is to position a robot, sensory-controlled, to a desired pose relative to an object, as characterized by nominal sensory patterns (Section 3.1). Redundant multisensory information (e.g. from cameras and range finder arrays) may be used to increase the performance of the servoing system as well as the robustness against failing sensors. Contrary to conventional methods which need camera and hand-eye calibration, we have developed two approaches which do not need calibrations, but are based on *learning by showing*. In both cases the control law may be written as

$$v_C = \alpha \, C(s - s^*)$$

where $(s - s^*)$ is the vector-valued deviation between the actual and the nominal sensory pattern indicating the displacement of the actual robot pose x from the nominal pose x^*, v_C is the velocity command, α represents a scalar dynamic expression, at least a real constant, determining the closed loop dynamics, and C represents a *projection operator* used for mapping the sensor space onto the control space. In the first approach, a generalized projection operator

$$C = (J^T W_m J + W_v)^{-1} J^T W_m$$

is based on linear minimum variance **estimation theory**. Here J is the sensor-to-control Jacobian, W_m the input weight matrix, and W_v the output weight matrix. We find the elements of the Jacobian experimentally by moving the robot along and around all the degrees of freedom near the nominal pose in order to approximate the partial derivatives by difference quotients. Sensor errors, up to failing sensors, can easily be taken into account by managing W_m, and subspace control is possible by managing W_v, both without the need of relearning.

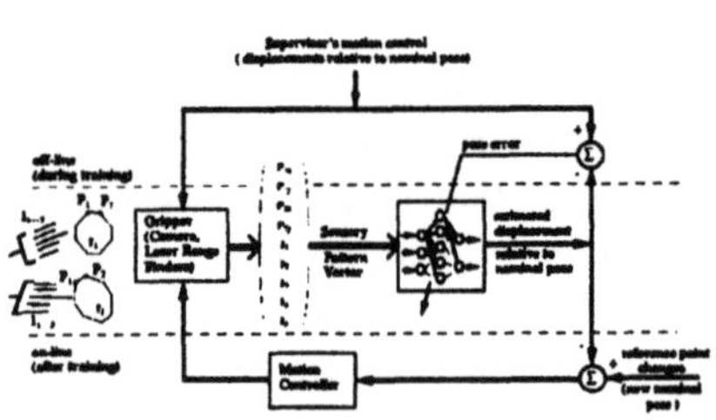

Fig. 22 Learning sensorimotion mappings from examples

Similarly the **neural network** approach (Fig. 22) learns the direct mapping C from the sensory pattern to the motion commands from a set of examples [9]. Besides the recall at the nominal position, we are able to modify the trained network such that other desired positions can be achieved without the need for retraining.

The dominant features of both learning approaches - successfully tested in a variety of experiments e.g. implying a stereo camera and four laser range finders - are: The controller design is straight forward. No analytical model is used, no partial derivatives are needed. Redundant informations of different sensor types are easily combined (fusion).

Model-based vision for robot servoing

The motivation for model-based vision approaches is to robustly analyze complex images with high performance, and to ensure at the same time that the methods developed are generally applicable. We assume that model information is available in the form of a polyhedral description of the 3-D object geometry, enhanced by information about circular features. The research reported here is devoted to *multisensory* eye-in-hand systems. The sensors considered are one or more video cameras and optical range sensors that yield an array of 3-D surface point measurements. The structure of the model-based multisensory tracking algorithm [10] is illustrated in Fig. 25. The task is to track the relative pose $\boldsymbol{x}$ of the target. Image feature vectors ${}^{Ci}\boldsymbol{f}$ along with the measurements ${}^{Sj}\boldsymbol{m}$ of the range sensing devices are the input to a least squares procedure that computes an estimate $\boldsymbol{x}$ of current pose and its covariance Σ_{xx}. Next, in order to compensate delay and to prepare feature extraction a prediction $\hat{\boldsymbol{x}}$ and its covariance are computed based on assumptions on the object motion, which is used to generate feature values ${}^{Ci}\hat{\boldsymbol{f}}$ expected in the next processing cycle. Thus, image processing can be limited to small regions of interest, potential occlusions can be predicted by hidden line removal. This structure is equivalent to that proposed by Dickmanns et al. [20] for monocular tracking.

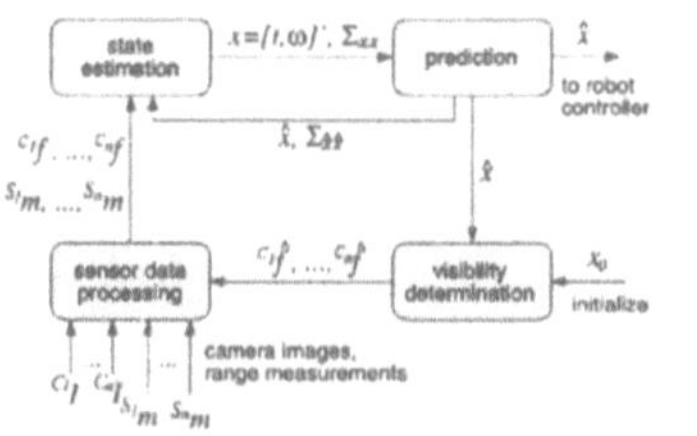

Fig. 23 Processing steps in model-based multisensory 3-D tracking

Based on the prediction $\hat{\boldsymbol{x}}$ expected features values are generated by projection of the model wire frame and subsequent removal of hidden lines. As the computing time required by estimation is low (about 1 msec), *complete analytic hidden-line elimination for non-convex objects* can be realized in one sensor cycle with our recursive pose estimation. Thus, not only self-occlusion can be handled but also occlusions of the target by other objects. Fig. 26 shows some examples of monocular tracking.

Fig. 24 Monocular, model-based real-time 3-D tracking: The wire frame of the object is projected into the image at the estimated pose. The good match indicates high accuracy. Minor occlusions do not affect accuracy.

Model-based pose estimation by registration

Novel algorithms [11] have been developed for registering a 3-D model to one or more of two-dimensional images without a-priori knowledge of correspondences. The method - a generalization of the iterative closest point algorithm (ICP) - handles full six degrees of freedom, without explicit 3-D reconstruction.

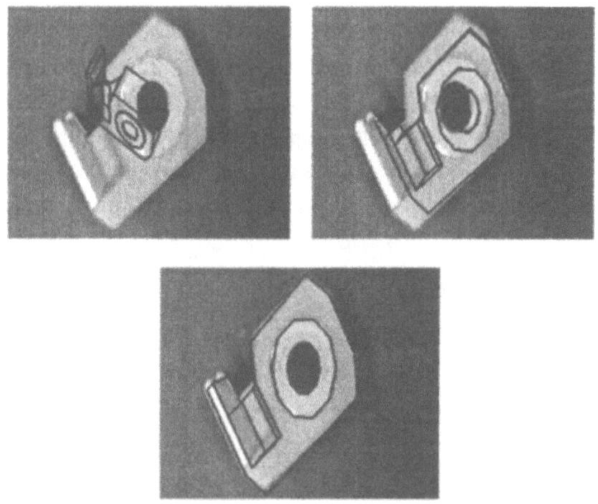

Fig. 25 An example of convergence of the registration algorithm. The initial rotational displacement in ZYX-Euler angles is {66.5°, 49.5°, 32.0°}

Experiments show that complex, three-dimensional CAD-models can be registered efficiently and accurately to images, even if image features are incomplete, fragmented and noisy (Fig. 25).

Shape from shading by neural networks

The problem of shape from shading is to infer the surface shape of an object based solely on the intensity values of the image. We proposed a new solution [12] of the shape from shading problem based on the optimization framework. We used a multilayer perceptron to parameterize the object surface.

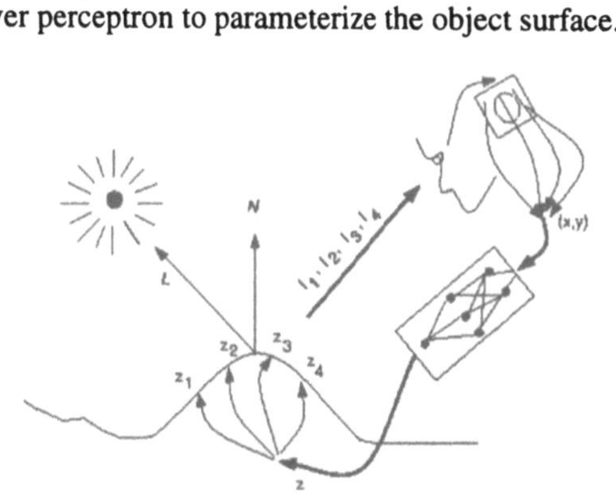

Fig. 26 Shape from shading: A learning scheme minimizes the intensity error

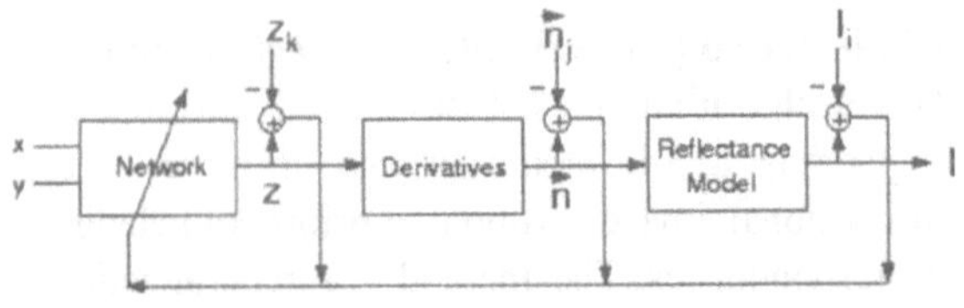

Fig. 27 A unified training framework

The weights of the network are updated so that the error between the given image intensity and the generated one is minimized. Fig. 26 and illustrate Fig. 27 the mechanism with which the shape is recovered from shading. We also showed that knowledge about object surface, e.g. known depths or orientations at some positions, can be easily incorporated into the shape from shading process so that errors due to lack of boundary conditions or self-shadows can be reduced. Results of the computational scheme are shown in Fig. 30.

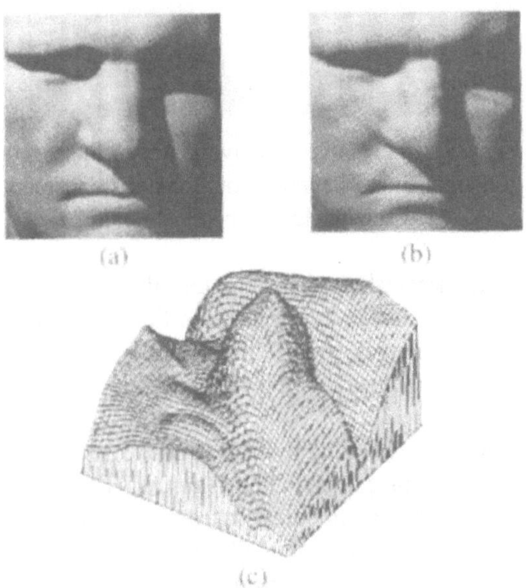

Fig. 28 Surface recovery of the Agrippa statue. (a) Input image; (b) Learned image; (c) Recovered 3D surface.

Typical applications

Automated laparoscope guidance in minimally invasive surgery

Unlike open surgery, where vision and action is concentrated at one person, in laparoscopic surgery the vision part is split up to two persons. A camera assistant has to move the endoscope to the surgeon's area of interest But a human normally cannot do this job without tremor and fatigue. Doing so he must react on the surgeons verbal advises as well as on small instrument motions. Beginners tend to react with delay and to overcompensate for delays by fast control motions. Tremor and trembling due to fatigue also disturb the success of operation. A robot automatically servoing the scope might be superior. We developed and patented a visual tracking system for mono- or stereo-laparoscopes, which is robust, simple and operates in real-time. [13]. We use color-coded instruments. The coding color is chosen different from the colors typical for laparoscopic images. With color image segmentation (Fig. 31), the color-coded instrument can be perfectly located in the image. Thus, with a mono-laparoscope the lateral motion can be controlled, with a stereo-laparoscope the transversal motion, too.

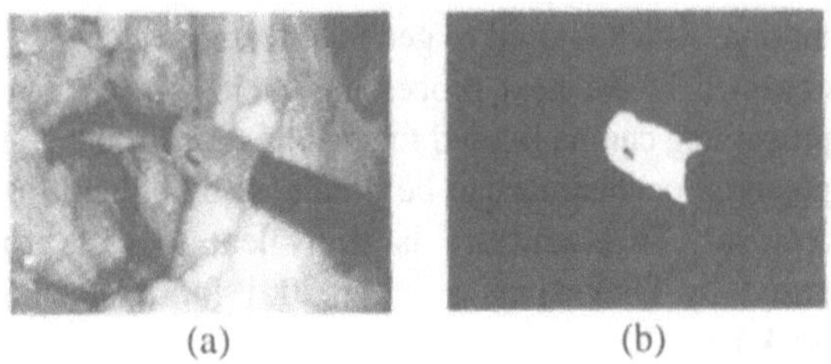

Fig. 29 (a) Laparoscopic image out of the abdomen with a color-marked instrument (b) post-processed segmentation

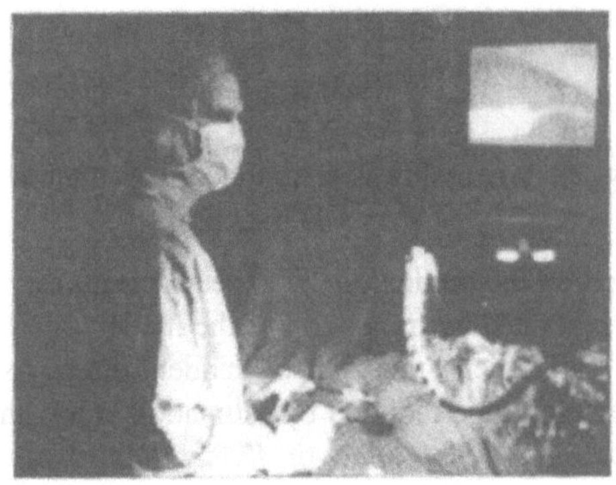

Fig. 30 Minimally Invasive Surgery using an Autonomous Robot Camera Assistant, for the first time tested on humans in September 1996 in the Munich hospital „Klinikum rechts der Isar"

This approach is very robust against occlusions, smoke, and image saturation. The control loop is designed such that the robot's motion is very smooth. Such a system, was successfully tested in the „Klinikum rechts der Isar" Munich hospital, on animals as well as on humans (Fig. 30). It was found that the surgeon's concentration onto the surgery is massively supported by this technology.

Model-based visual servoing in satellite capture

The capturing of a target satellite by a free-floating repair robot is intended to be automatically servoed by machine vision. As the geometry of the target satellite is known, a model-based vision approach is pursued, which also includes the well-known dynamics of a rigid body tumbling under zero gravity. Specifically, the prediction module of Fig. 23 is implemented as a Kalman filter that takes x and Σ_{xx} as an input (observation) and continuously estimates the parameters of the dynamic equations. From these estimates both a smoothed version of the object pose and a prediction are computed in each cycle.

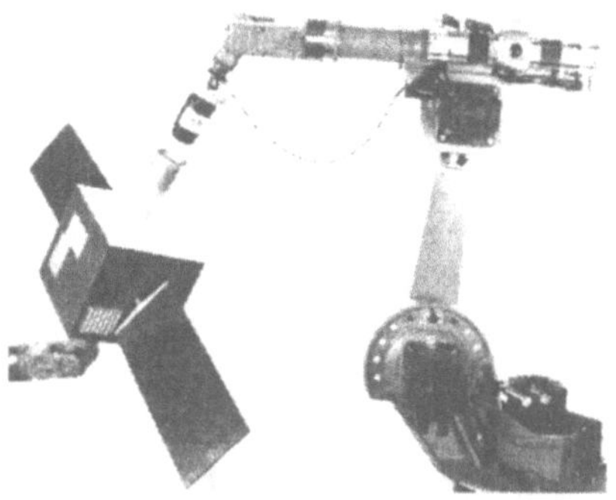

Fig. 31 A two robot system as testbed for a satellite repair project. The robot on the left carries a mockup of a satellite, that is tracked by a camera mounted in the tool of the tracking robot.

Image feature measurements are provided by a single camera mounted internally in the center of the capture tool. Fig. 31 shows results obtained with the testbed in the DLR lab. Recent video clips of the experiments can be downloaded from `http://www.op.dlr.de/FF-DR-RS/VISION`

3.4 Learning, skill-transfer and self-improvement

Introduction

Learning and adaptation are the core paradigms of intelligent control concepts which enable to increase skilled manipulation and to achieve higher levels of autonomy (Fig. 34). Data approximation and representation techniques like artificial neural networks enhance or may replace conventional model based approaches.

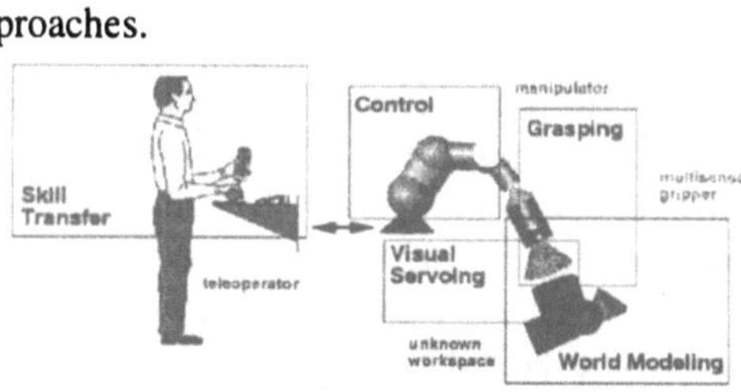

Fig. 32 Areas for learning and adaptation in advanced robotics

We use various kinds of neural network architectures for robot perception and manipulation. The multi-layer perceptron with sigmoid transfer functions is capable of solving non linearly separable classification problems. Another universal approximator is the Radial Basis Function network which uses Gaussian transfer functions. Furthermore multilayer networks may be structured to represent a fuzzy controller with adjustable parameters representing location and shape of the fuzzy sets and the weights of the truth value of rules. Recurrent networks have feedback connections within the network and are therefore dynamic systems themselves. Another important architecture is the self-organizing map, such as the Kohonen Feature map, which can be used to build discrete representations of data at optimal dimensions.

Neural network learning methods

For general approximation neural networks, second-order optimization methods have been shown to be superior to others. We have improved the learning algorithms to optimize the network weights [14],[15].

Learning problems remain when the controlled system is of high dimensionality, as is usually the case in robot/sensor systems. This is reflected in the conditioning of the Hessian of the feedforward network minimization task. In order to improve the condition of the Hessian matrix, we have proposed an extended network by adding an extra connection from each input unit to each output unit with a weight value coupled to the weight from the input to hidden unit. We call the new network the *linearly augmented feed-forward network* [14]. The well-known universal approximation theorems still apply to this type of network.

World modelling using Kohonen's feature map

Building a geometric description of a robot's workspace via unordered „clouds" of multisensory data out of the robot gripper is of crucial importance for robot autonomy.

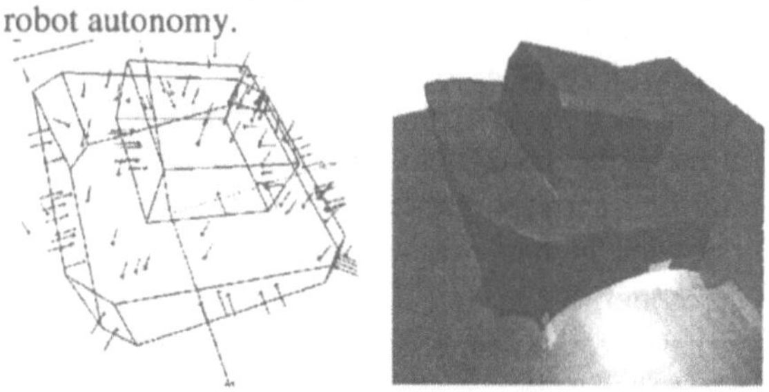

(a) (b)

Fig. 33 Self-organizing reconstruction of an object in the robot's work cell. Part (a) shows the wireframe representation of the object and the sampled 110 point and normal vector tuples, (b) the final surface after 3000 steps and a total time of 9 seconds.

To solve these kind of problems a surface reconstruction algorithm, based on Kohonen's self organizing feature map was developed and successfully implemented [16]. The two-dimensional array of neurons in Kohonen's algorithm may be considered as a discrete parametric surface description. To incorporate different types of surface information new training equations were developed for the Kohonen network. Successful surface reconstruction was performed for completely unordered *data clouds* of different surface information (Fig. 33). This method was used as a core for a complete world perception system for *Sensor-in-Hand* robots.

Shape from shading and visual servoing by neural networks

Application of multilayer perceptrons was outlined in Section 3.3.

Learning compliant motions by task demonstration

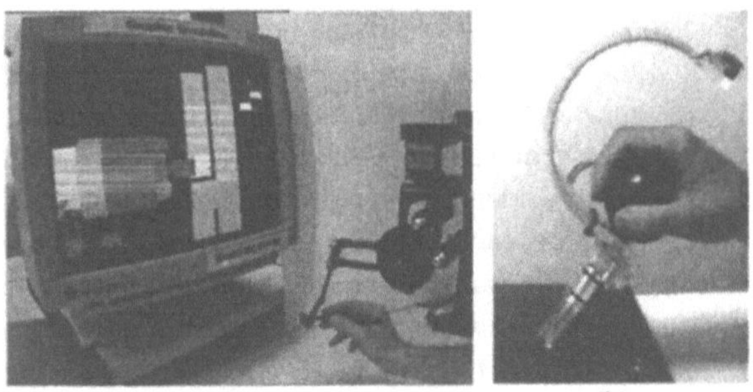

(a) (b)

Fig. 34 Demonstrating a compliant motion task in a natural way: (a) by haptic interaction in a telerobotic environment and (b) by direct manipulation using a teach device.

Neural networks are capable of representing nonlinear compliant motions. The training data preferably consist of the measured forces (input) and the human arm's motion (output). If sensorimotion data are directly recorded from a human demonstrating a task, the correspondence problem has to be solved, since the human executes an action u(t), due to a perception S(t-τ). In [21] we show that the correspondence problem is solved by approximating the function

$$u(t) = f\left[S(t), \dot{x}(t)\right]$$

with a neural network, taking the velocity $\dot{x}(t)$ as an additional input to the network.

With recent advances in virtual reality technology, haptics, and human interface systems, demonstrating compliant motion for execution on a robot can be performed by the sole process of human demonstration without the use of the actual robot system. We investigated two approaches: Interaction with a virtual environment using haptic interfaces [21] (Fig. 34(a)) and direct manipulation of the object using a teach device to measure force and motion of the human (Fig. 34(b)) [22]. Whereas the dynamics of the demonstration process is designed to support the human operator, the dynamics during execution depends on the robot system. By using dynamic manipulability analysis of motion and force, we investigate if a compliant motion trajectory obtained through teaching without a robot can be executed on a given robot.

Improving the accuracy of standard position controlled industrial robots

Learning control for standard industrial robots should be able to minimize the difference between the desired Cartesian path and the actual path. The optimal control input commands a fictive path which is deformed by the robot dynamics in such a way, that the resulting actual path coincides with the desired one. For convenience this task is splitted into subtasks which reduce the positional errors of the individual joints. The path deviations are results of static and dynamic effects. High speed errors due to static effects are compensated by sensory feedback. In this work dynamic effects, e.g. due to centrifugal or Coriolis forces are considered. The path errors due to the inertia of the robot mechanism can be minimized by independent linear feedforward learning controllers since they are almost decoupled and proportional to the accelerations. Neural networks may be used in addition to reduce the remaining effects, mainly the nonlinear couplings (see Fig. 35 (a) for a simple case). Common approaches for learning control focus on the improvement of fixed paths, thus, allowing accurate repeated execution of the desired trajectory, e.g. [23]. The advantage of our method comes into play if the path differs from the one used for training the controller, as it is valid for on-line planned paths which are modified in relation to unexpected sensor data.

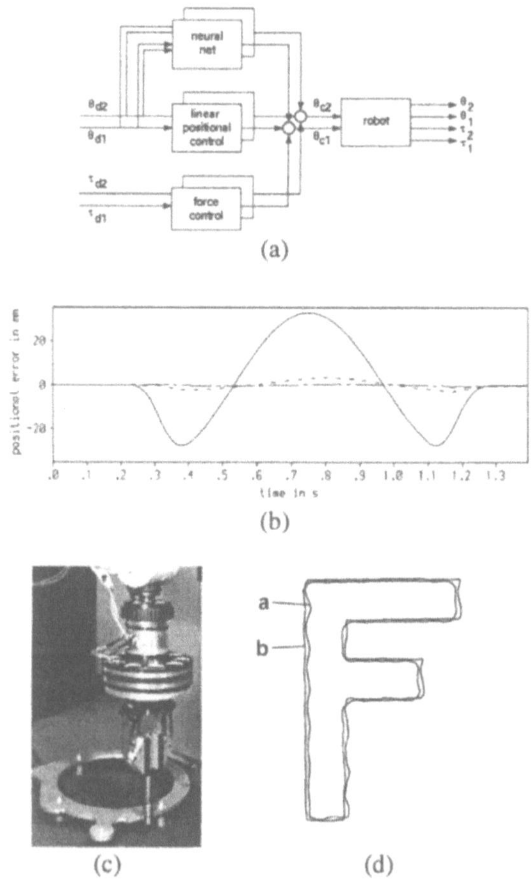

Fig. 35 Learning control structure (a) and contour following task experiment (c) Reduction of path errors to 20% (b). Distorted trajectory due to dynamic effects and compensated trajectory (d).

We use one network for each joint. The inputs are the positions of joints 2 to 5, and additional inputs, representing the time series of the desired accelerations of those joints which affect the motion of the actual joint. Multilayer perceptrons with sigmoid activation functions and two hidden layers are used. The training is performed with the Extended Kalman Filter algorithm mentioned in Section 3.3. The learning system was tested for high speed movements reducing path errors to 20 % and less in the first trial (Fig. 35 (c), [17]). Fig. 35 (b) shows a contour following task of a robot with conventional positional interface. First, the linear controllers are trained. Then, the neural networks are learned using training data from similar paths without contact. When applied to the untrained case of contact, the error of the desired force is reduced by 50% within the first execution of the path.
Standard industrial robots can be compensated by a simple Cartesian model of the robot dynamics. The trajectory „a" in Fig. 35 (d) shows the distortion of a high-speed motion path of a KUKA-Robot. Trajectory „b" shows the compensated path.

4 Resume

For us an interesting observation is that not only robotics research has made progress in the past, but that robot manufacturers are now accepting and integrating research results that have been achieved over the last twenty years. The coming years will bring fast sensory interfaces and feedback, more light-weight design and adaptability. Service robotics - be it for space or for terrestrial applications - is going to demonstrate what industrial robots should be capable of in the future.

References

[1] G. Hirzinger, B. Brunner, J. Dietrich, J. Heindl, „ROTEX - The First Remotely Controlled Robot in Space", IEEE Int. Conference on Robotics and Automation, San Diego, California, May 8-13, 1994

[2] G. Hirzinger, J. Butterfass, S. Knoch, H. Liu, DLR's multisensory articulated hand, ISER'95 Fifth International Symposium on Experimental Robotics, Barcelona, Catalonia, June 15-18, 1997

[3] G. Hirzinger, A. Baader, R. Koeppe, M. Schedl, "Towards a new generation of multisensory light-weight robots with learning capabilities", Prod. IFAC 1993 World Congress, Sydney, Australia, 1993.

[4] B. Gombert, G. Hirzinger, G. Plank, M. Schedl, „Modular concepts for a new generation of light weight robots", Prod. IEEE Int. Conf. on Industrial Elektronics, Instrumentation and Control (IECON), Bologna, Italy, 1994

[5] B. Brunner, K. Landzettel, B.M. Steinmetz, and G. Hirzinger, „Tele Sensor Programming - A task-directed programming approach for sensor-based space robots", Proc. ICAR'95 7th International Conference on Advanced Robotics, Sant Feliu de Guixols, Catalonia, Spain, 1995.

[6] G. Grunwald and G.D. Hager, „Towards Task-Directed Planning of Cooperating Sensors", Proc. SPIE OE/Technology: Sensor Fusion V, Conf. 1828,, Boston, Mass., 1992.

[7] V. Senft, G. Hirzinger, „Redundant Motions of Non Redundant Robots - A New Approach to Singularity Treatment", 1995 IEEE Int. Conf. on Robotics and Automation ICRA'95, Nagoya, Japan, vol 2, pp. 1553-1558, 1995

[8] J. Shi, Y. Lu, „Chatter Free Variable Structure Perturbarion Estimator on the Torque Control of Flexible Robot Joints with Disturbance and

Parametric Uncertainties", Proc. IEEE Conf. on Industrial Electronics, Control and Instrumentation, IECON'96, Taipei, 1996

[9] G. Wei, G. Hirzinger, B. Brunner, „Sensorimotion coordination and sensor fusion by neural networks", Proc. IEEE Int. Conf. on Neural Networks, San Francisco, USA, 1993

[10] P. Wunsch, G. Hirzinger, „Real-time visual tracking of 3D-objects with dynamic handling of occlusion", IEEE Int. Conf. on Robotics and Automation, Albuquerque, NM, 1997

[11] P. Wunsch, G. Hirzinger, „Registration of CAD-models to images by iterative inverse perspective matching", ICPR'96 Int. Conf. on Pattern Recognition, Vienna, Austria, 1996

[12] G.Q. Wei, G. Hirzinger, „Learning shape from shading by a multilayer network", IEEE Transactions on Neural Networks, vol. 7, pp. 985-995, 1996

[13] G.Q. Wei, K. Arbter, G. Hirzinger, „Real-time visual servoing for laparoscopic surgery", IEEE Engineering in Medicine and Biology, vol. 16, 1997

[14] P.v.d.Smagt, G. Hirzinger, „Local minima in ill-conditioned feed-forward neural networks", Workshop on ANNs and Continuous Optimization at NIPS 1996", Snowmass, Colorado, 1996

[15] P.v.d.Smagt, G. Hirzinger, „Optimization in feed-forward neural networks: On conjugate gradient, network size, and local minima", Workshop on Tricks of the Trade at NIPS, Snowmass, Colorado, 1996

[16] A. Baader, G. Hirzinger, „A self-organizing algorithm for multisensory surface reconstruction", Proc. IROS'94 IEEE Int. Conf. on Intelligent Robots and Systems, München, 1994

[17] F. Lange, G. Hirzinger, „Learning of a controller for non-recurring fast movements", Advanced Robotics, pp. 229-244, 1996

[18] M.W. Walker and R.P.C. Paul, „Resolved-Acceleration Control of Mechanical Manipulators", IEEE Trans. Automatic Control, vol. 25, pp. 468-474, 1980

[19] O. Khatib, „A Unified Approach for Motion and Force Control of Robot Manipulators: The Operational Space Formulation", IEEE J. Robotics and Automation, vol. 3, pp.43-53, 1987.

[20] E. D. Dickmanns and V. Graefe. Dynamic monocular machine vision. Machine Vision and Applications, 1:223-240,1988.

[21] R. Koeppe, G. Hirzinger, „Learning Compliant Motions by Task-Demonstration in Virtual Environments", Proc. of the Fourth International Symposium on Experimental Robotics, ISER'95, Stanford, CA, pp. 188-193, 1995.

[22] R. Koeppe, A. Breidenbach, and G. Hirzinger, "Skill representation and acquisition of compliant motions using a teach device," presented at Proc. IEEE/RSJ Int. Conference on Intelligent Robots and Systems IROS'96, Osaka, Japan, 1996.

[23] F. Arai, L. Rong, and T. Fukuda. Asymptotic convergence of feedback error learning method and improvement of learning speed. In IEEE Int. Conf. on Intelligent Robots and Systems (IROS) , Yokohama, Japan, July 26-30 1993

[24] F. Lange and G. Hirzinger, "Application of multilayer perceptrons to decouple the dynamical behaviour of robot links," presented at Proc. International Conference on Artificial Neural Networks, ICANN'95, Paris, France, 1995.

[25] E. Ralli and G. Hirzinger, "Fast path planning for 6-dof robot manipulators in static or partially changing environments," presented at Proc. ICAR'95 7th International Conference on Advanced Robotics, Sant Feliu de Guixols, Catalonia, Spain, 1995.

PART 6
NEW COMPONENTS
SESSION SUMMARY

Shigeo Hirose
Tokyo Institute of Technology
Department of Mechano-Aerospace Engineering
2-12-1 Ookayama Meguro-ku, Tokyo 152 Japan
e-mail: hirose @ mes.titech.ac.jp

Robots consist of multiple components and in order to realize a good robotic system we first have to strive to introduce innovative components. All the papers presented in the New Components session challenged this fundamental and important task. The papers presented were the following three:

1. Takeo Omichi, Shigetoshi Shiotani, and Reizo Miyauchi; The design of a serial communication link for built-in servo driver and sensors in a robot

2. Shree K. Nayar; Omni-directional Vision

3. Kurt Konolige; Small Vision System: Hardware and Implementation

The first paper by Takeo Omichi presents their trials to introduce standardized hardware for communication links of intelligent robot systems. It consists of AC servomotors, servo drivers, sensors, power supply links and circuit-saving-type communication links. The important point of this paper is that it is motivated through real experiences of developing more than 100 robot models at Mitsubishi Heavy Industry. Therefore, in all parts of the system design, we observe practical viewpoints typical of robot designers. They try to simplify the communication protocol for the ordinary robot engineer while opening the system structure, down sizing the system by reducing the wire numbers and introducing easy to use compact PC cards. Furthermore, they paid special attention to eliminate noise in compact systems that occurs in the combination of information and power lines. In short, this paper can be regarded as one of the trials to realize a "Mechatronic IC", where mechanical components are integrated together with an electronics system, for easy to use robotic components.

The second paper by Shree K. Nayar introduced a newly developed practical omni-directional vision. Nayar first reviewed former studies on omni-directional vision and then he pointed out that most of the former studies intended to introduce a single viewpoint on image sensors to simplify the mapping from image coordinates to world coordinates. While to the contrary, Nayar' s omni-directional vision makes the single viewpoint image first by using the paraboloidal mirror and then the reflected image is sensed by video camera with telecentric lens. As is often the case with practical systems, Nayar' s omni-directional vision is very simple in principle and construction but is apparently very effective. The device will be widely used for surveillance and navigation of mobile robot in the near future.

The third paper by Kurt Konolige also presented an extremely practical vision, a compact and easy to use stereovision system called "Small Vision Module" , or SVM. As SVM is extremely compact and self sufficient, we can expect that installation of the SVM into a robot system will directly enhance its sensing capability. It is reported that the SVM makes area correlation in stereo with the characteristics of LOG transform, L1 norm (absolute difference) correlation, 16-disparity search, and post-filtering.

I was extremely pleased to be the chairman of this session with such important and exciting papers. I also sincerely hope that the results of these researches will soon be commercialized and everyone in the robotic community can be permitted to use them freely for their robotic systems.

The Design of a Serial Communication Link for Built-in Servo Driver and Sensors in a Robot

Takeo Omichi, Shigetoshi Shiotani, Reizo Miyauchi
Takasago Research and Development Center, Mitsubishi Heavy Industries, Ltd.
2-1-1 Shinhama Arai-cho, Takasago, 676 (Japan)
e-mail: oomichi@trdc.mhi.co.jp

Abstract

The serial bus communication link was developed for servo drivers built in an AC servo motor, which has large share for a motor, upper control equipment, and a driver. This serial bus consist of two layers: the main bus that has three circuits based on the packet transmission formula and the local bus, made of a single circuit based on the polling method. It was confirmed that this development would make the external driver for a robot unnecessary, allow the facilitation and reliability enhancement of internal wiring, and thereby contribute toward the design of high-function robots.

1 Introduction

An actuator with a servo motor/driver has not yet materialized, although it has been a long-time dream of robot designers. The problem in meeting this requirement is that, because the servo driver is larger than a motor or equivalent in size, its installation as it is would enlarge the robot itself. Accordingly, placing the driver outside the manipulator, for instance, has been continually adopted. The shortcoming of this method is that a number of electric wires for the motor and servo driver pass through the robot, necessitating steps to keep smooth path at joints of the robot and provide for possible wire breaking. As such, this is a factor for the increase of robot costs.

Meanwhile, we proposed the open system architecture of intelligent robots ahead of others, engaging in the planning and development of the downsizing of servo drivers[1)] and circuit-saving of robot wiring amid progressing trends toward hierarchy and modules. The downsizing of servo drivers depends on the application of the generalized servo control theory, conversion of the control device into a single chip, lower heat generation of the power unit, and efficiency improvement of heat radiation, which can be attained through implementation of accumulated design ideas. Because the reliance of motors on hardware is substantial, we strove to attain steady improvements. Also, circuit-saving in the robot mechanism can be considered to be based on shifting to serial communications, but the current real-time communication method has merits and demerits of its own, as such, various ideas are necessary to cover a wide range of robots, which require diversified control. Fortunately, we have executed the development and practical use of more than 100 models of robots, and based on this experience, we judged that it was possible to extract more general communication protocols or driver control methods by classifying and analyzing common and special factors of such robot development.

Since we designed the circuit-saving-type serial link within robot mechanisms, judged to be highly practical, we hereby report on the outline of the design.

2 Standard Hardware Design of Intelligent Robots (SDIR)

Fig.1 shows the standard function module of an intelligent robot. As can be seen in Fig.1, the design of an intelligent robot requires technologies from a variety of realms, and it is not easy for a single designer to conceive and design all of robot architecture. For this reason, it becomes necessary to lighten the burden of the designer by stratifying functions of a robot and opening their

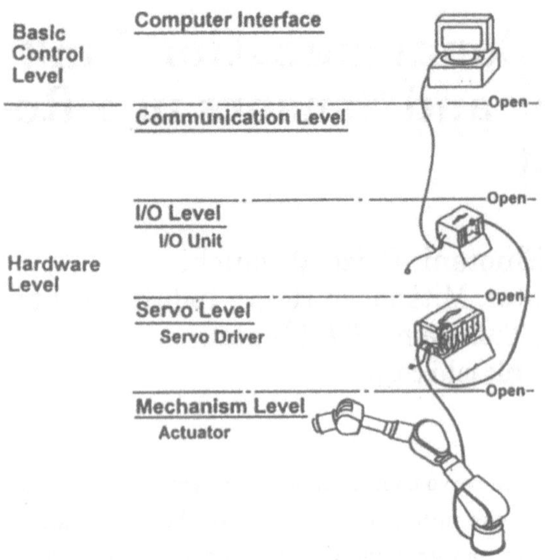

Fig. 1 Hierarchical Opened System on Hardware

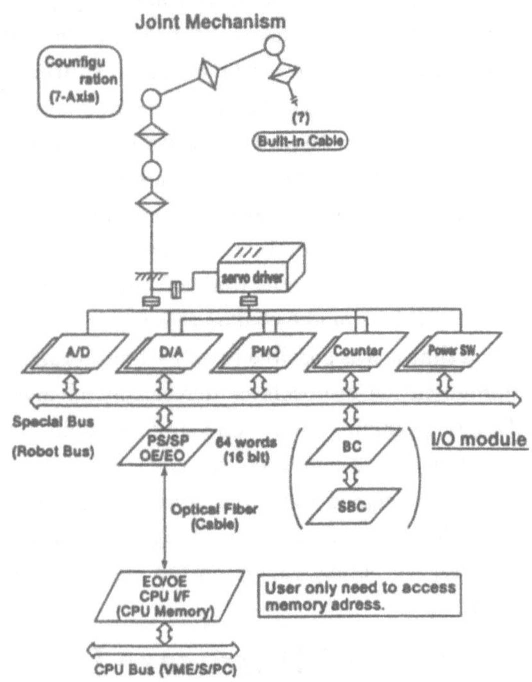

Fig. 2 Manipulator System by SDIR

interfaces. This is also true regarding hardware, so we carried out trial design shown in Fig.2. In this design, functions are stratified, while input/output related to sensors are handled as mere data groups in the communication system. Depending on the functions of hardware to be connected, therefore, no changes of upper-rank hardware for the communication system are required. However, because the communication system in Fig.2 is an interface board placed externally,

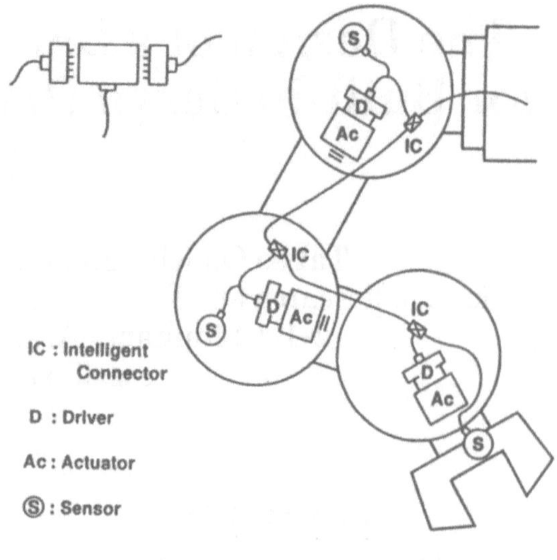

Fig. 3 Concept of New SDIR

parallel wiring is required within the robot body. Therefore, the aforementioned problem has not been solved.

Fig.3 shows our attempt at seeking circuit-saving with consideration given to such thinking. With the actuator driver downsized and placed near the motor, the interface device is designed to have a similar constitution. Here, the apparent composition of Fig.2 and Fig.3 may seem different, but the interface from the control unit only requires modification of similar addresses, and therefore the current program can be applied as it is. In this kind of development, ideas for standardization and opening the system are quite important, and they are considered to be essential for technological development that cover a wide rang of realms.

3 Basic concept for circuit-saving

Communications system are essential for circuit-saving. Regarding communications, various methods have so far been announced, and systematization, including stratification, has been deeply discussed. As such, the use of these methods may be one approach to the solution of the problem. Even so, these communication systems are not specifically designed with robots control communication. In other words, they are certainly well suited for the transmission of various kinds of information, but generally they are considered to be:

(1) Too difficult for ordinary robot engineers.

(2) So designed as to have much logic to deal with a variety of situations, making it difficult for the simultaneous attainment of the downsizing of devices and securing of the real-time basis.

Accordingly, while giving consideration to ease in achieving circuit-saving in the course of design, we set the following as the basic principles:

(1) Reducing wire number is possible for both communication and power supplying lines.
(2) The designer of upper control system doesn't need the knowledge of communications.
(3) Even if knowledge is necessary, application of the communication system can be easily designed, based on a quite simple protocol.
(4) Reliability and extensibility applicable to robots are ensured.

For the realization of these principles, we intended to conduct designing, based on the following thinking:

(1) The non-control-type protocol should be the basic rule.
(2) To seek the co-existence of the non-control formula and the real-time robot control, communication speed must be as fast as possible.
(3) Simplification of the communication protocol should be sought.
(4) For improvement of reliability, (a) The main communication system should have a redundancy system. (b) As a results, an increase in communication speed can be expected. (c) Simple wire-breaking and grounding countermeasures should be incorporated in the protocol.
(5) The communication system should have simple information to maintain and support the real-time control of the system.
(6) As for linkage with the upper control system, a package data input/output formula should be adopted.

4 Basic concept of the communication system

4.1 System design of communication system

The system concept, designed in accordance with the basic thinking, is shown in Fig.4. As the robot controller, we assumed a personal computer. Regarding interface with the communication

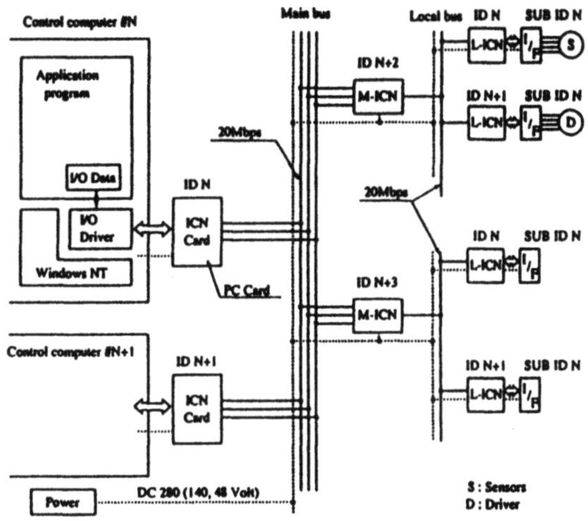

Fig. 4 Communication system concept

system, we adopted a PC card (called ICN CARD) in consideration of the need to be compatible with various kinds of computer hardware. The communication system has a two-tiered composition, made up of a multiplex system (3 lines) main bus and a local bus designed to be connected with the main bus, and a speed of about 20 Mbps is considered for the line of each bus. The interface module called M-ICN for the connection of the main bus and the local bus, servo driver, sensor, and the interface module called L-ICN for linkage with the local bus comprise the core of this system, and we intended to realize downsizing and higher speed by containing these functions in a single chip. On the ICN CARD, the M-ICN will be mounted to effect communications between M-ICNs.

Communications between M-ICNs are implemented based on interactive communication protocol, and communication is carried out using an empty line of the three lines. Communications between the M-ICN and the L-ICN are executed based on polling from M-ICN, with output data and input data always paired. The number of channels that can be handled in each ICN is:

(1) Number of I/O for linkage with the L-ICN (Number of SUB ID) : 16 bit × 16 channels Output, 16 bit × 16 channels Input
(2) Number of L-ICNs (L-ICN ID) : 255
(3) Number of M-ICNs (M-ICN ID) + Number of ICN CARD (IC CARD ID) : 255

SUB ID basically has parallel input and output. As such, if its number is increased, the signal lines will grow. Therefore, the number was determined in consideration of practically mounted wiring.

The ID number of ICN is determined based on the

number of terminal information (sensor and driver), as well as transmission speed and control cycles time. Judging from the designed speed, 8 bit (about 255) should be the benchmark. Also, considering that the method of combining the M-ICN and L-ICN can be freely chosen. Because the interface between the M-ICN on the ICN CARD and the main bus is the same as the interface between the M-ICN and the main bus, the total of both can be set at 255. In the design of actual mounting, therefore, it will be better if the ICN combination with allowance for the necessary control cycle is considered.

4.2 Evaluation of functions

(1) Circuit-saving for both communication and power supplying lines

Circuit saving in these units should be materialized by producing power sources for individual devices out of the motor-driving power source line. However, if there are other modules nearby, the power source should be used jointly.

(2) Easy I/O and simplified protocol

Because a control-system engineer can do his work only by describing the I/O area in package within a program, he can work out a control program without knowing communication system arrangements, except for special cases. Fig.5 shows the simplest example. It suggests that, with regard to the control system, it is sufficient only by inputting or outputting, as needed, output data and input data composed of the M-ICN ID (8 bit), L-ICN ID (8 bit), SUB ID (4 bit), and data (16 bit).

(3) Reliability/extendibility

Improvement of reliability should be sought by multiplying the main bus, which must be able to execute longer-distance transmission. In this case, detection of the breaking of wires will be

data output zone

8 bit	8 bit	4 bit (in 8 bit)	16 bit
M-ICN ID,	L-ICN ID,	SUB ID,	output data
......	......	......	......
......	......	......	......

data input zone

M-ICN ID,	L-ICN ID,	SUB ID,	output data
......	......	......	......
......	......	......	......

time zone

input date	transmission time
output date	transmission time

maintenance zone

ICN CARD ID etc.

Fig. 5 I/O example of control program

necessary, but because physical checking is difficult, the identification mode of the breakage point should be added to the protocol in line with the checking of transmission data on the ICN CARD. Because signal transmission under the polling has to be adopted for the local bus, a delay time should be used for the send/receipt of data, and if the communication time is over, special codes should be registered in data. By observing these, it is possible to identify the corresponding wire breakage point or L-ICN that failed.

Also, based on the ID number of ICN, the physically responsive terminal I/O will become as follows:

$$2^8 \times 2^8 \times 2^4 = 2^{20} \quad \cdots\cdots (1)$$

In a general robot, the practical I/O number is considered to be about 2^8. In actual application, however, various application can be conceived.

(4) Non-control-type protocol and real-time communication

Between M-ICNs, transmission of packets bearing the M-ICN ID on the receiving side, as well as the M-ICN ID on the sending side, should be sent from the sending side on an empty line. On the receiving side, when the M-ICN recognizes transmitted ID as its own ID, this packet will be taken in. Furthermore, at the M-ICN on the transmission side, the M-ICN ID of the sending side should be replaced with the M-ICN ID of the receiving side, and the packet should be transmitted back, after the completion of the data processing to be described later. On this occasion, time for the transmission in addition to L-ICN ID, SUB ID, and data should be set in the packet. Because this arrangement helps identification of the time for the transmission of the packet, it allows interpolation of prior or subsequent data. Thus, when more accurate data sampling is necessary, this function can be utilized.

As explained above, it is possible to satisfy basic requirements of design on the basic system functions.

5 Device functions and their application

5.1 Transmission format

The transmission format is shown in Fig.6. Broadly speaking, related information consists of synchronized transmission information, data transmission/receipt information, and concluding

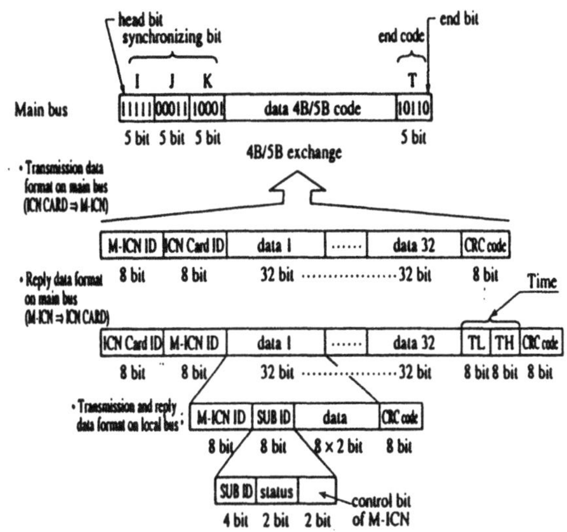

Fig. 6 Communication data format on the main bus

information. Here, the format for synchronized and concluding information is the same for the main bus and local bus, and only the format for the data transmission information is different.

(1) Data transmission/receipt signal format of the main bus

Transmission/receipt signals shall be composed of the transmitter, receiver M-ICN ID, data package (16 bit x 32 data max.), and CRC for transmission reliability. At the same time, the total frame shall be subject to 4 bit/5 bit exchanges. As such, the total frame length will be:

$$\left.\begin{array}{l} N_D \times (32+3\times 8)+20 = 56N_D+20 \text{ (from CARD)} \\ N_D \times (32+3\times 8)+20 = 56N_D+28 \text{ (to CARD)} \end{array}\right\} \cdots\cdots (2)$$

N_D : Number of data in the packet

(2) Data transmission/receipt signals of the local bus

The transmission data shall total 40 bit, made of L-ICN ID, SUB ID, and data CRC code. Because information for communication control information is added, the frame length will be:

$$(N_{DL}+8)\times\frac{5}{4}+20 = 70\text{bits / frame} \cdots\cdots (3)$$

N_{DL} : Local bus transmission data (excluding CRC) = 32 bit

Here, 8 bits shall be allotted to the SUB address because of ease in data processing. These bits shall be divided, and the 4 upper-rank bits shall be used for the sub address, and 2 following-rank bits shall be used for data status (effective / ineffective, transmission / receipt classification). The 2 lowest rank bits on the local line shall be empty bits.

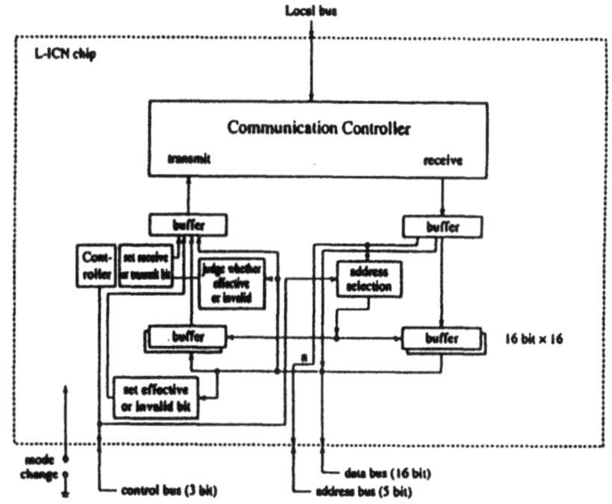

Fig. 7 Function block diagram of L-ICN

5.2 L-ICN

Functions of the L-ICN are:

(1) Interface with the local bus
(2) Receiving data from the local bus and transmitting data of the I/O unit
(3) Interface with I/O unit (CPU bus connection)
(4) Interface with I/O (direct connection)

As shown in Fig.7, the L-ICN has a buffer amounting to the SUB ID number (16 channels x 16 bits). Signals from the local bus shall be written in the receiving buffer, and after the completion of the write operation, data in the transmission buffer shall be sent back. At this time, the data shall be made effective if there is a corresponding input signal from I/O and made ineffective there isn't. Because this processing has a priority over access on the I/O side, it allows maintaining the speed of transmission without being affected by the processing speed of the I/O side. On the I/O side, the command value is issued from the receiving buffer at the possible timing, and the input value is entered in the transmission buffer (Fig.7). If the direct input of signals from the A/D converter etc., is desired, attach a chip selector to the A/D converter, etc., thereby directly controlling output and input.

The ICN logic unit shall be turned into one chip and contained in a small-sized package measuring 9 x 9 mm. Through this step, the downsizing of the whole L-ICN module unit shall be sought. Fig.8 shows the example of I/O that adopts an L-ICN chip. Further, because the transmission

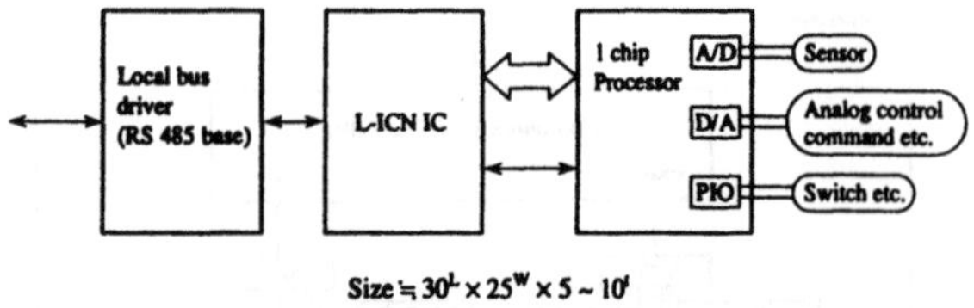

Fig. 8 L-ICN application example using 1 chip processor with I/O Port

and receipt are carried out at the same ratio, the arrangement will help improve the input and output process of data, while it will be possible to send the status communication monitoring data on the I/O side without executing new transmission / receipt by utilizing an empty buffer.

5.3 M-ICN

As shown in Fig.9, the M-ICN has the following functions:

(1) Communications between M-ICNs (empty line detection and transmission/receipt of data)

(2) Input/output of polling data about the L-ICN and during the time-over, corresponding data shall be made full-bit data.

(3) Parallel data input/output functions for ICN CARD.

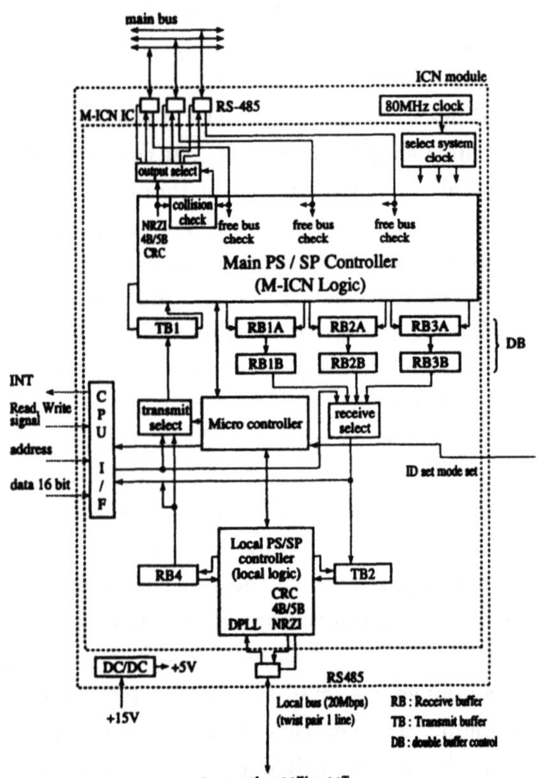

Fig. 9 Function block diagram of M-ICN

(4) Function of searching main bus wire-breakage points.

For each bus line, a transmission driver (high-speed RS485) shall be employed for the improvement of transmission reliability, as well as the securing of a certain distance (approximately 15m). If the common use of the power source unit is considered, it is desirable to arrange the common utilization of the 15V power source by placing the M-ICN near another module.

5.4 ICN CARD

An ICN CARD has the following four functions:

(1) Communication with the M-ICN

(2) Temporary storage of M-ICN communication information in the memory

(3) Interface between the memory and the computer bus

(4) Monitoring communication condition

By studying the use of a PC CARD in the linkage with the computer bus, consideration was given to the connection with notebook-type computers. In addition, based on the use of a dual-port memory as the storage memory, the M-ICN IC side and the computer bus side were separated in hardware. We considered system development into a more-diversified computer bus (without any need for changing the hardware/software design on the M-ICN IC side), based on these arrangements.

The processing between the M-ICN IC and the memory basically represents the function of sorting, so it is not difficult. However, choice by users is left, based on the following conditions:

(1) Data transmission, without reliance on the order of data, shall be made possible.

(2) If ICN IDs or SUB IDs are put in defined order, higher speed transmission will be-

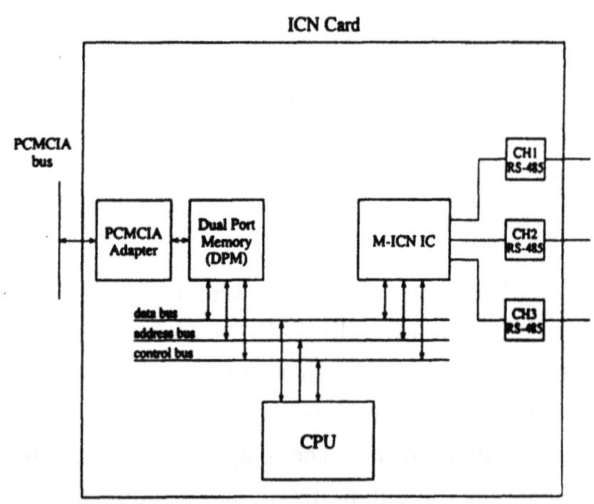

Fig. 10 Function block diagram of ICN CARD

come possible.

With regard to the communication monitoring function, we basically studied an arrangement for making it unnecessary by software but we determined that the ICN CARD should have the functions of confirming transmission and receipt to check the breaking of the bus line, of starting up the wire breakage line identification mode on the M-ICN, and of inputting information for prohibiting the use of the breaking line with the M-ICN. Fig.10 is a functional block diagram of the ICN CARD.

5.5 Transmission time

The transmission speed naturally deepens on the number of data. In the case of an ordinary robot, the speed can presumably be covered with one transmission data packet (output 32, input 32) of the main bus. Therefore, the time required for the input/output of the time for 32 channels is evaluated as follows:

(1) Local bus (Speed S_L = 20 Mbps)

$$t_L = \frac{N_{bL}}{S_L} \times 2 \times N_D + N_D \times t_{pL} = 360\mu\text{sec} \quad \cdots \text{(4)}$$

N_{bL} (the frame length of the local bus) = 70 bits

N_D (number of data) = 32

t_{pL} (time of L-ICN-based processing) < 5 μsec

(2) M-ICN polling

$$t_{MP} = t_{MP} \times 2 \times N_D = 160\mu\text{sec} \quad \cdots\cdots\cdots\cdots \text{(5)}$$

t_{MP} (polling processing time) = 2.5μsec/data

(3) M-ICN to M-ICN (ICN CARD)

Based on the formula (2)

$$t_{MM} = \frac{(56N_D + 20) + (56N_D + 28)}{S_L} = 170\mu\text{sec} \text{ (6)}$$

Accordingly, the time T_A required for the transmission system and transmission time t_a for one both of data shall be :

$$T_A = t_L + t_{MP} + t_{MM} \fallingdotseq 700m\sec/32data \quad \cdots \text{(7)}$$

$$t_a = \frac{t_A}{N_D} = 22\mu\text{sec/data I/O} \quad \cdots\cdots\cdots\cdots \text{(8)}$$

Further, processing time will be added to the above on the ICN CARD. Because this time depends on the kind of processor and program formula to be employed, it is impossible to make unified description. Even so, since it is possible to consider it to be blow 100 μsec, it is considered that the required time in this system is 25 μsec per data I/O. This, it is judged that a fairly wide range of application is possible.

6 Power source

Problems related to the power source voltages increase in the number of power splaying line wires, depending on the type of voltage and the disturbance of noise through power lines. The best way to solve these problems is to use completely isolate power sources for each ICN module. Generally speaking, however, the voltage of motor-driving power is greatly higher than that of the IC-driving voltage (+5V). As such, the design of isolation coils is difficult, and downsizing also poses problems (the height becomes larger). For this reason, we propose the following design concept.

The motor-driving power is used as the primary power source. In view of the generally used AC servo motor, the voltage is about 280V or 140V DC in Japan, but the level of about 48V will be considered, depending on each case. As for the power source, the lower the voltage is, the smaller the power unit can be made. On the other hand, shifting to lower voltage results in an increase in electric current, necessitating the thickening of power wires. Due to consistency with the design trade-off, therefore, the level of 140V is preferable. The upsizing of the ICN should be prevented by conceiving the isolation power system (secondary power source) of the highest voltage required for the signal system and by installing a suitable number of secondary power sources that cannot be downsized. Assuming the PWM driving of power elements for motor driver use, ±15V are considered as secondary power voltage.

Incorporation of ICN power sources (+5V, isolation type) in each ICN module shall be the basic rule. The secondary power source of +5V system will be used, however, if this is unavoidable because of placing priority on smaller sizes.

This design image is shown in Fig.11. It is possible to put overall wires in order by locally increas-

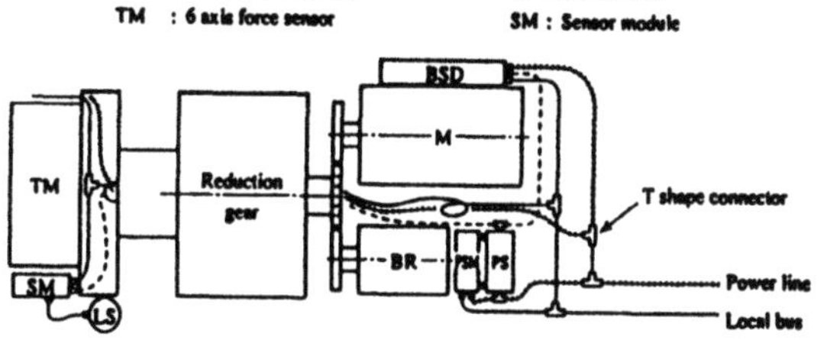

Fig. 11 Layout example of power supply module

ing the number of power wires. Also, noise resistance is expected to grow because of the adoption of isolation power sources for each module.

7 ICN application example

7.1 Servo motor driver

A representative linkage method is shown in Fig.12, although the method of linkage with the L-ICN differs, depending on the servo driver I/O. Because the system shown in Fig.12 employs microprocessors, it is basically the same as the system in Fig.8.

If output and input data number are the same, necessary data should be written as they are. If the number of input data is larger than that of output data, virtual data should be written as output data to the output SUB ID buffer concerned and output as ineffective data. If the number of output data is larger than that of input data virtual data will be transmitted as ineffective data from the L-ICN side to the input SUB ID concerned. As such, discard the transmitted input data, and after setting the SUB ID value concerned, output it as effective data.

7.2 Power Switch Module

Fig.13 shows the basic block diagram for a power switch module. In this case, we considered combining the L-ICN, 1-chip microprocessor, and the MOSFET, and turning on-off power switches, such as brakes. Here, because various kinds of power sources can be conceived as power sources to be used for brakes, voltage of PWM output will be produced with the processor, thereby seeking further standardization of the module. Thus, by using the ICN, it is possible to produce a variety of I/O families.

7.3 Application for autonomous guided vehicle

Fig.14(a) represents a research and development platform vehicle that has the ability to travel to all directions. Because the kind and place of appli-

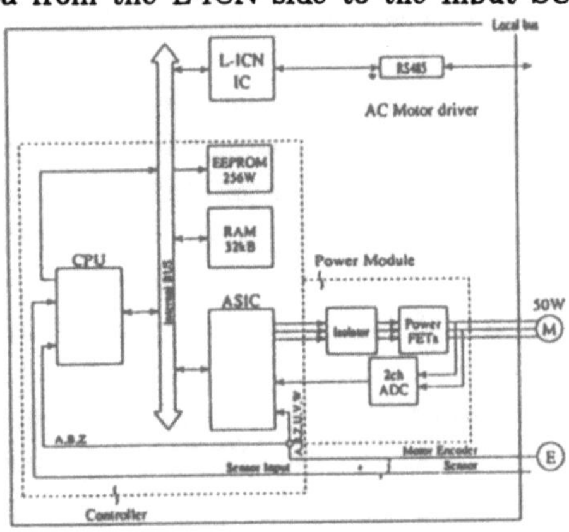

Fig. 12 Application for the servo motor driver

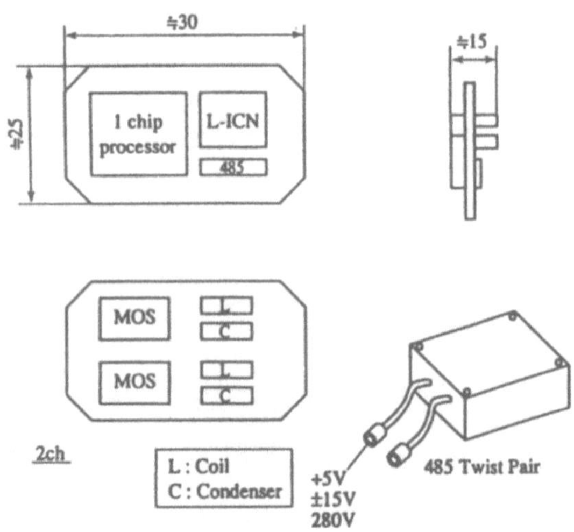

Fig. 13 Power switch module image

(a) Vehicle outview

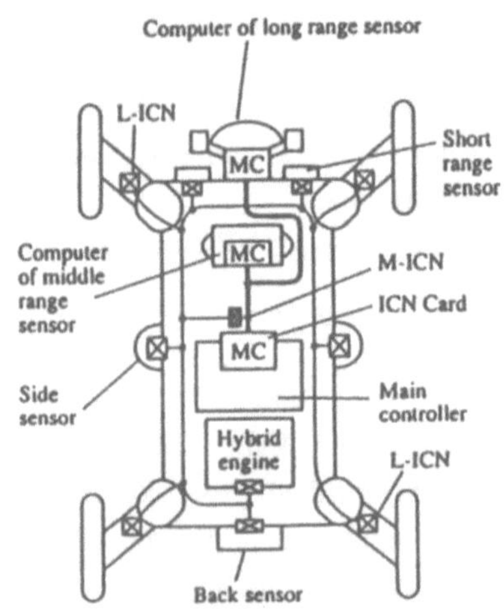

(b) Layout of the ICNs

Fig. 14 Omnidirectional vehicle designed by ICN system

cable sensors differ, depending on research purposes, and because the vehicle has much freedom, a great deal of effort is required for circuit wires design. Also, it was presumed that problems stemming from circuit wiring, such as mutual noise arising from vast wiring and breaking of wires, cannot be disregarded.

This being the case, we designed to install the L-ICN modules in the steering wheel unit and the sensor module unit, introducing this circuit-saving system. As a result, each function was completely turned into modules. This is expected to shorten the time for disassembly and assembly and to facilitate addition or expansion of vehicle function devices on an ICN basis. If the high-speed communications is utilized, in addition, data exchanges between individual modules can be executed in a blackboard method which does not need interrupting command for communication, so the simplification of the control system management program can be expected at the same time.

7.4 Application for a Manipulator

Fig.15 shows an example of introducing the ICN system into the manipulator manufactured based on the first standard design (SDIR 1: parallel wiring). The effect of wiring simplification is the same as that for vehicles, but in addition, the hardware independence of the manipulator (through the elimination of the driver unit) has greatly multiplied and improved the combination with other equipment.

Fig. 15 Manipulator designed by ICN system

8 Conclusion

(1) We proposed a circuit-saving system in the robot mechanism combining the main bus, composed of three redundant circuits, and the local bus, made of one polling-type circuit.

(2) A communication system that can be contained in the robot body was realized by turning the computer interface unit, main/local bus connection unit, and the communication logic unit for the local bus and I/O connection portions into a single chip.

(3) We see the prospect for applying the system to the input/output use of ordinary robots by ensuring high-speed characteristics based on shifting to IC and hardware.

(4) We evaluated the practical usability of the system by promoting the design that has introduced the ICN system.

9 Remarks and acknowledgments

As part of the hardware standardization design to promptly manufacture the hardware of intelligent robots, we proposed a circuit-saving system for the interior of a robot mechanism and evaluated the practical usability of the system based on the downsizing of devices. Standardization design can be an effective means of hardware manufacturing by itself, but the introduction of this method will enable complete independence at the functional module level of robots. This is considered to make much tasks previously required for robot function changes or addition unnecessary and to phenomenally help promote wider development of the research or improve its efficiency. In research fields of robots, proposed design and developed modules exactly matches the platform supply-type development suggested recently. As such, we are sure that it can make a great deal of contribution to robotics research and development .

However, the progress of electronic devices is unceasing, and further function and speed upgrading in line with such a progress is possible. In this perspective, it will also be necessary in the future to continue improvements by introducing this concept at an early stage and adopting various opinions based on diversified usage methods in a wide range of robotics fields.

For this reason, we earnestly hope for the application of the concept to various cases and would like to hear criticisms.

Finally, some portions of this research were enforced as part of the international standard creation research under the Ministry of International Trade and Industry, and we deeply appreciate research task force members who provided much cooperation in the formulation of specifications, etc.

10 References

1) Omichi, Ibe, Kawauchi, Hamana, and Isozaki: "Hardware Standard" Design Methods with Open System Architecture for Intelligent Robot" : Journal of Robotics Society of Japan, Vol. 14, No. 2, pp. 225-233, 1996.

2) T. Omichi, T. Ibe, and T. Susaki: "Hardware Design Methods and Proposals for Open System Architecture of Intelligent Robots," Robotics Research, Springer, pp. 626-633, 1996.

Omnidirectional Vision

Shree K. Nayar
Dept. of Computer Science, Columbia University, New York City, New York 10027, (USA)
email: nayar@cs.columbia.edu

Abstract

Conventional video cameras have limited fields of view that make them restrictive in a variety of vision applications. There are several ways to enhance the field of view of an imaging system. However, the entire imaging system must have a single effective viewpoint to enable the generation of pure perspective images from a sensed image. A camera with a hemispherical field of view is presented. Two such cameras can be placed back-to-back, without violating the single viewpoint constraint, to arrive at a truly omnidirectional sensor. The implications of such a sensor for computational vision are explored. Results are presented on the software generation of pure perspective images from an omnidirectional image, given any user-selected viewing direction and magnification. The paper concludes with a discussion on the spatial resolution of the proposed camera.

1 Introduction

Conventional imaging systems are quite limited in their field of view. Is it feasible to devise a video camera that can, at any instant in time, "see" in all directions? Such an *omnidirectional* camera would have an impact on a variety of applications, including autonomous navigation, video surveillance, video conferencing, virtual reality and site modelling.

Our approach to omnidirectional image sensing is to incorporate reflecting surfaces (mirrors) into conventional imaging systems. This is what we refer to as *catadioptric* image formation (Nayar 1997). There are a few existing implementations that are based on this approach to image sensing (see (Rees 1970), (Charles 1987), (Nayar 1988), (Yagi and Kawato 1990), (Hong 1991), (Goshtasby and Gruver 1993), (Yamazawa, Yagi, and Yachida 1995), (Bogner 1995), (Murphy 1995), (Nalwa 1996), (Southwell, Basu, Fiala, and Reyda 1996), (Nayar 1997) and (Nene and Nayar 1998)). As noted in (Rees 1970), (Yamazawa, Yagi, and Yachida 1995) and (Nalwa 1996), in order to compute pure perspective images from a wide-angle image, the catadioptric imaging system must have a single center of projection (viewpoint). In (Nayar and Baker 1997) (Baker and Nayar 1998), the complete class of catadioptric systems that satisfy the single viewpoint constraint is derived. Since we are interested in the development of a practical omnidirectional camera, two additional conditions are imposed. First, the camera should be easy to implement and calibrate. Second, the mapping from world coordinates to image coordinates must be simple enough to permit fast computation of perspective and panoramic images.

We begin by reviewing the state-of-the-art in wide-angle imaging and discuss the merits and drawbacks of existing approaches. Next, we present an omnidirectional video camera (Nayar 1997) that satisfies the single viewpoint constraint, is easy to implement, and produces images that are efficient to manipulate. We have implemented several prototypes of the proposed camera, each one designed to meet the requirements of a specific application. The single viewpoint constraint allows us to compute, distortion-free (perspective) images of the scene for any user-selected viewing direction, focal length and image size. Results on the mapping of omnidirectional images to perspective ones are presented. The resolution and enhancement of the acquired omnidirectional and computed perspective images are discussed. In (Peri and Nayar 1997), a software system is described that generates a large number of perspective and panoramic video streams from an omnidirectional video input. The proposed omnidirectional camera has been used for robust computation of egomotion (Gluckman and Nayar 1998). Also, planar and curved mirrors have been used to develop compact binocular stereo systems (Nene and Nayar 1998).

2 Single Viewpoint

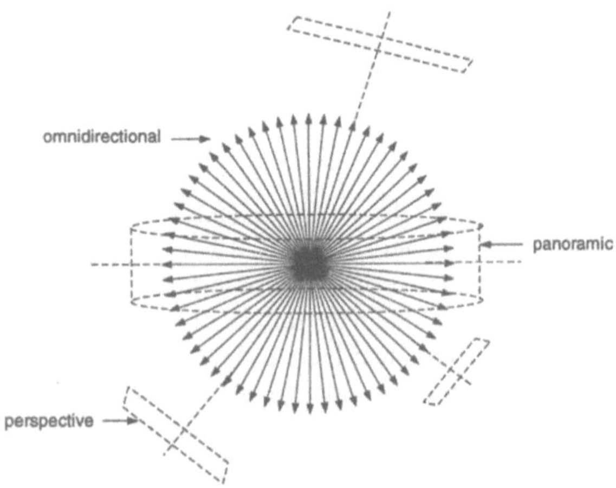

Figure 1: A truly omnidirectional image sensor views the world through an entire "sphere of view" as seen from its center of projection. The single viewpoint permits the construction of pure perspective images (computed by planar projection) or a panoramic image (computed by cylindrical projection). Panoramic sensors are not equivalent to omnidirectional sensors as they are omnidirectional only in one of the two angular dimensions.

The merits of having a single center of projection (viewpoint) have been emphasized by Rees (Rees 1970), Yamazawa et al. (Yamazawa, Yagi, and Yachida 1995) and Nalwa (Nalwa 1996). Consider an image acquired by a sensor that can view the world in all directions from a single effective pinhole (see Figure 1). From such an omnidirectional image, pure perspective images can be constructed by mapping sensed brightness values onto a plane placed at any distance (effective focal length) from the viewpoint, as shown in Figure 1. Any image computed in this manner preserves linear perspective geometry. Images that adhere to perspective projection are desirable from two standpoints; they are consistent with the way we are used to seeing images, and they lend themselves to further processing by the large body of work in computational vision that assumes linear perspective projection. Needless to say, a single viewpoint omnidirectional image facilitates the construction of not only perspective but also panoramic images by mapping the sensed data onto a cylinder, as shown in Figure 1.

3 State of the Art

Before we present our omnidirectional camera, a review of existing imaging systems that seek to achieve wide fields of view is in order. An excellent overview of some of the previous work can be found in (Nalwa 1996).

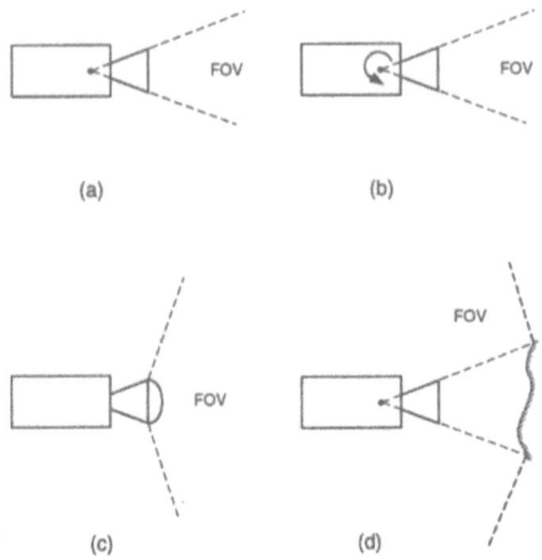

Figure 2: (a) A conventional imaging system and its limited field of view. A larger field of view may be obtained by (b) rotating the imaging system about its center of projection, (c) appending a fish-eye lens to the imaging system, and (d) imaging the scene through a mirror.

3.1 Traditional Imaging Systems

Most imaging systems in use today comprise of a video camera, or a photographic film camera, attached to a lens. The image projection model for most camera lenses is perspective with a single center of projection. Since the imaging device (CCD array, for instance) is of finite size and the camera lens occludes itself while receiving incoming rays, the lens typically has a limited field of view that corresponds to a small cone rather than a hemisphere (see Figure 2(a)). At first thought, it may appear that a large field can be sensed by packing together a number of cameras, each one pointing in a different direction. However, since the centers of projection reside inside their respective lenses, such a configuration proves infeasible.

3.2 Rotating Imaging Systems

An obvious solution is to rotate the entire imaging system about its center of projection, as shown in Figure 2(b). The sequence of images acquired by rotation are "stitched" together to obtain a panoramic view of the scene. Such an approach has been proposed by several investigators (see (Chen 1995), (McMillan and Bishop 1995), (Krishnan and Ahuja 1996), (Zheng and Tsuji 1990), for examples). Of these the most novel is the system developed by Krishnan and Ahuja (Krishnan and Ahuja 1996) which uses a camera with a non-frontal image detector to scan the world.

The first disadvantage of any rotating imaging system is that it requires the use of moving parts and precise positioning. A more serious drawback lies in the total time required to obtain an image with enhanced field of view. This restricts the use of rotating systems to static scenes and non-real-time applications.

3.3 Fish-Eye Lenses

An interesting approach to wide-angle imaging is based on the fish-eye lens (see (Wood 1906), (Slater 1932), (Miyamoto 1964)). Such a lens is used in place of a conventional camera lens and has a very short focal length that enables the camera to view objects within a hemisphere or more (see Figure 2(c)). The use of fish-eye lenses for wide-angle video has been advocated in (Oh and Hall 1987), (Kuban, Martin, Zimmermann, and Busico 1994) and (Xiong and Turkowski 1997), among others.

It turns out that it is difficult to design a fish-eye lens that ensures that all incoming principal rays intersect at a single point to yield a fixed viewpoint (see (Nalwa 1996) for details). This is indeed a problem with commercial fish-eye lenses, including, Nikon's Fisheye-Nikkor 8mm f/2.8 lens. In short, the acquired image does not permit the construction of distortion-free perspective images of the viewed scene (though constructed images may prove good enough for some visualization applications). In addition, to capture a hemispherical view, the fish-eye lens must be quite complex and large, and hence expensive. Furthermore, in our quest for a truly omnidirectional sensor, we are physically restricted in placing two fish-eye imaging systems back-to-back to image the complete sphere of view; the two viewpoint loci reside inside their respective lenses and hence cannot be brought close to one another. However, in applications where a single viewpoint is not critical, a back-to-back configuration such as the one implemented by Slater (Slater 1996) can be used.

3.4 Catadioptric Systems

As shown in Figure 2(d), a catadioptric imaging system uses a reflecting surface to enhance the field of view. However, the shape, position and orientation of the reflecting surface are related to the viewpoint and the field of view in a complex manner. While it is easy to construct a configuration which includes one or more mirrors that dramatically increase the field of view of the imaging system, it is harder to keep the effective viewpoint fixed in space. Examples of catadioptric image sensors can be found in (Rees 1970), (Charles 1987), (Yagi and Kawato 1990), (Hong 1991), (Yamazawa, Yagi, and Yachida 1995), (Bogner 1995), (Murphy 1995) and (Nalwa 1996). A recent theoretical result (Baker and Nayar 1998) reveals the complete class of catadioptric imaging systems that satisfy the single viewpoint constraint. This general solution has enabled us to evaluate the merits and drawbacks of previous implementations as well as suggest new ones (Baker and Nayar 1998).

Here, we will briefly review previous approaches. In (Yagi and Kawato 1990) and (Bogner 1995), a conical mirror is used in conjunction with a perspective lens. Though this provides a panoramic view, the single viewpoint constraint is not satisfied. The result is a viewpoint locus that hangs like a halo over the mirror. In (Hong 1991), (Bogner 1995) and (Murphy 1995), a spherical mirror was used with a perspective lens. Again, the result is a large locus of viewpoints rather than a single point. Hyperboloidal, paraboloidal and ellipsoidal mirrors have been used in the implementation of "all-sky" photographic cameras dating back to the late 1950's (examples can be found in (Charles 1987)). In most of these implementations, the single viewpoint constraint does not seem to have been a major consideration.

Rees (Rees 1970) appears to have been the first to use a hyperboloidal mirror with a perspective lens placed at its external focus to achieve a single viewpoint video camera system. A similar implementation was recently proposed in (Yamazawa, Yagi, and Yachida 1995). The hyperboloidal solution is a useful one. However, the sensor must be implemented and calibrated with care. More recently, Nalwa (Nalwa 1996) has proposed a panoramic sensor that includes four planar mirrors that form the faces of a pyramid. Four separate imaging systems are used, each one placed above one of the faces of the pyramid. The optical axes of the imaging systems and the angles made by the four planar faces are adjusted so that the four viewpoints produced by the planar mirrors coincide. The result is a sensor that has a single viewpoint and a panoramic field of view of approximately 360° × 50°.

4 Omnidirectional Camera

While all of the above approaches use mirrors placed in the view of perspective lenses, we approach the problem using an orthographic lens. It is easy to see that if image projection is orthographic rather than perspective, the geometrical mappings between the image, the mirror and the world are invariant to translations of the mirror with respect to the imaging system. Consequently, both calibration as well as the computation of perspective images is greatly simplified.

There are several ways to achieve orthographic projection. Most of these are described in (Nayar 1997). Here, we shall mention a few. The most obvious of these is to use commercially available telecentric lenses (Edmund 1996) that are designed to be orthographic. It has also been shown (Watanabe and Nayar 1996) that precise orthography can be achieved by simply placing an aperture (Kingslake 1983) at the back focal plane of an off-the-shelf lens. The approach that we have adopted in many of our implementations is the use of an inexpensive relay lens in front of an off-the-shelf perspective lens. The relay lens not only converts the imaging system to an orthographic one but can also be used to undo more subtle optical effects such as coma and astigmatism (Born and Wolf 1965) produced by curved mirrors.

Since orthographic projection is rotationally symmetric about the optical axis, all we need to determine is the cross-section $z(r)$ of the reflecting surface. The mirror is then the solid of revolution obtained by sweeping the cross-section about the axis of orthographic projection. A detailed derivation of the mirror shape for orthographic projection is given in (Nayar 1997). Not surprisingly, the mirror that guarantees a single viewpoint is a paraboloid with cross-section:

$$z = \frac{h^2 - r^2}{2h}, \tag{1}$$

where, $h > 0$ is the *parameter* of paraboloid.

Paraboloidal mirrors are frequently used to converge an incoming set of parallel rays at a single point (the focus), or to generate a collimated light source from a point source (placed at the focus). In both these cases, the paraboloid is a concave mirror that is reflective on its inner surface. In our case, the paraboloid is reflective on its outer surface (convex mirror); all incoming principle rays are orthographically reflected by the mirror but can be extended to intersect at its focus, which serves as the viewpoint. Note that a concave paraboloidal mirror can also be used. This solution is less desirable to us since incoming rays with large angles of incidence could be self-occluded by the mirror.

As shown in Figure 3, the parameter h of the paraboloid is its radius at $z = 0$. The distance between the vertex and the focus is $h/2$. Therefore, h determines the size of the paraboloid that, for any given orthographic lens system, can be chosen to maximize resolution.

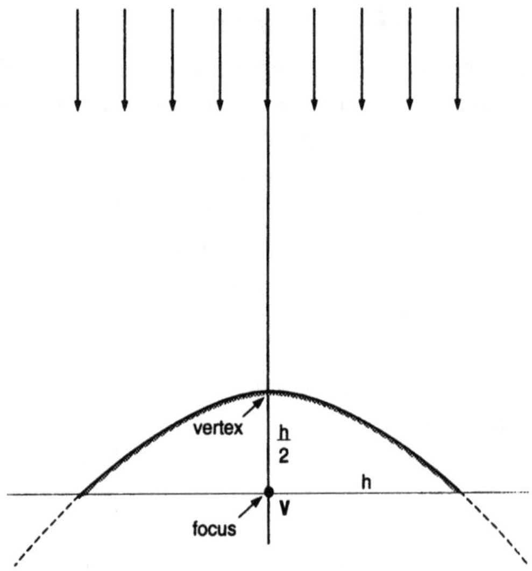

Figure 3: For orthographic projection, the solution is a paraboloid with the viewpoint located at the focus. Orthographic projection makes the geometric mappings between the image, the paraboloidal mirror and the world invariant to translations of the mirror. This greatly simplifies calibration and the computation of perspective images from paraboloidal ones.

5 Field of View

As the extent of the paraboloid increases, so does the field of view of the catadioptric sensor. It is not possible, however, to acquire the entire sphere of view since the paraboloid itself must occlude the world beneath it. This brings us to an interesting practical consideration: Where should the paraboloid be terminated? Note that

$$\left| \frac{dz}{dr} \right|_{z=0} = 1 . \tag{2}$$

Hence, if we cut the paraboloid at the plane $z = 0$, the field of view exactly equals the upper hemisphere (minus the solid angle subtended by the imaging system itself). If a field of view greater than a hemisphere is desired, the paraboloid can be terminated below the $z = 0$ plane. If only a panorama is of interest, an annular section of the paraboloid may be obtained by truncating it below and above the $z = 0$ plane.

In our prototypes, we have chosen to terminate the parabola at the $z = 0$ plane. This proves advantageous in applications in which the complete sphere of view is desired, as shown in Figure 4. Since the paraboloid is terminted at the focus, it is possible to place two identical catadioptric cameras back-to-back such that their foci (viewpoints) coincide. Thus, we have a truly omnidirectional sensor, one that is capable of acquiring an entire sphere of view at video rate.

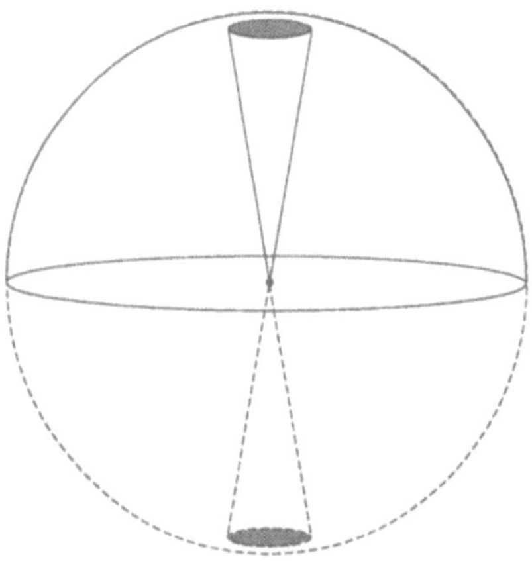

Figure 4: If the paraboloid is cut by the horizontal plane that passes through its focus, the field of view of the catadioptric system exactly equals the upper hemisphere. This allows us to place two catadioptric sensors back-to-back such that their foci (viewpoints) coincide. The result is a truly omnidirectional sensor that can acquire the entire sphere of view. The shaded regions are parts of the field of view where the sensor sees itself.

6 Implementation

Several versions of the proposed omnidirectional sensor have been built, each one geared towards a specific application. The applications we have in mind include video teleconferencing, remote surveillance and autonomous navigation. Figure 5 shows and details the different sensors and

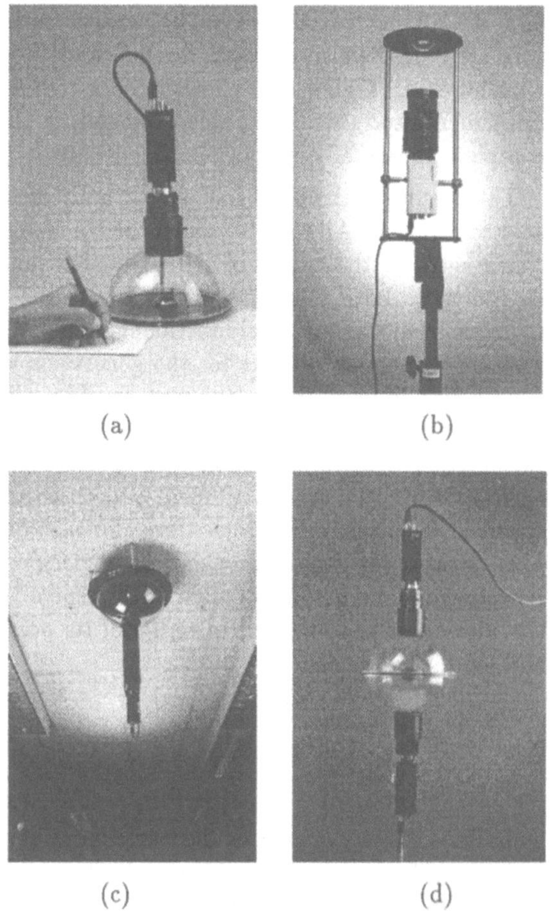

Figure 5: Four implementations of catadioptric omnidirectional video cameras that use paraboloidal mirrors. (a) This compact sensor for *teleconferencing* uses a 1.1 inch diameter paraboloidal mirror, a Panasonic GP-KR222 color camera, and Cosmicar/Pentax C6Z1218 zoom and close-up lenses to achieve orthography. The transparent spherical dome minimizes self-obstruction of the field of view. (b) This camera for *navigation* uses a 2.2 inch diameter mirror, a DXC-950 Sony color camera, and a Fujinon CVL-713 zoom lens. The base plate has an attachment that facilitates easy mounting on mobile platforms. (c) This sensor for *surveillance* uses a 1.6 inch diameter mirror, an Edmund Scientific 55mm F/2.8 telecentric (orthographic) lens and a Sony XR-77 black and white camera. The sensor is lightweight and suitable for mounting on ceilings and walls. (d) This sensor is a back-to-back configuration that enables it to sense the entire sphere of view. Each of its two units is identical to the sensor in (a).

their components. The basic components of all the sensors are the same; each one includes a paraboloidal mirror, an orthographic lens system and a CCD video camera. The sensors differ primarily in their mechanical designs and their attachments. Figure 5(d) shows a back-to-back implementation that is capable of acquiring the complete sphere of view.

The use of paraboloidal mirrors virtually obviates calibration. All that is needed are the image coordinates of the center of the paraboloid and its radius h. Both these quantities are measured in pixels from a single omnidirectional image. We have implemented software for the generation of perspective images. First, the user specifies the viewing direction, the image size and effective focal length (zoom) of the desired perspective image (see Figure 1). Again, all these quantities are specified in pixels. For each three-dimensional pixel location (x_p, y_p, z_p) on the desired perspective image plane, its line of sight with respect to the viewpoint is computed in terms of its polar and azimuthal angles:

$$\theta = \cos^{-1} \frac{z_p}{\sqrt{{x_p}^2 + {y_p}^2 + {z_p}^2}}, \quad \phi = \tan^{-1} \frac{y_p}{x_p}. \tag{3}$$

This line of sight intersects the paraboloid at a distance ρ from its focus (origin), which is computed using the following spherical expression for the paraboloid:

$$\rho = \frac{h}{(1 + \cos\theta)}. \tag{4}$$

The brightness (or color) at the perspective image point (x_p, y_p, z_p) is then the same as that at the omnidirectional image point

$$x_i = \rho \sin\theta \cos\phi, \quad y_i = \rho \sin\theta \sin\phi. \tag{5}$$

The above computation is repeated for all points in the desired perspective image. Figure 6 shows an omnidirectional image (512x480 pixels) and several perspective images (200x200 pixels each) computed from it. It is worth noting that perspective projection is indeed preserved. For instance, straight lines in the scene map to straight lines in the perspective images while they appear as curved lines in the omnidirectional image. A video-rate version of the above described image generation is detailed in (Peri and Nayar 1997).

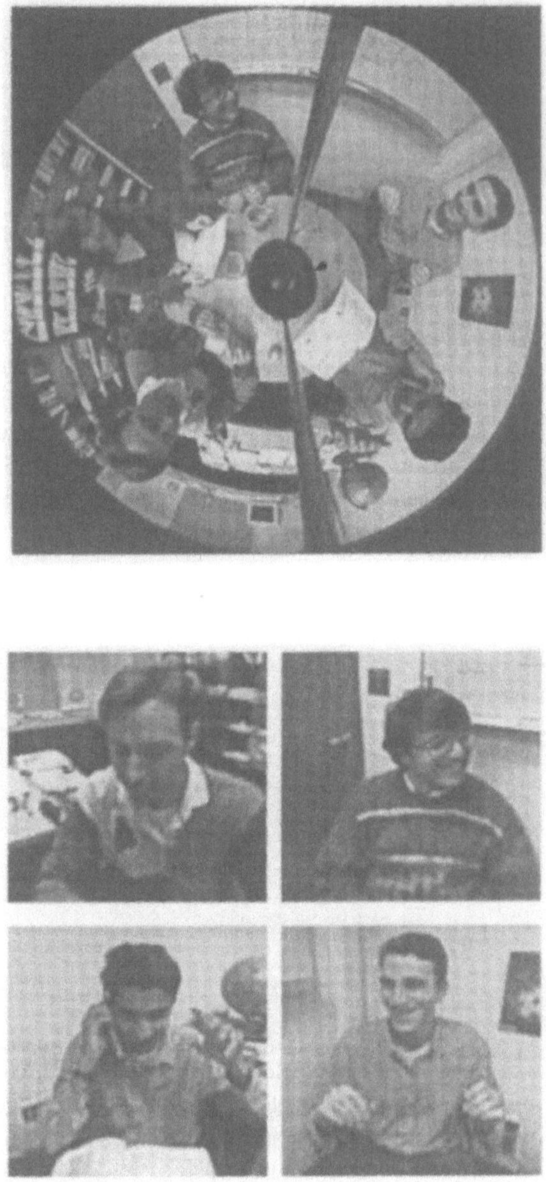

Figure 6: Software generation of perspective images (bottom) from an omnidirectional image (top). Each perspective image is generated using user-selected parameters, including, viewing direction (line of sight from the viewpoint to the center of the desired image), effective focal length (distance of the perspective image plane from the viewpoint of the sensor), and image size (number of desired pixels in each of the two dimensions). It is clear that the computed images are indeed perspective.

7 Resolution

Several factors govern the resolution of a catadioptric sensor. The most obvious of these is the spatial resolution due to finite pixel size. Let us assume that the area projected by a pixel along its line of sight is da, as shown in Figure 7. Note that for orthographic projection da is a constant, while for perspective projection it is easily computed from the distance of the corresponding point on the mirror and the focal length of the imaging system. Given that all reflections are specular, the reflecting surface area occupied by da is $ds = da/\cos\phi$. The foreshortened surface area as seen by the viewpoint $\mathbf{v}$ is $(da/\cos\phi)\cos\phi = da$. The solid angle subtended by the reflecting surface element is $d\omega = da/t^2$, where t is the distance of the surface element from the viewpoint. Hence, the spatial resolution for any catadioptric sensor can be written as

$$\frac{da}{d\omega} = t^2 = z^2 + r^2 . \tag{6}$$

In the case of our paraboloidal mirror, the resolution increases by a factor of 4 from the vertex ($r = 0$) of the paraboloid to the fringe ($r = h$). In many applications, this turns out to be a benefit of using a curved mirror instead of a fisheye lens; often, the panorama is of greater interest than the rest of the field of view. If a uniform resolution over the entire field of view is desired, it is of course possible to use image detectors with non-uniform resolution to compensate for the above variation. It should also be mentioned that while all our implementations use CCD arrays with 512x480 pixels, nothing precludes us from using detectors with 1024x1024 or 2048x2048 pixels that are commercially available at a higher cost.

More intriguing are the blurring effects of field curvature, coma and astigmatism that arise due to the aspherical nature of the reflecting surface (Born and Wolf 1965). Since these effects are linear but shift-variant, a suitable set of deblurring filters could be explored. Alternatively, these effects can be significantly reduced using inexpensive corrective lenses. We have recently used such corrective lenses to obtain excellent results.

Finally, the perspective images presented in Figure 6 are somewhat blocky because there is no interpolation scheme employed to account for the fact that pixel locations in the perspective images correspond to points that could lie between pixel coordinates in the paraboloidal image. We are currently exploring a variety of interpolation and deblurring schemes to enhance the quality of computed perspective images.

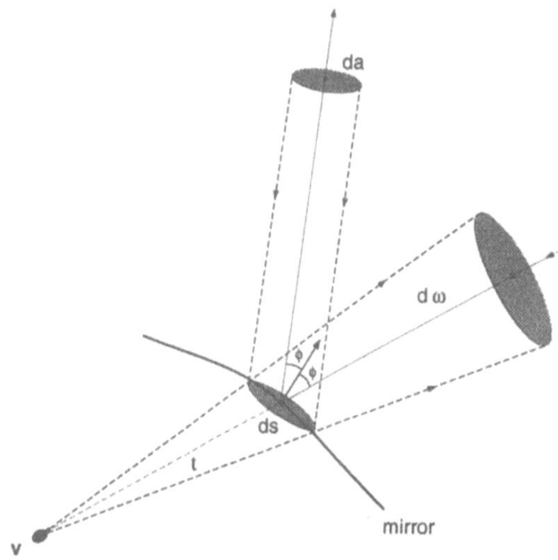

Figure 7: Geometry used to derive spatial resolution ($da/d\omega$) for any catadioptric sensor.

Acknowledgements

This work was inspired by the prior work of Vic Nalwa of Lucent Technologies. I have benefitted greatly from discussions with him. I thank Simon Baker and Venkata Peri of Columbia University for their valuable comments on various drafts of this paper. This project was supported in parts by an NSF National Young Investigator Award, an ONR MURI Grant (No. N00014-95-1-0601), and a David and Lucile Packard Fellowship.

References

Baker, S. and S. K. Nayar (1998, January). Catadioptric Image Formation. *Proc. of International Conference on Computer Vision.*

Bogner, S. (1995, October). Introduction to Panoramic Imaging. *Proc. of IEEE SMC Conference*, 3100–3106.

Born, M. and E. Wolf (1965). *Principles of Optics.* London:Permagon.

Charles, J. R. (1987, April). How to Build and Use an All-Sky Camera. *Astronomy Magazine.*

Chen, S. E. (1995, August). QuickTime VR - An Image Based Approach to Virtual Environment Navigation. *Computer Graphics: Proc. of SIGGRAPH 95*, 29–38.

Edmund (1996). *1996 Optics and Optical Components Catalog*, Volume 16N1. Edmund Scientific Company, New Jersey.

Gluckman, J. and S. K. Nayar (1998, January). Egomotion and Omnidirectional Cameras. *Proc. of International Conference on Computer Vision.*

Goshtasby, A. and W. A. Gruver (1993). Design of a Single-Lens Stereo Camera System. *Pattern Recognition 26*(6), 923–937.

Hong, J. (1991, May). Image Based Homing. *Proc. of IEEE International Conference on Robotics and Automation.*

Kingslake, R. (1983). *Optical System Design.* Academic Press.

Krishnan, A. and N. Ahuja (1996, June). Panoramic Image Acquisition. *Proc. of IEEE Conf. on Computer Vision and Pattern Recognition (CVPR-96)*, 379–384.

Kuban, D. P., H. L. Martin, S. D. Zimmermann, and N. Busico (1994, October). Omniview Motionless Camera Surveillance System. *United States Patent* (5,359,363).

McMillan, L. and G. Bishop (1995, August). Plenoptic Modeling: An Image-Based Rendering System. *Computer Graphics: Proc. of SIGGRAPH 95*, 39–46.

Miyamoto, K. (1964, August). Fish Eye Lens. *Journal of Optical Society of America 54*(8), 1060–1061.

Murphy, J. R. (1995, October). Application of Panoramic Imaging to a Teleoperated Lunar Rover. *Proc. of IEEE SMC Conference*, 3117–3121.

Nalwa, V. (1996, February). A True Omnidirectional Viewer. Technical report, Bell Laboratories, Holmdel, NJ 07733, U.S.A.

Nayar, S. K. (1988, November). Sphereo: Recovering depth using a single camera and two specular spheres. *Proc. of SPIE: Optics, Illumumination, and Image Sensing for Machine Vision II.*

Nayar, S. K. (1997, June). Catadioptric Omnidirectional Camera. *Proc. of IEEE Conf. on Computer Vision and Pattern Recognition.*

Nayar, S. K. and S. Baker (1997, May). Catadioptric Image Formation. *Proc. of DARPA Image Understanding Workshop.*

Nene, S. A. and S. K. Nayar (1998, January). Stereo Using Mirrors. *Proc. of International Conference on Computer Vision.*

Oh, S. J. and E. L. Hall (1987, November). Guidance of a Mobile Robot using an Omnidirectional Vision Navigation System. *Proc. of the Society of Photo-Optical Instrumentation Engineers, SPIE 852*, 288–300.

Peri, V. and S. K. Nayar (1997, May). Generation of Perspective and Panoramic Video from Omnidirectional Video. *Proc. of DARPA Image Understanding Workshop.*

Rees, D. W. (1970, April). Panoramic Television Viewing System. *United States Patent* (3,505,465).

Slater, D. (1996). Panoramic photography with fish eye lenses. *Panorama, International Association of Panoramic Photographers 13*(2,3).

Slater, J. (1932, October). Photography with the whole sky lens. *American Photographer.*

Southwell, D., A. Basu, M. Fiala, and J. Reyda (1996, August). Panoramic Stereo. *Proc. of International Conference on Pattern Recognition.*

Watanabe, M. and S. K. Nayar (1996, April). Telecentric optics for computational vision. *Proc. of European Conference on Computer Vision.*

Wood, R. W. (1906). Fish-eye views, and vision under water. *Philosophical Magazine 12*(Series 6), 159–162.

Xiong, Y. and K. Turkowski (1997, June). Creating Image-Based VR Using a Self Calibrating Fisheye Lens. *Proc. of IEEE Conf. on Computer Vision and Pattern Recognition.*

Yagi, Y. and S. Kawato (1990). Panoramic Scene Analysis with Conic Projection. *Proc. of International Conference on Robots and Systems (IROS).*

Yamazawa, K., Y. Yagi, and M. Yachida (1995, May). Obstacle Avoidance with Omnidirectional Image Sensor HyperOmni Vision. *Proc. of IEEE International Conference on Robotics and Automation*, 1062–1067.

Zheng, J. Y. and S. Tsuji (1990, June). Panoramic Representation of Scenes for Route Understanding. *Proc. of the Tenth International Conference on Pattern Recognition 1*, 161–167.

Small Vision Systems: Hardware and Implementation

Kurt Konolige
Artificial Intelligence Center, SRI International
333 Ravenswood Avenue, Menlo Park, CA 94025
email: konolige@ai.sri.com

Abstract

Robotic systems are becoming smaller, lower power, and cheaper, enabling their application in areas not previously considered. This is true of vision systems as well. SRI's Small Vision Module (SVM) is a compact, inexpensive realtime device for computing dense stereo range images, which are a fundamental measurement supporting a wide range of computer vision applications. We describe hardware and software issues in the construction of the SVM, and survey implemented systems that use a similar area correlation algorithm on a variety of hardware.

1 Introduction

Realtime stereo analysis, until recently, has been implemented in large custom hardware arrays (Kanade 1996, Matthies 1995). But computational power and algorithmic advances have made it possible to do such analysis on single processors. At the same time, increased density, speed and programmability of floating-point gate arrays (FPGAs) make custom hardware a viable alternative. In this paper, we discuss the implementation of area-based stereo algorithms on microprocessors, and describe in detail one such implementation, the SRI Small Vision Module (SVM), which achieves realtime operation at low power in a small package. We also survey area-based implementations on microprocessors and FPGAs, comparing speed and efficiency.

In a final section, we briefly describe some experiments with the SVM related to mobile robotics and human-computer interaction. These experiments make use of realtime stereo to segment interesting objects from the background.

2 Area-correlation Stereo

Stereo analysis is the process of measuring range to an object based on a comparison of the object projection on two or more images. The fundamental problem in stereo analysis is finding corresponding elements between the images. Once the match is made, the range to the object can be computed using the image geometry.

Matching methods can be characterized as *local* or *global*. Local methods attempt to match small regions of one image to another based on intrinsic features of the region. Global methods supplement local methods by considering physical constraints such as surface continuity or base-of-support. Local methods can be further classified by whether they match discrete *features* among images, or correlate a small *area* patch (Barnard and Fischler 1982). Features are usually chosen to be lighting and viewpoint-independent, e.g., corners are a natural feature to use because the remain corners in almost all projections. Feature-based algorithms compensate for viewpoint changes and camera differences, and can produce rapid, robust matching. But they have the disadvantage of requiring perhaps expensive feature extraction, and yielding only sparse range results.

Area correlation compares small patches among images using correlation. The area size is a compromise, since small areas are more likely to be similar in images with different viewpoints, but larger areas increase the signal-to-noise ratio. In contrast to the feature-based method, area-based correlation produces dense results. Because area methods needn't compute features, and have an extremely regular algorithmic structure, they can have optimized implementations. The SVM and other systems discussed in this paper all use area correlation.

2.1 Area Correlation Method

The typical area correlation method has five steps (see Figure 3):

1. Geometry correction. In this step, distortions in the input images are corrected by warping into a "standard form".
2. Image transform. A local operator transforms each pixel in the grayscale image into a more appropriate form, e.g., normalizes it based on average local intensity.

3. Area correlation. This is the correlation step, where each small area is compared with other areas in its search window.
4. Extrema extraction. The extreme value of the correlation at each pixel is determined, yielding a *disparity image*: each pixel value is the disparity between left and right image patches at the best match.
5. Post-filtering. One or more filters clean up noise in the disparity image result.

In the rest of this section we describe each of these operations in more detail.

2.2 *Epipolar Geometry*

The fundamental geometry for stereo correlation is the epipolar curve, illustrated in Figure 1. Consider any point p in one image; the ray from p through the focal point intercepts objects in the scene at different depths. The projection of this ray in the second image is the epipolar curve associated with p. The geometric significance of this curve is that, for any object imaged at p, the corresponding point in the second image must lie on the epipolar curve. Thus, it suffices to search along this curve when doing stereo matching.

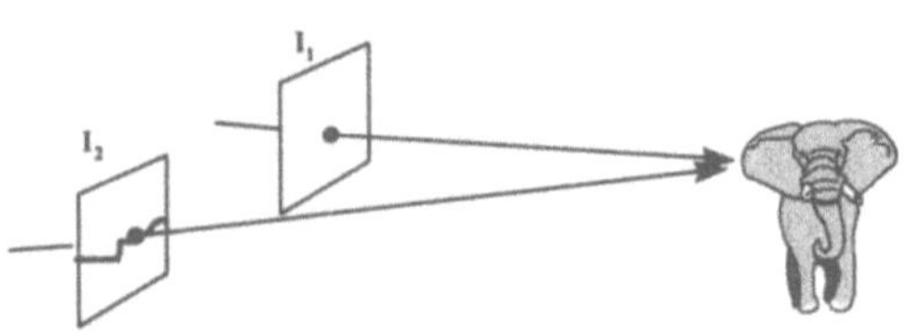

Figure 1 An epipolar curve is the image in I_2 of a ray projected from a point of I_1.

If the images are ideal projections of a pinhole camera, then the epipolar curves will be straight lines. For efficient processing, it is best to align epipolar lines in a regular pattern, usually along the scan lines of the camera. To accomplish this, the images must be in the same plane, with their scan lines and image centers horizontally aligned, and with the same focal length. Stereo images with this geometry are said to be in standard or scan-line form. In practice, it is difficult to achieve this physical geometry precisely, and in some cases vergence of the images is necessary to view close objects. However, if all internal and external camera parameters are known, then it is possible to warp each image so that it appears in standard form. In practice, it is difficult and time-consuming to completely calibrate a stereo setup, and repeated calibration necessitated by vibration or other mechanical disturbances is not practical. A compromise is to position the cameras in approximately the correct geometry, and to calibrate the internal camera parameters. The remaining offset parameters between the cameras can be dealt with by auto-calibration, discussed in Section 3.2.

2.3 *Image Transforms and Correlation Measures*

Correlation of image areas is disturbed by illumination, perspective, and imaging differences among images. Area correlation methods usually attempt to compensate by correlating not the raw intensity images, but some transform of the intensities. Table 1 lists some of the correlation measures found in the literature. There are three basic types of transforms:

1. Normalized intensities. Each of the intensities in a correlated area is normalized by the average intensity in the area.
2. Laplacian of gaussian (LOG). The laplacian measures directed edge intensities over some area smoothed by the gaussian. Typically the standard deviation of the gaussian is 1-2 pixels.
3. Nonparametric. These transforms are an attempt to deal with the problem of outliers, which tend to overwhelm the correlation measure, especially using a square difference.

Each of these transforms can be applied with different correlation measures. For example, LOG correlations include correspondence of zero crossings and their signs, and two measures of the magnitude: square and absolute difference (also called the *L1 norm*).

The census method computes a bit vector describing the local environment of a pixel, and the correlation measure is the Hamming distance between two vectors. This measure, as with other nonparametric measures, addresses the problem of outliers, which the square difference handles poorly. The L1 norm tends to weight outliers less than the square difference, and is more amenable to implementation on standard microprocessors, which usually do not have the bit-counting hardware necessary for computing Hamming distance efficiently.

There has been relatively little work comparing the quality of results from the different transforms. Partly this results from the difficulty of getting good ground truth test sets. Some recent work (Zabih and Woodfill 1997) comparing census and normalized intensities on synthesized and real images shows that census does better, in that it can

Normalized intensities square difference	Fua 1993
Laplacian of gaussian zero crossings sign square difference absolute difference	 Marr and Poggio 1979 Nishihara 1984 Matthies 1995 Kanade 1996
Nonparametric rank census	 Zabih and Woodfill 1994

Table 1 Correlation measures for area-based stereo

interpret more of the image correctly, especially at depth discontinuities. From the author's experience, the best quality results appear to come from absolute difference of the LOG, and from census.

Another technique for increasing the signal-to-noise ratio of matches is to use more than two images (Faugeras 1993). This technique can also overcome the problem of viewpoint occlusion, where the matching part of an object does not appear in the other image. The simple technique of adding the correlations between images at the same disparity seems to work well (Kanada 1996). Obviously, the computational expenditure for multiple images is greater than for two.

2.4 Filtering

Dense range images usually contain false matches that must be filtered, although this is less of a problem with multiple-image methods. Table 2 lists some of the post-filters that have been discussed in the literature. The correlation surface shape can be related to the uncertainty in the match. An interest operator gives high confidence to areas that are textured in intensity, since flat areas are subject to ambiguous matches. The left/right check looks for consistency in matching from a fixed left image region to a set of right image regions, and back again from the matched right region to a set of left regions. It is particularly useful at range discontinuities, where directional matching will yield different results.

Finally, a disparity image can be processed to give sub-pixel accuracy, by trying to locate the correlation peak between pixels. This increases the available range resolution without much additional work.

Correlation surface: peak width, peak height, number of peaks	Matthies 1993, Nishihara 1984
Mode filter	
Left/Right check	Fua 1993, Bolles and Woodfill 1993
Interest Operator	Moravec 1984
Interpolation	Nishihara 1984

Table 2 Range filtering operations

3 Algorithm and Implementation

Although area correlation is computationally expensive, good algorithm design coupled with current processors enable video rate performance for small frame sizes. In this section we describe the results of two optimized implementations of the same basic algorithm. The Small Vision Module (SVM), takes advantage of the instruction set and memory design of digital signal processors (DSPs) to achieve high performance at low power. The Small Vision System (SVS) is an implementation of the same algorithm in C on general-purpose microprocessors, and an optimized version for Pentium microprocessors that takes advantage of the MMX instruction set.

3.1 Algorithm

The algorithm we have implemented (Figure 3) has the following features:

- LOG transform, L1 norm (absolute difference) correlation;
- Variable disparity search, e.g., 16, 24, or 32 pixels;
- Postfiltering with an interest operator and left/right check;
- x4 range interpolation.

The LOG transform and L1 norm were chosen because they give good quality results, and can be optimized on standard instruction sets available on DSPs and microprocessors. The census method is an alternative, but requires fast bit-counting, and is more suitable for custom hardware.

Figure 2(b) shows a typical disparity image produced by the algorithm. Higher disparities (closer objects) are indicated by brighter green (or white, if this paper is printed without color). There are 64 possible levels of disparity; in the figure, the closest disparities are around 40, while the furthest are about 5. Note the significant errors in the upper

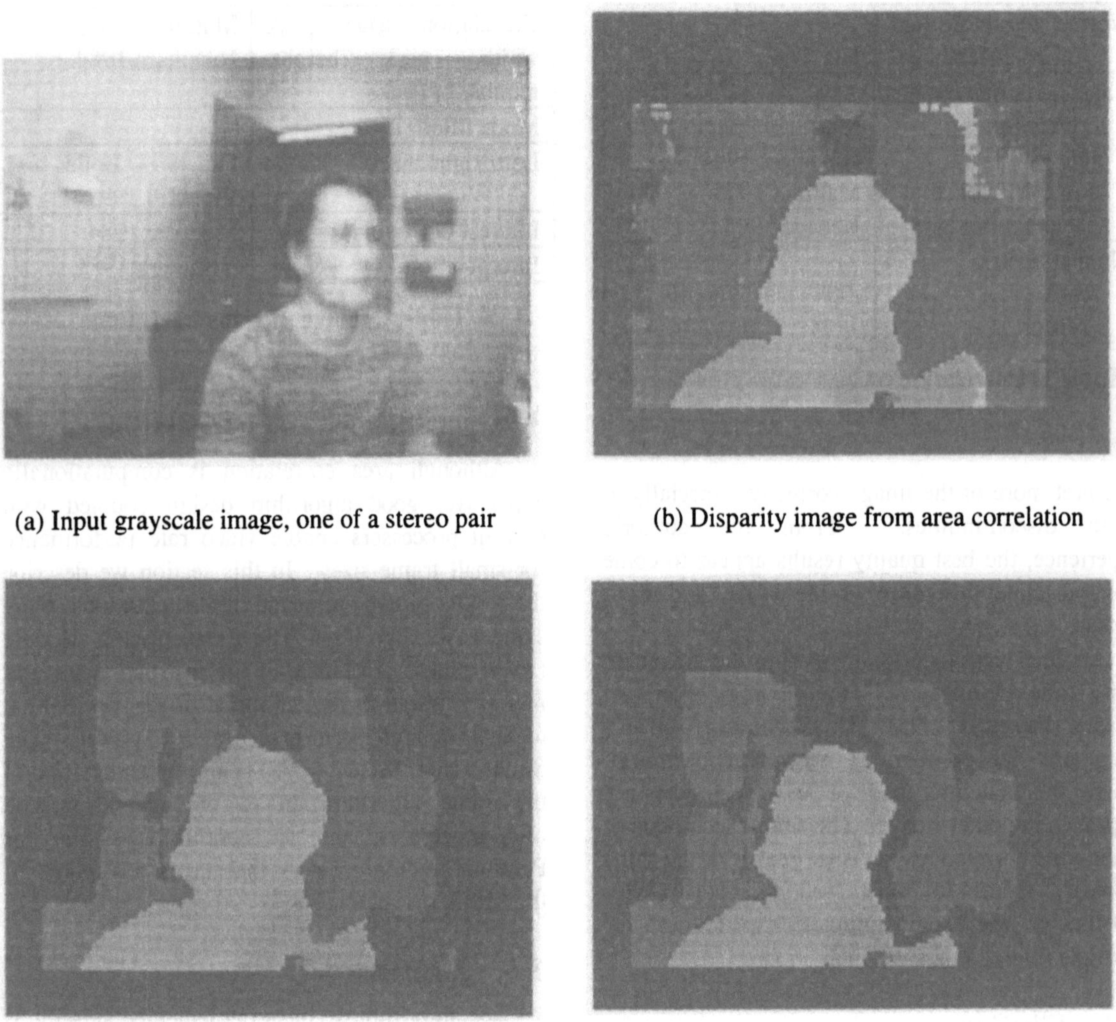

(a) Input grayscale image, one of a stereo pair

(b) Disparity image from area correlation

(c) Texture filter applied

(d) Left/right and texture filter applied

Figure 2 A grayscale input image and the disparity images from the SRI algorithm.

left and right portion of the image, where uniform areas make it hard to estimate the disparity.

In Figure 2(c), the interest operator is applied as a postfilter. Areas with insufficient texture are rejected as low confidence: they appear black in the picture. We've found that the best interest operator is one that measures texture in the direction of epipolar lines (Matthies 1993).

Although the interest operator requires a threshold, it's straightforward to set it based on noise present in the video input. Showing a blank gray area to the imagers produces an interest level related only to the video noise; the threshold is set slightly above that.

There are still errors in portions of the image with disparity discontinuities, such as the side of the subject's head. These errors are caused by overlapping the correlation window on areas with very different disparities. Application of a left/right check can eliminate these errors, as in Figure 2(d). The left/right check can be implemented efficiently by storing enough information when doing the original disparity correlation.

In practice, the combination of an interest operator and left/right check has proven to be the most effective at eliminating bad matches. Correlation surface checks, in our experience, do not add to the quality of the range image, and can be computationally expensive.

We have been careful in designing the algorithm to consider implementation issues, especially storage efficiency. Buffering of intermediate results, a critical part of the algorithm, is kept to a minimum by calculating incremental results for each new scanline, and by recalculating rather than storing some partial results. We have been especially careful about storage of correlation results, and require only space of size (number of

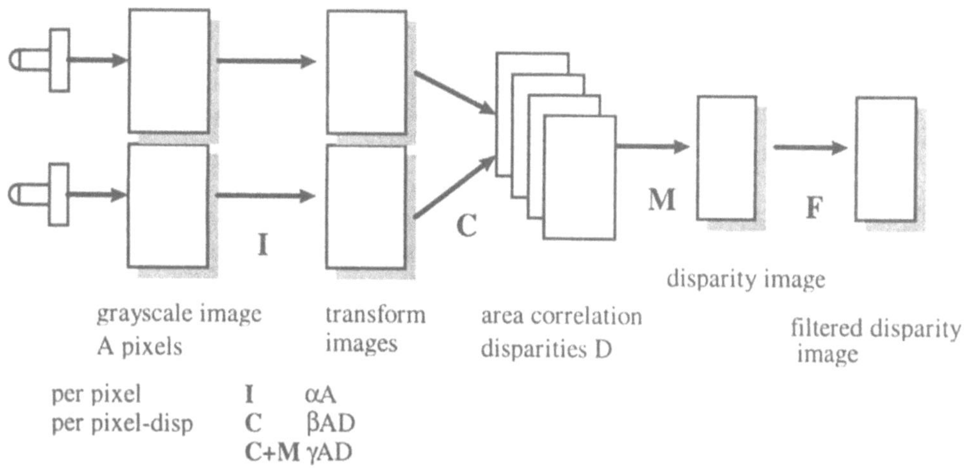

Figure 3 Area correlation algorithm diagram. *A* is the area of an image, and *D* is the number of disparities searched over. Image warping and post-filtering costs are not considered here. The cost of the algorithm is divided into per pixel and per pixel-disparity contributions.

disparities x line width). Storage efficiency is critical for DSP implementations, since DSPs have limited onboard high-speed memory. It also helps to make microprocessor implementations faster, since storage buffers can be contained in high speed cache.

3.2 Calibration

As described in Section 2.2, the standard form for stereo processing assumes that the two images are from pinhole cameras of the same focal length, and are co-planar, with scanlines and focal centers aligned horizontally. The difficulty of acquiring and maintaining this alignment has led us to consider strategies for automatic calibration. We assume that the internal camera parameters are already known, so that lens distortion can be corrected (Devernay 1995). We also assume that the cameras are in approximately the correct position, although there may be small errors in orientation. To a first approximation, orientation errors can be considered as generating horizontal, vertical, and rotational offsets of one image to another. Additionally, vergence may make the epipolar lines non-horizontal.

To calibrate these external offsets, we consider the appearance of disparity images under calibrated and noncalibrated conditions (Figure 4). The difference between these two is visually obvious, and we have found two image measure that correlate well with calibrated images. The first measure is simply the sum of all left/right consistency matches in the area correlation algorithm. If the input images are not calibrated, there will be few valid stereo matches, and the left/right check will reject many of these.

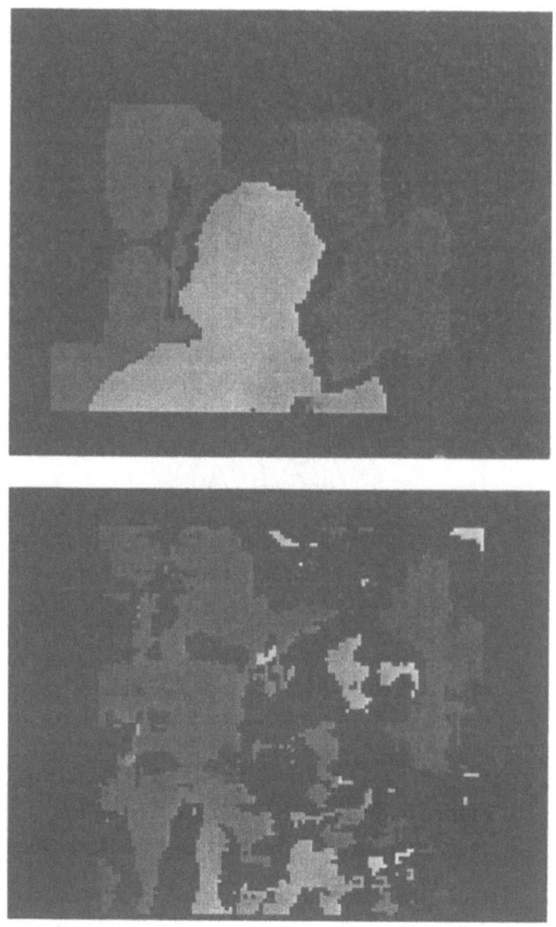

Figure 4 Calibrated and uncalibrated disparity images. The stereo input to the bottom image was offset 2 vertical pixels from the top.

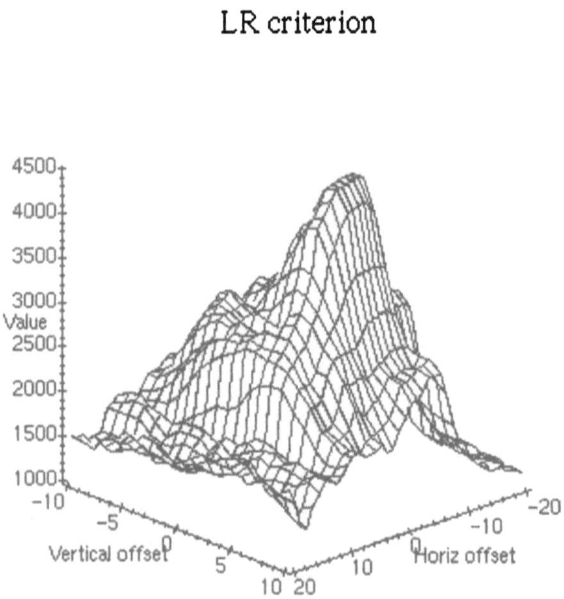

Figure 5 Number of left/right matches as a function of horizontal and vertical offset.

The second measure looks at the smoothness of the disparity image; it is the number of disparity values at the mode for a small area, summed over the entire image.

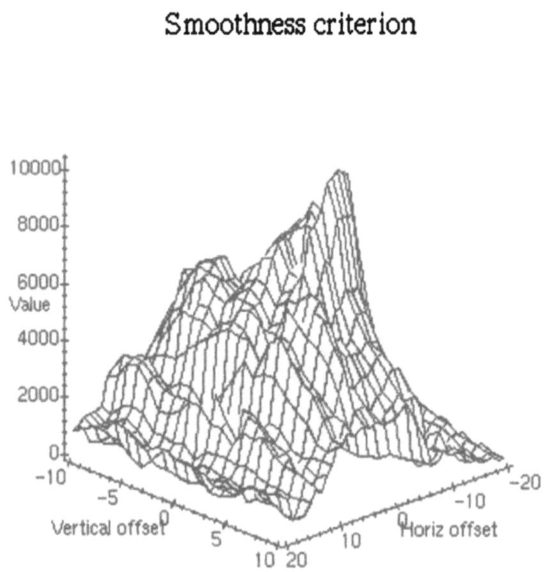

Figure 6 Smoothness of disparity image as a function of horizontal and vertical offset.

To use these measures, we first capture a stereo pair of a scene with objects at different distances, and compute disparity images at different offsets. Plots of the measures against vertical and horizontal offsets are shown in Figure 5 and Figure 6. Note that there is a peak at the calibrated offsets. The peak is very sharp in the vertical offset direction, because even small changes in vertical offset destroy the horizontal alignment of the images. The horizontal offset is more forgiving, since different offsets still give enough leeway to capture the depths of all objects in the image.

Unfortunately, the surfaces show that a hill-climbing algorithm is not sufficient to find the best offsets, since there are local maxima away from the peak. Instead, we can use a hierarchical search strategy where we first search at a coarse resolution, then use a fine resolution at a few of the highest points from the first search. This method takes on the order of 100 disparity image calculations, which is a few seconds at video rates.

Once horizontal and vertical offsets are calibrated, the measures can be used to compensate for vergence and rotation, using hill-climbing and perhaps iterating the offset calculation over a small space.

3.3 Small Vision Module

The Small Vision Module is a hardware and firmware implementation of the SRI area correlation algorithm. The hardware consists of two CMOS 320x240 grayscale imagers and lenses, low-power A/D converters, a digital signal processor and a small flash memory for program storage. All of these components are commercially available. The SVM is packaged on a single circuit board measuring 2" x 3" (Figure 7). Communication with a host PC for display and control is through the parallel port. During operation, the DSP and imaging system consume approximately 600mW.

Figure 7 The SRI Small Vision Module

In its current implementation, the SVM uses an ADSP 2181 running at 33 MHz. Figure 3 shows the general algorithm, with the data structures produced at each stage. One of the surprising aspects of the SVM is that the entire algorithm, including buffering for two 160x120 images, fits into the DSP's internal memory of 80 Kbytes.

Based on instruction set timings, the SVM can perform 8 fps on 160x120 images; in practice, communication overhead with the host reduces this to a maximum of 6 fps, enough for realtime operation of some range applications. Currently some 15 SVMs have been distributed to research labs for evaluation.

We have designed a second-generation SVM, the SVM II, using a new DSP from Texas Instruments (TMS320C60x), running at 200 MHz. In simulation, the SVM II has a performance improvement of x30 over the SVM. Performance of these devices relative to other implementations of area correlation is given in the next subsection.

One area of improvement for the SVM is in the imaging system. Current CMOS imagers are still an order of magnitude noisier and less sensitive than corresponding CCDs. On the other hand, they offer a degree of on-chip integration that cannot be achieved with CCDs: automatic exposure control, clock and control signals, and even A/D conversion. Future generations of CMOS imagers will narrow the gap in video performance with CCDs, and yield better stereo results. We have recently started using new imagers from Omnivision Technologies (Kempainen 1997), which achieve S/N ratios of 46 dB even with automatic gain, and which have sensitivities down to 1 lux.

3.4 Small Vision System

For development and experimentation on standard hardware, we have implemented the SRI area correlation algorithm in C on standard microprocessors. Further optimization of the algorithm is possible by using the Single-Instruction, Multiple Data (SIMD) instructions available on the Pentium (MMX) and several other microprocessors, which enable a parallel computation of the transform and correlation operations. For the SVM algorithms, these instructions increase speed by at least a factor of 4 over comparable μPs without SIMD.

4 Benchmark results

We survey some implementations of area correlation algorithms, comparing them to the SVM hardware, as well as software implementations of the SRI algorithm on general-purpose microprocessors. Figure 3 diagrams a generic area correlation algorithm, showing the operations and costs associated with each (Fua 1993). For concreteness, we take a benchmark problem to consist of input 320x240 images, with a search window of 32 disparities. The performance of an implementation is measured as a frame rate on the benchmark problem.

In implementations with multiple processors or distributed computation, algorithm operations can be performed in a pipelined fashion to increase throughput. The largest cost is in the correlation and extrama-finding operations, which are proportional to the number of pixels times the number of disparities. Note that the area correlation window size is not a factor, since algorithms can take advantage of redundancy in computing area sums, using a sliding sums method (Fua 1993).

Table 3 gives performance measurements for various implemented area correlation systems.[1] The columns list the type of correlation, the processor(s), and the operation coefficients, where available. Finally, as an overall measure of performance, the estimated results on the benchmark problem are given. From these results, it's possible to draw some general conclusions about the performance of the various types of processors: general-purpose microprocessors (μPs), digital signal processors (DSPs), and custom hardware (typically field-programmable gate arrays, or FPGAs).

First, it makes little difference which of the three correlation methods is chosen: all of the correlation methods can be implemented on all types of hardware. The census operator, because of its heavy reliance on bit-counting, is more suited to FPGA hardware, since DSPs and microprocessor generally don't have bit-counting instructions. But, algorithm design can make a big difference: both the Point Grey and SRI systems use the same correlation measure, but the latter has double the performance (considering Point Grey uses three images).

Second, the highest level of performance is provided by FPGAs and DSPs. General-purpose μPs, even when running at comparable clock rates, are at least three times slower on the benchmark problem. The best performance of μPs comes from using the Single-Instruction, Multiple Data (SIMD)

---

[1] These figures are the author's best estimate based on published reports and his own experiments.

System	Method	Processor(s)	α β γ	Bench fps	Remarks
Matthies JPL 1995	LOG square diff	Datacube MV200, 68040, Sparc		0.5	
Nishihara Teleos 1995	LOG sign Hamming	Pentium 166 MHz		0.5	
Woodfill Stanford 1995	Census Hamming	Sparc 10-51 50 MHz		1.0	
Faugeras INRIA 1993	Normalized correlation	PeRLe-1 FPGA array		2.5	
SVM+ SRI 1997	LOG abs diff	2 x ADSP 218x 50MHz	800 160 200	2.5	
CYCLOPS Point Grey 1997	LOG abs diff	Pentium II 233 MHz		3.0	3 images est. for P II
SVM algorithm SRI 1997	LOG abs diff	Pentium II 233 MHz	70 20 33	12	
Kanade CMU 1996	LOG abs diff	Custom hardware, TMS320C40 array	66 33 33	15	5 images α, χ pipelined
Dunn CSIRO 1997	Census Hamming	Occam FPGA array	15	20	
SVM II SRI 1997	LOG abs diff	TMS320C60x 200 MHz	50 10 12.5	30+	
Woodfill Interval 1997	Census Hamming	PARTS FPGA array		30+	Est. >100 fps with upgrade

Table 3 Performance of some implementations of area correlation stereo.

instructions, which enable a parallel computation of the transform and correlation operations. For the SRI algorithms, these instructions increase speed by a factor of 4 over comparable μPs without SIMD (e.g., the R10000).

Despite performance limitations, μPs offer the most flexible development environment. DSPs, while somewhat less flexible than μPs, can be programmed in C, with critical inner loops in assembly. At this point they probably offer the best compromise between performance and programmability. FPGAs, because they can be configured to pipeline successive operations and parallelize individual operations, offer the best performance. However, programming FPGAs can be difficult, and any changes in the algorithm can take significant development time (Woodfill and Von Herzen 1997).

4.1 Machine cycles and efficiency

A more critical examination of the same algorithm running on different processors reveals the tradeoffs in hardware and software design, and helps to quantify the efficiency of a hardware implementation. This measure is especially critical in some areas such as space missions or portable applications. To compare power ratings, selected processors of each type were rated based on how much power they would take to achieve the benchmark.[2] The results are in Table 4, which also details the number of instructions used per pixel per disparity in the correlation calculation. All of these processors ran the same SVM algorithm; the MMX and FPGA array implementations were optimized for parallel computation.

Without the SIMD MMX instructions, μPs do poorly. The Pentium MMX is fairly efficient, since it processes four to eight pixels at once. Since there are two MMX pipelines, theoretically it can process 2 instructions at once, which would make it almost as fast as the TMS320C6x DSP on a MIPS basis. However, because of cache miss limitations, the MMX instructions are issued at below their theoretic maximum rate. In DSPs, the programmer has control over fast cache memory, and can buffer critical information in this area as needed. The

[2] These power ratings are the author's estimates rather than measured results, and are based on processor data sheets and published results. Since processor characteristics change quickly, these results may not be valid in the future.

Processor	Number of Instructions	β	Bench Power
SGI R10000 200 MHz	24	100	280W
Sparc Ultra 200 MHz	24	90	180W
FPGA array Dunn	NA	16	40W
Pentium II MMX 233 MHz	5 (20 / 4)	20	25W
ADSP 218x 50 MHz	8	160	10W
TMSC6x 200 MHz	2 VLIW, 8FU	10	4W

Table 4 Processor power ratings for the benchmark problem.

general caching scheme of μPs means that the processor will idle while waiting for cache fills. With tuning of the code, and some attention to caching issues, it may be possible to double the performance of the MMX implementation. Since most of the major μP manufacturers plan versions of their chips with SIMD instructions, μPs are a viable platform for realtime stereo, except in low-power applications.

It is somewhat surprising that the DSPs are more efficient than FPGAs, since the latter have a speed advantage based on their highly-parallel implementation of the stereo algorithm. Because of their fixed functional units, however, DSP designs can be optimized for high performance at low power. The number of instructions for a standard design such as the ADSP 21xx is 1/3 that of a non-SIMD μP. The TMS320C6x achieves its low instruction count by having parallel functional units (in contrast to the parallel data concept of SIMD), achieving 12 operations per cycle in the correlation loop.

5 Applications

Practical realtime stereo has been available recently, and applications are just being developed. Kanade's group has implemented a Z-keying method, where a subject is placed in correct alignment with a virtual world (Kanade 1996). At SRI, we are experimenting with several applications involving mobile robotics and human interaction. In collaboration with Gaetan Marti and Nicola Chauvin of the Swiss Federal Institute of Technology in Lausanne, we have developed a *contour method* for fast segmentation of range images based on 3D areas. This method can be used, for example, to distinguish terrain height in front of a moving vehicle. The important part of the method is that results are computed within the range image, rather than in 3D space, making it suitable for realtime implementation.

A second application is the detection of people within the camera's range, in collaboration with Chris Eveland of the University of Rochester (Eveland 1997). Here the solution is to segment the range image by first learning a background range image, and then using statistical image comparison methods to distinguish new objects that appear in the field of view. A simple shape analysis is used to find the head and shoulder area of the person. The interesting part of this application is that the camera can pan and tilt to track the subject.

More details of these applications can be found at http://www.ai.sri.com/~konolige/svm..

References

Barnard, S. T. and M. A. Fischler (1982). Computational stereo. *Computing Surveys* **14**(4), 553-572.

Bolles, R. and J. Woodfill (1993). Spatiotemporal consistency checking of passive range data. *Proc ISRR93.*

Devernay, F. and O. Faugeras (1995). Automatic calibration and removal of distortion from scenes of structured environments. *Proc SPIE95.*

Dunn, P. and P. Corke (1997). Real-time stereopsis using FPGAs. *FPL97*, London.

Eveland, C., K. Konolige and R. C. Bolles (1997). Background Modeling for Segmentation of Video-Rate Stereo Sequences. Submitted to *CVPR98.*

Faugeras, O. et al. (1993). Real time correlation-based stereo: algorithm, implementations and applications. *INRIA Technical Report 2013.*

Fua, P. (1993). A parallel stereo algorithm that produces dense depth maps and preserves image features. *Machine Vision and Applications* **6**(1).

Kanada, T. et al. (1996). A stereo machine for video rate dense depth mapping and its new applications. *Proc. CVPR96.*

Kempainen, S. (1997). CMOS image sensors: eclipsing CCDs in visual information? *EDN*, October 9, 1997.

Marr, D. and T. Poggio (1979). A computational theory of human stereo vision. *Proc. Royal Society* B-204, 301-328.

Matthies, L. (1993). Stereo vision for planetary rovers: stochastic modeling to near realtime implementation. *IJCV* **8**(1), 71-91.

Matthies, L., A. Kelly, and T. Litwin (1995). Obstacle detection for unmanned ground vehicles: a progress report. *Proc. ISRR.*

Moravec, H. P. (1979). Visual mapping by a robot rover. *Proc. IJCAI*, Tokyo, 598-600.

Nishihara, H. K. (1984). Practical real-time imaging stereo matcher. *Optical Engineering* **23**(5), 536-545.

Point Grey Research (1997) *http://www.ptgrey.com.*

Woodfill, J. and B. Von Herzen (1997). Real-time stereo vision on the PARTS reconfigurable computer. *IEEE Workshop on FPGAs for Custom Computing Machines.*

Zabih, R. and J. Woodfill (1994). Non-parametric local transforms for computing visual correspondence. *Proc. ECCV94.*

Zabih, R. and J. Woodfill (1997). A non-parametric approach to visual correspondence. Submitted to *IEEE PAMI.*

PART 7

MOBILE ROBOTS

SESSION SUMMARY

Robert C. Bolles
SRI International, EJ254, 333 Ravenswood Avenue, Menlo Park, CA 94025 (U.S.A.)
email: bolles@ai.sri.com
Yoshiaki Shirai
Dept. of Computer-Controlled Machine Systems, Osaka University 2-1, Yamadaoka, Suita, 565 (Japan)
email: shirai@mech.eng.osaka-u.ac.jp

The field of mobile robots is in transition – vehicles are starting to leave the laboratory and enter day-to-day applications, such as cleaning floors in commercial buildings, driving underground mining machines, and warning drivers of dangers along the highway. These applications require a level of reliability that is substantially higher than is typically tolerated in laboratory environments.

A key question, then, is "How can we develop reliable robots?" Or as Hugh Durrant-Whyte, et al, ask in their paper, "How can we develop ultra-high integrity robots?" The authors of that paper are careful to make a distinction between reliability and integrity. A *reliable* robot operates correctly all the time. A *high-integrity* robot monitors its own behavior and shuts down if something is wrong. As a result, a high-integrity robot may not operate continuously, but when it does, it works correctly.

The distinction between reliability and integrity is critical to the mobile robot community at this time of transition, because it relaxes the requirements for a robot to be "trusted." Or put another way, since it is easier to make ultra-high integrity robots than it is to develop ultra-reliable robots, focusing on integrity will lead to practical robots sooner.

So, how do we develop ultra-high integrity robots? The papers in this section explore several approaches, but all of them involve multiple sensor types. Dillman and Weckesser use sonar sensors, a laser range scanner, stereo vision, a structured-light sensor, odometry, and a gyroscope. Thorpe, et al, work with long-range radar, short-range radar, and forward-looking video sensors. Durrant-Whyte, et al, use GPS sensors, an inertial measurement unit, steering encoders, and millimeter-wave radar. Everyone uses multiple sensors.

The application domains, on the other hand, determine how the authors approach the following three choices associated with mobile robots:

- interactive (with a person in the loop) vs. automatic
- self-contained and autonomous vs. an instrumented site
- statistical methods vs. structural (or symbolic) methods

Mori and Kotani, for example, are developing a robot system to help a visually impaired person navigate along sidewalks. They view their system as an interactive one in which the person plays a key part. In fact, the authors, in their design of an interface, are careful not to interfere with the person's hearing – they want the person to be able to use their hearing as another sensor as they move down the sidewalk. Durrant-Whyte, et al, however, have the goal of developing autonomous mining robots, so they want to maximize autonomy and minimize human interaction.

With respect to instrumenting the site, Mori and Kotani are exploring anything that might help a visually impaired person, including mounting cameras at intersections, while Thorpe, et al, have explicitly undertaken the task of making self-reliant vehicles. Their reasoning is that the development of self-reliant vehicles will make it possible to introduce automated vehicles earlier than vehicles requiring instrumented lanes, because self-reliant vehicles can be mixed with manually driven cars. In addition, the development of self-reliant cars can begin with systems that warn the driver of potential dangers, such as a vehicle in their blind spot, and work up to fully autonomous systems. Yang, et al, also focus on the development of an autonomous robot, but in their case, a robot to operate in a dynamic indoor environment, such as a room full of people walking around. Their system uses an enhanced topological map, an explicit representation of locational uncertainty, and a planner to provide the navigational information necessary to support the behavior modules.

Durrant-Whyte, et al, have developed a statistically based approach to navigation. By focusing on location determination, which is a critical function of all robots, they are able to formulate the navigational integrity problem as a generic problem in frequency analysis, which they solve by combining redundant pairs of sensors. Dillman and Weckesser, on the other hand, focus on the problem of dynamic mapping, where they use multiple sensors to construct symbolic descriptions of the local environment. Their goal is to build models that include such features as walls, doors, and corners.

In summary, the field of mobile robots is at an exciting time when several factors are coming together to place us at the beginning of the introduction of robots into everyday life. The sensors are maturing, the computing power is increasing dramatically, and the focus is shifting to evaluation and integrity!

Exploration of Unknown Environments with a Mobile Robot using Multisensorfusion

R. Dillmann P. Weckesser
Institute for Real-time Computer Systems & Robotics
Department for Computer Science
University of Karlsruhe
Kaiserstr.12, 76128 Karlsruhe, Germany
email: dillmann@ira.uka.de wecke@ira.uka.de

Abstract

The exploration of an unknown environment is an important task for the new generation of mobile service robots. These robots are supposed to operate in dynamic and changing environments together with human beings and other static or moving objects. Sensors that are capable of providing the quality of information that is required for the described scenario are optical sensors like digital cameras and laserscanners. In this paper sensor integration and fusion for such sensors is described. Complementary sensor information is transformed into a common representation in order to achieve a cooperating sensor system. Sensor fusion is performed by matching the local perception of a laserscanner and a camera system with a global model that is being built up incrementally. The Mahalanobis distance is used as matching criterion and a Kalman filter is used to fuse matching features. A common representation including the uncertainty and the confidence is used for all scene features . The system's performance is demonstrated for the task of exploring an unknown environment and incrementally building up a geometrical model of it.

1 Introduction

In this paper an approach to fuse sensor information from complementary sensors is presented. The mobile robot PRIAMOS (figure 1) was used as an experimental testbed. A multisensor system supports the vehicle with odometric, sonar, visual and laserscanner information. This work is part of a research project with the goal of making robot navigation safer, faster, more reliable and more stable under changing environmental conditions. An architecture for active and task-driven processing of sensor data is presented in [15]. With this architecture it is possible to control the sensor system according to environmental conditions, perceived sensor information, a-priori knowledge and the task of the robot.

1.1 Mobile robot navigation

The navigation tasks of a mobile robot can be subdivided into three subproblems.

1. *collision avoidance:* this is the basic requirement for safe navigation. The problem of collision avoidance is solved for dynamic environments with different kinds of sensors like sonars or laserscanners.
2. *mobile robot positioning:* if geometrical a priori information about the environ-

Figure 1: PRIAMOS & KASTOR

ment is available to the robot the following questions can be asked. 'Where am I?', Where am I going?' and 'How do I get there' [11]. With today's sensors these questions can be answered for static environments [8]. Though the problem is not solved in general for dynamic and changing environments.

3. *exploration and environmental modelling:* The problem of exploring an unknown environment was approached by various groups [5, 1, 12] but is not yet solved. Most approaches aim at generating a 2-dimensional map of a static environment. In this paper a 3-dimensional map of the environment is built up with an integrated sensor system consisting of a laserscanner and a trinocular vision system. The laserscanner only provides 2-dimensional information. The vision system is capable of perceiving the environment 3-dimensionally.

The purpose of this paper is to present and to discuss sensor fusion techniques in order to improve the system's performance for 'mobile robot positioning' and 'exploration of unknown environments'. The approach is capable of dealing with static as well as dynamic environments. On different levels of data-processing geometrical, topological and semantical models are generated *(exploration)* or can be used as a priori information *(positioning)*. The architecture for processing sensor data is displayed in figure 2. The system's performance is demonstrated for the task of building a geometrical model of an unknown environment.

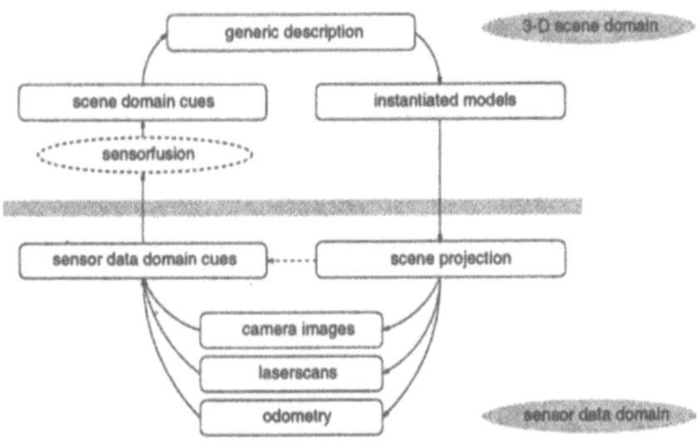

Figure 2: interpretation loop for the evaluation of sensor data

2 Sensors and processing of sensor data

The experiments described in this paper were carried out on the mobile robot PRIAMOS. PRIAMOS [4] (figure 1) is an omnidirectional, holonome Mecanum-wheel-driven platform for navigation experiments. In the following the sensors PRIAMOS is equipped with are described.

- **odometry** provides an estimation for the position and orientation of the robot.
- a **gyroscope** is used to measure the robot's orientation very accurately. The gyro is used to increase the accuracy of the odometry for rotational movements. The gyro is not described in this paper.
- 24 **sonar sensors** are mounted on PRIAMOS. Sonar sensors are capable of pro-

viding collision avoidance ability for a mobile system. This system is described in [13] and not content of this paper.

- a **structured-light** sensor is also used for collision avoidance and especially to detect descending stairs.

- the **stereo vision system** KASTOR (see figure 1) consists of three digital cameras which have nine motorized optical and mechanical degrees of freedom. The image-processing system is able to produce a symbolic contour description consisting of edges extracted from the images in video real-time. Additionally grey-value image-processing is performed in parallel on a programmable imaging computer. This enables the high performance of the system because edge-detection and grey-value processing are possible in parallel. The image-processing system is able to produce a symbolic contour description by edge segments extracted from the stereo images in real-time.

  The use of digital cameras advances image-processing because the quality of the images is much higher than the images of standard video cameras. The noise in the images is reduced by 30%. This is very important for edge-detection because edges are extracted from the images in a much more stable way. It also increases processing speed because smoothing is not necessary before edge-extraction any more. The full-frame capability of the camera also increases the accuracy of image-based measurements.

- a commercially available **laserscanner** (black box mounted on the front of PRIAMOS in figure 1) provides planar 180^0 scans at a rate of $25\,Hz$. 720 distance measurements are taken per scan. The used laserdiode of $10\,W$ is pulsed and so still eye-safe. The high power output of the laser enables measuring distances of up to $10\,m$ with an accuracy of $\pm 2cm$. Because of the high power output of the scanner there is no problem in detecting black, light-absorbing objects.

3 Obtaining 3D descriptions of the scene

In this section the reconstruction of the 3-dimensional scene with the trinocular vision system and the laserscanner is described. As scene features linear edge segments are used. These edge segments are represented by midpoint, direction-vector and half-length. The uncertainty of the segments is represented by a covariance matrix [3]. The xz-plane is the ground-plan of the coordinate system and the y-axis represents the height. As the laserscanner only provides 2-dimensional data the y-coordinate is always equal to the height in which the sensor is mounted on the robot.

3.1 3D reconstruction from trinocular stereo

In order to reconstruct a line segment the endpoints of the line $\mathbf{p_1}$ and $\mathbf{p_2}$ are reconstructed. The equations 1 to 5 describe the representation of a line segment by midpoint, normalized direction-vector and halflength and the corresponding covariance matrices for the representation of the uncertainty. For a minimal representation there is no uncertainty represented for the halflength.

$$\mathbf{x_m} = \frac{\mathbf{p_1}+\mathbf{p_2}}{2} \quad \text{midpoint} \quad (1)$$

$$\mathbf{r} = \frac{\mathbf{p_1}-\mathbf{p_2}}{\|\mathbf{p_1}-\mathbf{p_2}\|} \quad \text{normalized direction} \quad (2)$$

$$l = \frac{\|\mathbf{p_1}-\mathbf{p_2}\|}{2} \quad \text{halflength} \quad (3)$$

$$\Sigma_{\mathbf{x_m}} = \frac{\Sigma_{\mathbf{p_1}}+\Sigma_{\mathbf{p_2}}}{4} \quad \text{covariance of midpoint} \quad (4)$$

$$\Sigma_{\mathbf{r}} = \frac{\Sigma_{\mathbf{p_1}}+\Sigma_{\mathbf{p_2}}}{\|\mathbf{p_1}-\mathbf{p_2}\|^2} \quad \text{covariance of direction} \quad (5)$$

The state vector for a line segment in robot coordinates is given by $\mathbf{S}^{\mathbf{r}} = (\mathbf{x_m}, \mathbf{r}, l)$ with the covariance matrix

$$\mathbf{\Sigma}_{\mathbf{S}}^{\mathbf{r}} = \begin{pmatrix} \mathbf{\Sigma}_{\mathbf{x_m}} & 0 \\ 0 & \mathbf{\Sigma}_{\mathbf{r}} \end{pmatrix}.$$

Figure 3 shows one camera image from a stereo triple with the extracted line segments overlayed as white lines.

Figure 3: camera image

Figure 4 shows a projection of the corresponding trinocular stereo-reconstruction with the CAD-model of the environment overlayed.

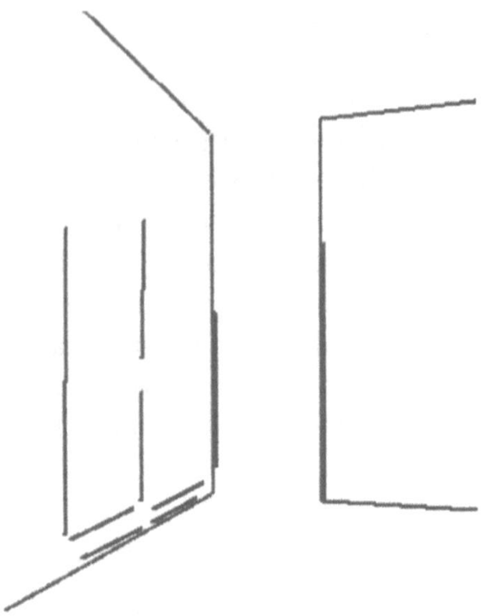

Figure 4: reconstruction from stereo camera

The accuracy of the stereo reconstruction Δx depends on the distance from the camera system to the scene-feature x, the distance of the cameras d, the focal length c and the uncertainty of the stereo matching Δu.

$$\Delta x = \frac{x^2 \Delta u}{c d} \tag{6}$$

For the values $x = 3m$, $d = 0.3m$, $c = 0.01m$ and $\Delta u = 1\,pixel = 10^{-5}m$ the stereo-reconstruction can be computed with an accuracy of $\Delta x = \pm 0.03m$.

3.2 3D descriptions obtained by a laser scanner

Figure 5 shows a laserscan that is acquired in a corridor environment. The sensor data provided by the laserscanner are 2-dimensional (ground plan) so y-coordinate is always 0. In order to show the quality of the laserscan the CAD-model of this environment is overlayed in grey lines.

The next step of processing is the extraction of linear edge segments from the laserscan. This is done by using the *iterative end-point fit* [16] algorithm for the determination of points belonging to a line segment. In order to obtain the same representation for the line segments of the laserscanner as for the line segments from the stereo-reconstruction a least-square approximation is computed and the extracted line segments are represented according to equations 1 to 5.

Figure 5: raw data from laserscan

3.3 Transformation to world coordinates

The line segments are so far represented in robot coordinates. The state vector of the robot is given by $\mathbf{x_r} = (x_r, z_r, \phi)$. For a common robot independent representation the

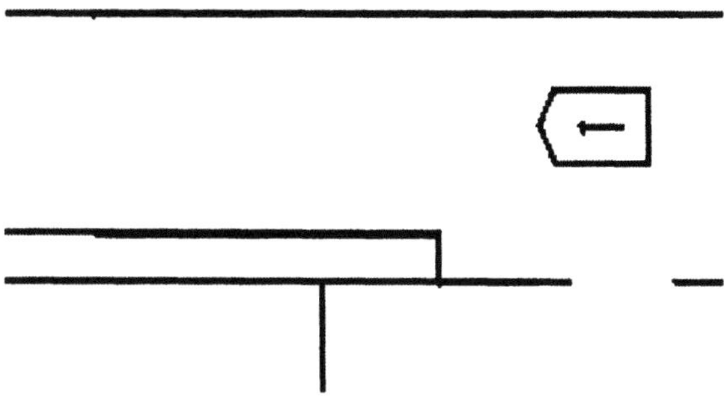

Figure 6: example for edge extraction by *iterative end point fit* combined with least-square approximation

transformation to world coordinates is necessary. This transformation is given by the rotation

$$\mathbf{R} = \begin{pmatrix} \cos(\phi) & 0 & \sin(\phi) \\ 0 & 1 & 0 \\ -\sin(\phi) & 0 & \cos(\phi) \end{pmatrix} \quad (7)$$

and the translation

$$\mathbf{T} = \begin{pmatrix} x_r \\ 0 \\ z_r \end{pmatrix} . \quad (8)$$

The transformation of the state vector in robot coordinates $\mathbf{S^r} = (\mathbf{x_m}, \mathbf{r}, l)$ to the state vector in world coordinates $\mathbf{S^w}$ becomes

$$\mathbf{S^w} = (\mathbf{R}\,\mathbf{x_m} + \mathbf{T},\ \mathbf{R}\,\mathbf{r},\ l)\,. \quad (9)$$

The propagation of the uncertainty of a random vector $\mathbf{x}$ with covariance matrix $\mathbf{\Sigma_x}$ under the transformation $\mathbf{y} = \mathbf{f}(\mathbf{x})$ is in a first order approximation given by

$$\mathbf{\Sigma_y} = \left.\frac{\partial \mathbf{f}(\mathbf{x})}{\partial \mathbf{x}}\right|_{\mathbf{x}=\bar{\mathbf{x}}} \mathbf{\Sigma_x} \left.\frac{\partial \mathbf{f}(\mathbf{x})^T}{\partial \mathbf{x}}\right|_{\mathbf{x}=\bar{\mathbf{x}}} . \quad (10)$$

With this equation the uncertainty of the state vector in world coordinates [7] becomes

$$\mathbf{\Sigma^w_{x_m}} = \mathbf{R}\,\mathbf{\Sigma^r_{x_m}}\,\mathbf{R}^T + \left(\frac{\partial \mathbf{R}}{\partial \phi}\mathbf{x_m}\right) \mathbf{\Sigma_R} \left(\frac{\partial \mathbf{R}}{\partial \phi}\mathbf{x_m}\right)^T + \mathbf{\Sigma_T} \quad (11)$$

and

$$\mathbf{\Sigma^w_r} = \mathbf{R}\,\mathbf{\Sigma^r_r}\,\mathbf{R}^T + \left(\frac{\partial \mathbf{R}}{\partial \phi}\mathbf{r}\right) \mathbf{\Sigma_R} \left(\frac{\partial \mathbf{R}}{\partial \phi}\mathbf{r}\right)^T , \quad (12)$$

with $\mathbf{\Sigma_R}$ and $\mathbf{\Sigma_T}$ beeing the uncertainties in $\mathbf{R}$ and $\mathbf{T}$.

4 Exploration

In order to explore an environment the robot is provided with a topological model (figure 7) and a certain mission (direction and distance to travel) is specified. The geometrical world model is built up incrementally. The local perception is matched with the global model and, if possible, fused according to the following section. The described technique is capable of dealing with environments that can be modelled by linear edge segments, which is the case for most indoor environments. The environment does not necessarily have to be rectangular though.

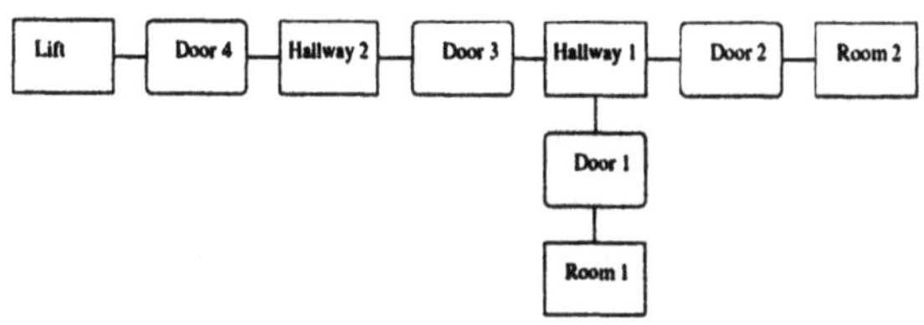

Figure 7: topological model

4.1 Fusion of symbolic edge segments

In this work linear edge segments are represented by midpoint, normalized direction vector and halflength because this representation is advantageous for the fusion of segments. This means an edge segment is defined by a state vector $\mathbf{S}(x_m, y_m, z_m, r_x, r_y, r_z, l)$ and the uncertainty is given by the covariance matrix $\mathbf{\Sigma_S}$.

In order to find corresponding scene features (nearest neighbor matching) in the local perception and the global model the Mahalanobis distance is applied. The Mahalanobis distance is a distance criterion for two state vectors normalized by the sum of their covariance matrices. The squared Mahalanobis distance is given by

$$d^2(\mathbf{S_0}, \mathbf{S_1}) = (\mathbf{S_0}-\mathbf{S_1})^T(\mathbf{\Sigma_{S_0}}+\mathbf{\Sigma_{S_1}})^{-1}(\mathbf{S_0}-\mathbf{S_1}).$$

The squared Mahalanobis distance defines a χ^2-distribution [2] if the means $\bar{\mathbf{S}}_0 = \bar{\mathbf{S}}_1$. For a segment in the local perception the nearest neighbor in the global model is defined by the minimal Mahalanobis distance. The two segments can be fused using a Kalman filter, because the measurements have a Gausian distribution and the system can be described by linear equations.

$$\begin{aligned} \mathbf{K} &= \mathbf{\Sigma_0}\,(\mathbf{\Sigma_0}+\mathbf{\Sigma_1})^{-1} && \text{Kalman-G.}(13) \\ \mathbf{S} &= \mathbf{S_0} - \mathbf{K}\,(\mathbf{S_0}-\mathbf{S_1}) && \text{fused segm}(14) \\ \mathbf{\Sigma} &= (\mathbf{I}-\mathbf{K})\,\mathbf{\Sigma_0} && \text{covariance }(15) \end{aligned}$$

Every time an edge segment is fused with a segment from the local perception its confidence C is incremented by one [3]. For multiple observations it is so possible to reject segments from the world model which have only been observed once and could be dynamic objects or artefacts. Futher details on the multi-sensor-fusion can be found in [14].

4.2 Ground plan exploration

In the following example the robot drove about 50 meters through a hallway environment and generated a geometrical map using the laserscanner and the above described techniques.

In figure 8 all percepted edges are displayed where in figure 9 only the edges are displayed which have a confidence $C \geq 5$. It is obvious that the number of edges can be reduced drastically without loosing important information about the modeled environment.

The accuracy of the generated map is basically defined by the odometry of the robot. This means that this map could be used as a priori information for later navigation tasks in order to enable position correction.

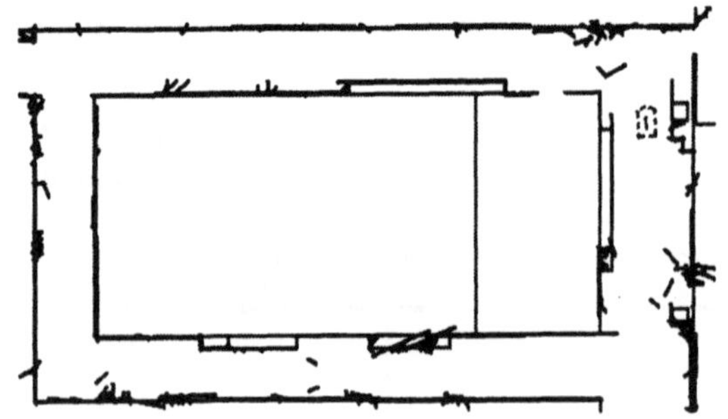

Figure 8: sensor fusion with laserscanner (confidence $C \geq 0$)

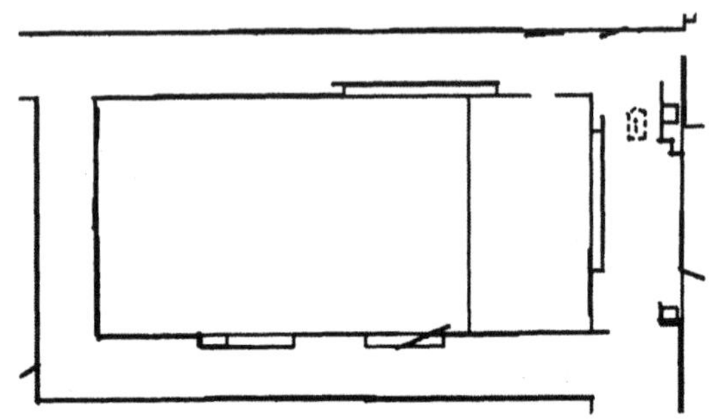

Figure 9: sensor fusion with laserscanner (confidence $C \geq 5$)

4.3 3-dimensional map building

In this second example an approach to fusing 3-dimensional stereo-reconstructions is presented. In this case the robot drives through a lab-environment. In figure 10 to 12 the fused reconstructions from a sequence of stereo triples and laserscans are displayed with different confidence values. Again the exact CAD-model of the environment is overlayed. Here the 3-D reconstruction is fully automatically computed with an accuracy of $\pm 3cm$ for the vertical edges.

Figure 10: reconstruction from a sequence of stereo-triples (confidence $C \geq 0$)

Figure 11: reconstruction from a sequence of stereo-triples (confidence $C \geq 1$)

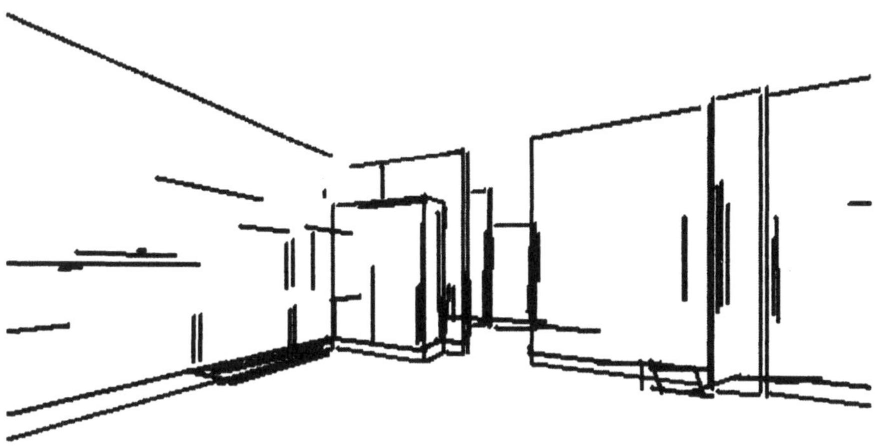

Figure 12: reconstruction from a sequence of stereo-triples (confidence $C \geq 4$)

The stereo vision system is the most powerful sensor that is currently used on PRIAMOS. It requires more effort to reconstruct scenes by stereo than with a laserscanner though.

5 Evaluation of Depth-Images

In the future the planar laserscanner will be replaced by the Range Imaging Camera (EBK[1]) (figure 13) manufactured by Dornier GmbH [6]. The EBK System provides depth images with an resolution of 128×64 pixels with a frame-rate of 5Hz with a field of view of $60° \times 30°$. The imaging range reaches from 5 to 50 meters. The EBK performs time of flight measurements of a 10W laserbeam. The depth-images are comparable to the depth-images acquired by the Video-Rate Stereo Machine [10]. This step will again improve the performance of the presented system, especially for 3-D modelling.

Figure 13: EBK

In order to evaluate the acquired sensor information an adequate segmentation of the depth-images has to be performed. The specific segmentation algorithm has to be selected according to the application. In this paper the given problem is to build a geometrical model of an indoor-environment which finally can be used for robot navigation. As robot navigation is a time-constrained problem the segmentation algorithm needs to real-time capable. In an indoor-environment it is usually possible to use planar surface patches for the geometrical description. For the given reasons a segmentation algorithm was developed which is based on the *iterative end point fit* of every single line of the acquired depth-image. This means that an extraction of linear edge segments is performed in every line of the depth image. Parallel line segments are then fused over the columns of the image. The result of this algorithm is the determination of the pixels that belong to common planar surface patches [9]. Finally a least square approximation is applied to these pixels to derive a symbolic description for the surface patches.

Figure 14: camera-image

In figure 14 to 17 an example for the evaluation of a depth-image is presented. Figure 14 shows an camera-images of an indoor-scene to be analyzed. The scene is in a lab-environment containing various objects. Figure 15 shows the depth-image which is acquired by the EBK. In this example it is the goal to model the walls of the displayed scene. The segmentation algorithm described above detects the planar patches which are overlayed in figure 15 and again displayed in figure 16. A least square approximation

[1] Entfernungs-Bild-Kamera

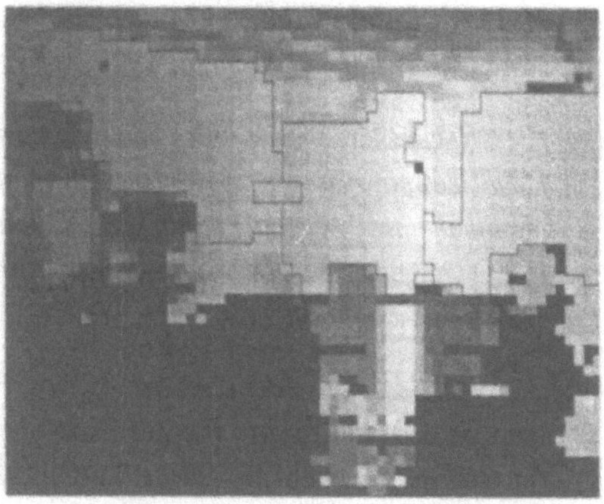

Figure 15: depth-image

is computed for the detected planar patches which results in a symbolic description of the surface patches. In figure 17 the projection of these patches into the ground-plan is displayed. It becomes obvious here that it is possible to detect and model the walls even in a very complex indoor-scene. It was experimentally shown that is possible to build a geometrical model of the navigation-relevant features of in indoor environment with the techniques described above.

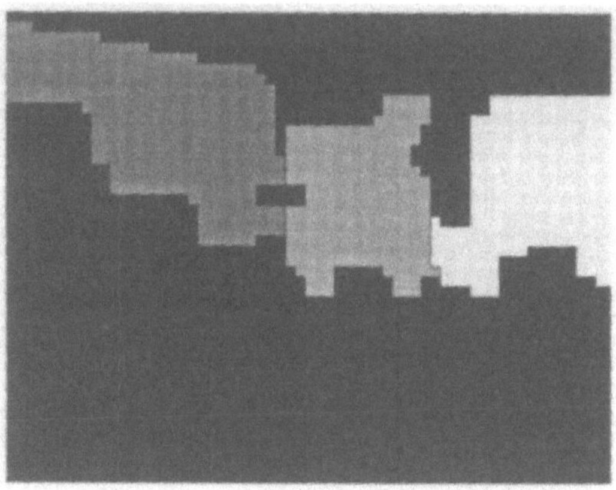

Figure 16: segmented depth-image

6 Conclusion

In this paper it was shown that stereo-reconstructions and planar laserscans can be represented in a common format. This enables the fusion of complementary sensor information. Experimental results were presented for 3-dimensional map building of an unknown environment. The accuracy of the generated model enables using this model for position correction during later on-line navigation tasks. It will also be the goal to derive topological and semantic knowlege from the geometrical reconstructions using generic object models.

Figure 17: ground-plan representation

Acknowledgement

This work was performed at the Institute for Real-Time Computer Control Systems & Robotics, Prof. Dr.-Ing. U. Rembold and Prof. Dr.-Ing. R. Dillmann, Department of Computer Science, University of Karlsruhe, 76128 Karlsruhe, Germany.

References

[1] F. Arman and J.K. Aggarwal. Model-based object-recognition in dense range images - a review. *ACM Computing Surveys*, 25(1):5 – 43, 1993.

[2] J.L. Crowley. *Principles and Techniques for Sensor Data Fusion*, volume 99 of *NATO ASI, Multisensor Fusion for computer Vision*, chapter 1, pages 15–36. J.K. Aggarwal, 1989.

[3] J.L. Crowley, P. Stelmaszyk, and P. Skordas, T. Puget. Measurement and integration of 3-D structures by tracking edge lines. *International Journal of Computer Vision*, 8(1):29 – 52, 1992.

[4] R. Dillmann, J. Kreuziger, and F. Wallner. The control architecture of the mobile system priamos. In *Proceedings of the 1st IFAC International Workshop on Intelligent Autonomous Vehicles. Southampton*, 1993.

[5] T. Edlinger and G. Weiß. Exploration, navigation and self-localization in an autonomous mobile robot. In *Autonome mobile Systeme*, pages 142 – 151. Springer-Verlag, 1995.

[6] M. Eibert, H. Hopfmüller, C. Schaefer, and K.R. Schulz. Laserkamera zur 3D-Vermessung. Technical report, Dornier, 1995.

[7] O. Faugeras. *Three-Dimensional Computer Vision*. MIT-Press, 1995.

[8] L. Feng, J. Borenstein, and H.R. Everett. *Where am I? Sensors and Methods for Autonomous Mobile Positioning*. University of Michigan, 1994.

[9] M. Fiegert. Experimentelle Untersuchungen zur Umweltmodellierung mit einem 3D-Laserscanner. Master's thesis, Institut für Prozeßrechentechnik und Robotik, Universität Karlsruhe, Deutschland, 1996.

[10] T. Kanade, H. Kano, S. Kimura, A. Yoshida, and O. Kazuo. Development of a video-rate stereo machine. In *Intelligent Robots and Systems*, pages 95–100. IROS, 1995.

[11] J.J Leonard and H.F. Durrant-Whyte. *Directed sonar sensing for mobile robot navigation*. MIT Press, 1992.

[12] E. Triendl and D. J. Kriegman. Stereo vision and navigation within buildings. In *International Conference on Robotics and Automation*. IEEE, 1987.

[13] F. Wallner, R. Graf, and R. Dillmann. Real-time map refinement by fusing sonar and active stereo-vision. In *IEEE International Conference on Robotics and Automation, Nagoya*, 1995.

[14] P. Weckesser. *Aktiver Einsatz eines Multisensorsystems zur Exploration der Umwelt mit einem mobilen Roboter*. PhD thesis, University of Karlsruhe, 1997.

[15] P. Weckesser, G. Appenzeller, A. von Essen, and R. Dillmann. Exploration of the environment with an active and intelligent optical sensor system. In *International Conference on Intelligent Robots and Systems*. IROS, November 1996.

[16] P. Weckesser and R. Dillmann. Sensorfusion of intensity- and laserrange-images. In *Proceedings of IEEE International Conference on Multisensor Fusion and Integration for Intelligent Systems*, pages 501–508, 1996.

Integration of Topological Map and Behaviors for Efficient Mobile Robot Navigation

Hyun S. Yang Byeong-Soon Ryu Jiyoon Chung*
Center for Artificial Intelligence Research / Dept. of Computer Science
KAIST, 373-1 Kusong-dong, Yusong-ku, Taejon, 305-701, Korea
email: hsyang@cs.kaist.ac.kr

Abstract

A navigation system for an autonomous mobile robot working in indoor environments is presented. The system takes advantage of *deliberativeness* and *reactivity*. In the reactive part of the system are local *behaviors* that are independent, action-generating entities. We also provide high-level deliberative modules so that the interactions become manageable and the system can accomplish more meaningful tasks. The deliberative modules control the activation and deactivation of each local behavior based on the current *situation*, which represents the current position of the robot and the path to the goal.

The situation is determined by a mapping subsystem that consists of an *enhanced topological map*, a localization module and a planning module. The use of an explicit world model, especially in a topological manner, makes it possible to reliably localize the robot and plan an efficient path.

1 Introduction

In order for an autonomous mobile robot to accomplish given tasks in a real environment, it must make decisions and execute actions in real-time as well as cope with various uncertainties. These uncertainties arise for various reasons. Knowledge of the environment is partial and approximate; it is changed dynamically and the changes can only be partially predicted; the robot's sensors are imperfect and noisy; and the robot's control is not precise.

Traditional *deliberative approaches* to planning and controlling mobile robots have been criticized for not being able to adequately cope with uncertain and complex environments. As an alternative, a number of *reactive approaches* have been proposed [1, 2]. These approaches handle uncertainty and unpredictable changes well by giving up the idea of modeling and reasoning about the environment and the future consequences of actions. To be successful, however, such systems should provide a mechanism to make the cooperation of the reactive modules manageable and meaningful.

One way to solve this problem is to limit interactions by adding top-down constraints that take advantage of regularities and knowledge in the operational domain in order to coordinate actions. A number of systems realizing this idea have been proposed since the end of the 1980s. Among them are Arkin[3], Mataric[4], and Saffiotti[5].

Although the previous work shows much improvement, they still have the weaknesses in utilizing world knowledge. Most of their systems use very simple representations of the world and provide simple localization algorithms. Even if the brief representation is sufficient, since the detail can be covered by the reactive modules, the localization can't be overlooked. This paper also focuses on developing a navigation system that utilizes both deliberativeness and reactivity. However, we have put special emphasis on the world model and the reliable localization algorithm.

2 Navigation System Architecture

Figure 1 depicts a block diagram of the proposed navigation system architecture. The system consists of three functional subsystems; the mapping subsystem, the control subsystem, and the perception subsystem. We have designed each subsystem to be independent from the others, so that

*He is currently with Communication System Division, Hyundai Electronics Industries Co.

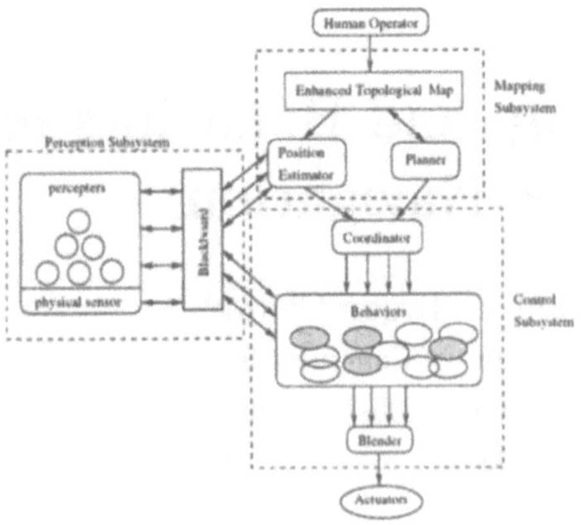

Figure 1: A block diagram of the proposed navigation system architecture.

we can modify one of them without changing the others.

Most of all the tasks to be accomplished by mobile robots are navigational ones; i.e., the robots must navigate autonomously from one place to other place. Thus, a human operator gives his robot the instructions with a world model. In this paper, we propose an *enhanced topological map* as a world model. The map represents the world in a topological manner with rough metrical information and provides information to two modules; a *position estimator* and a *planner*. Those modules determine the current situation of the robot, and they are capable of coping with the environmental and sensory uncertainties.

The control subsystem is composed of various modules called *behaviors*, each of which has its own goals and runs independently from others. As mentioned earlier, the behaviors are good at coping with uncertain and dynamic environments, but some additional mechanisms to manage the behaviors are required to be really useful in real world applications. For this reason, the proposed system provides two modules. One is a *coordinator* that controls the activation and deactivation of certain behaviors according to the current situation determined by the mapping subsystem, and the other is a *blender* that combines the multiple action commands produced by the activated behaviors into composed ones for the actuators.

The perception subsystem also consists of independent computational modules called *perceptors*. They are structured hierarchically from raw sensory data to high-level abstract information. Since each of them has its own thread of computation, delay in a perceptor does not make other perceptors delay. Use of a global storage called *blackboard* for communications among the perceptors and other two subsystems also makes the system more manageable and extensible.

3 Mapping Subsystem

3.1 Enhanced Topological Map

As mentioned earlier, we have proposed an enhanced topological map (ETMap) as a world model. The basic structure of the ETMap is a topological model consisting of nodes and links. However, it is different from traditional topological maps in that it is designed with special emphasis on efficient and reliable localization and navigation of mobile robots even with imprecise sensors and simple perception modules. As a result, it has the following three characteristics:

- It is provided with a topological structure with rough metrical information.
- It is provided with a reliable localization algorithm based on dead reckoning.
- It is provided with a planning algorithm for reliable localization and navigation, which considers sensory uncertainties.

Although this paper focuses mainly on the ETMap, we like to present our previous work using the pure topological map, too. It is because each algorithm (indeed the localization algorithms) has its own advantages and disadvantages. We are currently studying integrating the advantages of those two approaches.

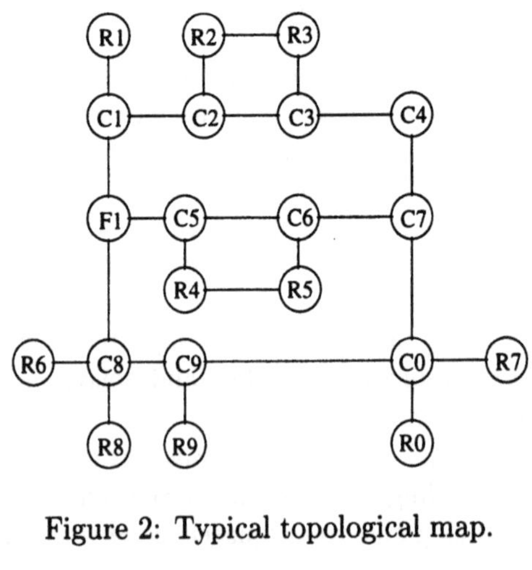

Figure 2: Typical topological map.

Figure 2 shows a typical topological map. The same representation of the map as the figure was given in the third international mobile robot competition sponsored by AAAI in 1994. The circles in the figure denote the nodes of a topological map and the line segments connecting the circles denote the links. The first character of each node classifies the node: e.g., 'R' stands for 'Room', 'C' stands for 'Corridor' and 'F' stands for 'Foyer'. The number written behind the classification letter identifies each node among the nodes of the same class. The map is used in explaining the first localization algorithm, localization using fuzzy automata, while the ETMap shown in Figure 3 is used in explaining the second algorithm, localization based on dead reckoning.

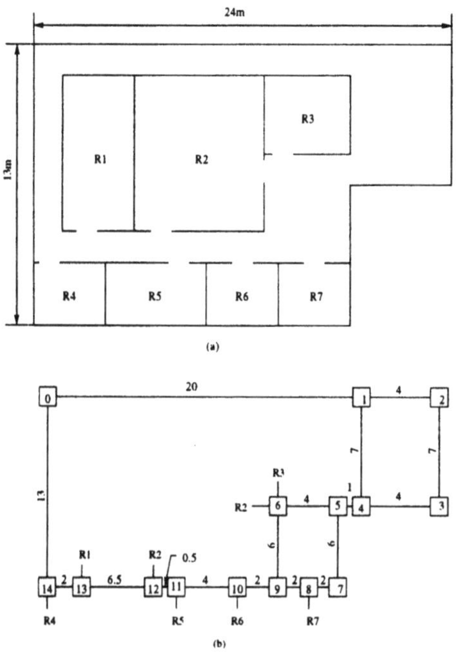

Figure 3: (a) A typical indoor environment and (b) its corresponding ETMap.

Figure 3 shows a typical indoor environment around our laboratory and a corresponding ETMap. As the figure shows, the ETMap has a topological structure, where nodes are *landmark places* (LPs) and arcs are *adjacency links* (ALinks). The LPs are drawn in rectangles and the ALinks are drawn in lines. In addition to a pure topological structure, the ETMap has metrical information (the length and orientation of the ALinks) although they need not be precise. Indeed, in our implementation, the length is given with the precision of 10cm and can be oriented in four primary directions; North, South, East, and West.

Besides ALinks, the ETMap has auxiliary links called *neighborhood links* (NLinks). While ALinks denote physical adjacency, NLinks denote logical adjacency, which we call *neighborhood.* We define the neighborhood of an LP (starting LP) as a set of LPs that can be reached by the robot from the starting LP without a drastic change of orientation. As the robot moves along the NLinks skipping several LPs, the robot can move at maximum speed.

3.2 Position Estimator based on Fuzzy Automata

Let's consider that there is a robot in a room that can get a lot of topological features necessary to confirm its position on a topological map while it is moving to another room. The navigation thus is the same as graph traversing, so we have expanded a topological map into a pseudo sequential machine in order to consider navigation as the operation of such a sequential machine. It can be exactly the same as deterministic finite automata if all possible percepts can be collected. Unfortunately, the inputs are determined uncertainly and they can be lost or duplicated, so we apply neither nondeterministic nor deterministic sequential machines for modeling navigation. William G. Wee introduced fuzzy automata to handle fuzzy values and fuzzy states [6]. We adopted fuzzy automata to estimate the position of the robot.

A *fuzzy automaton* is defined as a quadruple

$$\tilde{A} = (Q, \tilde{q}_0, I, \delta)$$

where
Q is a finite set of states,
$\tilde{q}_0$ is a fuzzy subset of Q called the fuzzy initial state,
I is a finite set of input symbols, and
$\delta : Q \times I \to Q$ is a transition function that maps states and inputs onto states.

The finite set of states, Q, corresponds to the set of nodes of the topological map, while a fuzzy subset of Q at time t, $\tilde{q}_t$, represents the possibility distribution over the nodes. This means every node has the possibility that the robot is located at the node. Since this position estimator keeps every possibility even it is very small, it is able to cope with environmental uncertainties and sensory noises. The fuzzy initial state, $\tilde{q}_0$, corresponds to the initial distribution of the possibilities about the robot's initial position. If the robot's initial position is known, $\tilde{q}_0$ is set correspondingly; i.e., assigns 1 to the node in which the robot initially is and 0 to the others. Otherwise, the values are assigned with uniform distribution or certain distribution according to *a priori* information. We have observed that the initial distribution is not critical to overall system performance.

Once the initial state, $\tilde{q}_0$, is set, the position estimator observes the environment and update the states according to the result of the observation. The state at time t, $\tilde{q}_t$, is estimated by the transition function, δ, as defined in the above. The definition implies that the state is estimated from the previous state and a set of input symbols. More precisely, we define the transition function as follows:

$$\tilde{q}_{t+1} = \delta(\tilde{q}_t, i) \tag{1}$$

where the membership of the state (or the possibility distribution) is defined as follows:

$$\mu_{t+1}(q) = \max_{q' \in N_q} \{(1-\rho)u(q', q, i) + \rho\mu_t(q')\} \tag{2}$$

where ρ is a coefficient, which determines the weight between new information and the history, and the N_q denotes a set of neighboring nodes of node q. In the transition function, certainties from all previous states (history) are not explicitly considered but implicitly. Since the certainties of states from 0 to $t-1$ affects the certainty of state at t recursively to some extent, we may not have to consider all history explicitly again to compute the certainty at $t+1$. Furthermore, if we provide too much weight to all previous ones, we might lose the advantage of the fuzzy automata in estimator, making system deterministic. From simulation and real world experiment, we confirmed our rationale.

The $u(q', q, i)$ represents an input function, which maps the set of input symbols to the possibility that the robot have moved from node q' to node q. We define the input function as follows:

$$u(q', q, i) = \sum_{l \in L_{q'q}} \{w_l C(l|i)\} \tag{3}$$

where the $L_{q'q}$ denotes a set of landmarks observable in the path between node q' and node q, and w_l is a coefficient that determines the importance of landmark l comparing to the other landmarks included in $L_{q'q}$. The $C(l|i)$ represents the *likelihood* that landmark l exists when the sensors have detected a landmark as it is landmark i. The likelihood is determined by the certainty matrix as shown in Table 2.

3.3 Position Estimator based on Dead Reckoning

Although various localization schemes for topological maps (including the algorithm described in the previous subsection) have been proposed, most previous works on topological approaches have inherent limitations as they have concentrated only on finding distinctive places to localize the robots but have seldom used metrical information. Even if they provide probabilistic or heuristic tools to select the best possible one among similar places, they have inherent limitations since they ignore metrical information. Suppose a robot is navigating down a corridor that has two adjacent and similar doors as shown in Figure 4. Since the robot's sensors are imperfect, one of the doors can be missed. For example, if the robot missed door 1 and detects a door in front of door 2, it might be unable to decide whether it is door 1 or door 2 without reference to such metrical information as the distance traveled.

Figure 4: An illustration of the proposed localization scheme.

In contrast, the algorithm proposed in this paper can solve this problem by utilizing dead reckoning. It first calculates the *Estimated Interval* (EI) of the target location and then navigates until the estimation of position using dead reckoning falls into the EI. Finally, it fine-tunes the robot's position using sensor-based landmark detection methods. We will explain how the EI and other intervals are used for localization and path planning in the following sections.

The idea of localization was motivated by the observations of human activities. When a person goes to a place where he does not know the way well, he depends on a map and looks for landmarks represented in the map. He seems not to evaluate the probabilities but rather to try to find distinctive landmarks. In addition he seems to estimate the possible distance to where the landmark can be found so that he realizes when he goes too far without detecting any landmark.

Generally speaking, it is hard to use dead reckoning alone because it has a large cumulative error. However, we have found that the *projected traveled distance* (PTD), which is the distance traveled projected to the major axis of the robot's motion, can be estimated quite reliably within the bound of small error when the motion of the robot is linearly constrained. The motion of the robot is linearly constrained when the robot navigates along a linear corridor, along a linear wall, along a linear road, or continuously aiming at a static target. Such situations are common in the robot's working environments.

In this position estimator, we assume most indoor navigation, especially office environment navigation, includes linear type motion to large extent without the loss of generality. In that case, the PTD is measured rather reliably. Of course, in case robot does not follow the linear path, we may not rely on dead reckoning. However, using the digital compass or gyro, we can adjust orientation very accurately, resulting in the decrease of distance error.

Below is a brief description of the proposed algorithm:

Step 1 We assume that the initial configuration of the robot is known although it need not be very precise.

Step 2 It selects the next place from the path planned by the planner. The planner plans the path so any path between adjacent places is linear.

Step 3 It extracts information on the EI of that place. The EI is pre-calculated with a constant error ratio for every NLink when the robot plans the path.

Step 4 It resets the orientation of the robot so the direction of the path to be followed becomes zero. Since we know the initial configuration, there is no problem in resetting the orientation at the initial place. However, at other places, the orientation of the robot should be estimated before resetting it because the configuration of the robot at that time is only estimated with dead reckoning and thus has a lot of error in its orientation. We have developed an algorithm to estimate the orientation using ultrasonic range finders and the Hough transform.

Step 5 While the robot navigates along the path, it estimates the PTD and moves to Step 6 when the PTD gets into the EI.

Step 6 It invokes the landmark detection modules related to the place.

Step 7 If any landmark detection module detects a landmark, it means that the target place is recognized. If the target place is the goal place, quit navigation. Otherwise, go to Step 2.

Step 8 If the PTD exceeds the EI without detecting any landmark, the localization module signals the robot to invoke the failure recovery module, which is described later.

3.3.1 *Distance Intervals*

To plan the path and localize the robot, the system introduces some notions of *distance intervals* as shown in Figure 5. The *Target Interval* (TI) denotes the interval where the robot can detect landmarks. The other two intervals are wider than the TI due to the dead-reckoning error. The *Estimated Interval* (EI) is the interval within which the estimation lies when the robot is actually in the TI. Since the robot knows only the estimation of its position, it uses the EI to localize itself. The *Planning Interval* (PI) is the interval within which the robot actually lies when the estimation of its position lies within the EI. The PI refers to the actual position of the robot when it estimates that its position is within the EI. So the PI is used by the path planner.

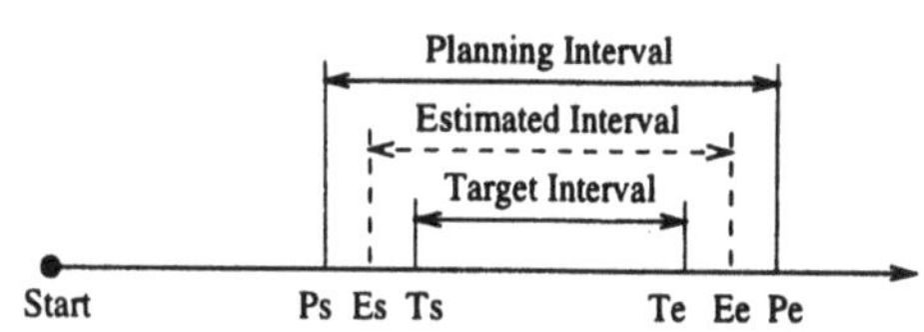

Figure 5: Distance Intervals.

3.3.2 *Projected Traveled Distance*

We are now going to show that the PTD can be estimated quite reliably. We prove it analytically and then show some experimental results.

The kinematics of a robot with two driving wheels is given by:

$$\begin{aligned} \dot{x} &= ((v_r + v_l)/2)\cos\theta &&= v\cos\theta \\ \dot{y} &= ((v_r + v_l)/2)\sin\theta &&= v\sin\theta \\ \dot{\theta} &= (v_r - v_l)/b &&= w \end{aligned} \tag{4}$$

where the state of the system (4), $q = [x, y, \theta]^T$, is the position of the wheel axis center, (x, y), and the orientation of the robot, θ, with respect to the x-axis. The velocities v_l and v_r are the tangent velocities of each wheel at its center of rotation and b is the distance between the wheels. Suppose the state of the robot at time t_0 is $q_0 = [x_0, y_0, \theta_0]^T$ and the robot moves with certain linear and angular velocities, v_1 and w_1, for the time interval Δt. Then the new state of the robot, $q_1 = [x_1, y_1, \theta_1]^T$, is obtained as follows:

$$\begin{aligned} x_1 &= x_0 + \int_{t_0}^{t_1} v_1 \cos\theta dt \\ &= x_0 + v_1(\sin\theta_1 - \sin\theta_0)/w_1 \\ &= x_0 + v_1 \tfrac{\sin\theta_1 - \sin\theta_0}{\theta_1 - \theta_0}\Delta t \\ y_1 &= y_0 + \int_{t_0}^{t_1} v_1 \sin\theta dt \\ &= y_0 - v_1(\cos\theta_1 - \cos\theta_0)/w_1 \\ &= y_0 - v_1 \tfrac{\cos\theta_1 - \cos\theta_0}{\theta_1 - \theta_0}\Delta t \\ \theta_1 &= \theta_0 + w_1\Delta t. \end{aligned} \tag{5}$$

When the robot moves along a wall, it seldom changes its direction except when it avoids obstacles. Generally, the number of obstacles is relatively small so the influence of the obstacles can be ignored. Therefore, $\theta_1 \approx \theta_0$ and the equation becomes:

$$\begin{aligned} x_1 &= x_0 + v_1 \cos\theta_1 \Delta t \\ y_1 &= y_0 + v_1 \sin\theta_1 \Delta t \\ \theta_1 &= \theta_0 \end{aligned} \tag{6}$$

Then, the position of the robot after k time steps, (x_k, y_k), is:

$$\begin{aligned} x_k &= x_0 + \textstyle\sum_{i=1}^{k} v_i \Delta t \cos\theta_i \\ y_k &= y_0 + \textstyle\sum_{i=1}^{k} v_i \Delta t \sin\theta_i \end{aligned} \tag{7}$$

In this equation, we implicitly included the orientation error term and derived the distance error as a function of maximum orientation error.

Let ε_i be a proportional error in velocity and $\hat{q}_i = [\hat{x}_i, \hat{y}_i, \hat{\theta}_i]^T$ be an estimation of the state of the robot at time i. Then the estimation of the position of the robot at time k is:

$$\begin{aligned} \hat{x}_k &= \hat{x}_0 + \textstyle\sum_{i=1}^{k} v_i(1+\varepsilon_i)\Delta t \cos\hat{\theta}_i \\ \hat{y}_k &= \hat{y}_0 + \textstyle\sum_{i=1}^{k} v_i(1+\varepsilon_i)\Delta t \sin\hat{\theta}_i \end{aligned} \tag{8}$$

When the initial position is given, $\hat{q}_0$ becomes identical to q_0. The positional error at time k, $E^k = (E_x^k, E_y^k)$, is then given as follows:

$$\begin{aligned} E_x^k &= |\hat{x}_k - x_k| \\ &= |\textstyle\sum_{i=1}^{k} v_i(1+\varepsilon_i)\Delta t \cos\hat{\theta}_i \\ &\quad - \textstyle\sum_{i=1}^{k} v_i \Delta t \cos\theta_i| \\ &= |\textstyle\sum_{i=1}^{k} v_i \Delta t(\cos\hat{\theta}_i - \cos\theta_i) \\ &\quad + \textstyle\sum_{i=1}^{k} \varepsilon_i v_i \Delta t \cos\theta_i| \\ E_y^k &= |\hat{y}_k - y_k| \\ &= |\textstyle\sum_{i=1}^{k} v_i(1+\varepsilon_i)\Delta t \sin\hat{\theta}_i \\ &\quad - \textstyle\sum_{i=1}^{k} v_i \Delta t \sin\theta_i| \\ &= |\textstyle\sum_{i=1}^{k} v_i \Delta t(\sin\hat{\theta}_i - \sin\theta_i) \\ &\quad + \textstyle\sum_{i=1}^{k} \varepsilon_i v_i \Delta t \sin\theta_i| \end{aligned} \tag{9}$$

Let D be the total distance traveled, so that it satisfies $D = \sum v_i \Delta t$. Several experiments show that the ε_i's are small enough to ignore, $\theta_i \approx 0$ and $\hat{\theta}_i \leq \alpha$ for a quite small constant α in most cases. Therefore, equation 9 can be simplified as follows:

$$\begin{aligned} E_x^k &\leq D(1-\cos\alpha) \\ E_y^k &\leq D\sin\alpha. \end{aligned} \tag{10}$$

Considering the characteristics of the sine and cosine functions around the origin, E_x^k is relatively smaller than E_y^k, and E_x^k is tolerable for qualitative indoor navigation. For instance, suppose the directional error, α, is 10° then E_x^k and E_y^k become $0.015D$ and $0.174D$ respectively; when the robot moves 10m, E_x^k is only 15cm while E_y^k is 174cm. Therefore, the PTD can be quite reliably estimated as long as the robot moves along a linear path.

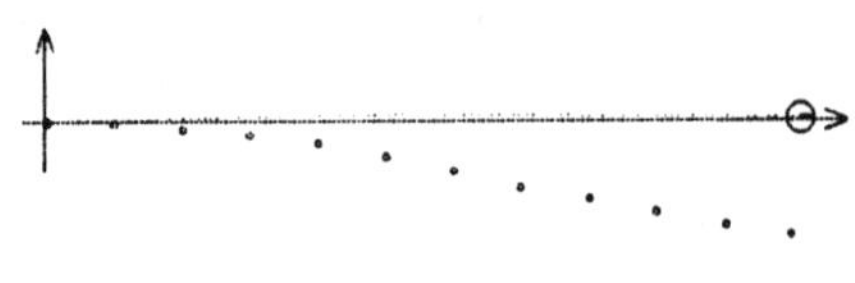

Figure 6: Sim1: a simulation in which a robot moves along the x-axis.

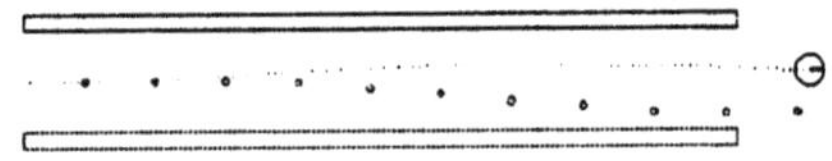

Figure 7: Sim2: a simulation i which a robot navigates down a corridor.

Figure 6 and Figure 7 show the simulated results. The robot moves along the x-axis in Figure 6 and along a corridor in Figure 7. In those figures a dotted line denotes the robot's trail while the small circles denote the estimated positions.

Table 1: Results of the simulations. (The errors are represented in %)

	Actual	Estimated	E_x	E_y
Sim1	(1647,3)	(1634,-235)	0.79	14.45
Sim2	(1632,12)	(1643,-48)	0.67	3.68

Table 1 shows the actual and estimated positions and positional errors of those experiments.

3.3.3 *Failure Detection and Recovery*

The planning algorithm described in the next subsection selects a path and the corresponding landmarks so that only one location has selected landmarks. So, the localization algorithm doesn't have to distinguish among the locations. Instead, it should be concerned with the situation where it doesn't detect any landmarks in the EI. When this situation occurs, the localization module signals 'Failed to Detect' and invokes the recovery module.

There may be many possible approaches to recovery. In our implementation, the recovery module makes the robot turn around and search for the landmarks again more carefully within the EI.

3.4 Planning with Sensory Uncertainty

As explained before, we plan the optimal path on the NLinks not on the ALinks for more efficient navigation. Efficiency is achieved by skipping several LPs. However, at the same time, the planner should be concerned with reliable sensing. The proposed localization algorithm first finds the EI using dead reckoning and then detects the landmarks within the EI. In order for the algorithm to be successful, the landmarks have to be unique in the EI. In other words, there should be only one door in the EI when we want to find a door. However, because of cumulative dead reckoning error, the EI becomes wider as the distance of the NLink becomes longer. Therefore, we should select the longest NLink where the EI has at least one landmark unique in the EI and reliably detectable.

Figure 5 defines three distance intervals. In order to check how reliable the NLinks are, the planner compares the landmarks included in the PI. The PI is divided into three sub-intervals: pre-TI, TI and post-TI. Let the sub-intervals be denoted simply by A, B and C, respectively. Then they

Table 2: A certainty matrix for the landmarks used in this paper.

	Door	Open	Front Wall
Nothing Detected	0.1	0.1	0.1
Door Detected	0.7	0.25	0
Open Detected	0.2	0.65	0
Front Detected	0	0	0.9

are defined as follows:

$$\begin{aligned} A &= \{x \mid P_s \le x < T_s\} \\ B &= \{x \mid T_s \le x < T_e\} \\ C &= \{x \mid T_e \le x < P_e\} \end{aligned} \tag{11}$$

An NLink is reliable if at least one landmark included in the sub-interval B is not included in A and C. Since the sensors are not perfect, we represent the sensory uncertainties with probabilities.

The probabilities that the robot detects or does not detect the target LP, or detects a wrong LP are

$$\begin{aligned} P(Success) &= P_A(\bar{D}) \cdot P_B(D) \\ P(Fail) &= P_A(\bar{D}) \cdot P_B(\bar{D}) \cdot P_C(\bar{D}) \\ P(False) &= 1 - P(Success) - P(Fail) \end{aligned} \tag{12}$$

, where the $P_*(D), P_*(\bar{D})$ denote the probabilities that the robot detects or doesn't detect the landmarks included in the TI when the robot is actually in sub-interval '*'. *Fail* occurs when the robot passes over the EI without detecting any landmarks and *False* occurs when the robot detects the landmarks in the wrong sub-interval. The latter is more critical and has to be avoided because if the robot goes the wrong way, it is hard to recover. Therefore, we check the false situation at first and discard it if $P(False)$ exceeds a certain amount.

In order to evaluate the $P_*(D)$ and $P_*(\bar{D})$, we first evaluate the probabilities related to each of the landmarks. We use a certainty matrix such as the one Nourbakhsh, Powers and Birchfield used for their mobile robot in [7]. The certainty matrix for ultrasonic range finders is shown in Table 2. The value of a certainty matrix at (i, j) represents the probability that the robot detects landmark l_i when the landmark is actually l_j. The probability can be written as $P(d_i|l_j)$.

Suppose a sub-interval, *, having a set of landmarks, L_*. The probability, $P_*(d_i)$, that the robot detects a landmark, l_i, within the sub-interval is $P_*(d_i) = \bigvee_{l_j \in L_*} P(d_i|l_j)$, where $\vee$ denotes a probabilistic 'or' operation. When the planner selects a set of landmarks, L_B, to be detected for localization, the probabilities for each landmark included in the L_B should be combined. Therefore, the probability, $P_*(D)$, that the robot detects any landmark included in the L_B within the sub-interval * is obtained as follows:

$$\begin{aligned} P_*(D) &= \bigvee_{l_i \in L_B} P_*(d_i) \\ &= \bigvee_{l_i \in L_B} \bigvee_{l_j \in L_*} P(d_i | l_j) \end{aligned} \quad (13)$$

And the probability, $P_*(\bar{D})$, that the robot doesn't detect any landmarks within the sub-interval is $P_*(\bar{D}) = 1 - P_*(D)$.

The probabilities given in equation (12) provide the reliability of the NLinks. For the sake of efficient navigation, they should be combined with other costs such as the length of the NLinks and the sensory cost. We define the cost in terms of time. Each of the landmarks restricts the speed of the robot by a certain amount. Therefore, the maximum speeds of the robot within PIs are determined by the selected landmarks. In other places outside the PIs, the robot can go at maximum speed. Therefore, the cost, C, is defined as follows:

$$C = \begin{cases} C(Success) + C(Fail) & \text{if } P(False) \leq \beta \\ \infty & \text{otherwise} \end{cases} \quad (14)$$

where $C(Success)$ and $C(Fail)$ are defined as follows:

$$\begin{aligned} C(Success) &= P(Success)\left(\frac{E_s}{V_m} + \frac{E_e - E_s}{V_s}\right) \\ C(Fail) &= P(Fail)\left(\frac{E_s}{V_m} + 3\frac{E_e - E_s}{V_s}\right) \end{aligned} \quad (15)$$

where the E_s and E_e are the starting and end point of the EI as depicted in Figure 5, and V_m is the maximum velocity and V_s is the minimum of constrained maximum velocities related to the selected landmarks; i.e., the robot can not exceed V_s to detect the selected landmarks. The definition of $C(Fail)$ implies that the robot scans the EI once more. The false situation is pre-checked with small constant β.

With the definition of the cost, the planner selects an optimal path and the corresponding sets of landmarks. First, it selects a set of landmarks for each of the NLinks so that the $P(Success)$ and V_s are high and thus eventually the cost of the NLink becomes lower. It then applies Dijkstra's shortest path algorithm to find the lowest cost path.

4 Control Subsystem

4.1 Reactive Behaviors

In order to cope with a complex, uncertain and dynamically changing environment, we have designed the control subsystem to be reactive. Many independent and reactive behavior modules cooperate to make the overall behaviors of the robot feasible. The most important factors considered by us in designing the behaviors are *extensibility* and *information loss*. The advocates of the behavior-based controller have argued that complex behavior emerges from cooperation among multiple local behaviors. However, the real world is very complex, so a lot of behaviors are needed to cope with such an environment. Therefore, the controller should be easily extensible. In the previous works, Arkin's *motor schemas* and Saffiotti's *fuzzy behaviors* are more adequate in this sense than the others. Their systems have no inter-connections among behaviors. Thus, their systems are easy to add/remove a behavior into/from the controller.

Those systems having no inter-connections among behaviors should provide an additional mechanism to blend the action commands produced by the multiple independent behaviors. Such a mechanism helps a system to be extensible, but the system may suffer from information loss. Since the behaviors use almost direct information, they exhibit little information loss. However, the blending module gets the information only indirectly via the output of the behaviors. Hence, the blending module can't know exactly why certain behavior produces such an output, so it may produce an unwanted result.

Such a problem can be found in the *potential field* approach, which is very popular and attractive because of its elegance and simplicity. Koren and Borenstein[8] have pointed out the inherent limitations of the approach, as illustrated in Figure 8. The first problem, the local minima, is not the problem of the potential field approach but the problem of purely reactive systems. Such a problem can be resolved by the integration of the deliberative approach. The remaining three problems are due to the information loss between the behaviors and the blending module. For example, in (b) of Figure 8, both obstacle 1 and obstacle 2 produce repulsive force F_r1 and F_r2, respectively, which are combined into F_r. Then, this repulsive force is combined with the attractive force F_t into a resultant force R, which makes the robot turn right. As a result, the robot can't pass between the obstacles despite the adequate width of the passage. Such a mistake is due to the repulsive forces. The F_r1 can be translated into a statement as "Go in the direction opposite to obstacle 1", while the expected statement is "Don't go in the direction of obstacle 1." Therefore, the information that the robot can go in any direction except toward the obstacle 1 is distorted and lost.

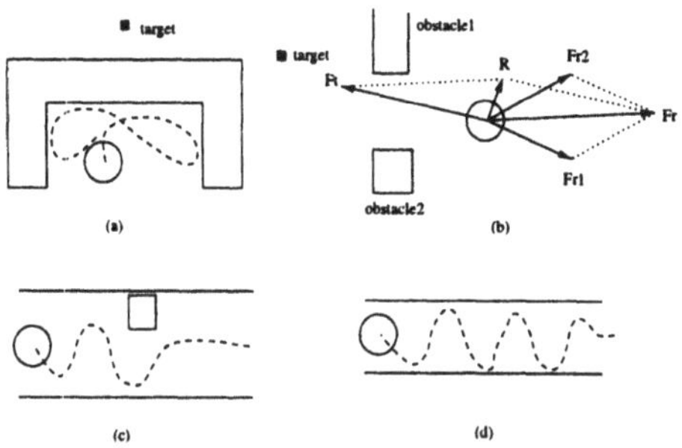

Figure 8: The inherent limitations of the potential field approach are: (a) trap situations due to local minima (b) no passage between closely spaced obstacles (c) oscillations in the presence of obstacles (d) oscillations in narrow passages.

In order to take "extensibility" and "no information loss", we have designed behaviors as computational modules that run independently without any inter-connections with other behaviors and produce action commands in desirability functions. More precisely, we say that each behavior B is implemented, as in Figure 9, and produces an action command in a desirability function as follows:

$$Des_B : Control \rightarrow [0,1] \tag{16}$$

Each behavior module has two parts: the initialization part and the control body. The initialization part is executed once the behavior is executed for the first time. In the initialization part, several state variables are initialized. Once the initialization is completed, the behavior acquires sensory information and then produces desirability as shown in Equation (16) repeatedly.

Saffiotti[5] has used the notion of the desirability, too. However, our work is different from his in that our behaviors are computation modules while his are fuzzy rules. We have decided to build the behaviors computationally rather than logically (with fuzzy rules) because it is more efficient and flexible in implementation.

```
BH_BEGIN
    initialization;

    loop forever
        gather sensory information;
        produce desirability function;
    end loop
BH_END
```

Figure 9: A frame of a behavior.

4.2 Blender

As many behaviors can be simultaneously active in the controller, each aiming at one particular goal, many desirabilities are produced. All these desirability functions are merged into a composite. Then the defuzzification module converts the resulting tradeoff desirabilities into one crisp control decision.

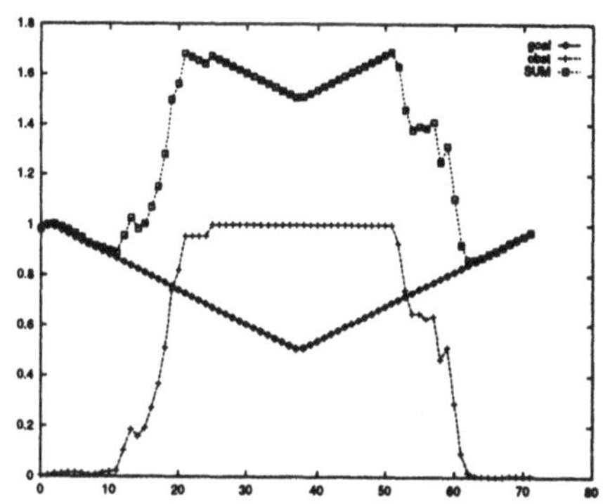

Figure 10: Action commands of *Move to Goal*, *Avoid Obstacle* and their composition.

Figure 10 illustrates how multiple action commands are merged into a composite one. In that figure, the x-axis represents the orientations; a value x denotes $5x$ degree. The 'goal' denotes the action command produced by the *Move to Goal* behavior – its current task is to orient the robot toward the goal which is about 10 degrees left, while the 'avoid' is the action command produced by the *Avoid Obstacle* behavior. The *Avoid Obstacle* behavior has found a small obstacle at the front. The desirability produced by the *Avoid Obstacle* behavior denotes that the robot should not go forward, but it doesn't prevent it from going in other directions.

In the current implementation, the desirabilities are combined by a simple summation. Though this summation makes the action command exceed the maximum of the probability value, it is not serious in this case because we need only a direction which has a maximum desirability. The composite action command is represented as 'SUM' in Figure 10.

The composite action command is then defuzzified to get an actual action command used to control the actuators. There are also many kinds of defuzzifing methods. In this paper, we take an action command having 'maximum desirability' as an input to the actuators. Another possible method is the 'center of mass'. In this method, the center of mass of the desirability distribution is selected. However, this method may produce undesirable results. Consider the case illustrated in Figure 10. The composite action command has two peaks and the center of mass is the center of these two peaks; this can make the robot run into the obstacle. In contrast, the 'maximum desirability' method does not suffer from such a problem, however it may cause oscillations between the two peaks. Some additional mechanism, such as momentum, is required to solve this problem.

4.3 The Coordinator

We hypothesize that many behavior modules are needed to make the system more intelligent and reliable. However, they need not operate all the time. Only one subset of behaviors is sufficient at one time. For example, a behavior "Pass the doorway" is not needed when the robot explores an open space. In this paper, a module called "coordinator" is provided for this purpose. It manages the activation of the behaviors according to the *situation*.

As the robot navigates, the situation continuously changes. In order to cope with such changes, the robot should recognize the situation. The system proposed in this paper recognizes the situation using the localization module and the planning module, which are described in the previous section. Since we use a topological map, the situation can be determined only roughly; i.e., one provable descriptive example of the situation is "the last identified location is Room #1024 and the robot is navigating toward Room #1025." However, it is sufficient for the coordinator because more precise variations in the situation can be handled by the behaviors.

The coordinator selects a set of behaviors according to the current and next locations. In fact, the sets of the behaviors for each NLinks are determined in the initialization time when the map is prepared. As mentioned earlier, the robot navigates along the NLinks so the robot's situations correspond to the NLinks. Table 3 shows some example correspondences between the NLink types and the behaviors. The NLink type is defined by the types of the current and next locations.

Table 3: Correspondences between the NLink types and the behaviors.

From	To	Behaviors
Corr.	Room	Pass doorway Avoid Obstacle in doorway
Corr.	Corr.	Follow a corridor Avoid Obstacle

When the robot recognizes that it has reached a sub-goal node and it is going to navigate to the next sub-goal along the NLink, the coordinator deactivates the previously activated behaviors and activates the behaviors corresponding to the NLink.

5 Experimental Results

Table 4 shows the simulation results of the localization algorithm based on the *fuzzy automata*. We used the same instruction "move to C1 from R0" in three different scenarios of the simulation, but used different simulator parameters.

In *Scenario A*, the simulator did not make any noise, so the input measure function u could be 1.00. In *Scenario B*, R7's door was sensed as a corridor, R6's door was not detected, and the other features were obtained with 20% noise. In *Scenario C*, we used the same parameters as in *Scenario B* except R9's door was not detected. R9's door was a very important feature since missing this feature caused failure to recognize place C9. However, the simulator shows that overall performance has not degenerated even if the door was missed.

Figures 11 and 12 show the results of the experiments using the ETMap and the localization algorithm based on dead reckoning. In those experiments, the robot navigated in the environment shown in Figure 3. The robot is commanded to navigate from place 0 to place 10 in the first experiment and from place 0 to place 14 in the other. In the second experiment, we cut the link between place 0 and place 14 intentionally to test the localization algorithm in the long run. The robot successfully reached its goal position in both experiments. It is noteworthy that, in the

Table 4: Indoor navigation under simulated environment

T	Scenario A			Scenario B			Scenario C		
	Decision	u	CV	Decision	u	CV	Decision	u	CV
1	r8 ⇒ c8	1.00	0.70	r8 ⇒ c8	0.98	0.68	r8 ⇒ c8	0.98	0.68
	r9 ⇒ c9	1.00	0.70	r9 ⇒ c9	0.98	0.68	r9 ⇒ c9	0.98	0.68
	r0 ⇒ c0	1.00	0.70	r0 ⇒ c0	0.98	0.68	r0 ⇒ c0	0.98	0.68
	r4 ⇒ c5	0.85	0.60	r4 ⇒ c5	0.82	0.58	r4 ⇒ c5	0.82	0.58
2	c0 ⇒ c9	1.00	0.91	c0 ⇒ c9	0.82	0.78	c0 ⇒ c9	0.82	0.78
	c8 ⇒ r6	0.57	0.61	c7 ⇒ c6	0.75	0.57	c7 ⇒ c6	0.75	0.57
	c9 ⇒ c8	0.50	0.56	c8 ⇒ r6	0.40	0.48	c8 ⇒ r6	0.40	0.48
	c5 ⇒ f1	0.50	0.53	c9 ⇒ c8	0.32	0.43	c9 ⇒ c8	0.32	0.43
3	c9 ⇒ c8	1.00	0.97	c9 ⇒ c8	0.95	0.90			
	c5 ⇒ f1	1.00	0.86	c6 ⇒ c5	0.95	0.84			
	c6 ⇒ c5	1.00	0.82	c5 ⇒ f1	0.95	0.79			
	c8 ⇒ r6	0.57	0.57	c3 ⇒ c2	0.60	0.54			
4	c8 ⇒ f1	1.00	0.99	c8 ⇒ f1	0.77	0.81	c8 ⇒ f1	0.77	0.67
	f1 ⇒ c1	0.50	0.61	f1 ⇒ c1	0.68	0.71	f1 ⇒ c1	0.68	0.59
	c0 ⇒ c7	0.57	0.49	c1 ⇒ r1	0.55	0.53	c1 ⇒ r1	0.55	0.46
	c1 ⇒ r1	0.32	0.37	c0 ⇒ c7	0.50	0.44	c0 ⇒ c7	0.50	0.41
5	f1 ⇒ c1	1.00	1.00	f1 ⇒ c1	0.90	0.87	f1 ⇒ c1	0.90	0.83
	c1 ⇒ r1	0.82	0.76	c1 ⇒ r1	0.76	0.74	c1 ⇒ r1	0.76	0.71
	c8 ⇒ f1	0.68	0.55	c8 ⇒ f1	0.57	0.48	c8 ⇒ f1	0.57	0.48
	c7 ⇒ c4	0.50	0.50	c7 ⇒ c4	0.45	0.45	c7 ⇒ c4	0.45	0.44

second experiment, the robot overcame a large positional error that occurred in the corridor linking place 0 and place 1. The results prove that the proposed localization scheme is reliable even with a somewhat large positional error and imprecise sensors.

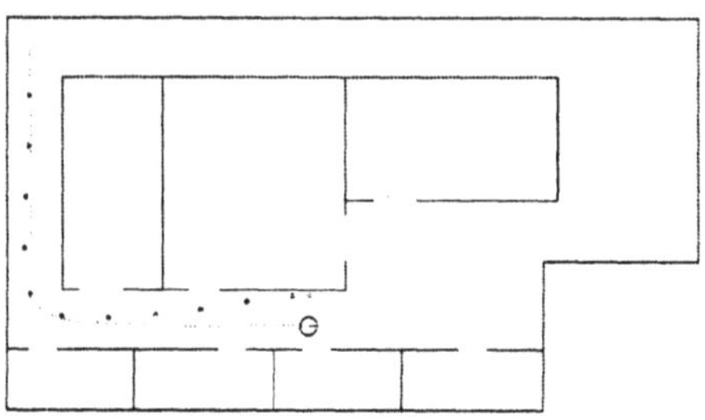

Figure 11: Experiment 1 : navigate from place 0 to place 10.

In each experiment, the planner selected the paths "0 → 14 → 10" and "0 → 1 → 4 → 6 → 9 → 12 → 14", respectively. The robot went to place 10 directly skipping all places from place 14 in experiment 1, but it visited place 12 when it went from place 9 to place 14. Because the distance between place 9 and place 14 is long, place 13 can be sensed instead of place 14.

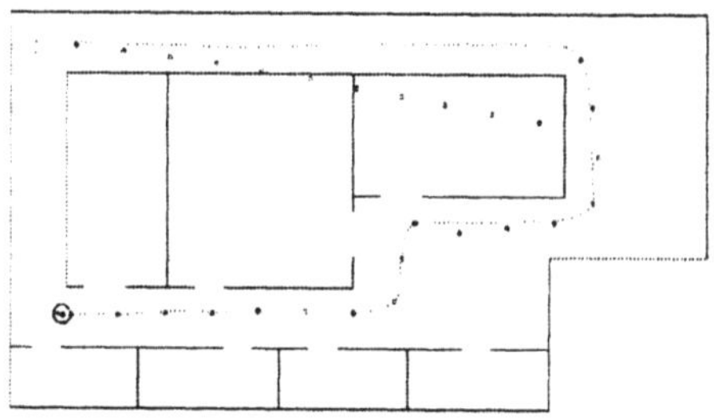

Figure 12: Experiment 2 : navigate from place 0 to place 14.

Besides, the effectiveness of the proposed control architecture has been proven in several real world experiments. Two major experiments were in the '93 Taejon World Exposition and the 4th Annual Mobile Robot Competition sponsored by IJCAI and AAAI. In the exposition, our mobile robot CAIR-2 reliably demonstrated reactive behaviors such as "Avoid obstacles and spectators" and "Track a moving visual landmark" in an outdoor environment for three months. And in the competition, CAIR-2 won the first place in the "Office Delivery" event that tested the ability of autonomous navigation in an office-like indoor environment. CAIR-2 used almost the same control architecture as the one proposed in this paper (we improved it after the competition) and proved the feasibility of the control architecture. A detailed description of the competition can be found in [9] and [10].

6 Conclusion and Future Work

We have described a new control architecture for an autonomous mobile robot. The proposed navigation system integrates the advantages of *reactivity* and *deliberateness*. In order to integrate them, we have provided two modules; one is for selecting the relevant behaviors and the other is for blending multiple action commands.

We have also concentrated on describing the mapping system, on which the behavior selection modules highly depend. The mapping system represents the environment in a topological map called the *enhanced topological map* (ETMap) and provides a localization module and a planning module for efficient and reliable navigation. In this paper, we have proposed two algorithms for position estimation. The first algorithm based on fuzzy automata provides great recovery capability but it has a weakness in distinguishing adjacent similar places. In contrast, the other algorithm based on dead reckoning is simple, efficient and reliable, though it is weak in recovery from failure. We are currently studying integrating the advantages of both algorithms.

The proposed system was implemented and tested on a real mobile robot, CAIR-2, and a simulator. The experiments show that the proposed system can accomplish the tasks even with such imprecise sensors as ultrasonic range finders.

References

[1] Rodney A. Brooks. A robust layered control system for a mobile robot. *IEEE J. of Robotics and Automation*, 2(1):14–23, March 1986.

[2] Ronald C. Arkin. Motor schema based navigation for a mobile robot: An approach to programming by behavior. In *Proc. of IEEE Int. Conf. on Robotics and Automation*, pages 264–271, 1987.

[3] Ronald C. Arkin. Integrating behavioral, perceptual, and world knowledge in reactive navigation. *Robotics and Autonomous Systems*, 6:105–122, 1990.

[4] Maja J. Mataric. Integration of representation into goal-driven behavior-based robots. *IEEE Trans. on Robotics and Automation*, 8(3):304–321, 1992.

[5] Alessandro Saffiotti. Some notes on the integration of planning and reactivity in autonomous mobile robots. In *Proc. of the AAAI Spring Symposium on Foundations of Automatic Planning*, pages 122–126, 1993.

[6] Eugene S. Santos and William G. Wee. General formulation of sequential machines. *Information and Control*, (12):5–10, 1968.

[7] Illah Nourbakhsh, Rob Powers, and Stan Birchfield. Dervish: An office-navigating robot. *AI Magazine*, 16(2):53–60, 1995.

[8] Yoram Koren and Johann Borenstein. Potential field methods and their inherent limitations for mobile robot navigation. In *Proc. IEEE Int. Conf. on Robotics and Automation*, pages 1398–1404, 1991.

[9] Hyun S. Yang, Jiyoon Chung, Byeong-Soon Ryu, and Juho Lee. Cair-2: Intelligent mobile robot for guidance and delivery. *AI Magazine*, 17(1):47–53, 1996.

[10] Jiyoon Chung, Byeong-Soon Ryu, and Hyun S. Yang. Integrated control architecture based on behavior and plan for mobile robot navigation. *Robotica*, to appear.

A Robotic Travel Aid for the Blind
-Attention and Custom for Safe Behavior-

Hideo Mori & Shinji Kotani
Dept. of E.E. and C. S., Yamanashi University Takeda-4, Kofu 400(Japan)
forest@koihime.esi.yamanashi.ac.jp

Abstract

We have been developing Robotic Travel Aid(RoTA) "HARUNOBU" to guide the visually impaired in the sidewalk or campus. RoTA is a motor wheel chair equipped with vision system, sonar, differential GPS system, dead reckoning system and a portable GIS. We estimate the performance of RoTA in two viewpoints, the viewpoint of guidance and the viewpoint of safety. RoTA is superior to the guide dog in the navigation function, and is inferior to the guide dog in the mobility. It can show the route from the current location to the destination but can not walk up and down stairs. RoTA is superior to the portable navigation system in the orientation, obstacle avoidance and physical support to keep balance of walking, but is inferior in portability.

1 Introduction

Among 307,000 visually impaired in Japan 65,000 are the complete blind. Most of them lost their sight in the elderly age. It is very difficult for the aged to learn to walk with the long cane or the guide dog, because they are not so rich in the auditory and haptic sensing and have not good memory for cognitive map.

We have been developing Robotic Travel Aid(RoTA) "HARUNOBU" since 1990 to guide the visually impaired in the sidewalk or campus[1]. RoTA is a motor wheel chair equipped with vision system, sonar, Differential GPS system, dead reckoning system and a portable GIS. In designing the RoTA, we add a guidance function and a safety function to the conventional mobile robot functions.

MoBIC Project (the mobility of blind and elderly people interacting with computers) was carried out from 1994 to 1996 with support of the Technology for the Integration of Disabled and Elderly People (TIDE) program of the Commission of the European Union[2]. MoBIC travel Aid(MoTA) which consists of MoBIC Pre-Journey System (MoPS) and MoBIC Outdoor System (MoODS).

MoPS is a simulator which facilitates the exploration of an previously unknown area and the selection and preparation of a route before actual walk out. For this purpose several presentation media are used, including verbal descriptions of routes and virtual exploration facilities. Natural language descriptions can be obtained for routes with varying detail enabling the user to customize the output to his preference.

MoODS is a portable system which gives assistance during the walk. It consists of a small wearable PC kernel of 16 x 11 x 7 cm in the size, a GPS, an electronic campus and a pair of special earphones which prevent masking the ambient sound essential for echo location. The system provides on-route information about the current position of the traveller and his/her environment (e.g. "You are in Harborn Road near a bakery"). On pressing the designated button of the small keypad, the system responds with "Go ahead 50 meters to the next junction and turn right" for example. The system informs the traveller automatically when they are leaving the chosen route or if the accuracy of the system has degraded. A mute button has been added to key pad to allow a sudden stop of the speech when reaction to the environmental sounds becomes necessary.

A testbed of MoTA was developed and estimated

through a field test and found the design philosophy was useful in the human navigation.

RoTA is superior to the guide dog in the navigation function, and is inferior to the guide dog in the mobility. It can show the route from the current location to the destination but can not walk up and down stairs. RoTA is superior to MoTA in the orientation, obstacle avoidance and physical support to keep balance of walking, but is inferior in portability. The functional comparison of RoTA, the guide dog and MoTA is shown in Table 1.

2 Sensing strategy

Conventional methods for the car and pedestrian detection seem to simulate the perception of the human beings. We get the idea of objects discrimination from the study of ethologists especially Nobel winner Tinbergen [3]. He showed that the animal behavior was represented by a chain of fixed action patterns even if the behavior was an advanced and complex one. To explain the mechanism of the behavior Tinbergen proposes three concepts: sign stimulus, Central Excitatory Mechanism (CEM) and Innate Releasing Mechanism (IRM). The animal dose not recognize objects as human beings does, it makes a response not to the whole of the object but to the part inherent in the object. The part of the object which activate the fixed action pattern is called sign stimulus. CEM is similar to the modern multi-tasking system of the computer OS. All the fixed action patterns are in the dormant state, and when a sign stimulus appears IRM activate one of the fixed action pattern corresponding to the sign pattern. We think that Tinbergen's concepts are useful for the configuration of our vision based mobile robot. We use sign pattern as the technical term of the sign stimulus. Sign pattern is different from the landmark in three factors as shown in Table 2.

3 The guidance

A Geographic Information System(GIS) is required as the base of the navigation system of RoTA. The GIS of RoTA has to include the robot guide information and the human guide information. The robot guide information should give the sensor system of the robot the information about the environment.

Table 1 RoTA, MoTA/MoODS and guide dog

	Obstacle avoidance orienta-tion	Mobility (Physical support)	Mobility (Up down stairs)	Navi-gation
RoTA	○	○	×	○
MoTA	×	×	○	○
Guide dog	○	○	○	×

Table 2 Comparison of land mark and sign pattern

	Landmark	Sign pattern
Purpose	To verify the current position	To activate or control the fixed action pattern
Object	Permanent objects	Permanent & temporal objects
Represen-tation	2-D model, 3-D model	Simple feature, rhythm, shadow

3.1 Sign pattern

The robot dose not recognize the total environment but it recognize only two kinds of signals which are required to guide in the environment. One is sign pattern and the other is landmark. 'Sign pattern' is synonym of 'sign stimulus' defined by Tinbergen. In the engineering sense it is defined as a signal which is used to correct the position and heading error of the dead reckoning system. As the sign pattern of RoTA we use such elongated feature on the road as road boundary, lane mark, tacktile block, etc.. The landmark is defined as a feature which is used to verify the current location. The difference between the sign pattern and the landmark is shown in Table 2.

3.2 Human guide information

Petrie et al.[4] studied user requirements by the deep interview in which the potential users of the system were asked about their current levels of mobility, problems which they currently encountered in traveling independently and their opinion and ideas to design MoTA. The interviews were conducted with mobility officers and others involved in the training and rehabilitation of the visually disabled people. The information to be provided by the MoTA should be accessible at different levels of detail. The requirements of MoPS is shown in Table 3 and Table 4.

The opportunity to add personal comments and reminders to the route description(e.g. shopping to collect as well as the name or the room number of person to visit) was highly favored by all interviewees.

For the MoODS system, all the potential users expressed their desire to be able to control the flow of information and to be able to request as they felt they needed at the time. If unsure about current location, more detail may be required, wheres if a user is quite confident about the location and the route, a constant flow of instructions may be disracting. The three levels of interaction as shown in Table 5 were suggested in MoODS.

Table 3 Information to be presented at route planning in MoPS

Essential: Basic maps, routes including street names, distance information, known hazards and road gradings.

Desirable: Shops, public buildings, Road crossings, public transport, nearest Church and ticket office.

Ideal: Up-to-date information concerning transient obstacles, e.g. road works scaffolding and diversions.

Table 4 Required interaction levels in MoPS

Essential: User should dictate the level of detail incorporated in any particular journey plan.

Desirable: User should be able to request area maps or specific routes, shortest route and safest route.

Ideal: User should be able to edit the journey plan with own comments and memos.

Table 5 The suggested control levels in MoODS

Basic: Information to check you are still on course such as direction of travel, grade of road and nearest crossing, any known obstacles.

Detailed: Information to include the type of detail selected for inclusion in the pre-journey plan, for example shops, public buildings etc.

Transport: Information nearest bus stop, rail or tube station, taxi rank and telephone.

We have been developing the human navigation system of RoTA which follows the idea of MoTA. RoTa has a synthesized voice maker and tells the blind the desired information through it.

3.3 The robot guide information

To keep safe and to follow the Japanese traffic regulation RoTA should move right or left sidewalk and zebra crossing marks. For this reason we define the path on which RoTA and the blind can move safely. When the road has the sidewalk of enough width for RoTA to move along, the path is specified on it. When the road has not a sidewalk, the path is specified right or left roadside which is free from falling into creek, downstairs and depression. The digital map of RoTA includes a road network, a path network, sign patterns and land marks, The road network includes road information such as the type, the distance, the direction and absolute location of the street and the junction. After route searching the GIS feeds to the locomotion control system the robot guide information along the route.

3.4 Input device

The blind should input RoTA the destination and the requirements described in 3.2 in the textual form. We think the input device and its usage should be changed depending on the ability and the level of disablement. We think a voice-controlled portable PC is useful for the intelligent user to make dialogue with RoTA .

4 Moving obstacle detection

To perform the safe behavior, moving obstacles as well as stationary obstacles should be detected. Among them vehicles and pedestrians are very serious obstacles.

4.1 Conventional strategy

Many models have been proposed to identify the car. As an example of 2-D model, Regensberg and Graefe developed the rectangular model in which the right and left and the upper and bottom side of a vehicle are detected horizontally and vertically by a local edge operation[5]. The problem of extracting the upper edge of the vehicles is complicated by the fact that it is difficult to distinguish the upper edge from the structured background of the road. To represent the rear vehicle Kuehhnle presented three symmetry criteria: contour symmetry, gray level symmetry and horizontal line symmetry[6]. The problems are the unsymmetrical image of the car caused by the slantwise illumination and the covering by the other car. Optical flow is another approaches in object detection. Frazier and Nevatia employed Complex Logarithmic Mapping(CLM) method to detect simply the desired movement. Moving objects move in angular direction, while stationary objects remain constant[7]. The problem is that the correct optical flow is obtained only when the car is in the homogeneous background in the image intensity. The stereo vision based method is a recent trend in obstacle detection. Saneyoshi et al. applied stereoscopy for detecting objects on the road environment. The system currently can detect two lane marks and less than seven obstacles on the road every 100msec[8]. The problem is the cost of the stereo vision system.

The recognition of an object in motion is one of the topics of psychophysics. The moving light display(MLD) method proposed by Johansson[9] is capable of perceiving the body and the motion of a walker. Kozlowsky and Cutting have demonstrated that MLD can tell the sex a walker by the pattern of the motion[10]. These facts suggest that the motion of the object could be the key to its recognition. Conventional human motion tracking includes the modeling of human body and the finding of the correspondence between model and real data. The stick figure model is a well-known model of human body[11], but it should be modified by the distance between video camera and the person and the clothes which will be changed by man or woman, summer or winter.

4.2 Vehicle detection by shadow

The sun light and the diffused sky light don't reach the part of the road underneath a car. The image intensity of the underneath part is almost noise level in the video image. Its intensity is lower than any other part such as a wet part or a patched part repaired by new asphalt even in the cloudy day. To apply this phenomena, a window is set up in the lane as shown in Fig.1(a), and the vertical projection

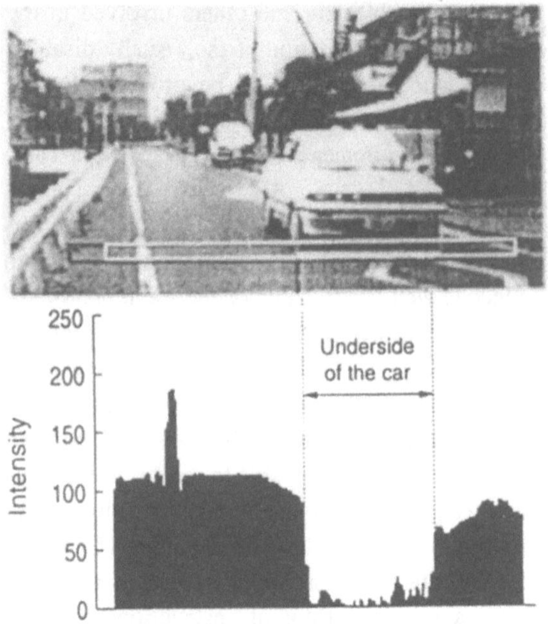

Fig.1 Intensity in the window underneath a vehicle

of the window is obtained as shown in Fig.1(b). When the projection curve shows a bottom of a certain width with cliffs at the right and left end, we define the bottom as the sign pattern of the car. This method is implemented on a vision system with a simple program code of real time processing. It shows almost 100% success in no shadow or entirely shadow scene in fine day and in cloudy day, but less than 1% of error happens in sunny and shadow mixed scenes[4]

4.3 Pedestrian detection by rhythm

When one walks in the sidewalk, the rhythm of the walk is almost constant. The rhythm can be seen in the alternate motion of feet and hands and in the up down motion of the head and shoulder. Among these motion the feet motion is the most detectable by the computer vision, because their rhythm is clearer than those of the head and the shoulder, the background of the feet is simpler than those of head and shoulder, and the clothes and another part of the body don't cover the feet in the image. The rhythm of the feet is a good sign pattern as it is easy to detect by the computer vision. The difficult process of scaling to fit the object image to the model is not needed. It is free from the distance, the clothes and the weather. The implemented method is as follows[12].

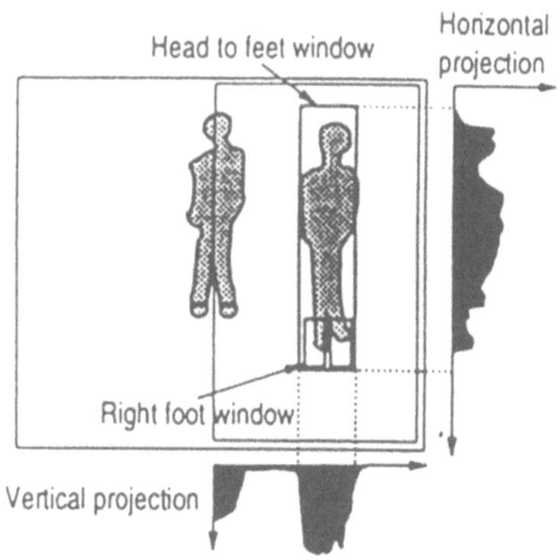

Fig.2 Setting of three windows W, W_r and W_l

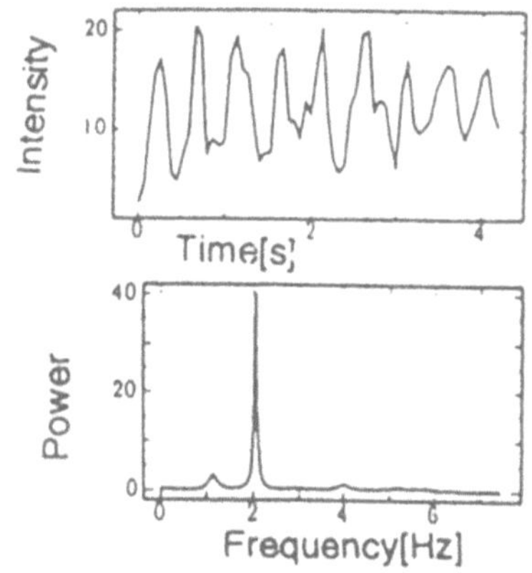

Fig.3 An Example of the time series of the area of W_L and its power spectral

4.3.1 *Motion segmentation*

The frame subtraction is applied to detect moving objects as shown in Fig.2. So this method is effective when the video camera is in the stationary state. A horizontal projection is operated after binarizing the subtracted image. The horizontal projection is sliced by a threshold to obtain H segment which may represent the height of a person. A vertical projection is operated and sliced by another threshold to obtain V segment which may represent the width of the person. If V segment satisfies the threshold of width, window W of $H \times V$ in size is assumed to be the head to feet window of the person. Then the right and the left foot window, W_R and W_L, each of which is 1/5 of H segment and 1/2 of V segment are set up in the lowest part of W. This process is called finding process, and is followed by tracking process as follows. Window W_R and W_L of the last frame are a little enlarged in length and width to trace the feet in the next frame, and the horizontal and vertical projections are operated on the new binarized subtracted image. New W_R and W_L are obtained by the same slicing operation as the finding process.

4.3.2 *Location matching and data acquisition*

The most significant features of W_R and W_L are the ordinate of the bottom of the windows and the area of the binarized subtracted image in the windows, as the ordinate shows the distance of the person, and the periodic change of the area of the sliced image depends on the rhythm of walking. they are recorded in each step.

4.3.3 *Rhythm matching*

Fast Fourie Transformation is operated on time series of the area of W_R and W_L. When the primary components of the power spectrum of the two time series are satisfied with 2σ of the mean rhythm of walking , W_R and W_L are judged as the feet of a person. An example of time series of the area in W_L and its power spectral are shown in Fig 3.

5 The safety

The safety means 1) the safe behavior not to be damaged from the other and 2) the safe behavior not to damage the other. The first is performed by avoiding depression and moving heavy obstacles such as the vehicle and the motor cycle. The second is performed by avoiding pedestrian and street furniture such as litters and trees. The vehicle and the pedestrian keep the traffic regulation and the custom of the community. To avoid the collision with the vehicle, we should formulate and implement the traffic regulation and the custom which may be different by the nation and the region.

5.1 Danger estimation at an intersection

When the driver and the blind keep the traffic regulation perfectly, they will not meet with any accident. However as they often pay less attention to the right and the left sides of an intersection, they will occasionally have an accident. According to the statistics of the traffic accidents on the road in Japan, about 50 % of them occur at or near intersections. To avoid collision we should estimate danger level of vehicles at or near the intersection[13].

5.2 Danger levels

Three levels of danger are defined in this work.

0: *Safe* 1: *Warning* 2: *Dangerous*

The robot detects the location *si* and moving direction *rj* of the vehicle by its sign pattern and predicts its future path based on the Japanese traffic regulation. The danger coefficient *dij* for a vehicle at (*si*,*rj*) is defined as follows. When the future paths of the vehicle and the robot do not cross(*dij*=0), possibly cross(*dij*=1), surely cross(*dij*=2).

5.3 Japanese traffic regulation

We formulate the traffic regulation including the behavior of the careless driver as follows.

JTR1: Vehicles move along the left lane mark.

JTR2: Vehicles follow a typical path.

JTR3: When the driver moves straight, he will only pays attention to the front. When he turns left, he will pay attention to the front and the left. When he turns right, he will wait until all the straight moving and right turning vehicles pass by.

JTR4: When the blind starts moving across the intersection, the vehicle must not obstruct his/her way.

5.4 Robot's traffic regulation

We consider that the robot follows the same traffic regulation as the guide dog.

RTR1: The robot moves along the left side of road.

RTR2: When the danger estimate value is *safe*, the robot sends the blind the permission message for start crossing.

RTR3: After the blind receives the permission message, he gives the robot a start command.

RTR4: The has an intelligent disobedience function. The robot dose not follow the blind's command before the danger estimate value becomes *safe*.

Based on the traffic regulation of the vehicle and the robot, danger matrixes *dij* are given as shown in Fig.4.

6 Implementation and results

We have implement the concept of RoTA on a mobile robot "HARUNOBU 6" as shown in Fig.5 HARUNOBU 6 has a motor wheel chair (SUZUKI Co.) as the undercarriage part, a video camera(Sony) and a real time image processing board (HITACHI, IP-2000) as the vision module, a dead reckoning system with an optical gyroscope(HITACH WIRE Co.) and a differential GPS system.

We set up three kinds of test courses to evaluate the function of RoTA. The first is in a small zone of our university campus of 50m by 50m. In this course RoTA changes 360 degrees in its heading.

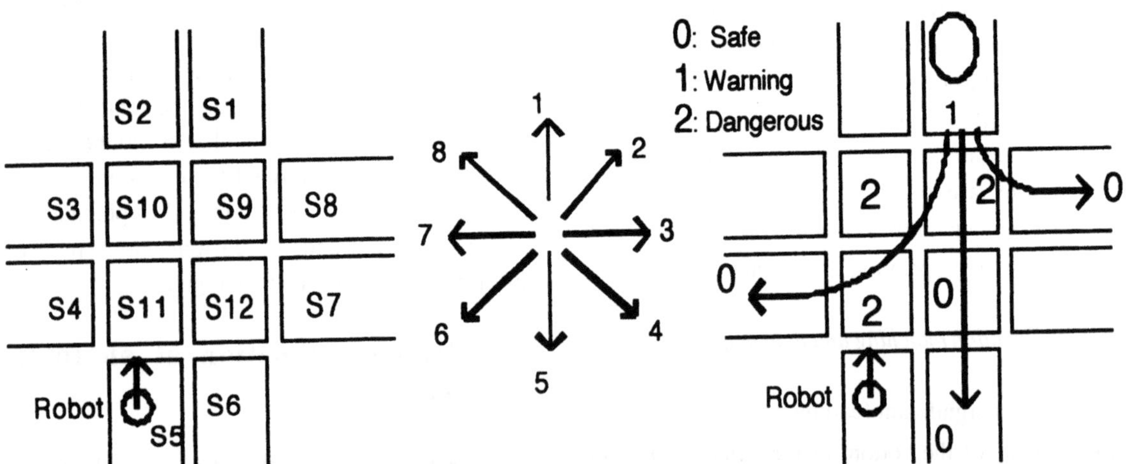

(a) Sections at the Intersection (b) Quantizes directions (c) Danger estimates for a front vehicle

Fig. 4 Danger estimates of a vehicle at a cross

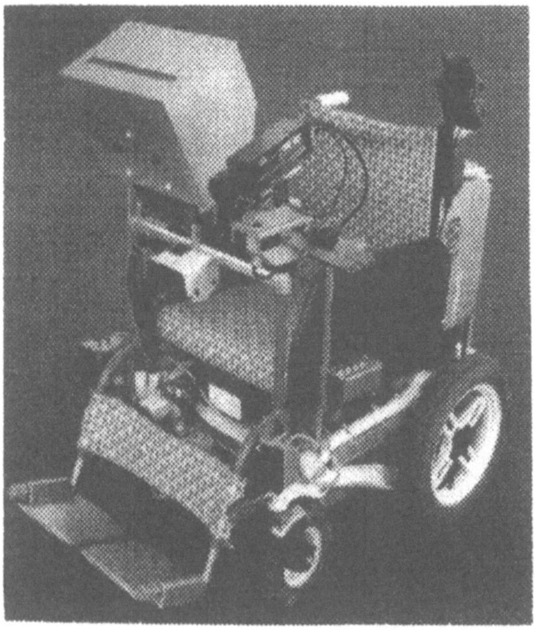

Fig. 5 Robotic Travel Aid "Harunobu-6"

The illumination of sunlight changes from back light to counter light. Through the test we can study the problem of iris control. The second is in the open field of a sport park in our city. RoTA can move anywhere in the park. We can compare cruising toward land mark and cruising along sign pattern. The third is in the sidewalk which lies between a road and a building or shop. In addition obstacles such as street trees, poles and bicycles are scattered in the sidewalk.

Through the experiment we found HARUNOBU 6 is enough to move in the campus and open field.

6.1 Portable GIS

For the navigation system of RoTA a portable geographic information system(GIS) with a data base is required. At first we implement a GIS and database on sale (Town Map developed by ZENRIN Co.), because it is 1/2,000 in the reduced scale, shows the shape of the road and the square, covers 95 % of Japanese cities and towns and is gradually transformed from the paper to the CD-ROM. However it is too big to store in the memory and too complex to build the navigation system. We are developing a simple GIS and a small data base for the navigation of RoTA. Town Map is used to make the panning of measurement and verify the measurement by drawing measured sign patterns and land marks on Town Map.

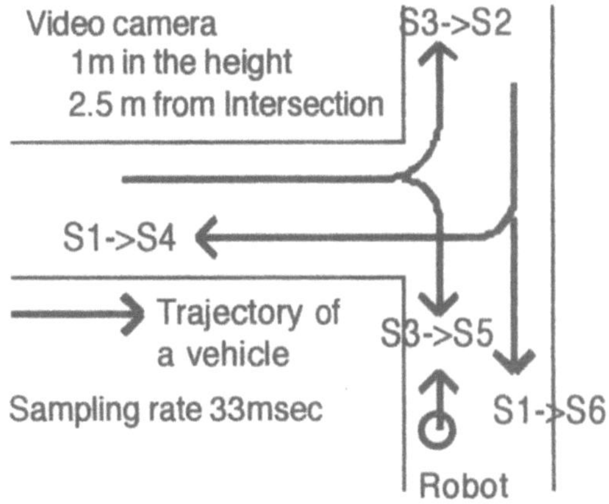

Fig.6 Experimental set up at a T shape intersection

Table 6 Results of danger estimation at a T shape intersection

Course	No. of vehicles	Correct	False
S1->S6	36(100%)	30(83%)	6(17%)
S1->S4	16(100%)	15(93%)	1(7%)
S3->S2	23(100%)	22(95%)	1(5%)
S3->S5	30(100%)	28(93%)	2(7%)
Total	105(100%)	95(90%)	10(10%)

6.2 Results of danger estimation

The vehicle detection algorithm is implemented on a one-board image processor(MC68040 25Mhz). The vehicle tracking is performed at video rate(33ms). The danger level is estimated while a vehicle passes an intersection within 3 - 4 sec. A video camera was fixed 1m in the height and 2.5m apart from a T shape intersection as shown in Fig.6. Images of 105 vehicles were recorded on a video tape. Fig.7 shows examples of three sampled scenes. In the right side of the display six parameters are shown: "TIME" indicates the processing time of one frame. "Moving car" shows the result of the moving object identification process. "DIST" is the measured distance of the vehicle from the video camera in meters. "Speed" shows the measured velocity of the vehicle in KM/h. "AP_TM" indicates the predicted arrival time at the intersection. The trajectory of the vehicle is shown at the bottom. Finally, the danger coefficient of the vehicle shows the intersection status, *safe*, *warning* and *risky*.

By looking at the display we judged the status

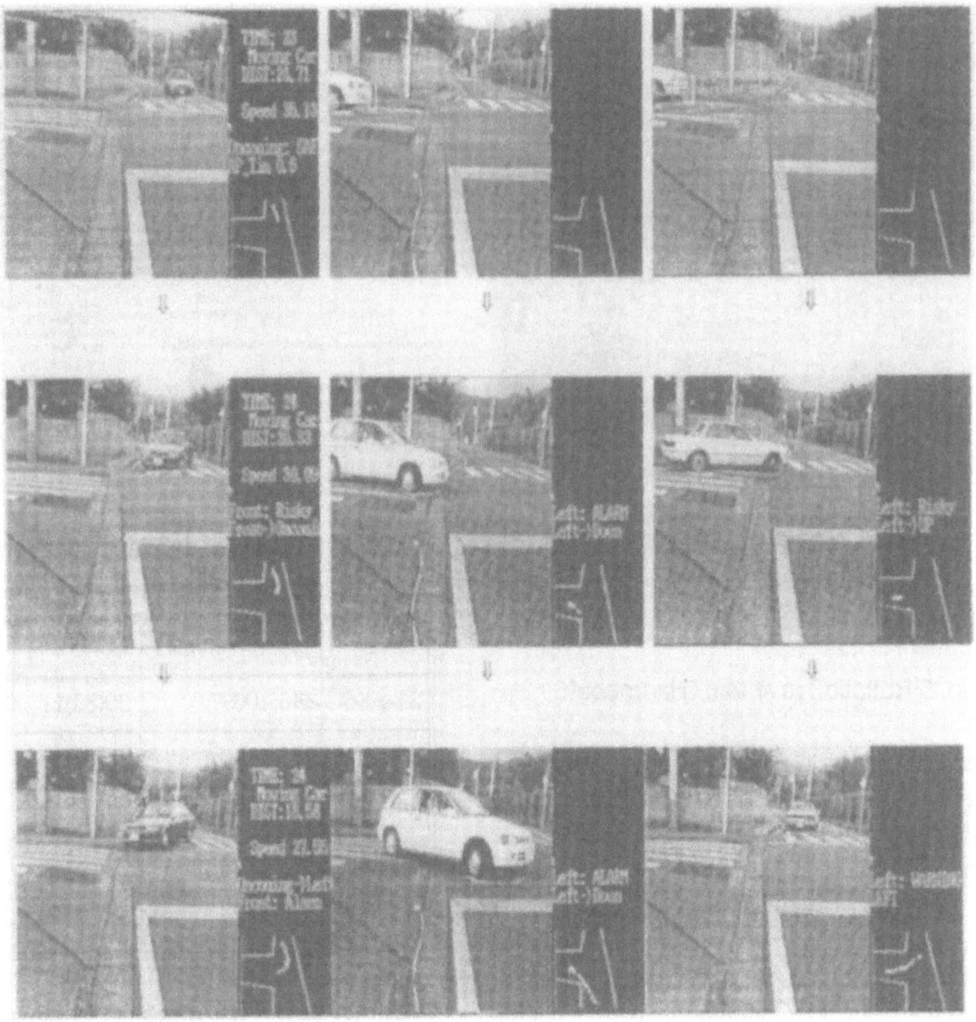

(a) Entered from ahead, turned left (b) Entered from left, come here (c)Entered from left, go ahead
Fig.7 The result of danger estimation for vehicles:

given by the system. The performance of the system for 105 vehicles on the video tape shows 90 % of success as shown in Table 6. Among 10 misjudgments, one was caused by a trajectory other than typical ones. Another was caused by the mistracking of an image of a vehicle too large to process at video rate. The remaining eight misjudgments were caused by two or three vehicles in successive running at less than 20 m apart. The vision system is successful in tracking the first vehicle, but often fails in detecting the second vehicle.

6.3 Results of pedestrian detection

Pedestrian detection algorithm for stationary camera is implemented on a monochromatic image processing system(ICS400 XM9, ANDROX Co.) . It samples a moving object every 67ms and takes 64 samples(4.3sec) to judge by the rhythm. A video camera was fixed 1m in the height and 15° in the depression angle. 407 pedestrians and 106 non-pedestrian including bicycles and dogs were recorded on a video tape in a sidewalk in a cloudy day at our university campus. Among the pedestrians 82% of them wear pants, 11% of them wear short pants or short skirt and 7% of them wear long skirt. The experimental results on the video tape are shown in table 7. Five % of errors are caused by 1) noise of video signal which makes jitters on the image,
2) swaying of trees and grass which make the same rhythm as that of pedestrian, 3) The almost same color of shoes as that of the asphalt paved road.

Table 7 Results of pedestrians detection

	Samples	Correct	False
Pedestrians	407	94.9%	5.1%
Non-pedestrians	106	96.2%	3.8%
Total	513	95.0 %	5.0%

7 Concluding remarks

We have a plan to develop several testbeds of RoTA in corporation with Japanese companies to make field tests by two kinds of the blind. The first is the blind who can walk with the guide dog and the second is the diabetic who lose his sight recently and cannot walk without a helper.

The guide dog user will want to walk in the crowded streets for visiting and shopping. We think he can use PC with voice interface to communicate with RoTA. The difficult problem in this case is how to make the map data base.

The blind of diabetes will want to learn waking in a safe place such as the campus of a hospital or a park. The difficult problem is the human interface because the diabetic loses not only the vision but also auditory sense and the haptic sense. He will not be able to use PC.

This work is supported by Grant-in-aid for Scientific Research on Priority Areas No.07245105, Grant-in-Aid for Scientific Research(B)(2) No.07555428 by the Ministry of Education, Science, Sports and Culture, The Mechanical Industry Development & Assistance Foundation (MIDAF).

References

1) S.Kotani, H.Mori & N. Kiyohiro,"Development of the robotic travel aid "HITOMI"", Robotics and Autonomous Systems, No.17,1996, pp.119-128

2) Mobility of Blind and Elderly People Interacting with Computers Final Report, Royal Natinal Institute for the Blind on behalf of the MoBIC Consoetium, 1997

3) N. Tinbergen, "The study of instrinct", The Claredon Press Oxford, 1969

4) H.Mori, N.M.Charkari and T.Matushita,"On-line Vehicle and Pedestrian Detection Based on Sign Pattern", IEEE Trans. On Industrial Electronics, Vol.41, No.4, 1994, pp384-391

4) Pertrie H., Johnson V., Strothotte T., Rabb A., Fritz S. And Michel R.,"MOBIC: Designing a Travel Aid for Blind and Elderly People", J. Of Navigation, Vol.49, No.1, 1996, pp45-49

5) Regensburg U. And Graefe V. "Visual Recognition of Obstacles on Road", IROS'94, 1994

6) Kuehnell A. "Symmetry based Recognition of Vehcle Rears", Pattern Recognition letters 12. North Holland, 1991, pp.249-258

7) Fraizer J. & Neatia R."Detecting Moving Objects from a Moving Platform", Int"l Conf. On Robotics and Automation, 1992, pp.1627-1633

8) Saneyoshi K. et al."3-D Image Recognition System for Driver Assist", Intelligent Vehicle 93 Symposium, 1993, pp.60-65

9) Johanson G."Spatio-temporal differentiation and integration in visual motion perception", Psychological Research. 38,1976, pp.379-383

10) Kozlowski L.T. and Cutting J.E. "Recognition the sex of a walker from dynamic point-light Display", Perception & Psychophysics Vol. 21,No.6, 1977, pp.575-580

11) Akita K. "Image Sequence Analysis of Real World Human MotionPattern Recognition, Vpol.17,No.1,1984,pp.73-83

12) S. Yasutomi, H. Mori & S.Kotani, "Finding pedestrians by estimating temporal-frequency and Spatial-period of the moving objects "Robotics and Autonomous Systems, No.17,1996, pp.25-34

13) S.Kotani, H.Mori & N.M.Charkari, "Danger estimation of the Robotic Travel Aid(RoTA)at intersection", Robotics and Autonomous Systems, No.18,1996,pp.235-242

Automated Highways And The Free Agent Demonstration

Chuck Thorpe, Todd Jochem, Dean Pomerleau
Robotics Institute, Carnegie Mellon University
{thorpe,pomerleau,jochem}@ri.cmu.edu

1 Abstract

In August of 1997, The US National Automated Highway System Consortium (NAHSC) presented a proof of technical feasibility demonstration of automated driving. The 97 Demo took place on car-pool lanes on I-15 in San Diego, California. Members of the Consortium demonstrated many different functions, including:

Vision-based road following
Lane departure warning
Magnetic nail following
Radar reflective strip following
Radar-based headway maintenance
Ladar-based headway maintenance
Partial automation and evolutionary systems
Close vehicle following (platooning)
Cooperative maneuvering
Obstacle detection and avoidance
Mixed automated and manual driving
Mixed automated cars and buses
Semi-automated maintenance operations

Carnegie Mellon University (CMU) led the effort to build the Free Agent Demonstration (FAD). The FAD involved two fully-automated cars, one partially-automated car, and two fully-automated city buses. The scenario demonstrated speed and headway control, lane following, lane changing, obstacle detection, and cooperative obstacle avoidance maneuvers.

This paper describes the demonstration itself, the technology that made the demonstration possible, and the current efforts to turn the demonstration system into a practical prototype.

2 Introduction

In August of 1997, the National Automated Highway Systems Consortium (NAHSC) organized a public demonstration of automated cars, trucks, and busses. This demo was requested by the US Congress in the 1991 ISTEA legislation. The legislation read in part: "The Secretary of Transportation shall develop an automated highway and vehicle prototype from which future fully automated intelligent vehicle-highway systems can be developed ... The goal of this program is to have the first fully automated roadway or an automated test track in operation by 1997." Unfortunately, Congress didn't fund research on AHS until 1995. But the US Department of Transportation (USDOT) still requested the demo in 1997.

This paper reviews the motivation for AHS, and a variety of interesting technologies. It goes into more detail on two of the demonstration Scenarios: Platooning, led by the University of California at Berkeley PATH program, and Free Agents, led by CMU.

3 Motivations

The most important motivation for building automated vehicles and highways is improved safety. In the US alone, accidents cost 40,000 lives, and 150 billion dollars, every year. The number of fatalities peaked in the 1960s. Since then, a combination of safer cars (e.g., designed crumple zones), safer roads (the Interstate highway system), and policies (mandatory seat belt use, tougher drunk driving laws) has eliminated many of the mechanical causes of accidents and made collisions more survivable. At this point, the dominant cause of accidents is human error: in 90% of all accidents, the driver is at least partly to blame. The next step, to make further significant safety improvements, requires eliminating driver error, either by offering driver assistance or by automating the vehicles.

Congestion is also an increasing problem. Vehicle miles traveled have steadily increased in the US at 4% per year, much faster than the rate of growth of highway miles. The Interstate Highway System is now complete. Adding new lanes in congested urban areas can cost as much as $100 million per lane-mile. Automation is an attractive solution to many congestion problems. On today's roads, with manual driving, the maximum capacity is about 2000 vehicles per lane per hour. If traffic were evenly spaced, this would translate, at 100 kph, to an average spacing of 50 meters per vehicle. But traffic is not evenly spaced; there is bunching and

Figure 1: Westrack automated pavement test vehicle

gapping, and lane changing and weaving. Automated vehicles, communicating with each other and with the infrastructure, should be able to maintain much closer and more even spacing, and to double or triple roadway capacity.

The two motivations, of improved safety and improved traffic flow, are neither directly competing nor directly complementary. Different research groups have developed different technologies and different architectures, partly in response to different emphases on the two main motivations.

4 Alternative Technologies

The NAHSC is not the first group to invent automated driving, nor the only currently active group working on the problem. Serious research efforts began in the late 1950s at the GM Research Center. Their Firebird 2 followed a buried cable. Interestingly, one of the junior engineers on the Firebird 2, Bill Spreitzer, just retired after the 97 demo from his position as head of GM efforts worldwide on intelligent vehicles. The Firebird 2 was brought out of retirement and restored to running order to be shown at the 97 demo.

Figure 2: Westrack pickup coil

4.1 Westrack

Some of the most practical current efforts focus on special-purpose roadways dedicated to automated vehicles. Westrack, at the Nevada Automotive Testing Center, is a pavement test site. They have 26 different types of asphalt pavement along a 2.9 km oval test track. In order to do load testing of the pavement, they have four semi-tractors, each pulling triple trailers around the oval. Since driving in circles is boring, fatiguing, and dangerous for humans, they decided to automate the trucks. The centerpiece of their work is automated steering based on following a buried cable. They put an audio-frequency signal on the cable and sense its lateral position with pickup coils mounted under the front bumper of the trucks. That signal is used for steering control.

The Westrack trucks have logged over 800,000 km of automated driving. The designers took great care to ensure system safety: there are dual pickup coils on the front of the vehicle, dual wires operating at different frequencies, dual windings on the steering motor, triply redundant shaft encoders, and redundant controllers with no computers in the critical loop. There are still some shortcomings in the system, though. Since there is no preview information in the steering system, the trucks overshoot the entrance and exit of each corner. Where there was a small error in laying the original cables, the vehicles tracked the cable and oscillated on each lap until they wore grooves in the pavement. Eventually, vehicle control became difficult, and that section of track had to be repaved and the cable had to be replaced. Finally, since asphalt is flexible, the passage of the trucks "milked" the cable through the conduit with each lap, eventually pulling out the 10" service loop and snapping the cable.

Overall, the system has performed its design task admirably, but it may not be suitable for mass deployment.

4.2 O-Bahn

Several systems around the world use mechanical guideways. The O-bahn system uses concrete rails on both sides of the road. Buses are equipped with horizontal rollers near the front wheels. These buses can run normally on city streets. Then they enter the O-bahn through a centering system, similar to a car wash.

The horizontal rollers run along the concrete rails and directly control the steering.

The O-bahn installation in Essen, Germany allows buses to drive through the very narrow streets at the old city center. The O-Bahn in Adelaide, Australia provides a narrow busway elevated over an ecologically sensitive wetland.

As with Westrack, O-Bahns do a remarkably effective job at providing a service, but are probably not easily extensible to mass use. In particular, they work well for situations where there is one entrance, one exit, and a fleet of well-maintained vehicles that are not likely to break down and block traffic.

4.3 FSS

The Ohio State University showed their Frequency Selective Strip (FSS) technology at Demo 97. The FSS is lane marking tape with an aluminum foil layer in the middle. The foil has slots punched in it, spaced to provide a strong return from an automotive radar at a particular shallow angle. This way the same sensor used to see other vehicles on the roadway can also be used to see the road position, at a distance of several meters in front of the vehicle. The system shown in August 97 used a 10 GHz radar and a strip placed down the center of the lane. Current work is designing a strip for the 77GHz frequency that is now the standard for automotive radars, and is designing strips that are visible from a range of azimuth angles, so they can be placed as lane markers.

4.4 Vision

Besides the CMU group, many other groups around the world are also working on road-following using vision, including: Professor Dickmanns at the Military University of Munich[1], Daimler Benz[2], BMW,[3] and several of the Japanese auto companies.[4] Two of those, a Honda system and a Toyota system, were shown as part of the 97 Demo.

5 97 Demonstration

The 1997 AHS Demo included seven demonstration scenarios, designed to showcase different technologies and different functions:

Platoons, with closely-spaced vehicles following buried magnets (coordinated by UC Berkeley PATH program) [5]

Free agents, with cars and busses using vision and radar (CMU)

Evolutionary, showing how this technology can be introduced incrementally for driver assistance (Toyota)[6]

Control transition, using both vision and buried magnets (Honda) [7]

Alternative technology, using a radar-reflective strip for lateral control (Ohio State) [8]

Infrastructure diagnostic, checking the accuracy of the magnets (Lockheed Martin) [9]

Heavy trucking, using radars for smart cruise control and driver warning (Eaton Vorad)

Of these seven demos, the Platoon and Free Agent demos were the largest and the most distinctive, and are therefore the best examples to explain.

5.1 Platoon Demo

The Platoon demo used two interesting technical approaches: lateral control by magnetic nail following, and longitudinal control in tightly spaced groups of vehicles, or platoons.

Magnetic nails are permanent magnets. The markers were installed every 1.4 m by surveying the location, drilling a hole, placing the magnets, then sealing them with epoxy. Each vehicle had three magnetometers mounted beneath the front and rear bumpers. As they drove over each magnet, they sensed its location, and servoed the steering to follow the markers. The magnets can be installed either North Pole up or South Pole up. This creates a simple binary code which can be used to signal upcoming curves or intersections.

The motivation for platoons is that packing vehicles very closely can add to safety. In the unlikely event that a computer-controlled vehicle has an abrupt failure in its velocity regulation, there may be a collision with a leading or trailing

Figure 3: Platoon Demo

vehicle. But since the space is so small, any collision will happen quickly, before a large relative velocity can build up. Generally, platoons run at inter-vehicle spacing of a few meters down to one meter.

In order to provide the tight control needed to maintain these spacings, platoons need good vehicle range sensing, an accurate dynamic model, high performance actuators, and good inter-vehicle communications to provide control preview information from leading vehicles. For the 97 demo, UC Berkeley used a specially-modified Delco radar, identical Buick LeSabres with modified and instrumented transmissions, and digital radios provided by Hughes for communicating the state of the lead vehicle to the rest of the platoon at 50 Hz.

The platoon demo ran a string of eight vehicles, with 6.5 meter inter-vehicle spacing. They engaged automated control from a stop, and ran completely automatically up to highway speeds and back to a stop. During the run, the second vehicle requested a lane change. The platoon automatically separated to provide maneuvering room, the second vehicle changed into the right lane, and the platoon reformed. After the second vehicle dropped back to the end, it changed back into the left lane, and rejoined the platoon.

5.2 Free Agent Demo

The Free Agent demo included five vehicles: two fully-automated Pontiac Bonneville sedans, a partially automated Oldsmobile Silhouette minivan, and two fully automated New Flyer city busses. The vehicles are named Navlab 6 and 7 (the Bonnevilles), 8 (minivan), and 9 and 10 (busses). Each of the vehicles in the scenario demonstrated slightly different functions. As an example, the following is the trace of a run on one of the sedans.

The Navlab 7 enters the AHS lane following a bus, a sedan, and another bus, and trailed by the minivan. All vehicles start under manual control. As the vehicles pick up speed to 50 mph, the lead vehicles drift off the road under manual control, to demonstrate the lane departure warning system. The warning system beeps, the drivers

Figure 4: Navlabs 6 through 10, from front to back

note they are drifting off the road, and they steer safely back onto the roadway. Once the vehicles are all safely back in their lanes, the Navlab 7 driver engages auto control by pressing the cruise control engage switch. A gentle voice says "automatic control on", a confirming display appears on the interface screen and on the HUD, and the driver takes his hands off the wheel and his feet off the pedals.

In a real AHS system in an urban environment, there could be a Traffic Management Center sending speed commands to the vehicles. The demo did not include a real TMC, so the vehicles simulated receiving a command to increase speed. The computer communicates with the cruise control, and the car increases speed to 55 mph automatically, passing the lead busses and car.

The minivan, driving manually, approaches from the rear at 65 mph. The minivan driver receives a warning that he is closing quickly on the vehicle ahead, triggered by his forward-looking radar. On Navlab 7, the rear looking ladar detects the approaching van. The human interface announces "high speed vehicle approaching". Navlab 7 checks its vision system to see if there is a lane to the right, checks its side looking sensors to confirm that the lane is clear, and checks its rear looking sensor for vehicles approaching in the right lane. If it is not clear of the busses, Navlab 7 holds its position. Once it is safe to change lanes, the voice and displays indicate "changing to right lane", and the vehicle smoothly changes lanes, allowing the minivan to pass.

Later, the second sedan pulls in behind the Navlab 7. The two vehicles communicate by digital radio to establish that they are both automated. The trailing sedan tracks Navlab 7 by radar, and maintains a comfortable 1.5 second gap using the throttle and brake actuators. Navlab 7 detects an obstacle in its lane, in this case an orange plastic construction barrel. Inside the vehicle, the interface indicates "obstacle detected - swerving to left", and the Navlab moves to the side. Since the radar has high angular accuracy, the vehicle only moves over far enough to clear the obstacle. It also communicates the location of the obstacle to the trailing sedan and busses, which automatically and safely change lanes even before their own sensors have spotted the obstacle.

The trailing sedan passes Navlab 7, and pulls back into the right lane. The driver of Navlab 7 wishes to re-pass the other sedan, without disengaging automated control. He presses the "change lanes left" button, presses the "increase speed" cruise control button, and the Navlab 7 changes lanes, speeds up, and passes the other sedan. Similarly, he requests a slowdown and a return to the right lane, and, once the spacing is clear, the Navlab 7 changes back.

Eventually, the Navlab 7 detects obstacles completely blocking its lane. For this part of the scenario, a simulation is set up indicating that there is traffic in the left lane, so it is impossible to change lanes. Navlab 7 brakes to a safe halt, and through a combination of radio communication and radar sensing, the trailing sedan also comes to a halt, followed by the busses.

5.3 Underlying Technology

Much of the underlying technology in the Free Agent Demo is new, built specifically for the Demo. Other components have been adapted from previous work. To as great an extent as possible, all systems on the three passenger cars and the two busses are identical. Components include the following:

RALPH: The vision system on all 5 vehicles is the RALPH system, built by Pomerleau[10]. This system uses a forward-looking video camera, mounted behind the rear view mirror of the cars and on the inside of the bus windshield, to image the road. The image is re-sampled to produce an overhead view of the road. The overhead view is processed to find the road curvature, by looking for the swept arc that maximizes the sharpness of edges along the swept line segment. This effectively finds the curve that most closely follows all visible road features. This was especially important for the 1997 Demo, since highways in California use raised dots instead of painted lines, so vision systems that rely on continuity of lines may have difficulty with this course. RALPH uses the raised dots, but also uses pavement joints and the edge of the shoulder and other parallel linear features, in order to find and track the road. This system is now commercially available through Assistware Technology Inc.

Radar: Headway maintenance (keeping a consistent gap from the lead vehicle) relies on a radar. Our partner Delco electronics supplied a 77GHz mechanically scanned radar with software for detecting and tracking targets within a 12 degree field of view, out to a range of 150 meters. It is important to measure both target range and bearing; commercially available automotive radars usually have no measurement of bearing, and therefore cannot properly track targets on curved roads. We have integrated the radar output with RALPH to register detected targets with detected road position. This lets our vehicles classify targets as to whether they are in the current lane, in an adjacent lane, or off the road. The sensors used on the busses are commercially available radars from Eaton Vorad that report range but not bearing.

Side-looking sensors: Each vehicle is equipped with four side-looking short-range radars from Eaton Vorad for detecting objects adjacent to the vehicles.

Rear-looking sensors: The rear-looking sensors are scanning ladars from Riegel. They have a field of view of approximately 20 degrees.

Lane changing: The logic requesting a lane change is based on desired speed, speed of preceding vehicles, and locations of vehicles in adjacent lanes. For the demonstration, the scenario was constructed so lane changes were easily executed when expected. In the more general case, deciding on a lane is an example of "tactical driving", the subject of a recent thesis by Rahul Sukthankar[11], a member of our group. His SAPIENT simulated vehicles do careful analyses of upcoming exits, velocities as well as positions of surrounding vehicles, and other factors, all combined in a distributed behaviorist framework.

Actuators: The car brake and steering actuators were custom provided by our partners at General Motors. The bus air brake and steering actuators were custom built by K2T, Inc. For all vehicles, the throttle actuation is through the existing cruise control. The Free Agent philosophy is to have large enough separations between vehicles that high-bandwidth throttle and brake servos are not needed. Using the existing cruise controls shows that low-bandwidth speed control is sufficient. As an added benefit, it reduces cost, provides commonality of interface between buses and cars, and increases safety by using tested commercial components.

Safety circuit: There are several safety checks in the system, to maximize safety on the demo vehicles. First, at the lowest level, any actuator can be overridden by the human safety driver. The steering motors and amplifiers are deliberately torque-limited to be easily overpowered by a person. The driver can similarly drive the throttle or brakes, and the computer controls have no way to backdrive the pedals. As a last hardware check, an independent safety board can at any time cut power to all actuators. The safety board continually monitors computer heartbeat, lateral acceleration, and state of emergency kill switches. In addition, the vehicle driving behaviors in the Free Agent philosophy are designed to keep safe space around vehicles, and to provide opportunity for defensive driving.

6 Comparison

At each major design point, CMU and Berkeley made different choices:

	PATH	CMU
Obstacles	Exclude	Sense
Driver role	Exclude	Cruise control
Grouping	Platoons	Free agents
Lanes	Dedicated	Mixed traffic
Deployment	Revolutionary	Evolutionary

6.1 Choices

For obstacle handling, CMU chose to give the vehicles forward-looking obstacle detection sensors. PATH did not demonstrate obstacle detection in this demo, and instead emphasized obstacle prevention through inspection of vehicle loads and construction of barriers.

For the role of the driver, in the PATH platoons spacing is so tight that the slow reflexes of a person would lead to unacceptable danger, so the human is locked out of control. When a driver wants to take control, he must request control, and wait until the vehicle has been moved out of the platoon and out of the dedicated lane. In the Free Agent scenario, the automated system is treated like a cruise control, that can be engaged, disengaged, and overridden by the driver.

The grouping strategy used by PATH, the platoon, generates high traffic density. The platoon strings are so long that they block merge lanes for significant distances, so any merging requires coordination to make sure vehicles enter only in the gaps between platoons. This probably relegates platoons to dedicated lanes, on which only automated vehicles can drive. The Free Agents usually run as individual vehicles, but can communicate to share information on emergency braking or obstacle avoidance.

PATH demonstrated a scenario in a dedicated lane. CMU demonstrated driver assist, mixed manual and automated traffic, and finally full automation with communicating vehicles[12].

A dedicated lane system requires a revolution in transportation. The free agents are designed to run in mixed traffic, and to be deployed one vehicle at a time. The dedicated lanes are an easier engineering problem in some ways; but building new lanes is a larger societal challenge.

6.2 Discussion

These different approaches to AHS are appropriate. First, it makes more sense for the NAHSC to explore alternatives rather than to build duplicate systems. But more fundamentally, the different demo systems emphasize the different goals of AHS, and perhaps the different geographic imperatives of the two developers. Berkeley is located in the crowded San Francisco Bay area. Traffic congestion is a major problem. Many freeways are at least three lanes wide in each direction. That encourages them to work on designs that dedicate a lane to automated vehicles, and pack as many vehicles as possible into that lane. For those circumstances, platoons are a reasonable alternative to consider. And for the relatively few miles of urban freeway, infrastructure investments such as placing magnets or building barriers to separate a lane may make sense.

CMU, in contrast, is in the rural western end of Pennsylvania. Few miles of expressways are congested and the only 3-lane wide roads are for climbing lanes on hills, near some interchanges, and for a very few miles of the newest freeway. In those circumstances, safety is more of a market force than congestion. Since accidents occur on rural roads as well as interstates, and since there are 4 million miles of road (vs. 100,000 miles of limited access roads), it is infeasible to consider major infrastructure investments for safety upgrades. This leads to a solution where vehicles are sensor-rich and independent, and where the computing and sensing can be used for driver safety assistance even when the vehicles are driven manually on rural roads.

So which system is better, the Platoons or the Free Agents? Perhaps there is no one right answer; each system has its niche in which it may be more appropriate. A useful deployment strategy may embrace Free Agent technology sooner, at least for driver warning and assist, and later selectively deploy platooned systems in dedicated lanes, after market penetration justifies that lane usage.

7 Conclusion and the Future

The 97 AHS Demo was designed to be an intermediate checkpoint on the way to building a prototype AHS. According to the original schedule, the next two years of the AHS program were supposed to be spent building three separate approaches to AHS, followed by a downselect, followed by designing and building a final prototype and conducting tests up to the year 2002.

The NAHSC program was on schedule and within budget, and the AHS Demo was widely regarded as an outstanding success. But politics and funding priorities change. The current emphasis in the US Department of Transportation is on nearer-term results, and particularly on safety and driver warning devices. The new program, to kick off in 1998, is called the Intelligent Vehicle Initiative. Much of the work on full automation will be reduced or postponed. Increased emphasis will be placed on vehicle-centered technology, and on human factors and driver interfaces.

These new priorities are useful and will generate interesting research questions. But in the longer term, driver warnings will only achieve limited results. The imperative remains for increasing automation, both for congestion relief and for increased safety.

8 Acknowledgements

The CMU AHS group includes Parag Batavia, Michelle Bayouth, Tunde Balvanyos, Frank Dellaert, Dave Duggins, John Hancock, Martial Hebert, Todd Jochem, Katsumi Kimoto, Phil Koopman, John Kozar, Bala Kumar, Dirk Langer, Sue McNeil, Illah Nourbakhsh, Dean Pomerleau, Rahul Sukthankar, Toshi Suzuki, Todd Williamson, and Liang Zhao. Assistware Technology Inc. provided copies of the vision system, and Tad Dockstader, of K2T Inc, did the mechanical engineering on the buses.

Primary funding for this work comes from the US Department of Transportation, agreement DTFH61-94-X-00001, "Automated Highway System", from the National Highway Traffic Safety Administration (NHTSA) under contract DTNH22-93-C-07023, and from Houston Metro. Other support comes from the Ben Franklin of Pennsylvania Technology Partnership through contract 95W.CC005OR-2, and from DARPA through TACOM in contracts DAAE07-90-C-R059 and DAAE07-96-C-S075, "CMU Autonomous Ground Vehicle".

Core members of the NAHSC are: Bechtel, Caltrans, CMU, Delco, General Motors, Hughes, Lockheed Martin, Parsons Brinkerhoff, the USDOT, and the University of California. Other groups providing demonstrations were Ohio State, Honda, Toyota, and Eaton Vorad.

Figure 3 is courtesy of California PATH Publications.

Portions of this paper have appeared in the Proceedings of ITSC 97 and in the Proceedings of FSR 97.

9 References

---

[1] Dickmanns E. D. "Vehicles Capable of Dynamic Vision," in 15th International Conference on Artificial Intelligence, Nagoya, Japan, August 1997.

[2] Zomotor, Z, and Franke, U., "Sensor Fusion for Improved Vision Based Lane Recognition and Object Tracking with Range Finders" in Proceedings of IEEE Conference on Intelligent Transportation Systems, Boston MA, November 1997.

[3] Lerner G. and Steinkohl, F. (1997) "Driver Assistance: Concepts and Systems" Traffic Technology International, Oct/Nov. 1997

[4] Proceeding of 1996 Intelligent Vehicles Symposium, Sept. 19-20 1996, Tokyo Japan.

[5] Zhang, Wei-Bin, "National Automated Highway System Demonstration: A Platoon System" in Proceedings of IEEE Conference on Intelligent Transportation Systems, Boston MA, November 1997.

[6] Andrews, Scott, "Steps to Deployment: An Outline of Toyota's U.S. AHS Program" in Proceedings of IEEE Conference on Intelligent Transportation Systems, Boston MA, November 1997.

[7] Shladover, Steve and Empey, Dan, "Characteristics of the Honda-PATH AHS Team Vehicles for the 1997 AHS Feasibility Demonstration" in Proceedings of IEEE Conference on Intelligent Transportation Systems, Boston MA, November 1997.

[8] Ozguner, Umit, et. al. "The OSU Demo '97 Vehicle" in Proceedings of IEEE Conference on

Intelligent Transportation Systems, Boston MA, November 1997.

[9] McKenzie, Patrick, "Infrastructure Diagnostic Vehicle" in Proceedings of IEEE Conference on Intelligent Transportation Systems, Boston MA, November 1997.

[10] Pomerleau, D. and Jochem, T. (1996) Rapidly Adapting Machine Vision for Automated Vehicle Steering. IEEE Expert, Vol. 11, No. 2

[11] Sukthankar, Rahul. "Situational Awareness for Driving in Traffic", PhD thesis, Carnegie Mellon University, January 1997.

[12] Bayouth, M., and Thorpe, C. "An AHS Concept Based on an Autonomous Vehicle Architecture", in Proceedings of 3rd Annual World Congress on Intelligent Transportation Systems, 1996.

A. Appendix: Overall Demo Statistics

The 97 Demo was a very large public undertaking, perhaps the largest public participation demonstration of robotics and automated vehicles.

Attendees	
Total	3500
Public	1000
Press	100
VIP	1400
Industry	1000
I-15 Demo	
Total vehicle	26
Total automated	21
Automated types	1 truck, 2 busses, 18 cars
Automated makes	Freightliner, New Flyer, Buick, Oldsmobile, Pontiac, Toyota, Honda
Demo runs	20 each
Trial runs	8 each
Total automated vehicle runs	588
Total automated miles during demo	4468
Dress rehearsal runs	22 each
Automated miles during dress rehearsal	3511
Mini demo on short track	
Total vehicles	5
Total automated runs	144
Mini-demo total miles	180
Passengers	
I-15 during demo runs	1350
I-15 during demo trials	500
I-15 during dress rehearsals	1000
Minidemos	1400
Total passengers	4250

The Design of High Integrity Navigation Systems

H. Durrant-Whyte, E. Nebot, S. Scheding, S. Sukkarieh and S.Clark

Australian Centre for Field Robotics,

Department of Mechanical and Mechatronic Engineering,

University of Sydney, NSW. 2006, Australia

E-mail [hugh,nebot,scheding,sala,clark]@tiny.me.su.oz.au

Abstract— This paper describes some key elements of our current work to design high-integrity navigation systems for use in large autonomous mobile vehicles. A frequency domain model of sensor contributions to navigation system performance is introduced and used to demonstrate the manner in which conventional navigation loops fail. On this basis, a new navigation system structure is introduced which is capable of detecting faults in any combination of navigation sensors. The main elements of the proposed structure are analysed. Two example implementations of this structure are described; a system for land-vehicle navigation employing twin GPS/Inertial and radar/encoder navigation loops, and an underground vehicle navigation system employing twin laser/sonar/inertial and laser/encoder navigation loops.

1 INTRODUCTION

This paper addresses the problem of developing ultra-high integrity navigation systems large outdoor autonomous guided vehicles (AGVs). Navigation, the ability to locate a vehicle in space, is a fundamental competence of any AGV system. To achieve any useful purpose, to guide itself from one place to another, an AGV must know where it is in relation to its environment. If this position information is inaccurate, unreliable, or in error, the AGV system will fail in a potentially catastrophic manner.

No system can be made 100% reliable; every sensor, all electronic components and any mechanical design will fail in some (possibly unknown) way, in some circumstance at some time. Consequently, navigation system design must always focus on integrity: if there is a failure it should be detected, identified and dealt with in a fail-safe manner. A system which is constructed from high reliability components may not itself have high integrity. However, a system constructed of relatively low reliability components may be designed to have high integrity.

The key to the development of a high-integrity, and thus useful, navigation system is through the use of a number of physically different sensors whose information is combined in such a way that all possible failures in all possible sensor combinations are detectable. This is the rational for the proposed design described in this paper.

This paper begins in Section 2 by describing and justifying an overall navigation system architecture which achieves this goal of high integrity. We describe how different sensor and system faults are detected and show that with the proposed design it is possible to quantify system integrity *a priori* in terms of fault rejection sensitivity. Section 3 describes one specific installation if this architecture which we are currently developing which employs both a DGPS/Inertial navigation loop together with a millimetre-wave radar/encoder navigation loop. Preliminary results of trials with these sensors are described are described in Section 4.

2 THE STRUCTURE OF A HIGH-INTEGRITY NAVIGATION SYSTEM

The overall structure of the proposed navigation system is shown in Figure 1. Fundamentally, It consists of two physically independent "navigation loops", a third independent sensor that monitors differences between these two loops and an arbitration unit which compares and combines the results from these navigation different systems.

The essential principle encapsulated by this structure is the explicit recognition that all sensors fail at some point in time in a number of (possibly unknown) ways. The structure is designed such that any one failure in any one sensor can be detected and identified on-line. A failure in any two or more sensors may also be detected but not identified. The only way that this navigation structure will fail catastrophically is if any two sensors fail at the same time in exactly the same way. As the sensors are specifically chosen so that each operates on completely different physical principles, the probability of this common mode failure is as low as it can be.

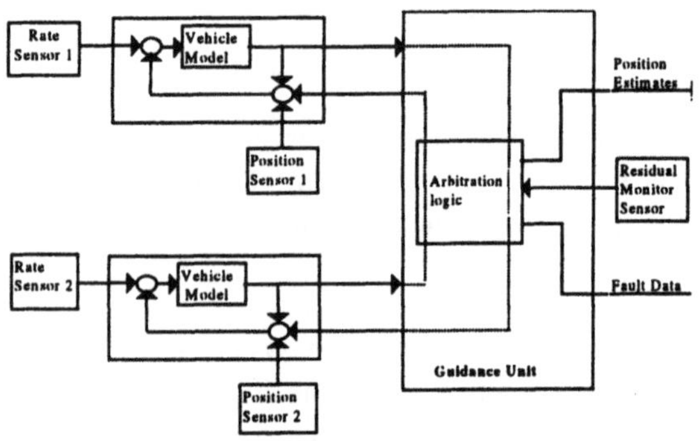

Figure 1: Proposed Navigation System Architecture

2.1 Navigation Loops

Any navigation system or navigation loop must contain at least i) a low frequency or position measurement sensor; ii) a high-frequency or rate measurement sensor; and iii) a model of the vehicle or process whose position and velocity is to be determined [9]. The reason for this is as follows:

A position measurement sensor is clearly essential to provide position estimates. In the position sensing process, the required position measurement is physically always a band-limited signal embedded in broad-band noise. Thus the position estimate must be constructed by passing this measurement through a low-pass filter (a smoother, weighted time average or a Kalman filter), with fixed cut-off frequency dictated by the position signal bandwidth. Beyond this cut-off frequency, position measurements are assumed dominated by signal noise. To provide position estimates beyond this cut-off, a sensor sensitive to high-frequency position changes must be employed. In particular, sensors that provide measurements of velocity or acceleration, once integrated to provide position, yield estimates of rapid changes in position. However, such sensors are not good at providing estimates of long term or slow changes in position as integration of the noise component in the signal leads to Brownian-like divergence of estimates. Thus the "low frequency" information provided by a position measurement sensor must be combined with "high frequency" information provided by a rate measurement sensor in a complimentary manner. Each sensor provides different information (they are not in any sense redundant), both sensors are necessary in any one navigation loop to provide essential vehicle navigation information.

In Figure 1, each navigation loop consists of this low-frequency/high frequency sensor combination arranged in a feedback mode (known as a feed-back complimentary filter [9]). Rate information is fed forward through the model of vehicle dynamics to provide an estimate of position. This estimate is compared with a position measurement and the error between the two is fed back to correct the rate sensor position estimates at source. This ensures that high frequency information from the rate sensor is fed straight through to the loop output, while low frequency information is corrected and filtered. The essential role of the vehicle model is now clear, it provides the means of relating rate information, obtained from wheel rotation and steer angle or from acceleration and rotation rates in an inertial package, to position information through a model of vehicle motion.

Thus, a position measurement sensor, a rate measurement sensor and a vehicle model are essential to any one navigation loop. Each is complimentary and there is little or now redundancy between components.

2.2 Resilience to Faults

Each filter in each navigation loop acts as a low-pass filter with respect to position observations and as a high pass filter with respect to rate observations. The cut off frequency of the filter is determined by the relative noise levels in the two sensors but is otherwise always first order (with a 20db/decade roll-off) [10]. This is discussed further in the example described in Section II.3. With this information it is not difficult to quantify fault rejection capability both in each loop and for the overall system. Consider the following four (exclusive and exhaustive) scenarios:

- Rejection of low frequency faults (drift, bias, offset) in low frequency (position) sensors: In any one navigation loop these faults are passed straight through the low-pass filter to the output and are thus undetectable. For the combined system, such faults are detected with a discrimination power equal to the ratio of the covariance in each loop; if the two loops are of essentially equal accuracy then the discrimination power is 3db.
- Rejection of high-frequency faults (jump fault, mis-match, sudden failure) in low frequency sensors: Such faults are naturally detected and attenuated by the low-pass characteristic of each navigation loop with a discrimination power that increases as 20db/decade past the filter cut-off frequency. Such faults are thus readily detected and rejected in any well design navigation loop.
- Rejection of low-frequency faults (drift, bias offset) in high frequency (rate) sensors: Again, such faults are naturally detected and attenuated by the high-pass (with respect to rate sensing) characteristic of each navigation loop with a discrimination power that decreases as 20db/decade up to the cut-off frequency.
- Rejection of high-frequency faults (jump fault, sudden failure) in high-frequency (rate) sensors: In any one navigation loop these faults are passed straight through the high-pass filter to the output and are thus undetectable. For the combined system, discrimination power is dependent on both the sensor error rates and the vehicle model employed but is again, typically, 3db beyond the cut-off frequency.

An analysis of the fault discrimination capabilities of any given suite of navigation sensors is additionally a powerful tool in actually determining what combination of sensors as best fitted to a given navigation task.

2.3 Quantifying Fault Rejection Capability: An Example

The innovations of a filter (or observer) are the only internal performance metric of the navigation system, having known statistics when no fault has occurred. The difficulty in determining whether a fault has occurred in a sensor from the innovation sequence arises when the filter tracks a fault rather than rejecting it, resulting in a state estimate that diverges from the true state, without changing the statistics of the innovations. This shows, through a simple linear example, how fault detectability may be quantified as a function of frequency. It is shown that for a single sensor system, low frequency faults are undetectable.

Consider a linear system represented in standard state space form by the equations,

$$\dot{\mathbf{x}}(t) = \mathbf{F}\mathbf{x}(t) + \mathbf{w}(t)$$
$$\mathbf{z}(t) = \mathbf{H}\mathbf{x}(t) + \mathbf{v}(t) \qquad (1)$$

where $\mathbf{x}(t)$ is the state of the system at time t, $\mathbf{F}$ is the continuous time process model and $\mathbf{H}$ is the matrix mapping the state to observation space $\mathbf{z}(t)$. The variables $\mathbf{w}(t)$ and $\mathbf{v}(t)$ are the process noise and sensor noise vectors respectively and are assumed Gaussian with zero mean.

For a system described in this form, the continuous time Kalman filter update equations are given by

$$\dot{\hat{\mathbf{x}}}(t) = \mathbf{F}\hat{\mathbf{x}}(t) + \mathbf{K}[\mathbf{z}(t) - \mathbf{H}\hat{\mathbf{x}}(t)] \qquad (2)$$

where $\mathbf{K}$ is the steady state Kalman gain. This may be described in the frequency domain as,

$$s\hat{\mathbf{x}}(s) = \mathbf{F}\hat{\mathbf{x}}(s) + \mathbf{K}[\mathbf{z}(s) - \mathbf{H}\hat{\mathbf{x}}(s)] \quad (3)$$

giving the transfer between the estimate and the output as,

$$\frac{\hat{\mathbf{x}}(s)}{\mathbf{z}(s)} = [s\mathbf{I} - \mathbf{F} + \mathbf{K}\mathbf{H}]^{-1}\mathbf{K} = \mathbf{G}(s). \qquad (4)$$

Therefore, under no-fault conditions,

$$\hat{\mathbf{x}}_{nf}(s) = \mathbf{G}(s)\mathbf{z}_{nf}(s) \qquad (5)$$

where the subscript nf denotes the no-fault condition.

Consider an observation corrupted by a fault vector $\mathbf{f}(s)$

$$\mathbf{z}_f(s) = \mathbf{z}_{nf}(s) + \mathbf{f}(s). \qquad (6)$$

From equation (5) the state estimate becomes

$$\hat{\mathbf{x}}_f(s) = \mathbf{G}(s)[\mathbf{z}_{nf}(s) + \mathbf{f}(s)]. \qquad (7)$$

The no-fault filter innovations are defined as

$$\nu_{nf}(s) = \mathbf{z}_{nf}(s) + \mathbf{f}(s) \qquad (8)$$

so, under fault conditions

$$\begin{aligned} \nu_f(s) &= \mathbf{z}_f(s) - \mathbf{H}\hat{\mathbf{x}}_f(s) \\ &= [\mathbf{z}_{nf}(s) + \mathbf{f}(s)] - \mathbf{H}[\mathbf{G}(s)[\mathbf{z}_{nf}(s) + \mathbf{f}(s)]] \\ &= [\mathbf{z}_{nf}(s) - \mathbf{H}\hat{\mathbf{x}}_{nf}(s)] + [\mathbf{f}(s) - \mathbf{H}\mathbf{G}(s)\mathbf{f}(s)] \\ &= \nu_{nf}(s) + [\mathbf{1} - \mathbf{H}\mathbf{G}(s)]\mathbf{f}(s) \end{aligned}$$

(9)

Therefore, a fault is detectable in the innovation sequence when the term $[\mathbf{1} - \mathbf{HG}(s)]$ is non-zero. It is interesting to note that this has a direct analogy with the no-fault innovation which may be written as $[\mathbf{1} - \mathbf{HG}(s)]\mathbf{z}(s)$. Consider now a system consisting of a constant velocity (linear) vehicle together with a sensor directly observing position. Figure 2(a) shows the filter transfer function, between observation and position or velocity estimate, for this system. Figure 2(b) shows the detectability of a fault in the position sensor – found by plotting the term $[\mathbf{1} - \mathbf{HG}(s)]$. This shows that low frequency faults such as bias and drift are completely undetectable for this system. If a second position sensor is introduced, the detectability plot for one sensor at fault is shown in Figure 3. This shows that by using redundancy, all faults are captured in the innovation sequence, allowing for relatively simple fault detection schemes to be used such as whiteness tests. Furthermore, by tracking the transmittance of the fault, the magnitude of the fault may be estimated.

2.3 Arbitration and Residual Monitoring

The two navigation loops must each provide estimates of vehicle location and heading derived from physically different sensor operating principles. The output from these two navigation loops may be compared. If there is statistically no difference between the two estimates, then they may be combined to provide an improved overall estimate of vehicle location. If there is a significant difference between the two estimates a fault can be declared. If, at any one time, only one navigation loop is in fault, a comparison can be made with a third (residual monitoring sensor), to determine which loop is at fault and potentially why the fault has arisen. In the unlikely event that both loops fail simultaneously a fault can still be detected and declared on the basis of a difference in estimate output from each other and the residual sensor. However, in this case, it is not possible to identify the fault. The only case in which a fault can not be detected is if both navigation loops and the residual monitoring sensor all fail at exactly the same time in exactly the same way. Given the physical dissimilarity between the sensors employed in each loop, such a common mode failure will essentially never occur.

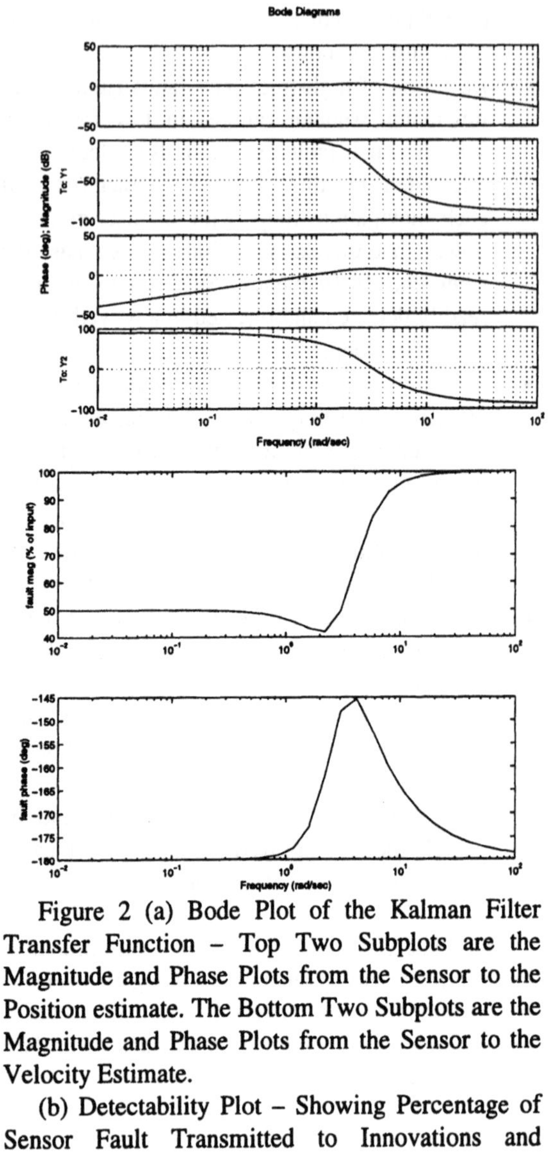

Figure 2 (a) Bode Plot of the Kalman Filter Transfer Function – Top Two Subplots are the Magnitude and Phase Plots from the Sensor to the Position estimate. The Bottom Two Subplots are the Magnitude and Phase Plots from the Sensor to the Velocity Estimate.

(b) Detectability Plot – Showing Percentage of Sensor Fault Transmitted to Innovations and Corresponding Phase

2.4 Decentralised Loop Structure

We briefly describe the structure of the arbitration unit. The algorithms used for combining loop information are based on the use of decentralised data fusion methods employing the information filter as described in [4]. The main advantage of these methods are that they cater directly for the correlation inherent between the navigation loops, and thus allow simple integration of navigation tracks from any number of navigation loops.

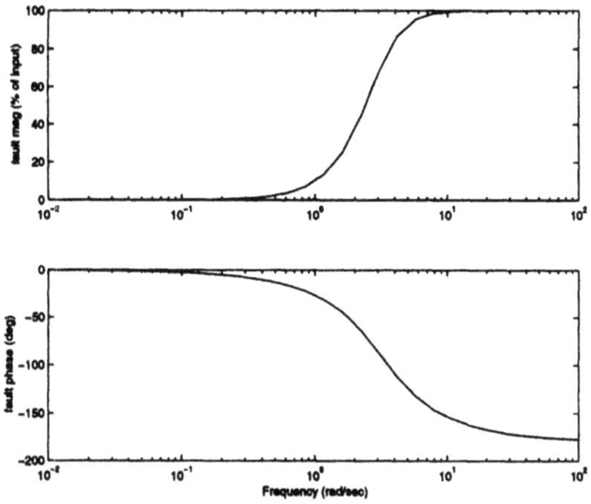

Figure 3 Detectability Plot – the detection of a faults in a single position sensor.

The decentralised estimation system implemented is shown in Figure 4. It consists of a group of local navigation loops operating with independent local filter and sensor inputs.

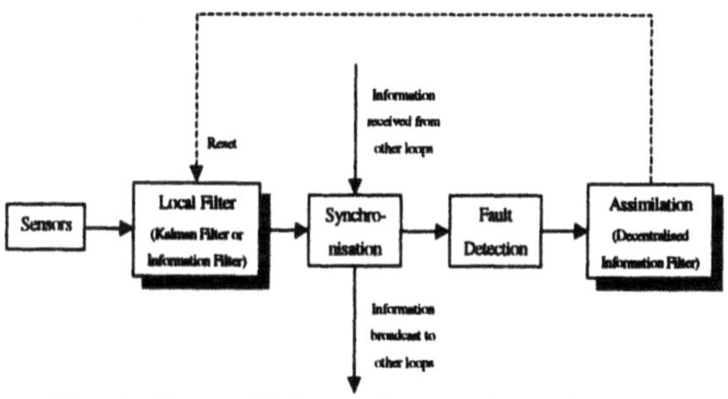

Fig. 4 Flow of information in a local loop

Each loop has a local Kalman filter in which prediction, observation and estimation takes place independent of the other loops. Each loop has a prediction time interval and observation rate dictated by their particular sensory system. The Kalman filters implemented in the local loop will make predictions of states and covariance ($\hat{\mathbf{x}}$ and $\hat{\mathbf{P}}$), at constant interval of times. When new observations become available, the filter will fuse the data to obtain new estimations for the states and covariance matrices ($\tilde{\mathbf{x}}$ and $\tilde{\mathbf{P}}$). At this point the loop has gathered new information that can be broadcast to the rest of the system.

To fuse the information obtained from the different loops in the assimilation stage, it is required that all of the information corresponds to the same time. Therefore, considering that the sensors observe at different times, it is required to synchronise all the information obtained from the sensors before the assimilation stage.

To obtain the new information at the assimilation time, $\hat{\mathbf{x}}$, $\hat{\mathbf{P}}$, $\tilde{\mathbf{x}}$ and $\tilde{\mathbf{P}}$ are propagated forward to the next assimilation time. Denote the values corresponding to the next assimilation time by $\hat{\mathbf{x}}_a$, $\hat{\mathbf{P}}_a$, $\tilde{\mathbf{x}}_a$ and $\tilde{\mathbf{P}}_a$. The information form of the above vectors are obtained as:

$$\begin{aligned}\hat{\mathbf{y}}_a &= \hat{\mathbf{P}}_a^{-1}\hat{\mathbf{x}}_a\\ \hat{\mathbf{Y}}_a &= \hat{\mathbf{P}}_a^{-1}\\ \tilde{\mathbf{y}}_a &= \tilde{\mathbf{P}}_a^{-1}\tilde{\mathbf{x}}_a\\ \tilde{\mathbf{Y}}_a &= \tilde{\mathbf{P}}_a^{-1}.\end{aligned} \tag{10}$$

The new information obtained from the observation is then calculated

$$\begin{aligned}\mathbf{i}_a &= \tilde{\mathbf{y}}_a - \hat{\mathbf{y}}_a\\ \mathbf{I}_a &= \tilde{\mathbf{Y}}_a - \hat{\mathbf{Y}}_a.\end{aligned} \tag{11}$$

where $\mathbf{i}_a$ and $\mathbf{I}_a$ are information state vector and information matrix corresponding to the assimilation time, respectively.

In the assimilation stage, all information obtained from the various loops (after having been synchronised) are fused with the local loop prediction through the equations:

$$\begin{aligned}\tilde{\mathbf{y}}_l(k\mid k) &= \hat{\mathbf{y}}_l(k\mid k-1) + \sum_{j=1}^{N}(\mathbf{i}_a)_j\\ \tilde{\mathbf{Y}}_l(k\mid k) &= \hat{\mathbf{Y}}_l(k\mid k-1) + \sum_{j=1}^{N}(\mathbf{I}_a)_j,\end{aligned} \tag{12}$$

where the index "l" is the loop number, N is the total number of loops and $(\mathbf{i}_a)_j$ and $(\mathbf{I}_a)_j$ are the propagated values of the new information obtained from the loop number "j" to the assimilation time.

Once $\tilde{\mathbf{y}}(k/k)$ and $\tilde{\mathbf{Y}}(k/k)$ have been updated with the new information provided by all the other loops of the system, the local state vector $\tilde{\mathbf{x}}_a$ and covariance $\tilde{\mathbf{P}}_a$ are finally updated in the local loop using the relation described in Equation (12).

It is interesting to note that a system that assimilates all information present will be optimal with respect to the estimate covariances but will not be fault tolerant [2,7]. There are low frequency faults that individual loops will not be able to detect and will consequently broadcast to other loops. It is important to include low frequency fault detection capabilities here, since it is the only stage where these faults can be detected. There will be a trade off between the assimilation rate and the lower frequency of the fault that the system will be able to detect.

3. THE DESIGN OF A HIGH-INTEGRITY NAVIGATION SYSTEM FOR LAND VEHICLE APPLICATIONS

This section describes a current project to develop and demonstrate a high-integrity navigation system for a number of land vehicle projects. The system consists of two navigation loops, one based on the use of GPS as a position sensor and an inertial measurement unit (IMU) as a rate sensor, the other based on millimetre-wave radar as a position sensor and encoders (wheel rate and steer angle) as rate sensors. This is shown in Figure 5. We now explain and justify the design of each proposed navigation loop.

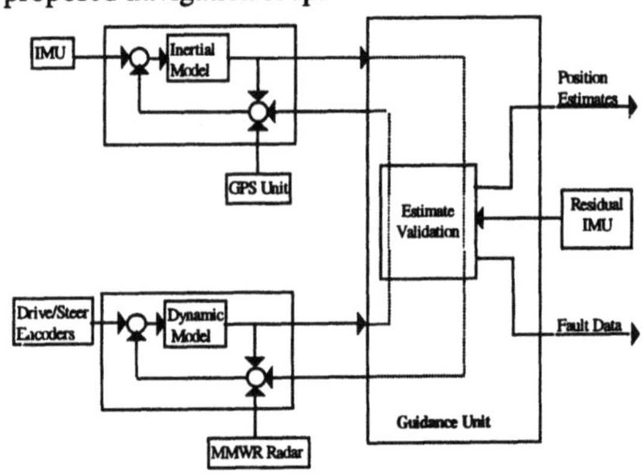

Figure 5: Land Vehicle Navigation System Design

3.1 The GPS/IMU Loop

The Global Positioning System (GPS) is now a widely used source of position measurement information in tasks such as vehicle tracking and dispatch, and surveying. The operating principles of GPS and also DGPS are well known and thus not discussed here [8].

GPS is relatively cheap to implement and is there for the using. It therefore makes sense to incorporate it in any proposed land vehicle navigation system. However, while GPS technology improves daily, as with any sensor it is important to understand the ways it may fail. Failures in GPS are predominantly caused by either constellation or local environment effects. Constellation faults, poor geometry, satellite availability, ephemeris error are generally abrupt and detectable with modern receivers. Environment effects, shadowing, EM interference and particularly multi-path propagation, are more problematic, leading to failures which may be both of abrupt and gradual (drift) types, many of which are unpredictable and undetectable.

In the design, position information from the GPS unit is put in feedback configuration with rate information from an inertial measurement unit (IMU) [10,12]. The unit employed is a Watson triaxial IMU system using vibrating structure gyros, solid state accelerometers, and two pendulum sensors to provide full three dimensional rate, acceleration and 2-D attitude information. The use of the pendulum sensors is critical in providing accurate initial alignment information. Errors of the order of 0.1 degree in attitude will introduce drift of up to 0.5 meters after only 1 minute of operation. This is due to the incorrect compensation of the gravity vector.

The "vehicle model" used in this navigation loop consists of a standard model of three-dimensional point kinematics relating accelerations and angular velocities to point position.

The GPS/IMU loop employs an indirect feedback filter. In this case high-frequency IMU information is still able to propagate through to the output without attenuation, and is compensated with low frequency corrections. The main feature of this filter is that the IMU may be reset from the filter estimates. This approach has the potential advantage of including other IMU internal parameters, such as platform misalignments, gyro and accelerometer bias, in the estimated state vector. The indicated position/velocity information derived from the INS is subtracted from the position/velocity GPS information. An error propagation model is used to fuse this observation and predict position/velocity and some parameters to correct the inertial measurement unit information. These parameters are fed back to the IMU unit.

3.2 The MMWR/Encoder Loop

Over the past five years, millimetre wave radar (MMWR) has found increasing use as automotive sensors, in guidance and collision detection tasks, as part of a number of international "intelligent vehicle highway" systems projects [5]. As an outdoor all weather navigation sensor, MMWR is almost ideal. The high operating frequency (77GHz) and the relatively short wave-length (4mm), provide a number of key operating characteristics. The radar itself can be made relatively small (100mm antenna is typical) while maintaining high levels of directional accuracy (1o is common). In frequency modulation (FM) range measurement, very high swept band-widths (600MHz) can be obtained, translating in to very accurate range measurements (of order 10-25cm). MMWR has typical operating ranges of 2-300m and significantly is largely unaffected by heavy rain and dust. The MMWR unit used in this work is shown in Figure 6.

Figure 6: 77GHz MMW Radar Unit and Beacon used in the navigation system.

The MMWR/encoder loop is based on the same technology and principles previously employed by the authors in [3]. Position information is obtained by mechanically scanning the radar beam through 360°, obtaining range and bearing measurements to a sparse number of special passive reflectors placed along the haul route. Position information from the MMWR is combined with encoder (wheel rotation rate and steer angle) information using a detailed kinematic vehicle model [6], together with a Kalman filter, in standard feedback configuration. The integrated MMWR/encoder system is capable of providing position accuracies of a few centimetres.

3.3 Preliminary Results from the Land Vehicle System

We focus here on describing results from the GPS/IMU system. The sensors were retrofitted in a Holden Ute vehicle. The car was driven in the neighbourhood of the University where buildings and other structures were present obstructing in some cases the direct path of the GPS satellites. Figure 7 shows the estimates produced by the filter running on GPS and IMU data. Figure 8 shows an enlargement of the position estimate after the first turn. In this case the DGPS generates a position solution which is affected by multipath. Since the filter is fusing all DGPS data the estimate becomes erroneous. Fault detection capabilities were included in the run shown in Figure 9. In Figure 10 it can clearly be seen that the previous multipath fault is detected and the filter navigates with IMU alone during the period that the DGPS is in fault. There is no noticeable degrading of performance during the GPS fault.

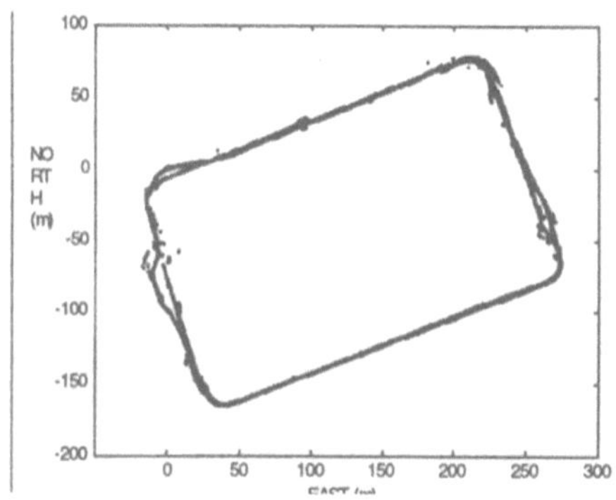

Figure 7 Trajectory without Fault Detection

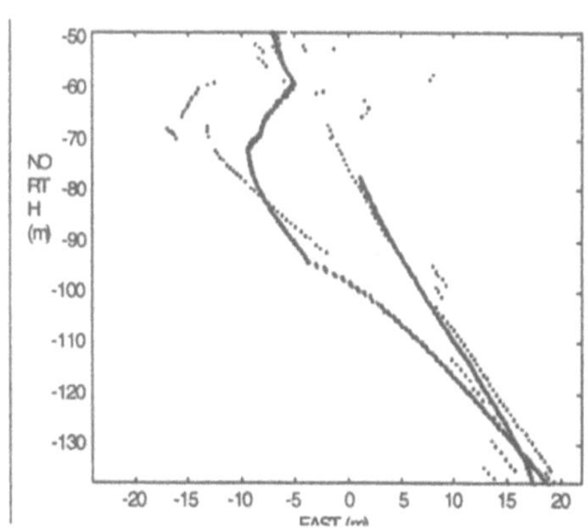

Figure 8: Enlarged Trajectory in Multi-Path area.

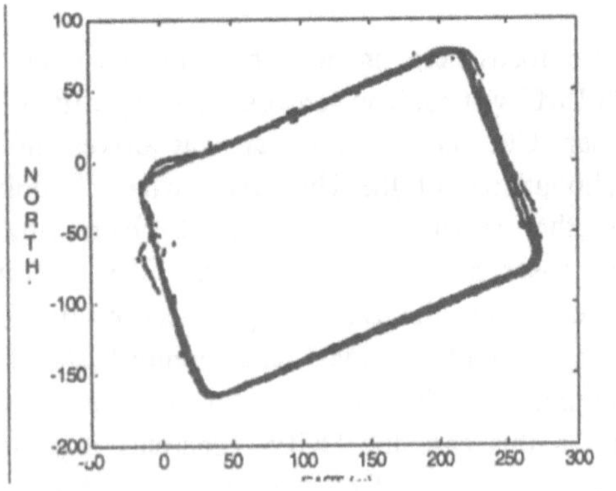

Figure 9: With GPS and Fault Detection

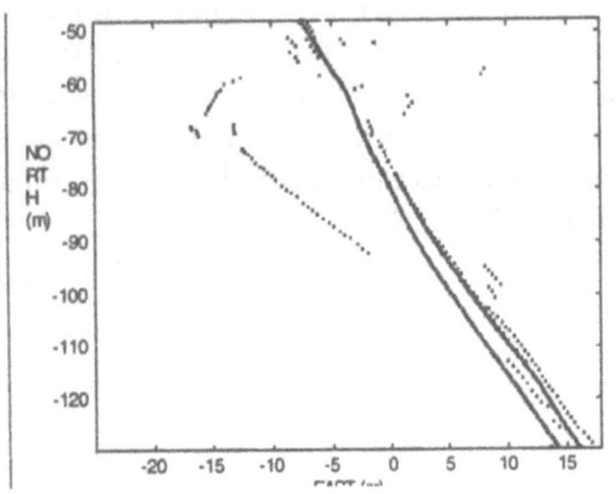

Figure 10. Enhanced GPS with Fault detection

4 Underground Vehicle Navigation

The Load Haul and Dump (LHD) truck (see Figure 11) is the workhorse of the underground mining industry. A typical use for the LHD is in moving ore from the rock face to a centralised dumping point. There is a strong case for automation of these vehicles, for reasons of both safety and productivity. The safety issue has been addressed with several working implementations of teleoperated systems. Teleoperation, however, offers little to increase productivity, as these systems tend to run at speeds slower than conventional manned systems, resulting in significantly lesser productivity levels with the additional overhead of the infrastructure required to teleoperate. The next step to increasing productivity whilst maintaining safety is therefore to fully automate the vehicle, using as little in-mine infrastructure as possible.

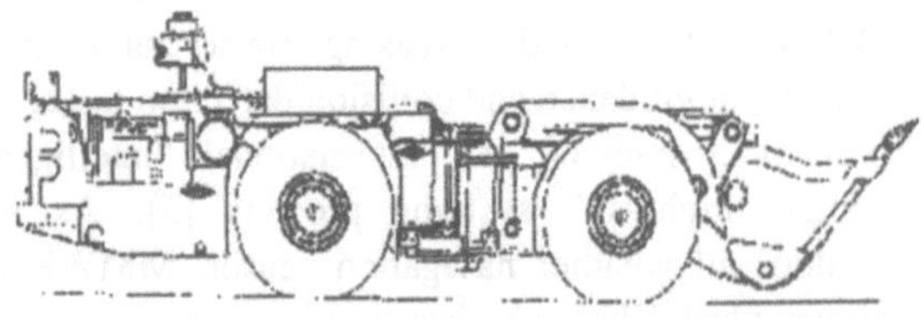

Figure 11 A Load Haul and Dump (LHD) truck.

This section presents data obtained during field trials at an operational underground mine conducted to determine which sensors or combinations of sensors offer a solution to providing high-integrity navigation systems for underground mining vehicles. Figure 12 shows part of the sensor array used in the field trials.

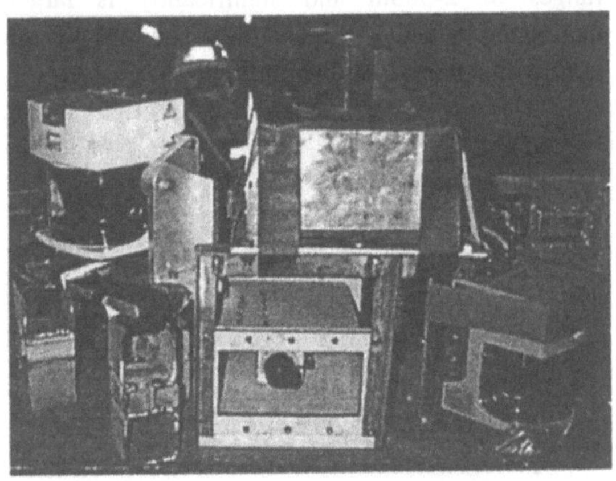

Figure 12 The Sensor Array

4.1 Localisation System A: Using Artificial Landmarks

Perhaps the simplest way to build a localisation system is to rely on dead reckoning information coupled with artificial landmark detection to determine vehicle location. Dead reckoning sensors in this case include odometer, articulation angle encoders, and a heading gyro. Dead reckoning sensors have the advantage of being simple and relatively cheap. Artificial Landmarks may include beacons, reflective stripes and radio tags.

The artificial landmark navigation system uses the sensors:

Bearing-only laser: This type of sensor is now commonly used for navigation of industrial autonomous guided vehicle (AGV) systems (see top-center of Figure 12). The basic principle of operation is very simple. A pencil laser beam is mechanically scanned in azimuth around the environment. Passive reflectors (occasionally bar-codes) are placed around the environment at known (surveyed) locations. When a scanner mounted on a vehicle moves through the environment it detects the presence of these reflectors or beacons. An encoder mounted on the scanner is used to record the azimuth angle or bearing at which a reflector is detected. Thus as the vehicle moves through the environment, a sequence of bearing measurements, to a number of fixed and known locations, are made. If the locations of these reflectors are known to the vehicle navigation system, then the absolute location of the vehicle may in principle be computed from the sequence of bearing measurements made.

Odometry: An inductive proximity sensor located above a gear wheel inside the transmission of the LHD was used to determine the angular velocity of the wheels. To know the direction the articulated vehicle was heading, the articulation angle was measured with a rotary position transducer.

4.2 Localisation System B: Using Natural Landmarks

Just as it is possible to infer position information using artificial landmarks, natural landmarks can also be used. Natural landmarks have an advantage over artificial landmarks because no added infrastructure is required (such as tags, stripes, etc). The position of natural landmarks would, however, have to be known in some form.

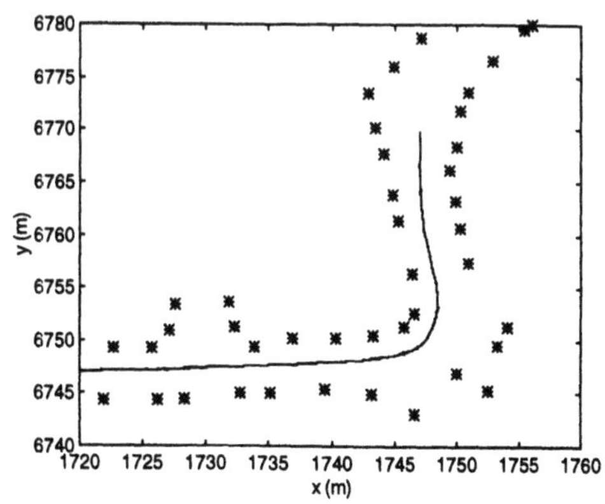

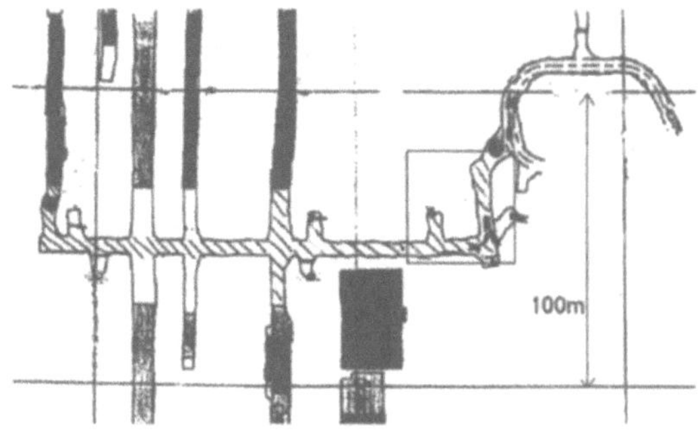

Figure 13 (a) Estimated path of the LHD (stars represent beacon positions) and (b) Trial Area – Small Square Indicates Area for which Results Presented.

The natural landmark navigation system uses the sensors:

Sick PLS Laser Ranger \ Ultrasonics: The PLS is a laser ranger manufactured by Sick optoelectronics of Germany. It does not require reflectors and works on the time-of-flight principle. The PLS produces 361 range estimates (one for each half a degree of its 180 degree field-of-view) at approximately 5Hz (communicating at 38400 baud). Side tunnels can be reliably detected using the Sick PLS laser ranger (see Figure 14(a)). Ultrasonic sensors also measure distance using the time-of-flight principle, but are prone to producing misleading information due to specular/multi-path reflections. However, the ultrasonic sensors were also found to reliably detect side tunnels (see Figure 14(b)).

Inertial Navigation System (INS): The INS system used in the trials program is a strap-down system consisting of a single triaxial accelerometer and three gyros. Theoretically, the accelerometer and gyro outputs may simply be integrated to provide 3D position. In practise however, the INS drifts with time and must be periodically reset with the landmark sensors.

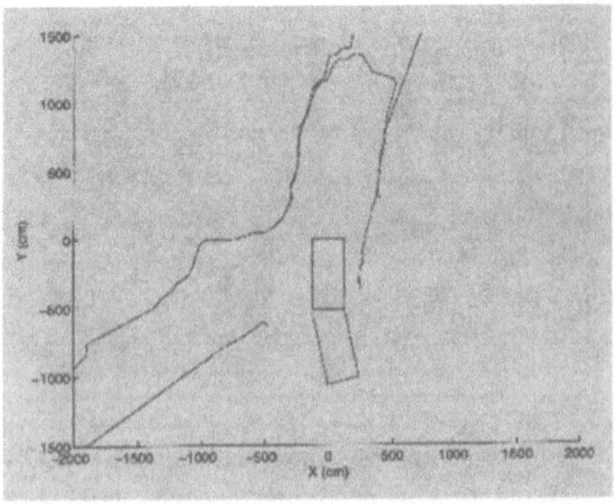

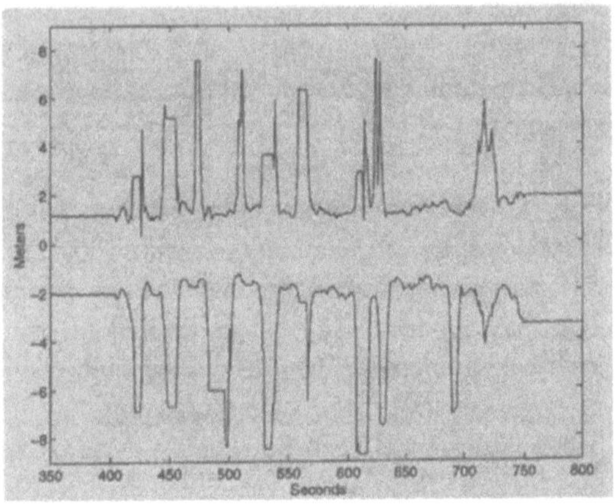

Figure14: Side Tunnel Detection using (a) SICK Laser and (b) Ultrasonics

5 Conclusions

This paper has described on-going work in the development of high-integrity navigation systems for large autonomous vehicles. Some preliminary data from two application areas has been presented. Current work is directed at providing quantitative measures of performance for these navigation systems and in evaluating these in production field applications.

References

[1] B. Barshan and H.F. Durrant-Whyte, "Inertial Sensing for Mobile Robotics", IEEE Trans. Robotics and Automation, Vol 11, No 3, 1995.

[2] N. A. Carlson, "Federated filter for computer-efficient, near-optimal GPS integration," IEEE Position Location and Navigation Symposium (PLANS), pp. 306-314, 1996.

[3] H.F. Durrant-Whyte "An Autonomous Guided Vehicle for Cargo Handling Applications", Int. J. Robotics Research, 15(5), p407-441, 1996.

[4] S. Grime and H. F. Durrant-Whyte, "Data fusion in decentralized sensor networks," Control Engineering Practice, vol. 2, pp. 849-863, 1994.

[5] M. Herbert, C. Thorpe and T. Stenz, (editors) "Autonomous Navigation Research at Carnegie Mellon", Kluwer Academic Publishers, 1997.

[6] S. Julier and Durrant-Whyte H.F. "Navigation and Parameter Estimation of High Speed Road Vehicles", Proc. IEEE Int. Conf. Robotics and Automation, 1995.

[7] P. J. Lawrence, Jr. and M. P. Berarducci, "Navigation sensors, filter, failure mode simulation results using the distributed Kalman filter simulator (DKFSIM)," Proc. IEEE Position Location and Navigation Symposium (PLANS), 1996.

[8] A. Leick, "GPS Satellite Surveying, Second Edition", Wiley Interscience, 1995.

[9] P. Maybeck, "Stochastic Models, Estimation and Control Vol I", Academic Press 1979.

[10] E.M. Nebot, H.F. Durrant-Whyte, S. Scheding, "Frequency Domain Modeling of Aided GPS with application to High Speed Vehicle Navigation Systems", Proc. IEEE Int. Conf. Robotics and Automation, Alberquerque, April 1997.

[11] E.M. Nebot, H.F. Durrant-Whyte, S. Scheding, G. Dissanayake, "Slip Modeling and Aided Inertial Navigation of an LHD", Proc. IEEE Int. Conf. Robotics and Automation, Alberquerque, April 1997.

[12] E.M. Nebot., S. Sukkarieh, H.F. Durrant-Whyte, "Inertial Navigation Aided with GPS Information", Proc. of M2VIP-97, Toowoomba, Australia, September 1997.

[10] A. Stevens, M. Stevens and H.F. Durrant-Whyte, "OxNav: Reliable Autonomous Navigation", Proc. IEEE Conf. Robotics and Automation, Japan, 1995.

PART 8

HAPTICS

SESSION SUMMARY

Paolo Dario

Scuola Superiore Sant'Anna
Pisa, Italy
dario@arts.sssup.it

The terms *haptics* and *haptic perception* were coined about 35 years ago and then used quite extensively by the psychophysics research community. According to the terminology adopted in psychophysics, *touch* is a general term, including the study of passive cutaneous sensitivities. But it may also include the *active* exploratory and manipulative use of the skin, and hence stimulation of receptor systems in the muscles, tendons, and joints – the kinesthetic system. The term *haptics* is often used to indicate exploratory and manipulative touch, in contrast to the tactile «sensations» resulting from stimulation of passive skin receptors. In general, the term *tactile* is used primarily in referring to passive touch («being touched»); the term *haptics* is used primarily in referring to active exploratory and manipulative touch.

In the robotics research community, haptics became popular in the '80 as a model to imitate for providing (autonomous) robots with perceptual capabilities.

More recently, haptics has been investigated also with a different aim: to design devices («haptic displays») which can induce artificially to a human user sensations which correspond as faithfully as possible to those he/she would experience when manipulating real objects.

The first problem to address is the following: how, and how well, do people sense and resolve intensive, spatial and temporal variations in mechanical (and thermal) inputs to the skin (the «cutaneus» system) and to muscles, tendons and joints (the «kinesthetic» system)? In this contest psychophysical research can help the design considerations and the selection of appropriate physical parameters and associated values for the interface system. Psychophysics can also provide the set of formal experimental tools that are needed for evaluating how well operators perform with the haptic interface.

The second problem to consider is the engineering design of haptic interfaces, and in particular of haptic displays. Tactile display devices stimulate the skin to generate sensations of contact. It is useful to distinguish between systems for vector force feedback and devices that convey distributed sensations, including vibrations, small-scale shape or pressure distribution and thermal display.

Current research on tactile display has much in common with previous work on sensory substitution for the disabled. On the other hand, research on vector force feedback has much in common with previous work on teleoperation. Both teleoperation and virtual environments require rich and self-consistent sensory feedback.

In teleoperation the loop between the operator's motions «inputs» and forces applied by the haptic device is closed via a mechanization and kinematics of teleoperation.

Details on the evolution of haptics research and technology are available in the recently established "Haptic Community Web Page" (http://haptic.mech.nwu.edu/).

This session includes three papers addressing important aspects of current research on haptics.

In the first paper, entitled *Tactile Displays for Increased Spatial and Temporal Bandwidth in Haptic Feedback*, R.D. Howe, D.A. Kontarinis, W.J. Peine and P.S. Wellman observe that most research on haptic interfaces has focused on force feedback, whereas conveying high frequency vibratory information and small-scale object shape is also very important and requires tactile feedback modalities. Cutaneous feedback is produced by distributed sensations across the skin including both high temporal frequencies (up to at least 1 kHz) and high spatial frequencies (up to at least few cycles/mm). The authors have used for their experiments a telemanipulation system equipped with a vibration sensor on the remote robot's hand and a vibrotactile display on the user's finger tip. Task performance is compared with and without vibrotactile feedback. Experimental work demonstrates that tactile feedback enables telerobotic execution of a number of new tasks, and can increase productivity and decrease error rates in many others. Small-scale tactile shape feedback is still in an early stage of development. Development of this type of tactile feedback, however, promises significant benefits in a number of application areas, particularly in medicine.

The second paper, entitled *Design of an Anthropomorphic Haptic Interface for the Human Arm*, by M. Bergamasco and G.M. Prisco, discusses the development of an exoskeleton system with seven rotational joints whose axes are aligned with those of the user arm in order to preserve freedom of movement. The objective pursued is to reproduce in a realistic way the pattern of forces on the human body which arise during activities such as manipulation and haptic exploration. The interface can apply forces and torques at the level of the arm, forearm and palm; hand force feedback systems can be easily integrated with the arm interface. The design of the interface was particularly challenging. The authors discuss how they were able to meet the requirements of workspace, wearability, adjustability and dynamics posed by the anthropomorphic approach.

The objective of the third and final paper of the session, entitled *Testing a Visual Phase Advance Hypothesis for Telerobots* by R.W. Daniel and P.R. McAree is to investigate the extent to which visually derived information improves the performance of a high fidelity force reflecting teleoperation system. Experimental evidence has been reported in the literature that a human operator is able to generate phase advance when performing pursuit tasks, and sophisticated models of the human operator when carrying out tracking tasks have been developed. However, no mechanisms have been identified in the physiology literature for the generation of phase advance. In their paper Daniel and McAree explore the hypothesis that with vision an operator is able to generate phase advance and inject this into an outer feedback loop and so stabilize a marginally stable system. Based on experiments, the authors claim that vision has the capacity to generate anticipative information that can be used directly by the neuromuscular control system, and argue that quantification of this effect is possible.

At the end of each paper, many questions were posed to the presenting authors on technical and fundamental aspects of haptics.

The paper by Howe *et al.* raised questions on the importance of vibrations for haptic perception and on the role of stimulation frequency on sensitivity. The answer was that human response to vibration is frequency dependent. Questions were also asked on the possibility of using different stimulating devices for eliciting haptic sensations, including local electrical stimulation. A comment was that, though possible and desirable, electrical stimulation may not be reliable for practical and long term use, since fatigue and pain are observed after prolonged stimulation.

The paper by Bergamasco and Prisco presented interesting design considerations and quite sophisticated mechanical solutions for haptic interfaces, which attracted specific questions from the audience on such important aspects like the stiffness of cable transmission, the arrangement of some d.o.f. at the wrist, and the solution adopted for weight compensating during operation.

The paper by Daniel and McAree raised questions on absolute performance and emphasis on bandwidth: the solution adopted by the authors for their experiments was to use limited bandwidth for the master-to-slave control loop, and much larger bandwidth for the slave-to-master reflection loop. Questions were also asked on the importance of vision cues versus "pure" force feedback in some practical applications of teleoperation, for example in minimally invasive surgery. The answer was that the actual importance of vision is higher than the non professional teleoperator could say. In fact, the authors believe that quantifying the extent to which vision cues can help an operator is possible, but so far they have only demonstrated plausibility.

In conclusion, the papers included in this session, though limited in number, cover some among the most critical and important aspects of current research on haptics, and provide many useful hints for further investigation in this exciting and promising area of robotics research.

frequency dependent. Questions were also raised on the robustness of some different stimulating devices for eliciting rapid sensations, including local electrical stimulation. A comment was that though possible and desirable, electrical stimulation may not be reliable for practical [illegible]

[illegible]

The paper by Hammond and Pryor presented interesting [illegible]

Tactile Displays for Increased Spatial and Temporal Bandwidth in Haptic Feedback

Robert D. Howe Dimitris A. Kontarinis William J. Peine Parris S. Wellman
Division of Engineering and Applied Sciences, Harvard University
Pierce Hall, Cambridge, MA 02138 USA
email: howe@deas.harvard.edu

Abstract

We are investigating tactile feedback modalities designed to convey high frequency vibratory information and small-scale object shape. Vibrotactile feedback systems transmit information about textures and events like contact and slip that reveal the mechanical state of the remote environment. Shape display devices consist of regular arrays of pin elements that rest against the user's finger tip. Each pin is raised and lowered to approximate the desired surface shape on the skin. Experiments with prototype tactile feedback systems help improve our understanding of human tactile information requirements and to determine the relationship between task characteristics and the benefits of tactile feedback.

1 Introduction

Most research on haptic interfaces has focused on force feedback. Spatial information is limited to a single force vector applied to the end of the finger or arm, and time variation is usually limited to a few cycles per second (Sheridan 1992). Human manipulation, however, also relies on cutaneous feedback, produced by distributed sensations across the skin (Johansson and Westling 1987). This includes both high temporal frequencies (up to at least 1 kHz) and high spatial frequencies (up to at least a few cycles/mm) (Boff and Lincoln 1988, Johansson and Vallbo 1983).

These high frequencies are the basis for a wide range of our perceptual and manipulative abilities. Fast variations in contact force are perceived as vibrations, which are produced by transient events such as contact and slip that indicate the progress of manipulation tasks. Vibrations are also used to distinguish textures by stroking a finger over a surface. High frequency spatial information is useful for both object recognition and for manipulation control. Local object curvature determines the kinematics of the finger-object contact; robotic analysis has shown that this geometric information is essential for effective grasping and manipulation. Small-scale spatial information is also used in perceptual tasks like medical palpation to distinguish hidden lumps and other anatomical landmarks.

We are developing new methods for providing both types of tactile feedback in teleoperated robotics and virtual environments (Figure 1). For teleoperation, sensors built into the remote robot's finger tips measure transient contact forces and spatial distribution of pressure. To relay these sensations to the human operator, tactile display devices are integrated with conventional force reflection interfaces to increase the effective bandwidth of information transfer. In the following sections, we describe our work in each of these areas, note important application areas, and outline some of the notable challenges.

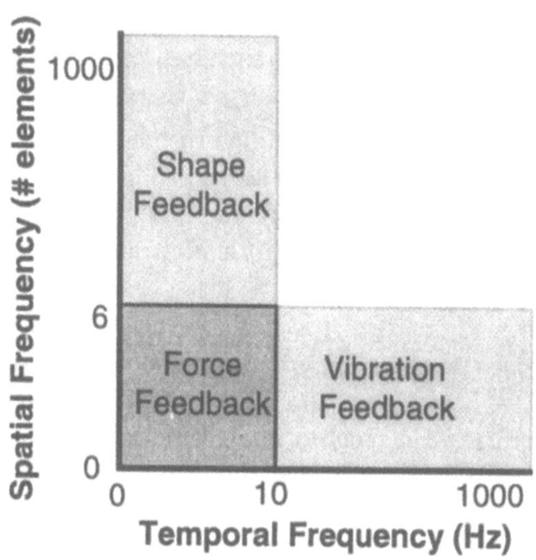

Figure 1. Frequency ranges of force, vibrotactile, and shape feedback.

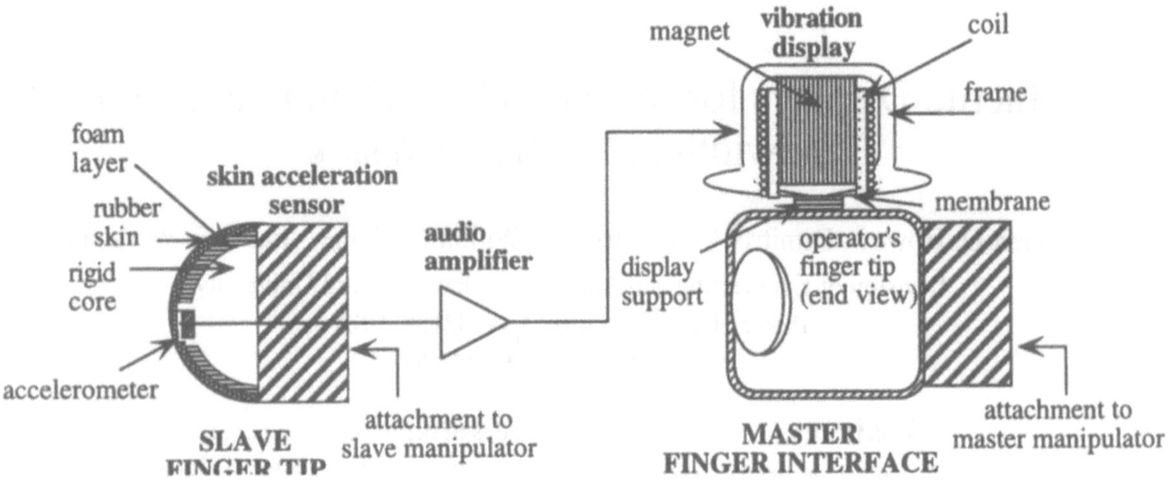

Figure 2. Vibrotactile sensor and display.

2 Vibrotactile Feedback

People make use of high frequency vibrations in many common manipulation tasks. We feel with our fingers for a rattle indicating a loose screw, or for a gritty sensation indicating a damaged ball bearing. Vibrations are often designed into the function of manufactured items, such as snap closures on clothing or calculator keys with detents. In many tasks, contact between hard surfaces is accompanied by copious vibrations; a typical example is the clatter of aligning a steel wrench with a bolt. Texture perception also relies on high frequency vibrations, and professional machinists determine the roughness of machined surfaces by comparing the "feel" of a newly-cut surface to calibration surfaces of known roughness.

Despite their importance in manipulation, vibrations have received little attention in haptic interface research. Some work has appeared on sensory substitution aids for the blind such as the Optacon (Bliss et al. 1970), which use vibrating pin elements to stimulate the skin. Here the vibratory stimulus is designed to provide information about another physical parameter such as shape or optical intensity, and no attempt is made to relay information about vibrations that occur as part of a task. Similarly, Massimino and Sheridan (1993) used a vibrotactile display to relay force information in teleoperation, rather than to portray the vibrations at the remote manipulator.

In this research, we have examined the use of tactile displays for conveying task-related vibratory information in teleoperation and virtual environments. This is a low-cost, open loop display modality that can be easily added to many existing systems to improve performance. Our initial goal was to demonstrate the utility of this type of feedback, and to delineate the kinds of tasks where high frequency vibratory feedback is important (Kontarinis and Howe 1995). In more recent work, we have created algorithms for generating realistic vibrations in virtual environments (Wellman and Howe 1995) and developed systems for practical teleoperation applications (Dennerlein, Millman and Howe 1997).

2.1 Vibrotactile feedback hardware

From human factors considerations (Kontarinis and Howe 1995), a high frequency vibration tactile display device should produce mechanical vibrations in the range of 40-1000 Hz with variable amplitude and frequency. Since this is in the audio frequency range, miniature loudspeakers may be readily modified for this application. For example, we have used 0.2 watt loudspeakers mounted "upside down" on the force reflection system, with the outer cones and metal frames removed (Figure 2). The remaining structure containing the magnet, coil, and central diaphragm is then attached to the master manipulator near the operator's fingers. The base containing the permanent magnets is thus free to move in space. Passing current through the coil generates a force against the magnet, which accelerates the 35 gram base and produces an inertial reaction force against the manipulator.

This inverted mounting results in a higher moving mass compared to the usual audio configuration, providing larger inertial forces. The range of motion is 3 mm, and the displays can, for example, produce up to 0.25 N peak force at 250 Hz. Vibrations

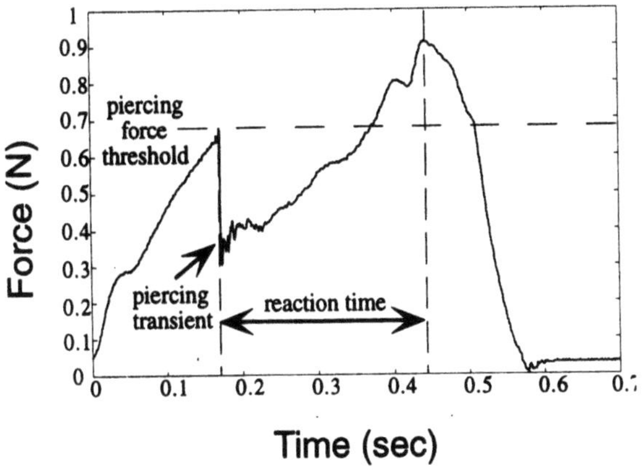

Figure 3. Slave robot fingertip force sensor waveform generated as grasped needle punctures membrane.

are transmitted to the fingers of the human operator through aluminum bracket "handles" mounted at the ends of the master manipulator finger links, as shown in Figure 2.

A number of different tactile sensors can detect task-related vibrations at the remote robot finger tips (Howe 1994). We use skin acceleration sensors, which consist of miniature instrumentation accelerometers mounted on the inner surface of the rubber finger tip skin of the slave fingers, shown in Figure 2 (Howe & Cutkosky 1989). A layer of foam rubber beneath the skin provides passive compliance to improve grasp stability and isolate the sensor from vibrations in the robot mechanism. In the mounting configuration described here, the skin acceleration sensors have their greatest sensitivity to vibrations in the vertical direction; however, the compliant skin and foam readily couple vibrations in other directions to create an omnidirectional sensor.

One advantage of the skin acceleration sensor is its excellent sensitivity to vibrations at the frequencies we are concerned with here. Acceleration sensing is also appropriate because certain human mechanoreceptors that are particularly important in this context (Pacinian corpuscles, the FAII receptors) appear to respond to acceleration of the finger tip skin. Furthermore, the vibration display produces inertial forces which will in turn accelerate the operator's finger tip skin, reproducing the same physical quantity that the sensor is measuring at the remote manipulator. In our experimental system, the measured response amplitude from sensor mechanical input to display mechanical output is flat to within 7 dB across the frequency range of interest.

2.2 Tasks and vibrations

To investigate the role of vibrotactile feedback, we have used a telemanipulation system equipped with a vibration sensor on the remote robot's hand and a vibrotactile display on the user's finger tip (Kontarinis and Howe 1995). Task performance is compared with and without vibrotactile feedback, and correlated with task properties. Experimental results suggest that tasks may be broadly divided into three categories: perceptual tasks where detection of vibrations is the fundamental goal of the task; manipulation tasks where vibrations indicate the state of the task; and tasks where vibrations are not directly important. In the first case, many inspection tasks require the detection of vibrations for their successful completion. Touch inspection differs from visual inspection in that it requires mechanical interaction with the object under inspection. Examples here include tasks where vibrations are produced as parts are manipulated, such as detection of looseness in an assembly of parts or damage to a ball bearing set. Also in this category is texture perception, where the objective is to determine the roughness of a surface, or detecting the presence of surface contaminants such as dirt – the sorts of tasks where we stroke a finger over a surface to generate vibrations. In all of these tasks, vibrations are by definition essential.

In the second category, vibrations can enhance performance by indicating the mechanical state of the hand-object system. This can reduce reaction times or permit minimization of forces. Vibrations can indicate that contact has occurred between a grasped tool and a surface in the environment; examples include bringing a wrench into contact with a bolt, or lowering a grasped object into contact with a table top. Rapid detection of contact permits the human operator to stop the motion, which reduces contact forces and prevents dislodging the object. This can be particularly important in delicate probing tasks, where detection of the vibrations that indicate the earliest instant of contact can be essential. One example from medical applications is needle biopsy, where vibrations signal the puncture event (Figure 3), and vibration feedback can reduce reaction times and permit minimization of forces.

Finally, in some tasks vibration information may be unimportant. In simple positioning tasks, or tasks that are limited by precise control of forces, vibration feedback may not improve performance, even though vibrations are generated by the task. However, it may still be useful to provide vibratory information, as it adds to the subjective "feel" of the system for the user. This is particularly true in virtual environments, where vibrations may significantly contribute to the sense of remote presence, even though they may not improve a particular performance measure for a given task.

2.3 Vibrotactile feedback for industrial telemanipulators

A key advantage of vibrotactile feedback is its simplicity. Since human vibrotactile sensation has poor spatial resolution, a single vibration display is often adequate for an entire finger or even an entire hand. This decreases system complexity and increases reliability. We are now working with Schilling Robotics, a leading manufacturer of teleoperated robot arms for the offshore oil industry, to develop a practical vibrotactile feedback system for their subsea manipulators.

In this demanding application, force feedback has not proved practical, due to increased cost and decreased reliability. This means operators must rely exclusively on visual information to determine the contact state of the remote robot. With vibration feedback from the robot's gripper, the operator can detect contact, slipping, and other mechanical events that signal changes in the remote environment. Initial tests show that vibrotactile sensors can be robust to the extremely high loads of this application, and can improve operator performance in various contact tasks (Dennerlein, Millman and Howe 1997).

2.4 Virtual vibrations

Vibrotactile sensations can also enhance the perceived reality of virtual environments. A pioneering example is Minsky's (1995) system for generating vibrations appropriate to stroking fine surface textures in virtual environments. In our work, we examined the problem of surface stiffness perception, where people distinguish the hardness of an object surface by tapping it with a hand-held probe. To determine the appropriate waveforms to use in virtual environments, we first measured the waveforms that are produced when subjects tap on real surfaces of varying stiffness with an instrumented stylus. Generalizing from these results, a simple exponentially damped sinusoidal model captures the essential characteristics of these vibrations. This model is then used to produce waveforms for an experiment that assessed the effectiveness of vibrotactile stimulation in conveying stiffness information in a virtual environment. The results are compared with a similar experiment in which subjects tapped directly on different real surfaces using a stylus. Using these simple waveforms for vibrotactile display in our test system gives a relatively accurate description of the phenomenon and allows subjects to accurately distinguish surface stiffness (Wellman and Howe 1995).

3 Small-scale shape feedback

Our everyday experience shows that distributed sensations across the skin of our fingers is essential for dexterous manipulation (Johansson and Westling 1987). This type of information can be characterized by its high spatial frequency content compared to force feedback, where only a single force vector is presented. To convey this type of information in teleoperation and virtual environments requires a set of sensors and actuators distributed across the contact surface. The usual approach to sensing this type of information is a tactile array sensor, a regular grid of sensors that measure local pressure or displacement. For tactile feedback, it is straightforward to use arrays of pins that rest against the finger tip, where each pin is individually raised and lowered to approximate the measured shape.

The development of a tactile relay system draws on a number of disparate areas of research, including robotic tactile sensing, human tactile display interfaces, and dexterous telemanipulation. In robotic tactile sensing research the major focus has been the development of new array sensor devices, which can measure pressure or displacement at a large number of locations across the contact area; see Howe (1994) for a review. These devices are now a relatively mature technology, and a number of designs have demonstrated good performance.

Tactile shape display, however, presents far more difficult challenges. The few reported studies of tactile array sensing in teleoperation often use visual feedback to convey remote touch information to the human operator. To convey this information in tactile form requires small and light tactile display devices that can be easily mounted on force feedback interfaces, with sufficient force output and temporal bandwidth to match human voluntary motion. We have developed a series of systems for tactile shape feedback. In the following sections, we will describe the hardware developed for these systems, as well as some of human perceptual and motor behaviors we have measured.

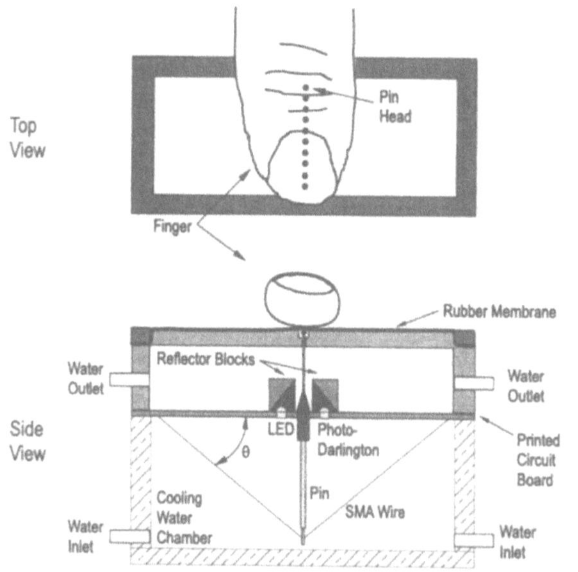

Figure 4. Tactile shape display with 10x1 pin configuration and water cooling.

3.1 Tactile shape displays

Tactile shape displays consist of an array of pins that are raised and lowered against the finger tip. Our goal is to recreate skin deformations that might be produced by object features such as corners, raised edges, and surface textures. One important application area for these displays is minimally invasive surgery, where the display can reproduce tactile information sensed by instruments within the patient's body. A key example is the localization of tumors, which often appear as hard lumps embedded in soft tissues, such as the lung and the liver.

Our recent experiments (Peine et al. 1997) show that surprisingly high bandwidth is required for effective tactile display. These experiments used an optical tracking system to measure finger speeds while experienced surgeons located simulated tumors in soft rubber models. They found that maximum finger speeds for 90 percent of the population tested were approximately 120 mm/second.

This measurement allows us to determine the temporal bandwidth required for effective shape display. Given a shape display that has pins spaced 2.0 mm apart, the maximum spatial frequency that can be created (in the Nyquist limit, with pins alternating up and down) is 0.25 cycles/mm. If the maximum spatial frequency is scanned across the display at 120 mm/sec, each pin must travel up and down in 33.3 milliseconds. This means that each pin must achieve a temporal bandwidth of at least 30 Hz.

If the tactile display device cannot raise and lower the pins in this time interval, then the tactile information will not be correlated with the gross motion of the finger. The lack of correlation between finger motion and tactile information can make it difficult to quickly and accurately locate tactile features. Very low bandwidths can lead to a "move and wait" strategy, where the user conducts a search using discrete moves followed by pauses to allow the display to catch up.

In addition to high bandwidth, high spatial resolution is required. For example, Johnson and Phillips (1981) have shown that humans can reliably distinguish between two points that are separated by as little as 0.9 mm. In addition to the small spacing requirement, we have found that human finger pads have an indentation stiffness as high as 3.5 N/mm at 1.2N indentation force (Pawluk and Howe 1997). Since we are interested in displaying shape, rather than pressure, we require pins with high stiffness and high force capability. This is especially true when designing displays to be used on force reflecting devices, because the pins will need to support all of the force produced by the user's fingers during manipulation.

A number of researchers have constructed tactile shape displays (e.g. Cohn, Lam, and Fearing 1993, Hasser and Weisenberger 1993, Kontarinis and Howe 1993). These displays have been constructed in many configurations including single pin and multi-pin matrix displays. Unfortunately, these previous displays have all been limited by low bandwidth, low stiffness or lack of static response. Thus, in order to effectively portray small scale tactile shapes a display must have small pin-to-pin spacing, produce large forces, have high pin stiffness and relatively high bandwidth.

3.1.1 Mechanical design

Our previous prototype shape displays have included arrays with 3x3 and 6x4 pin resolutions. None of these devices attained the required 30 Hz bandwidth. In our most recent prototype design, we have concentrated on increasing bandwidth at the expense of spatial resolution, with a resulting 10x1 pin configuration. Using a single row of pins simplifies packaging and mechanical design constraints, yet retains much of the functionality of a matrix of pins because textures can be easily perceived along the line of the pins. The user can also sweep the display to perceive curvature perpendicular to the line of pins.

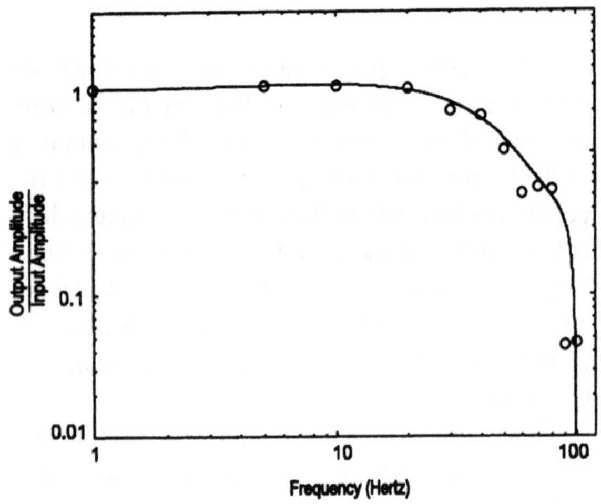

Figure 5. Frequency response of the prototype shape display.

Figure 4 shows that the line of pins runs along the finger pad rather than across it. The size of various components limits the pin spacing to 2.0 mm on center. The small pin-to-pin spacing, compact size and the high force and stiffness requirements, led us to choose shape memory alloy (SMA) wires to actuate each of the pins. This material is more suited to this application than electromagnetic actuators because of its high force-to-weight ratio and high intrinsic stiffness. It also provides much larger displacements than piezoelectric or magnetostrictive materials. There is a price to be paid for these benefits and SMA has its own set of design challenges which must be addressed.

SMA typically undergoes less than 5% strain during the phase transition that actuates the pin motion. In order to compactly package the display we have configured the wires in a V shape, as shown in Figure 4. In order to facilitate the construction of the display, it uses a layer-based modular design that includes all electrical connections on a single printed circuit board. This also minimizes the number of parts that are required and makes for an overall size of 78 mm by 35 mm by 57 mm. The position of each of the pins is measured by an infrared light-emitting diode (LED) and photoDarlington transistor pair. As seen in Figure 4, the pair is mounted in the printed circuit board, and the light from the LED is reflected past the pin to the photoDarlington by a pair of reflectors. Dividers separate each pin to prevent crosstalk. Each pin is shaped so that the amount of light it blocks varies with its position, which causes the collector current in the photoDarlington to vary as a monotonic function of pin position.

The circuit board also serves as a bearing surface for one end of each pin, while the top block on the display provides the second bearing surface. Together these constrain the pins to move linearly with minimal side loading. The top of each pin is recessed into the surface of the top block to provide a nearly flush surface for the user's finger tip to press against. Finally, the necessary spring return force for each pin is provided by a latex rubber membrane which also serves as a seal.

3.1.2 Thermal Design

In this application, the phase transition of SMA is thermally driven, and because it is always possible to apply more electrical current to cause the wire to contract faster through resistive heating, the fundamental limitation is cooling. Because the rate of heat transfer depends on the ratio of surface area to volume, we use two small (75 micron) wires rather than one larger one to actuate each pin. We also use SMA wire with a transition temperature of 90C because the rate of cooling is linearly related to the temperature difference between the wires and the cooling medium.

In previous work, we found that ambient air cooling resulted in a bandwidth of perhaps 1 Hz, while forced air cooling increased this to 5-6 Hz. To obtain the required 30 Hz bandwidth, the present design uses a slowly recirculating bath of water. Water has been chosen as a cooling fluid because of its high thermal conductivity and tremendous heat capacity, which allows for a low recirculation rate within the display. This ensures more uniform cooling of all of the wires in the display, and thus more uniform pin-to-pin performance.

Because of the success of the mechanical and thermal design of the display, a simple linear controller has proved adequate to drive each pin. To obtain maximum performance, it is desirable to always keep the wires operating between the minimum transition temperature and the maximum transition temperature. We use a proportional controller, with a constant offset current. The current offset is used to heat the wire to near the transition temperature and its magnitude and the gains are determined experimentally; for this design, they were 0.5 amps and 4 amps/mm, respectively.

3.2 Performance Characterization

Characterizing the bandwidth of this nonlinear system is problematic because the fall time is ef-

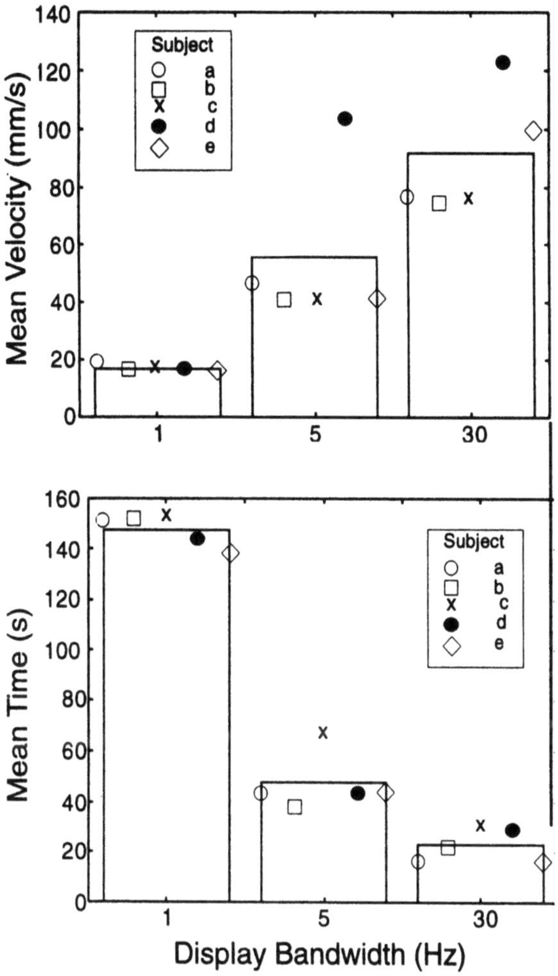

Figure 6. Mean finger velocity and search time for all subjects (bars) and for each individual subject (symbols) at each test frequency.

fectively slew rate limited by the wire cooling rate. One useful performance measure is the output amplitude at each frequency, in response to a triangle wave command at the maximum static displacement. Figure 5 shows the frequency response of the display when driven with a 3 mm triangle wave position command for one second with a finger pressed firmly (total force approximately 5 N) against the display. The frequency at which the output amplitude is 3 dB below the commanded input amplitude is approximately 40 Hz. Although the maximum frequency shown in the figure is 100 Hz, detectable output can still be felt at frequencies approaching 150 Hz, suggesting that the display can be used for vibrotactile feedback as well as shape feedback.

3.3 Bandwidth vs. Performance

To quantify the utility of high bandwidth shape display on task performance, we asked subjects to perform a prototypical search task while a digital filter with a cutoff frequency of 1, 5 and 30 Hz was applied to the commanded positions for the display. These frequencies were chosen because they correspond to the frequency response for SMA shape displays cooled with still air, forced air, and water respectively.

The display was mounted to the mouse pointer of a digitizing tablet. Subjects grasped the display with the dominant hand and rested the index finger tip on the pins. The tablet measured finger position as subjects moved within a test area. The computer sampled the display location and raised and lowered the display pins to represent virtual small-scale shapes at fixed locations. Subjects were asked to search a 150 mm diameter circle as quickly as possible to find two small dots (3 mm diameter by 2 mm height cylinders). They were asked to locate each dot in the search space and center the display on the dot and press a button. Once they found both dots, they were asked to press a second button to end the experiment. The time subjects took to search the space and their average velocity while searching were recorded.

Five subjects (three male, two female, 24 to 29 years of age) were asked to search each of eight virtual environments with two randomly placed dots as described above for the two highest bandwidths, and four dots for the slowest bandwidth. Each subject was presented with the same eight (or four) virtual textures as the other subjects, and these were presented in random order.

Figure 6 shows the results of the experiment. The average search velocity more than doubled, going from 18 mm/sec to 55 mm/sec when the filter cutoff frequency was increased from 1 Hz to 5 Hz. The effect of going from 5 Hz to 30 Hz was somewhat less pronounced, as the velocity only increased to about 90 mm/second. The search times showed a steady decrease from approximately 150 seconds, to 50 seconds, to 25 seconds when the filter cutoff frequency was changed from 1 to 5 to 30 Hz respectively.

3.4 Discussion

This prototypical experiment shows, that at least for some tasks, high bandwidth is essential for good task performance. Subjects described "extreme frustration" when using the display to search the space with the filter set at the lowest cutoff frequency. It is interesting to note that although subject d moved at a much higher average velocity than any of the other 4 subjects during the 5 Hz frequency trials she took nearly the same amount of time to complete the task as the other subjects. This suggests that she wasted a large amount of time searching at high speeds when the display could not move the pins up and down quickly enough. During experimentation she was observed to pass directly over the top of the dots apparently without noticing on more than one occasion. This is a real concern in tasks like tumor localization where finding tumors accurately and quickly is a priority, because it will be very difficult to maintain a regular search pattern and ensure that no tumors will be missed. Thus, it is important that any tumors that may be encountered are felt the first time.

4 Conclusions and future work

Experimental work has demonstrated that tactile feedback enables telerobotic execution of a number of new tasks, and can increase productivity and decrease error rates in many others. As with force feedback, suitability of tactile feedback for particular applications depends on a complex combination of task characteristics, manipulator capabilities, and cost considerations. Further research can help to elucidate the relationship between task characteristics and feedback requirements, which will facilitate design of effective tactile sensing and display systems.

Vibrotactile display has proved to be an easy and inexpensive means of providing contact information in teleoperation, either with or without force feedback. Vibrations are particularly effective in transmitting texture information, and in conveying events like contact and slip that reveal the mechanical state of the gripper-object system. Both sensor and display design is straightforward, and the feedback system functions independently of the manipulator controller, minimizing design complexity and impact on reliability. Vibrotactile feedback for virtual environments is also practical, and more sophisticated algorithms should be developed for tactile rendering and partitioning feedback between force and vibration channels.

Small-scale tactile shape feedback, in contrast, is still in an early stage of development. While tactile sensors are reasonably effective, display devices are relatively crude. Progress in shape feedback is still largely limited by device design, and none of the devices that have appeared in the literature are entirely satisfactory. Much also remains to be learned about the pertinent aspects of the human haptic sensing system. Development of this type of tactile feedback, however, promises significant benefits in a number of application areas, particularly in medicine.

Acknowledgments

This work is supported by National Science Foundation Young Investigator Award No. IRI-9357768, Office of Naval Research grants No. N00014-92-J-1887 and N00014-96-C-0325, and by a Whitaker Foundation Biomedical Engineering Research Grant.

References

Bliss, J. C., et al. (1970). "Optical-to-Tactile Image Conversion for the blind," *IEEE Trans. Man-Machine Systems*,. **MMS-11**(1):58-65.

Boff, K.R., and Lincoln, J.E., eds. (1988). *Engineering Data Compendium: Human Perception and Performance*. Wright-Patterson AFB, Ohio: Armstrong Aerospace Medical Research Laboratory.

Cohn, M. B. , M. Lam, and R. S. Fearing (1992). "Tactile feedback for teleoperation," in Telemanipulator Technology, SPIE.

Dennerlein, J.T., P. Millman, and R.D. Howe (1997). "Vibrotactile Feedback for Industrial Telemanipulators," Symp. Haptic Interfaces for Virtual Env. and Teleop. Sys., ASME Intl. Mech. Eng. Congress, Dallas, Nov. 15-21,.

Hasser, C. and Weisenberger J.M. (1993). Preliminary evaluation of a shape memory alloy tactile feedback display. *Symp. Haptic Interfaces Virtual Env. Teleop. Sys., ASME Winter Annual Meeting*, New Orleans.

Howe, R. D. (1994). "Tactile sensing and control of robotic manipulation," *Journal of Advanced Robotics*, **8**(3):245-261.

Howe, R.D., Peine, W.J., Kontarinis, D.A., and Son, J.S. (1995). Remote Palpation Technology, *IEEE Engineering in Medicine and Biology*. **14**(3):318-323, May/June.

Johansson, R. S., & Vallbo, Å. B. (1983). Tactile sensory coding in the glabrous skin of the human hand. *Trends in Neuroscience* **6**(1), 27-32.

Johansson, R. S., & Westling, G. (1987). "Signals

in tactile afferents from the fingers eliciting adaptive motor responses during precision grip." *Experimental Brain Research*, **66**:141-154.

Johnson, K.O. and Phillips, J.R. (1981). Tactile Spatial Resolution. I. Two-Point discrimination, Gap Detection, Grating Resolution, and Letter Recognition. *J. Neurophys.* **46**(6):1177-1191.

Kontarinis, D.A. and R.D. Howe (1995). "Tactile display of vibratory information in teleoperation and virtual environments," *Presence* **4**(4):387-402.

Minsky, M.D.R. (1995). Computational haptics : The Sandpaper system for synthesizing texture for a force-feedback display. Ph.D. Thesis, MIT, Program in Media Arts & Sciences.

Pawluk, D. T.V. and R. D. Howe (1997). "Contact pressure distribution on the human finger pad," presented at the 26th Congress of the International Society of Biomechanics, Tokyo, August 25-29.

Peine, W. J., K. C. Foucher, and R. D. Howe (1997). "Finger speed during single digit palpation," in press, *Human Factors.*

Sheridan, T. B. (1992). *Telerobotics, Automation, and Human Supervisory Control.* Cambridge, MA: MIT Press.

Wellman, P. and R.D. Howe (1995)."Towards realistic vibrotactile display in virtual environments," Symp. Haptic Interfaces for Virtual Env. and Teleop. Sys., Proc. ASME Intl. Mech. Eng. Congress, San Francisco, Nov. 12-17, T.E. Alberts, ed., DSC-Vol. 57-2, p. 713-718.

Design of an Anthropomorphic Haptic Interface for the Human Arm

Massimo Bergamasco Giuseppe Maria Prisco
PERCRO,
Simultaneous Presence, Telepresence and Virtual Presence
Scuola Superiore S.Anna,
Via Carducci 40, I-56127 Pisa, Italy
email: massimo@percro.sssup.it

Abstract

In this paper, methods and technical issues dealing with the design of an anthropomorphic haptic interface for the upper limb are presented. The device features seven rotational joints whose axes are aligned with the user arm degrees of freedom (d.o.f.) in order to fully preserve the freedom of movement of the human arm. Its links adhere just to the external side of the user arm, without fully wrapping it. The interface can apply forces and torques at the level of the medium part of the arm, forearm and palm of the user. Highly intuitive operation, universal applicability, large workspace and the possibility of being used together with an hand force feedback system, are the most remarkable consequences of its coherent anthropomorphic design. The system requirements and the design choices are discussed; the innovative mechanical solutions are described.

1 Introduction

This paper develops the concept of anthropomorphic haptic interface for the human arm, which has evolved from previous studies on arm exoskeleton devices for force feedback on the human operator hand.

The key feature of an anthropomorphic device is that it has seven rotational joints whose axes are aligned with the user arm d.o.f., in order to fully preserve the freedom of movement of the human arm. The links of such device adhere just to the external side of the user arm, without fully wrapping it as in a typical exoskeleton-like design.

According to this concept, we have designed a 7 DOF anthropomorphic haptic interface for the human arm. The design of an haptic interface with such a high number of degrees of freedom and with the further constraints given by the choice of an anthropomorphic kinematics, has required a considerable effort in research and technical development because such a design combines the complexity of "classical" robot design with the challenge to meet the specific performance required to achieve effective haptic feedback.

In our view, the goal of having an anthropomorphic haptic device is very attractive because such device has the potentiality to reproduce in a realistic way the patterns of forces on the human body which arise during activities such as manipulation and haptic exploration. Our device can apply forces and torques at the level of the medium part of the arm, forearm and palm of the user. Its distinctive qualities are highly intuitive operation, universal applicability, large workspace and the possibility of being used together with an hand force feedback system.

The present paper analyses the design methods and choices which have been required to implement the concept of anthropomorphic haptic interface for the human arm.

Several studies in the field of haptics focus on determining the requirements for master arms and force feedback haptic interfaces (Fischer and Daniel 1990); others focus on quantitative measurements of their performance (Hayward and Astely 1995).

On the contrary, when coming to the problem of designing haptic interfaces, most of the analyses available at the state of the art are, unfortunately, of little use. In fact, even if the optimal device performance is known in advance (and we believe the state of the art has not reached this point yet), an aware design process requires to put in quantitative relation the design alternatives with the final performance. As a consequence, models representing a suitable abstraction of the system, models specifically intended to evaluate each design alternative, must be developed and implemented. Often their predictions need to be validated experimentally on simplified hardware setups (Hayward and Cruz-Hernandez 1995).

The paper is organized as follows. In Section 2, some existing designs of master arms and exoskeletons with 5 to 7 d.o.f. are briefly reviewed; then in Subsection 2.1, the concept of anthropomorphic design is analysed; finally in Subsection 2.2, the precedent work which we have done on the topic is critically reviewed to find guidelines for a better implementation of the concept of anthropomorphic haptic interface. In Section 3, the principal strategic objectives of the design are summarized. In Section 4, the key innovative mechanical solutions which are at the base of the whole design are discussed. In Section 5, the design tools which have made possible the development are presented. The final design of the haptic interface is presented in Section 6; finally, in Section 7, the proposed design is quantitatively analysed.

2 State of the Art

There are several arm masters designed to provide force feedback from a virtual environment (an up-to-date state of the art can be found in (Burdea 1996)). A small set of very different devices can be identified by the common idea of resembling in their design the kinematics of the human arm. It is interesting to review the approaches underlying the design of the SARCOS Dextrous Arm Master and the FREFLEX Master, which can be both regarded as attempts towards a no compromise solution to feedback forces on the human arm.

The SARCOS Dextrous Arm Master, by SARCOS Co. (Jacobsen, Smith, and Backman 1991), is composed of a 7 d.o.f. arm and of a 3 d.o.f. hand. The kinematic chain of the arm has three perpendicular, intersecting d.o.f. forming a shoulder, then an elbow joint and, at the end, a spherical wrist. The joints ranges of movement are 180° at the shoulder, 105° at the elbow, 105° × 180° × 100° at the wrist. The SARCOS Master shoulder is positioned close to the operator shoulder; during operation, the elbow and the wrist of the device remain only approximatively next to the human arm. Hydraulic actuators, equipped with servo valves and potentiometers, are used for joint level position control. The force exchanged with the user hand is measured with load cells. The hydraulic pipes and electric wiring are routed inside the links. The maximum torque available at the joints ranges from 98 Nm, at the shoulder, to 22 Nm, at the wrist.

The FREFLEX (Force REFLecting EXoskeleton), developed by Odetics Co. for the Wright-Patterson Air Force Base (Burke 1994), is an electrically actuated, seven d.o.f. device, whose kinematics is similar to that of the SARCOS Master. Direct current servomotors are located on the ground and the joints are driven by bidirectional cable transmissions. Each transmission cable is routed over idlers to reach the joint. The axes of the idlers and of the joints are designed to be either perpendicular or parallel to each other. Gearboxes are present at the motors. The maximum exertable force at handgrip is 25 N.

2.1 Anthropomorphic Devices

Human joints are not lower-pair joints but are composed of non-conforming smooth surfaces that are constrained in their motion by elastic ligaments and compliant tissues. Even just the kinematic simulation of such mechanisms is mathematically quite complex (Ellis, Zion, and Tso 1997). On the other side, mechanics is, at present, mainly confined to the construction of mechanisms with lower-pair joints. Moreover in robotics, there is undoubtedly a preference for the usage of pure rotary joints which, for several reasons, guarantee a better performance in motion control with respect to translational ones.

These considerations led to the chioce of building an anthropomorphic haptic interface with a kinematics composed of seven rotary joints, which approximates the kinematics of the arm of the operator (see Figure 2.1). The first three joints are intersecting in a point in the operator shoulder and account for the shoulder adduction/abduction movement, the shoulder flexion/extension movement and the arm rotation. The fourth joint is associated to the elbow flexion/extension; the fifth to the forearm pronation/supination movement. The third, fourth and fifth joint axes intersect in a point in the operator elbow. The two last joints account for the wrist abduction/adduction and wrist flexion/extension. In order to better approximate

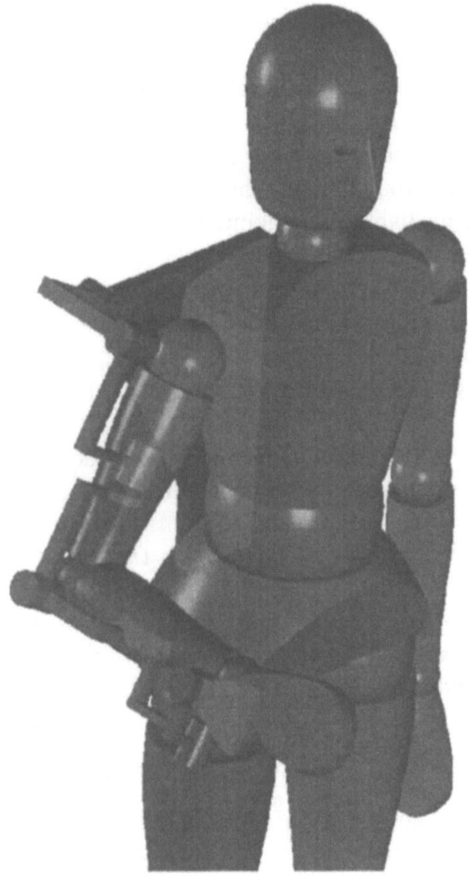

Figure 1: Anthropomorphic kinematics

the wrist movements these two axes are some millimeters apart.

An anthropomorphic design, which is "coincident" with the human kinematics in the sense just specified, offers several advantages and peculiarities which are discussed in the following. It is worth noticing that, in this sense, the SARCOS and the FREFLEX devices are not really anthropomorphic.

2.1.1 *Device-Body Interface*

Matching the kinematics of the haptic interface to that of the human makes the arm, forearm and palm links of the device always integral with the respective human arm links. It becomes then possible to have three points of attachment between the device and the human arm, using fixtures positioned at the center of the arm, forearm and palm. In other words, the interface braces the whole arm of the user letting his fingers free. In this way, by adopting an appropriate control strategy, the interface can exert forces on the three main segments of the user arm.

2.1.2 *Monitoring of User Body Posture*

The measurements of joint rotation angles of an anthropomorphic interface offer an estimation of the posture of the whole user arm; such estimation is coherent with the approximated kinematic model chosen for the human arm. The ability to record the whole human arm posture, instead of recording just the hand position, has several advantages both for teleoperation applications and virtual environments. In teleoperators, the operator can control not only the slave "end-effector" position but also its posture, especially the elbow position which in many arm manipulators is not uniquely determined by the end effector position. In Virtual Environment applications, the monitoring of the human arm posture is necessary to represent the artificial body of the user in the VE.

2.1.3 *Set of Reproducible Forces*

An anthropomorphic haptic interface for the arm with the kinematics depicted in Figure 2.1 can exert forces applied on the arm, forearm and hand of the user. An arbitrary wrench can be applied by the interface on the user palm; four independent components of a wrench can be applied on the user forearm, while only three independent components of a wrench can be applied on the arm link. For example, forces aligned with the arm link rotation axe cannot be applied on the arm and forearm links. Moreover, since the device has in all seven degrees of freedom, only seven components of wrench can be controlled independently at the same time. For instance, a wrench can be applied on the user palm and at the same time a single force component can be applied on the user elbow.

Forces on the arm and forearm arise in many cases during manipulation procedures in complex environments. When the operator is working in a cluttered space, it can collide with obstacles in any point of the upper limb. During manipulative operations, the operator can deliberately lay his arm/forearm on surfaces or edges. Inserting the arm in objects having deep hollows gives also rise to patterns of forces applied on the arm. It is clearly very interesting to handle these cases in the simulation of a realistic Virtual Environment.

In teleoperation, the ability of an anthropomorphic master to feedback forces not only at the level of the operator hand, but also on his arm and forearm, makes possible to think of master-slave systems which can perform "whole arm" manipulation procedures (i.e. manipulation with several link surfaces in contact with the environment or manipulandum, instead of interactions exclusively at the level of the end-effectors). Moreover, anthropomorphic master arms, which can constrain the motion of the whole operator arm and not only of his hand, can be used to enforce the avoidance of known objects in the remote environment.

2.1.4 *Human-like Workspace*

In order to preserve the user arm dexterity, the workspace available when wearing the device must match that of the free human arm. The available workspace is the intersection between the human arm workspace and the device workspace. An anthropomorphic interface is in the best conditions to maximize such intersection because the two workspaces tend to coincide if the device's joints ranges of movement are matched to the human joints ranges of movement.

2.1.5 *Mechanical Design Challenges*

The mechanical design of an anthropomorphic interface presents several challenges.

First of all, the kinematics of an anthropomorphic interface is determined by anthropometric considerations. From statistical data expressing the lengths of human upper limb as percentiles, the links lengths have been computed to ensure, on the 95 percentage of population, a maximum telerated joint axes misalignment. Kinematic parameters are not degrees of freedom in the design but rather constraints. As a consequence, the stiffness and the inertial properties of the interface cannot be improved by optimizing kinematics parameters.

Secondly, an anthropomorphic device is like a robot which the human operator wears on his body (on his arm in this specific case) and therefore it must be tailored on the operator body. The elementary requirement of avoiding interferences between the device and the human arm and trunk in every configuration gives rise to big difficulties about "where to place the links" with respect to the human body. The link design is greatly affected by the problem of wearability too. The links should be slender and close to the human limb so that both the operator and the space around him is as unencumbered as possible. For

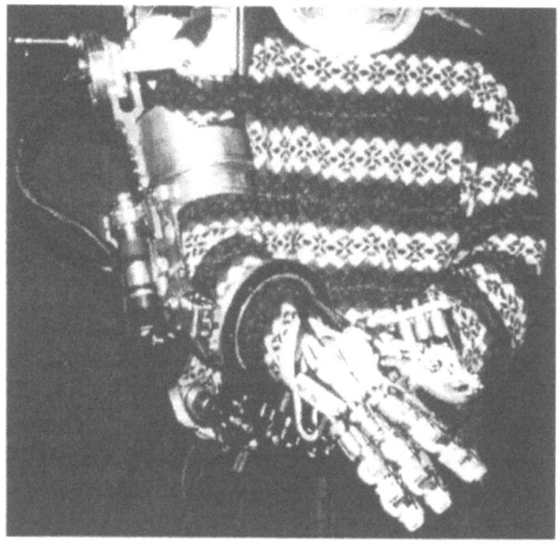

Figure 2: Arm and Hand Exoskeletons

the same reason, the links should have their surfaces free of protrusions, which could hurt the operator coming in contact with his arm or trunk during movements.

Finally, aligning the joint axes of the device with the approximated anthropomorphic kinematics is particularly difficult from the mechanical design point of view. In fact, it is necessary to design joints whose axes intersect *inside* the human arm at the level of the shoulder, elbow and wrist. The case of the arm and forearm rotation joints is even more critical, because they must have their rotation axes completely *inside* the human arm and forearm (see Figure 2.1).

2.2 Precedent Work

At PERCRO, Scuola Superiore S. Anna of Pisa, Italy, a Force Display Device, consisting of two exoskeleton systems devoted to replicate forces at the level of the palm and of the operator's fingers, has been designed and realized (Bergamasco 1995a) (Bergamasco, Allotta, Bosio, Ferretti, Prisco, Parrini, Salsedo, and Sartini 1994). In particular the two systems are:

- an Arm Exoskeleton, or External Force Feedback (EFF) system, with 7 d.o.f. which are coincident with the principal joint axes of the human arm. The EFF is attached to the user arm at the level of the palm (see Figure 2.2);
- an Hand Exoskeleton of Hand Force Feedback (HFF) system, consisting of 4 independent finger exoskeletons wrapping up four

fingers of the human hand (little finger excluded) and each one possessing 3 actuated d.o.f. in correspondence of the finger flexion axes, while fingers abduction movements are just sensed. The HFF is connected to a base plate located on the user's metacarpus and corresponding to the end -point of the Arm Exoskeleton (see Figure 2.2).

The EFF and the HFF are both anthropomorphic haptic interfaces in the sense specified in the precedent section. The present work has its origins in the analysis and the experimental work done on the EFF (Bergamasco and Prisco 1995) (Bergamasco 1995a) (Bergamasco 1995b)(Prisco and Bergamasco 1997). Some limitations of the Arm Exoskeleton mechanical design have been highlighed.

First of all, the workspace available to the operator is limited by the reduced range of motion available at the shoulder joints: the maximum shoulder adduction and extension are respectively 73° and 49°. These limitations don't allow to take full advantage of its anthropomorphic design.

Secondly, the exoskeleton structure, wrapping completely the human arm, forces the user to wear the system inserting his arm through the shoulder hollow. This has proved to be cumbersome to the point that unexperienced or impaired users may require external help. Moreover a exoskeleton around the arm, even if compact, is anyway an hindrance to bringing the arm laterally close to the trunk, a position which we have noticed to be the most natural for resting.

Thirdly, although the EFF is fully anthropomorphic, it does not exploit the possibility, discussed in Subsection 2.1.3, to replicate forces also on the users arm and forearm link.

The actuation system of the EFF is dimensioned to apply a $20N$ force on the user palm positioned still at the center of the workspace with the HFF (weighting around 1.4 kilos) mounted on the seventh link. Such performance is realized using small DC servo motors, designed for linear torque output, gearboxes with reduction ratio of 66, and a cable transmission allowing for a further reduction ratio of around 3. The presence of gearboxes causes friction which affect negatively joints backdrivability. The reduction of friction by means of closed loop joint torque control has proved successful but it has been implemented only on the wrist joints, where torque sensors are available.

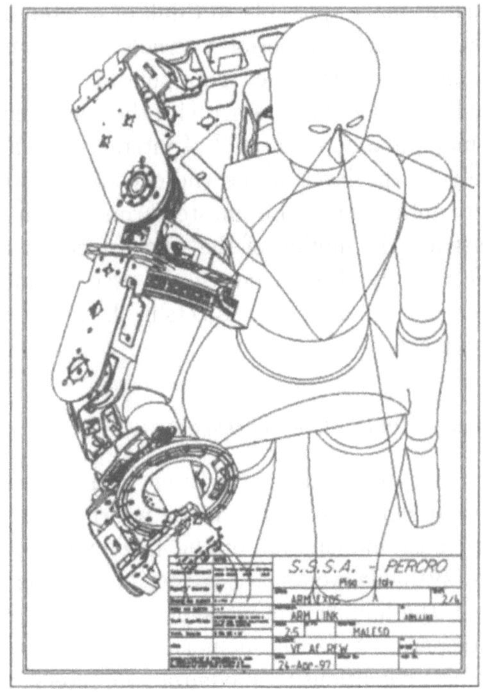

Figure 3: Anthropomorphic haptic device for the upper limb in reference posture

The joint position sensing obtained with optical encoders on the motor axes gives a satisfactory resolution, also because of the reduction rate, but it doesn't allow to derive a clean velocity signal by numerical differentiation. The need of a noise free velocity signal, especially for the implementation of interactions with virtual rigid surfaces, is experimentally demonstrated in (Bergamasco and Prisco 1995).

The EFF control strategy is based on driving the joints with a 7-dimensional torque vector $\tau_{control}$ which compensates the weight of the device and balancea the wrench $\mathbf{F}_{react}$ desired on the user's hand. The torque vector $\tau_{control}$ is computed according to the relation:

$$\tau_{control} = \hat{G} + J^T(\mathbf{q})\mathbf{F}_{react}$$

where $\hat{G}$ indicates an estimate of the gravity effects at the joints and $J(\mathbf{q})$ is the jacobian matrix.

3 Objectives of the Research

The scope of the present research is to proceed towards the realization of a satisfactory implementation of the concept of anthropomorphic haptic interface, as it has been outlined in Section 2.1, by designing a device which exploits the advantages of the anthropomorphic approach and at the same time overcomes some of the limitations of the Arm Exoskeleton described in Section 2.2.

Costs and development time have played no role in the system specifications; reliability has been considered of secondary importance at this stage of the project. In fact, our effort has aimed at demonstrating the system feasibility and its usefulness; we rely on technology advance and industrial re-engineering to cut on costs and realization times and to increase system reliability.

Apart from the requirements associated to the anthropomorphism of our device, our design has been influenced by two other functional requirements:

- adjustability. It is particularly important for an anthropomorphic haptic interface to be adjustable to accommodate to different sizes of the user arm (different arm and forearm lengths and circumferences);
- wearability. From our experience with the Arm Exoskeleton, easiness of wearing is a key issue for the acceptance of a device among users which are not specifically trained.

3.0.1 *Dynamic Requirements*

The dynamic properties, which we have chosen as design requirements for our device, are the exertable peak force (transient and continuous value) and the force resolution (both specified at the palm). The maximum velocity and acceleration have been specified defining a set of reference arm trajectories, as discussed in Section 5.0.1.

The backdrivability, i.e. the impedance of the controlled device measured at the palm, and the force bandwidth have been judged very important requirements too. It is very difficult to translate into design indications such requirements since they depend also on the joint torque control laws. Therefore the open loop joint torque bandwidth has been used as reference criteria for the dynamic properties of the final controlled system.

4 Design Innovations

The main innovative design solutions which we have introduced in our system are related to the design of the tendon transmission and to the mechanical solution which permits lateral wearing of the device.

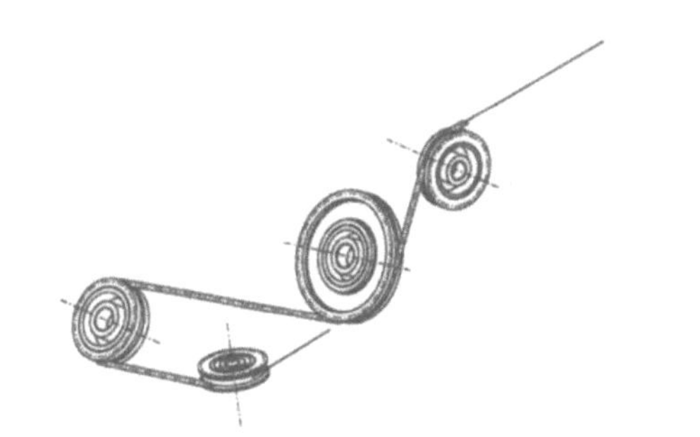

Figure 4: Spatial placement of idlers

4.1 Tendon transmission

We have designed a tendon based driving system which permits to place the motors away from the joints. Our goals have been to achieve reduced link weight (especially for the distal joints), increased joint compactness, reduced encumbrance in the workspace of the moving parts of the device. Moreover, such solution allows the usage of grounded actuators with high peak torque and consequent reduction or elimination of gearboxes (with the associated friction and torque transmission problems).

The main drawbacks of tendon transmissions are associated to the tendon elasticity and routing. The elasticity between the motor and the link, which is inevitably introduced by tendons, tends to lower the stiffness and the mechanical bandwidth of the device. Therefore these values have been kept under control during the design as described in Section 7.

The routing of the tendons from the motor to the joint becomes complex in multi d.o.f. systems. If the tendons are guided by sheaths the routing is simplified but severe problems arise in force control due to dry friction. In our design, the tendons are routed over idlers mounted on ball bearings. Such a method has been used in several other robotic structures (Hirose 1993) (Burdea 1996) (Bergamasco 1995a). Usually such type of transmission is planar, that is to say, all the idlers lay in a common plane. A variant to this approach has been proposed in (Hirose 1993) and it allows for a small skew angle between an idler and the following. In other designs, such as the WAM from MIT and the Arm Exoskeleton from PERCRO, the axes of two successive idlers

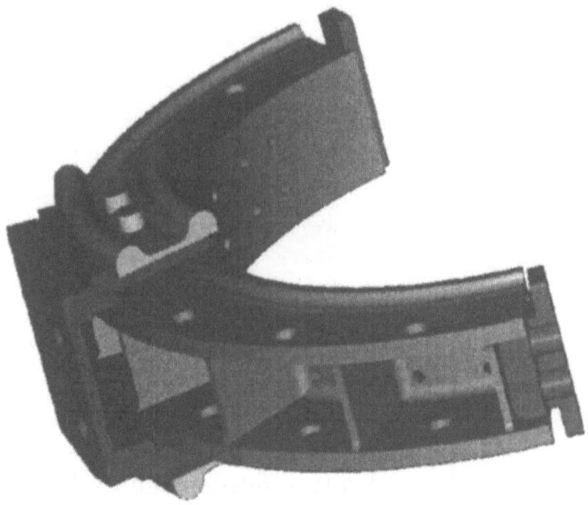

Figure 5: Anthropomorphic kinematics

can be perpendicular in order to route the tendon between two orthogonal planes.

The innovative design solution introduced inthe anthropomorpic haptic interface is that the idlers of tendon transmissions are arbitrarily placed in the space; just the constitutive condition that two successive idlers share a common tangent line, as shown in Figure 4, is respected. This is the most general type of tendon transmission guided over pulleys and, to our knowledge, it has never been adopted in other designs. In this case, it has been a key technology which has allowed to route the tendons around the human arm and forearm.

4.2 Lateral wearing

In the aim of achieving a comfortable wearability, 'open' links for the arm and forearm have been designed. This means that the links do not wrap completely the user limb like in an exoskeleton, but instead just adhere to the external part of the arm. This solution allows easiness of wearing, since the user arm enters laterally into the device (while an arm exoskeleton has to be worn just like a sleeve) and it adapts to a broader range of user arm circumferences. Fast lateral wearing/unwearing gives also a greater intrinsic safety and a broader acceptance among users. Moreover such solution allows the user to bring his arm very close laterally to his trunk.

The design of open links around the user's arm and forearm, while meeting the constraint of coincident anthropomorphic kinematics, has been possible thanks to an on purposely developed a partial (semicircular) rolling ball bearing. Such a mechanical component, depicted in Figure 4.2, has the same performance in terms of stiffness, weight and friction of a precision ball bearing of the same diameter.

The lateral wearing is only partially allowed by the present design, in fact only the arm link is open (while the forearm link wraps the user forearm as can be seen in Figure 3). This is only due to time/budget constraints.

5 Design tools

The mechanical design of the anthropomorphic haptic interface for the human arm has been possible only thanks to a set of software tools, which have given to the mechanical designers a better control of their work.

First of all, a key element has been the adoption of a fully associative, 3D, parametric CAD environment, integrated with a structural analysis tool. This has allowed to draw and analyse intrinsically 3D solution, such as the transmission system described in Subsection 4.1. Moreover, it has been possible an iterative process of design, by changing the dimensional parameters of a part in consequence of the results of the analyses performed on the whole system.

Secondly, a software tool has been developed in order to evaluate the lowest resonant mode associated to the multi DOFs, coupled transmission system. Such tool implements the dynamic model of such type of tendon driven robot according to (Prisco and Bergamasco 1997).

Finally, a dynamical simulation software application has been developed to evaluate quantitatively the performance required to the actuators (see Section 5.0.1.

5.0.1 *Dynamical simulation tool*

The derivation of quantitative requirements for the actuators has been based on the following specifications:

A the kinematics and dynamics of the device, in terms of joint variables;

B the transmission structure, in terms of the relationship it introduces between joint variables and motor variables;

C a description of the arm free movements workspace, in terms of joint positions, velocities, accelerations during the expected usage of the device. Since the haptic device has no repetitive usage, a quite large set of typical

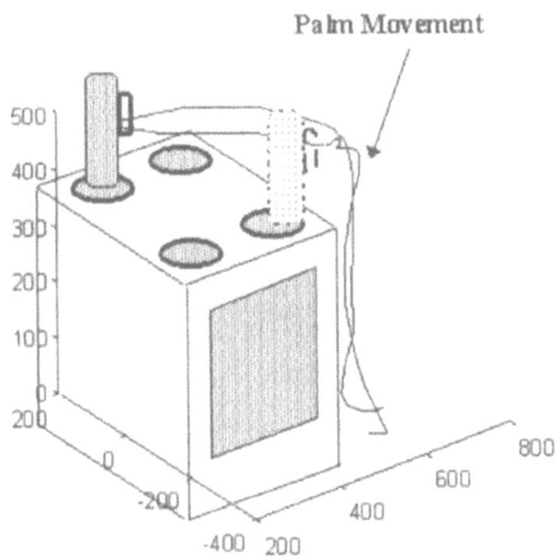

Figure 6: An example of arm trajectory recorded to derive quantitative requirements for the actuation system

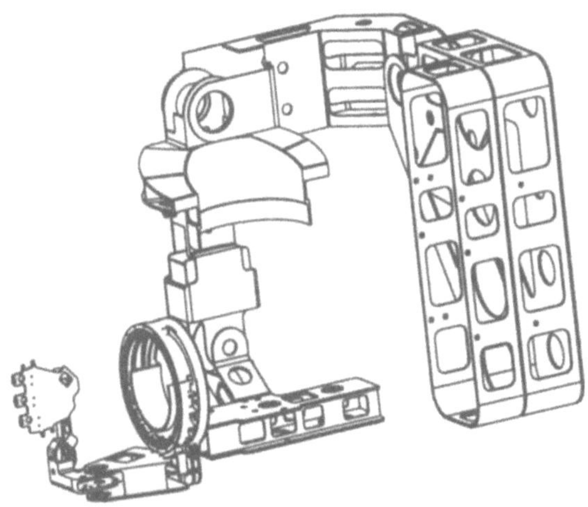

Figure 7: Anthropomorphic haptic device for the upper limb in reference posture

operation has been chosen and arm trajectory acquisition has been done using infrared markers attached on user arm (see Figure 6 for an example trajectory).

D a specification of the peak forces the haptic device is requested to apply on the human arm.

Given the complexity of the anthropomorphic haptic device, the first two items have been determined with confidence only thanks to the adoption of a 3D CAD environment capable of computing masses and inertia of the complex assemblies of mechanical parts, which forms the links of the device.

Items A,B,C and D have been used as inputs to the dynamical simulation software tool (see the precedent Section) which solves the inverse dynamic problem along all the arm trajectories and computes the associated motor trajectories, under the hypothesis of rigid transmissions. From the motor trajectories, a set of indexes, such as peak continuous and rms values of motor torque, velocity, acceleration and mechanical power output are computed. These values have been used to select the motors (see Table 2).

6 Design solutions

The device closely follows the human arm from the shoulder to the palm; it features 7 rotational joints in order to fully preserve the freedom of movement of the user arm (see Figure 7 for an global view).

The choices, already experienced in the Arm Exoskeleton, of a totally anthropomorphic kinematics and of the usage of electrical actuation, cable transmission and joint torque control approach, have been renewed.

The kinematic congruence of the device to the user arm kinematics is guaranteed by means of a regulation system for the arm and forearm links. The lengths of the arm link and the forearm link are continuously adjustable within a range of $0.05m$ and $0.03m$. The adjustments induce no rotation on the actuators and they can be performed with the device powered and controlled by a special software.

The device is fixed to the user upper link at the level of the medium part of the arm, of the forearm and of the palm (see Figure 3).

The wearability, for users with arm circumferences of a wide range of values, is guaranteed because the arm link of the device just surrounds half of the user arm (see Figure 8). This has been realized using the semicircular rolling ball bearing described in Subsection 4.2, and guiding the transmission tendons around the joint 3 using a spatial transmission, as explained in Subsection 4.1. The same solution could be implemented for the forearm link.

6.1 Kinematics

Figure 8: Link 2

The joints ranges of movement with respect to the arm reference position of Figure 7 are reported in Table 1.

Joint	#1	#2	#3	#4	#5	#6	#7
θ_{min}	0	−180	−60	−90	−60	−70	−45
θ_{max}	90	0	60	20	60	60	15

Table 1: Joint rotation limits (in degrees)

The shoulder d.o.f. have a range of movement which is wider with respect to the Arm Exoskeleton described in Section 2.2. This allows the user to raise his hand over his shoulder and head. The range of movement of the arm and forearm rotations are limited by the presence of the semicircular ball bearings.

6.2 Actuators

Motor	#1	#2	#3	#4	#5	#6	#7
τ_{cont}	13	13	13	2	0.8	0.8	0.8
τ_{peak}	21	21	21	3.5	1.5	1.5	1.5

Table 2: Available torques at motors' axes (Nm)

Current controlled DC electric motors, designed for high peak torque (i.e. pancake shaped motors), with one stage gear boxes or no gear boxes at all, have been used. The actuators for the first

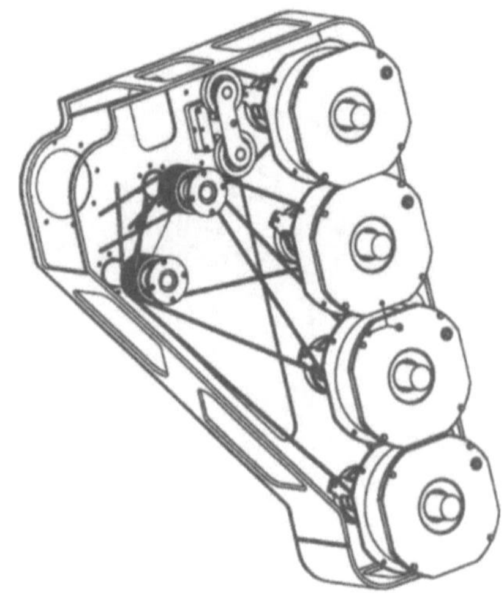

Figure 9: Link 0 with 4 motors

4 joints are placed on the back of the user (see Figure 9), while the 3 actuators for the wrist joints are placed under the forearm with a similar mechanical design.

6.3 Transmissions

The basic design choice for the transmission system is that of using a $2N$ tendon configuration, in which a pair of opposed tendons drives each single joint and one rotary actuator is used to drive a pair of tendons (see Figure 10). Multistage transmissions are used whenever the motor is not placed near the joint (see Figure 10). In such cases, the fact that the tendons driving a joint are routed over the preceeding joints, causes a coupling in the relationship between motor variables (angular displacements and torques) and joint variables (see (Prisco and Bergamasco 1997) for a detailed discussion). Multi strands, steel cables, with a diameter of 2 mm, are used as tendons; the tendons are routed over pulleys mounted on ball bearings. Pretensioning is obtained with a regulation mechanism located at the driven pulleys.

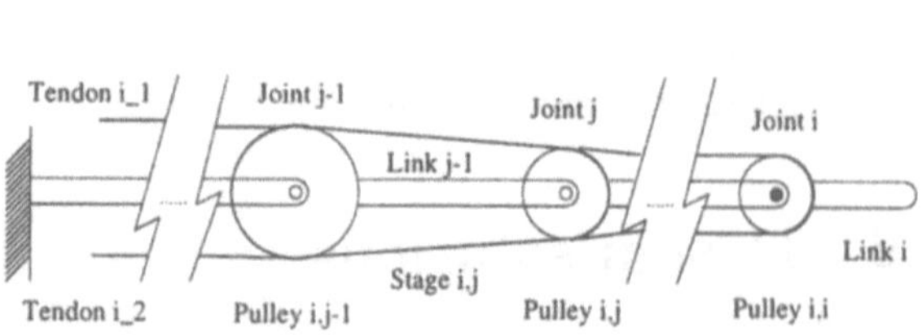

Figure 10: A multi stages transmission with pairs of opposed tendons

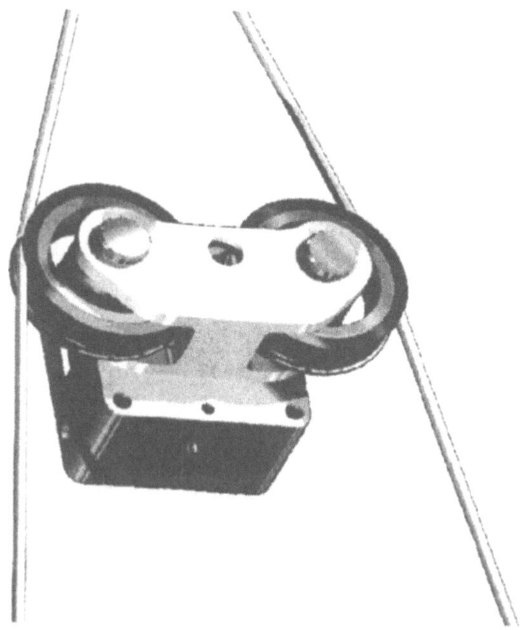

Figure 11: Differential tendon tension sensor

6.4 Sensors

Resolvers, mounted on the motor shafts, are used in order to have high resolution in the position measurement and in order to have a direct velocity measurement available to implement damping control.

On purposely developed, differential tendon tension sensors are employed, in order to measure joint torques (see Figure 11). Such sensors use strain gauge full bridges as sensing elements. They are equipped with local electronics for signal conditioning in order to improve signal to noise ratio. The performance of the sensor module is summarized in Table 3. The sensing resolution is adapted to the requirement of each joint, by varying the incidence angle of the cables on the sensor frame.

Transversal force	Max	200 N
	Min	-200 N
Transversal torque	Max	2 Nm
	Min	-2 Nm
Overload factor	1.5	
Force resolution	0.1 N	
Force accuracy	2.0 N	

Table 3: Performance of the torque sensor module

7 Design analysis

An analysis of the global results of the design has been performed iteratively during the design process, in order to evaluate the impact of some solutions on the most critical global performance indexes. Three sets of global properties of the haptic interface have been kept under control:

- the masses and inertia tensors of the links;
- the worst-case end point stiffness of the device;
- the worst-case mechanical bandwidths of the joints.

7.1 Link inertial properties

The estimation of the mass and inertia tensors of the links has been obtained directly by the 3D CAD design environment. In this way, every single mechanical part, up to the screws, has been taken into consideration; we have also been able to evaluate the gain in mass which is realizable by using special materials, such as composite material or carbon fiber, instead of aluminium, for the structural parts.

The results of the analysis have shown that the greatest impact on link masses is due to the actuator modules, each integrating an unhoused pancake rotor/stator, a resolver and a mechanical frame. In fact, from Table 4, it is noticeable the high masses of links 0 and 4, which house respectively four and three actuator modules. A second significant contribution to the link masses is represented by the steel parts, such as torque sensor frames, ball bearing, idlers and pulleys.

7.2 Stiffness

The estimation of the worst-case end point stiffness of the device has required a special care, because of the complexity of the mechanism. From a methodological point of view, the correct computation of the end-point stiffness has required to take into account the contribution of both the link structural parts and the tendons. The worst case has been identified as occurring when the

Link	#0	#1	#2	#3	#4	#5	#6	#7
Mass	8.1	2.6	3.2	1.4	5.1	0.64	0.28	0.07

Table 4: Link masses (with structural parts made of aluminium)

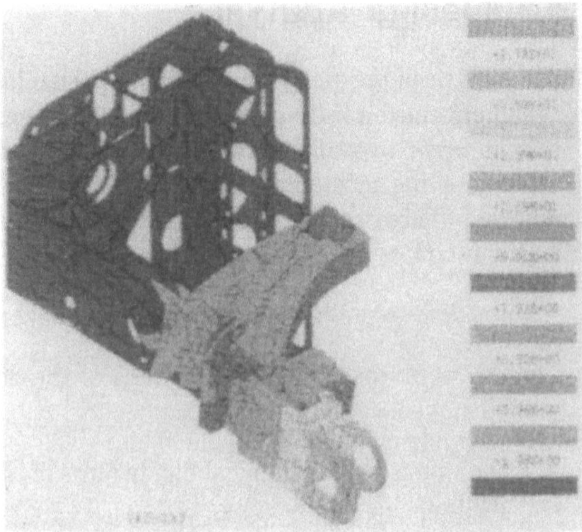

Figure 12: Displacement magnitude of the assembly composed of the first four links, including cable transmission contributions, under a generic load condition

Figure 13: Displacement magnitude analysis of the assembly composed of the last 3 links, including cable transmission contributions, under a generic load condition

human arm is completely stretched in the front (the joint angles values in such case are reported in Table 5). In Figure 12 and 13, the diagrams representing the displacement magnitudes, under a general load condition, are reported. The elastic constant along z, which is by far the worst direction, is $1.8N/mm$.

8 Conclusions

The ideation and the design of an anthropomorphic haptic interface for the himan arm has been presented. The methodologies and technical innovations developed to accomplish the design have been outlined. The presented design has been successful in meeting the requirements of workspace, wearability, adjustability while the dynamic performance has been kept under control. It is noticeable that the tendon transmission and link length regulation mechanisms have tripled the design time of the basic mechanical solution.

The realization of the first prototype will allow to evalute the technical and functional performance of the device.

Joint	#1	#2	#3	#4	#5	#6	#7
θ	0	−90	0	−90	0	0	0

Table 5: Joint angle in test configuration(in degrees)

9 Acknowledgements

The authors wish to thank the PERCRO team of mechanical engineers, which was involved in the presented work in the years 1994-1997, and in particular Fabio Salsedo who has leaded the team to the ideation and design of many mechanical solutions and innovations.

The present work has been partially funded under the European Commission RTD Programme TIDE VETIR n.1216.

References

Bergamasco, M. (1995a). Force replication to the human operator: the development of arm and hand exoskeletons as haptic interfaces. *ISRR International Symposium of Robotics Research*.

Bergamasco, M. (1995b). Haptic interfaces: The study of force and tactile feedback systems. *Proc. of IEEE Workshop on Robot and Human Communication ROMAN '95*.

Bergamasco, M., B. Allotta, L. Bosio, L. Ferretti, G. M. Prisco, G. Parrini, F. Salsedo, and G. Sartini (1994). An arm exoskeleton system for teleoperation and virtual environments applications. *Proc. 1994 IEEE International Conference on Robotics and Automation*.

Bergamasco, M. and G. M. Prisco (1995). Virtual surfaces countur following: An experimental approach exploiting and arm exoskeleton as haptic interface. *Proc. 1995 ASME Winter Annual Meeting-Session on Haptic Interfaces for VirtualEnvironment and Teleoperation.*

Burdea, G. C. (1996). *Force and Touch Feedback for Virtual Reality* (1 ed.). Wiley-Interscience Publication.

Burke, J. (1994). *Exoskeleton Master Arm, Wrist and End Effector Controller with Force-Reflecting Telepresence.* Technical Report AL/CF-TR-1994-0146, Odetics Inc.,Anaheim, CA.

Ellis, R. E., P. Zion, and C. Y. Tso (1997). Interactive visual and force rendering of human-knee dynamics. *Fifth International Symposiom on Experimental Robotics.*

Fischer, P. and R. Daniel (1990). Specification and design of input devices for teleoperation. *Proc. 1990 IEEE International Conference on Robotics and Automation.*

Hayward, V. and O. Astely (1995). Performance measures for haptic devices. *ISRR International Symposium of Robotics Research.*

Hayward, V. and J. M. Cruz-Hernandez (1995). Parameter sensitivity analisys for design and control of tendon transmissions. *Fourth International Symposium on Experimental Robotics.*

Hirose, S. (1993). *Biologically inspired robots* (1 ed.). Oxford University Press.

Jacobsen, S., F. Smith, and D. Backman (1991). High performance, dextrous telerobotic manipulator with force reflection. *Intervention/ROV'91.*

Prisco, G. M. and M. Bergamasco (1997). Dynamic modelling of a class of tendon driven manipulators. *International Conference on Advanced Robotics.*

Testing A Visual Phase Advance Hypothesis for Telerobots*

R. W. Daniel and P. R. McAree
Department of Engineering Science
Oxford University, Parks Road, Oxford OX1 3PJ, UK.
email: rwd@robots.ox.ac.uk

Abstract

We report on the capacity of an operator to utilize visual feedback to stabilize an otherwise marginally stable force reflecting system. We use a simple model of the force reflecting system to argue that such a stabilising influence can only arise from phase advance generated by the operator's view of the task representation.

1 Introduction.

Our work in teleoperation focusses on providing high fidelity control over a teleoperated arm when in a hazardous environment. Over the last ten years, the Oxford laboratory has developed a very high performance input device (the Bilateral Stewart Platform or BSP) that is used by the nuclear industry for decommissioning activities and is distinguished from other teleoperation systems by its capacity to transmit high frequency 6-D force information (up to 80 Hz) to an operator. The BSP is subject to the usual stability problems seen in force reflecting systems and the whole system is unstable if too high a force reflection ratio is used. The objective in this paper is to investigate the extent to which visually derived information improves the performance of our force reflecting teleoperation system. We ask, in particular, whether the human visual system can facilitate the generation of *phase advance* within the human neuro-muscular control system when carrying out contact tasks.

The study of human operator dynamics has a long history. During the 1960s and 1970s McRuer and his co-workers at STI developed a sophisticated model of the human operator when carrying out tracking tasks; this work is summarized in (McRuer 1980). Underlying these investigations was the need to understand the response of aircraft pilots to visual inputs. McRuer demonstrated conclusively that a human operator is able to generate phase advance when performing pursuit tasks. He also demonstrated other forms of filtering based on visual information, but for our purposes it it the generation of phase-lead that is of principal interest. This phase-lead is evidenced by the first order roll-off of his operator models which are at variance with the expected second-order roll-off due to the arm's mass. This expected second-order behaviour has been confirmed in several studies of the dynamics of the human arm when subjected to exogenous physical disturbances, e.g. the study of elbow stiffness by (Bennett, Hollerbach, Xu, and Hunter 1992) which shows the operator's neuromuscular control system acts as a lightly damped second order system with second order roll-off. Our own studies (Daniel and McAree 1997) on the limits of force reflection support Bennett's findings.

We might interpret McRuer's findings as confirmation that an operator is able to anticipate future events. Several studies in the physiology literature have explored this theme. Johansson and Edin (Johansson and Edin 1993), for example, have studied the capacity of the CNS to use prediction when carrying out a 'reach and grasp' task. They identify two mechanisms: i) anticipatory parameter control, i.e. a strategy tuned to the anticipated characteristics of an object to be grasped, its mass, its roughness, etc. and ii) predictive feedforward control using tactile information, which they conjecture imposes a sequence of learnt compensatory feedback controllers appropriate to each stage of a manipulation. They do not, however, identify any mechanisms for the generation of phase advance.

Many researchers have reported on the desir-

*Supported by EPSRC grant GR/15005 in collaboration with UK Robotics Ltd. and BNFL P LC.

ability of high fidelity force reflection and many have reported on how such feedback often leads to instability. Here we ask whether the human capacity to generate phase advance when carrying out visual servoing tasks can be used to improve the stability of a force reflecting system.

2 A phase advance hypothesis.

Our experience of teleoperation indicates that if the operator has a good view of a task then the system is more stable than with a poor view. Is it possible that the provision of visual cues enables an operator to generate phase advance and hence extend the limits of stability of a marginally stable force reflecting teleoperator? Certainly we think McRuer's work supports this hypothesis, but as there is a latency of approximately 0.2 seconds associated with the generation of phase advance, see (McRuer 1980), the bandwidth must be limited to below 1.25 Hz. Above 1.25 Hz the phase delay from the latency would exceed that of the phase advance of the anticipation.

McRuer's findings and our experience prompts the following hypothesis:

With vision an operator is able to generate phase advance and inject this into an outer feedback loop and so stabilize a marginally stable system.

3 Physics of teleoperation and stability.

The physical characteristics of the remote slave arm imposes a fundamental limit on the loop gain of a teleoperator system which manifests itself most obviously as a limit in the extent to which force can be reflected to the operator. We have developed a description of the limits on performance based on the dynamics of the slave when coupled to the operator which is matched to the dynamic characteristics of the operator's mechanoreceptors (Daniel and McAree 1997). This is achieved by separating the transmitted signals into two components: i) a low frequency energetic channel; and ii) a high frequency information channel. The energetic channel passes around the loop and must satisfy the passivity constraints for stability. The information channel is present only in the force feedback

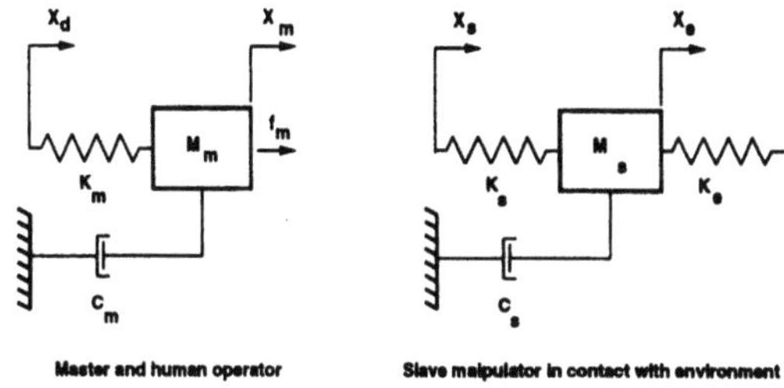

Figure 1: Simple model of master and slave contacting the environment

section and is matched to the physical characteristics of the operator's vibration receptors in the hand, in particular the the Pacinian or PC system which is responsible for transmission of high frequency events to the CNS. The two channels are separated using a notch filter, the shape of the notch depending on the physical limits imposed by the system. The model predicts the modes of failure of system stability, the likely strategy followed by an operator at the margins of stability (stiffening of the arm), and the limits on the force reflection ratio for a given reflection bandwidth.

The arguments presented in this paper are based on examining the behaviour of a pure force reflecting system which has been stripped of its channel separating notch filter. This filter is removed to reduce the complexity of possible interacting effects. Our simple model is illustrated in Figure 1.

In this model M_m is the mass of the master combined with that of the operator's hand and arm, C_m is the net damping provided by the master and operator, K_m is the operator's muscular stiffness. The input demand to the system, generated by the operator's central nervous system (CNS), is X_d, the force reflection system imposes a force f_m on the master and the resulting position of the master is X_m. The slave is driven under PD control and has mass M_s, servo stiffness K_s, damping C_s. The position demand sent by the master is X_s, which is equal to X_m, the environmental stiffness is K_e and the position of the slave is X_e.

The equations of motion of the resulting system are

$$M_m \ddot{X}_m = f_m - K_m(X_m - X_d) - C_m \dot{X}_m,$$

$$M_s \ddot{X}_e = K_s(X_s - X_e) - K_e X_e - C_s \dot{X}_e,$$

and these form the starting point of the analysis described in (Daniel and McAree 1997). Our model predicts a maximum force reflecting ratio,

λ, of

$$\begin{aligned}\lambda &= \frac{C_s C_m^2}{K_s(C_m M_s + C_s M_m)} \\ &+ \frac{2C_s C_m M_m}{K_s(C_m M_s + C_s M_m)^2}[\{(K_m M_s - K_s M_m)^2 \\ &+ (C_m M_s + C_s M_m)(C_m K_s + C_s K_m)\}^{\frac{1}{2}} \\ &- (K_m M_s - K_s M_m)],\end{aligned}$$

where the critical environmental stiffness (the stiffness for marginal stability) K_{ec} is

$$\begin{aligned}K_{ec} &= \frac{1}{M_m}[(K_m M_{xs} - K_s M_m)^2 \\ &+ (C_m M_s + C_s M_m)(K_m C_s + K_s C_m)]^{\frac{1}{2}}\end{aligned}$$

Using typical values ($M_m = 1$ kg, $C_m = 21.2$ Ns/m, $K_m = 600$ N/m, $M_s = 16$ kg, $C_s = 603$ Ns/m, $K_s = 5.6 \times 10^3$ N/m) for the human arm and the PUMA 560 slave arm, the model predicts that i) the maximum force reflection ratio is 0.14,ii) critical stability occurs for an environmental stiffness of 2.2×10^4 N/m and iii) the marginally stable system will have a pair of lightly damped poles at $\pm j31.6$ rad/sec.

A straightforward test of the phase advance hypothesis based on the model is confounded because the operator has independent control over both the net force that his muscles apply and the effective stiffness of the elbow. This ability was demonstrated in (Hogan 1984) where it is shown that such control is possible given the physics of an agonist-antagonist pair of muscles. The operator thus has two channels of control available:

- By changing the sum of excitations to the muscles while keeping the difference in excitation constant the operator may change the effective stiffness K_m in our simple model.
- By changing the difference in excitation while keeping the sum constant the operator may change X_d in our simple model.

To remove the ambiguity as to the cause of any change in observed performance, the above operating conditions for our model were chosen to be at the extreme end of the operator's ability to stabilize the system by stiffening his arm. The stable regime would require full excitation of both agonist and antagonist muscles in order to achieve maximum stiffness. The only means of control available is to back off the stiffness slightly and to manipulate the difference in excitation, that is, we establish experimental conditions such that only X_d is under control by the operator and any attempt to reduce K_m significantly makes the system unstable.

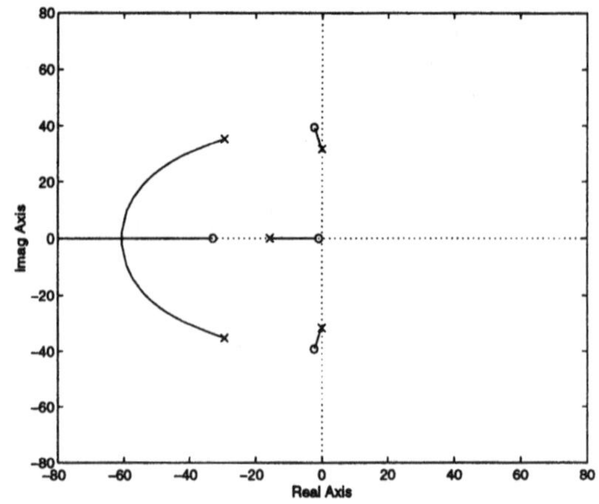

Figure 2: Pole-zero locations at marginal stability; root locus showing phase advance.

Let f_r be the net force applied by the operator to the input device (f_r is easily available on our system for display purposes and is obtained from a force sensor mounted on the input device). The transfer function between X_d and f_r is

$$\frac{f_r}{X_d} = -\frac{K_m\left(C_m s\left[M_s s^2 + C_s s + K_s + K_e\right] + \lambda K_e K_s\right)}{(M_m s^2 + C_m s + K_m)(M_s s^2 + C_s s + K_s + K_e) + \lambda K_s K_e}$$

and the resulting pole zero locations are shown in Figure 2

If the operator has access to a visual display of the applied force, then it is plausible that the CNS could complete a proportional loop using this additional information. Not surprisingly, this loop would result in the oscillatory poles migrating up the imaginary access before moving towards the zeros in the left-half plane. A simple angle of departure calculation, as illustrated in 2, confirms that for a stable angle of departure, any loop connected by the CNS via the operator's visual cue must have phase advance. The angle shown is for a phase advance compensator with a maximum phase advance of 60 degrees at 0.625 Hz, which is a reasonable upper bound on the phase advance predicted by McRuer once the effects of intellectual latency are included.

Theory thus predicts that *if* the operator uses vision as an extra sense within the loop, and *if* this extra sense helps to stabilize a marginally stable force reflecting system, then the operator *must* use phase advance.

Complications arise from the operator's 'higher intellectual functions'. One could argue that the operator is able to use his intelligence to identify the presence of a lightly damped pole and thus pre-filter the demand to the system. We can identify any pre-filtering that the operator performs by deconvolving the measured response of the system with the system model to arrive at the operator's pre-filtered input. (It is most unlikely that an intelligent operator would use a step input into such a lightly damped system). It could be argued that the operator could use vision information to 'super-smooth' the input demand over that used without vision. To overcome this objection we will recover the smoothed operator input for both systems, with and without vision, and then use the recovered inputs to drive models of the system with and without vision used as feedback. We assert that if the model without vision feedback has large oscillations when driven by the 'smoothed' input, then this argues that pre-filtering is not a good strategy for removing unwanted oscillations. If the model with vision does not display oscillations when driven by the input derived from the system *without* vision, this argues that it is the feedback properties of the extra sense that are generating the improvement in performance.

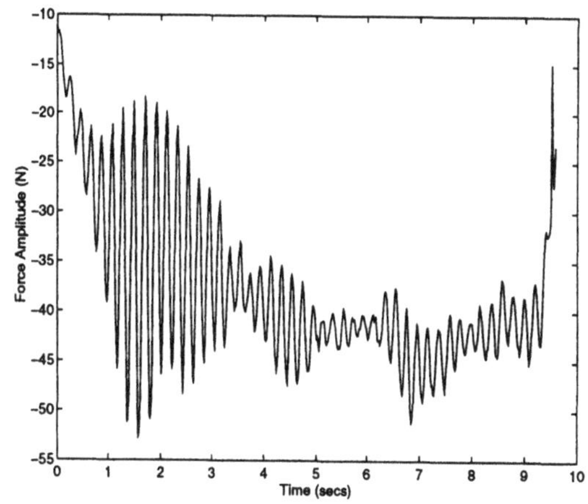

Figure 3: Force response with no visual feedback.

4 The experiments.

The experiment consisted of asking the operator to impose a step change in force, with the slave arm in contact with the environment, without going unstable. Only forces in the z-direction were reflected in order to remove interactions with the operator's wrist stiffness. The operator was first asked to carry out this task without access to any visual cues to the force being applied. In the second stage of the experiment a bar chart representation of the force being applied to the input device was provided, and the operator asked to repeat the step change in force task.

The force applied by the operator was recorded for the two situations, with and without the visual force cue. In both situations the operator had clear and direct view of the slave arm's interaction with its environment.

Figure 3 shows a typical force response of the master when the slave is in contact with a stiff environment and the operator asked to apply a step force without any visual indication of the fed-back force he is feeling.

Figure 4 shows a typical force response when the operator is asked to perform the same task but with a visual representation of the force being applied to his wrist through the master.

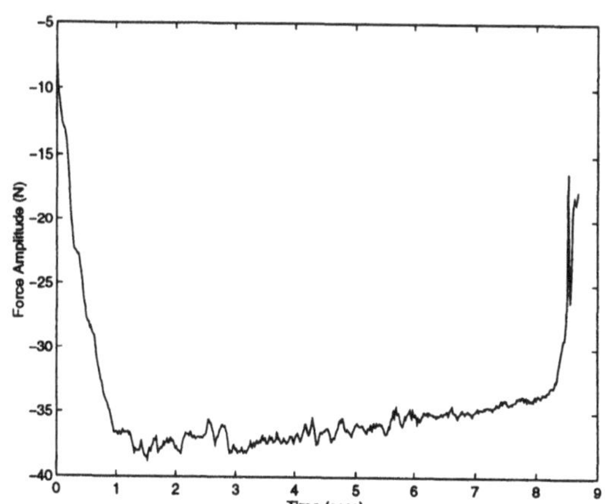

Figure 4: Force response with extra visual information.

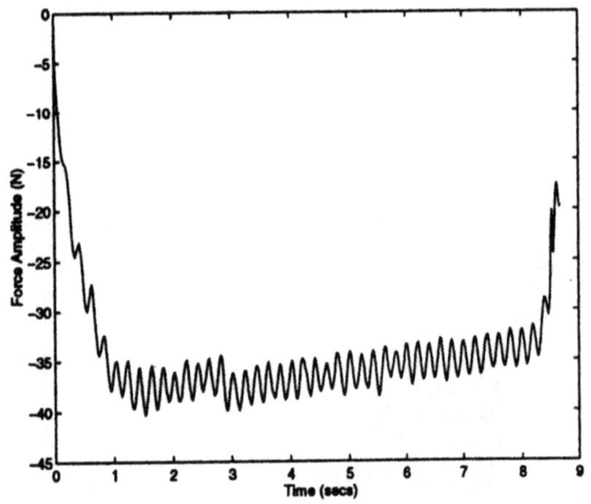

Figure 5: Simulation of the operator force (no vision is available) when: driven by the deconvolved 'with vision' input.

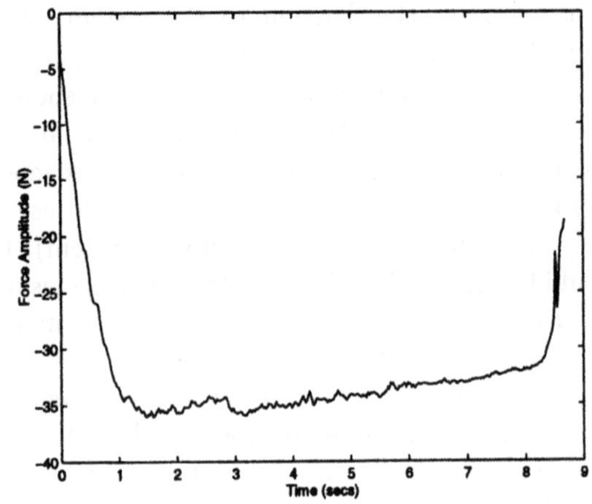

Figure 6: Simulation of the operator force (vision available) when: driven by the deconvolved 'with vision' input.

Note that the response without visual feedback has an oscillatory component which was significantly reduced when the operator has a visual representation of force. The operator reported that with force information, the system was easier to control and improved his confidence in making contact with the environment. We thus claim that the vision feedback improves matters and now discuss the likely mechanism for explaining the improvement.

5 Analysis and discussion.

The operator's 'intellectual demand' X_d was reconstructed using a simple band-limited inverse of the transfer function X_d/f_r. As expected, the inputs for both the system with vision and the system without vision gave a pre-filtered step function, indicating that the operator was using his intelligence to minimize the excitation of resonant modes. The deconvolved inputs are omitted for brevity, but simply consist of smoothed ramps to the desired force-level. Figure 5 shows the response of the model of 'no vision' feedback to the reconstructed 'with vision' input. Note that the model, as expected, is highly oscillatory, that there was nothing 'magical' done by the operator in smoothing the demand generated by the CNS. In fact the operator's demand would have to be remarkably free of noise not to excite the undamped poles using open loop compensation.

Figure 6 shows the response of the model with vision feedback. A phase advance network with a maximum advance of 60 degrees at 0.625 Hz has been included in the loop to model the expected upper limit on operator performance. The phase advance network has been chosen to represent the maximum possible reasonably achievable performance given the time delay suggested by (McRuer 1980) and is based on engineering judgement alone. It would require a more detailed experimental regime than reported here to quantify the compensation generated by the operator. It can be seen that the oscillation has been significantly reduced and the response is very like that seen in Figure 4.

This completes the demonstration of plausibility of phase advance being the most likely mechanism for improving performance of a force reflecting system.

6 Conclusions and future work.

Much of the work carried out to date on visual displays for non time-delayed systems has concentrated on the generation of a sense of 'immersion' and an investigation into the effects of low bit-rate transmission of images. These studies are based on the not unreasonable premise that visual feedback is a 'good thing', and thus follow a line of investigation which compares the performance of gross tasks, such as pick and place under a range of conditions. Implicit in much of the work on the use of vision is that it is only used for *planning* and that it has little or no dynamic role in manipulation (Klatzky, Purdy, and Lederman 1996). The community's attention is firmly focussed assessing the capacity of humans to plan based on visual information.

In this paper we asked whether there is any evidence of the operator using vision *during* manipulation, and have concentrated on the capacity of vision to stabilize an otherwise marginally stable system. Our findings indicate that it is possible to quantify the gain in performance that vision provides, particularly in any linkage of vision's anticipatory capacity with the neuro-muscular control functions. We have only demonstrated that such a function is plausible but our investigations are certainly not definitive. We invite comment from the physiology research community as to how the existence of such a mechanism may be confirmed.

We claim that the phase advance hypothesis is proven and that vision has the capacity to generate anticipative information that can be used directly by the neuro-muscular control system in situations, where, left to its own devices, it would be unstable. The effect is not large, but may have implications for the design of teleoperation systems where 'operator comfort' and 'ease of use' are nebulous artifacts from the engineer's point of view, but are improved by the provision of the correct information to the operator. We hope that this paper has gone some way to providing some engineering measurements to back up what is usually a psycho-physical problem.

Our current work involves the use of computer vision derived object models to optimally anticipate collision and to use this information to help the operator generate a soft impact with the environment. We propose to use the results reported in this paper to derive limits on the capacity of the operator to use a 'good view' to improve his performance, useful engineering measures of such information being notoriously difficult to come by.

References

Bennett, D. J., J. M. Hollerbach, Y. Xu, and I. W. Hunter (1992). Time-varing stiffness of human elbow joint during cyclic voluntary movement. *Experimental Brain Research 88*, 433–442.

Daniel, R. W. and P. R. McAree (1997). Fundamental limits of performance for force reflecting teleoperation. *To appear in the International Journal of Robotics Research 17.*

Hogan, N. (1984). Adaptive control of mechanical impedance by coactivation of antagonist muscles. *IEEE Transactions on Automatic Control 29*(8), 681–690.

Johansson, R. S. and B. B. Edin (1993). Predictive feedforward sensory control during grasping and manipulation in man. *Biomedical Research 14*(4), 95–106.

Klatzky, R. L., K. A. Purdy, and S. J. Lederman (1996). When is vision useful during a familiar manipulatory task? *ASME Dynamic Systems and Control 58*, 561–566.

McRuer, D. (1980). Human dynamics in man-machine systems. *Automatica 16*, 237–253.

PART 9

MEDICAL

SESSION SUMMARY

Gerd Hirzinger

DLR (German Aerospace Center), Institute of Robotics and System Dynamics

berpfaffenhofen, D-82234 Wessling, Germany

e-mail: Gerd.Hirzinger@dlr.de

The session on medical robots kind of reflected the most interesting developments under way in this area. Indeed medical robotics presently help to accept robot technology in the public opinion quite a bit, as it has been acknowledged meanwhile that sensorized, intelligent mechanisms in many cases can do a better job than the human hand holding e.g. a rigid endo-scope or surgical tool.

Ikuta's talk was partly philosophical, posing questions about the merits of medical robots and their safety. However it contained also interesting technical concepts, e.g. the description of a cybernetic actuator with piezoelectric elements and electromagnetic coils. Indeed endoscopic investigation (and may be surgery) in colon and intestine with devices that allow to smoothly follow the curvature of tubes without hurting or piercing the inner surface seems to be one of the most challenging problems in medical diagnostics and therapy. Ikuta also outlined the de-velopment of force-reflection forceps in 4 dof, usable for remote surgery as well as for virtual training. The RAT (robot assisted technology) is stronger than the CAT (computer assisted technology) was one of his remarkable statements.

An impressive description of the state of the art in virtual surgical training was given by Ni-colas Ayache. Indeed the more pressure is exerted on doctors by patients and medical assur-ances to operate minimally invasive, the more important is it to avoid failures by extensive training; indeed virtual training may help to reduce training with animals, however it requests realistic simulation of deformations caused by the forceps and by force feedback. It has turned out meanwhile that distributed mass-spring systems for simulating the elastic behaviour of tissue and organs is inadequate for providing a realistic force feeling. Instead finite element techniques computable in real time are requested. N. Ayache demonstrated such a system, with an artificial liver composed from geometric data as taken from the "visible human". Re-alistic cutting of tissue seems achievable in the near future as well as blooding and other ef-fects needed to make surgical simulations realistic.

The paper given by Jaqueline Troccaz tried to shed light into the various types of mechanical and mechatronic devices for surgery by classifying them on function rather than mechanism. She discriminated between localizers (i.e. devices which measures the coordinates of a tool or a pointer but do not control the location), robots and socalled synergistic devices. These latter kind of devices are particulary interesting, as they allow to share a task with a surgeon who e.g. cooperatively controls motion in some subspace while the device control motion or forces in the remaining degrees of freedom. Thus preoperative planning may be automatically taken in account, relieving the surgeon form the burden of executing highly precise motions not adequate for human performance.

All three talks in the medical session were an impressive demonstration of how smart mechatronic devices and intelligent mechanisms in combination with virtual reality and 3D world modelling are going to revolutionize surgery and therapy in the future, thus making the bene-fits of robot technologies visible for everyone.

Robot Assisted Surgery and Training for Future Minimally Invasive Therapy

Koji Ikuta

Department of Micro System Engineering
Graduate School of Engineering
Nagoya University

Furocho, Chikusa-ku, Nagoya, Aichi 464-01, Japan
e-mail: ikuta@mech.nagoya-u.ac.jp
TEL: +81 52-789-5024
FAX: +81 52-789-5027

Abstract

Advanced medical robotics based on the micro and macro mechatronics has been investigated by the author. Firstly, advantages of the medical robotics and unsolved issues for practical application are introduced. Secondly, safety "cybernetic actuator", "hyper endoscope" , "remote minimal invasive surgery system with force sensation", "virtual endoscope system with force sensation for clinical training" and "active micro forceps for retina micro surgery" are reported. Finally, prospect for medical robotics in future are discussed.

1. Introduction

There is a very fundamental question on medical robotics as described below.

"Can robot contribute to medicine? "

My answer is "Yes, it can. But, please don't expect too much. All the tasks of medical doctor are not fully replaced by the robot. It is important to utilize only merits of the robot. This is a very similar relationship between computer and human in today. "

There are so many questions and interests on medical robotics as follows;

"What kind of merits should be expected for medical robot now ?"

" Are there any safety problems on medical robot?"

"Is it possible to keep human safety during clinical operation by robot system ?" so on.

I will describe both merits and demerits of medical robot and several trials in my laboratory toward really useful medical robots.

2. Merits of Robots in Medicine

As everybody know, robot has following strong points that human doesn't have.

1. Micro and narrow space task
2. High precision task
3. Repeatable task

The "Active endoscope" (SMA-ACM II) developed by Ikuta,Tsukamoto and Hirose [1] , "Hyper endoscope" [2] , "Micro active forceps "(MAF) by the author's group [3][4] and "Active catheters" been developed by several companies in Japan are utilizing advantage on micro and narrow space task of micro medical robots.

Cutting/drilling bone robot for hip replacement by Russell Taylor et al.,[5], the "Neuronavigator" for navigational task during brain surgery developed by Watanabe et al[6] and the "Automatic endoscope manipulation" (EASOP) by Y.Wang are taking advantage both merits of high precision and repeatability. It is very important to investigate medical robot so as to expand strong points.

On the other hand, following issues exist in medical robotics now.

1. Safety issue
2. Low task efficiency
3. Contamination and cleaning
4. Miniaturization technology
5. Man-machine interface (vision & force)

Although all of above requirements are extremely essential, they completely new for conventional robotics research. To solve each issue is not only indispensable for medicine but also new challenge for robotics. Furthermore, above issues are almost similar needs for future service robots working nearby human in the home. Based on above reason, the author has been conducting systematic researches for these ten years.

3. Remote Surgery Robotics

Before discussing remote surgery robotics which is the one of exciting subject in medical robotics, the meaning of " remote" should be deeply considered.

Author has pointed out that the word of "remote" in medicine has following two meanings.

1. Physical distance
2. Functional distance

Of course, the first one is "physical distance" where doctor makes therapy under far away from patient.

The second one is "functional distance" where access for therapy is difficult even if patient is sitting in front of the doctor. For example, a tumor in deep site of the brain is not physically far but functionally far from doctor, because it is very hard to approach it. It is much more needed for ordinary cases to shorten functional distance than physical one, except for risky therapy at war.

Based on above consideration, the research theme on remote medical robotics in author's laboratory are classified into following two categories.

Physical distance based research:

1. Remote rapaloscopic surgery system with force sensation

Functional distance based research:

1. Active endoscope
2. Hyper redundant active endoscope (Hyper Endoscope)
3. Variable stiffness endoscope
4. Micro active forceps for retinal micro surgery
5. Remote rapaloscopic surgery system with force sensation
6. Virtual endoscopic surgery system with force sensation for training

Detail of each research is described in later sections.

4. Hyper Endoscope driven by Cybernetic Actuator

4.1. Cybernetic Actuator

A tiny linear actuator so called "Cybernetic Actuator" to satisfy both safety and miniature requirement has been developed by the authors. The definition of cybernetic actuator is the actuator having four driving states such as free, lock, increasing and decreasing was proposed by Tomovic and McGee in 1966.[7] Unfortunately the prototype has been never developed so far.

The author is interested in cybernetic actuator from safety view point. The state of free is extremely important for safety feature while the medical robot is off drive. Because, the robot can be easily back driven by human if the actuator is completely free. Most of the actuators such as geared motor should be locked during no electric input. If human makes a accidental collision to medical robot, human would be injured in this case. The cybernetic actuator with free mode can avoid above mentioned tragedy.

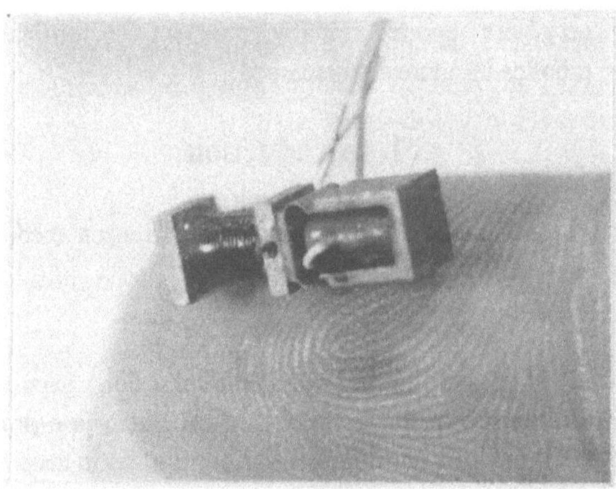

Fig.1 Miniature cybernetic actuator (**Cybernator**)

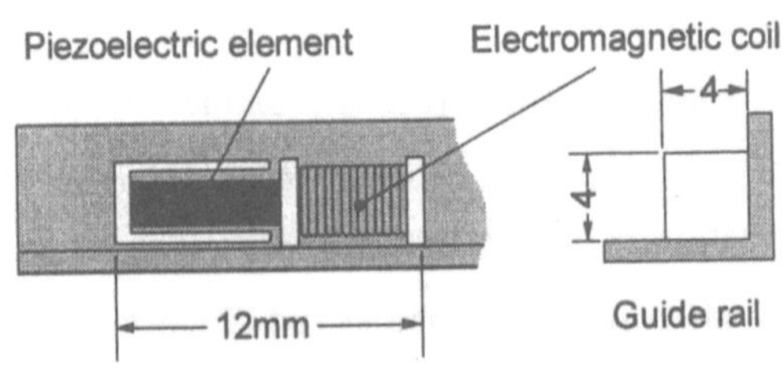

Fig.2 Basic design of cybernetic actuator

Fig.1 shows the prototype of miniature cybernetic actuator. The size is 12 x 4 x 4 [mm] and the weight is only 1[g]. The maximum producing force is 200 [gf]. This device consists of small piezoelectric device and electric magnetic coil as shown in Fig.2. Unlike conventional inchworm mechanism, the cybernetic actuator can run based on new driving principle. Since the piezoelectric device produces can fairly large impact force by applying special input voltage pattern, maximum speed is about 35 [mm/sec]. Though the impact drive principle was invented by Higuchi,[8] it has been used as slow positioning mechanism.

4.2. Hyperendoscope

Fig.3 shows the "active universal joint mechanism" driven by a couple of cybernetic actuators. Two degrees of freedom motion was obtained in spite of small diameter. (ϕ10 [mm]) Hyper redundant active endoscope named "Hyper Endoscope" was made by using active universal joints. Fig.4 is the first prototype.

The first goal is to develop Hyper Redundant Active Endoscope (Hyper Endoscope) with highly multi degrees of freedom composed by many actively bending segments. Ikuta et al have already developed the first "active endoscope (SMA-ACMII)" driven by miniaturized Shape Memory Alloy servo actuator in 1986.[1] This unique endoscope was focused on various kind of inspections in digestive canals such as intestine and colon.

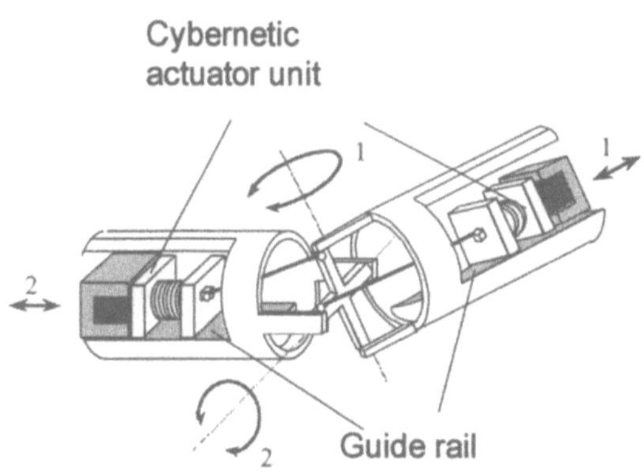

Fig.3 Active universal joint driven by a couple of cybernetic actuators

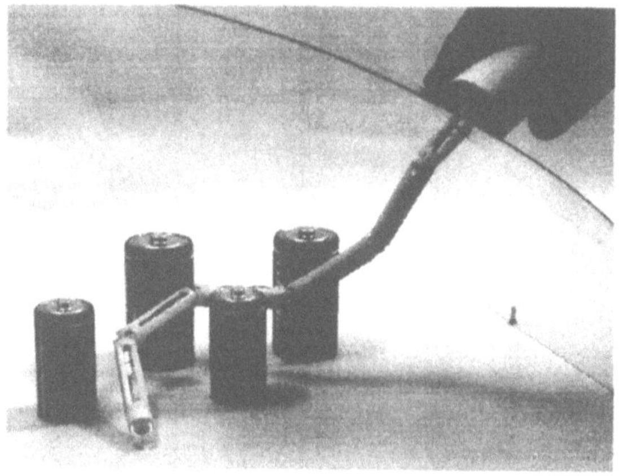

Fig.4 Hyper redundant active endoscope (Hyper endoscope) with ten degrees of freedom

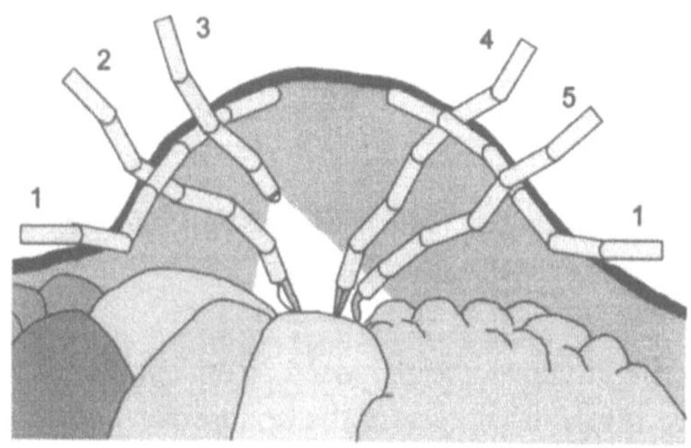

Fig.5 Basic concept of minimally invasive surgery using several types of Hyper endoscope
1. Expander 2. Bistoury 3.Active Endoscope
4. Forceps 5. Hook

On the other hand, the purpose of Hyper Endoscope is not only inspections but also remote surgery in the abdominal cavity. Therefore, new requirements for Hyper Endoscope are small diameter and three dimensional motion with "free" and "position lock" ability of whole segments. The state of "free" for the active joints is significant to pull out safely from the abdominal cavity.

Moreover, the state of "lock" to suspend active joints is necessary for the "Expander" as described before and precise maneuver.

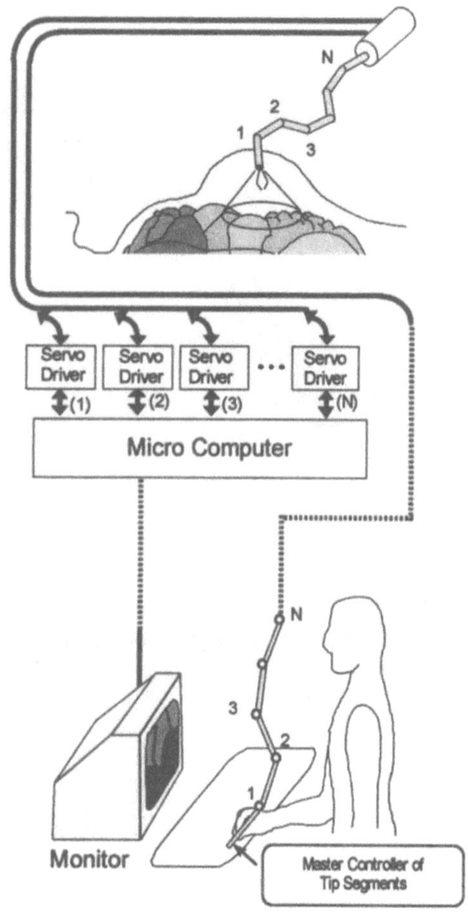

Fig.6 Total control system of hyper endoscopic minimally invasive surgery

Fig.5 shows the proposed concept for minimum invasive surgery in the abdominal cavity in future. There are five types of active endoscope working together or alone as follows,

1."Expander" keeps suspending the skin on the abdominal cavity
2."Bistoury" cuts the internal organs by the scalpel on its tip
3."Endoscope" provides light and view for the operator controlling from out side.
4."Forceps" can handle and pick up tissue.
5."Hook" can hook the organ

Fig.6 shows the total control system with force feedback. Bending motion of each active joint is controlled from outside. In order to realize above system, active endoscope useful for various purposes should be developed. Especially active joint in miniature size is one of most important parts.

5. Remote Minimally Invasive Surgery System by Servo Forceps

5.1. Basic Concept

Another type of remote surgery system is proposed here. Unlike hyper endoscopic minimally invasive surgery as mentioned before, this system doesn't need any special training for surgeon to maneuver , because real surgical tools such as forceps is the master controller. This new system aims to provide almost similar environment for surgeon to make a remote minimally invasive surgery.

Moreover, this system can be easily applied to virtual training system for young doctor as describe later. Surgeon have to learn new surgical skills in order to carry out endoscopic surgery, because he can get limited information for surgery in comparison with ordinal open surgery. The visual image of internal abdominal cavity form the endoscope and position/force information of forceps are available.

On the contrary, this limitation of information becomes advantage for building remote surgery system. It is enough to transmit small bits of information for mutual communication between surgeon and patient.

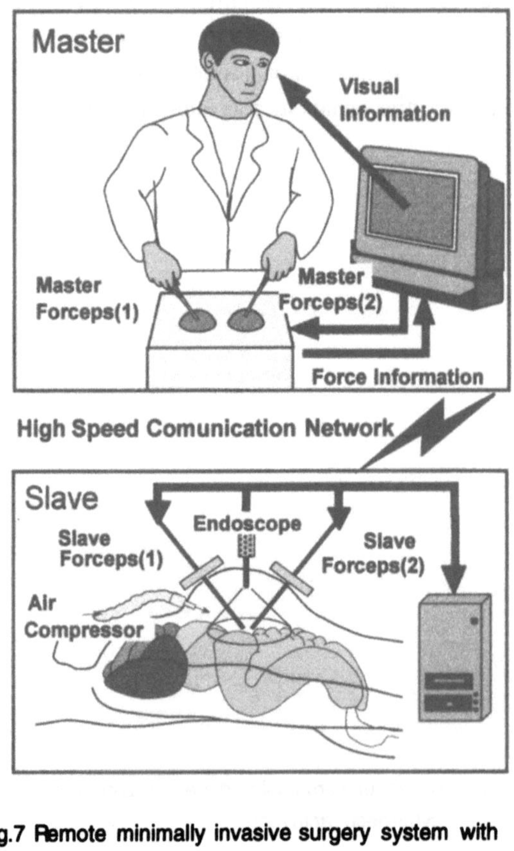

Fig.7 Remote minimally invasive surgery system with force sensation by using master/slave forceps

5.2. Servo Forceps

In order to present same maneuverability of forceps, special "servo forceps" useful for both master and slave system was designed as shown in Fig.8. The servo forceps consists of gimbals with linear/ rotational motion mechanisms to provide 4 D.O.F. pivot motion around the small hole opened on the abdominal membrane of the patient. Each motion has both position(angle) sensor and force sensor to achieve force sensation for all degrees of freedom.

Fig.9 shows developed prototype. Since the both of master and slave servo forceps uses same mechanism, bilateral control is easy to make. Pre-clinical test is under going new successfully now.

As I mentioned before, it necessary for surgeon to make a retraining to pursuit newly developed minimally invasive surgery. However, surgical training needs fairly long term, since surgeon must learn only from real surgery under supervisor surgeon.

5.3. Expand to Virtual Minimally Invasive Surgery

Our servo forceps can be easily applied to virtual minimally invasive system as shown in Fig.10. Instead of real slave hardware, virtual slave forceps and virtual organs should be implemented in the computer. We have already built up system and basic performance was verified successfully.

It should be noticed that dynamic model of various kind of organ in real time formula is the key issue to obtain realistic force sensation. Since conventionally investigated dynamic models in biomechanics are based on FEM (finite element method) or non-real time one, it is hard to utilize. Basic research to measure dynamic response under large deformation of soft organs is also been carrying out in our laboratory.

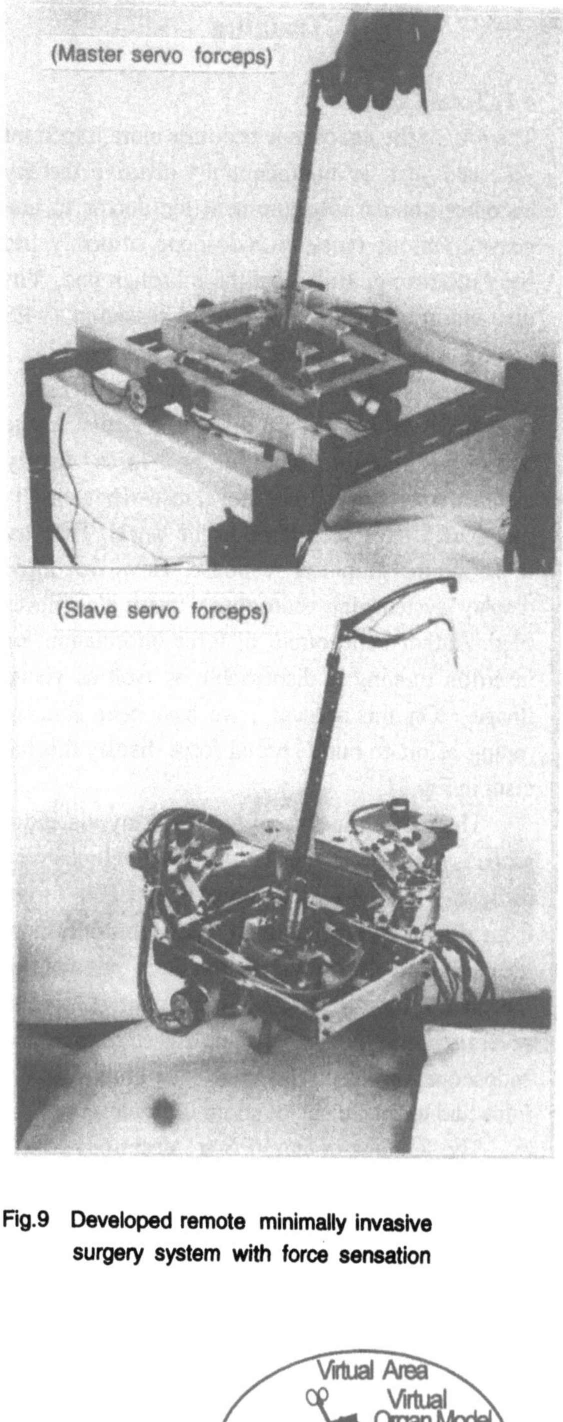

Fig.9 Developed remote minimally invasive surgery system with force sensation

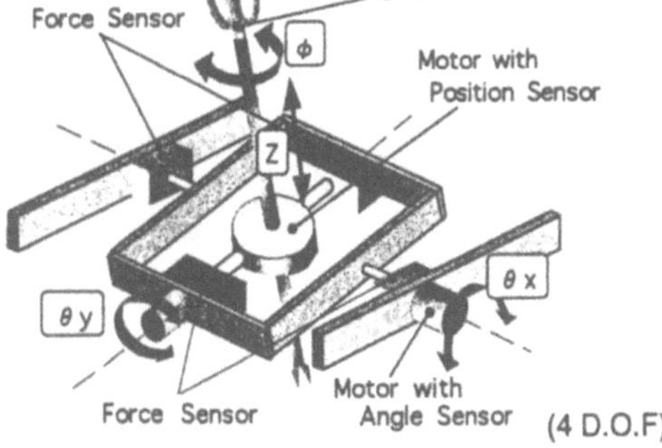

Fig.8 Servo mechanism of master/slave forceps

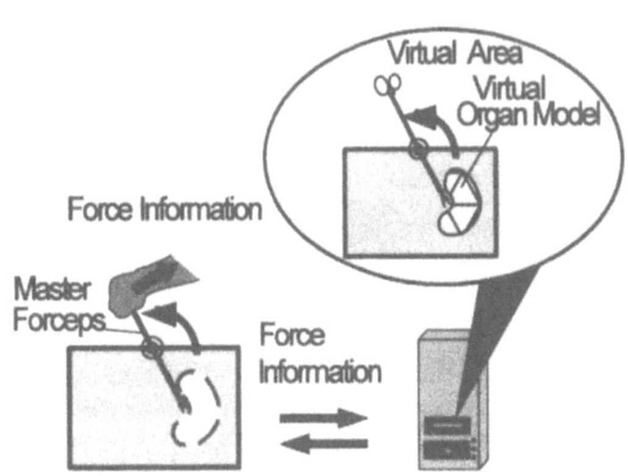

Fig.10 Basic concept of virtual minimally invasive surgery system with force sensation

6. Virtual Endoscope for Insertional Training

6.1. Total Concept

The role of the endoscope becomes more important year and year, as non/minimally invasive therapy becomes popular. Requirement for doctor to maneuver various types of endoscope smoothly has been increasing. Based on these background, Virtual endoscope system with force sensation (VES project) has been running in my laboratory.

Total system of our VES shown in Fig.11 consists of not only visual display of virtual image inside the digestive tube but also force display mechanism. Several high level researches on CG based VES have been made in the world. They try to show the graphical response. However, force display system for endoscopic training is never tried. Author believe role of force information for insertion training is dispensable as well as visual image. On this account , we have been concentrating effort to build special force display mechanism in Fig.11

The basic principle of force display via endoscope is described in Fig.12. Although it looks very difficult to display complicated reaction forces from digestive tube through the endoscope, new idea solves it. According to the basic principle of mechanics, both the frictional and contact force at several contact points between digestive tube and endoscope can be transferred into a total thrust force and torque at one point on the endoscope.

The features of our VES are described as follows:

1. Reaction force display ability
2. Real endoscope is available
3. Compatible for various diameter
4. Infinite thrust and rotational motion

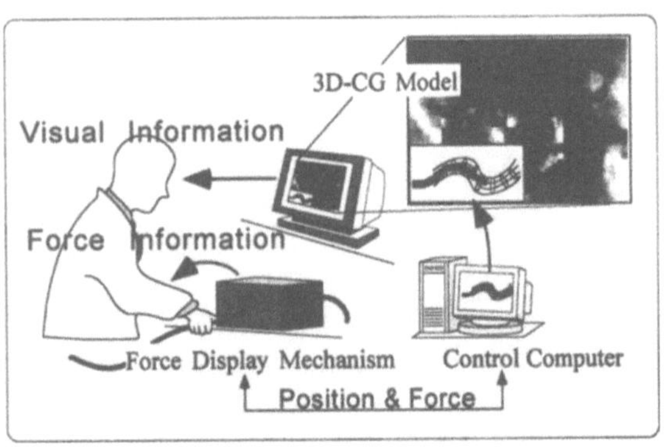

Fig.11 Virtual endoscope system with force sensation for insertion training (VES)

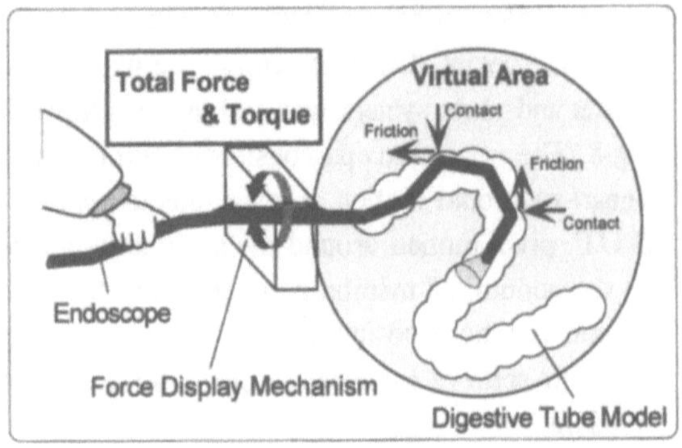

Fig.12 Principle of force display

6.2. Force Display Mechanism

The force display mechanism as shown in Fig.13 was developed to satisfy above features. Two pairs of rubber roller produce thrust and rotational motion at the same time. Although the internal forces between rollers and endoscope are coupling, pure thrust and rotational force can be made by controlling input to DC servo motors. Advantage of this mechanism is realizing infinite thrust/rotational drive under force feedback.

Another important feature is introducing differential gears which contribute not only reducing number of motor but also making easier to control by decoupling. The roller shafts are connected via differential gear so that complicated coupling of control is decoupled mechanically.

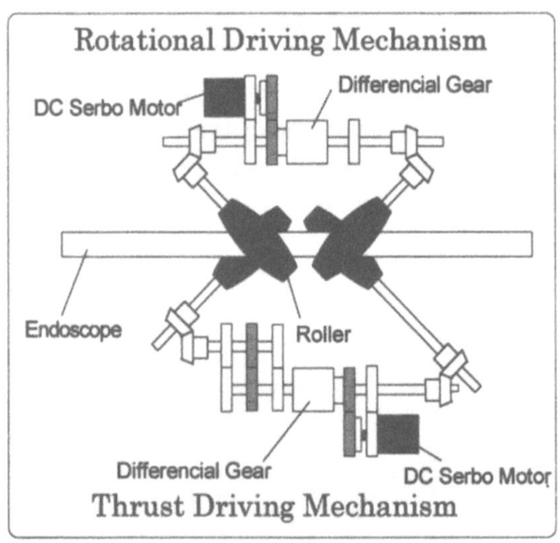

Fig.13 Force display mechanism with rollers and differential gears

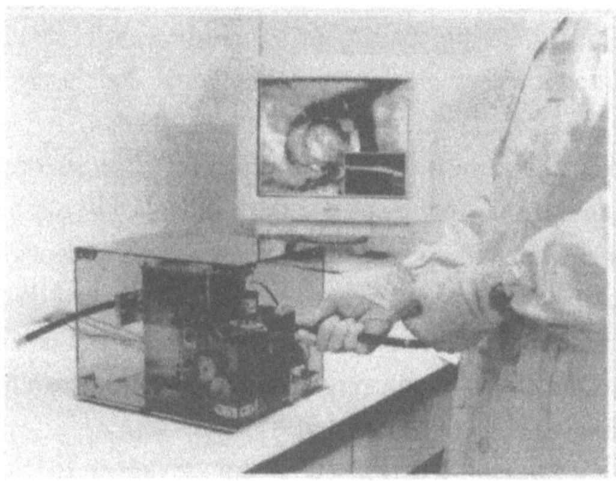

Fig.14 Prototype of force display mechanism

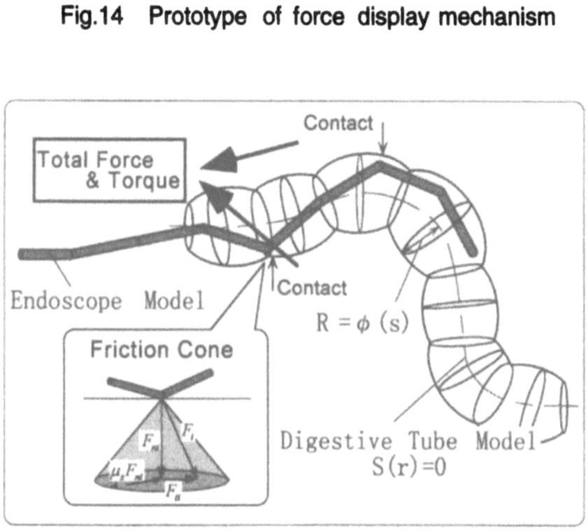

Fig.15 Interactional model of endoscope and digestivetube

The prototype of force display mechanism in Fig.14 works successfully as a first step. Fig.15 shows dynamic model of endoscope and digestive tube to calculate contact frictional force from digestive tube. The endoscope is modeled as a serially connected linkage via viscoelastic joints. The digestive tube is simplified to elastic tube with various diameter around curved center line. The frictional force is calculated under friction cone assumption. All of calculation should be done in real time manner to make force feedback. Next step of the project is combining CG system.

7. Micro Active Forceps with Optical Fiber Scope for Retinal Micro Surgery

Microsurgery is the most attractive but most difficult treatment at present. Many field in medicine has been developing special tools along with special skill of the surgeon. Now a days, the great demand from various kind of microsurgery for the micro machine technology has been claimed.

Especially demand from eye microsurgery is very easy to understand by the engineer. Eye microsurgery on the retina needs extremely difficult and special skill. The issues in this area have general aspect for all kind of microsurgery. Therefore, technology obtained for this subject is considered to be easily generalized to other medical field.

7.1. Problems of Today's Eye Microsurgery on the Retina

7.1.1. Present Method of Eye Microsurgery

Let's start from the brief introduction of eye microsurgery. Fig.16 shows today's microsurgery in the eyeball to operate retina. A skillful operator handles forceps under binocular microscope while an assistant controls light guide forceps to light up the small surgical area.

7.1.2 . Problems

This surgical method brings following two issues.

1) **Limited visual information**

Only top view image through the cornea is available. It is extremely difficult to see the depth and the distance between tip of forceps and retina. Sometime the retina has been seriously damaged by penetration of forceps by mistake.

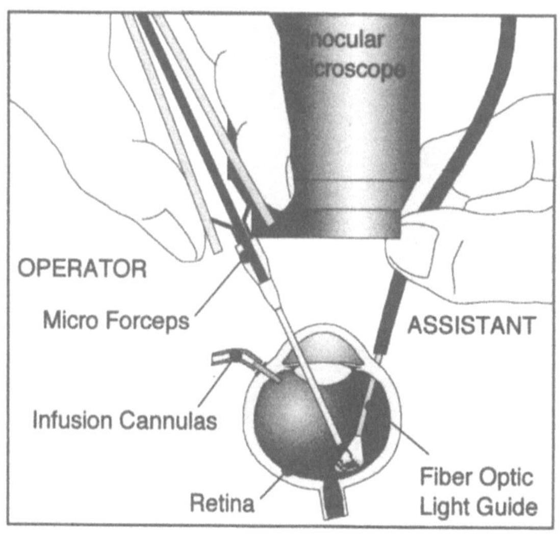

Fig.16 Today's retina micro surgery by using micro forceps

2) **Limited operational area**

Since the conventional forceps is straight bar type as shown in Fig.17, working space in the eyeball is very small and only bottom half of eyeball can be reached. This reason can be explained in Fig.18.

Eye surgeon can maneuver forceps via small hole on the upper half of the cornea and this hole acts as a pivot for forceps motion. Therefore micro surgery on the upper half from inside eyeball has been seriously demanded.

7.2. Development Concept and Tactics

Before describing technological articles, we would like to mention about the concept of research approach and tactics.

Our final goal is not only improving surgical tools but also establishing new microsurgery system applicable today's most difficult microsurgery at the bottom of eyeball. In this kind of research, the engineering researcher should be careful about "interface between new technology and users". Of course, the users are medical doctors.

Unfortunately there has been a lot of misunderstanding research results in the medical engineering field. Even though the technology itself is enough excellent, that would never be used if medical doctor does not like it. It is obvious that the main reason of above unhappy cases lay on the miscommunication between engineer and doctor.

Another big reason is conservatism in the medicine. Most of the surgeon want to keep their own skill obtained through long term training. It is extremely difficult to break it out. One solution, authors think, is new development concept based on " **drastic improvement under the continuity of operational skill** ".

Based on above considerations, we decided to start from improvement of tools such as micro forceps as a first step. And the second step is total system as shown in Fig. 19 and Fig.20 explained later.

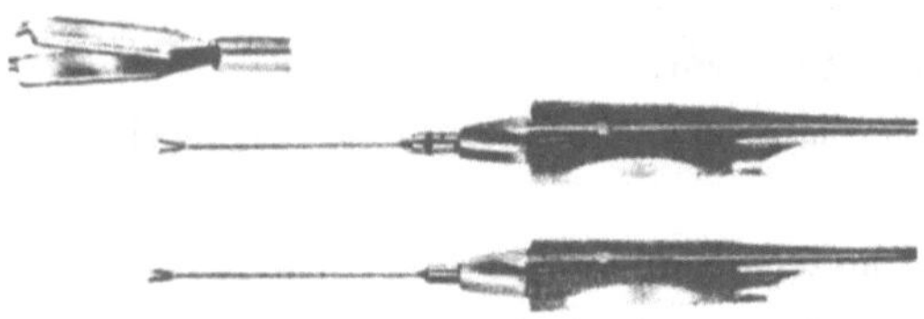

Fig.17 Conventional micro forceps

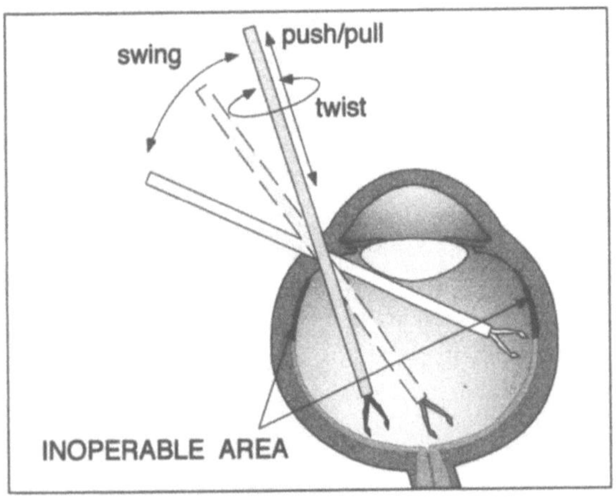

Fig.18 Limited operational area by conventional forceps

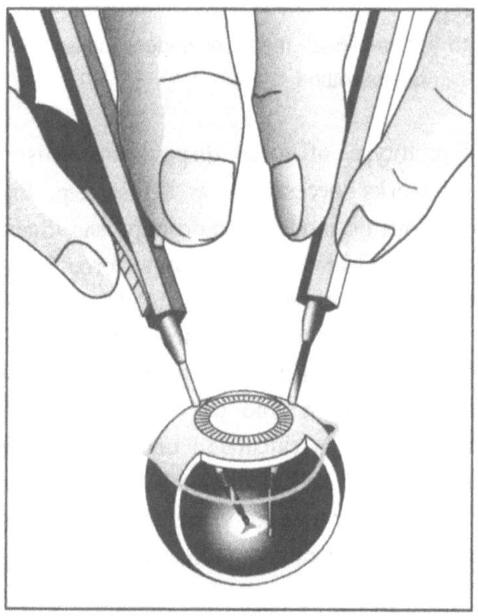

Fig.19 Proposed two hand operation using two actively bending micro forceps with built-in fiber scope

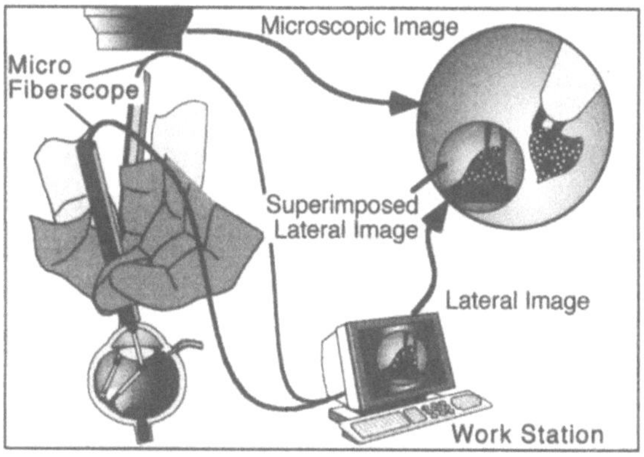

Fig.20 Total system of new retina micro surgery

7.3. Micro Active Forceps

7.3.1. Basic Design and Prototype

Fig.21 shows the overview of prototyped micro active forceps and Fig.22 shows the basic design.

An active bending joint and micro gripper can be operated by one hand using the micro cable running through inside of micro forceps. Bending angle of the active joint is controlled by a forefinger and the micro gropper is controlled by a thumb.

A thin optical fiber scope (diameter is 0.4mm with 2000 pixels) set at the tip of forceps provides us lateral image near the operation area on the retina.

Basic features of the micro active forceps are summarized as follows;

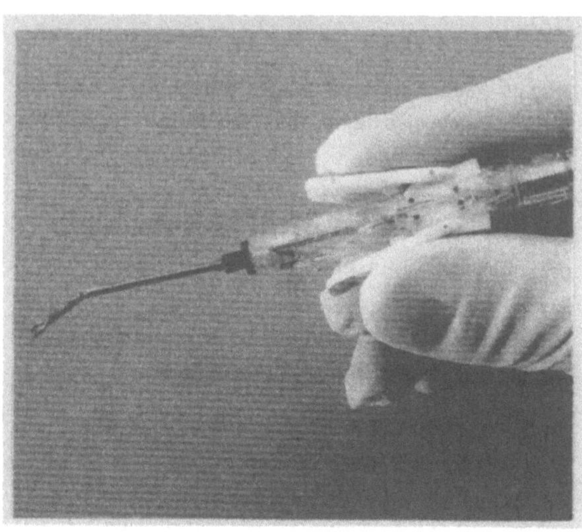

Fig.21 Developed micro active forceps (MAF-II)

1) **Active bending joint with lock mechanism** (Fig.21 and 22)

An active bending joint enables us to address upper part of eyeball from the inside. Bending angle of active joint is controlled and adjustable depending on operational area in the eyeball after first insertion. And then, bending angle is fixed by micro lock mechanism prior to real micro operation.

2) **Micro gripper**

A micro gripper on the tip of active forceps was made to grasp fine tissue on the retina. Both the open and close operations are manueved by finger of the surgeon via tiny cable system running through the stem.

3) **Built-in optical fiber scope with light guide**

In order to obtain lateral view during operation, one thin optical fiber scope was built in the stem. Since authors assume two hands operation by one surgeon as described later, it is possible to see the tip of micro forceps in another hand. Simply speaking, operational area by left hand forceps can be seen by the fiber scope of right hand.

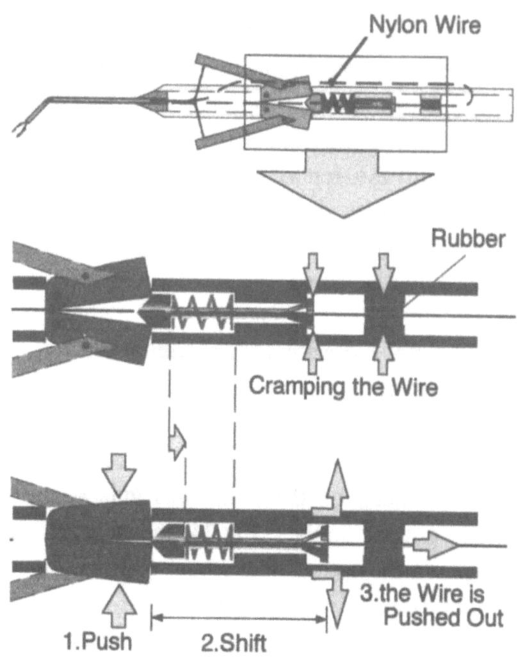

Fig.22 Internal mechanism of MAF-II

4) **One hand operation**

All of the task is possible to do by fingers in one hand without any assistance of another hand. Therefore, a surgeon can handles both of hands like a famous Japanese samurai "Musashi Miyamoto" who could handle both of arms dexterously and finally became the highest great swordsman.

5) **No assistant is needed**

Because an eye surgeon can make two arm operation using two micro active forceps, an assistant to maueve right guide during operation is not required any more. This merits contributes to decrease clinical cost due to man power also.

6) **Introducing micro mechatronic technology**

The IH Process based on micro stereo lithography to fabricate any shape of three dimensional miniature and micro mechanical parts was introduced. This unique process was developed by the authors in 1992 and it can shorten prototyping time. [9][10]

Since the IH process can fabricate any shape of micro structure quickly, design improvement and optimization are much easier and faster.

7.3.2. Total System

Fig.19,20 show the total micro surgery system. Eye surgeon can easily grasp and pick up fine tissue on the retina by using two micro active forceps. Surgeon can see the superimposed lateral micro scopic image provided from optical fiber scope along with the top view image by binocular microscope. Three dimensional position of the forceps's tip and distance between retina can be easily noticed by this system.

7.4. Animal Experiment for Evaluation

Preliminary experiment using pig's eyeball was conducted as shown in Fig.23. The micro active forceps was inserted through the cornea. The blood capillary on the retina was grasped by the micro gripper softly. Fig.24 shows real lateral image via the optical fiber scope. (Micro grippe is handling thin blood capillary)

One hand operation under no assistant was also verified successfully.

Feasibility and maneuverability were verified by eye micro surgeon. Further miniaturization and improvement are under going.

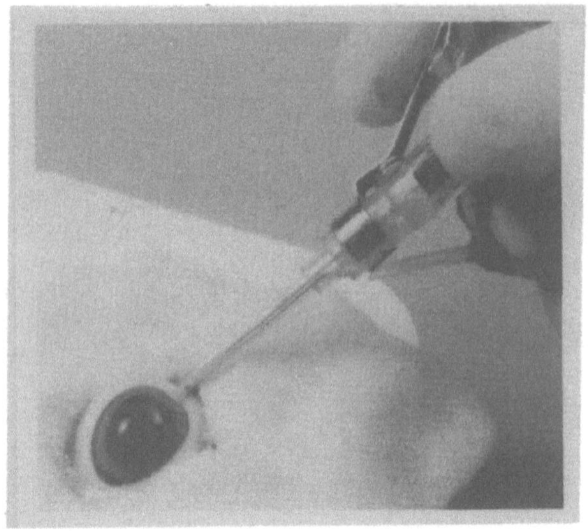

Fig.23 Experiment of retinal surgery by using pig's eye ball

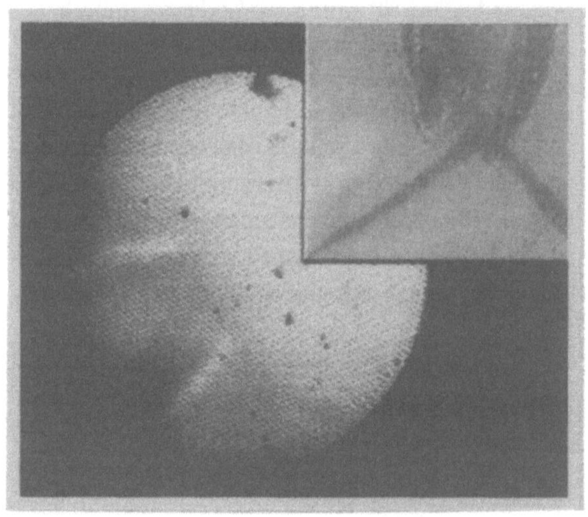

Fig.24 Super imposed images of retina surface from binocular (right) and built-in fiber scope (left)

8. Prospect for Medical Robotics in Future

Following subjects should be investigated in robotics and mechatronics in near future.

1. Micro endoscopic surgery with high dexterity
2. Tools for non invasive inspections
3. Tools for minimally invasive surgery
4. New mechatronics for medical and welfare
5. Robot assisted training (**RAT**)
6. Real-time model of human organs and tissue
7. Welfare robotics
8. Safety strategy for medical and welfare robotics

Not limited to above subjects, more contributions from mechanical engineering including robotics and mechatronics should be made. As well known, electronics contributes to medicine for these thirty years.

On the other hand, the contributions from the mechanical engineering has been not sufficient except biomechanical analysis in orthopedics.

Unlike medicine in thirty years ago, the non and minimally invasive therapy are demanded greatly in today. Beyond the stage of analysis, various types of robotic systems are expected to be utilized for both insepctional and surgical applications.

" The RAT is stronger than the CAT"

One example of advantage of robotics in medicine is the robot assisted training (RAT) for beginner surgeon. The CAT (computer assisted training)can use mainly CG technique. On the other hand, RAT can handle position and force indispensable for the surgical training. Frankly speaking, the robot has greater advantage than computer graphics approach. This fact encourages searchers in robotics.

Safety issue on both medical and welfare is extremely important. No systematic research has been done yet.. Software and hardware approach should be carried out.

Ikuta and Nokata have been engaged in safety problems by theoretical and practical approaches. As mentioned before in chapter 4, the cybernetic actuator was developed as a mechanical safety solution. Another trial with simplicity is the safety transmission mechanism by using " Non-contact magnetic gear" developed authors. [11]

Fig.25 shows the schematic diagram of non-contact magnetic gear. It is unique that the magnetic gear has no tooth and gap space between gears. Torque can transmit through the gap, because a lot of magnetic poles are made around the magnetic gear made of magnetic material such as Ferrite and rare earth metal. Silent torque and adjust free transmission suitable for welfare application were also verified.

Moreover, since the magnetic gear is torque limited feature, an excessive torque from outside can be released by racing and suitable for simple safety solution without any feedback control. Significance of the " soft mechatronics" constructed by hard mechanism will increase more and more.

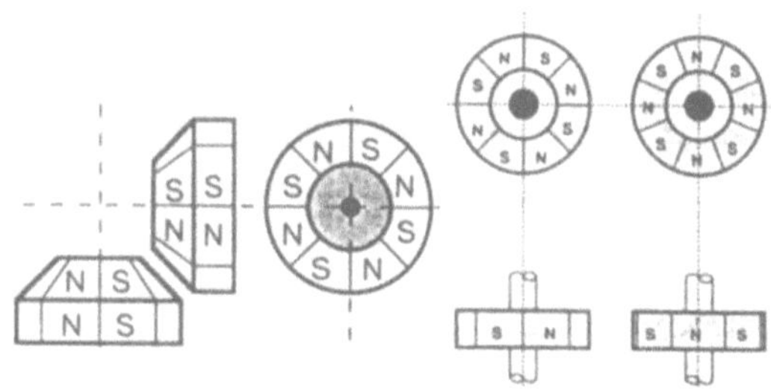

Fig.25 Schematic diagram of Non-contact magnetic gear

As a conclusion, the author stress that one of key concept for medical and welfare robotics is a design concept of " **Simple with neat solution** " . I hope a lot of young and able researchers enter this field.

Acknowledgment

Author would like to thank all of co-authors for each research introduced in this paper.

References

[1]K.Ikuta, M.Tsukamoto, S.Hirose, " Shape Memory Alloy Servo Actuator System with Electric Resistance Feedback and its Application to Active Endoscope", Proc. of IEEE International Conference on Robotics and Automation, pp.427-430, (1998)

[2]K.Ikuta, M.Nokata, S.Aritomi, "Biomedical Micro Robots Driven by Miniature Cybernetic Actuator", Proc.of IEEE International Workshop on Micro Electro Mechanical systems (MEMS-94), pp.263-268, (1994)

[3]K.Ikuta, T.Kato, S.Nagata, "Micro Active Forceps with Optical Fiber Scope for Intra-Ocular Microsurgery", Proc. of IEEE International Workshop on Micro Electro Mechanical Systems (MEMS'96), pp.456-461 (1996)

[4] K.Ikuta, T.Kato, S.Nagata, "Development of Micro Active Forceps with Built-in Fiber Scope for Intra-

Ocular Microsurgery", Proc. of Computer Assisted Radiology and Surgery (CAR'97), pp.914-919, Excerpta Medica International Congress Series 1134, ELSEVIER, (1997)

[5] L.Joskowicz, R. H. Taylor, B.K.Williamson, R.Kane, A.Kalvin, A. Gueziec, G. Taubin, J. Funda, S. Gomory, L. Brown, J. McCarthy, R. Turner," Computer Integrated Revision Total Hip Replacement Surgery : Preliminary report.", Proc. of International Symposiumu on Medical Robotics and Computer Assisted Surgery (MRCAS'95), pp.193-202, (1995)

[6] E.Watanabe,Y.Mayanagi, Y. Kosugi, S. Manaka, K. Takakura, " Open Surgery Assisted by the Neuronavigator, articulated, sensitive arm", Neurosurgery 28, pp.792-800, (1993)

[7]R.Tomovc, R.B.McGee, " A Finite State Approach to the Synthesis of Bioengineering Control Systems ", IEEE Trans. on Human Factor in Electronics, Vol.7, No.2. p65, (1966)

[8] T.Higuchi, et al, Proc. of IEEE International Workshop on Micro Robot and Teleoperation, (1987)

[9]K.Ikuta, K.Hirowatari, T.Ogata, ""Three Dimensional Micro Integrated Fluid System (MIFS) Fabricated by Stereo Lithography", Proc. of IEEE International Workshop on Micro Electro Mechanical Systems (MEMS'94), pp.1-6, (1994)

[10]K.Ikuta, Plenary Talk Paper "3D Micro Integrated Fluid System Toward Living LSI",Proc. of Fifth International Conference on Artificial Life (A-Life V),MIT Press, pp.17 - 24 , (1996)

[11] K.Ikuta, S.Makita, S.Arimoto, " Non-contanct Magnetic Gear for Micro Transmission Mechanism", Proc. of International Workshop on Micro Electromechanical Systems (MEMS'91), pp.125-130, (1991)

Surgery simulation with visual and haptic feedback

N. Ayache S. Cotin H. Delingette
Epidaure Project, INRIA
2004 Route des Lucioles, 06902 Sophia-Antipolis (France)
email: [ayache,cotin,delingette]@sophia.inria.fr

Abstract

This article reports the ideas presented in (Cotin, Delingette, and Ayache 1996; Cotin, Delingette, and Ayache 1997) for developing a real-time surgery simulation system. This system allows the interaction with volumetric deformable models of organs, and provides visual and haptic feedback in real-time. The geometry of organs is acquired from medical images. The physical properties are based on linear elasticity, and deformations are computed with finite elements. A pre-processing technique allows real-time computation of deformations and forces. The method has been extended to introduce a non linear behavior closer to the biomechanical behavior of soft tissues, while preserving real-time. We present the basic principles of the approach and results obtained with our experimental system (figure 1).

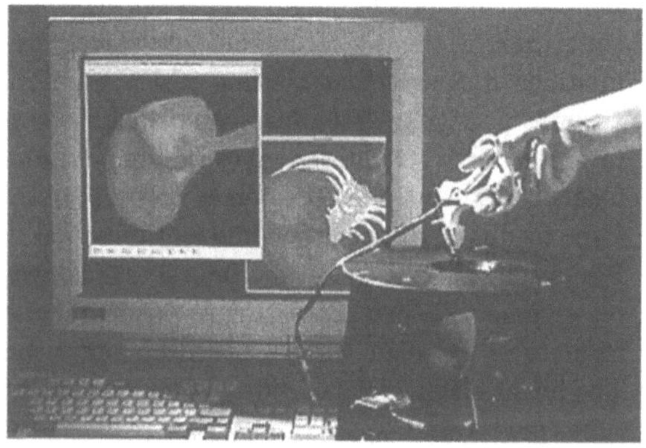

Figure 1: The surgeon manipulates the force feedback device. If a collision is detected with the surface of the virtual organ, the mesh is deformed in real-time and a non-linear reaction force is computed and sent back to the force feedback system.

1 Simulation in Surgery

With the development of laparoscopic techniques which reduce operating time and morbidity, surgical simulation appears to be an essential element of tomorrow's surgery. Indeed, most unexperienced surgeons and students have to practice this novel technique. Nowadays they practice on an endotrainer, on living animals or cadavers. The lack of realism in the first solution and the ethical problems linked to the other solutions show a real need for simulated surgery.

A lot of research has focused on spring-mass methods, because of their simplicity of implementation and their relatively low computational complexity (Baumann and Glauser 1996; Meseure and Chaillou 1997; Joukhadar and Laugier 1996). For example, (Kuehnapfel and Neisius 1993) present a simulation of endoscopic surgery based on a surface spring-mass model. Although in this case the interactions are driven by instruments with sensors, no force feedback is used. The simulation environment at MERL (Gibson, Samosky, Mor, Fyock, Grimson, Kanade, Kikinis, et al 1997) takes into account the volumetric nature of the organs with a deformation law derived from a spring-mass model. (Cover, Ezquerra, and O'Brien 1993) have developed a model based on thin plates for endoscopic gall bladder surgery simulation.

Finite element models are less widely used due to the difficulty of their implementation and the large computing time (Bainville, Chaffanjon, and Cinquin 1995). Nevertheless, (Sagar, Bullivant, Mallinson, Hunter, and Hunter 1994) proposed a method for simulating features of the human eye with a complex behavior (large incompressible 3D elastic deformations). Another example of eye surgery is given by (Le Tallec, P. and Rahier, C. and Kaiss, A. 1993), but still very far from real-time applications.

Reduction of computing time was studied by (Bro-Nielsen and Cotin 1996) using a condensation technique. With this method, the computation time required for the deformation of a volumetric model can be reduced to the computation time of a model only involving the surface nodes of the mesh. (Song and Reddy 1995) described a technique for cutting linear elastic objects defined as finite element models. This technique was only applied to very simple two dimensional objects. One can also cite a method for free-form cutting in tomographic volume data (Pflesser, Tiede, and Hoehne 1995) based on voxel operations for cutting and visualization.

In our approach, we tried to integrate all the requirements for a realistic simulation. The static equations of the elastic model are solved by a modified finite element method which takes into account particular boundary conditions. The solution of these equations gives not only the deformed mesh but also the forces to be sent to a force feedback device according to the actual deformation. Finally, real-time interaction is possible thanks to a pre-processing of elementary deformations coupled with a speed-up algorithm. The linear elastic deformations, computed in real-time, give a first approximation of reality as reported in (Cotin, Delingette, Clément, Tassetti, Marescaux, and Ayache 1996; Cotin, Delingette, and Ayache 1996). This linear model was then improved to introduce non-linear biomechanical properties of soft tissues as reported in (Cotin, Delingette, and Ayache 1997).

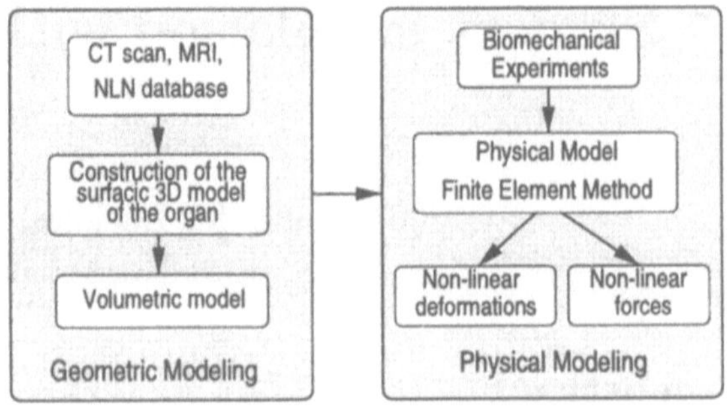

Figure 2: Creation of the deformable model. A geometric model is extracted from a 3D medical image. Then a deformation law is added.

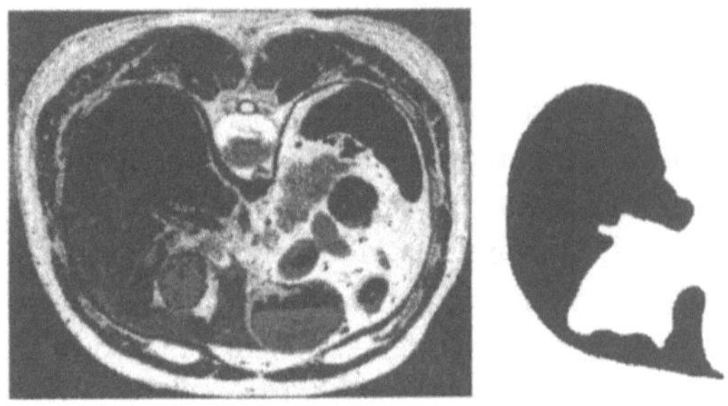

Figure 3: Segmentation of the liver, slice by slice. The initial data (left) is high resolution photography of an anatomical slice of the abdomen. The binary image (right) corresponds to the segmented liver cross-section.

2 Geometric Modeling

The choice of a specific geometric representation is of prime importance for simulators. This choice should be governed by the resulting trade-off between realism and fast interaction. In our method, the realistic aspect of the organs is linked, to an accurate segmentation of a volumetric medical image and also to an appropriate representation of the surface of the segmented organ.

We decided to start modelling the *liver*,a deformable organ requiring complex surgical procedures, in collaboration with the medical team of Pr. Marescaux at IRCAD (cf. acknowledgments).

In order to produce a model of the liver with anatomical details, a dataset provided by the *National Library of Medicine* was used. This process is illutrated by figures 2, 3, and 4.

This approach is used for building a *generic* liver model. It is also possible to build *specific* models corresponding to the geometry of a given patient. For this purpose, one can use the CT images of this patient, as shown in (Montagnat and Delingette 1997).

3 Physical modeling

Once the geometry of the organ is acquired, it is necessary to model its physical properties. Our physical model is initially based on elasticity theory.

3.1 Force feedback

Recent work (Barfield and Hendrix 1995) has shown that, in virtual environments, the sense of presence is highly correlated with the degree of immersion in that environment. In particular, the sensation of forces (haptic feedback) is very important in medical applications.

The information flow in the simulator should form a closed loop (Jackson and Rosenberg 1995):

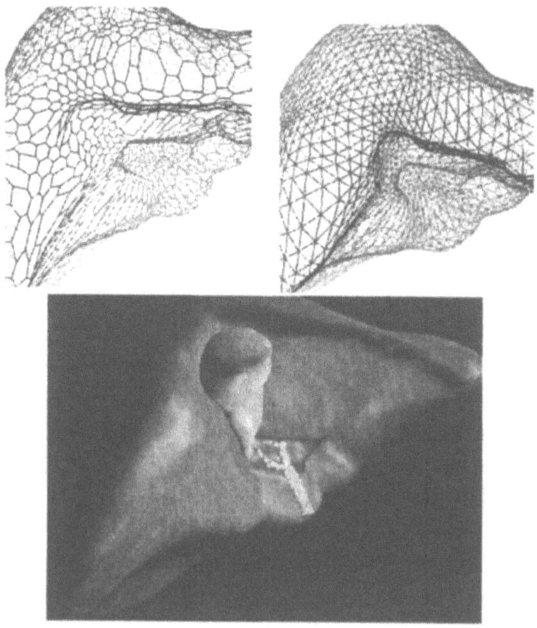

Figure 4: Different representations of the geometric liver model. The simplex mesh fitting the data (top-left) with a concentration of vertices in areas of high curvature, the triangulated dual surface (top-right) and a texture-mapped model with anatomical details (gall bladder and ducts) from an endoscopic viewpoint (bottom).

the model deforms according to the motion induced by the surgeon with the force feedback device. This deformation allows us to compute the contact force and finally the loop is closed by generating this force through mechanical actuators.

When adding force feedback in a simulation, the main difficulty is related to the real-time constraint imposed by such a system: to remain realistic, the forces must be computed at a very high frequency, at least equal to 300 Hz. Deformations and graphics rendering need to be performed at about 24 Hz. Finite element methods or spring-mass models are computationally expensive. To guarantee interactive rates with large meshes, we use a speed-up algorithm presented in section 3.4.

3.2 Biomechanical behavior of soft tissues

It is hard to quantify the realism of a deformable model since very little information is available regarding the deformability of human tissue. Recent publications in the field of surgical simulators make use of thin-plate deformable surfaces, spring-mass models, or volumetric linear elasticity. With such models, the real-time constraint is hardly reached to the detriment of realism. Conversely, research in the field of biomechanics has shown that the fairly realistic model for soft tissues is probably a visco-elastic non-linear model (Fung 1993; Chinsei and Miller 1997; Sagar, Bullivant, Mallinson, Hunter, and Hunter 1994). The major disadvantage of such a model remains its high computational complexity.

3.3 3D linear elasticity

In a previous work (Cotin, Delingette, and Ayache 1996), we showed the interest of linear elasticity as a reasonable approximation for soft tissue deformation. Here we recall the statement of the elasticity problem. Let Ω be the configuration of an elastic body before deformation. Under the action of a field of volumetric forces f_Ω and surface forces f_Γ, the elastic body is deformed and takes a new configuration Ω^*. The problem is then to determine the displacement field u which associates with the position $\boldsymbol{p_0}$ of any particle of the body before deformation, its position $\boldsymbol{p}$ in the final configuration. In order to solve this problem numerically, we use a classical finite elements approach, i.e., with Lagrange elements of type P_1. Using this class of elements implies the decomposition of the domain Ω into a set of tetrahedral elements. Through variational principles, the elasticity theory problem's solution becomes equivalent to the solution of a linear system:

$$[\boldsymbol{K}]\,[\boldsymbol{X}] = [\boldsymbol{F}] \tag{1}$$

where $[\boldsymbol{K}]$ is the stiffness matrix and is symmetrical, positive definite, and sparse. $[\boldsymbol{X}]$ is the unknown displacement field and $[\boldsymbol{F}]$ the external forces.

In a general approach, a set of external forces are applied to the surface of the solid while some mesh nodes are fixed. When using a force feedback device for the interactions with the deformable model it is impossible to measure the forces exerted by the operator. The position of the tip of the surgical tool is the only information transmitted by the device. Consequently, the deformation must be driven by specific constraints *in displacement* and *not in force*. Hence, our boundary conditions are mainly the contact points displacements between the surgical tool and the body. We can then deduce both the forces exerted on the end effector of the tool and the global deformation by applying the linear elasticity equations. The computed forces are finally set into the force feedback system in order to provide a mechanical resistance to the surgeon's hands: the loop is closed.

3.4 Reduction of computing time

The degree of realism needed in surgical simulation requires a complex model of the organ. The number of mesh vertices has a direct impact on the size of the matrices involved in the linear system $[\boldsymbol{K}][\boldsymbol{X}] = [\boldsymbol{F}]$. The computation time required for solving the system is too high for real-time deformation of the mesh. In order to insure real-time computation, we take advantage of the following properties: linearity and superposition principle. We introduced a pre-computation of elementary deformations which are stored beforehand and can be accessed during the actual use of the simulator to insure real-time interaction with the deformable organ.

The pre-processing stage can take between a few minutes and several hours depending on the size of the model and on the desired precision. For example, the pre-processing time required for a mesh with 193 vertices and 725 tetrahedra takes 30 mn. When the mesh size increases to 1735 vertices and 8124 tetrahedra, the computation time reaches about 36 hours. Finally, these results are stored in a file (typically a few megabytes). Since they depend on the fixed nodes and elasticity parameters, it is possible to generate a different set of files that correspond to various elastic behaviors, with specified boundary values for a given geometry.

3.5 Non-linear elasticity

We have also extented this approach to non-linear elasticity behavior, based on the properties of human soft tissues. Due to the lack of space, we refer the interested reader to (Cotin, Delingette, and Ayache 1997). For a bibliography on soft tissue modeling in medical simulation, one can consult (Delingette 1998).

4 Experiments & validation

A set of modules including the geometrical and physical models, a collision detection algorithm, and the deformation and force feedback algorithm has been integrated in our prototype of laparoscopic surgery simulator (figure 1). These modules run on a distributed architecture, based on a PC and a Dec Alpha station (figure 5). The force feedback device (*Laparoscopic Impulse Engine*[1]) is connected to a PC (Pentium 166 Mhz). The collison detection is performed on the PC as well as the force evaluation since it is possible, thanks to our pre-processing algorithm, to split the computation of forces from the computation of the deformation.

The deformation is computed on the Alpha station (233 Mhz with 3D graphics hardware) as well as the display of the different parts of the scene. Typical computing times corresponding to the computation of the deformations on a model of the liver with 1400 nodes and 6500 tetrahedra are reported in table 1. The size of the matrix $[\boldsymbol{K}]$ is 4200×4200. The pre-computing step takes approximatively 24 hours.

	normal	*accelerated*
linear elasticity	50000 msec.	1 msec.
non-linear elast.	> 100000 msec	2 msec.

Table 1: CPU times to deform a model of the liver.

The communication between the two workstations is performed via an ethernet connection. The data transmitted between the two computers is limited to the five degrees of freedom of the surgical tool plus some information issued from the collision detection algorithm. Consequently, we have a very high frequency (> 300 Hz) in the simulation loop with a very little latency between force feedback and visual feedback.

Two typical sequences are shown in figures 6 and 7.

5 Conclusion & Perspectives

The simulator has been tested by surgeons, specialized in laparoscopic surgery. It appears that the generated sensations are very close to reality, probably due to the addition of haptic feedback. Of course, the evaluation of the results is essentially qualitative, but our goal is not to compute an exact deformation of an organ. Actually, this objective seems very difficult to achieve given the actual knowledge in the field of soft tissues and the complexity of the interactions between the organs of the abdomen.

[1] The Laparoscopic Impulse Engine is a product from Immersion Coorporation.

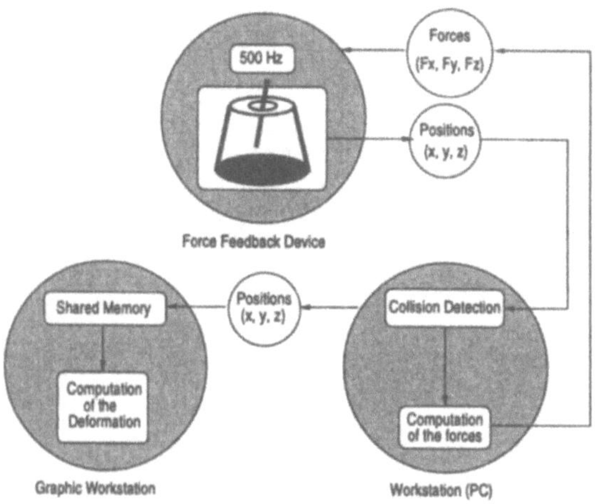

Figure 5: Description of the simulator architecture.

Nevertheless, several improvements can be made in the simulation process. One is adding a visco-elastic behavior (since the speed of the instrument has an importance in the deformation) and taking large deformations into account. Another important improvement is based on the possibility to simulate tissue cutting. Since the pre-computations allowing real-time interactions are dependent on the geometry of the mesh, it seems impossible to use only finite element models in the simulator. Consequently, we are now working on a dynamic tissue cutting simulation technique that could be merged with our actual model.

Acknowledgements

This project was partially funded by a contract with IRCAD (Institut de Recherche sur les Cancers de l'Appareil Digestif, Strasbourg, France) with the collaboration of professor J. Marescaux, director, and the help of J.M.Clément and V. Tassetti. The authors are grateful for their collaboration in this project and their valuable advices. We also acknowledge Morten Bro-Nielsen for stimulating discussions.

References

Bainville, E., P. Chaffanjon, and P. Cinquin (1995, May). Computer generated visual assistance during retroperitoneoscopy. *Computers in biology and medecine 2*(25), 165–171.

Barfield, W. and C. Hendrix (1995). *Interactive Technology and the New Paradigm for Healthcare*, Chapter 4: Factors Affecting Presence and Performance in Virtual Environments, pp. 21–28. IOS Press and Ohmsha.

Baumann, R. and D. Glauser (1996, January). Force Feedback for Virtual Reality based Minimally Invasive Surgery Simulator. In *Medecine Meets Virtual Reality*, Volume 4, San Diego, CA.

Bro-Nielsen, M. and S. Cotin (1996). Real-time Volumetric Deformable Models for Surgery Simulation using Finite Elements and Condensation. In *Proceedings of Eurographics'96 - Computer Graphics Forum*, Volume 15, pp. 57–66.

Chinsei, K. and K. Miller (1997). Compression of Swine Brain Tissue - Experiment *In Vitro*. *Journal of Mechanical Engineering Laboratory*, 106–115.

Cotin, S., H. Delingette, and N. Ayache (1996). Real Time Volumetric Deformable Models for Surgery Simulation. In K. Hohne and R. Kikinis (Eds.), *Vizualisation in Biomedical Computing*, Volume 1131 of *Lecture Notes in Computer Science*, pp. 535–540. Springer.

Cotin, S., H. Delingette, and N. Ayache (1997). Real-time non-linear elastic deformations of soft tissues for surgery simulation. IEEE Transactions on Visualization and Computer Graphics, accepted for publication. (see also Ph.D Thesis of S. Cotin, Nov. 1997, available through INRIA (in French))

Cotin, S., H. Delingette, J.-M. Clément, V. Tassetti, J. Marescaux, and N. Ayache (1996, January). Geometric and Physical Representations for a Simulator of Hepatic Surgery. In *Medecine Meets Virtual Reality IV*, Interactive Technology and the New Paradigm for Healthcare, pp. 139–151. IOS Press.

Cover, S. A., N. F. Ezquerra, and J. F. O'Brien (1993). Interactively Deformable Models for Surgery Simulation. *IEEE Computer Graphics and Applications*, 68–75.

Delingette, H. (1998, March). Towards realistic soft tissue modeling in medical simulation. *Proceedings of the IEEE*. in press.

Fung, Y. C. (1993). *Biomechanics - Mechanical Properties of Living Tissues* (Second ed.). Springer-Verlag.

Gibson, S., J. Samosky, A. Mor, C. Fyock, E. Grimson, T. Kanade, R. Kikinis, H. Lauer, and N. McKenzie (1997, March). Simulating arthroscopic knee surgery using volumetric object representations, real-time volume rendering and haptic feedback . In J. Troccaz, E. Grimson, and R. Mosges (Eds.), *Proceedings of the First Joint Conference CVRMed-MRCAS'97*, Volume 1205 of *Lecture Notes in Computer Science*, pp. 369–378.

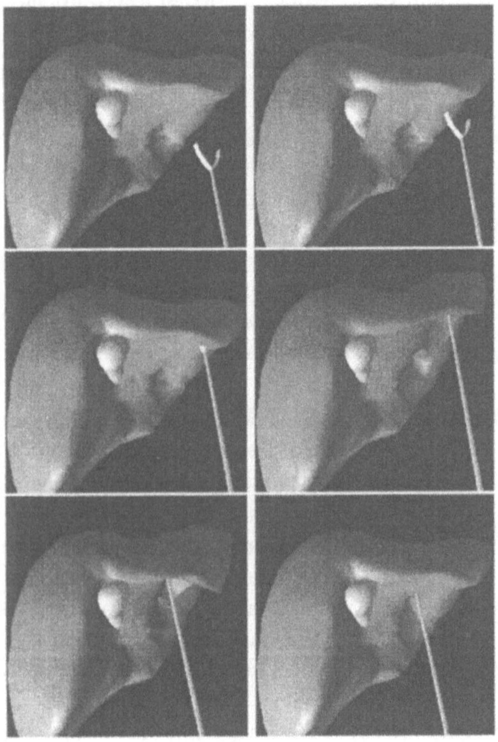

Figure 6: A first sequence of images issued from a simulation. The liver model contains about 1000 nodes and the gall bladder (in white) contains 200 vertices.

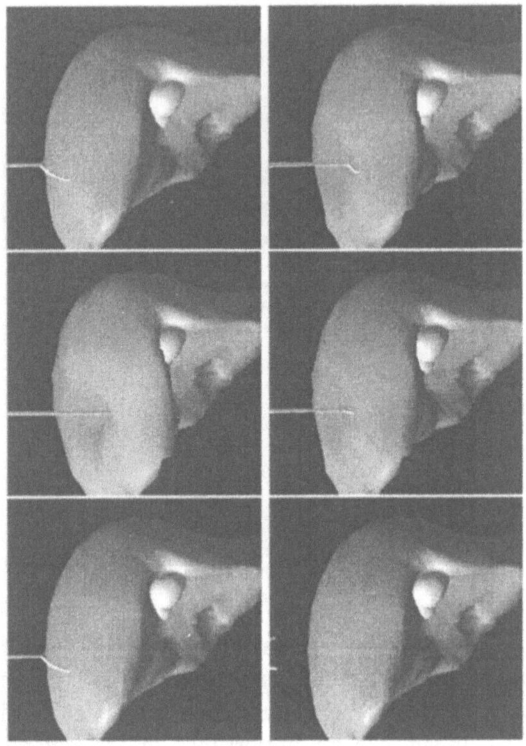

Figure 7: Second sequence of images. The collision detection and force computation are performed at about 300 Hz.

Jackson, B. and L. Rosenberg (1995). *Interactive Technology and the New Paradigm for Healthcare*, Chapter 24: Force Feedback and Medical Simulation, pp. 147–151. IOS Press.

Joukhadar, A. and C. Laugier (1996, November). Adaptive time step for fast converging dynamic simulation system. In *Proc. of the IEEE-RSJ Int. Conf. on Intelligent Robots and Systems*, Volume 2, Osaka (JP), pp. 418–424.

Kuehnapfel, U. and B. Neisius (1993). CAD-Based Graphical Computer Simulation in Endoscopic Surgery. *End. Surg. 1*, 181–184.

Le Tallec, P. and Rahier, C. and Kaiss, A. (1993). Three Dimensional Incompressible Viscoelasticity in Large Strains. *Computer Methods in Apllied Mechanics and Engineering 109*, 233–258.

Meseure, P. and C. Chaillou (1997, February). Deformable Body Simulation with Adaptative Subdivision and Cuttings. In *Proceedings of the WSCG'97*, pp. 361–370.

Montagnat, J. and H. Delingette (1997, March). Volumetric Medical Images Segmentation using Shape Constrained Deformable Models. In J. Troccaz, E. Grimson, and R. Mosges (Eds.), *Proceedings of the First Joint Conference CVRMed-MRCAS'97*, Volume 1205 of *Lecture Notes in Computer Science*. Springer.

Pflesser, B., U. Tiede, and Hoehne (1995, April). Towards realistic visualization for surgery rehearsal. In *Computer Vision, Virtual Reality and Robotics in Medecine*, Volume 905 of *Lecture Notes in Computer Science*, pp. 487–491. Springer.

Sagar, M. A., D. Bullivant, G. Mallinson, P. Hunter, and I. Hunter (1994). A Virtual Environment and Model of the Eye for Surgical Simulation. In *Computer Graphics*, Annual Conference Series, pp. 205–212.

Song, G. J. and N. P. Reddy (1995). Tissue Cutting In Virtual Environment. In *Medecine Meets Virtual Reality IV*, pp. 359–364. IOS Press.

Synergistic mechanical devices: a new generation of medical robots

J. Troccaz
TIMC/IMAG Laboratory, Faculté de Médecine de Grenoble
Domaine de la Merci
38706 La Tronche cedex - France
email: Jocelyne.Troccaz@imag.fr
M. Peshkin
Mechanical Engineering Dept., Northwestern University,
Evanston IL 60208 - USA
B. Davies
Mechatronics in Medicine Lab.,
Imperial College of London,
London SW7 2BX - UK

Abstract

There are many roles for electromechanical devices in image guided surgery. One is to help a surgeon accurately follow a preoperative plan. Devices for this purpose may be localizers, robots[1]*, or recently, synergistic systems in which surgeon and mechanism physically share control of the surgical tool. This paper discusses available technologies, and some emerging technologies, for guiding a surgical tool. Characteristics of each technology are discussed, and related to the needs which arise in surgical procedures. Three different approaches to synergistic systems, under study by the authors (PADyC, ACROBOT, and Cobots), are highlighted.*

1 Introduction

An electromechanical device of some sort is needed in image-guided surgery, in order to connect the "information world" of images, plans, and computers, to the physical world of surgeons, patients, and tools. That is the situation in which a surgical plan has been created based on diagnostic images, and it is the job of the surgical system to *guide* the surgeon in the accurate execution of his own preoperative plan. The surgeon is again in direct contact with the surgical tool, but an interface device must *also* be connected to that tool, so that the computer may in some way provide guidance. Thought of as human interfaces, the perceptual quality of such a device is often the most prominent factor in the performance of surgical systems. We appreciate a quality that is sometimes called transparency – the quality of being perceptually absent. One purpose of this paper to describe several classes of interface devices, with examples. Previous descriptions of such devices relied on a decomposition in passive, active and semi-active systems (1) in which the degree of passivity was often associated with a type of technology. We prefer to define a new classification based on function rather than mechanism including localizers, robots, and also a new class which we call *synergistic devices.* Synergistic devices are intended for direct physical guidance of a surgical tool which is also held and controlled directly by a surgeon. Each of the authors is pursuing a different approach to synergistic devices, and these approaches are outlined. The paper concludes with a discussion of the applicability of the technologies to various surgical purposes.

2 Classes of interface devices

2.1 Localizers

Localizers are devices that measure the coordinates of a tool or a pointer, but do not directly

[1] By *robot* we mean a mechanism with some level of autonomy, programmability and adaptivity

control that location. The location is controlled by the surgeon, by physically moving the tool or pointer, and is unconstrained by the localizer. Examples of localizers are passive arms with joint angle sensors, such as the Faro arm (2). Optical tracking systems perform a similar function, simply collecting coordinates. Localizers have the advantage that achieving transparent behavior is easier than for devices with actuators. In other words they cooperate easily with a surgeon, interfering little with his intended motion. Lacking actuators, however, they cannot offer guidance to the surgeon by providing physical constraint. Instead, the surgeon must explicitly observe and obey some other less immediate mode of guidance, usually a video display of some sort. An interesting variation is the addition of brakes to the joints of an otherwise unpowered localizer arm. In this way, if a surgeon can be guided visually to position the arm in a desired location according to a preoperative plan, the device can "lock" in that position and can subsequently be used as a physical guide. However the intrinsic physical mode of the localizer is passive, allowing the surgeon full mobility.

2.2 Robots

Most robots are fully actuated, having a motor driving each joint. Thus the position of the robot's end-effector is predominantly determined by how it runs its motors, and it intrinsically has little patience for physical "cooperation" with a surgeon. For some applications no cooperation is required; the robot works autonomously. An example is the Adler/Latombe radiosurgery system, in which a heavy payload is moved about a large workspace, both of which exceed human scale. No direct physical input from the surgeon is possible, or needed (see §3.2). In some circumstances, the robot needs some help from the operator, for instance for registration. In this case the "cooperation" problem can be addressed by adding a force sensor to the robot end-effector. The control computer is then aware of forces reflecting a surgeon's intended motion. It may direct the robot motors to comply with that intent. In practice it has so far been very difficult to acheive perceptually smooth cooperative motion in this way, but even primitive "force following" by the robot is useful (see §3.2). Another approach, however, leaves some of the joints of the robot unactuated but still equipped with sensors. Motion of these joints is naturally free and smooth. Since the decision to leave a joint unactuated is permanent, clever kinematic design of the robot is required so that the resulting "free" motions remain the appropriate ones even as the robot's configuration changes. An example of these mixed acturated/unactuated mechanisms is CMI's Aesop, which holds a laparoscope inserted through a trocar into the body (cf. §3.2). The intrinsic physical mode of actuated robots is active controlling position, and cooperating physically is not their natural mode.

2.3 Synergistic devices

Part of our purpose in this paper is to introduce the notion of synergetic devices, in contrast to localizers and autonomous robots. Synergistic devices are intended for cooperative physical interaction with a surgeon. Both the surgeon and the synergistic device hold the tool, apply forces to it and to each other, and impart motions. Under computer control, the synergistic device may allow the surgeon to have control of motion within a particular plane, while the device dictates motion perpendicular to that plane, for instance. As an example, suppose the surgeon and the synergistic device cooperatively hold a bone saw. The surgeon may maneuver the saw at will within the defined plane, cutting at any desired speed from any angle of approach, and avoiding anatomic structures that must not be damaged. At the same time, the synergistic device confines the blade of the saw to a defined plane based on a preoperative plan, so that the eventual resected surface is flat and corresponds to the plan. Arbitrarily shaped surfaces, with greater or fewer than two dimensions, can be defined based on preoperative plans, and enforced by the synergistic device. The surgeon is free to control the remaining degrees of freedom. Synergistic cooperation has the benefit that the robot can provide accurate, precise geometric motions whilst the surgeon holding the tool can feel the forces applied and modify them appropriately. It also has the psychological benefit that the surgeon is in direct control of the procedure. Several of us have realized the value of a synergistic control of motion. Several distinct approaches to achieving that goal will be described in §3.3.

3 Technologies

3.1 Localizers

Mechanical localizers were introduced for endonasal surgery in the early eighties (3; 4) and for neurosurgery (5; 6). They generally consist of man-powered mechanisms that have several degrees of freedom and encoders on each joint. The position and orientation of a tool attached to the end-effector of the mechanism is computed in real-time from the geometric model and the instantaneous values of the encoders. In small workspaces that are typical in surgery, an accuracy ranging from $0.1mm$ to $1mm$ can be achieved. However, constraints or large forces applied to the mechanism can deteriorate these values significantly. As compared to non mechanical navigators, these systems have the disadvantage of being cumbersome in the operative field. A major limitation is also that they can track only one object. However, they always give information, without any possibility of being obstructed, as can occur with the non-contact localizers. Another advantage is that they can be fixed in a definite position to hold an instrument (however, in some systems, the application of brakes can cause a small motion when they are operated). Such a mechanism has to be as light and balanced as possible to limit the efforts to be produced by the human operator especially when anthropomorphic mechanical architectures ("arms") are used. Therefore, the workspace and inertia have to be small and the "drag" on the various joints have to be similar. Transparency has to be as good as possible. This includes also a good visual interaction since all the topographic information given to the operator has to be rendered on displays. They must be as fast and ergonomic as possible. Because motions are generally man-powered, such mechanisms are intrinsically very safe.

3.2 Robots

ROBODOC The Robodoc system has been developped for machining of femoral bones in hip surgery (7; 8). Accurately machining the bone according to the shape of the prosthesis to be implanted allows perfect fit between the cavity and the implant and is intended to provide best bio-mechanical behaviour and long term stability of the implant. The robot is a SCARA based architecture which workspace has relatively limited interaction with the surgical field. In the Robodoc system, the robot control subsystem performs an extensive safety check and monitors cutting force to ensure that unnecessary force cannot harm the patient (9). The RoboDoc system uses force following to allow the surgeon to guide the robot into proximity of the surgical site, after which the robot performs the surgery autonomously.

Radiotherapy irradiation robot (10) developed a frameless system for neuroradiosurgery based on the use of an industrial six axes robot which carries the radiation device. The robot is rather big because it has to position very accurately a heavy payload. It has parallel elements in its structure to increase its rigidity. Its very large workspace intersects the patient area. This system allows position tracking of the patient head during irradiation. Tracking is allowed for small motions only to avoid potential collision with the environment. A spherical architecture would have certainly been more adapted to this kind of application. A number of watch dogs are used at the control level to increase safety.

AESOP The AESOP system from Computer Motion Inc. is dedicated to laparoscopic procedures (11). It is used to move the laparoscope and is controlled by foot pedals operated by the surgeon from video images. Mounted onto the surgical table, this SCARA-based architecture has a very limited workspace and a task-dedicated design. Indeed, it has 6 dofs, 4 actuated joints and 2 passive joints. The passive joints (no 4 and 5) are designed such that the laparoscope can rotate freely about the pivot point constraint imposed by the patient's abominal wall therefore describing only conical motions. This provides the system with very interesting safety characteristics.

3.3 Synergistic systems

Measures of perceptual quality for synergistic devices might focus on the question of transparency. In particular we may ask how unobtrusive the device can be when it wishes to allow the surgeon full control over position. We will call this its transparency in "free mode". Equally important is the smoothness with which the device can enforce a constraint surface. Optimally the surgeon would be able to use a software-defined constraint surface, as exhibited physically by a synergistic device, in much the same way that a surgeon normally uses a physical guide or jig. One wishes a guide to be smooth, preferably of low friction so that one may glide across it, and for it to be rigid and strong. We will refer to these characteristics as the smoothness of the device in "constraint mode".

Mechanical guide At the border of this classification, we can find systems such as (12) for which a six axes actuated mechanism autonomously positions a mechanical guide in stereotactic neurosurgery. This guide is used by the surgeon to guide a linear tool according a pre-planned trajectory. In this case, the constraint is simple (a linear trajectory) and rigid.

Moderated braking We mentioned above the possibility of superimposing brakes on the joints of a passive localizer arm. The surgeon thus has complete and free control of position, until a desired position is reached, at which point the localizer can be entirely "frozen" and subsequently used as a rigid guide. An extension of this idea, which approaches the function of a synergistic device, is to use brakes which can be fractionally activated rather than turned entirely on or off. One would hope that, with appropriate control of the brakes, such a device could constrain the motion of a surgical tool grasped jointly by the robot and the surgeon, and could for instance confine the surgeon's motion of the tool to with a region or on a plane. Such a device has in fact been explored in the CAS field by Taylor (13). The transparency of such a device in its free mode can be excellent, since it reverts to being a passive localizer when the brakes are off, and thus can be moved very easily. Unfortunately it turns out that brakes are extremely difficult to control smoothly. It is very difficult using brakes to exhibit a constraint surface at all, except in the fortuitous instances when one joint can be fully locked and another left fully free.

Passive Arm with Dynamic Constraints Exhibiting a smooth constraint surface requires the establishment of allowed, non zero, velocities, such that the end-effector can move parallel to the constraint surface freely. Such a mechanism is the core of the Passive Arm with Dynamic Constraints (PADyC). In PADyC, each joint is velocity-limited by two reference clutch plates, which thereby define an angular-velocity "window". The joint may turn only at an angular velocity which falls between the limits established by the two reference clutch plates. The angular velocities of these two reference plates are controlled by a computer. When PADyC is in free mode, the angular velocity windows of all joints are set wide open, and the device allows unrestricted freedom of motion naturally. As a constraint surface is approached however, the angular velocity windows are made narrower in some directions, such that ultimately the only velocities available to each joint are the one which move the device away from or parallel to the constraint surface. Constraints include "free", "position", "trajectories" and "regions" modes (14). Those are rigid and programmable constraints; nevertheless, some soft surface behaviours can be simulated by suitable velocity windowing. Because of its principle (no anticipation of next motion is made possible), PADyC natural modes are the free and region modes. The last one is particularly interesting for anatomic obstacle avoidance (neurosurgery or endonasal surgery for instance) and resection operations. The system is smooth and frictionless. A two link (see figure 1.a) and a three link PADyC have been built.

Cobots Armlike cobots with revolute joints are more difficult to describe than translational cobots, and interested readers are referred to (15; 16). Suffice it to say that the principle of operation is the same. Here in the interest of space we will describe the simplest possible translational cobot. As presented below it is a two degree of freedom device. Several higher degree of freedom cobots are under development, as well as an armlike cobot. The two degree of freedom translational cobot consists of a rolling wheel, free to roll on a flat working surface. A computer controlled motor steers the wheel, and a handle is attached to it as shown in figure

Figure 1: (top) *PADyC*: A two degrees of freedom laboratory prototype. The operator moves the PADyC in the plane and the system constrains the motion according to pre-planned strategy. (Courtesy of Dr Jocelyne Troccaz, TIMC Laboratory) (middle) *Cobot*: A two degrees of freedom prototype. The operator moves the cobot in the plane using the handle and the system automatically rotates the wheel in order to describe a given trajectory. (Courtesy of Dr Michael Peshkin, North Western University) (bottom) Close-up of *ACROBOT* mechanism and end-effector showing controlled degrees of freedom (Courtesy of Dr Brian Davies, Imperial College of London)

1.b. The user and the cobot interact through this handle, and the workspace of the cobot is the horizontal plane in which the user can move the handle. Note that the motor cannot make the handle move; only the user can do that. The motor only steers. It can however enforce a constraint surface, which in this example should be a understood as a constraint curve in the planar workspace. It can enforce this constraint simply by steering the wheel parallel to it. Because the rolling wheel can only be moved in the direction it is aimed at each moment, the user perceives an impenatrable boundary at the constraint surface. In practice this illusion is convincing. Since the constraint arises mechanically, it is smooth and frictionless. The instrinsic modes of PADyC are its free and region modes. In contrast, cobots have an instrinsic mode which is the trajectory mode. Free and region modes must be acheived through computer control. To allow the user full freedom of motion, the control computer uses a force sensor to detect which direction the user wishes to move the handle. It then steers the rolling wheel to coincide with the desired direction, much the way that a caster wheel under the leg of a piece of furniture aligns itself with the desired direction. The constraints are rigid and programmable.

ACROBOT As mentioned in section 4.3, a conventional robot may attempt to cooperate with a human user by measuring the the user's applied force and driving joint motors to comply with these forces. However the transparency is usually poor. ACROBOT (or Active Constraint ROBOT) is a robot specially designed for transparent cooperation with a human user, while nevertheless using actuated joints. ACROBOT uses backdrivable motors and transmissions, where conventional robots are usually made strong and stiff at the expense of backdrivability. Mechanically, ACROBOT places the human user and the robot's motor on a more nearly equal basis for controlling position. A conventional robot gives a great advantage to the robot's motors, as it is intended to be insensitive to externally applied forces. In "region" mode ACROBOT's motors are actively driven to comply with the user's force. Good tranparency can result due to mechanical sharing of forces between motor and human, made possible by backdrivability. As the user approaches and then contacts a constraint surface defined in the preoperative plan, the motors are actuated to gradually increases its resistance until, at the edge of the permitted region, it prevents further motion by the operator. Constraints may be deformable (elastic or plastic) and rigid. Following preliminary trials of a prototype a new 4 axis robot, mounted on passive structure, has been constructed and evaluated (see figure 1.c) Whilst the ACROBOT is currently being used for knee surgery, the system is also suited to a range of orthopaedic and soft tissue procedures.

4 Discussion

Synergistic devices are intended to cooperate physically with a surgeon. They must offer good transparency, but also be able to produce forces. These forces (based on a preoperative plan) can guide the surgeon physically, much in the way that a physical jig or guide offers guidance. In contrast to video guidance, direct physical guidance promises to be more efficient and accurate. Especially when the number of degrees of freedom (angles and positions) that the preoperative plan specifies is large, it becomes difficult and frustrating for a surgeon to follow video guidance. Three approaches to synergistic devices were described. In one, ACROBOT, a special purpose robot is used. It may bring speed and tracking properties that PADyC and Cobots cannot. We also described two other approaches to synergistic devices, both of which rely on novel joint mechanisms. Both of them are intrinsically safer than ACROBOT. PADyC's intrinsic modes are the "region" and "free" modes, whilst Cobots' intrinsic mode is the "trajectory" mode. ACROBOT has no preferred mode. It is debatable what will be the future of those CAS systems that perform surgery based on a preoperative plan. Is the computer's primary role to be to present the plan visually to the surgeon, and the surgeon works essentially freehand but with visual reference to the plan? If so, localizers offer the needed functionality with a minimum of interference to the surgeon's delicate work. Or perhaps CAS systems will increasingly execute surgical plans themselves, with the surgeon's direct touch becoming less and less important. If so, semi-autonomous robots with little opportunity for physical input from the surgeon may hold the future. Yet a third possibility, and the one to which the synergistic devices being developed by the authors is addressed, is that surgeon and computer will need to interact physically in a direct and cooperative and sensitive way. This is required if the surgeon is to remain responsible for some aspects of tool motion, while simultaneously the computer is responsible for others. Synergistic devices are designed for this intimate cooperation.

References

[1] P. Cinquin and al. Computer Assisted Medical Interventions at TIMC Laboratory: passive and semi-active aids. *IEEE Engineering in Medicine and Biology magazine, special issue Robots in Surgery*, 14(3):254–263, 1995.

[2] ISG Technologies. Viewing wand operator's guide. Technical report, ISG Technologies Inc., 1993.

[3] R. Mosges, G. Schlondorff, L. Klimek, D. Meyer-Ebrecht, W. Krybus, and L. Adams. Computer assisted surgery. an innovative surgical technique in clinical routine. In H.U. Lemke, editor, *Computer Assisted Radiology, CAR 89*, pages 413–415, Berlin, June 1989. Springer-Verlag.

[4] L. Adams, W. Krybus, Meyer-Ebrecht D., R. Rueger, J.M. Gilsbach, R. Moesges, and G. Schloendorff. Computer assisted surgery. *IEEE Computer Graphics and Applications*, pages 43–51, May 1990.

[5] Y. Kosugi, E. Watanabe, and J. Goto. An articulated neurosurgical navigation system using MRI and CT images. *IEEE Transactions on Biomedical Engineering*, 35(2):147–152, 1988.

[6] H.F. Reinhardt. Neuronavigation : a ten years review. In R. Taylor, S. Lavallee, G. Burdea, and R. Mosges, editors, *Computer-integrated surgery*. MIT Press (Cambridge, MA), 1995.

[7] R.H. Taylor, H.A. Paul, B.D. Mittelstadt, E. Glassman, B.L. Musits, and W.L. Bargar. A robotic system for cementless total hip replacement surgery in dogs. In *2nd Workshop on Medical & Healthcare Robotics*, pages 79–89, Newcastle, UK, 1989. D.T.I.

[8] R.H. Taylor, H.A. Paul, C.B. Cutting, B. Mittelstadt, W. Hanson, P. Kazanzides, B. Musits, Y.Y. Kim, A. Kalvin, B. Haddad, D. Khoramabadi, and D. Larose. Augmentation of human precision in computer-integrated surgery. *ITBM (Innovation and Technology in Biology and Medicine) , Special Issue on Robotic Surgery*, 13(4):450–468, July 1992.

[9] Kazanzides P., Zuhars J., Mittlestadt B., and Taylor R.H. Force sensing and control for a surgical robot. In *IEEE Conference on Robotics and Automation*, pages 612–617, May 1992.

[10] J. Adler, A. Schweikard, R. Tombropoulos, and J-C. Latombe. Image-guided robotic radiosurgery. In *MRCAS 94, Medical Robotics and Computer Assisted Surgery*, pages 291–297, Pittsburgh, PA, 1994.

[11] Uecker D.R., Lee C., Wang Y.F., and Wang Y. A speech-directed multi-modal man-machine interface for robotically enhanced surgery. In *1st Symposium on Medical Robotics and Computer-Assisted Surgery*, 1994.

[12] J. Lavallee, S. Troccaz, L. Gaborit, P. Cinquin, A.L. Benabid, and D. Hoffmann. Image guided robot: a clinical application in stereotactic neurosurgery. In *IEEE Int. Conf. on Robotics and Automation*, pages 618–625, Nice , France, 1992.

[13] R.H. Taylor, C.B. Cutting, Y.Y. Kim, A. Kalvin, D. Larose, B. Haddad, D. Khoramabadi, M. Noz, R. Olyha, M. Bruun, and D. Grimm. A model-based optimal planning and execution system with active sensing and passive manipulation for augmentation of human precision in computer-integrated surgery. In *2nd International Workshop on Experimental Robotics*, Toulouse(France), June 1991. Springer Verlag.

[14] J. Troccaz and Y. Delnondedieu. Semi-active guiding systems in surgery. A two-dof prototype of the passive arm with dynamic constraints (PADyC). *Mechatronics*, 6(4):399–421, June 1996.

[15] M.A. Peshkin J.E. Colgate and W. Wannasuphoprasit. Nonholonomic haptic displays. In *IEEE International Conference on Robotics and Automation*, 1996.

[16] J.E. Colgate M. Peshkin and C. Moore. Passive robots and haptic displays based on nonholonomic elements. In *IEEE International Conference on Robotics and Automation*, 1996.

PART 10
LEARNING FROM HUMAN
SESSION SUMMARY

Professor R.A. Jarvis
Intelligent Robotics Research Centre
Monash University
Wellington Road, Clayton 3168
Victoria, Australia

This session, entitled "Learning from Human", is made up of four fairly disparate presentations linked by their common focus on human behaviour as an inspiration, a source of design data and as a working system which is a challenge to emulate. The recent unveiling of the achievements of the Honda humanoid walking robot has amazed us, made some of us feel a tingle in the spine provoked by the sensation (real though falsely based) of being in the presence of an intelligence not unlike our own and others of us felt our hackles rise as if we had a new rival. Such research should continue with renewed energy and support and not be suppressed through the false public perception that all important aspects of humanoid research have already been demonstrated. It is clear that we are only at the beginning of this fascinating journey along the way to realising the dream shared by many in the robotics research community, to create artificial beings that can behave like humans in many different ways. For some, this quest has become an obsession - such obsessions, guided by good science and engineering, can lead to excellent research and we hope it does.

The first paper in this session presented by Minoru Asada, is concerned with the emergence of robot intelligence in co-operative tasks through vision based learning and focuses on the Robocup soccer challenge that has had growing support in recent times and is inspiring robotics research communities around the world to participate in it as a means of developing new co-operative robotic systems that can demonstrate their value by competitive performance rather than speculative hopes. The importance of the physical environment and the tight coupling of perception and action are emphasised. That a complex environment, appropriately accommodated, can lead to complex behaviour through systems involving state estimation and reinforcement learning is clearly demonstrated. The interaction between the static environment (field and goal), a passive agent (the ball), and active agents (team members and competitors) represents a very complex game environment which is a very rich learning domain in which to "train" robots. This type of problem domain can be treated as a member of the class of differential games within an optimality seeking framework. One would hope that the type of convergence to very narrow optimal behaviours the micro-mouse competition seems to have lead to will not be evident with the much richer environment of the Robocup challenge.

The second paper, presented by Rod Brooks, continues the saga on the education of COG, the legless humanoid robot who learns, through interaction in a real world of humans and moving objects, to be able (at this early stage) to foveate and reach for objects and play what seems to be copy-cat games, without the unnecessary baggage of complex three dimensional models of the world. The emphasis of the project is on making the system adaptable and self-calibrating through unsupervised learning forged by social interactions with humans. The ideal of building small pockets of human-like behaviour using only 2D maps (saccade and ballistic) for hand/eye co-ordination (foveating and reaching) is the backbone of this approach of "intelligence without reasoning". What remains unproven, at this stage, is whether combining a multitude of small working systems will lead to coherent behaviour or chaos. For example, it would be interesting to test whether vision guided reaching and grasping can be combined with obstacle avoidance as a first step towards answering this critically important question. Another important area of continuing development is the construction of motivation systems which can direct capabilities towards specific tasks within socially defined circumstances.

The third paper, presented by Jessica Hodgins, is not directly about humanoid robotics but rather on dynamic visual simulation of human motion using a tool box approach to implementing a hierarchy of control laws, with state machines at the top, control actions at the intermediate level and low level motion control at the bottom. Passive strategies are used where possible along with the synergistic use of joints and physical intuition. Readers will unfortunately not be able to fully appreciate the capabilities of the system developed without viewing the very impressive video clips of athletic human action shown during the presentation. Potential applications for the work include, in addition to computer animation in its own right, include the development of interactive environments, the support of sports training and the provision of specifications for controlling humanoid robots.

The final paper of the session, presented by Masa Inaba, is a genealogy of a species of robotic humanoid machines in the "remote brain" class developed over a number of years. There are machines that crawl, walk, kick, fall over and get up, swing and feel with convincingly life-like motions. Relieving the machines from the burden of carrying sophisticated computational support onboard but instead controlling the many degrees of freedom of their model servo driven joints via broadcast radio signals allows for much shorter development cycles, more freedom of mechanical construction, and improvements in speed, dexterity and physical robustness. Furthermore, the "brain" can be developed independently of the body and even replaced when better computational resources become available without a need to change the "body". The evolution of this species has been along the line of balanced growth taking the development of a human child as the model. Thus, this is a fabricated evolution without the destruction of the "unfit". Goals can be defined and developments carried out off-line with goal seeking behaviour activated on-line. The basic idea is powerful through simplicity and flexibility whilst economical in terms of the ready availability of most of the components common amongst the various "beings" created. A new "being" can be created by constructing the mechanics out of easily "bolted together" components and using the evolving software tools to define new control structures and action sequences.

Considering this session as a whole, there is a certain, if not obvious, balance amongst the four papers in it. Human-like behaviour is common amongst them, the first in respect of co-operative emergent behaviour, the second in terms of unsupervised learning through "social" interactions with humans, the third in terms of high fidelity visualisation of humans in athletic, flowing actions and the last as an example of non-biological evolution of a species of remote brained humanoids with ever improving capabilities emulating human child development.

Vision-based Behavior Learning and Development for Emergence of Robot Intelligence

M. Asada, K. Hosoda, and S. Suzuki
Dept. of Adaptive Machine Systems, Graduate School of Engineering
Osaka University, Suita, Osaka 565 (Japan)
email: asada@ams.eng.osaka-u.ac.jp

Abstract

This paper focuses on two issues on learning and development; a problem of state-action space construction, and a scaling-up problem. The former is mainly related to sensory-motor mapping and its abstraction, and we show two our methods for the state and action space construction for reinforcement learning. For the latter issue, we attempt to define the environmental complexity based on the relationships between observations and self motions. Based on this view, we introduce a method which can cope with the complexity of multi-agent environment by a combination of a state vector estimation process and a reinforcement learning process based on the estimated vectors. As example tasks in our work, we adopt the domain of soccer robots, RoboCup [1]. Computer simulations and real robot experiments are given.

1 Introduction

The ultimate goal of our research is to design the fundamental internal structure inside physical entities having their bodies (robots) which can emerge complex behaviors through the interactions with their environments. In order to emerge the intelligent behaviors, physical bodies have an important role of bringing the system into *meaningful* interaction with the physical environment – complex, uncertain, but with automatically consistent set of natural constraints. This facilitates the correct agent design, learning from the environment, and rich meaningful agent interaction. The meanings of "having a physical body" can be summarized as follows:

1. Sensing and acting capabilities are not separable, but tightly coupled.
2. In order to accomplish given tasks, the sensor and actuator spaces should be abstracted under the resource bounded conditions (memory, processing power, controller etc.).
3. The abstraction depends on both the fundamental embodiments inside the agents and the experiences (interactions with their environments).
4. The consequences of the abstraction are the agent-based subjective representation of the environment, and its evaluation can be done by the consequences of behaviors.
5. In real world, both inter-agent and agent-environment interactions are asynchronous, parallel and arbitrarily complex. The agent should cope with increasing complexity of the environment to accomplish the given task at hand.

In this paper, we focus on two issues on learning and development; a problem of state-action space construction, and a scaling-up problem. The former is mainly related to 2 and 3, and we show two our methods for the state and action space construction for reinforcement learning. One is based on an off-line learning method [2] and the other on-line one [3].

The latter issue is closely related to 4 and 5, and we attempt to define the environmental complexity based on the relationships between observations and self motions. Based on this view, we introduce a method which can cope with the complexity of multi-agent environment by a combination of a state vector estimation process and a reinforcement learning process based on the estimated vectors [4].

As example tasks in our work, we adopt the domain of soccer robots, RoboCup, which is an attempt to foster intelligent robotics research by providing a standard problem where a wide range of technologies can be integrated and examined [1].

The remainder of this article is structured as follows. We first give an explanation of the problem of state-action space construction along with our real robot experiments in the context of reinforcement learning. Then, we show our method to cope with more complicated tasks in multi-agent environment. Finally, we give a conclusion.

2 A Problem of State-Action Space Construction

Reinforcement learning [5, 6] has been receiving increased attention as a method for robot learning with little or no *a priori* knowledge and higher capability of reactive and adaptive behaviors. In such robot learning methods, a robot and an environment are generally modeled by two synchronized finite state automatons interacting in a discrete time cyclical processes. The robot senses the current state of the environment and selects an action. Based on the state and the action, the environment makes a transition to a new state and generates a reward that is passed back to the robot. Through these interactions, the robot learns a purposive behavior to achieve a given goal.

To apply robot learning methods such as reinforcement learning to real robot tasks, we need a well-defined state-action space by which the robot learns to select an adequate action for the current state to accomplish the task at hand. Traditional notions of state and action in the existing applications of the reinforcement learning schemes fit nicely into deterministic state transition models (e.g. one action is forward, backward, left, or right, and the states are encoded by the locations of the agent). However, it seems difficult to apply such deterministic state transition models to real robot tasks. In real world, everything changes asynchronously [7]. Therefore, the construction of state-action space is one of the most important issues in robot learning.

Generally, the design of the state-action space in which the necessary and sufficient information to accomplish a given task should be included depends on the capabilities of agent sensing and acting. The abstraction process from sensory information to a state seems to depend on the process from motor commands to an action, and *vice versa*. This resembles the well-known "chicken and egg problem" that is difficult to be solved (see Figure 1). Therefore, we need a sort of constraint to solve the problem.

Figure 1: The inter-dependence between sensor and motor spaces from a viewpoint of state-action space construction

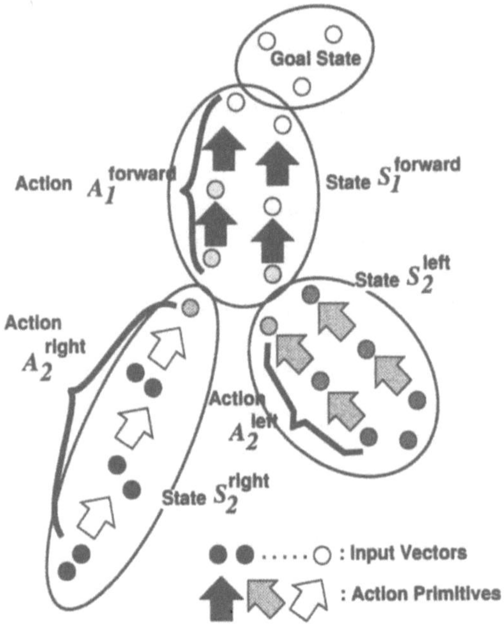

Figure 2: A basic idea of state-action space construction

2.1 An Off-Line Learning Method

Basic ideas of our first approach to cope with this problem are:

- we define an action primitive as a motor command to be executed during a fixed time interval, and an input vector as sensory data of the consequence of the action primitive, and

- we define a state as a cluster of input vectors from which the robot can reach the goal state (or the state already obtained) by a sequence of one kind action primitive regardless of its length, and one action as this sequence of action primitives.

Figure 2 shows the basic idea of the state-action space construction. The initial state space consisting of the goal state and the other is iteratively separated into several states.

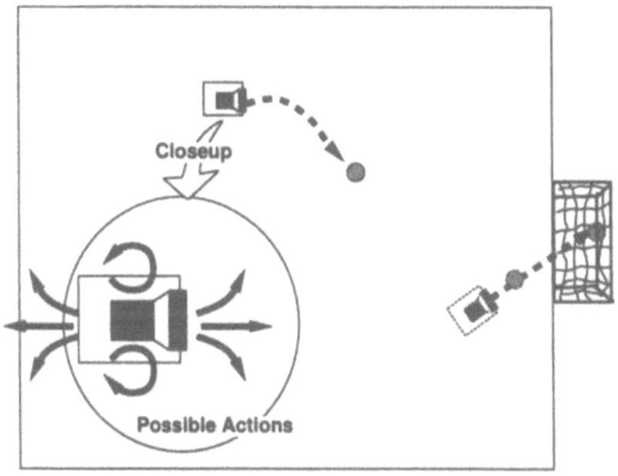

Figure 3: Task and environment

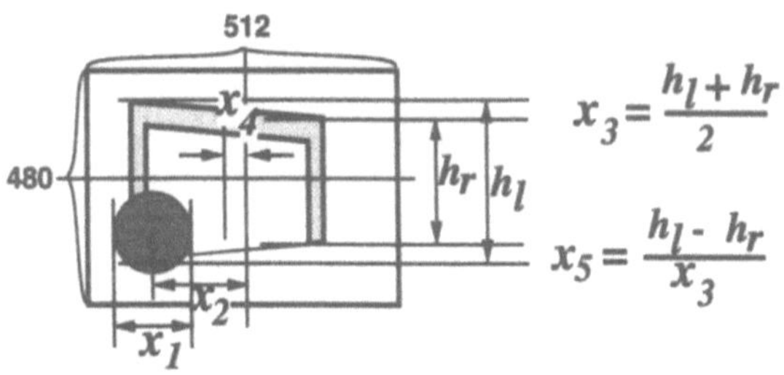

Figure 4: Input vector consisting of five parameters

The method is applied to a soccer robot which tries to shoot a ball into a goal (see Fig.3). The size of the observed image is 512 by 480 pixels, and the center of image is the origin of the image coordinate system (see Figure 4). An input vector x for a shooting task consists of:

- x_1: the size of the ball, the diameter that ranges from 0 to about 270 pixels,
- x_2: the position of the ball ranging from -270 to +270, considering the partial observation,

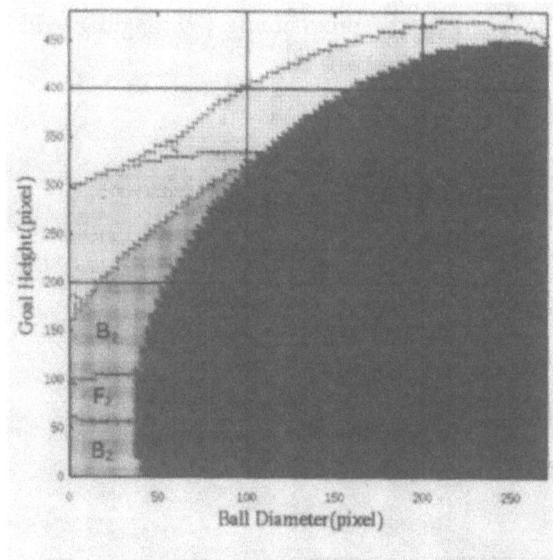

Figure 5: 2-D projection of the result of state space construction

- x_3: the size of the goal, the height average of the left and right poles (or ends in image) that ranges from 0 to 480 pixels,
- x_4: the position of the goal, the average of the positions of the left and right poles (or ends in image) that ranges from -256 to +256, and
- x_5: the orientation of the goal, the ratio of the height difference between the left and right poles (or ends) to the size of the goal x_3. x_5 ranges from -1.00 to +1.00.

Figure 5 shows the result in which the final state space is projected into two dimensional space in terms of the ball size and the goal size (when their positions are frontal and the orientation of the goal is horizontal). The intensity indicates the order of the division: the darker is the earlier. Labels "F" and "B" indicate the motions of forward and backward, respectively, and subscript shows the number of state transitions towards the goal. Grid lines indicate the boundaries divided by programmer in the previous work [8]. The remainder of the state space in Figure 5 corresponds to infeasible situations such as "the goal and the ball are observed at the center of image, and the size of the goal is large, but that of the ball is small" although we had not recognized such a meaningless state in the previous work. As we can see, the sensor space categorization by the proposed method is quite different from the one designed by the programmer (rectangular grids) in the previous work [8].

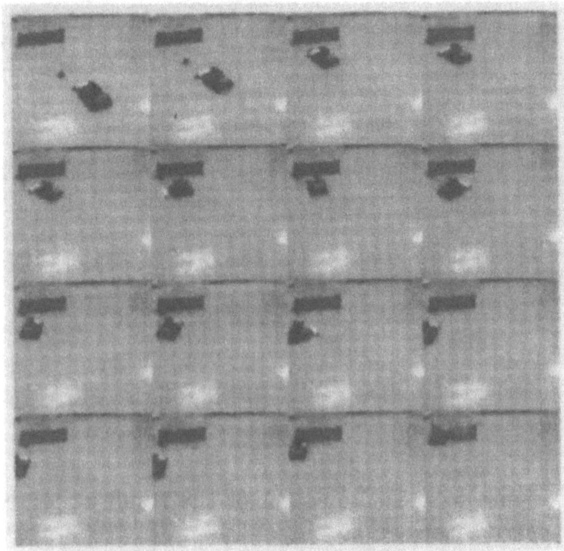

Figure 6: The robot succeeded in finding and shooting a ball into the goal

We applied the method to a real robot environment. The success ratio is worse than the simulation because of the disturbances due to several causes such as eccentricity of the ball centroid and slip of the tires that make the ball or the robot move into unpredictable directions. Figure 6 shows how a real robot shoots a ball into a goal by using the state and action map obtained by the method. 16 images are shown in raster order from the top left to the bottom right in every 1.5 seconds, in which the robot tried to shoot a ball, but failed, then moved backward so as to find a position to shoot a ball, finally succeeded in shooting. Figure 7 shows a sequence of images taken by the robot during the task execution shown in Figure 6. Note that the backward motion for retry is just the result of learning and not hand-coded.

2.2 An On-line Learning Method

The above method needs sufficient amount of uniformly sampled data to construct the state space suitable for the robot to perform the given task, and therefore, does not cope with dynamic changes happened in the environment. These problems are resolved by the second approach [3] which obtains a purposive behavior within less learning time by incrementally segmenting

Figure 7: Images taken by the robot during the task execution

the sensor space based on the experiences of the robot. The incremental segmentation is performed by constructing local models in the state space, which is based on the function approximation of the sensor outputs to reduce the learning time, and the reinforcement signal to emerge a purposive behavior. They applied their method to the same task as in [2]. The basic ideas are as follows:

1. Set up a state space consists of two states; the goal state and the other.
2. Apply function approximation to the changes of the input vectors caused by action primitives. If the function approximation cannot cope with these changes, then segment the states into two and apply the function approximation to a new state. This process might cause to merge a state with one of the separated states. These processes can reduce ineffective explorations.
3. Initialize the action-value for the new state, and apply the reinforcement learning. The learning time is very short because the number of states to be updated is small.
4. Apply stochastic action selection to cope with dynamic change of the environment.

Figures 8, 9, and 10 show the experimental results. Figure 8 shows a projection of the state

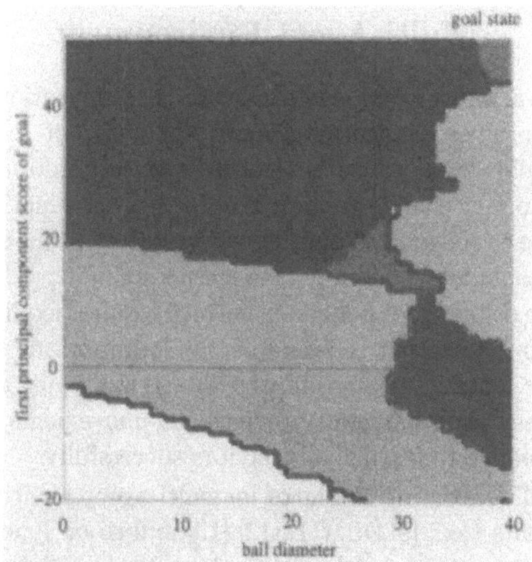

Figure 8: Obtained state space

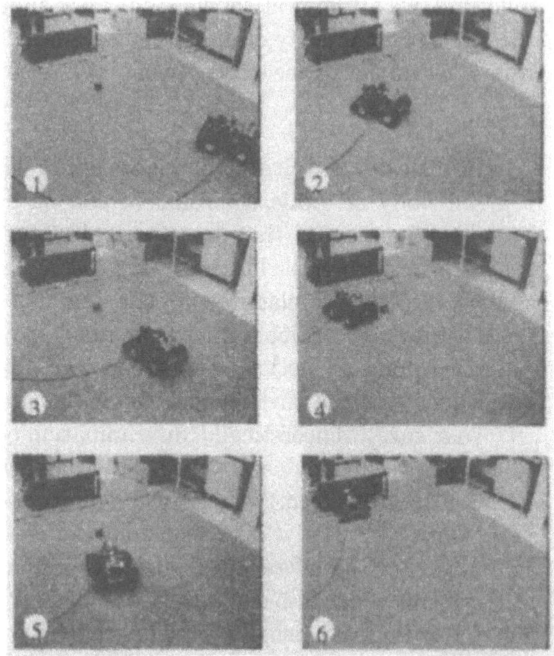

Figure 10: The robot succeeded in shooting a ball into the goal

space after 1,110 trials, where the state space in term of ball size and goal size is indicated when the position of the ball and the goal are center of the screen and the orientation of the goal is frontal. As we can see the shape of the resultant state space is complicated and quite different from the previous result (see Figure 5). Figure 9 indicates the changes of the success rate and the number of states in the case that the ball size is suddenly changed twice at the 500th trial. These suggest that the method cope with nonlinear mapping between states and actions and deal with dynamic change of the environment.

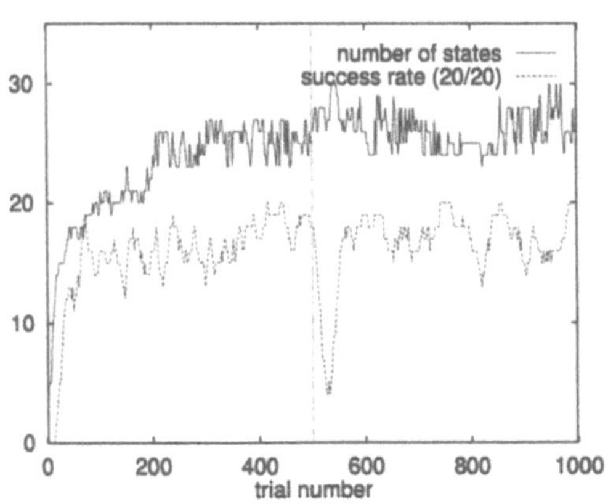

Figure 9: Success rate and the number of states

Figure 10 shows how the robot tries to shoot a ball into the goal. Because of the sensor noise and the uncertainty of the motor commands, the robot often misunderstands the states, and takes wrong actions, therefore it fails to do the task. ① indicates that the robot is going to shoot a ball into the goal and moves forward. But it fails to kick the ball at ② because the speed is too high to turn. Eventually, the ball is occluded by the robot in ②. Then, it goes back left so that it can shoot a ball at ③. But it fails again at ④. Then it goes back left again at ⑤. After all, the robot shoots the ball into the goal successfully at ⑥.

3 A Scaling-Up Problem

Since each species of animals can be regarded to have its own intelligence, difference of intelligence seems to depend on the agent (capabilities in sensing, acting, and cognition) and its environment. If agents have the same bodies, differences or levels in intelligence can occur in the complexity of interactions with their environments. In case of our soccer playing robot with vision, the complexity of interactions may change due to other agents in the field such as common side players, opponents, judges and so on. In the following, we attempt at showing our view about the levels of complexity of interactions, especially from a viewpoint of the existence of other agents.

1. **Self body and Static Environment:** The self body or static environment can be defined in a sense that the observable parts of which changes in the image plane can be directly correlated with the self motor commands (ex. looking at your hand showing voluntary motion, or observing an optical flow of the environment when changing your gaze). Theoretically, discrimination between "self body" and "static environment" is a hard problem because the definition of "static" is relative and depends on the selection of the reference (the base coordinate system) which also depends on the context of the given task. Usually, we suppose the natural orientation of the gravity, which provides the orientation of the ground coordinate system.

2. **Passive agents:** As a result of actions of the self or other agents, passive agents are movable or can be stopped. A ball is a typical one. As long as they are stationary, they can be categorized into the static environment. But, not so simple correlation with motor commands as the self body or the static environment can be expected when they are in motion.

3. **Active (other) agents:** Active other agents do not have a simple and straightforward relationship with the self motions. In the early stage, they are treated as noise or disturbance because of not having direct visual correlation with the self motor commands. Later, they can be found as having more complicated and higher correlation (coordination, competition, and others). The complexity is drastically increased.

According to the complexity of the environment, the internal structure of the robot should be higher and more complex to emerge various intelligent behaviors. We show one of such structure coping with the complexity of agent-environment interactions with real robot experiments and discuss the future issues.

3.1 A More Complicated Task in Multi-Agent Environment

In a multi-agent environment, the conventional reinforcement learning algorithm does not seem applicable because the learner's sensory information may change regardless of the learner's motion due to the motion of other active agents in the environment. Therefore, the learner cannot predict the other agent behaviors correctly unless explicit communication is available. It is important for the learner to discriminate the strategies of the other agents and to predict their movements in advance to learn the behaviors successfully.

The existing methods in multi agent environments (ex., [9],[10],[11],[12],[13] and so on.) need state vectors in order for the learning to converge. However, it is difficult to obtain a reasonable analytical model in advance. Therefore, the modeling architecture is required to make the reinforcement learning applicable.

Here, we show a method which estimates the relationship between the learner's behaviors and the other agents through interactions (observation and action) using the method of system identification. In order to construct the local predictive model of other agents, we apply Akaike's Information Criterion(AIC) [14] to the result of Canonical Variate Analysis(CVA) [15], which is widely used in the field of system identification. The local predictive model is constructed based on the observation and action of the learner (observer).

We apply the proposed method to a simple soccer-like game in a physical environment. The task of the agent is to shoot a ball which is passed back from the other agent. Since the environment consists of the stationary agents (the goal and the line), a passive agent (the ball) and an active agent (the passer), the learner has to construct the adequate models for these agents. After constructing the models and estimating their parameters, the reinforcement learning is applied in order to acquire purposive behaviors. The proposed method can cope with the moving ball because a state vector for learning is selected appropriately so as to predict the successive steps.

Figure 11 shows an overview of the proposed method consisting of local predictive models and reinforcement learning architecture. At first, the learning agent collects the sequence of sensor outputs and motor commands to construct the local predictive models. By approximating the relationship between inputs (learner's action) and

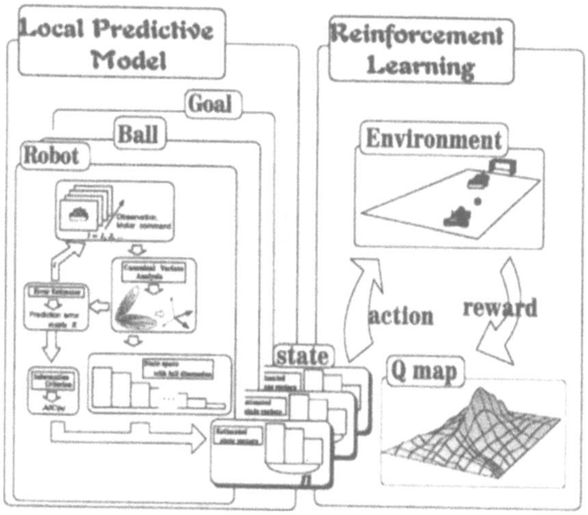

Figure 11: An overview of the proposed method

outputs (observation), the local predictive model gives the learning agent not only the successive states of the agent but also the priority of state vectors, which means that first a few vectors might be sufficient to predict the successive states.

The flow of the proposed method is summarized as follows:

1. Collect the observation vectors and the motor commands.

2. Estimate the state space with the full dimension directly from the observations and motor commands (Section 3.1.1).

3. Determine the dimension of the state vectors which is the result of the trade off between the error and the complexity of the model.

4. Apply the reinforcement learning based on the estimated state vectors.

3.1.1 *Canonical Variate Analysis*

A number of algorithms to identify multi-input multi-output (MIMO) combined deterministic-stochastic systems have been proposed [16]. In contrast to 'classical' algorithms such as PEM (Prediction Error Method), the subspace algorithms do not suffer from the problems caused by a priori parameterizations. Larimore's Canonical Variate Analysis (CVA) [15] is one of such algorithms, which uses canonical correlation analysis to construct a state estimator

Let $\boldsymbol{u}(t) \in \Re^m$ and $\boldsymbol{y}(t) \in \Re^q$ be the input and output generated by the unknown system

$$\begin{aligned} \boldsymbol{x}(t+1) &= \boldsymbol{A}\boldsymbol{x}(t) + \boldsymbol{B}\boldsymbol{u}(t) + \boldsymbol{w}(t), \\ \boldsymbol{y}(t) &= \boldsymbol{C}\boldsymbol{x}(t) + \boldsymbol{D}\boldsymbol{u}(t) + \boldsymbol{v}(t), \end{aligned} \quad (1)$$

with

$$E\left\{\begin{bmatrix} \boldsymbol{w}(t) \\ \boldsymbol{v}(t) \end{bmatrix} \begin{bmatrix} \boldsymbol{w}^T(\tau) & \boldsymbol{v}^T(\tau) \end{bmatrix}\right\} = \begin{bmatrix} \boldsymbol{Q} & \boldsymbol{S} \\ \boldsymbol{S}^T & \boldsymbol{R} \end{bmatrix} \delta_{t\tau},$$

and $\boldsymbol{A}$, $\boldsymbol{Q} \in \Re^{n\times n}$, $\boldsymbol{B} \in \Re^{n\times m}$, $\boldsymbol{C} \in \Re^{q\times n}$, $\boldsymbol{D} \in \Re^{q\times m}$, $\boldsymbol{S} \in \Re^{n\times q}$, $\boldsymbol{R} \in \Re^{q\times q}$. $E\{\cdot\}$ denotes the expected value operator and $\delta_{t\tau}$ the Kronecker delta. $\boldsymbol{v}(t) \in \Re^q$ and $\boldsymbol{w}(t) \in \Re^n$ are unobserved, Gaussian-distributed, zero-mean, white noise vector sequences. CVA uses a new vector $\boldsymbol{\mu}$ which is a linear combination of the previous input-output sequences since it is difficult to determine the dimension of $\boldsymbol{x}$. Eq.(1) is transformed as follows:

$$\begin{bmatrix} \boldsymbol{\mu}(t+1) \\ \boldsymbol{y}(t) \end{bmatrix} = \boldsymbol{\Theta} \begin{bmatrix} \boldsymbol{\mu}(t) \\ \boldsymbol{u}(t) \end{bmatrix} + \begin{bmatrix} \boldsymbol{T}^{-1}\boldsymbol{w}(t) \\ \boldsymbol{v}(t), \end{bmatrix}, \quad (2)$$

where

$$\hat{\boldsymbol{\Theta}} = \begin{bmatrix} \boldsymbol{T}^{-1}\boldsymbol{A}\boldsymbol{T} & \boldsymbol{T}^{-1}\boldsymbol{B} \\ \boldsymbol{C}\boldsymbol{T} & \boldsymbol{D} \end{bmatrix}, \quad (3)$$

and $\boldsymbol{x}(t) = \boldsymbol{T}\boldsymbol{\mu}(t)$. We follow the simple explanation of the CVA method.

1. For $\{\boldsymbol{u}(t), \boldsymbol{y}(t)\}$, $t = 1, \cdots N$, construct new vectors

$$\boldsymbol{p}(t) = \begin{bmatrix} \boldsymbol{u}(t-1) \\ \vdots \\ \boldsymbol{u}(t-l) \\ \boldsymbol{y}(t-1) \\ \vdots \\ \boldsymbol{y}(t-l) \end{bmatrix}, \quad \boldsymbol{f}(t) = \begin{bmatrix} \boldsymbol{y}(t) \\ \boldsymbol{y}(t+1) \\ \vdots \\ \boldsymbol{y}(t+k-1) \end{bmatrix},$$

2. Compute estimated covariance matrices $\hat{\boldsymbol{\Sigma}}_{pp}$, $\hat{\boldsymbol{\Sigma}}_{pf}$ and $\hat{\boldsymbol{\Sigma}}_{ff}$, where $\hat{\boldsymbol{\Sigma}}_{pp}$ and $\hat{\boldsymbol{\Sigma}}_{ff}$ are regular matrices.

3. Compute singular value decomposition

$$\hat{\boldsymbol{\Sigma}}_{pp}^{-1/2}\hat{\boldsymbol{\Sigma}}_{pf}\hat{\boldsymbol{\Sigma}}_{ff}^{-1/2} = \boldsymbol{U}_{aux}\boldsymbol{S}_{aux}\boldsymbol{V}_{aux}^T, \quad (4)$$

$$\boldsymbol{U}_{aux}\boldsymbol{U}_{aux}^T = \boldsymbol{I}_{l(m+q)}, \quad \boldsymbol{V}_{aux}\boldsymbol{V}_{aux}^T = \boldsymbol{I}_{kq},$$

and $\boldsymbol{U}$ is defined as:

$$\boldsymbol{U} := \boldsymbol{U}_{aux}^T\hat{\boldsymbol{\Sigma}}_{pp}^{-1/2}.$$

4. The n dimensional new vector $\mu(t)$ is defined as:

$$\mu(t) = [I_n \; 0]Up(t), \qquad (5)$$

5. Estimate the parameter matrix Θ applying the least square method to Eq (2).

Strictly speaking, all the agents do in fact interact with each other, therefore the learning agent should construct the local predictive model taking these interactions into account. However, it is intractable to collect the adequate input-output sequences and estimate the proper model because the dimension of state vector increases drastically. Therefore, the learning (observing) agent applies the CVA method to each (observed) agent separately.

3.1.2 Determine the dimension of other agent

It is important to decide the dimension n of the state vector x and lag operator l that tells how long the historical information is related in determining the size of the state vector when we apply CVA to the classification of agents. Although the estimation is improved if l is larger and larger, much more historical information is necessary. However, it is desirable that l is as small as possible with respect to the memory size. For n, complex behaviors of other agents can be captured by choosing the order n high enough.

In order to determine n, we apply Akaike's Information Criterion (AIC) which is widely used in the field of time series analysis. AIC is a method for balancing precision and computation (the number of parameters). Let the prediction error be ε and covariance matrix of error be

$$\hat{R} = \frac{1}{N-k-l+1} \sum_{t=l+1}^{N-k+1} \varepsilon(t)\varepsilon^T(t).$$

Then $AIC(n)$ is calculated by

$$AIC(n) = (N-k-l+1)\log|\hat{R}| + 2\lambda(n), \qquad (6)$$

where λ is the number of the parameters. The optimal dimension n^* is defined as

$$n^* = \arg\min AIC(n).$$

Since the reinforcement learning algorithm is applied to the result of the estimated state vector to cope with the non-linearity and the error of modeling, the learning agent does not have to construct the *strict* local predict model. However, the parameter l is not under the influence of the $AIC(n)$. Therefore, we utilize $\log|\hat{R}|$ to determine l.

1. Memorize the q dimensional vector $y(t)$ about the agent and m dimensional vector $u(t)$ as a motor command.

2. From $l = 1\cdots$, identify the obtained data.

   (a) If $\log|\hat{R}| < 0$, stop the procedure and determine n based on $AIC(n)$,

   (b) else, increment l until the condition (a) is satisfied or $AIC(n)$ does not decrease.

3.2 Reinforcement Learning

After estimating the state space model given by Eq. 2, the agent begins to learn behaviors using a reinforcement learning method. Q learning [17] is a form of reinforcement learning based on stochastic dynamic programming. It provides robots with the capability of learning to act optimally in a Markovian environment. In the previous section, appropriate dimension n of the state vector $\mu(t)$ is determined, and the successive state is predicted. Therefore, we can regard an environment as Markovian.

3.3 Task and Assumptions

We apply the proposed method to a simple soccer-like game including two agents (Figure 12). Each agent has a single color TV camera and does not know the location, the size and the weight of the ball, the other agent, any camera parameters such as focal length and tilt angle, or kinematics/dynamics of itself. They move around using a 4-wheel steering system. As motor commands, each agent has 7 actions such as go straight, turn right, turn left, stop, and go backward. Then, the input u is defined as the 2 dimensional vector as

$$u^T = [v \;\; \phi], \quad v, \phi \in \{-1, 0, 1\},$$

where v and ϕ are the velocity of motor and the angle of steering respectively and both of which are quantized.

The output (observed) vectors are shown in Figure 13. As a result, the dimension of the observed vector about the ball, the goal, and the other robot are 4, 11, and 5 respectively.

Figure 12: The environment and our mobile robot

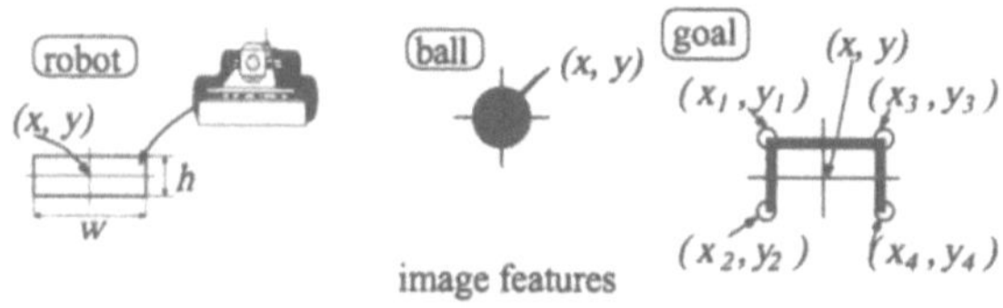

robot	ball	goal
area	area	area
center position	center position	center position
height	radius	4 corners
width		

Figure 13: Image features of the ball, goal, and agent

3.4 Experimental Results

3.4.1 *Simulation Results*

Table 1: The estimated dimension (computer simulation)

agent	l	n	$\log \lvert \boldsymbol{R} \rvert$	AIC
goal	1	2	−0.001	121
ball	2	4	0.232	138
random walk	3	6	1.22	232
move to the ball	3	6	−0.463	79

Table1 shows the result of identification. In order to predict the successive situation, $l = 1$ is sufficient for the goal, while the ball needs 2 steps. The motion of the random walk agent can not be correctly predicted as a matter of course while the move-to-the-ball agent can be identified by the same dimension of the random agent, but the prediction error is much smaller than that of random walk.

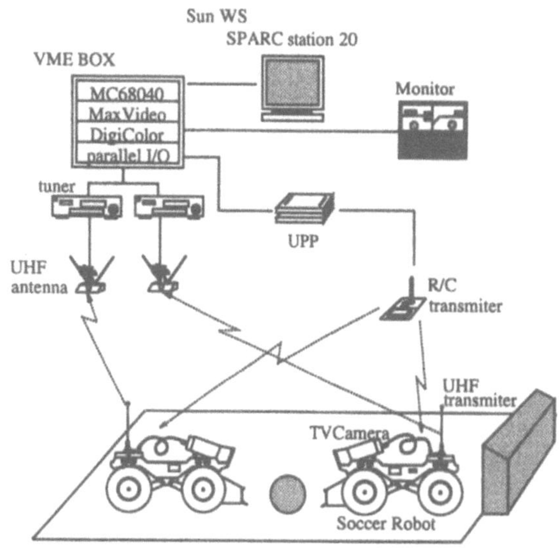

Figure 14: A configuration of the real system.

Table 2 shows the success rates of shooting and passing behaviors compared with the results in our previous work [?] in which only the current sensor information is used as a state vector. We assign a reward value 1 when the robot achieved the task, or 0 otherwise. If the learning agent uses the only current information about the ball and the goal, the leaning agent can not acquire the optimal behavior when the ball is rolling. In other words, the action value function does not become to be stable because the state and action spaces are not consistent with each other.

Table 2: Comparison between the proposed method and using current information

state vector	success of shooting (%)	success of passing (%)
current position	10.2	9.8
using CVA	78.5	53.2

3.4.2 *Real Experiments*

We have constructed the radio control system of the robot, following the remote-brain project by Inaba et al. [19]. Figure 14 shows a configuration of the real mobile robot system. The image taken by a TV camera mounted on the robot is transmitted to a UHF receiver and processed by Datacube MaxVideo 200, a real-time pipeline

Table 3: The estimated dimension (real environment)

from the shooter				
	l	n	$\log \lvert \boldsymbol{R} \rvert$	AIC
ball	4	4	1.88	284
goal	1	3	−1.73	−817
passer	5	4	3.43	329
from the passer				
	l	n	$\log \lvert \boldsymbol{R} \rvert$	AIC
ball	4	4	1.36	173
shooter	5	4	2.17	284

video image processor. In order to simplify and speed up the image processing time, we painted the ball, the goal, and the opponent red, blue, and yellow, respectively. The input NTSC color video signal is first converted into HSV color components in order to make the extraction of the objects easy. The image processing and the vehicle control system are operated by VxWorks OS on MC68040 CPU which are connected with host Sun workstations via Ether net. The tilt angle is about −26 [deg] so that robot can see the environment effectively. The horizontal and vertical visual angles are about 67 [deg] and 60 [deg], respectively.

The task of the passer is to pass a ball to the shooter while the task of the shooter is to shoot a ball into the goal. Table 3 and Figure 15 show the experimental results. The value of l for the ball and the agent are bigger than that of computer simulation, because of the noise of the image processing and the dynamics of the environment due to such as the eccentricity of the centroid of the ball. Even though the local predictive model of the same ball for each agent is similar ($n = 4$, and slight difference in $\log |\boldsymbol{R}|$ and AIC) from Table3, the estimated state vectors are different from each other because there are differences in several factors such as tilt angle, the velocity of the motor and the angle of steering. We checked what happened if we replace the local predictive models between the passer and the shooter. Eventually, the large prediction errors of both side were observed. Therefore the local predictive models can not be replaced between physical agents. Figure 16 shows a sequence of images where the shooter shoots a ball which is kicked by the passer.

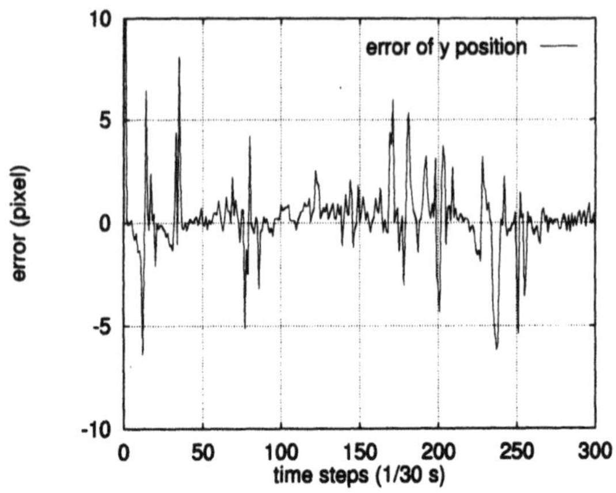

(a) y position of the ball

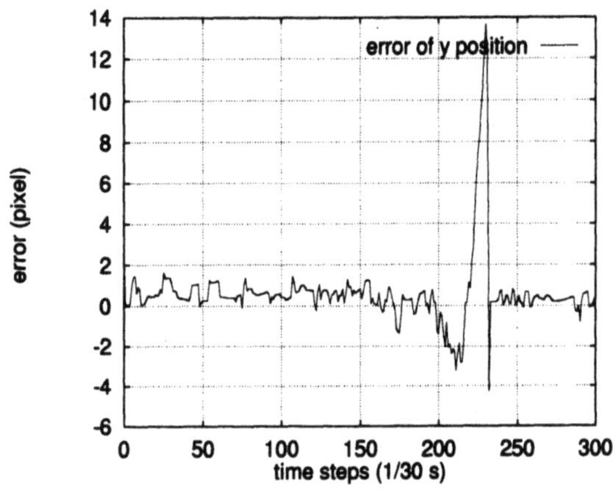

(b) y position of the left-upper of the goal

Figure 15: Prediction errors in the real environment

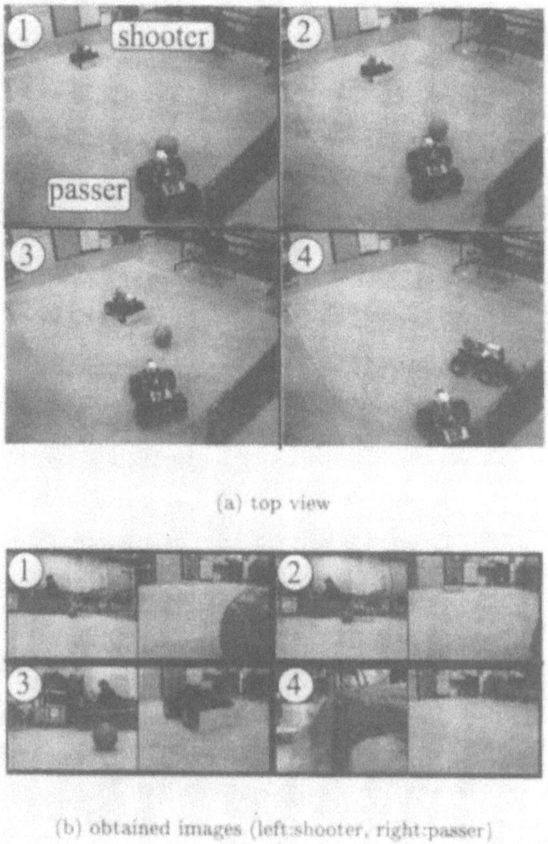

Figure 16: Acquired behavior

4 Concluding Remarks

Along with examples of soccer robots, we have claimed the importance of the design of the internal structure which reflects the complexity of the interactions with the agent's environment. Although the task and the environment seem simple and limited, the design of the soccer robots includes a variety of the fundamental and important issues as a standard problem in robotics and AI [1]. We expect that more agents in the field cause much higher interactions among them, which emerges a variety of more complex behaviors.

Acknowledgment

We like to thank Eiji Uchibe, Yasutake Takahashi, Masateru Nakamura, and Chizuko Mishima for their supports of our work described in this paper.

References

[1] H. Kitano, M. Asada, Y. Kuniyoshi, I. Noda, E. Osawa, and H. Matsubara. "robocup: A challenge problem of ai". *AI Magazine*, 18:73–85, 1997.

[2] M. Asada, S. Noda, and K. Hosoda. Action-based sensor space categorization for robot learning. In *Proc. of IEEE/RSJ International Conference on Intelligent Robots and Systems 1996 (IROS '96)*, pages 1502–1509, 1996.

[3] Y. Takahashi, M. Asada, and K. Hosoda. Reasonable performance in less learning time by real robot based on incremental state space segmentation. In *Proc. of IEEE/RSJ International Conference on Intelligent Robots and Systems 1996 (IROS96)*, pages 1518–1524, 1996.

[4] E. Uchibe, M. Asada, and K. Hosoda. Vision based state space construction for learning mobile robots in multi agent environments. In *Proceedings of 6-th European Workshop on Learning Robots, EWLR-6*, pages 33–41, 1997.

[5] C. J. C. H. Watkins and P. Dayan. "Technical note: Q-learning". *Machine Learning*, 8:279–292, 1992.

[6] R. S. Sutton. "Special issue on reinforcement learning". In R. S. Sutton(Guest), editor, *Machine Learning*, volume 8, pages –. Kluwer Academic Publishers, 1992.

[7] M. Mataric. "Reward functions for accelerated learning". In *Proc. of Conf. on Machine Learning-1994*, pages 181–189, 1994.

[8] M. Asada, S. Noda, S. Tawaratumida, and K. Hosoda. Purposive behavior acquisition for a real robot by vision-based reinforcement learning. *Machine Learning*, 23:279–303, 1996.

[9] Peter Stone and Manuela Veloso. Using machine learning in the soccer server. In *Proc. of IROS-96 Workshop on Robocup*, 1996.

[10] E. Uchibe, M. Asada, and K. Hosoda. Behavior coordination for a mobile robot using modular reinforcement learning. In *Proc. of IEEE/RSJ International Conference on Intelligent Robots and Systems 1996 (IROS96)*, pages 1329–1336, 1996.

[11] Michael L. Littman. Markov games as a framework for multi-agent reinforcement learning. In *Proc. of Conf. on Machine Learning-1994*, pages 157–163, 1994.

[12] Tuomas W. Sandholm and Robert H. Crites. On multiagent Q-learning in a semi-competitive domain. In *Workshop Notes of Adaptation and Learning in Multiagent Systems Workshop, IJCAI-95*, 1995.

[13] Long-Ji Lin. Self-improving reactive agents based on reinforcement learning, planning and teaching. *Machine Learning*, 8:293–321, 1992.

[14] H. Akaike. A new look on the statistical model identification. *IEEE Trans. AC-19*, pages 716–723, 1974.

[15] W. E. Larimore. Canonical variate analysis in identification, filtering, and adaptive control. In *Proc. 29th IEEE Conference on Decision and Control*, pages 596–604, Honolulu, Hawaii, December 1990.

[16] Peter Van Overschee and Bart De Moor. A unifying theorem for three subspace system identification algorithms. *Automatica*, 31(12):1853–1864, 1995.

[17] C. J. C. H. Watkins and P. Dayan. Technical note: Q-learning. *Machine Learning*, pp. 279–292, 1992.

[18] E. Uchibe, M. Asada, and K. Hosoda. Behavior coordination for a mobile robot using modular reinforcement learning. In *Proc. of IEEE/RSJ International Conference on Intelligent Robots and Systems 1996 (IROS96)*, pp. 1329–1336, 1996.

[19] M. Inaba. Remote-brained robotics : Interfacing AI with real world behaviors. In *Preprints of ISRR'93*, Pitsuburg, 1993.

Using Human Development As A Model For Adaptive Robotics

Rodney A. Brooks
MIT AI Lab, 545 Technology Square, Cambridge, MA 02139 (USA)
email: brooks@ai.mit.edu

Abstract

Over the years an approach to robotics developed that relied on the notion of accurate control; position accuracy was the ultimate measure of success, and in the service of that goal kinematics, dynamics, and path planning became dominant, with some aspects of accurate three dimensional recovery receiving attention. More recently two new threads have developed. In one, the structure of the task is analyzed and the robot is coupled into the environment in such a way that uncertainty is reduced as a consequence of action in the world. The second operates in much less structured environments and uses collections of simple behaviors to guarantee success. In this paper we suggest yet another approach to robotics. Here we suggest combining many theories from artificial intelligence, cognitive sience, physiology and neuroscience in order to build a robot which discovers the structure of the world for itself through a development process in the world.

1 Introduction

Robotics has gotten stuck in a difficult place.

In order for our robots to manipulate the world we need to have exact kinematic models of the robots, very precisely calibrated dynamic models, and need to reconstruct extremely accurate three dimensional models of the world. Most effort goes into dealing with these accuracy issues.

This is the same place we were with mobile robots twelve years ago. Behavior-based robotics has changed that to various degrees although there are the more theoretical papers which continue to treat the old issues as being of prime importance. Mobile robotics has gotten to the point of having a robot which is capable of operating totally autonomously on another planet, and robots which can routinely drive along normal freeways crowded with cars and trucks driven by people, by rejecting the idea of explicit three dimensional models. The successful robots operate without explicit kinematic models, nor with any three-dimensional reconstruction of the world about them.

It is time that we tried to do the same for robots that are to manipulate the world around them. We believe that a fruitful way to tackle this problem is to look at animal models—living systems can adapt in seconds to very large kinematic changes when required and routinely adapt to widely varying dynamic changes.

2 The Cog Project

The Cog project is an attempt to explore these issues. The humanoid robot is meant to operate in the world with people—certainly these are things which can not be modelled geometrically in any simple way. Furthermore Cog is meant to cooperatively carry out manipulation tasks with people.

There are two main ideas driving the design of the robot Cog as a humanoid. The first is based on the ideas of Lakoff (1987) and Johnson (1987) that all human representations are ultimately grounded in sensory motor patterns. Thus to build an artificial system with similar grounding to a human system it is necessary to build a robot with human form. An open question remains as to whether our decision to use metal and silicon rather than flesh and neurons disqualifies our robot from having enough similarity with humans for this to be true. The second reason that our robot has humanoid form is so that people will interact with the robot in social ways, just as they would with another person.

Our major conclusions to date are (1) that much of what people perceive as being highly characteristic of human behavior can be achieved

Figure 1: The Cog robot built at the MIT Artificial Intelligence Laboratory. The robot is approximately adult human in size, but it has no legs.

with rather low level sets of sensory motor processing, and (2) that people, even aware observers, naturally fall into modes where they social lead a system that is less than fully human, so that to a third observer it appears to be having human like interactions with the person.

3 The Physical Robot

The Cog robot is a human sized torso with a head and two arms. Cog has no legs. The details of Cog have changed over time as we have repeatedly re-engineered its subsystems as a result of experience with earlier versions of them.

Cog has three degree of freedom hips. Its torso has a touch sensitive chest plate. The hip motors all have temperature sensors mounted on them so that the system gets a kinesthetic sense of how hard it is working.

The neck has three degrees of freedom. The eye system has one shared tilt degree of freedom, and two independent pan degress of freedom. All motors are DC motors with built in gearheads and 16 bit encoders. The tilt and pan motors for the eyes use tendon drive systems for further geardown. Every motor in the robot has its own servo processor, a Motorola 6811. The servo loops run at 1000Hz. Usually the position, velocity, and/or stiffness commands for a motor are updated at at most 30Hz.

The two arms are each six degree of manipulators. Each joint is a DC motor with a gearhead and a tendon drive system. Additionally there is a spring in series with each arm motor. It has strain gauges mounted on it, and potentiometers in the load bearing end. With these two sensors it is possible to measure the torque in each joint. These series elastic actuators are controlled as virtual springs, with equilibrium point control layered on top of that (Williamson 1996).

Each eye consists of two cameras. These are mounted one above the other, with one have a 120 degree field of view, and one having a 23 degree field of view. These correspond to peripheral and foveal fields of vision in a human, although the peripheral field is smaller than that of a human, and the foveal is wider. In both cases the actual resolution is much less than that of a human. There are also two microphones, in the ear positions with elementary pinnae, which are used for simultaneous localization of multiple sound sources.

The head has a vestibular system based on gyroscopes and tilt sensors.

The processing system for Cog has evolved over time. The oldest part of processing system is a network of Motorola 68332s. Each of these runs a multi-threaded subset of Common Lisp. The nodes of the network communicate point to point via shared dual-ported RAMs. The topology of the network is established by plugging in cables joining each node to up to 8 different shared memories. We have a "hot-wiring" system in place so that cables can be moved while the system is powered up and the processors are running. There is an additional slower network connecting the 68332s via their serial ports to a front end Macintosh computer, where there is an interface written in Macintosh Common Lisp. This allows listeners to be open on each 68332 and for us to dynamically reload debugged program fragments while the system continues to run. It also gives us a number of graphical tools such as digital oscilloscopes that can track sensor and motor values in real time. The 68332's can talk to slave Motorola 6811 processors which act as servo processors for Cog's motors (one processor per motor, servoing at 1000Hz), and exchange packets with their slaves at an aggregate rate of 400Hz per 68332.

A newer visual cortex now runs on Cog (although older vestigial vision code runs on the 68332 network). It is a fixed topology network of Texas Instrument C40 digital signal processors. They run C code with very little timing flexibility—we arrange them to simply run as fast as possible, streaming visual and aural information through at maximum possible rates from one DSP to another via their standard communication ports (up to six per processor). They interface to a host PC and to the 6811 processors via ISA cards. The host PC can be used to monitor them, but there are no facilities for dynamically re-linking subparts of the overall programs.

A third, newer network has recently been added. It has high performance single board PCs as its nodes, and modulates the DSP network, while receiving outputs from many nodes within that network. It is used for higher level functions within Cog.

Figure 2: Cog entraining to a slinky toy. It has no kinematic or dynamic models, but it repsponds to the forces it feels as it moves its limbs.

4 Building Many Behaviors

The original outline of the Cog project was given in Brooks & Stein (1994). During the beginning of the project we worked on isolated subsytems, demonstrating the ability to mimic many basic capabilities of the human system. The system is organized in much the same way as the subsumption architecture (Brooks 1986) has been used to build navigation systems for mobile robots. Simple low level behaviors are recruited by higher level behaviors, not by explicitly invoking them, but by letting them run, and modulating them when appropriate.

Early behaviors included making the arms compliant and safe for interaction with people (Williamson 1996), localizing sounds (Irie 1995), and learning to saccade to motion and sound cues (Ferrell 1996). More complex behaviors were built on top of this, such as learning to orient towards motions and sounds, and then learning to reach out a hand towards the source (Marjanović, Scassellatti & Williamson 1996). All of these simple behaviors were based on the notion that there should be no external calibration—the system should operate with unsupervised learning, and should adapt to local variations. In fact when learning to orient and reach we changed the kinematics of the eyes, and more recently the arms, and have not had to change a single line of code—Cog simply relearns how to operate with new arrangements of its sensors and actuators.

Over the last year we have built up a large repertoire of slightly more complex behaviors. These include visual smooth pursuit, a vestibular occular reflex, segmentation by occular vergence and zero disparity detection, finding face pop outs, localizing human voices, two armed manipulation of toy slinkies, pendulums, and cranks, coordinate eye and neck motions, and simple mechanisms of attention.

Figure 2 show's Cog playing with a slinky. Two neuron oscillators control its arm joints and they entrain to the small forces produced by the slinky as the mass shifts from side to side. When proprioception is not present the two arms oscillate at different natural frequencies and the slinky does not move in a rythmic manner. With proprioception, no matter how the arms are moved about in their workspace (and there is no feedback of just where they have been manually placed) the two arms very quickly entrain and cooperate in playing with the slinky. A stable rythmic motion is produced.

Exactly the same neural circuits have been used to let the robot play with pendula of different natural frquencies. Without proprioception the robot waves them about randomly. With proprioception the robot quickly entrains to the characteristic frequency of the pendulum and excites it at that frequency.

The fundamental idea behind thus controlling the arms is that it is based on the dynamics they *experience* in the world. It is not based on detailed *a priori* models of what will be encountered. Rather, just as with animals and in par-

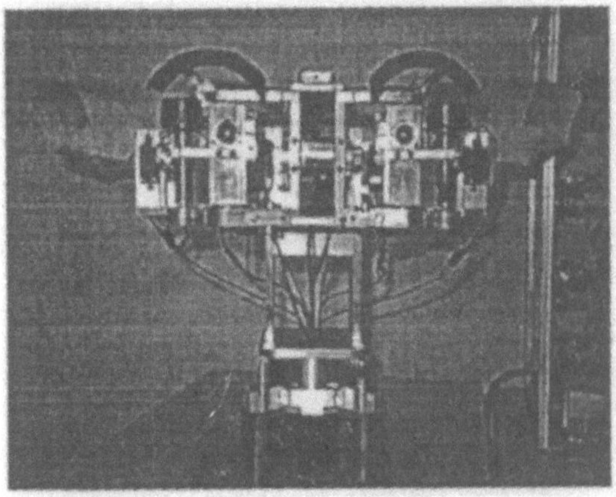

Figure 3: A new version of Cog's head. We are adding expressive elements to the face so that humans interacting with the robot can get feedback from the robot as the robot itself tries to get feedback from the human.

ticular people, the system adapts to the detected dynamics of the world in order to carry out tasks.

This is a significantly different approach to that used for conventional manipulation. It is early in the development history of this approach and it remains to be seen how far it can be pushed.

5 Human Interaction

Over the course of operating the Cog robot we have become aware of just how powerful is the human form of Cog in encouraging people to interact with it in human like ways. As it saccades from person to person as they speak during a conversation it appears to be following the conversation. As people engage in play with it they unconsciosly capitalize on pieces of its dynamic with the world and engage it in play for which it does not have intentions built in—the person playing with it projects their intentions on to the robot and they manipulate it into seemingly cognitive behavior.

This has brought about a significant realization for the Cog research team. We can capitalize on this natural behavior of people to provide the robot with structured training and bias during interaction. Figure 3 shows a new version of Cog's head which now includes expressive elements such as eyebrows and moveable ears. They are rather exaggerated in form to make it easier to interpret the emotional states that can be displayed with the system. In the future we will add lips, eyeballs, and eyelids.

In order to extend the interactions between Cog and people it is necessary to add a lot of perceptual machinery (and this process is underway) so that it can determine both the emotional state and the focus of attention of its human teachers.

In the meantime the following, taken from Turkle (1995), describes the experience of one skeptic in interacting with Cog in a much early instantiation of its software:

> Cog "noticed" me soon after I entered its room. Its head turned to follow me and I was embarrassed to note that this made me happy. I found myself competing with another visitor for its attention. At one point, I felt sure that Cog's eyes had "caught" my own. My visit left me shaken–not by anything that Cog was able to accomplish[1] by by my own reaction to "him". For years, whenever I had head Rodney Brooks speak about his robotic "creatures", I had always been careful to mentally put quotation marks around the word. But now, with Cog, I had found that the quotation marsk disappeared. Despite myself and despite my continuing skepticism about this researhc project, I had behaved as though in the presenced of another being.

The humanoid form is thus a dangerous intellectual weapon that researchers may accidentally use against themselves. It is very easy to slip into the role of maximizing the illusion of success, as people are predisposed to interpret the actions of a humanoid as being driven by great intelligence. We must avoid such showmanship. At the same time there are subtle forces to consider in the other direction. Rather than being worried about observers over anthropomorphizing our robots, perhaps we should realize the possibility that we routinely over anthorpomorphize people.

References

Brooks, R. A. (1986), 'A Robust Layered Control System for a Mobile Robot', *IEEE Journal of Robotics and Automation* **RA-2**, 14–23.

Brooks, R. A. & Stein, L. A. (1994), 'Building Brains for Bodies', *Autonomous Robots* **1**, 7–25.

[1] Turkle was completely cognizant of the rather low level behaviors running on Cog during her visit.

Ferrell, C. (1996), Orientation Behavior Using Registered Topographic Maps, *in* 'Fourth International Conference on Simulation of Adaptive Behavior', Cape Cod, Massachusetts, pp. 94–103.

Irie, R. E. (1995), Robust Sound Localization: An Application of an Auditory Perception System for a Humanoid Robot, Master's thesis, Massachusetts Institute of Technology, Cambridge, Massachusetts.

Johnson, M. (1987), *The Body in the Mind: The Bodily Basis of Meaning, Imagination, and Reason*, The University of Chicago Press, Chicago, Illinois.

Lakoff, G. (1987), *Women, Fire, and Dangerous Things: What Categories Reveal about the Mind*, University of Chicago Press, Chicago, Illinois.

Marjanović, M., Scassellatti, B. & Williamson, M. (1996), Self-Taught Visually-Guided Pointing for a Humanoid Robot, *in* 'Fourth International Conference on Simulation of Adaptive Behavior', Cape Cod, Massachusetts, pp. 35–44.

Turkle, S. (1995), *Life on the Screen*, Simon and Schuster.

Williamson, M. (1996), Postural Primitives: Interactive Behavior for a Humanoid Robot Arm, *in* 'Fourth International Conference on Simulation of Adaptive Behavior', Cape Cod, Massachusetts, pp. 124–131.

Developmental Processes in Remote-Brained Humanoids

Masayuki INABA
Dept. of Mechano-Informatics, University of Tokyo
7-3-1 Hongo, Bunkyo-ku, Tokyo 113 (Japan)
email: inaba@jsk.t.u-tokyo.ac.jp

Abstract

Across the ages, one dream of intelligent robotics has been humanoid robots. To realize a real humanoid requires long developmental cycles. If we can understand the developmental cycles and explore new ideas on common operations for the developmental processes, it may help to generate the next developmental stages. To do this, we have built an environment with remote-brained humanoids and progressed through several stages in development. This paper introduces the idea of research on the developmental processes and describes the several development stages of remote-brained humanoids.

1 Introduction

Across the ages, one dream of intelligent robotics has been humanoid robots(Kato, Ohteru, Kobayashi, Shirai, and Uchiyama 1974; Sugano, Tanaka, Ohoka, and Kato 1985; Brooks and Stein 1994; Inoue and Sato 1994). Building a humanoid is a long process. The available technology during the process decides the design and the goal. The researcher who designs the goal may take long time to achieve the goal and learn many things from the building process. Once the goal is achieved, the developed humanoid will be recorded as the result of long time research. The researcher publish the ideas learned and may make the developed humanoid available in public.

Because developing a humanoid usually takes a long time, the next generations take over from the ealier researchers. They design new goals and try to achieve them using new technologies available for the next generations. As the available technologies are new, the goal for the new humanoid becomes a more advanced level and the process of development starts from the new stage different from the former development done by the former researcher. The humanoid requires many generations in nature. This makes the developmental research harder since the researchers can hardly inherit the former ideas in development.

Although these many development cycles over generations may produce an advanced humanoid which may be usable in the real world, what can we learn about the principles of development from these development cycles? If we can learn about common or essential process from the series of developmental cycles, we may apply it for the next steps of further development.

In order to learn about development of humanoid, we have aimed to build a research environment where we can realize several generations of humanoids and overview the developmental processes done over generations. As one of the big problems in the real integration of a humanoid is to combine a large scale brain and a limber body into the humanoid system, we have taken a methodology of the 'remote-brained robot' approach(Inaba 1993; Inaba, Kagami, Kanehiro, Takeda, Oka, and Inoue 1996).

This paper introduces the research on developmental processes in humanoid and describes the processes done in our remote-brained humanoids.

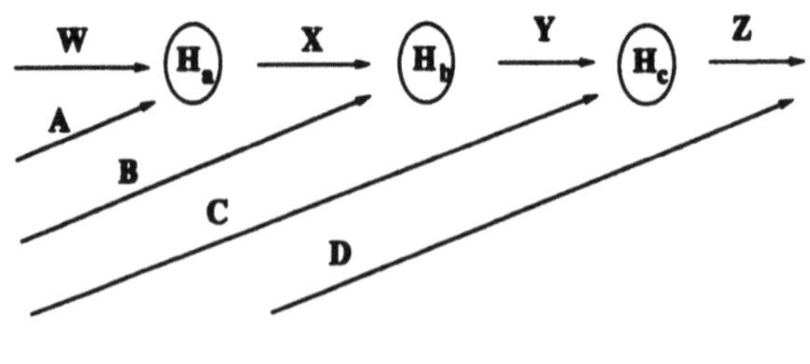

Figure 1: Overview of Research on Developmental Processes

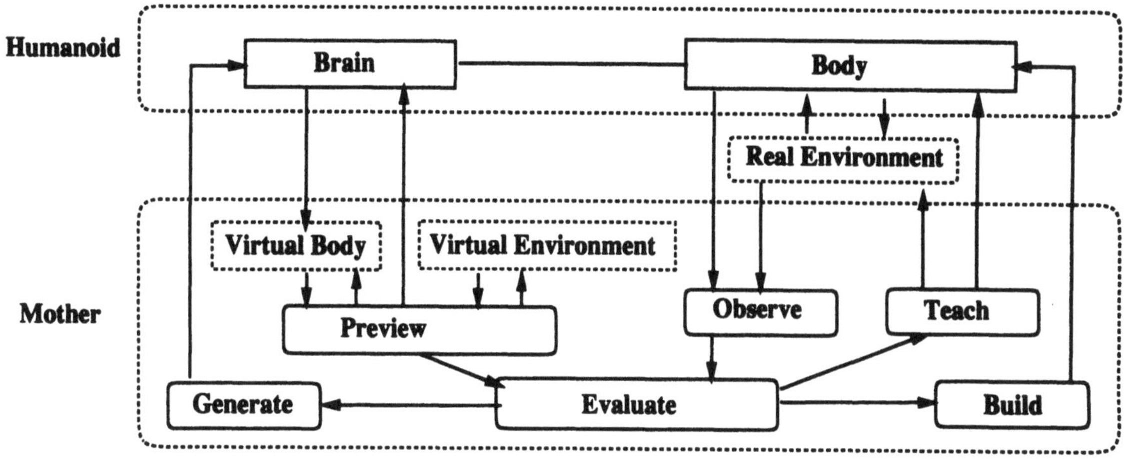

Figure 2: Mother Operations

2 Research on Developmental Processes

In order to see the developmental processes done over generations for building humanoids, we have to clarify the research framework. Figure 1 is shown to explain the idea of the framework. An arrow is a developmental process to a stage corresponding to a design goal of the research. H_a, H_b, H_c are some of the stages. The processes A,B,C,D are done by the researchers who can not share or inherit the former developmental processes. The processes W,X,Y,Z forms the phased stages. Each of these processes has the begining stage and the goal stage. It is the differential process that can compare the difference between the two stages.

We aim at research to build those differential developmental processes and learn from them about the essentials of development. As the reality of humanoids makes the development harder, we need a good methodology to advance real humanoids and an usable environment to do the research. The questions in the research of the developmental processes are: What is the motivation to make the differential change of the current development? What did the change make possible? What was the essential operation to advance the current change. Is there commonness in the series of the differential developmental processes? Is it possible to perform the next developmental process with the same strategy of the former process? What is necessary for the change in each processes to be expressed by the common one?

3 Mother Operations to Build Developmental Processes

In order to start the research in developmental processes, we have tried to investigate what processes are required to engender and grow an embodied robot and to build a model of a mother(Inaba, Kagami, Kanehiro, Nagasaka, and Inoue 1996). The structure of the mother model is proposed in Figure 2 which shows the relationship between a robot and a mother and the operations done by the mother. This model is based on the remote-brained design of a robot. In this model, a mother has six operations: **Generate**, **Preview**, **Observe**, **Teach**, **Build** and **Evaluate**. So far, the interface between **Body** and **Real Environment** has been emphasized in robotics and the new wave in AI. However, this model stresses the importance of the interface between a robot and its mother.

Preview simulates the situation of the real environment and **Body** behaviors according to action values given from **Brain**. It shows simulated results to the **Brain**. **Observe** monitors behaviors of the **Body** through the real sensors of the mother. **Evaluate** receives observed behaviors from **Preview** and **Observe** and selects appropriate operations from **Teach**, **Generate**, **Build**. In order to pursue evolutional and physically grounded intelligence, it is crucial to investigate how to build these mother operations.

In the remote-brained approach, **Brain** is placed in the mother side. The advantages of

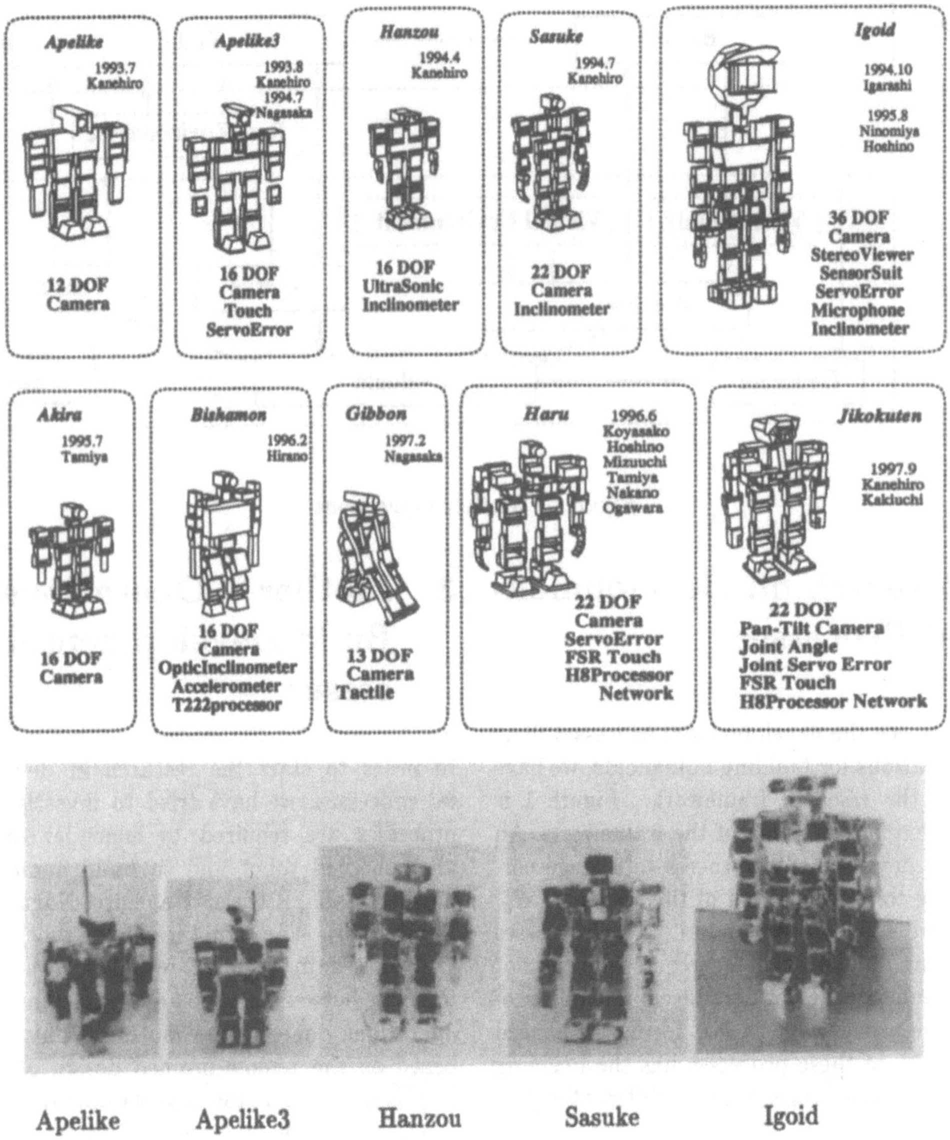

Figure 3: A series of humanoid robots

remote-brained approach can be explained by this structure. One advantage is that the mother operations are divided into three independent operations; **Generate** is a software operation, **Build** is hardware construction, **Teach** is interaction such as teaching-by-showing(Kuniyoshi, Inaba, and Inoue 1989). A mother can employ experts to suit each operation.

Second, the operation **Generate** is simplified because the brain is not placed inside the robot body but in the mother. This means that it is possible to build a loop from **Preview** to **Generate** by software on the same computer. If the mother has real eyes to perform an **Observe** process, the system can have parallel evaluation processes through both **Preview** and **Observe** because the **Brain** is shared by both virtual body and real body. This allows the mother to compare the result with those paths and utilize them to perform **Evaluate**.

4 The Remote-Brained Humanoids

The operation **Build** is done by a human serving as mother for the present. We have designed and built several remote-brained humanoids in differ-

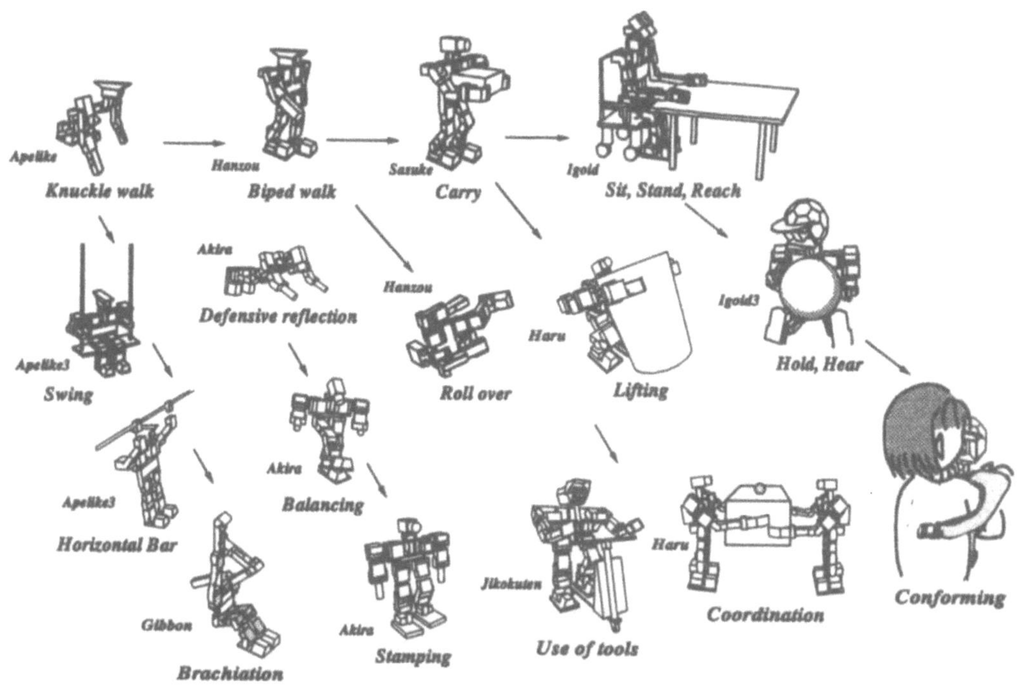

Figure 4: Developed actions

ent developmental stages. Figure 3 shows some of them. The date of build, the human mother and the key features of each body are described in it.

Apelike(Inaba 1993) is the first model of our humanoid; it has 12 joints and a camera. It has 3DOF per leg and 3DOF per arm. It performs knuckle walk and vision-based walking. As it has no joints for a head, the view direction is fixed to the body.

Apelike3(Inaba, Kanehiro, Kagami, and Inoue 1995b) has 2DOF in neck and a yaw axis in the shoulder. It can control the view direction by the neck and turn the walking direction by using the arms. It could climb curved stairs by using visual feedback. Apelike3 is used for experiments of swing motion and horizontal bar.

Hanzou(Inaba, Kanehiro, Kagami, and Inoue 1995a) has 4DOF per leg. A ankle joint is added to Apelike. It allows Hanzou to roll the body and stand on a single leg. Hanzou can perform bipedal walk. As it does not have the yaw axis in the legs, it is hard to change the direction of the walk. Hanzou has 2DOF shoulder. It can roll over by coordinating the arms and the legs. Hanzou can stand up even when it falls down in biped walking.

Sasuke(Kanehiro, Inaba, and Inoue 1996) has 5DOF per leg. A joint of yaw axis is added to the leg of Hanzou. Sasuke can perform biped turns and walk on a curved line. As it does not need arms to change the walking direction, it can carry an object with two arms and walk in any position.

Igoid(Inaba, Igarashi, Kagami, and Inoue 1996) has 6DOF per leg. It has performed sitting on a chair, standing up using arms for support, and crawling. As it has 6DOF per arm and 4DOF per hand, it can manipulate an object. When it sits on a chair, it can reach an object on a table using vision. Igoid3 has the same body as Igoid but with the stereo viewing adapter as the head. The stereo viewer provides a binocular view to a single camera (Inaba, Hara, and Inoue 1993).

Akira has 2DOF neck, 3DOF arm, 4DOF leg. It does not have grippers. It is designed to study the generation of whole-body motions. A puppet which has the same structure as Akira is designed for the master controller to teach whole body motions.

Bishamon is the first body which has a microprocessor on board. The processor filters the radio transmitted signals and transmits the sensor information of 3D acceleration sensor to the brain.

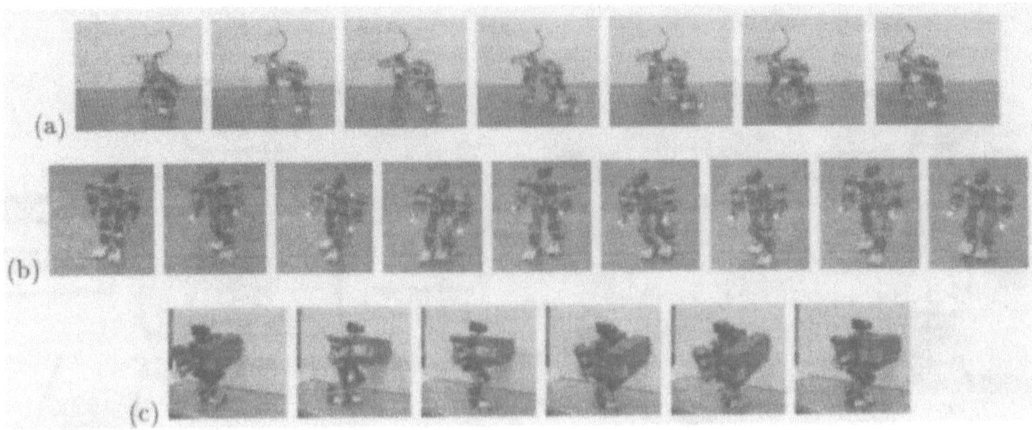

Figure 5: Development of Locomotion: (a) Knuckle Walking, (b) Biped Walking, (c) Carrying an Object

Gibbon is specially designed for experiments in brachiation. It has 3DOF leg, 2DOF arm, 1DOF gripper and 1DOF neck.

Haru is designed as a platform for sensor based behaviors. It has onboard network for information channels to receive all the sensors: load torque and angle of all joints, 3D accelerometer, and force sensor under the feet. We have developed four Harus to share sensor-based software that is developed by different mothers independently.

Figure 4 shows the flow of the developed typical actions of all the remote-brained humanoids. The development of the horizontal series of the top is based on the body structure. The downward arrows relate the development of the sensors on the body and the structure of the brain software.

5 Behavior Development in Locomotion

The development in locomotion has been advanced through the **Observe** and **Generate** operations by human mothers so far. Figure 5 shows real walking sequences developed by the Apelike3, Hanzou, Sasuke respectively. A walking sequence is generated through **Observe** and **Generate** loop based on the structural differences in them. If the mother has only the virtual body simulator inside, it is able to build a teaching loop through the **Preview**. This makes the **Generate** operation interactively. The next stages are to make methods for **Teach**, to make **Observe** by computer vision and to make **Preview** by computer program.

In order to connect an automatic loop through the **Build** operation, the mother can make use of the humanoids explained above. That is, if the mother has computer vision-based **Observe**, it can change the body structure by selecting one from already developed humanoids. As the remote-brained system does not place the brain inside body, the mother system can continue raising the same brain after switching the body. This is another feature of the remote-brained system.

The goal here is to know how the brain for Apelike3 should be raised to accept the body of Hanzou and so on. Another goal is to know whether there is a brain which accepts all body structures, and whether there is a general method to build such brain. In order to approach these questions, we have tried to share the software environment in the mother environment. The software of next humanoids such as Igoid and Haru is designed over the software developed for Apelike3, Hanzou, and Sasuke.

6 Development in Brain Generation

Figure 6 shows the swing motion generated by a brain program written by a human developer. The robot controls the legs and the upper body to change the center of gravity of the robot body. In this experiment the mother is a human developer. The human mother analyzed the set of the experiment and wrote a brain program that mea-

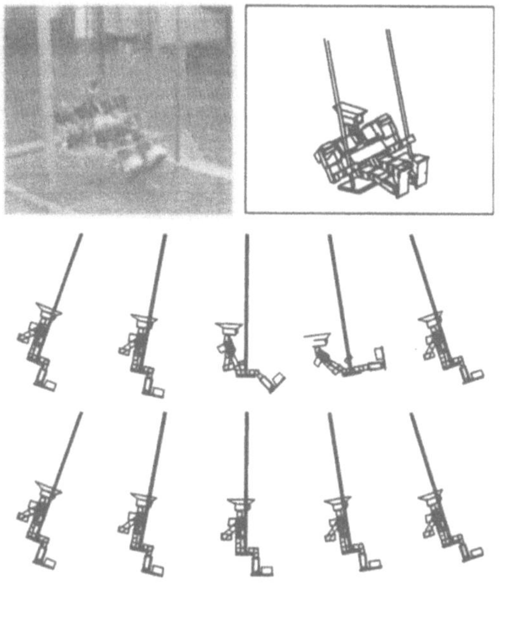

Figure 6: Swing motion done by a control program written by a human mother

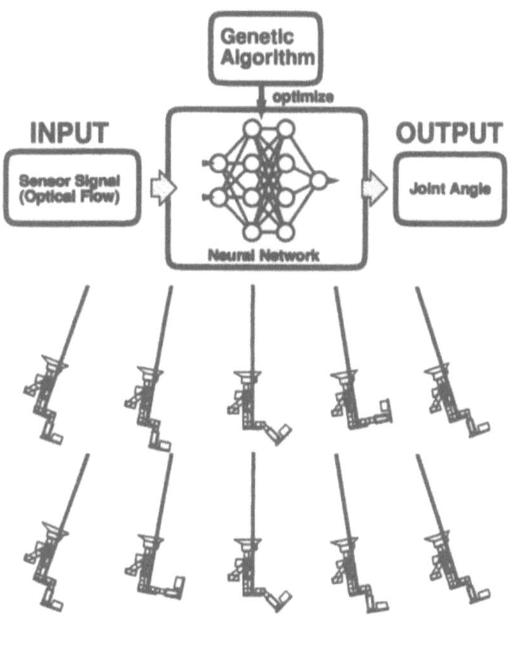

Figure 7: Swing motion done by a neural network brain generated by genetic algorithm

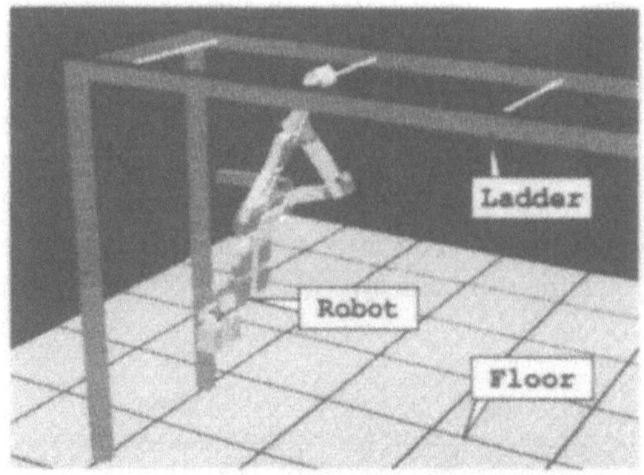

(a) The environment for brachiation

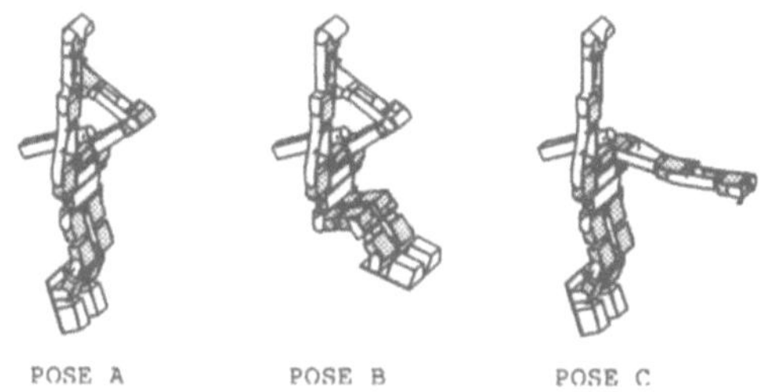

(b) Given poses

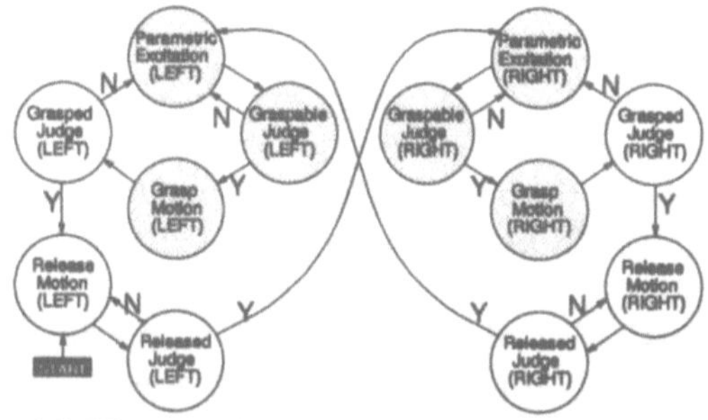

(c) The transition network for brachiation

Figure 8: Brachiation Experiment

sures the parameters such as phase and cycle time of the swing motion at initial motion and issues motion commands using those parameters(Inaba, Nagasaka, Kanehiro, Kagami, and Inoue 1996). The **Generate** operation is to provide the parameter measure program (PMP) and the parameterized execution program (PEP). The robot executes them. There is two ways to execute them. One is execute PMP before PEP. Another is execute PEP during executing PMP.

Figure 7 shows the case where the brain is generated by a computer mother. The brain is a neural network that receives the optical flow in the vision and controls the knee joint. The neural network is generated by a computer program using genetic algorithm. The computer mother is the program that simulates the system of the swing and the robot and evaluates the brain behavior in the simulated world(Nagasaka, Inaba,

Inoue, and Konno 1997) . In this case the human developer prepares the simulation world and the brain generator. The brain is generated based on the evaluation rule of the task. The computer mother **Preview** the body motion in the simulated world and **Evaluate** the actions executed by the brain. The fitness to **Evaluate** is given by a human in this system.

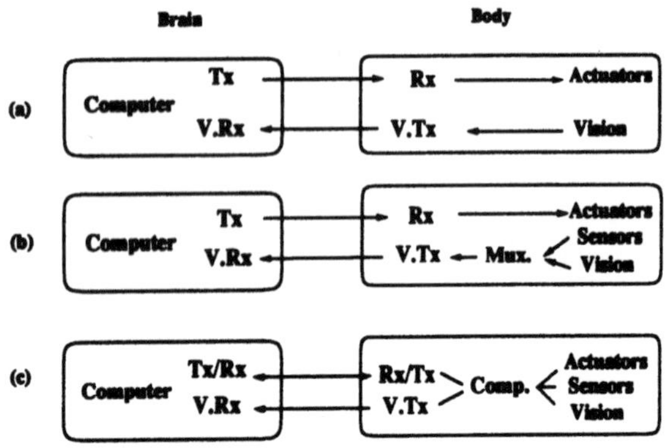

Figure 9: Development in brain-body interface

In this GA-based swing experiment, the output of the brain is 1DOF of the knee. In order to extend the output DOFs, another experiment was done in performing vision-based swing with horizontal bar. In this experiment, the output has 2DOF, crotch angle and knee angle. The GA-based mother generates the neural networks which has 2DOF output.

As the next stage, brachiation is designed. Brachiation requires sequential execution including conditional branches. If the robot can't hold a bar, it has to accelerate its swing motion by a single arm and try reaching the bar again. Gibbon has succeeded to performd the brachiation motion using vision and gripper sensors. Figure 8 shows the brachiation experiment. The poses A and B in (b) are given by a human developer to make swing. The pose C is given for grasping the bar. The brain is a compound program generated by the human coded program and the GA-based learning. The brain is described as the state transition network as shown in Figure 8 (c). Each node of the network shows an action. The node modules are (1) Parametric excitation, (2) Graspable judge, (3) Grasp motion, (4) Grasped judge, (5) Release motion, (6) Released judge. In this experiment, the shaded nodes ((1), (2), (3)) are generated by using GA and the others are generated by the human programmer. (K.Nagasaka,

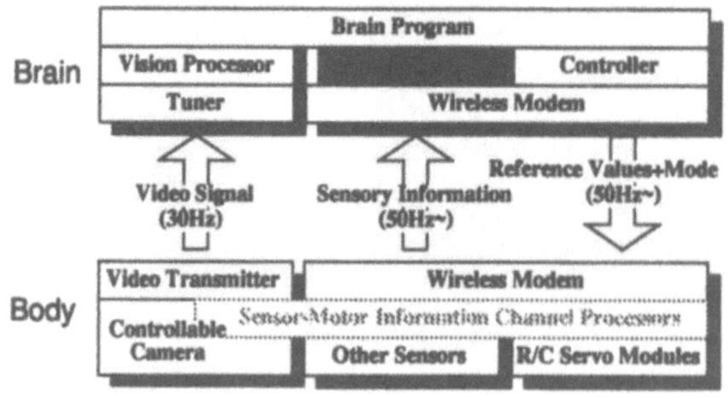

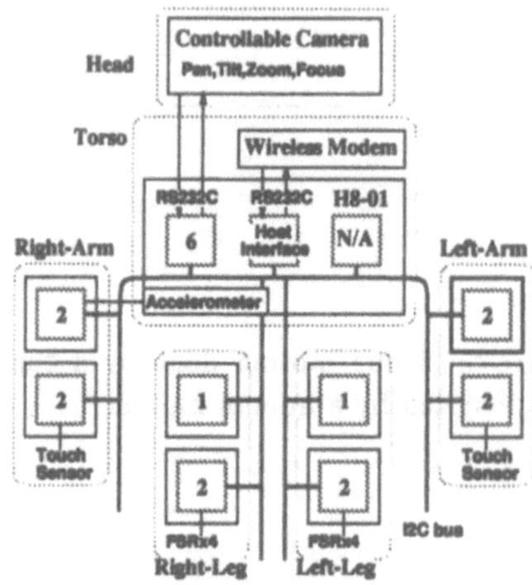

Figure 11: The new humanoid Haru, Jikokuten, with the information channel processors on body

A.Konno, M.Inaba, and H.Inoue 1997).

7 Development on Brain-Body Interface

The physical connection between the brain and the body is important in the remote-brained approach. It dominates the possible behaviors and

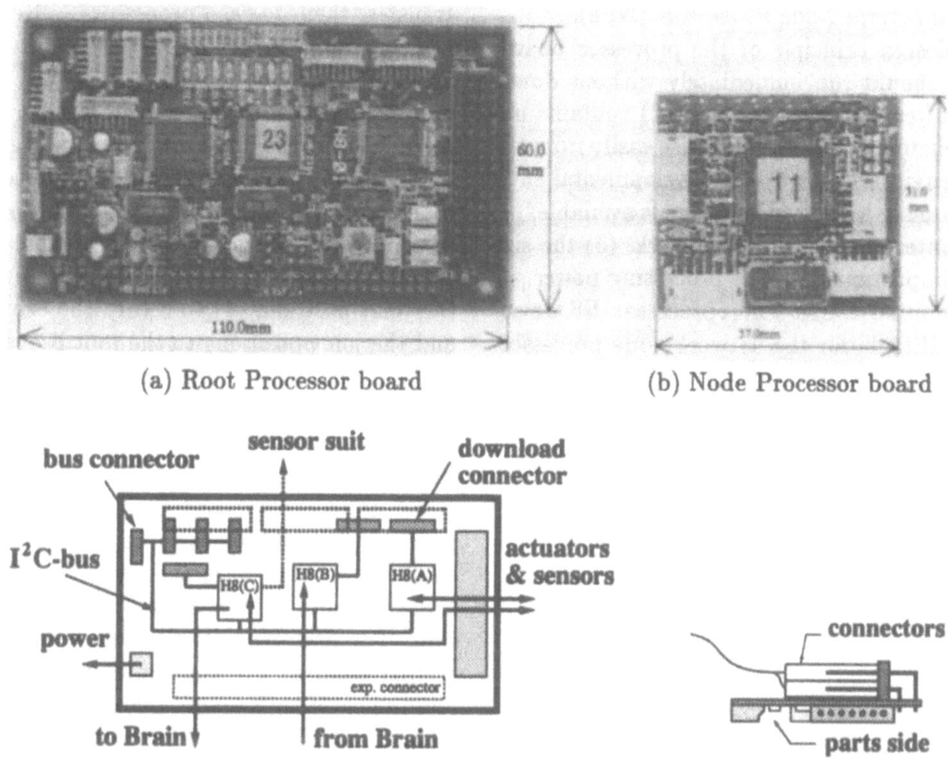

(a) Root Processor board (b) Node Processor board

(c) Interfaces of Root Processor (d) Side view of Node Processor

Figure 10: Processor Boards for Information Channels on Body

the developable stages. We have made three stages in the connection.

Figure 9 shows several ways to implement remote-brained robots. The left is the brain part and the right is body part in the figure. Tx and Rx mean transmitter and receiver. V.Tx and V.Rx mean video transmitter and video receiver. Comp. is a computer set in the diagram.

The connection (a) was adopted at the first remote-brained system(Inaba 1993; Inaba, Kanehiro, Kagami, and Inoue 1995b). The sensor was just vision. In order to put the other sensors on body, we adopted the connection (b), which we called the image-based approach(Inaba, Kagami, Sakaki, Kanehiro, and Inoue 1994), which multiplexes the sensor signals onto the vision signal and forms the sensor image which describes all the input to the brain from the body. The key to simplify the configuration for multisensor integration is the method for multiplexing different kinds of sensor signals. We have developed two different multiplexers: mechanical method and electrical method(Inaba 1996). Apelike3 sends back the joint error signal and the tactile sensor signals through the mechanical method using LEDs put in front of the camera head. Igoid uses the electrical method to merge the tactile signals of the sensor suit onto the stereo image of the camera. Multiplexing via the video signal gives a convenient method for practical integration.

The system based on the connection (c) has a computer to support the information connection between the brain and the body. Bishamon is the first remote-brained adopting the connection (c). It has two different radio modules connected to an onboard micro processor. The processor filters the reference values from the brain and sends back the sensor signals to the brain. The video signal uses the different channel. The new humanoids using the onboard network use the connection (c).

8 Body Development with Information Channel Processors

In order to put more sensors distributed on the body, we have designed and developed the modules to build an information channel system on the body. The modules are shown in Figure 10. The channel system consists a Root Processor

(RP) and several Node Processors (NPs).

The design principle of the processor boards are: (1) should run immediately without downloading after power-on or reset, (2) contains unvolatile memory that we can revise easily from the mother environment, (3) the developmental environment using high level language is available, (4) has the interface to build a network, (5) the size is more important than the processing power.

We adopted a 8bit microprocessor H8 developed by Hitachi for the network node processor. It has 32KB EEPROM, 1KB RAM, eight 10bit A/D ports, two 8bit D/A ports, some 8bit and 16bit timers, about 50 bit digital I/O ports and the interface of the serial bus I^2C.

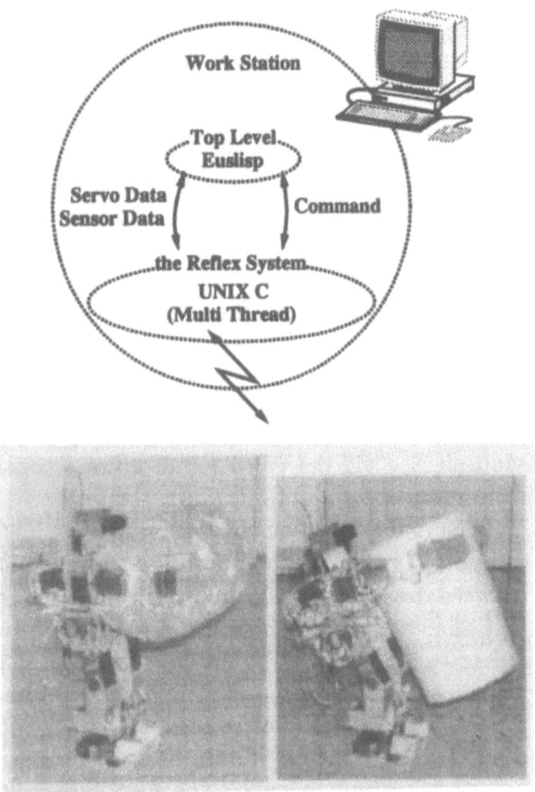

Figure 12: Holding Unknown Object

Figure 11 shows Jikokuten, the new model of Haru body which contains the information channel system. It contains a root processor and eight node processors. The head is a commercially available camera, EVI-G20, which has the mechanism pan-tilt, zoom and focusing inside. The function of the camera is controllable by the message from the brain. The root processor decodes the head messages and the joint messages and transfers them to each processors.

Eight node processors are distributed on the arms and the legs as shown in figure 11. The number in the box shows the number of servo modules that the node processor manages.

Each node processor repeats the processes: (1) to receive the reference value of the joint servo from the root processor, (2) to provide it to the servo modules, (3) to read the servo errors and the joint position, (4) to return the error signals and the joint positions to the root processor.

The two images of the new Haru body in Figure 11 show that the number of cables on body are reduced. The white box on the back contains the root processor and the radio modem.

By using the onboard channel processors, the general system becomes to have the architecture as shown in the top of the figure 11. The brain contains the real time control processing part inside. The first generation of the remote-brained system adopted the transputer network for the real time system(Kagami, Inaba, and Inoue 1996; Oka, Takeda, Inaba, and Inoue 1996). The higher level of the brain is written in Euslisp/MT developed at ETL(Matsui and Sekiguchi 1995).

9 Towards Whole-Body Adaptation

Once the humanoid has enough sensors for a certain stage of development sequence, we can concentrate on the whole-body behaviors that performs adaptive interactions with the real environment.

Figure 12 shows our ongoing research on the reflex system to get autonomous adaptation in picking up an unknown object using the whole body. The brain is divided into the higher level and the reflex level. The reflex level controls the body in local feedback control. The higher level can control the feedback control of the reflex level depending on the situation.

The experiment shown in the figure 12 is balancing the body when it holds an unknown object using the joint servo error signals. The system can estimate the joint load torque from the joint servo signal.

Figure 13 shows another experiment on whole-body adaptation with Igoid wearing the sensor suit(Inaba, Hoshino, Nagasaka, Kagami, and Inoue 1996; Inaba, Ninomiya, Hoshino, Nagasaka, Kagami, and Inoue 1997). The task is to have the humanoid respond to being held by a human.

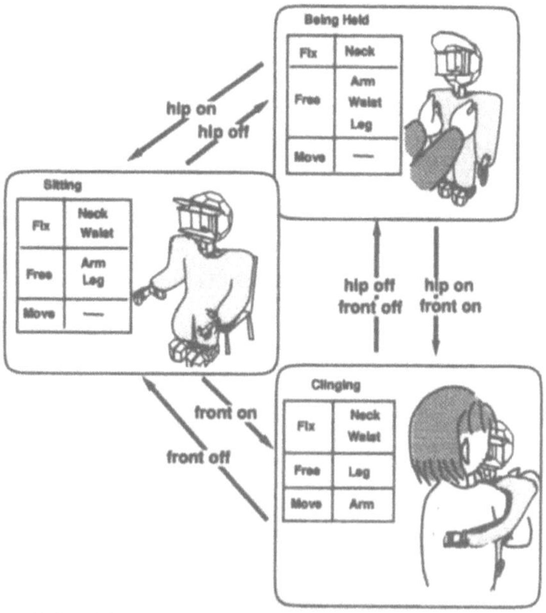

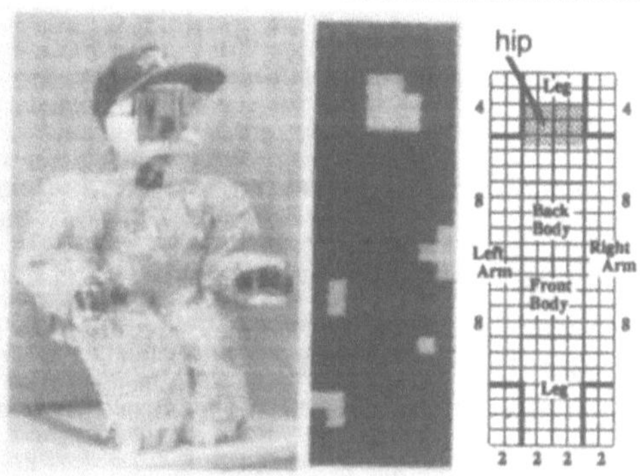

(a) The pattern of the sensor suit in sitting

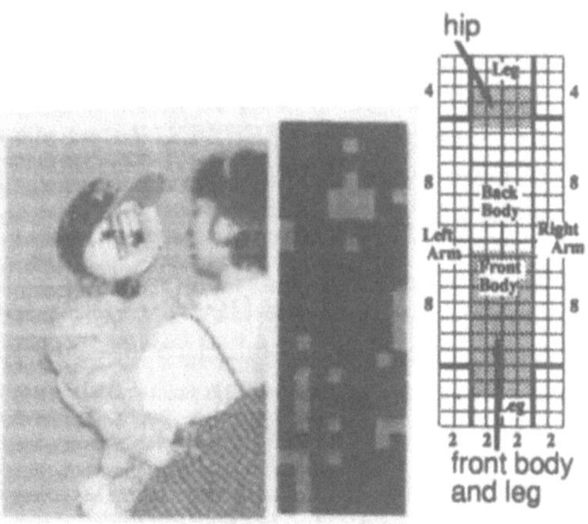

(b) The pattern of the sensor suit in being held

Figure 13: Behavior of conforming to human shape

In order to perform such task, the humanoid has the ability to conform to human shape. The experiment was done the situation containing three phases: sitting on a chair, being held, and clinging. There is three control modes in a joint: fix, free and move. The fix mode is to continue servoing with compliance. The free mode is that the joint does not perform servoing. In the move mode, the humanoid makes a reaction and controls the joint adaptively using local sensing information. As the first experiment, the three stages are distinguished by the sensor suit. The figure 13 (a) and (b) shows the experimental result of the sensor suit signals. The center image is the real sensor image of the sensor suit. The right image shows the map of the sensor image. The shaded region of the map shows the attention region to monitor the situation.

10 Concluding Remarks

This paper describes the idea of research on developmental processes and our research that tries to recognize the developmental processes for the remote-brained humanoids. We aim at the research to build those differential developmental processes and learn from them about the essential way on development.

The remote-brained approach allows us to develop many levels of real humanoids and the tools to develop the behaviors using the bodies in short periods. When we find new ideas for development, we can start quickly to test them in this environment. In this environment we can discuss with real experiments on all the differential processes and test again by sharing the developed environment of behavior software.

The mother operations are the first trial to denote the operation that advances the differential developmental processes. The mother is usually human but we have tried to build an artificial mother for the cyclic actions such as swing and horizontal bar. When the task contains different states such as brachiation, the mother operations were done by both humand and computer. The motion controller such as swinging its body by an arm can inherit the mother facility used in the tasks of swing and horizontal bar. Thus the transition net of behaviors and their sequences is described by human and control parameters to use within some nodes of that net are learned.

The body governs the levels of the behavioral development. Especially the sensors take an important role in autonomous development. The onboard network system developed for the information channels allows us to build a body which has rich sensors. The stages of developmental processes on behaviors are decided by these sensor systems. We are looking at the next stages for the whole-body adaptations based on multiple sensor coordination such as conforming to human shape when it is held.

The connection level between the brain and the body changes the aspects on dynamics. One of the next connection level in the future will be what the processor on body performs reactive processing. Once we put a processor on body, we have to provide the environment to develop the software for the onboard processors. The robustness of the processor and the turn around time for modification of the program become critical. That is the reason why we insist on the remote-brained approach. In order to prepare the next connection level, we design the brain architecture in two layers: the real time reflex part and the higher part. The reflex layer developed in the remote brain will be shifted onto the body in the near future after we learned much about that reflex behaviors.

Acknowledgements

The author would like to note his great gratitude to Professor H. Inoue for supporting and encouraging the group for this research and the contributors to develop the series of humanoids. Dr. S. Kagami developed the first generation of remote-brained environment and supported the development of the remote-brained humanoid systems. F. Kanehiro has built Apelike, Apelike3, Hanzou, Sasuke, and Jikokuten, and developed all the software needed for the humanoid series. K. Nagasaka did the experiments of swing, horizontal bar and brachiation, and developed the dynamic control system in the brain system. Y. Hoshino has designed and developed the sensor suit for Igoid and adaptive behaviors with it. Y. Tamiya has developed Akira with whole-body balancing and designed the body of the new Haru series. I. Mizuuchi has developed the onboard channel system and developed the reflex architecture. K. Koyasako has designed and built the onboard processors. My sincere thanks to them all. This research has been partly supported by several grants of Grant-in-Aid for Scientific Research, and JSPS-RFTF96P00801.

References

Brooks, R. A. and L. A. Stein (1994). Building Brains for Bodies. *Autonomous Robots 1*(1), 7–25.

Inaba, M. (1993). Remote-Brained Robotics: Interfacing AI with Real World Behaviors. In *Robotics Research: The Sixth International Symposium*, pp. 335–344. International Foundation for Robotics Research.

Inaba, M. (1996). Extended Vision with Robot Sensor Suit: Primary Sensor Image Approach in Interfacing Body to Brain. In G. Giralt and G. Hirzinger (Eds.), *Robotics Research: The Seventh International Symposium*, pp. 499–508. Springer.

Inaba, M., T. Hara, and H. Inoue (1993). A stereo viewer based on a single camera with view-control mechanisms. In *Proceedings of IEEE/RSJ International Workshop on Intelligent Robots and Systems IROS '93*, pp. 1857–1864.

Inaba, M., Y. Hoshino, K. Nagasaka, S. Kagami, and H. Inoue (1996). A Full-Body Tactile Sensor Suit Using Electrically Conductive Fabric and Strings. In *Proceedings of the IEEE/RSJ International Conference on Intelligent Robots and Systems*, pp. 450–457.

Inaba, M., T. Igarashi, S. Kagami, and H. Inoue (1996). A 35 DOF Humanoid that can Coordinate Arms and Legs in Standing up, Reaching and Grasping an Object. In *Proceedings of the IEEE/RSJ International Conference on Intelligent Robots and Systems*, pp. 29–36.

Inaba, M., S. Kagami, F. Kanehiro, K. Nagasaka, and H. Inoue (1996). Mother operations to evolve embodied robots based on the remote-brained approach. In *Preprint of Artificial Life V*, pp. 278–285.

Inaba, M., S. Kagami, F. Kanehiro, K. Takeda, T. Oka, and H. Inoue (1996). Vision-Based Adaptive and Interactive Behaviors in Mechanical Animals using the Remote-Brained Approach. *Robotics and Autonomous Systems 17*(1–2), 35–52.

Inaba, M., S. Kagami, K. Sakaki, F. Kanehiro, and H. Inoue (1994). Vision-Based Multisensor Integration in Remote-Brained Robots. In *1994 IEEE International Con-*

ference on Multisensor Fusion and Integration for Intelligent Systems, pp. 747–754.

Inaba, M., F. Kanehiro, S. Kagami, and H. Inoue (1995a). Two-armed bipedal robot that can walk, roll-over and stand up. In *Proceedings of the IEEE/RSJ International Conference on Intelligent Robots and Systems*, Volume 3, pp. 297–302.

Inaba, M., F. Kanehiro, S. Kagami, and H. Inoue (1995b). Vision-Equipped Apelike Robot Based on the Remote-Brained Approach. In *Proceedings of the IEEE International Conference on Robotics and Automation*, pp. 2193–2198.

Inaba, M., K. Nagasaka, F. Kanehiro, S. Kagami, and H. Inoue (1996). Real-Time Vision-Based Control of Swing Motion by a Human-Form Robot Using the Remote-Brained Approach. In *Proceedings of the IEEE/RSJ International Conference on Intelligent Robots and Systems*, pp. 15–22.

Inaba, M., T. Ninomiya, Y. Hoshino, K. Nagasaka, S. Kagami, and H. Inoue (1997). A Remote-Brained Full-Body Humanoid with Multisensor Imaging System of Binocular Viewer, Ears, Wrist Force and Tactile Sensor Suit. In *Proceedings of the IEEE International Conference on Robotics and Automation*, pp. 2497–2502.

Inoue, H. and T. Sato (1994, June). Humanoid Autonomous System. *RWC Technical Report*, 109 - 110.

Kagami, S., M. Inaba, and H. Inoue (1996). Design of Real-Time Large Scale Robot Software Platform and its Implementation in the Remote-Brained Robot Project. In *Proceedings of the IEEE/RSJ International Conference on Intelligent Robots and Systems*, pp. 1394–1399.

Kanehiro, F., M. Inaba, and H. Inoue (1996). Development of a Two-Armed Bipedal Robot that can Walk and Carry Objects. In *Proceedings of the IEEE/RSJ International Conference on Intelligent Robots and Systems*, pp. 23–28.

Kato, I., S. Ohteru, H. Kobayashi, K. Shirai, and A. Uchiyama (1974). Information-Power Machine with Senses and Limbs(Wabot 1). In *First CISM - IFToMM Symposium on Theory and Practice of Robots and Manipulators*, Volume 1, pp. 11–24. Springer-Verlag.

K.Nagasaka, A.Konno, M.Inaba, and H.Inoue (1997). Motion Acquisition and Control Method of Vision-Equipped Two-Armed Bipedal Brachiation Robot. In *Proc. of 15th Annu. Conf. of RSJ*.

Kuniyoshi, Y., M. Inaba, and H. Inoue (1989). Teaching by showing: Generating robot programs by visual observation of human performance. In *Proc. of 20th International Symposium on Industrial Robots and Robot Exhibition*, pp. 119–126.

Matsui, T. and S. Sekiguchi (1995). Design and Implementation of Parallel EusLisp Using Multithread. *Journal of Information Processing Society of Japan 36*(8), 1885 - 1896.

Nagasaka, K., M. Inaba, H. Inoue, and A. Konno (1997). Acquisition of Visually Guided Swing Motion Based on Genetic Algorithms and Neural Networks in Two-Armed Bipedal Robot. In *Proceedings of the IEEE International Conference on Robotics and Automation*, pp. 2944–2949.

Oka, T., K. Takeda, M. Inaba, and H. Inoue (1996). Designing Asynchronous Parallel Process Networks for Desirable Autonomous Robot Behaviors. In *Proceedings of the IEEE/RSJ International Conference on Intelligent Robots and Systems*, pp. 178–185.

Sugano, S., Y. Tanaka, T. Ohoka, and I. Kato (1985). Autonomic Limb Control of the Information Processing Robot. *Journal of Robotics Society of Japan 3*(4), 81–95.

Animating Human Athletes

J. K. Hodgins and W. L. Wooten
College of Computing, Georgia Institute of Technology Atlanta, GA (USA)
email: [jkh|wlw]@cc.gatech.edu

Abstract

This paper describes algorithms for the animation of male and female models performing three dynamic athletic behaviors: running, bicycling, and vaulting. We animate these behaviors using control algorithms that cause a physically realistic model to perform the desired maneuver. For example, control algorithms allow the simulated humans to maintain balance while moving their arms, to run or bicycle at a variety of speeds, and to perform two vaults. For each simulation, we compare the computed motion to that of humans performing similar maneuvers. We perform the comparison both qualitatively through real and simulated video images and quantitatively through simulated and biomechanical data.

1 Introduction

People are skilled at perceiving the subtle details of human motion. We can, for example, often identify friends by the style of their walk when they are still too far away to be recognizable otherwise. If synthesized human motion is to be compelling, we must create actors for computer animations and virtual environments that appear realistic when they move. We use dynamic simulation coupled with control algorithms to produce natural-looking motion.

In particular, this paper describes algorithms that allow a rigid-body model of a man or woman to stand, run, and turn at a variety of speeds, to ride a bicycle on hills and around obstacles, and to perform gymnastic vaults (Fig. 1). The rigid-body models of the man and woman are realistic in that their mass and inertia properties are derived from data in the biomechanics literature and the degrees of freedom of the joints are chosen so that each behavior can be completed in a natural-looking fashion.

Although the behaviors are very different in character, the control algorithms are built from a common toolbox: state machines are used to enforce a correspondence between the phase of the behavior and the active control laws, synergies are used to cause several degrees of freedom to act with a single purpose, limbs without required actions in a particular state are used to reduce disturbances to the system, inverse kinematics is used to compute the joint angles that would cause a foot or hand to reach a desired location, and the low-level control is performed with proportional-derivative control laws.

We have chosen to animate running, bicycling, and vaulting because each behavior contains a significant dynamic component. For these behaviors, the dynamics of the model constrain the motion and limit the space that must be searched to find control laws for natural-looking motion. This property is most evident in the gymnastic vault. The gymnast is airborne for much of the maneuver, and the control algorithms can influence the internal motion of the joints but not the angular momentum of the system as a whole. The runner, on the other hand, is in contact with the ground much of the time and the joint torques computed by the control algorithms directly control many of the details of the motion. Because the dynamics do not provide as many constraints on the motion, much more effort went into tuning the motion of the runner than into tuning the motion of the gymnast.

Many computer animations and interactive virtual environments require a source of human motion. The approach used here, dynamic simulation coupled with control algorithms, is only one of several options. Two alternatives, motion capture and keyframing, have been demonstrated to be practical and are widely available in commercial systems. The difficulty of designing control algorithms has prevented the value of simulation from being demonstrated for systems with internal sources of energy such as humans. However, simulation has several potential advantages over other approaches. Given robust control algorithms, simulated motion can easily be computed to produce similar but different motions while

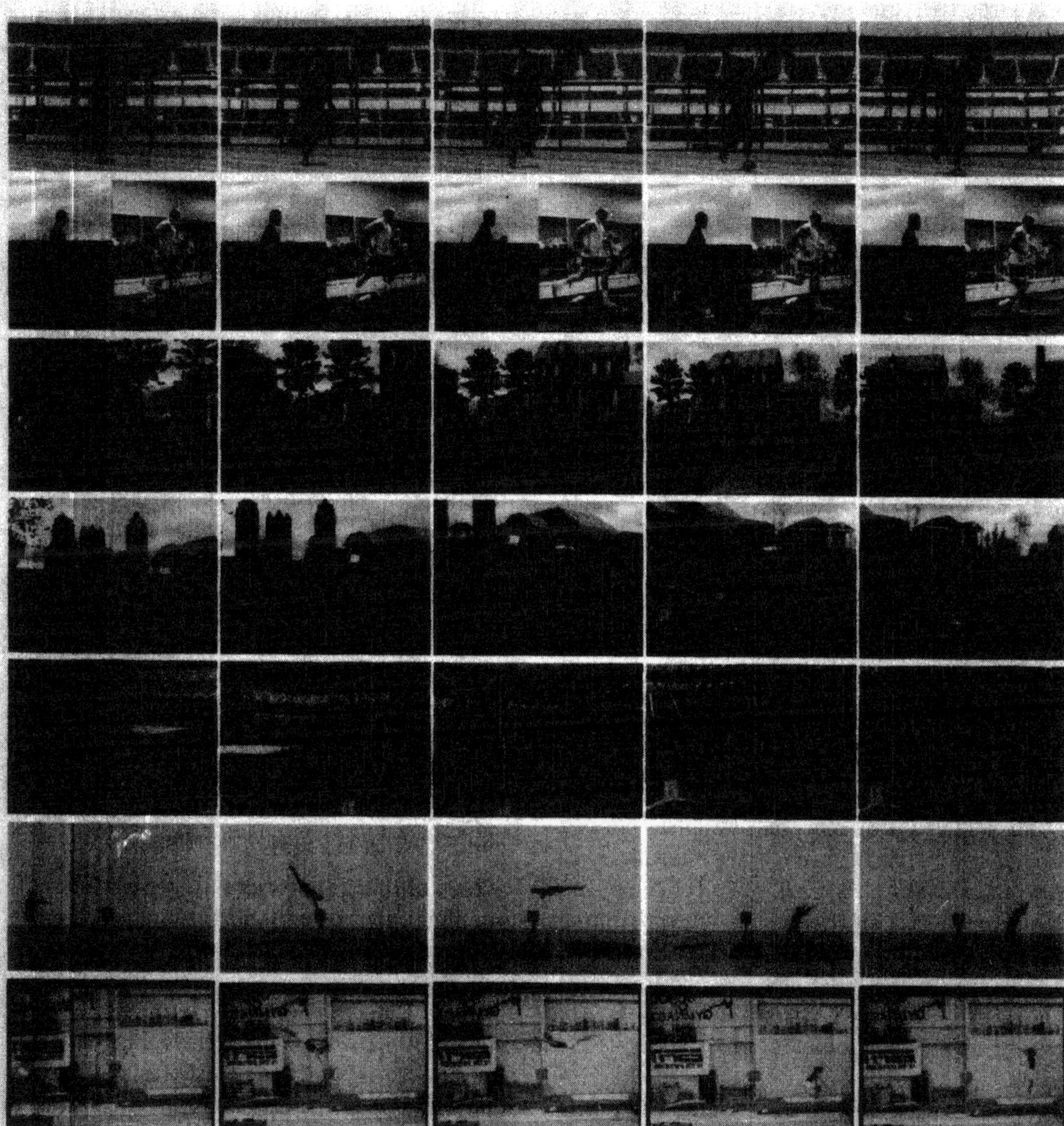

Figure 1: Images of an athlete running on a quarter mile track in the 1996 Olympic Stadium, comparison of the simulated runner and a runner on a treadmill, a bicyclist jumping off the curb, a group of bicyclists riding around a corner during a race, a gymnast performing an arabian vault, and comparison of an handspring vault and a simulated performance. In each case, the spacing of the images in time is equal with the runners at intervals of 0.066 s, the single rider and the group of bicyclists at 0.33 s, and the gymnasts at 0.5 s.

maintaining physical realism (running at 4 m/s rather than 5 m/s for example). Real-time simulations also allow the motion of an animated character to be truly interactive, an important property for virtual environments in which the actor must move realistically in response to changes in the environment and in response to the actions of the user. And finally, when the source of motion is dynamic simulation we have the opportunity to use multiple levels of simulation to generate either secondary motion such as the movement of clothing and hair or higher-level behaviors such as obstacle avoidance and grouping.

2 Background

Research in three fields is relevant to the problem of animating human motion: robotics, biomechanics, and computer graphics. Researchers in robotics have explored control techniques for legged robots that walk, run, balance, and perform gymnastic maneuvers. While few robots have been built with a complexity similar to that of the human body, control strategies for simpler machines provide basic principles that can be used to design control strategies for human models.

Raibert and his colleagues built and controlled a series of dynamic running machines, ranging from a planar machine with one telescoping leg to three-dimensional machines that ran on two or four legs. These machines walked, jumped, changed gait, climbed stairs, and performed gymnastic maneuvers ([22], [12], [13], [11], [23], and [21]). The control algorithms for human running described in this paper build on these control algorithms by extending them for a system with many more controlled degrees of freedom and more stringent requirements on the style of the motion.

Biomechanics provides the data and hypotheses about human motion required to ensure that the computed motion resembles that of a human performing similar maneuvers. The biomechanics literature contains motion capture data, force plate data, and muscle activation records for many human behaviors. These data were used to tune the control algorithms for running, bicycling, vaulting, and balancing. Cavagna presents energy curves for walking and running as well as studies of energy usage during locomotion[5]. McMahon provides graphs of stance duration, flight duration, and step length as a function of forward speed[18]. Gregor surveys biomechanical studies of bicyclists[10]. Takei presents biomechanical data of elite female gymnasts performing a handspring vault and relates the data to the scores that the gymnasts received in competition[27].

Many researchers in computer graphics have explored the difficult problems inherent in animating human motion. The Jack system developed at the University of Pennsylvania contains kinematic and dynamic models of humans based on biomechanical data[1]. It allows the interactive positioning of the body and has several built-in behaviors including balance, reaching and grasping, and walking and running behaviors that use generalizations of motion capture data[15]. Jack has been used extensively for ergonomic analysis and human factors engineering as well as distributed simulation applications.

Bruderlin and Calvert used a simplified dynamic model and control algorithms to generate the motions of a walking human[4]. The leg model included a telescoping leg with two degrees of freedom for the stance phase and a compound pendulum model for the swing phase. A foot, upper body, and arms were added to the model kinematically, and were made to move in an oscillatory pattern similar to that observed in humans. Pai programmed a walking behavior for a dynamic model of a human torso and legs in a high-level fashion by describing a set of time-varying constraints, such as, "maintain ground clearance during leg swing," "lift and put down a foot," "keep the torso vertical," and "support the torso with the stance leg"[20].

None of these approaches to generating motion for animation are automatic because each new behavior requires significant additional work on the part of the researcher. In recent years, the field has seen the development of a number of techniques for automatically generating motion for new behaviors and new creatures. Witkin and Kass[31], Cohen[7], and Brotman and Netravali[3] treat the problem of automatically generating motion as a trajectory optimization problem. Another approach finds a control algorithm instead of a desired trajectory ([30], [29], [19], [25], [26] and [16]). In contrast, the control algorithms described in this paper were designed by hand, using a toolbox of control techniques, our physical intuition about the behaviors, observations of humans performing the tasks, and biomechanical data. While automatic techniques would be preferable to hand design, automatic techniques have not yet been developed that can find solutions for systems with the number of controlled degrees of freedom needed for a realistic model of the human body. Furthermore, although the motion generated by automatic techniques is appealing, much of it does not appear natural in the sense of resembling the motion of a biological system. We do not yet know whether this discrepancy is because only relatively simple models have been used or because of the constraints and optimization criteria that were chosen.

3 Dynamic Behaviors

The motion of each behavior described in this paper is computed using dynamic simulation. Each simulation contains the equations of motion for a rigid-body model of a human and environment (ground, bicycle, and vaulting horse), control algorithms for balancing, running, bicycling, or vaulting, a graphical model for viewing the motion, and a user interface for changing the parameters of the simulation. The user is provided with limited high-level control of the animation. For example, the desired velocity and facing direction for the bicyclist and runner are selected by the user. During each simulation time step, the control algorithm computes desired positions and velocities for each joint based on the state of the system, the requirements of the task, and input from the user. Proportional-derivative servos compute joint torques based on the desired and actual value of each joint. The equations of motion of the system are integrated forward in time taking into account the internal joint torques and the external forces and torques from interactions with the ground plane or other objects. The details of the human model and the control algorithm for each behavior are described below.

The human models were constructed from rigid links connected by rotary joints with one, two or three degrees of freedom. The dynamic models were derived from the graphical models shown in Fig. 2. The mass and moment of inertia of each body part was calculated by computing the moment of inertia of a polygonal object of uniform density[17] using density data measured from cadavers[8]. The controlled degrees of freedom of the models are shown in Fig. 2. Each internal joint of the model has a very simple muscle model, a torque source, that allows the control algorithms to apply a torque between the two links that form the joint. The equations of motion for each system were generated using a commercially available package[24].

The points of contact between the feet and the ground, and the gymnast's hands and the vault are modeled using constraints. Interaction of the

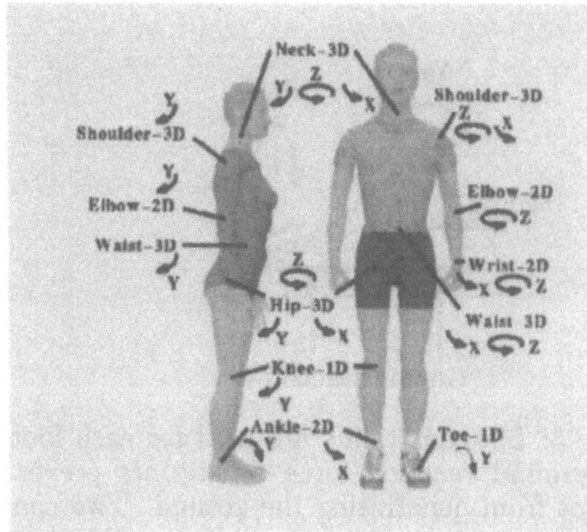

Figure 2: The controlled degrees of freedom of the human model. The gymnast represented in the figure has 15 body segments and a total of 30 controlled degrees of freedom. The runner has 17 body segments and 30 controlled degrees of freedom (two-part feet with a one degree-of-freedom joint at the ball of the foot and only one degree of freedom at the ankle), The bicyclist has 15 body segments and 22 controlled degrees of freedom (only one degree of freedom at the neck, hips, and ankles). The directions of the arrows indicates the positive direction of rotation for each degree of freedom. The polygonal models were purchased from Viewpoint Datalabs.

feet with the ground use six constraints for each foot: two keep the metatarsus and heel above the surface of the ground, two prevent the foot from sliding on the ground, and two prevent the foot from rolling or twisting about the center of mass of the foot (Fig. 3). The linear and rotational acceleration of the contact point of the foot with respect to the ground is the constraint error. The penetration of the foot into the ground and the velocity of the foot relative to the ground stabilize the constraint error[2]. Friction is infinite in our model. The ground contact forces computed with the constraint matrix are applied to the foot at the contact points to prevent the foot from penetrating or slipping on the ground. Torques are applied about the center of mass of the foot to prevent rolling or twisting. To allow the feet to leave the ground, the force is applied only when the foot has penetrated the ground and the vertical velocity of the foot is negative. Contact between the hands and the environment are modeled in the same manner as the feet with contact points at the base of the palm and the tips of the fingers.

The points of contact between the bicycle wheels and the ground are also modeled using constraints. The constraints allow rolling with-

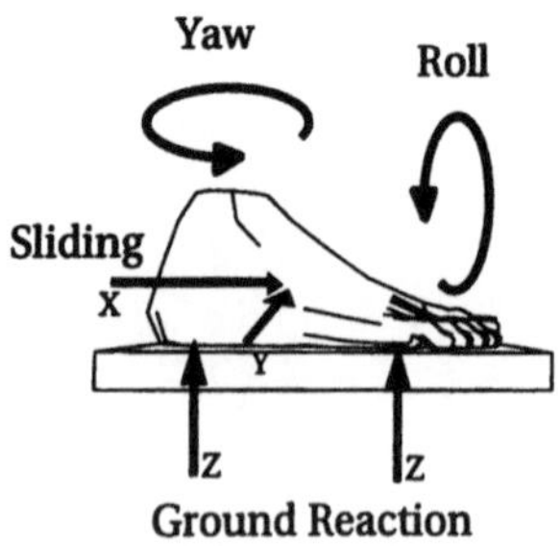

Figure 3: Six constraints are used for each foot. Two ground reaction force constraints prevent the foot from penetrating the ground. Two constraints prevent the foot from sliding on the plane in x and y. Two more constraints are used to prevent the foot from rolling or yawing about its center of mass.

out slipping. Yaw damping is applied to the wheel about the z axis and is proportional to the yaw rotation rate of the wheel up to a friction limit.

3.1 Running

Running is a cyclic behavior in which the legs swing fore and aft and provide support for the body in alternation. Because the legs perform different functions during the phases of the locomotion cycle, the muscles are used for different control actions at various times in the cycle. When the foot of the simulated runner is on the ground, the ankle, knee, and hip provide support and balance. During the flight phase, a leg is swung forward in preparation for the next touchdown. These distinct phases and corresponding changes in control actions make a state machine a natural tool for selecting the control actions that should be active at a particular time. The state machine and transition events used for the simulation of running are shown in Fig. 4. The main states correspond to the points of contact with the ground: flight, heel contact, heel and toe contact, and toe contact. Transitions between states are based on events that can be easily determined from the variables of the simulation (contact of the foot with the ground or movement of the knee joint, for example). The control laws for each state are described in Fig. 4.

To interact with the animation of the runner, the user specifies desired values for the magnitude of the velocity on the ground plane and the facing direction. The control laws for each state compute joint torques that move the velocity and facing direction toward these desired values while maintaining balance. The simulated runner can run at speeds between 2.5 m/s and 5 m/s.

We call the leg that is on the ground or actively being positioned for touchdown the *active leg*. The other leg is the *idle leg*. During flight, the active leg is swung forward in anticipation of touchdown. Using the degrees of freedom of the leg in a synergistic fashion, the foot is positioned at touchdown to correct for errors in forward speed and to maintain balance. Forward speed is controlled by placing the average point of support during stance underneath the hip and taking into account the change in contact point from heel to metatarsus during stance. At touchdown, the desired distance from the hip to the heel projected onto the ground plane is

$$x_{hh} = 1/2(t_s\dot{x} - \cos(\theta)l_f) + k(\dot{x} - \dot{x}_d) \qquad (1)$$

where t_s is an estimate of the period of time that the foot will be in contact with the ground (based on the previous stance duration), $\dot{x}$ is the forward velocity of the runner, $\dot{x}_d$ is the desired velocity, θ is the facing direction of the runner, l_f is the distance from the heel to the ball of the foot, and k is a gain for the correction of errors in speed. A similar equation holds for y_{hh}. The length of the leg at touchdown is fixed and is used to calculate the vertical distance from the hip to the heel, z_{hh}. The disturbances caused by the impact of touchdown can be reduced by decreasing the relative speed between the foot and the ground at touchdown. This technique is commonly called *ground speed matching*. In this control system, ground speed matching is accomplished by swinging the hip further forward in the direction of travel during flight and moving it back just before touchdown.

The equations for x_{hh}, y_{hh}, and z_{hh} and the kinematics of the model are used to compute the desired knee and hip angles at touchdown. The angle of the ankle is constant during flight and is chosen such that the toe will not touch the ground at the same time as the heel at the beginning of stance.

The idle leg plays an important role in locomotion by reducing disturbances to the body attitude caused by the active leg as it swings forward and in toward the centerline in preparation for touchdown. The idle leg is bent so that the toe does not stub the ground, and the hip angles mirror the motion of the active leg to reduce the net torque on the body:

$$\alpha_{x_d} = \alpha_{x_{lo}} - (\beta_{x_d} - \beta_{x_{lo}}) \qquad (2)$$

where α_{x_d} is the desired rotation of the idle hip with respect to the pelvis, $\alpha_{x_{lo}}$ is the rotation

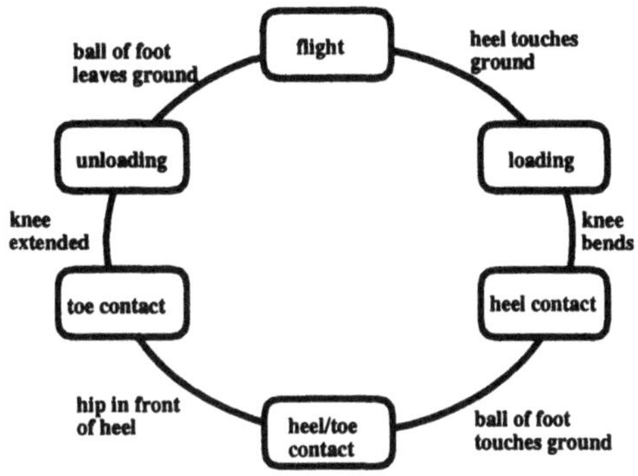

State	Action
All States	Neck: turn in desired facing direction Shoulder: mirror hip angle Elbow: mirror magnitude of shoulder Wrist: constant angle Waist: keep body upright
Flight	Active Leg swing leg forward for touchdown straighten knee Idle Leg mirror hip angles of active leg hold knee and ankle at flight angle
Heel Contact	Active Leg pitch control with hip allow ankle to extend knee acts as a spring Idle Leg mirror hip angles of active leg bend knee to prevent foot contact hold ankle at flight angle
Heel/Toe Contact and Toe Contact	Active Leg pitch control with hip extend ankle for thrust extend knee for thrust Idle Leg mirror hip angle of active leg bend knee to prevent foot contact hold ankle at flight angle

Figure 4: A state machine is used to determine the control actions that should be active for running given the current state of the system. The transition events are computed for the active leg. At liftoff the active and idle legs switch roles.

of the idle hip at the previous liftoff, β_{x_d} is the desired position of the active hip, and $\beta_{x_{lo}}$ is the position of the active hip at the previous liftoff. A similar equation holds for the y component of the hip rotation. The mirroring action of the idle leg is modified by the restriction that the legs should not collide as they pass each other during stance.

During stance, the knee acts as a passive spring to store the kinetic energy that the system had at touchdown. The majority of the vertical thrust is provided by the ankle joint. During the first part of stance, *heel contact*, the toe is passively moved toward the ground because the contact point on the heel is behind the ankle joint. Contact of the ball of the foot triggers the transition from *heel contact* to *heel and toe contact*. The transition from *heel and toe contact* to *toe contact* occurs when the hip has moved a certain distance in front of foot. After the transition, the ankle joint is extended, causing the heel to lift off the ground and adding energy to the system for the next flight phase.

Throughout stance, proportional-derivative servos are used to compute torques for the hip joint of the stance leg that will cause the attitude of the body (roll, pitch, and yaw) to move toward the desired values. The desired angle for roll is zero except during turning. The desired angle for pitch is inclined slightly forward, and the desired angle for yaw is set by the user. The x and y joints of the waist maintain the body attitude by allowing the upper part of the body to be more vertical than the pelvis. The z joint of the waist allows the upper body to turn to face in the direction specified by the user.

Throughout the running cycle, the shoulder and elbow joints swing the arms fore and aft in a motion that is synchronized with the motion of the legs: $\gamma_{y_d} = k\alpha_y + \gamma_0$ where γ_{y_d} is the desired fore/aft angle for the shoulder, k is a scaling factor, α_y is the fore-aft hip angle for the leg on the opposite side of the body, and γ_0 is an offset. The other two degrees of freedom in the shoulder (x and z) and the elbows also follow a cyclic pattern with the same period as γ_{y_d}. The neck is used to turn the head to look at the ground in front of the runner in the desired facing direction. The motion of the upper body is important in running because the counter oscillation of the arms reduces the yaw oscillation of the body caused by the swinging of the legs. However, the details of the motion of the upper body are not strongly constrained by the dynamics of the task and amateur athletes use many different styles of arm motion. Observations of human runners were used to tune the oscillations of the arms to produce a natural-looking gait.

The control laws compute desired values for each joint. Proportional-derivative servos are used to control the position of all joints. For each internal joint the control equation is

$$\tau = k(\theta_d - \theta) + k_v(\dot{\theta}_d - \dot{\theta}) \tag{3}$$

where θ is the angle of the joint, θ_d is the desired angle, and k and k_v are the proportional and derivative gains. The desired values used in the proportional-derivative servos are computed as trajectories from the current desired value of the joint to the new desired value computed by the control laws. Eliminating large step changes in the errors used in the proportional-derivative servos smoothes the simulated motion.

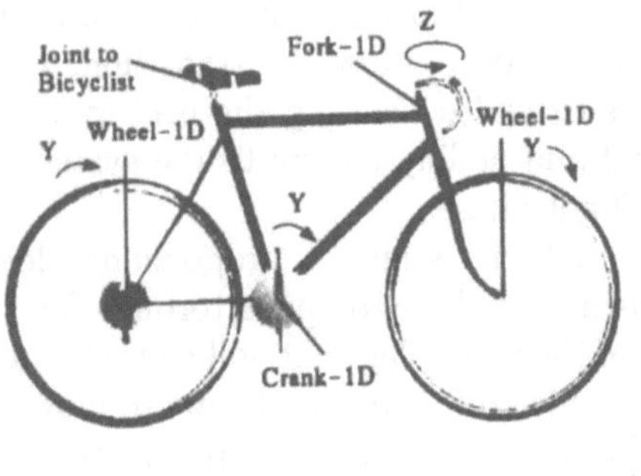

Figure 5: The four degrees of freedom of the bicycle model. The direction of the arrows indicates the positive direction of rotation for each degree of freedom. The polygonal model is a modification of a model purchased from Viewpoint Datalabs.

To generate running motion for a specific virtual environment or animated sequence, the user specifies a path through the environment and a desired velocity along that path. As the simulated runner moves along the path, the control system computes a desired facing direction that will cause the runner to face along a line between the runner's current location and the point on the path that he will reach in 2 seconds.

When the path is curved, the control algorithms are modified to allow the runner to turn more tightly. The timing and magnitude of thrust are functions of the yaw velocity in order to decrease the thrust from the inside leg and increase the thrust from the outside leg. This modification to the control laws maintains an approximately constant flight duration for the two legs. The idle leg is swung further to the outside to avoid a collision with the active leg during stance and the desired roll angle is computed to cause the torso to lean into the curve.

3.2 Bicycling

The bicyclist controls the facing direction and speed of the bicycle by applying forces to the handlebars and pedals with his hands and feet. The rider is attached to the bicycle by a pivot joint between the bicycle seat and the pelvis (Fig. 5). Spring and damper systems connect the hands to the handlebars, the feet to the pedals, and the crank to the rear wheel. The connecting springs are two-sided and the bicyclist is able to pull up on the pedals as if the bicycle were equipped with toe-clips and a fixed gear (no freewheel). The connection between the crank and the rear wheel includes an adjustable gear ratio. The bicycle wheels have a rolling resistance proportional to the velocity.

The control algorithm adjusts the velocity of the bicycle by using the legs to produce a torque at the crank. The desired torque at the crank is $\tau_c = k(v - v_d)$ where k is a gain, v is the magnitude of the bicyclist's velocity, and v_d is the desired velocity. The force applied by each leg depends on the angle of the crank because we assume that the legs are most effective at pushing downwards. For example, the front leg can generate a positive torque and the rear leg can generate a negative torque when the crank is horizontal. To compensate for the crank position, the desired forces for the legs are scaled by a weighting function between zero and one that depends on the crank position, θ_c:

$$w = \frac{\sin(\theta_c) + 1}{2}. \tag{4}$$

θ_c is zero when the crank is vertical and the right foot is higher than the left. If $\tau_c > 0$, the force on the pedal that the legs should produce is

$$f_l = \frac{w\tau_c}{l} \tag{5}$$

$$f_r = \frac{(1-w)\tau_c}{l} \tag{6}$$

where f_l and f_r are the desired forces from the left and right legs respectively, and l is the length of a crank arm. If τ_c is less than zero, then the equations for the left and right leg are switched. An inverse kinematic model of the legs is used to compute hip and knee torques that will produce the desired pedal forces.

To steer the bicycle and control the facing direction, the control algorithm computes a desired angle for the fork based on the errors in roll and yaw:

$$\theta_f = -k_\alpha(\alpha - \alpha_d) - k_{\dot\alpha}\dot\alpha + k_\beta(\beta - \beta_d) + k_{\dot\beta}\dot\beta \tag{7}$$

where α, α_d, and $\dot\alpha$ are the roll angle, desired roll, and roll velocity and β, β_d, and $\dot\beta$ are the yaw angle, desired yaw, and yaw velocity. k_α, $k_{\dot\alpha}$, k_β, and $k_{\dot\beta}$ are gains. The desired yaw angle is set by the user or high-level control algorithms; the desired roll angle is zero. Inverse kinematics is used to compute the shoulder and elbow angles that will position the hands on the handlebars with a fork angle of θ_f. Proportional-derivative servos move the shoulder and elbow joints toward those angles.

These control laws leave the motion of several of the joints of the bicyclist unspecified. The wrists and the waist are held at a nearly constant angle with proportional-derivative controllers. The ankle joints are controlled to match data recorded from human subjects[6].

3.3 Vaulting

To perform a vault, the gymnast uses a springboard to launch herself toward the horse, pushes off the horse with her hands, and lands on her feet on the other side of the horse. While airborne, the gymnast performs a combination of somersaults or twists. We have implemented two vaults, a handspring and an arabian. In the handspring vault, the gymnast performs a full somersault over the horse while keeping her body extended in a layout position. In the arabian vault, the gymnast performs a half-twist piked and lands facing the horse.

The handspring and arabian vaults are structured by a state machine with seven phases: *compression*, *decompression*, *flight1*, *horse*, *flight2*, *landing* and *balance*. The control actions required during each of the seven phases are presented in Fig. 6. The simulation of each vault begins in a flight state just before touchdown on the springboard. The initial conditions were estimated from video footage and confirmed with data measured from gymnasts[27] (forward velocity is 6.75 m/s and the height of the center of mass is 0.9 m).

The vaulting simulation uses the same control hierarchy as the running and bicycling simulations. A state machine selects the appropriate phase for the vault and middle level controllers calculate new desired angles for the joints. The lowest level of control uses proportional-derivative servos to compute the torque required to generate the desired motion. The next four sections will discuss the implementation of the middle level controllers for vaulting: leaping, tumbling, landing, and balancing.

3.3.1 *Leaping*

The leaping controller has two phases, *compression* and *decompression* and the control laws are similar to those used during the stance phase of running. During *compression*, the simulated gymnast lands on a springboard that is modeled as a board attached to the ground with a pivot joint. Using a technique called *blocking*, the control system positions the hips forward before the touchdown on the springboard so that some of the horizontal velocity at touchdown is transformed into rotational and vertical velocity at liftoff.

When the springboard reaches maximum deflection, the simulation enters the *decompression* phase. In this phase the control system extends the knees to push on the springboard and add energy to the system. As the springboard rebounds, the simulation extends the leg muscles rapidly and is launched into the air. When the simulated gymnast's feet leave the springboard, the *flight1* phase begins.

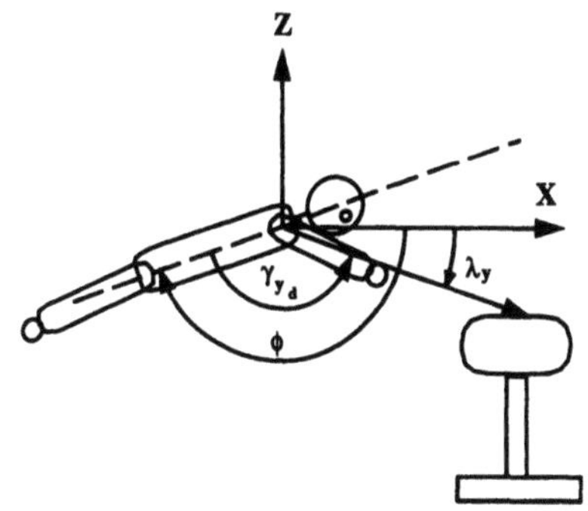

Figure 7: The control system positions the hands on the line between the shoulder and the horse. The desired shoulder angle is γ_{y_d}.

3.3.2 *Tumbling*

During the *flight1* phase, the control system prepares to put the gymnast's hands on the horse by positioning her hands on the line between the shoulders and the desired hand position on the vault: $\gamma_{y_d} = \lambda_y - \phi$, where γ_{y_d} is the desired shoulder angle relative to the body, λ_y is the angle between horizontal and a vector from the shoulder to the desired hand position on the vault, and ϕ is the pitch angle of the body with respect to horizontal (Fig. 7). This control strategy causes the hands to be nearly motionless with respect to the horse at contact, in a similar fashion to the ground speed matching strategy that reduces the impact at touchdown in the simulated runner. The wrists are servoed to ensure that the hands hit palm down and parallel to the surface of the horse: $\phi_{y_d} = \arccos(V_a \cdot V_x)$, where ϕ_{y_d} is the desired angle for the wrist, V_a is a vector from the palm to the shoulder, and V_x is a vector along the positive x-axis.

While her hands are in contact with the horse, the gymnast generates the angular momentum required for the maneuver. For the handspring vault, the gymnast maintains a layout position. During an arabian vault, the gymnast enters a pike position before her hands leave the horse in order to generate the required twist. The control system does not apply a torque at the shoulders or wrists for either vault, allowing the gymnast's angular and forward velocity to carry her over the

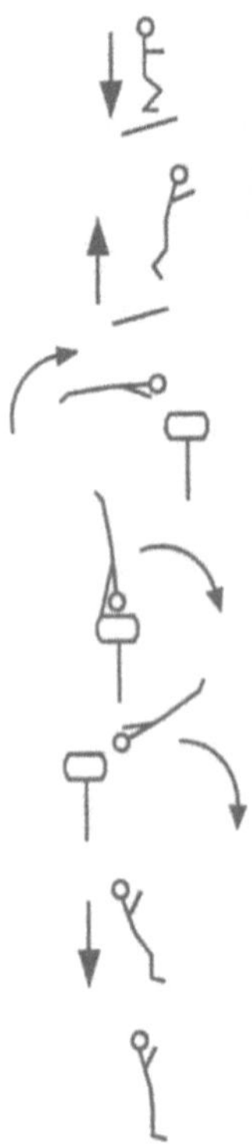

State	Handspring	Arabian
Compression	Place feet in front of COM: Bend knees $+y$ Bend hips $-y$ Bend waist $-y$ Compress springboard	Same as handspring
Decompression	Jump from springboard: Decompress springboard Straighten knees Straighten hips Straighten waist Extend ankles	Same as handspring
Flight1	Move hands to vault: Swing arms upward $-y$ Bend at hips slightly $+y$ Bend neck upwards $-y$	Same as handspring
Horse	Initiate somersault: Keep wrists flat Passively swing over vault	Initiate twist: Keep wrists flat Passively swing over vault Pike at hips $-y$ Pike at waist $-y$ Twist at hips $-z$ Twist at waist $-z$
Flight2	Prepare to land: Extend legs for wide stance $\pm x$	Prepare to land: Extend legs for wide stance $\pm x$ Straighten hips $+y$ Straighten waist $+y$ Untwist hips $+z$ Untwist waist $+z$
Landing	Recover balance: Straighten neck Bend at knees $+y$ Bend at hips $-y$ Bend at waist $-y$	Same as handspring
Standing	Stand up: Raise arms over head Straighten knees Straighten hips Straighten waist	Same as handspring

Figure 6: The state machine determines the control laws that are in effect at each phase of the vault.

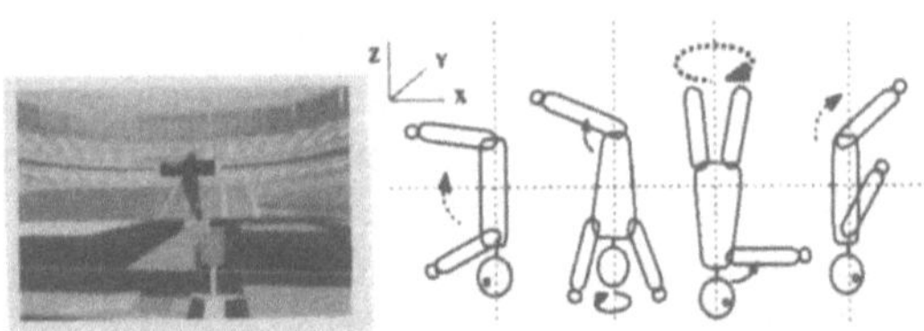

Figure 8: The gymnast performs a half twist using a ***swivel hips*** maneuver. As the gymnast leaves the horse she rotates her upper torso as far as possible while unpiking the lower torso. As the hips straighten, the body performs the twist about the z-axis. The gymnast then "swivels" her hips and ends the maneuver by pulling her arms down while returning to a piked position (figure adapted from Frohlich[9]).

horse.

During the *flight2* phase, the control system completes the specified maneuver and prepares to land. For the handspring vault, the control system maintains the layout position. For the arabian vault, the control system performs a half-twist by moving the limbs in a specified sequence to perform a ***swivel hips*** maneuver (Fig. 8). This maneuver allows the gymnast to rotate the body 180° about the z-axis even when the gymnast has little angular momentum. Often gymnasts will swing their arms to generate more twist, but the control system for the simulated gymnast did not use this strategy and all of the twisting velocity for the arabian vault was generated by piking the hips and twisting the body.

As the simulated gymnast's feet approach the mat, the control system prepares for landing. The feet are spread slightly to give a larger area of support. The angle of the legs with respect to the ground at touchdown is chosen so that the gymnast's forward velocity will decrease until it is nearly zero when her center of mass is above the support polygon. This strategy and the blocking strategy used for the springboard contact are essentially the same as the strategy used to control the forward speed of the runner.

3.3.3 *Landing*

When the feet hit the ground, the control system for the *landing* phase must remove the horizontal and rotational energy that was required for the maneuver and must establish an upright, balanced posture. At touchdown, the gymnast is not in a balanced state because the projection of the center of mass is outside the support polygon. But if the angle of the legs at touchdown was chosen correctly, the forward velocity will be near zero when the center of mass is over the center of the support polygon. Feedback control is used to servo the ankles to adjust the forward velocity of the center of mass as it moves forward. During landing, the knees and waist bend to absorb most of the energy.

The controllers for the handspring vault and arabian vaults are similar except that the orientation of the landing is reversed. We found the handspring vault easier to control than the arabian because it is easier to recover balance when the twisting angular velocities are near zero.

3.3.4 *Balancing*

When the gymnast's forward velocity is nearly zero, the *balance* phase begins. The balance controller straightens the hips and knees to cause the gymnast to stand up from the crouch used during landing. Changes in the ankle and hip angles compensate for disturbances to the location of the center of mass caused by motion of the upper body and arms.

The balance controller moves the projection of the center of mass of the gymnast on the ground, (C_x, C_y), towards the desired center of mass, (C_{x_d}, C_{y_d}). For balance, the center of mass should be in the center of the polygon formed by the four contact points of the feet, the *support polygon*. The desired angles for the ankle are

$$\rho_{x_d} = -(k_{ax}(C_{x_d} - C_x) - \dot{k}_{ax}\dot{C}_x) \quad (8)$$

$$\rho_{y_d} = k_{ay}(C_{y_d} - C_y) - \dot{k}_{ay}\dot{C}_y \quad (9)$$

where k_{ax}, k_{ay}, $\dot{k}_{ax}$, and $\dot{k}_{ay}$ are gains for the proportional-derivative controller. The same equations are used for both the left and right ankles. Similar equations are used to compute the desired hip angles, the only difference being the sign and the gains

$$\alpha_{x_d} = k_{hx}(C_{x_d} - C_x) - \dot{k}_{hx}\dot{C}_x \quad (10)$$

$$\alpha_{y_d} = -(k_{hy}(C_{y_d} - C_y) - \dot{k}_{hy}\dot{C}_y) \quad (11)$$

This controller not only allows the gymnast to stand after landing, but also compensates for disturbances resulting from the motion of other parts of the body while she is standing. For example, balance is maintained as the simulated gymnast throws her arms back in a gesture of success after completing the vault.

4 Discussion

This paper presents algorithms that allow an animator to generate motion for several dynamic behaviors. Elsewhere we have presented animations of other dynamic behaviors including platform diving, unicycle riding, and pumping a swing ([28, 14]). Taken together with this previous work, the dynamic behaviors described here represent a growing library. While these behaviors do not represent all of human motion or even of human athletic endeavors, an animation package with ten times this many behaviors would have sufficient functionality to be interesting to students and perhaps even to professional animators.

Several open questions remain before the value of simulation as a source of motion for animation and virtual environments can be conclusively demonstrated:

How can we make it easier to generate control algorithms for a new behavior? This paper partially addresses that question by presenting a toolbox of techniques that can be used to construct the control algorithms for a set of diverse behaviors. However, developing sufficient physical intuition about a new behavior to construct a robust control algorithm remains time consuming. We hope that these examples represent a growing understanding of the strategies that are useful in controlling simulations of human motion and that this understanding will lead to the development of more automatic techniques.

What can we do to reduce the number of new behaviors that need to be developed? One idea that has been explored by researchers in the domain of motion capture and keyframing is to perform transitions between behaviors in an automatic or semiautomatic fashion. Such transitions may be much more amenable to automatic design than the design of entire control algorithms for dynamic simulations.

What rules can we add to the system to improve the naturalness of the motion? The techniques presented here are most effective for behaviors with a significant dynamic component because the dynamics constrain the number of ways in which the task can be accomplished. When the gross characteristics of the motion are not constrained by the dynamics of the system, the task can be completed successfully but in a way that appears unnatural. For example, the simulated runner can run while holding his arms fixed at his sides, but an animation of that motion would be amusing rather than realistic. Humans are strong enough and dextrous enough that simple arm movements such as picking up a coffee cup can be completed in many different ways. In contrast, only good athletes can perform an arabian vault and the variations seen in their performances are relatively small. When the dynamics do not significantly constrain the task, the control algorithms must be carefully designed and tuned to produce motion that appears natural while matching the key features of the behavior when performed by a human. The tuning process might be aided by using data from psychophysical experiments to provide additional constraints on the motion.

Can human motion be simulated interactively? To be truly interactive, the motion of synthetic actors in virtual environments must be computed in real time (simulation time must be less than wall clock time). Our implementation of the bicyclist runs in real time on a Silicon Graphics Indigo[2] Computer with a R8000 processor. The runner is six times slower than real time on the same processor and the vaulter is twenty times slower than real time. We anticipate that with improved dynamic simulation techniques and the continued increase in workstation speed, a three-dimensional human simulation with a realistic number of degrees of freedom will run in real time within a few years.

Are the behaviors robust enough for use as interactive synthetic actors? The runner can run at a variety of speeds and change direction, but abrupt changes in velocity or facing direction will cause him to fall down. The planning or reactive response algorithms that lie between the locomotion control algorithms and the perceptual model of the simulated environment will have to take in account the limitations of the dynamic system and control system in order to provide robust but agile higher-level behaviors.

One goal of this research is to demonstrate that dynamic simulation of rigid-body models can be used to generate natural-looking motion. Fig. 1 shows a side-by-side comparison of video footage of a human runner and gymnast with images of the simulated runner and gymnast. This comparison represents one form of evaluation of our success in generating natural-looking motion. The simulated runner's leg motion is similar to that of the athlete on the treadmill although the simulated runner has a shorter stride and runs with a more upright body posture. The simulated and human vaults differ in that the simulated vaulter has a straighter position in the layout and less motion on the landing.

A second form of comparison involves the use of biomechanical data to measure how closely the simulated motion resembles natural-looking motion. Data available in the biomechanical literature such as force platform records illustrate some ways in which the motion of the simulated runner does not match the motion of a human. Fig. 9 compares data from the handspring vault performed by 24 gymnasts in the 1987 Pan American Games to our simulated handspring vault[27]. Most of the simulated data fell within the range of the measured data.

Variables	Human			Sim.
	Mean	Min	Max	
Mass (kg)	47.96	35.5	64.0	64.3
Height (m)	1.55	1.39	1.66	1.64
Board contact (s)	0.137	0.11	0.15	0.105
First flight (s)	0.235	0.14	0.30	0.156
Horse contact (s)	0.245	0.19	0.30	0.265
Second flight (s)	0.639	0.50	0.78	0.632
Horiz. vel. (m/s)				
Board touchdown	6.75	5.92	7.25	6.75
Board liftoff	4.61	3.97	5.26	4.01
Horse touchdown	4.61	3.97	5.26	4.01
Horse liftoff	3.11	2.48	3.83	2.83
Vert. velocity (m/s)				
Board touchdown	-1.15	-1.54	-.71	-1.13
Board liftoff	3.34	2.98	3.87	3.81
Horse touchdown	1.26	0.74	2.39	2.13
Horse liftoff	1.46	0.56	2.47	1.10
Aver. vert. force (N)				
Board contact	2175	1396	2792	5075
Horse contact	521	309	752	957

Figure 9: Comparison of velocities, contact times, and forces for a simulated vaulter and human data measured by Takei. The human data was averaged from 24 subjects. The simulated data was taken from a single trial.

A final form of evaluation would be a Turing test with direct comparison between simulated motion and human motion. The question we would like to ask is the following: if simulated data and human data were represented using the same graphical model, would the viewer sometimes choose the simulated data as the more natural motion? Unfortunately, the capture techniques for human data have characteristic flaws making this test more difficult than it might at first appear.

The animations described in this paper motion can be seen on the WWW at **http://www.cc.gatech.edu/gvu/animation**

5 Acknowledgments

Earlier versions of portions of this paper were presented at Siggraph '95 and the 1996 IEEE International Conference on Robotics and Automation. The authors would like to thank Debbie Carlson and Ron Metoyer for their help in developing our simulation and rendering environment and the CAD Systems Department at the Atlanta Committee for the Olympic Games for allowing us to use models of the Olympic venues. This project was supported in part by NSF NYI Grant No. IRI-9457621, Mitsubishi Electric Research Laboratory, and a Packard Fellowship. Wayne Wooten was supported by a Intel Foundation Graduate Fellowship.

References

[1] Badler, N. I., Phillips, C. B., Webber, B. L. 1993. *Simulating Humans*. Oxford: Oxford University Press.

[2] Baumgarte, J. 1972. Stabilization of Constraints and Integrals of Motion in Dynamical Systems. *Computer Methods in Applied Mechanics and Engineering* 1:1–16.

[3] Brotman, J. S., Netravali, A. N. 1988. Motion Interpolation by Optimal Control. *Proceedings of SIGGRAPH*, 309–315.

[4] Bruderlin, A., Calvert, T. W. 1989. Goal-Directed, Dynamic Animation of Human Walking. *Proceedings of SIGGRAPH*, 233–242.

[5] Cavagna, G. A., Thys, H., Zamboni, A. 1976. The Sources of External Work in Level Walking and Running. *Journal of Physiology* 262:639–657.

[6] Cavanagh, P., Sanderson, D. 1986. The Biomechanics of Cycling: Studies of the Pedaling Mechanics of Elite Pursuit Riders. In *Science of Cycling*, Edmund R. Burke (ed), Human Kinetics: Champaign, Ill.

[7] Cohen, M. F. 1992. Interactive Spacetime Control for Animation. *Proceedings of SIGGRAPH*, 293–302.

[8] Dempster, W. T., Gaughran, G. R. L. 1965. Properties of Body Segments based on Size and Weight. *American Journal of Anatomy* 120: 33–54.

[9] Frohlich, C. 1979. Do springboard divers violate angular momentum conservation? *American Journal of Physics* 47:583–592.

[10] Gregor, R. J., Broker, J. P., Ryan, M. M. 1991. Biomechanics of Cycling *Exercise and Sport Science Reviews* Williams & Wilkins, Philadelphia, John Holloszy (ed), 19:127–169.

[11] Hodgins, J. K. 1991. Biped Gait Transitions. In *Proceedings of the IEEE International Conference on Robotics and Automation*, 2092–2097.

[12] Hodgins, J., Raibert, M. H. 1990. Biped Gymnastics. *International Journal of Robotics Research* 9(2):115–132.

[13] Hodgins, J. K., Raibert, M. H. 1991. Adjusting Step Length for Rough Terrain Locomotion. *IEEE Transactions on Robotics and Automation* 7(3): 289–298.

[14] Hodgins, J. K., Sweeney, P. K, Lawrence, D. G. 1992. Generating Natural-looking Motion for Computer Animation. *Proceedings of Graphics Interface '92*, 265–272.

[15] Ko, H., Badler, N. I. 1993. Straight-line Walking Animation based on Kinematic Generalization that Preserves the Original Characteristics. In *Proceedings of Graphics Interface '93*.

[16] Laszlo, J., van de Panne M., Fiume, E. 1996. Limit Cycle Control and its Application to the Animation of Balancing and Walking. *Proceedings of SIGGRAPH*, 155–162.

[17] Lien, S., Kajiya, J. T. 1984. A Symbolic Method for Calculating the Integral Properties of Arbitrary Nonconvex Polyhedra. *IEEE Computer Graphics and Applications* 4(5):35–41.

[18] McMahon, T. A. 1984. *Muscles, Reflexes, and Locomotion*. Princeton: Princeton University Press.

[19] Ngo, J. T., Marks, J. 1993. Spacetime Constraints Revisited. *Proceedings of SIGGRAPH*, 343–350.

[20] Pai, D. 1990. Programming Anthropoid Walking: Control and Simulation. Cornell Computer Science Tech Report TR 90-1178.

[21] Playter, R. R., Raibert, M. H. 1992. Control of a Biped Somersault in 3D. In *Proceedings of the IEEE International Conference on Robotics and Automation*, 582–589.

[22] Raibert, M. H. 1986. *Legged Robots That Balance*. Cambridge: MIT Press.

[23] Raibert, M. H., Hodgins, J. K. 1991. Animation of Dynamic Legged Locomotion. *Proceedings of SIGGRAPH*, 349–356.

[24] Rosenthal, D. E., Sherman, M. A. 1986. High Performance Multibody Simulations Via Symbolic Equation Manipulation and Kane's Method. *Journal of Astronautical Sciences* 34(3):223–239.

[25] Sims, K. 1994. Evolving Virtual Creatures. *Proceedings of SIGGRAPH*, 15–22.

[26] Sims, K. 1994. Evolving 3D Morphology and Behavior by Competition. *Artificial Life IV*, 28–39.

[27] Takei, Y., 1990. Techniques Used by Elite Women Gymnasts Performing the Handspring Vault at the 1987 Pan American Games. *International Journal of Sport Biomechanics* 6:29-55.

[28] Wooten, W. L., Hodgins, J. K. 1996. Simulation of Human Diving. *Computer Graphics Forum* 15(1):3–13.

[29] van de Panne M., Fiume, E. 1993. Sensor-Actuator Networks. *Proceedings of SIGGRAPH*, 335–342.

[30] van de Panne M., Fiume, E., Vranesic, Z. 1990. Reusable Motion Synthesis Using State-Space Controllers. *Proceedings of SIGGRAPH*, 225–234.

[31] Witkin, A., Kass, M. 1988. Spacetime Constraints. *Proceedings of SIGGRAPH*, 159–168.

PART 11

FUTURE ROBOTS

SESSION SUMMARY

Hirofumi Miura
Department of Mechano Informatics
The University of Tokyo, 7-3-1 Hongo, Bunkyoku
Tokyo 113, Japan
e-mail: miura@leopard.t.u-tokyo.ac.jp

I had the pleasure of chairing the session entitled "Future Robots". Four papers were presented by Joel Burdick(CALTECH), James Trevelyan(The University of Western Australia), Hiroshi Ishiguro(Kyoto University) and Isao Shimoyama (The University of Tokyo).

At both ISRRs in 1993 and 1995, I fortunately chaired the session on futuristic topics of robotics. The titles of session were "Micro/Miniature Mechanisms" in 1993 and "Micro-robotics" in 1995 . In those days micromachine technology by which sub-millimeter sized mechanical parts can be fabricated was highlighted taking advantage of a NSF report " Small machines, Large opportunities--a report on the emerging field of microdynamics--". As one of interesting fields of micromachine technology, the researches on microrobots started at those days in several research groups inthe world. Consequently, several papers on microrobots were submitted and four papers were accepted at ISRR in 1993 and 1995 respectively. Micro-robots were one of the most futuristic topics in robotics research in those days and I was very happy to chair the above mentioned sessions. However, at the 8th ISRR in 1997, no papers on microtechnology were submitted. But the above first mentioned four authors' papers were considered dealing with new and futuristic topics in robotics, being grouped into the session "Future Robots".

J. Burdick reviews progress towards the development of more unifying principles for the analysis and control of biomimetic robotic locomotion. Biomimetic methods of robotic locomotion, which are analogous to patterns of movement found in nature, do not rely upon wheels, jets,or propellers. In an attempt to derive useful results for specific examples, prior studies have generally focused either on a particlar assumption (such as the quasi-static assumption) or on a particular robot morphology (such as a biped or quadruped). Unfortunately, results derived for one morphology often do not extend to other morphologies. To enable future widespread deployment of cheap and robust robotic locomotion platforms, he says that they must ultimately seek a more unifying engineer-ing frame-work for biomimetic robotic locomotion. The realization of this frame work could be based on the following sequence: (1) establish a general form for the equations of motion for locomotion systems; (2) develop a control theory for this class of nonlinear equations; and (3) abstract motion planning and feed back control algorithms from the control theory. Progress in each of these areas is summarized in Burdick's paper.

For mechanics and modelling, two interesting examples are described. The first one is the case of microorganism. Microorganisms such as paramecia orchestrate the methachronal flexing of their cillia to effect cyclic changes in their body envelope. Consider an artificial "amoeba": roughly circular or spherical robot that would swim by small deformations of its surface. They would like to know how such a swimmer should deform in order to propel and steer itself, and how effective this kind of swimming would be. The class of locomotors of the first example is likely to have limited practical applications. The second example is the fluid flow associated with most high performance aquatic propulsors involves significant vorticity. Consider "carangiform" swimmers (a tuna or a porpoise) whose propulsive body deformations are nominally planar and restricted to a rear body segment. The correlation between the model and experiment at steady state is quite good indicating that modeling assumpution are valid in practice.

The physics of biomimetic locomoton are also highly nonlinear. However, there do appear to be some common underlying principles. Hopefully, continued investigation of these principles will lead to a useful foundation for robotic locomotion engineering.

J. Trevelyan discussed robots for the land mine problem as described bellow. He says that many people have expressed the hope that robots can help clear the millions of anti-personnel and other land mines which litter over 70 countries. An effective robotics solution could help and relieve the efforts of thousands of "deminers" who are using techniques which have changed little since the 1940's. Several large research efforts have failed, so far, to develop an effective mine clearance alternative to the existing manual technique. Robots have been tried at great expense, but without sucsess. This paper argues that robots are not an appropriate solution for mine clearance. First, there is little likelihood of sensing improvements in short term. Second, the huge variety of mines and minefields defined any automated solution. Third, robotic solutions are likely to be too expensive to be practical for humanitarian demining operations in countries like Angola, Afganistan and Cambodia. The effort devoted to robotics solutions would be more help-ful if it were directed at simple equipment improvements and low-cost robotic devices might provide some useful improvements in safety and cost-effectiveness in the short to medium term.

To understand why these efforts have achieved so little, he says that they need to understand more about mines, minefields, and the huge and continuing effort to develop suitable sensors. Sadly, there is little likelihood of success if present approaches are followed. A robotics project requires several importantfactors to be sucessful. What is unusual about robotics research is that it requires the successfulintegration of number of disparate technologies, most of which must perform at or near ultimate limits. For example, there are no suitable sensors yet, and no avairable experimental results suggest that there is little likelihood in the next 2 - 3 years. As a result of discussions with deminers, he has been able to devise several simple equipment improvements which could help demining teams.

H. Ishiguro proposed a "Perceptual Information Infrastructure(PI^2)" for robot navigation in real world and showed, as an example, a distributed vision system for navigating mobile robots. He insists that the perceptual information infrastructure enables to develop a new research direction of robots which are integrated with environments. He says that there are two possible way in realizing the intelligent robot system. One is to develop autonomous robots which have flexible mechanisms like humans. Several research groups are challenging to develop the humanoid robots and discover fundamental principles of intelligent systems. In this approach, there are remained serious problems. One example is that robots do not have sophisticated sensor systems to observe external world as they like. Another approach is to prepare infrastructures for robots so that robots can obtain necessary information with cameras as a example. He calls this infrastructure "Perceptual Information Infrastructure". The infrastructure in his paper differs from the infrastructure for mobile robots which move in factories. The purpose is not to develop systems which support individual functions of the robots such as guide lined and landmarks for locomotion, but to develop an infrastructure which actively provides various information for real world agents, such as robots and humans. That is, the PI^2 monitors the environment, maintains the dynamic environment models, and provides information for the real world agents.

The experiment was done in the model town, of which reduced scale is 1/12, representing enough realities of an outdoor environment, such as shadows, textures of tree, lawns and houses. 16 Vision Agents(cameras) were established in the model town and used for navigating the mobile robot. He calls the research direction with the infrastructure "Social Robotics". In the case that robot tasks are complex and it is hard to design proper mechanisms and sensors, robots need to execute possible tasks with the infrastructure.

I. Shimoyama discussed bio-robotic systems based on insect fixed behavior by artificial stimulation. He selected the insects as the model of micro-robots. He says that a large complicated robot cannot be scaled down to a micro-system due to the scale effect. An insect is thought to be a good model for a microrobot because an insect has obtained mechanisms and functions suitable to its scale. Also, insect behavior is fascinating because the behavior emerges from a small number of neurons. The number of neurons of human is larger than 10^{10} and on the other hand the number of neurons of insect is about 10^5-10^6 . The purpose of this research is to investigate the relationship between stimulation and behavior. A silkworm moth and a cockroach are used in the experiment because both can be obtained easily even in winter. At first, pysical stimulation including wind, sound, light, heat and electricity was applied to several points of insects. As a result, it was found that heat stimulation and electric stimulation at antennae were the most effective for causing behavior. He also tried to develop a robot which behaves as a insect. He says that by extending his research there will be possibility to find out how behavior comes out from the interaction with the environment. These results will be useful for developing a microrobot controller.

Mechanics and Control of Biomimetic Locomotion

J. Burdick B. Goodwine R. Mason
Mechanical Enginering, CALTECH, Mail Code 104-44, Pasdena, CA 91125 (U.S.A.)
email: {jwb,goodwine,mason}@robby.caltech.edu

Abstract

Biomimetic locomotion refers to the movement of robotic mechanisms in ways that are analogous to the patterns of movement found in nature. This paper reviews progress towards the development of more unifying principles for the analysis and control of biomimetic robotic locomotion.

1 Introduction

Since mobility is essential for many autonomous systems, robotic locomotion has been actively studied for three decades. Most mobile robots use wheels, since they provide the simplest means for mobility. The assumption that these wheels do not slip provides nonholonomic kinematic constraints on a vehicle's motion, and these systems have been extensively studied. For robotic operation on rugged terrains *biomimetic* schemes, such as legs, have also been considered. Biomimetic methods of robotic locomotion are analogous to the patterns of movement found in nature. Practically speaking, biomimetic locomotors do not rely upon wheels, jets, or propellers.

Quasi-static legged locomotion has been the most extensively studied type of biomimetic locomotion, and several 4- and 6-legged robots have been successfully demonstrated (Hirose and et al 1985; Song. and Waldron 1989). Beginning with Raibert (Raibert 1986), legged hopping robots have received considerable experimental and analytical attention (Koditschek and Buhler 1991; M'Closkey and Burdick 1993). Bipedal walking and running has also been an active area of study (Kajita and Tani 1991; McGeer 1990). Others have implemented various forms of "snake-like" locomotion (Hirose and Umetani 1976; Chirikjian and Burdick 1995), as snake-like robots can potentially enter environments that are inaccessible to legged or wheeled vehicles.

In an attempt to derive useful results for specific examples, prior studies have generally focused either on a particular assumption (such as the quasi-static assumption) or on a particular robot morphology (such as a biped or quadruped). Unfortunately, results derived for one morphology often do not extend to other morphologies. To enable future widespread deployment of cheap and robust robotic locomotion platforms, we must ultimately seek a more unifying engineering framework for biomimetic robotic locomotion. This framework should have the following properties:

- the systems analysis, design, and control methodologies can be uniformly applied to a broad class of locomotory problems.
- significant aspects of the approach can be implementable in automated software tools.
- the methods is sufficiently rigorous so that system performance can be predicted.

The realization of this framework would allow non-specialists to design and deploy effective biomimetic locomotors. This realization could be based on the following sequence: (1) establish a general form for the equations of motion for locomotion systems; (2) develop a control theory for this class of nonlinear equations; and (3) abstract motion planning and feedback control algorithms from the control theory. The remainder of this paper summarizes progress in each of these areas. The references contain more detailed discussions.

2 Mechanics and Modeling

Biomimetic propulsion is typically generated by a coupling of periodic mechanism deformations to external constraints. The forces generated by these constraint interactions (e.g. pushing, rolling, or sliding) induce net robot movement. The creeping, sidewinding, and undulatory gaits of snakes rely upon no-slip, or non-holonomic, constraints. Slug and snail movement depends upon the viscous fluid constraint of slime trails, while amoebae and paramecia move via a constraint between their surfaces and the surrounding fluid. Fish use a variety of fluid mechanical constraint principles. As shown below, common principles underlie the modeling and mechanics of these seemingly different systems. Techniques for modeling locomotion mechanisms that employ continuous nonholonomic constraints are given in (Burdick and Ostrowski 1995), and are briefly reviewed below. Extensions to other types of constraints are considered by way of example.

For a given locomotor, we can assume that there exists a Lagrangian, $L(q,\dot{q})$, and a set of constraints equations. Q is the mechanism's configuration space, and $q \in Q$. The form of the constraint equations will vary with the type of constraint. Non-holonomic constraints take the form $\omega^j(q)\dot{q} = 0$, $j = 1,\ldots,n_c$, where n_c is the number of constraints. One can always determine the locomotor's equations of motion from the Euler-Lagrange equations, which for the case of nonholonomic constraints takes the form

$$\frac{d}{dt}\left(\frac{\partial L}{\partial \dot{q}^i}\right) - \frac{\partial L}{\partial q^i} + \lambda_j \omega_i^j - \tau_i = 0. \qquad (1)$$

where the τ_i are the generalized external forces acting on the system, the λ_j are the Lagrange multipliers associated with the constraints, and $i = 1,\ldots,dim(Q)$. Unfortunately, it is generally difficult to solve for the Lagrange multipliers, and the resulting equations are not practically useful for locomotion analysis and control. Fortunately, the mechanics of locomotion systems have additional structure that can be exploited.

It is always possible to divide a locomoting robot's configuration variables into two classes. The first set of variables describes the robot's *position*, which is the displacement of a coordinate frame attached to the moving robot with respect to a fixed reference frame. The set of frame displacements is the Lie group $SE(m), m \leq 3$, or one of its subgroups (hereafter denoted by G). The second set of variables defines the *shape* of the mechanism. We only require that the "shape space," M, define a manifold. The configuration space is $Q = G \times M$, and its interesting structure is described below.

If the robot's initial body fixed frame position is denoted by h (e.g., h is a homogeneous transformation matrix), and it is displaced by g, then its new position is gh. This displacement is a *left translation*, and can be thought of as a map $L_g : G \to G$ given by $L_g(h) = gh$. The left translation induces a *left action* of G on Q. A left action of a Lie group G on Q is a smooth mapping $\Phi : G \times Q \to Q$ such that for all $q \in Q$:

1. $\Phi_e(q) = q$, and
2. $\Phi_g(\Phi_h(q)) = \Phi_{gh}(q)$ for all $g, h \in G$.

The *lifted action*, which describes the effect of Φ_g on velocity vectors in TQ, is the linear map, $D_q\Phi_g : T_qQ \to T_qQ$.

An action is *free* if $\Phi_g(q) = q$ implies $g = e$ (the identity in G) for each $q \in Q$. A ***principal fiber bundle*** over M with Lie group G consists of a manifold Q and a free left action of G on Q satisfying the following: (1) $M = Q/G$, where the *canonical projection* $\pi : Q \to M = Q/G$ is differentiable, and (2) Q is locally *trivial*. The sets $\pi^{-1}(r) \subset Q$ for $r \in M$ are the *fibers*, and Q is the union over M of its fibers, which are diffeomorphic to G.

The importance of the principal fiber bundle structure of configuration space is related to the following facts. The shape and position variables are coupled by the constraints acting on the robot. By making changes in the shape variables, it is possible to effect changes in the position variables through the constraints. Hence, one would like to systematically derive an expression that answers the question: "if I wiggle the body, how does the mechanism move." Answering this question is one of the central goals of locomotion research A related and important fact is that every principal fiber bundle can be endowed with a *connection*. The connection can be used to encode the all-important relationship between shape and position changes.

Formally, a *connection one-form*, $\mathcal{A}(q): T_qQ \to \mathfrak{g}$, is a Lie-algebra valued one-form satisfying:

(i) $\mathcal{A}(q)$ is linear in its action on T_qQ.

(ii) $\mathcal{A}(q)\xi_Q = \xi$ for $\xi \in \mathfrak{g}$.

(iii) $\mathcal{A}(q)\dot{q}$ is equivariant, i.e. it transforms as: $\mathcal{A}(\Phi_h(q))T_q\Phi_h(q)\dot{q} = Ad_h\mathcal{A}(q)\dot{q}$

where $\mathfrak{g}$ is the Lie algebra of G, Ad is the adjoint action on $\mathfrak{g}$, and η_Q for any $\eta \in \mathfrak{g}$ is the *infinitesimal generator*:

$$\eta_Q(q) = \frac{d}{d\epsilon}\Phi_{\exp(\epsilon\eta)}(q)\Big|_{\epsilon=0}$$

It can be shown that in local coordinates $q = (g,r)$, where $g \in G$ and $r \in M$, that $\mathcal{A}(q)\dot{q}$ takes the form

$$Ad_g(g^{-1}\dot{g} + A(r)\dot{r})$$

where $A(r)$ is termed the "local" form of the connection, and only depends on the shape variables. The connection one-form is useful in the locomotion context because the kernel of the connection one-form defines the *horizontal space*:

$$H_qQ = \{z \in T_qQ |\ \mathcal{A}(q)z = 0\}.$$

When possible, *we will define the horizontal space to be the space of motions that satisfy the physical locomotion constraints.* Hence, constraints that satisfy the mathematical properties of a connection can always be expressed in the form of a linear and separable relationship between shape and position velocities:

$$g^{-1}\dot{g} = -A(r)\dot{r}. \tag{2}$$

We now relate the mathematical concept of a connection one-form to the physical constraints that arise in locomotion.

For locomotion problems, it will be convenient to use the constraints associated with the problem to define the connection. In our context, a set of constraints will be interpreted as constraint distribution.

Definition 1 *A* constraint distribution $\mathcal{D}$ *is a distribution on Q composed of the allowable directions of motion at each point $q \in Q$, written $\mathcal{D}_q$. A curve $c : [a,b] \to Q$ is said to* satisfy the constraints *if $\dot{c}(t) \in \mathcal{D}_{c(t)}$, $\forall t \in [a,b]$.*

Let's first consider a mechanism that is subject to a "sufficient" number of nonholonomic kinematic constraints. Kinematic nonholonomic constraints can be expressed in the form:

$$\omega^i(q)\dot{q} = 0 \qquad i = 1, \cdots, l \qquad \dot{q} \in T_qQ$$

These kinematic constraints define a constraint distribution,

$$\mathcal{D}_q = \{\dot{q} \mid \omega(q)\dot{q} = 0\}.$$

It is assumed that $\dim(\mathcal{D})$ is constant over Q. This restriction allows us to write $\mathcal{D}$ as the kernel of a set of one-forms over Q—i.e., the one-forms $\{\omega^i(q)\}$ in the equation above.

If the constraint distribution is invariant with respect to the group action, then it defines a connection: the horizontal space is itself the constraint distribution. The constraint distribution will be invariant if the constraint one-forms are invariant with respect to the lifted action. This can be seen as follows. Let $\mathcal{D}_q$ and $\mathcal{D}_{\Phi_h(q)}$ denote the constraint distributions at two configurations q and $\Phi_h(q)$, where $h \in G$:

$$\begin{aligned} \mathcal{D}_q &= \{v \mid w(q)v = 0\} & (3) \\ \mathcal{D}_{\Phi_h(q)} &= \{z \mid w(\Phi_h(q))z = 0\} & (4) \end{aligned}$$

If the vector fields $v \in \mathcal{D}_q$ are invariant, then it must be true that $z = T_q\Phi_h v$. Thus,

$$w(\Phi_h(q))T_q\Phi_h v = 0$$

However, it is true that $w(\Phi_h(q))T_q\Phi_h v = T_q^*\Phi_h w(\Phi_h(q))v = 0$. Hence, v must be a horizontal vector of the one-form $T_q^*\Phi_h w(\Phi_h(q))$, and thus a sufficient condition for a set of constraint one-forms to define a connection is that the one-forms are invariant with respect to the lifted action. Hence, if a mechanical system is subject to the same number of kinematic nonholonomic constraints as $dim(G)$, and if the constraints satisfy this invariance condition, then it is automatically true that the kinematic constraints define a connection for this system. By Equation (2), it must be true that the kinematic constraints equations can be rearranged into the form:

$$g^{-1}\dot{g} = A_{loc}(r)\dot{r}$$

Instead of taking the form in Eq. (1), the complete equations of motion "reduce" to the equations (Ostrowski 1995):

$$\begin{aligned} \xi = g^{-1}\dot{g} &= -A(r)\dot{r} \\ M(r)\ddot{r} &= \tau - B(r,\dot{r}) - C(r) \end{aligned} \tag{5}$$

The reduced dynamics (i.e., the second equation in Eq. (5)) are not important for understanding locomotion as they merely describe the dynamics of the robot's internal shape deformations. Thus, the equations which are important for understanding locomotion are reduced to a set of 1st-order equations (the connection) that make explicit how shape changes lead to robot motion.

Next consider a mechanism in the absence of any external constraints—for example, a free floating satellite with appendages. If the Lagrangian is *invariant* with respect to the group action (i.e., if $L(\Phi_g(q), D_q\Phi_g(q)\dot{q}) = L(q,\dot{q})$), then there must exist $dim(G)$ conservation laws, or "symmetries." These symmetries can be thought of as a type of constraint. When $G = SE(m)$ (which is always true for locomotion mechanisms), these conservation laws correspond to conservation of linear and angular momentum. Practically, we only require that the dynamical equations do not change if we displace the mechanism in its ambient space. This is nearly always true.

These conservation laws define a connection (known as the *mechanical* connection). That is, the conservation laws can roughly be described as a set of velocity constraints of the form Eq. (2). Instead of taking the form in Eq. (1), the complete equations of motion "reduce" to the equations:

$$\begin{aligned} \xi = g^{-1}\dot{g} &= -A(r)\dot{r} + I^{-1}(r)\mu \\ \dot{\mu} &= ad^*_{\xi}\mu \\ M(r)\ddot{r} &= T(r)\tau - B(r,\dot{r}) - C(r) \end{aligned} \tag{6}$$

where $g^{-1}\dot{g}$ is interpreted as the velocity (in body coordinates) of the body fixed moving reference frame, μ is the momentum as expressed in body coordinates, and τ are forces acting on the shape variables. The body coordinates momentum, μ, is not necessarily conserved. However, when expressed in in spatial coordinates, the momentum is conversed. In the case of zero initial momentum, the mechanics of the this class of systems looks identical to the kinematic nonholonomic systems.

In some mechanisms, such as the snakeboard, there are fewer kinematic constraints than symmetries, and the Lagrangian symmetries and nonholonomic constraints interact. In these cases one needs to define an extended notion of a connection and the reduced dynamical equations. As shown in (Burdick and Ostrowski 1995), the equations of motion in these cases take the form:

$$\begin{aligned} g^{-1}\dot{g} &= -A(r)\dot{r} + I(r)^{-1}p, \\ \dot{p} &= \tfrac{1}{2}\dot{r}^T\sigma_{\dot{r}\dot{r}}(r)\dot{r} + p^T\sigma_{p\dot{r}}(r)\dot{r} + \tfrac{1}{2}p^T\sigma_{pp}(r)p, \\ M(r)\ddot{r} &= \tau - C(r,\dot{r}) + N(r,\dot{r},p) \end{aligned} \tag{7}$$

where $p(t) = \frac{\partial L}{\partial \dot{q}^i}(\xi^c(t))_Q$ (for any $\xi^c \in \mathfrak{g}$ that satisfies the nonholonomic constraints) is termed the "nonholonomic momentum.". I.e. the Euler-Lagrange equations can be reduced to a system of $dim(G)$ first-order constraints (the connection), $(dim(G) - n_c)$ first-order momentum equations, and $dim(M)$ second-order base space equations. Only the first two equations are of interest in the analysis of locomotion behavior and control.

The applications of these ideas to the modeling of a "snakeboard" and Hirose's active cord mechanism can be found in (Burdick and Ostrowski 1995). These terrestrial systems rely upon continuous non-holonomic constraints. For an extension of these ideas to viscous fluid constraints, see (Kelly and Murray 1996). Two examples are given to show how these ideas extend to the realm of fluid mechanical constraints, and in Section 5 we discuss how to analyze systems with discontinuous constraints, such as legged systems.

If we assume that our locomotor operates in an incompressible fluid, then the fluid-mechanism system is subject to two symmetries. The locomotor is subject to the previously discussed $SE(m)$ symmetry. An incompressible fluid naturally has the symmetry group of all volume-preserving diffeomorphisms, $Diff_{vol}(\mathcal{D})$, of the fluid domain, $\mathcal{D}$. As before, let M denote the internal shape space of the aquatic propulsor. The configuration space is thus the infinite dimensional space $Q = SE(m) \times Diff_{vol}(\mathcal{D}) \times R$. The nature of the constraints, and the resulting reduction of the equations of motion will depend on the assumptions made about the fluid flow. The examples illustrate representative assumptions.

Example #1: Microorganisms such as paramecia orchestrate the metachronal flexing of their cilia to effect cyclic changes in their body envelope. Consider an artificial "amoeba": a roughly circular or spherical robot that would swim by small deformations of its surface. We would like to know how such a swimmer should deform in order to propel and steer itself, and how effective this kind of swimming would be. Since real amoebae are microscopically small, the surrounding fluid flow is governed by the very low Reynolds number equations of creeping flow. However, a macroscopic "robot" amoeba in water would experience a higher Reynolds number flow. The idealization that the macroscopic amoeba swims through an inviscid, incompressible, and irrotational fluid is a very good approximation.

Figure 1: The boundary deformation modes

Any practical robot amoeba would be constructed with a finite number of actuators—preferably with the fewest possible. Therefore, we consider a boundary deformation model incorporating a finite number of parameters. Fix a frame in the amoeba body, and let the polar coordinates of the amoeba's boundary be:

$$\begin{aligned} r(t,\theta) &= R(t,\theta) = r_0[1 + \epsilon(k_1(t)\cos(2\theta) \\ &\quad + k_2(t)\cos(3\theta) + k_3(t)\sin(3\theta))] \end{aligned} \tag{8}$$

where $\epsilon > 0$ is the amplitude of the boundary deformation, which is assumed to be "small," and r_0

is the nominal radius. The shape deformation parameters k_1, k_2, and k_3 correspond to three modes of deformation and are assumed to be time periodic (Fig. 1). Deformation modes one and two together yield motion in the x-direction. Modes one and three together yield motion in the y-direction.

To derive the equations of motion, one can start from a Lagrangian framework. For the flow assumptions made above, there must exist a potential function ϕ such that the velocity field is $u = \nabla\phi$ and that the potential ϕ must satisfy Laplace's equation, $\nabla^2\phi = 0$. The potential flow formulation leads to a total kinetic energy term of the form $K(g, r, \dot{g}, \dot{r}) = \frac{1}{2}\dot{r}^T M_r(r)\dot{r} + \frac{1}{2}\xi^T M_g(r)\xi$, where $\xi = g^{-1}\dot{g}$ is the body velocity of the vehicle. The appropriate symmetries can the be applied to show that the equations of motion reduce exactly to the form of Eq. (2). The actual calculations in the Lagrangian framework are difficult. Fortunately, the theory dictates that the equations resulting from any other approach can be rearranged in the form of Eq. (2).

In practice, the equations of motion can be more readily derived using a result from Kelvin:

$$m\dot{C} = K(\phi) = \rho \int_S \phi\, n\, dS \qquad (9)$$

where m is the amoeba mass, C is the location of the robot's center of mass, ρ is the fluid density, S represents the amoeba's boundary, n is the unit normal to S, ϕ is the appropriate potential function for the given geometry, and the ***fluid impulse*** $-K(\phi)$ is the time integral of the pressures acting on S. If in a world frame the body frame origin has coordinates (x, y) and velocity $(\dot{x}, \dot{y})$, then the centroid of a homogeneous amoeba has world-frame coordinates $C(t)$:

$$\begin{bmatrix} x(t) \\ y(t) \end{bmatrix} + \frac{1}{A(t)} \int_0^{2\pi} \int_0^{r(t,\theta)} \begin{bmatrix} r' \cos(\theta) \\ r' \sin(\theta) \end{bmatrix} dr'\, d\theta$$

Where $A(t) = \pi r_0^2[1 + \frac{1}{2}\epsilon^2(k_1(t)^2 + k_2(t)^2 + k_3(t)^2)]$ is the area of the body. Differentiation and expansion in terms of ϵ yields the centroid velocity:

$$\dot{C} = \begin{bmatrix} \dot{x} \\ \dot{y} \end{bmatrix} + \epsilon^2 r_0 \begin{bmatrix} k_1\dot{k}_2 + \dot{k}_1 k_2 \\ k_1\dot{k}_3 + \dot{k}_1 k_2 \end{bmatrix} + O(\epsilon^3) \quad (10)$$

The potential flow solution, ϕ, can be obtained by realizing that through the separation of variables method, any solution to Laplace's equation has the form: $\phi(r, \theta) = (a_0 \log(r) + b_0)(c_0\theta + d_0) + \sum_{n=1}^{\infty}(\frac{a_n}{r^n} + b_n r^n)(c_n \sin(n\theta) + d_n \cos(n\theta))$. The constants a_n, b_n, c_n, d_n are determined by the boundary conditions at the amoeba's surface, by the requirement that $u = \nabla\phi$ go to zero as r approaches infinity, and by the requirement that there be no circulation around the amoeba. We thus find an approximate solution for ϕ: $\phi = \phi_0 + \epsilon\phi_1 + \epsilon^2\phi_2 + O(\epsilon^3)$.

Substituting the approximate expressions for ϕ and $\dot{C}$ into Eq. (9) and solving to find expressions for $\dot{x}, \dot{y}$ as an expansion in ϵ, we obtain:

$$\begin{bmatrix} \dot{x} \\ \dot{y} \end{bmatrix} = \epsilon^2 r_0 \left(\frac{\alpha}{\beta} \begin{bmatrix} \dot{k}_1 k_2 \\ \dot{k}_1 k_3 \end{bmatrix} - \begin{bmatrix} k_1 \dot{k}_2 \\ k_1 \dot{k}_3 \end{bmatrix} \right) + O(\epsilon^3) \qquad (11)$$

where $\alpha = \rho\pi r_0^2 - m$ and $\beta = \rho\pi r_0^2 + m$. Notice that this equation is a connection!

Example #2: Eq. (11) shows that the amoeba moves at the disappointingly slow pace of $\epsilon^2 r_0$ per deformation period. Hence, this class of locomotors is likely to have limited practical applications. The fluid flow associated with most high performance aquatic propulsors involves significant vorticity. Consider "carangiform" swimmers (a tuna or a porpoise) whose propulsive body deformations are nominally planar and are restricted to a rear body segment. The forward and rear sections of the swimmer's body are joined by a region of decreased hydrodynamic aspect, called a "peduncle" (see Fig. 2).

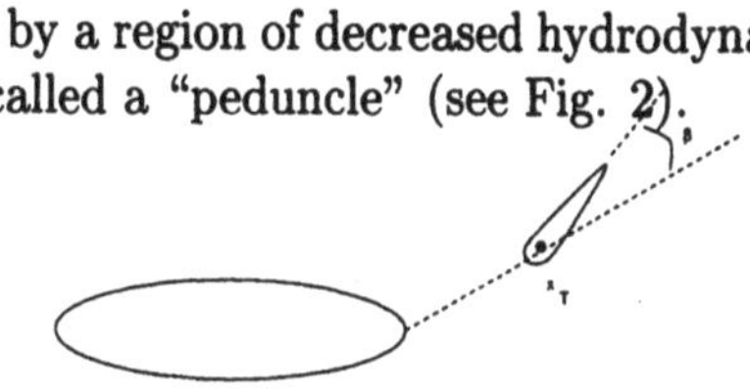

Figure 2: Carangid Schematic

The Lagrangian analysis starts with a Lagrangian whose state space includes the propulsor variables and the fluid particles, $L(q, \dot{q})$. However, the $SE(m)$ symmetries and fluid symmetries lead to a reduction of the Lagrangian to the form $l(g^{-1}\dot{g}, \dot{\psi}\psi^{-1}, r, \dot{r})$, where $\xi = g^{-1}\dot{g}$ is the velocity of the swimmer, as seen in it's body coordinates, ψ is a function that assigns position to a fluid particle, and $u(x) = \dot{\psi}\psi^{-1}$ is the fluid velocity field (in spatial coordinates).

Carangiform propulsion involves (and benefits from) significant vorticity generation and control. Vorticity in turn depends upon fluid viscosity. However, for carangiform motion, we can approximate the fluid mechanics by inviscid irrotational flow driven by a "substitution vortex" (Kelly, Mason, Anhalt, Murray, and Burdick 1998). The general form of the equations of motion then reduce to

$$\begin{aligned} \frac{d}{dt}\frac{\partial l}{\partial \xi} + ad^*_\xi \frac{\partial l}{\partial \xi} &= \rho U \Gamma \\ \nabla \times \left(\frac{d}{dt}\frac{\partial l}{\partial u} - ad^*_u \frac{\partial l}{\partial u}\right) &= \dot{\Gamma}\delta_{te} \\ \frac{d}{dt}\frac{\partial l}{\partial \dot{r}} - \frac{\partial l}{\partial \dot{r}} &= Bv \end{aligned} \tag{12}$$

where U is the free stream velocity and Γ is the strength of the vortex shed by the fish. The specific form will depend upon the carangid geometry and other assumptions. Generally, the first equation will reduce to a connection and possibly momentum equation. The second equation describes the fluid flow, and the third equation describes the internal shape deformation dynamics.

A detailed model of the experimental flapping foil model shown in Fig.s 3 and 4 is described in (Kelly, Mason, Anhalt, Murray, and Burdick 1998). This model is used to validate the modeling and control of carangiform locomotion.

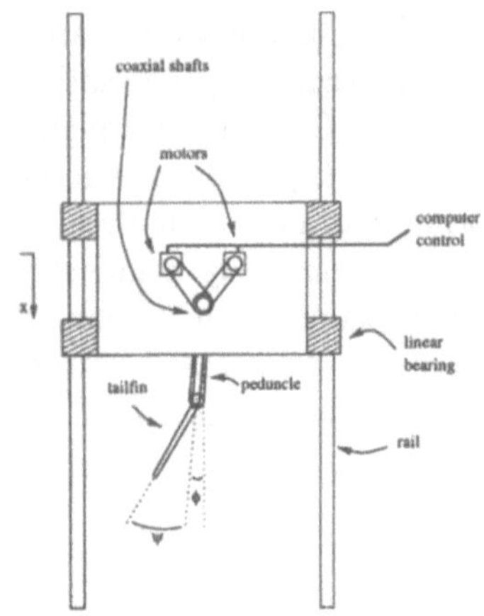

Figure 3: Top view of experimental apparatus

Figure 4: Photo of apparatus

Fig. 5 compares the experimentally determined carangiform location compared with a simulation of a steady state fluid mechanical model. The correlation between the model and experiment at steady state is quite good, indicating that our modeling assumptions are valid in practice.

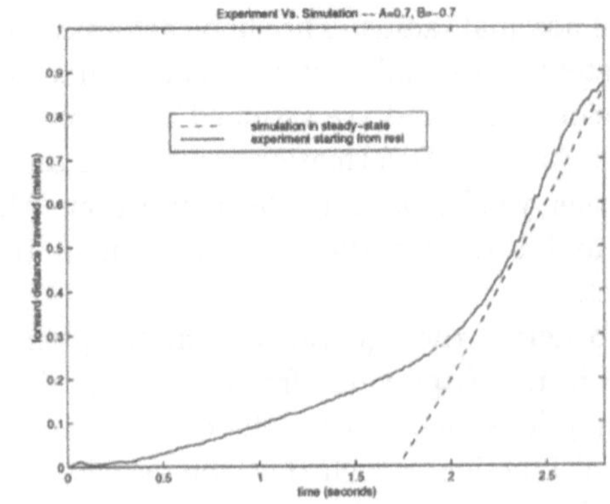

Figure 5: Comparison of theory and experiment

3 Controllability

A locomotion mechanism is *controllable* if there exists an admissible control which drives the system from the initial configuration, q_i, to the final configuration, q_f. The issue of controllability is a fundamental question in nonlinear control theory. As the example in Section 5 shows, it is not always obvious if a complicated locomotion mechanism is controllable. Historically, the determination of controllability is often an important first step in the development of a useful control theory for a given class of equations, since trajectory generation and control techniques typically are derived from the controllability test.

Recall that affine control systems have the form:

$$\dot{z} = f(z) + h_1(z)u_1 + \cdots + h_p(z)u_p \tag{13}$$

where z is the system state, the $\{u_i\}$ are the control inputs, the $h_i(z)$ are the control vector fields, and $f(z)$ is termed the "drift" vector field. The system is "driftless" if $f(z) = 0$. Classically, the controllability of nonlinear systems whose equations can be expressed in the form of Eq. (13) is considered by computing the *accessibility algebra*. Let $\Delta_0 = \text{span}\{f, h_1, \ldots, h_m\}$ and $\Delta_k = \Delta_{k-1} + \text{span}\{[X,Y] \mid X, Y \in \Delta_{k-1}\}$, where $[\cdot,\cdot]$ is the Lie bracket of vector fields. The sequence of Δ's terminates at some k_f. The system is *accessible* at z if $\dim(\Delta_{k_f}) = \dim(T_z N)$. For driftless systems, accessibility implies controllability. Unfortunately, accessibility is a necessary, but not sufficient condition, for a driftless system to be controllable.

For locomotion systems with nonholonomic constraints, controllability can be determined directly from the connection and momentum equations. For the case of kinematic systems, the kinematic connection takes the form of a driftless system with $z = [g\ r]^T$ and $h(z) = [-gA(r)\ I]^T$ In general, letting $z = (g, p, r, \dot{r})$, the connection and momentum equations can be put in the form of Eq. (13) where:

$$f(z) = \begin{bmatrix} g(-A\dot{r} + I(r)^{-1}p) \\ \frac{\dot{r}^T \sigma_{\dot{r}\dot{r}} \dot{r}}{2} + p^T \sigma_{p\dot{r}} \dot{r} + \frac{p^T \sigma_{pp} p}{2} \\ \dot{r} \\ 0 \end{bmatrix} \quad h_i(z) = \begin{bmatrix} 0 \\ 0 \\ 0 \\ e_i \end{bmatrix} \tag{14}$$

Controllability of kinematic, or driftless, case has been considered in (Kelly and Murray 1995). The mixed constraint case is discussed in (Ostrowski and Burdick). In summary, controllability is achieved if (1) Δ_∞ is full rank; (2) $\sigma_{\dot{r}\dot{r}}$ is an onto map; and (3) $(\sigma_{\dot{r}\dot{r}})_{ii} = 0$ for $i = 1, \ldots, m$. The interesting issue of controllability for legged systems is considered in Section 5.

4 Trajectory Generation

From a control perspective, locomotion systems are almost universally "underactuated." The control literature dealing with underactuated kinematic nonholonomic systems contains a number of schemes for determining the control inputs to approximately steer a system along a given trajectory. Since locomotion systems with kinematic connections can be put into the form of a driftless underactuated system, these techniques can be adapted for the purposes of trajectory planning. Unfortunately, these techniques have two problems. First, none of these techniques apply to systems with drift. Since the most interesting locomotion systems fall into this class, the nonlinear control literature is currently unable to provide useful results in this regard. Second, even for driftless systems, practical experience has shown that the trajectories that result from applying known techniques are quite unsatisfying. We are currently investigating three approaches.

In some instances, optimal control theory can provide useful trajectory planning schemes. For example, in the case of the amoeba, if the cost function is related to the square of the total boundary deformation, then the optimal control inputs are sinusoids: $k_1(t) = \cos(\omega t)$, $k_2(t) = -a(t)\sin(\omega t)$, and $k_3(t) = -b(t)\sin(\omega t)$, $a(t)$ and $b(t)$ functions determined from the boundary conditions and system parameters.. The motion of the amoeba's centroid due to these prescribed boundary deformations is:

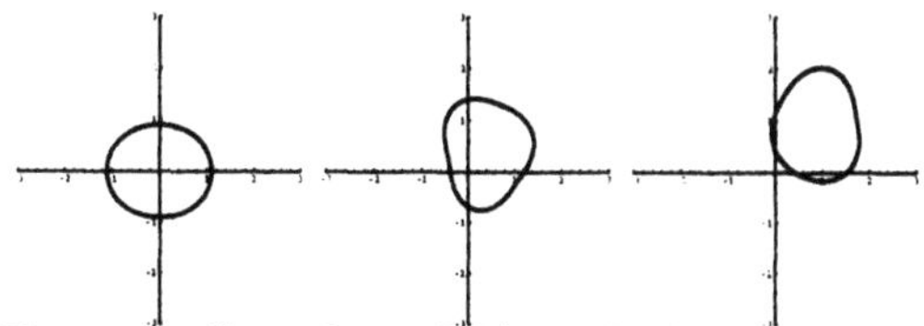

Figure 6: Snapshot of "Amoeba" Simulation

$$\dot{C} = \epsilon^2 r_0 \frac{2\rho\pi r_0^2}{\rho\pi r_0^2 + M} \begin{bmatrix} a(t)\omega \sin^2(\omega t) \\ b(t)\omega \sin^2(\omega t) \end{bmatrix} + \mathcal{O}(\epsilon^3)$$

Thus, moving the amoeba centroid along a given curve is remarkably easy. Fig. 6 shows snapshots of a an amoeba simulation model for the case $a = b = -1$, which causes the circle to move along a line slanted at a 45° angle to the horizontal.

Alternatively, at least for the case of kinematic connections, one can construct motion planning schemes by using a local approximation to the solution of the differential equation Eq. (2):

$$\begin{aligned} g(t) &= g(0)(I + A^i(0) \textstyle\int_{\alpha_i} dr^i + \\ &\quad (DA)^{ij}(0) \textstyle\int_{\alpha_i} dr^i \int_{\alpha_j} dr^j + \cdots) \end{aligned} \tag{15}$$

For systems with drift, we currently used a heuristic approach based on the relationship between "gaits" and Lie brackets in the controllability calculations (Burdick and Ostrowski 1995).

5 Legged Systems

A legged robot has discontinuous equations of motion near points in the configuration space where each of its "feet" come into contact with the ground, and it is precisely the ability of the robot to lift its feet off of the ground that enables it to move about. Consider the six–legged hexapod robot illustrated in Figure 7. Note that each leg has only two degrees of freedom, i.e. the robot can only lift its legs up and down and move them forward and backward. Such limited control authority may be desirable in practical situations because it decreases the mechanical complexity of the robot; however, the decreased complexity comes at the cost of requiring more sophisticated control. It is not immediately clear whether the robot can move "sideways," and even if it can, it is clear that conventional trajectory planning schemes based on foot placement can not be easily applied. Traditional nonlinear control analyses are inapplicable because they rely upon differentiation in one form or another. Yet it is the discontinuous nature of such systems that is often their most important characteristic.

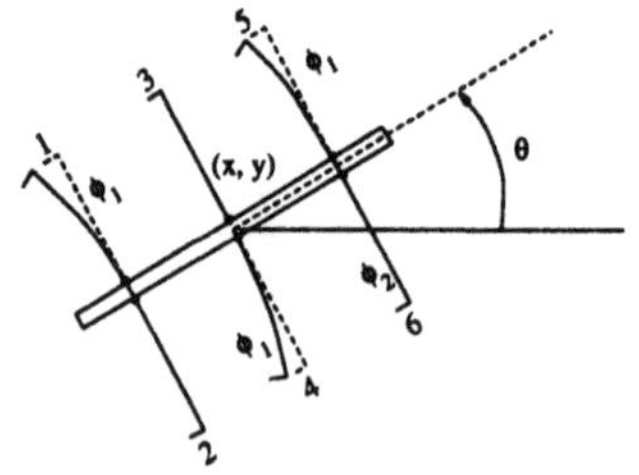

Figure 7: A Simple Hexapod Robot Model

By realizing that the configuration spaces of such systems are ***stratified***, they can be handled by a direct extension of the previously summarized techniques. Consider a biped robot. It will be subjected to constraints if one or more of its feet are in ground contact. The set of configurations corresponding to one foot in ground contact is a codimension one submanifold of the c-space. The same is true when the other foot contacts the ground. Similarly, when both feet are in ground contact, the system lies on a codimension 2 submanifold of the c-space formed by the intersection of the single contact submanifolds. The c-space structure for such a biped is abstractly illustrated in Fig. 8, and is said to be "stratified." Except for when a foot transitions from ground contact to free motion, the system's equations of motion are smooth on each strata, and they take the form Eq. (6) or Eq. (7).

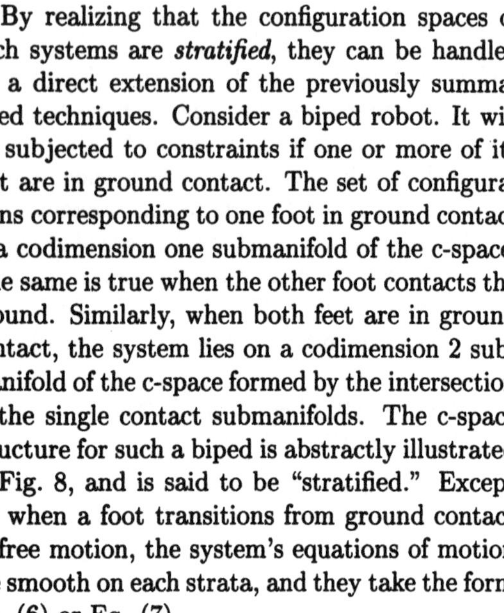

Figure 8: Schematic depiction of biped c-space

Classically, a ***regularly stratified*** set $\mathcal{X}$ is a set $\mathcal{X} \subset \mathbf{R}^m$ that can be decomposed into a finite union of disjoint smooth manifolds, called ***strata***, satisfying the Whitney condition. The dimension of the strata varies between zero, which are isolated point manifolds, and m, which are open subsets of $\mathbf{R}^m$. Let $S_i \subset Q$ denote the codimension one submanifold of Q that corresponds to all configurations where only the ith foot contacts the terrain. Denote, the intersection of S_i and S_j, by $S_{ij} = S_i \cap S_j$. The set S_{ij} physically corresponds to states where both the ith and j^{th} feet are on the ground. Further intersections can be similarly defined in a recursive fashion: $S_{ijk} = S_i \cap S_j \cap S_k = S_i \cap S_{jk}$, etc. In the classical definition of a stratification, stratum $\mathcal{X}_i$ consists of the submanifold S_i with all lower dimensional strata (that arise from intersections of S_i with other submanifolds) removed. However, we will refer to the submanifolds S_i, as well as their recursive intersections S_{ij}, S_{ijk}, etc, as strata.

For kinematic (i.e., quasi-static) legged systems on strata $S_I = S_{i_1,\cdots,i_k}$ the equations of motion take the form:

$$\dot{x} = g_{S_I,1}(x)u^{S_I,1} + \cdots + g_{S_I,p_I}(x)u^{S_I,p_I}$$

where $g_{S_I,j}$ is the jth control vector field and p_I is the number of control inputs on S_I. The controllability of kinematic legged systems can be determined as follows. Define $\Delta_{S_I} = \text{span}\{g_{S_I,1}, \ldots, g_{S_I,p_I}\}$. The stratified kinematic system is small time locally controllable if:

$$\overline{\Delta}_Q + \sum_i \overline{\Delta}_{S_i} + \sum_{i,j} \overline{\Delta}_{S_{ij}} + \cdots = T_q Q.$$

where $\overline{\Delta}_{S_I}$ is the involutive closure of Δ_{S_I}. A straightforward calculations shows that the hexapod of Fig. 7 is indeed controllable, and therefore it can move "sideways." However, the calculations show that direct sideways motions must involve complicated gyrations, as there is no direct sequence of foot placements that can move the mechanism in the sideways direction.

A general method for determining the controls inputs that will drive a quasi-static legged locomotors along a given trajectory is presented in (Goodwine and Burdick 1997). This method, which is an adaptation of prior nonholonomic motion planning methods to stratified geometry, does not rely upon foot placement concepts. In principle, the method is also independent of the number of legs. This method is useful for dealing with systems such as the simplified hexapod where foot placement methods are not applicable.

Fig. 9 shows an overhead view of the time history of the center of the robot's body as it follows and elliptical path with constant orientation. The inability of the mechanism to move directly sideways accounts for the unusual pattern. In order to illustrate the complexity of motion that arises due to the mechanical simplicity of this model, Fig. 10 shows the foot placements associated with a straight line path along which the body orientation uniformly increases.

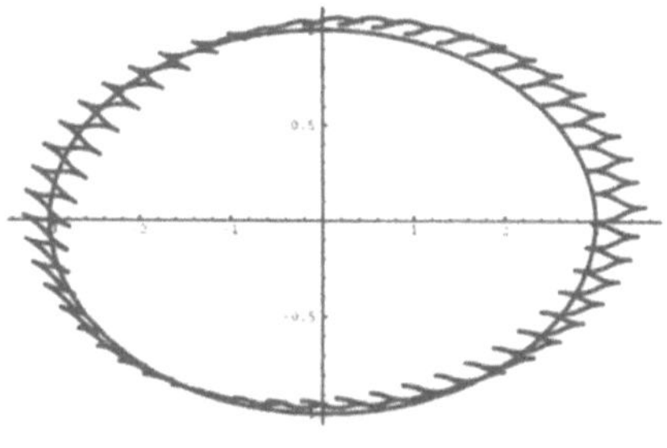

Figure 9: Elliptical path with fixed orientation

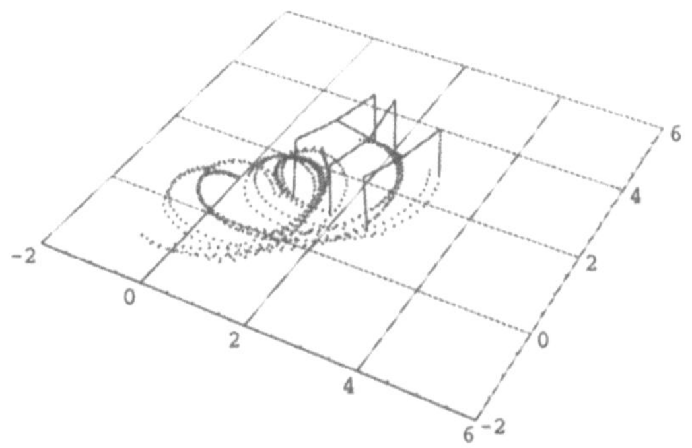

Figure 10: View of Foot Placements

6 Conclusion

The realm of biomimetic locomotion encompasses a large number of morphologies and physical principles of propulsion. The physics of biomimetic locomotion are also highly nonlinear. However, there do appear to be some common underlying principles. Hopefully, continued investigation of these principles will lead to a useful foundation for robotic locomotion engineering.

References

Burdick, J. and J. Ostrowski (1995). Geometric perspectives on the mechanics and control of robotic locomotion. In *Proc. 7th Int. Symp. Robotics Research*, Germany.

Chirikjian, G. and J. Burdick (1995). The kinematics of hyper-redundant robot locomotion. *IEEE Trans. on Robotics and Automation 11*(6), 781–793.

Goodwine, G. and J. Burdick (1997). Trajectory generation for kinematic legged robots. In *Proc. IEEE Int. Conf. Robotics and Automation*, Albuquerque, NM.

Hirose, S. and et al (1985). Titan III: A quadruped walking vehicle. In 2nd *Int. Symp. on Robotics Research*, Tokyo, Japan.

Hirose, S. and Y. Umetani (1976). Kinematic control of active cord mechanism with tactile sensors. In *Proc.* 2nd *Int. CISM-IFT Symp. on Theory and Practice of Robots and Manipulators*, pp. 241–252.

Kajita, S. and K. Tani (1991). Study of dynamic biped locomotion on rugged terrain. In *Proc. IEEE Int. Conf. Robotics and Automation*, Sacramento, CA, pp. 1405–1411.

Kelly, S., R. Mason, C. Anhalt, R. Murray, and J. Burdick (1998). Modeling and experimental investigation of carangiform locomotion for control. In *(submitted) Proc. Americal Control Conference.*

Kelly, S. and R. Murray (1995, June). Geometric phases and robotic locomotion. *Journal of Robotic Systems 12*(6), 417–431.

Kelly, S. and R. Murray (1996). The geometry and control of dissipative systems. In *Proc. IEEE Conf. on Decision and Control.*

Koditschek, D. and M. Buhler (1991, Dec.). Analysis of a simplified hopping robot. *Int. J. of Robotics Research 10*(6), 587–605.

McGeer, T. (1990). Passive dynamic walking. *Int. J. of Robotics Research 9*(2), 62–82.

M'Closkey, R. and J. Burdick (1993, June). On the periodic motions of a hopping robot with vertical and forward motion. *Int. J. of Robotics Research 12*(3), 197–218.

Ostrowski, J. (1995, Sept.). *The Mechanics and Control of Undulatory Robotic Locomotion.* Ph. D. thesis, California Institute of Technology, Pasadena, CA.

Ostrowski, J. and J. Burdick. Control of mechanical systems with symmetries and nonholonomic constraints. *J. of Applied Mathematics and Computer Science (to appear).*

Raibert, M. (1986). *Legged Robots that Balance.* MIT Press.

Song., S. and K. Waldron (1989). *Machines that walk: the Adaptive Suspension Vehicle.* MIT Press.

Robots:
A Premature Solution for the Land Mine Problem

James Trevelyan

Department of Mechanical and Materials Engineering,
The University of Western Australia,
Nedlands 6907, Western Australia
jamest@mech.uwa.edu.au

Abstract

Several large research efforts have failed, so far, to develop an effective mine clearance alternative to the existing manual technique. Robots have been tried at great expense, but without success. This paper argues that robots are not an appropriate solution for mine clearance. First, there is little likelihood of sensing improvements in the short term. Second, the huge variety of mines and minefields defies any automated solution. Third, robotic solutions are likely to be too expensive to be practical for humanitarian demining operations in countries like Angola, Afghanistan and Cambodia. The effort devoted to robotic solutions would be more helpful if it were directed at simple equipment improvements and low-cost robotic devices might provide some useful improvements in safety and cost-effectiveness in the short to medium term.

Understanding why "high tech" research efforts have failed, so far, may help to avoid similar mistakes in other ambitious robotics research programmes.

1 Introduction

Many people have expressed the hope that robots can help clear the millions of anti-personnel and other land mines which litter over 70 countries. An effective robotic solution could help and relieve the efforts of thousands of "deminers" who are using techniques which have changed little since the 1940's. The popular film "The English Patient" provides a graphic demonstration of this technique in an early scene.

Several different robotic solutions have been proposed and some are described below. Unfortunately, none of these solutions shows any prospect of making a significant contribution in the forseeable future.

Nicoud[11] proposed a small two-wheeled autonomous mine detector robot (Pemex) carrying a sensor package in an inverted dome which provides the third support point and skids over the ground. The robot is designed to be light enough not to detonate buried mines, and uses large diameter bicycle wheels for enhanced mobility. This was intended as a low-cost method for automatically locating mines. Several other similar concepts (using legs, rollers, balloon wheels, air cushions etc.) have been proposed but no reliable mine detecting sensors are available.

Several researchers have proposed armoured vehicles which can withstand the effects of anti-personnel mines which they would set off by ground pressure or tripwires. The vehicles either carry a mini-flail [12] to detonate mines or sensors to provide locations for subsequent manual removal. Both teleoperated and autonomous versions have been proposed. These vehicles are vulnerable to anti-tank mines and unexploded munitions (UXO's), even fragmentation anti-personnel mines (see later section for description of mines).

The author proposed a cable-suspended robot [15] which could be particularly useful over rough open terrain which vehicles cannot access. This has not proceeded beyond the paper concept stage. There are no suitable sensors for mine detection, and research on low-cost equipment improvements for manual demining is likely to me more effective, as this paper argues.

The British company ERA proposed a mine detection system consisting of a robotic arm carrying a ground penetrating radar detector and demonstrated a prototype [2]. However, tests have shown that ground penetrating radar, by itself, is not a reliable sensor.

Several research efforts have been directed at anti-vehicle mines which are potentially easier to detect.

Figure 1a: Bofors mine clearance machine.

Figure 1b: Vehicular Mine Detection System (VMDS) developed by US Defense Department programmes.

The US Defense Department funding has supported several ambitious projects to develop a useful anti-tank mine capability. Similar projects have been started in several other countries. Figure 1 (right) illustrates one concept which emerged from this research. This work was directed mainly at the need to maintain safe road communications in times of guerilla or civil war conflicts. Dirt roads are particularly vulnerable because mines can be laid in minutes by irregular forces, particularly in shallow potholes where the effects of excavating a hole to lay the mine cannot be distinguished easily. Frequently, the mines are laid to disrupt a road convoy in order to carry out an ambush attack to steal vehicles, weapons and valuables. The mines are either purchased or removed from other minefields. This may even be carried out by non-combatants who resort to such desperate measures to feed starving families or to obtain means to escape the conflict.

A mine detection vehicle would need to clear roads at a speed of at least 30 km/hr to be useful. However, the speeds reached in trials so far have not been much more than 1 m/sec. Apart from the problems posed by mine detection, the vehicle has to react quickly enough to stop before activating a mine. McMichael [9] proposed a light weight autonomous detection vehicle driving 50 - 100 m in front of the main vehicle to provide a greater stopping distance.

The South African company MECHEM have demonstrated a mine-proof vehicle. This has an armoured V-shaped underside and wheels which can be replaced if necessary. However, it is only safe against certain types of mines, and detection capabilities are limited. It is said to be more useful as a transport vehicle rather than an effective mine clearance technique.

One of the problems facing such efforts is the ingenuity of the people who lay the mines. In Afghanistan, for instance, anti-vehicle mines were laid in deep holes to prevent Russian forces from detecting them. Rocks laid carefully over the mines, or simple wooden blocks, which transmit the weight of the vehicle to the mine beneath. Road material was packed carefully around the rocks or stones so that the surface of the road would show no disturbance. Roads have had to be literally 'bulldozed' away to a depth of a metre to clear these mines. Sophisticated 'Western' mines were supplied to Mujahedeen forces specifically to counter Russian mine clearance machines.

To understand why these efforts have achieved so little, we need to understand more about mines, minefields, and the huge and continuing effort to develop suitable sensors.

1.1 Mechanised Clearance

Several attempts have been made to devise and build mine clearing machines to replace the well-known flail (see http://diwww.epfl.ch/lami/detec/rodemine.html#breaching) which has not proved to be satisfactory for humanitarian demining. A flail can clear as few as 70% of AP mines and can throw some mines several hundred metres away [19]. One development is a machine from Bofors (figure 1 - left) which grinds up mines and other buried objects to a depth of about 50 cm. This machine has been tested in Bosnia. However, it is too heavy to reach minefields in many countries (inadequate roads) and may be vulnerable to anti-tank mines and unexploded ordnance. Nicoud [12] describes the Krohn roller which has been used in southern Africa but is now impounded pending legal action. No machine has yet achieved anything approaching the required 99.96% clearance rate. Experience suggests that manual clearance after mechanised demining can be more dangerous.

2 Defining the Problem

2.1 Global landmine problem and its effects

More than 65 countries are affected to varying degrees by land mines. The Red Cross [13] estimated that mines and other unexploded ordnance are killing between 500 and 800 people and maiming 2000 others per month, mainly innocent civilians who had little or no part in the conflicts for which the mines were laid. Carefully targetted efforts to clear land mines can significantly reduce these statistics.

Some countries have banned the use of landmines and others are supportive of a complete ban. However, their low cost ($3 - $30) and the large numbers in existing stockpiles make them an attractive weapon for insurgency groups which operate in many countries with internal conflicts—the most common cause of wars today. Mines are often used by renegrade groups to harass civilians in order to extort money and food. They are also used for self-defence by villages and groups of people travelling in many districts where civil law and order provides little effective protection. Many countries retain massive landmine barriers on their borders or near military installations. Some of the most severe landmine problems exist in Egypt, Angola, Afghanistan, Rwanda, Bosnia, Cambodia, Laos, Kuwait, Iraq, Chechnya, Kashmir, Somalia, Sudan, Ethiopia, Mozambique, and the Falkland Islands.

Apart from deaths and injuries to people, the major effect is to deny access to land and its resources, causing deprivation among the affected populations. Millions of refugees cannot yet return home, and impose huge burdens on neighbouring countries (3–4 million Afghans in Pakistan, for instance) and require major international relief efforts. Regardless of any efforts to stop the further use of landmines, the safe restoration of productive land is an urgent issue in affected regions. While current clearance programmes measure results in terms of a 10–20 square kilometres per year, up to 60% of available agricultural land is unusable in several of the affected countries.

The widely quoted statistics suggesting that there are as many as 110,000,000 mines are now being questioned [5]. The original statistics were prepared in a great hurry by the US State Department in 1994. Mine clerance teams in the field were anxious not to under-estimate the problem, so they applied generous factors to cover uncertainties. Whereas 10,000,000 mines were thought to have been laid in Afghanistan, current estimates are less than 1,000,000. To further compound the problem, statistics have been quoted in terms of the cost to remove each mine and the rate of removal (up to $1,000 each, 20,000 mines and UXO's per year in Afghanistan cleared) rather than the cost to clear minefields (per quare metre). Thus one might deduce from published statistics that the problem in Afghanistan will cost $10,000,000,000 and require perhaps 500 years of clearance work. According to UN staff directing the mine clearance programme in Afghanistan, about 5 - 10 years work will clear most of the mines from high priority areas and enable the Afghan population to start rebuilding their country. Reconstructing roads, towns and other infrastructure will take longer, of course. Many mines will be left in or on the ground, but they are concentrated around military posts on hill-tops and in mountain areas which contribute little to economic activity and which can be left until other more important reconstruction is well under way.

Elsewhere, many millions of the land mines included in statistics lie along national borders or around vital installations such as airports where security is seen as a major local problem. Governments are not yet ready to remove these mines.

The main implication is that the land mine problem has been overstated in many places by perhaps a factor of 50 or more in terms of cost and time needed to clear the mines which are causing current problems. It is still a major problem, of course, but not as big, in relative terms, as the current publicity suggests. One should remember that perhaps 200,000 or more people die in road accidents each year around the world, and perhaps as many as 2,000,000 die from preventable Malaria.

Land mines are still a major problem. However, the problem can be fixed for what is. in international terms, a modest sum of money. About $2,000,000,000 spent over 10 to 20 years would deal with most of the land mine problems we now face.

2.2 Types of mines (figure 2)

Land mines are usually very simple devices which are readily manufactured anywhere. There are two basic types of mines: anti-vehicle or anti-tank (AT) mines, and anti-personnel (AP) mines[4]. AT mines are comparatively large (>4 kg explosive), usually laid in unsealed roads or potholes, and detonate when a vehicle drives over one. They are typically activated by force (>100 kg), magnetic influence, or remote control. AP mines (AP) are much smaller (80 - 250g explosive, 7 - 15cm diameter) and are usually activated by force (3 - 7kg) or tripwires. There are over 700 known types with many different designs and actuation mechanisms. There are two main categories of

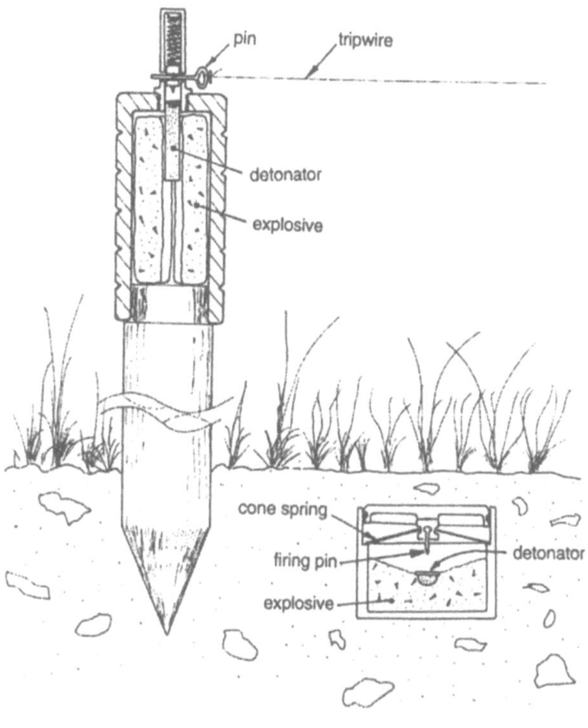

Figure 2. Two types of mine. Left a fragmentation mine which is activated by a trip wire which pulls out the firing pin, deton‑‑ng the main charge around the detonator. The outer shell breaks into lethal fragments which spread to a radius of 50 metres when the mine explodes. Right: a small blast mine (about 7 cm diameter) activated by foot pressure which causes the fibre-reinforced plastic cone shaped spring washer to snap down, firing the detonator which initiates the explosion of the surrounding explosive. The only metal parts are the firing pin and the detonator shell. Safety pins and time delay mechanisms have not been shown. The most common buried mine is the PMN (Russian) which has many more large metal components and is easy to detect and very sensitive, requiring as little as 1 kg force to set it off if disturbed

AP mine. A blast mine is usually small, and detonates when a person steps on it: the shoe and foot is destroyed and fragments of bone blast upwards destroying the leg. When a fragmentation mine explodes, metal fragments are propelled out at high velocity causing death or serious injuries to a radius of 30 or even 100 metres, and penetrating up to several millimetres of steel if close enough. Simple fragmentation mines are installed on knee high wooden posts and activated by tripwires (stake mines).

Another common type of fragmentation mine (a bounding mine) is buried in the ground. When activated, it jumps up to groin height before exploding. Mines of one type have often been laid in combination with another type to make clearance more difficult: stake mines with trip wires may have buried blast mines placed around them.

Humanitarian demining operations face many complications which need to be understood if we are to work towards useful improvements. The mines have often been in place for 5 or more years, may be corroded, waterlogged, impregnated with mud or dirt, and can behave quite unpredictably [19]. Stakes which carried fragmentation mines may have fallen over, and trip wires may be caught up in overgrown bushes, grass or roots. A wind gust may sway a bush enough to pull a trip wire and detonate a nearby mine. One cannot rely on mechanically activating old mines (or even new ones) to neutralise them. Many mines were laid by untrained personnel or civilians who had little idea how to lay them correctly, or who buried them too deep to stop more organised forces finding them with metal detectors. Hitting a mine may simply dislodge dirt which allows the mine to detonate the next time a person steps on it. Deeper mines may not detonate when the ground is hard, but later rain may soften the ground to the point where even a child's footstep will set them off.

2.3 Minefields

For an excellent review of what mine clearance is really all about, the reader is referred to Colin King's paper "Mine Clearance in the Real World" [6].

It is more helpful to think of areas of land which are suspected to be mined rather than "minefields" which may give the impression of mines laid in neat rows at regular intervals. Mined areas are often only found as a result of accidents to people or animals. A large area around the accident location is then declared as "probably mined".

A "typical" minefield is hard to find. Figure 3 illustrates some minefields in Afghanistan and helps to show how mines have been laid in every conceivable place in every type of environment.

Figure 3: Photographs of minefields in Afghanistan (kindly supplied by Afghan Technical Consultants, and Mine Clearance Planning Agency)

Residential area where hard ground has to be "dismantled" with a pick to find buried mines.

Open grazing land can be cleared much more quickly.

Ghazni Fort—a hill top surrounded by steep minefields which are littered with metal shrapnel from intense fighting.

A single mine (centre) on top of a ruined wall. This may be a solitary "nuisance" mine or the whole compound may be mined.

A PMN mine has been exposed on a steep rocky hillside

A single POM-Z fragmentation mine set up as a booby trap round a blind corner in an alley-way.

Often mines have been used as a defensive measure and are laid in large numbers in a concentrated area. Elsewhere just a single mine may have been laid in a highly visible spot to create the impression that a whole area has been mined. Many minefields are littered with unexploded shells, rockets or mortar rounds, often just below the surface lying unnoticed. Where there has been fighting, there will also be millions of small metal fragments left behind. And everywhere there will be the rubbish left behind by soldiers and civilian occupants alike. In come climates (eg Cambodia) thick scrub may have grown up in former fields and rice paddies.

The vegetation has to be completely removed by hand before metal detectors can be used, and this consumes as much or more effort than clearing the mines. In open country or river beds, storms or floods may have carried mines far from their original locations or have buried them under layers of soil and debris. Mines placed on, in, or near buildings may lie deep under fallen rubble, with yet more mines laid on top. Some mines may have been set up as booby traps, detonating (for example) when something of apparent value is picked up. Irrigation canals which had mines laid on the sides and bottom may have been deliberately filled in with bulldozers, burying the mines under a metre or more of soil.

2.4 Current demining methods

Before demining can start, surveys are needed to produce detailed maps of minefields to be cleared. The survey team may use specially trained dogs to narrow down the limits of a mined area, and normally verifies a 1 or 2 metre wide "safe lane" around each minefield (using the same demining method as below) to define the minefield which may be surrounded with unknown land or other minefields. Typical minefields are 100–200 metres across and 0.1–10 Hectares in area.
The manual clearance method is still the only one which works. In Afghanistan, a team of 30 deminers is assigned to clear each minefield. Two man clearance parties work on clearing parallel lanes, 1 metre wide, across the minefield with each lane about 25 metres from the next (considered to be a safe distance) [19].

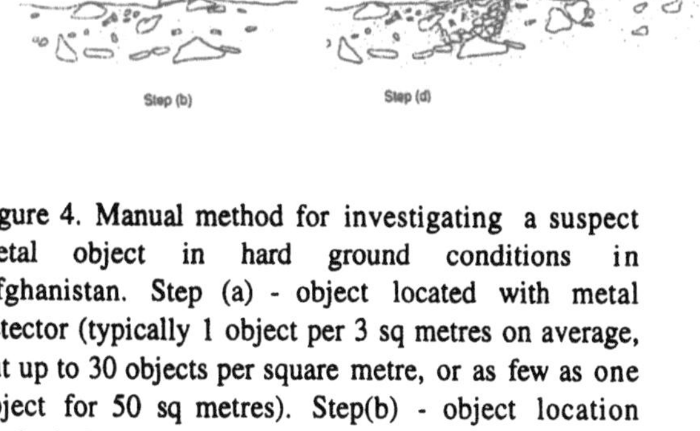

Figure 4. Manual method for investigating a suspect metal object in hard ground conditions in Afghanistan. Step (a) - object located with metal detector (typically 1 object per 3 sq metres on average, but up to 30 objects per square metre, or as few as one object for 50 sq metres). Step(b) - object location marked. Step(c) - deminer scrapes surface carefully to see if fragment is lying on the ground. Otherwise, he digs a trench 30cm behind the location. Step(d) - deminer works forward, dismantling the ground piece by piece with a prodder (usually a bayonet) until the object is found.

The three main tools are:

A whisker wire which is gently swung or lifted gently to check for tripwires.

A metal detector which is swung from side to side to check for metal objects.

A prodder (typically a bayonet, screw driver or knife) which is used to probe the ground at an angle of about 30° to the horizontal, and to excavate earth from around a suspect object. Usually a prodder is used to investigate a suspect metal object. However, when dealing with minimum metal mines or large numbers of metal fragments, then entire area has to be prodded by hand.

Dog teams are being introduced to help clear these minefields.

One must not forget the essential human skills deminers need. With experience and training, their eyes reveal vital cues such as slight depressions in the ground caused by settling after mines were buried, their ears can distinguish different sounds from the metal detector, and their hands develop a feel for different buried objects.
The principal limitation of this method is the time needed to investigate the many suspect metallic objects found in minefields.
Analysis of statistics from 2500 Afghan minefields [17] reveals that deminers typically find 1 to 1000 suspect objects for every 100 square metres, and clear 3 to 50 square metres per hour. On average for a 30 man team, between 1 and 15 hours pass between finding each mine or unexploded ordnance meaning that an individual deminer may work for months without finding a mine. Boredom and fatigue are key factors.

3 Sensors — recent research efforts

It is essential to review recent research results on sensing technology, because a robot which can sense landmines, even using human interpretation if necessary, would be an important step forward. However, there is no evidence yet that a new technique will be available in 2–3 years to replace the current range of metal detectors.

The sensing problem can best be understood in its inverse form. Rather than detecting mines, we need to reliably detect the *absence* of mines. We must be at least 99.99% sure that there is not a mine in the ground where we are about to step, or the vehicle is about to move.

With only a tiny number of exceptions, all anti-personnel mines currently in the ground contain metal components. Sensitive metal detectors can detect all these mines, but also detect thousands of pieces of buried metal junk. The challenge, therefore, is to find additional sensors to safely classify a suspect metal object as "not a mine".

Radar (ground penetrating, wideband, arrays, synthetic aperture radar), infra-red and microwave radiometry, explosive vapour sensors, acoustic sensors, electromagnetic induction, magnetometers, and electrical impedance tomography are some of the techniques which have been tried. Vast sums of money have been spent on this research, particularly in military research programmes in the USA, Britain, France and Germany. Trevelyan [16] presents a survey of some recent research results.

Ideally, a sensor would react to the presence or absence of explosives. Dogs can smell explosive vapours, but the most sensitive artificial detectors developed so far have orders of magnitude less sensitivity. Unfortunately, dogs are not 100% consistent, and treat their job as a game: they soon become bored. Also, their location accuracy is usually only about 5 metres. A consistent and reliable sensor with the sensitivity of a well-trained dog would be of great value.

Several research groups are hoping that sensor fusion techniques will provide major performance improvements. However, the best single sensor alternatives to metal detectors have 80% - 90% detection probability for minimum metal mines which means that the performance has to improve by orders of magnitude before sensor fusion could provide useful results.

4 Experts and Communication Problems

In this author's opinion, the main reason why so little useful progress has been achieved lies in human communication failure.

Deminers have excellent knowledge of minefields and practical problems in mine clearance. They know less about individual mines, treating all of them with great respect.

Military experts have excellent knowledge of explosives, mines and certain types of mine warfare.

Sensor experts (researchers and engineers) have excellent knowledge of their own sensing techniques, often in several different application areas.

Robotics experts have a broad knowledge of sensors, control systems, mechanisms and computer software. However they need sensor experts to tell them about sensors which are unfamiliar to most robotics people, and usually rely on domestic resident military experts for their knowledge of mines and minefields. While it would be logical to ask them about mines and minefields, they may have little experience of real mine clearance conditions. Perhaps because of this, very few people involved with sensor or robot developments have appropriate working knowledge of minefield conditions. Just one example will suffice.

The US Army recently evaluated several metal detectors for work with minimum metal mines (so called plastic mines). The trials were set up by military experts who removed the detonators from the mines to prevent accidents. However, without the metal encased detonators, the mines were much more difficult to detect than they would have been under operational conditions. The electronics experts did not know enough about mines to realise the implications so some metal detectors were incorrectly classified as unsuitable [7]. This illustrates the communication problem.

By working through the problem with many different people we realized that there were many simple and low cost improvements which would make an impact much quicker than a robotics research programme. Suprisingly, almost no one else seemed to have suggested these.

5 Future Prospects for Robots

Sadly, there is little likelihood of success if present approaches are followed. A robotics project requires several important factors to be successful. What is unusual about robotics research is that it requires the successful integration of a number of disparate technologies, most of which must perform at or near ultimate limits. Some of these factors are:

Incentive: there must be a large enough economic problem to justify the research expenses.

Timescale: the problem will still be there in a few years time.
Sensors: the robot must be able to perceive the problem.
Cost and availability: the robot solution must be affordable by users.

5.1 Sensors

As we have seen, there are no suitable sensors yet, and available experimental results suggest that there is little likelihood that there will be in the next 2 - 3 years.

5.2 Incentive

Mine clearance programmes are making a big impact and, at current rates of progress, will clear high priority land areas to enable displaced populations to return within a few years. In Afghanistan, for instance, US$75,000,000 over the next three years could achieve this. US$2 billion (the cost of just two major city office towers) would be sufficient to clear most of the urgent land mine problems around the world. Although most of the mines would still be left in the ground, the ones causing civilian casualties and keeping people from their homes and land would have been removed. The rest lie in concentrated areas around military installations, hill tops, and national borders and can be safely fenced off until local authorities decide how to deal with them.
The money needed to solve the urgent problem is not large by world standards. Even in Australia, a mining company recently wrote off US$750,000,000 in cost overruns on just one new minerals project, though that was regarded as a serious problem on a local scale!

5.3 Timescale

Significant progress is being made in many different countries, and while the prospects of an effective land mine ban seem remote, there is more awareness now, even among combatants, that land mines should not be used if possible. Even though progress is slow, another decade should see most of the urgent problems solved.

5.4 Cost and Availability

Humanitarian demining organizations already have access to metal detectors and dog teams. While better sensing devices would help, the cost would have to be competitive in the sense that the cost savings in detection would have to pay for the increased price of the detector. However, there are other factors in such decisions other than pure cost alone.
External funding for indigenous demining organizations provides a substantial cash boost to the local economy. When confronted with a decision on whether to import cost-saving equipment on a limited budget, they must balance the adverse impact on employment in their community. The objective for demining is to restore land and homes to people so they can become less reliant on external aid funding. If this is at the expense of their economy, then the effort may be counter-productive. Unemployed deminers have to feed their families, and this may mean returning to the armies of local war-lord or drug empires.
Any country which develops an effective sensor for detecting land mines and other buried explosive devices will immediately gain a significant military advantage over other countries. Therefore, irrespective of cost, it is unlikely that such technology would be released to non-state organizations.
This is not a problem for robots yet. The widespread belief that the global landmine problem can be solved using a combination of advanced robotics, sophisticated sensors, and powerful computing devices is simply a myth. The urgent problem is not large in financial terms. While there are simple and effective ways to solve current problems we should explore those first. The money now being spent on elaborate sensors and robotic solutions is very unlikely to lead to useful improvements. If it were to be spent on simple improvements, the same money could transform humanitarian demining operations almost overnight.

6 Appropriate Solutions

As a result of discussions with deminers, we have been able to devise several simple equipment improvements which could help demining teams [18].
These include:

- Support frame for demining helmet and visor (which weigh up to 2 kg) to reduce discomfort and neck fatigue
- Light weight, low cost blast protection aprons for deminers
- Ventilation fan for helmet (Pentium CPU fan)

Improved light weight and low-cost prodders and probes

- Portable manual powered excavators
- Depth sensing for conventional metal detector
- Water jets for excavating mines
- Low cost training mines
- Magnetic devices for collecting metal fragments

Some of my students have declined to participate in this research giving comments such as "I cannot see enough technology in this to write a thesis"! In fact, devising simple solutions has been a rewarding challenge, requiring some new or emerging materials technologies. We have also developed close collaboration with groups in less developed countries where the combination of low labour costs (particularly for writing software) and technical ingenuity has opened several interesting prospects for innovative robotics projects in the future. But not for clearing minefields!

7 Acknowledgements

Our work is supported principally by a contract from the US Army Night Vision and Electronic Sensors Directorate. There are many who deserve thanks for helping with this research, in particular, M. M. Iqbal (Hameed and Ali Research Centre, Islamabad), UNOCHA staff (Islamabad), Afghan Technical Consultants and Mine Clearance Planning Agency (photographs and information), Australian Army Engineers and the authors many colleagues and students who have participated in the project.

8 References

[1] Cain, B. and T. Meidinger (1996). 'The Improved Landmine Detection System.' (EUREL, 1996), pp. 188-192.

[2] Daniels, R. A. (1996), Videotape shown at EUREL conference Detecting Abandoned Landmines, Edinburgh, available from ERA plc, Britain.

[3] EUREL (1996) Proceedings of EUREL International Conference on the Detection of Abandoned Landmines, IEE Conference Publication 431, London, UK.

[4] Janes (1996). Janes Mines and Mine Clearance Techniques, King, C (Ed) Coulsdon (163 Brighton Rd, Coulsdon, Surrey CR5 2NH, UK) Jane's Information Group.

[5] Jefferson (1997). Mines, Damn Lies and Statistics, Manchester Guardian, September 1997 (available from http://www.mech.uwa.edu.au/jpt/demining/lies.html).

[6] King, C. (1997) Mine Clearance in the Real World. SusDem97: International Workshop on Sustainable Humanitarian Demining, Zagreb, October, pp S2.1-8.

[7] King, C. (1996) Personal communication. (Editor of Janes 'Mines and Mine Clearance).

[8] McGrath, R. (1994). *Landmines, Legacy of Conflict: A Manual for Development Workers*, Oxfam, Oxford, UK.

[9] McMichael, D.W. (1996) 'Data Fusion for Vehicle-Borne Mine Detection.' (EUREL, 1996), pp. 167-171.

[10] Nicoud, J .D. (1995) *Proceedings of Workshop on Anti-personnel Mine Detection and Removal WAPM '95*, Swiss Federal Institute of Technology Microprocessors and Interfaces Laboratory (EPFL-LAMI), Lausanne, Switzerland.

[11] Nicoud, J-D. (1996) 'A Demining Technology Project.' (EUREL, 1996), pp.37-41.

[12] Nicoud, J-D. (1997) Vehicles and Robots for Humanitarian Demining, Industrial Robot Vol 24, No. 2, pp 164-168.

[13] Red Cross (1995) *Landmines must be stopped: Chapter VI Mine Clearance.* Special Brochure, International Committee of Red Cross, Geneva, Switzerland.

[14] Trevelyan, J. P. (1992). Robots for Shearing Sheep: Shear Magic. Oxford Science Publications, UK.

[15] Trevelyan, J. P. (1996c) 'A suspended device for humanitarian demining.' (EUREL, 1996), pp.42-45.

[16] Trevelyan, J. P. (1997a) Robots and Landmines, Industrial Robot Vol 24, No. 2, pp 114-125.

[17] Trevelyan, J. (1997b). Modelling minefield clearance statistics. Technical Report, Department of Mechanical and Materials Engineering, University of Western Australia.

[18] Trevelyan, J. P. (1997c). Better tools for deminers: International Workshop on Sustainable Humanitarian Demining, Zagreb, October, pp S6.1-12.

[19] UNOCHA (1996) Notes on interviews with UN Office for Coordinating Humanitarian Aid to Afghanistan (UNOCHA), Islamabad. (Available from author).

Robots Integrated with Environments

- A Perceptual Information Infrastructure for robot navigation -

Hiroshi Ishiguro
Depertment of Information Science, Kyoto University, Sakyo-ku, Kyoto 606-01 (Japan)
email: ishiguro@kuis.kyoto-u.ac.jp

Abstract

This paper proposes a Perceptual Information Infrastructure which supports robots in a real world and realizes real-time and robust robot navigation. One of the examples is a distributed vision system consisting of multiple vision agents and a computer network for the agent communication. The design policies and experimental results for mobile robot navigation are shown. The perceptual information infrastructure enables to develop a new research direction of robotics in which robots are considered as integrated systems with environments. In addition to the infrastructure for robots, this paper discusses research issues which should be considered in order to develop the new intelligent robot systems.

1 Introduction

In order to realize intelligent robots, there are two possible directions. One is to develop autonomous robots which have flexible mechanisms like humans. Such humanoid robots enable to study complex relations between environments, robot mechanisms and information processing. Several research groups [3, 7] are tackling to develop the humanoid robots and discover fundamental principles of intelligent systems.

The approaches with the humanoid robots are very interesting and challenging. However, there are remained serious problems. Especially, development of the mechanisms is basically hard. Recently, HONDA Motor Co., Ltd. [7] have developed a humanoid robot which can walk with two legs. Although this is great progress in robotics, the robot lacks sophisticated sensor systems to observe external worlds.

Another approach is to prepare infrastructures for robots. That is, many sensors embedded in the environment provide various information for

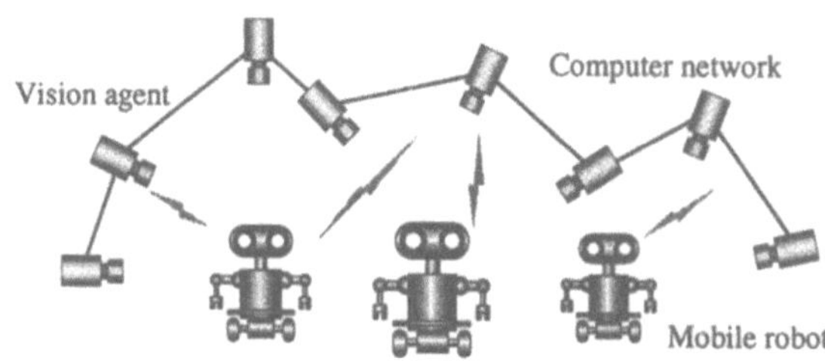

Figure 1: Distributed Vision System

the robots. In this approach, there is no difficulty in the development of the mechanisms. Fig. 1 shows an example of the PI^2 using cameras as the sensors. The robots can obtain necessary information with the cameras. The author calls the infrastructure *Perceptual Information Infrastructure* (PI^2) [9]. The PI^2 has several merits for fundamental research of robotics and real applications of robots comparing with the autonomous robots.

Let us overview previous robotics research approaches. Navlab project at Carnegie Mellon University [14] investigated possibilities of autonomous mobile robots. The Navlab vehicle detects obstacles with a ERIM laser range finder, recognizes road regions with color images and moves along detected roads by referring to a road map. The main purpose of Navlab project was development of vision algorithms for mobile robot navigation. Dickmanns and others [6] also developed an autonomous vehicle which can drive with a high speed in a highway environment. Before developing the robot, they have developed a parallel computer system which enables high speed visual feed back control by signal processing methods. The robot was an application of the computer system. Yamabico robots developed by Yuta and others [15] are compact autonomous robots which moves in a structured environment with ultrasonic sensors and laser range finders. An interesting point of Yamabico robots is its goal of the project, which is long distance navigation by the completely autonomous robots. Distance in which the robot can autonomously move is one of the reasonable system performance measures. Chatila and others [5] proposed and developed an open and modular architecture for autonomous robots. They applied the architecture to planet rovers and bay transportation systems. Brooks [4] proposed and developed a robust architecture of behavior based robots called *Subsumption architecture*. Arkin and others [1] also developed an architecture of behavior based robots called *Schema system*. Their purpose was to develop architectures for robust autonomous robots.

While surveying these previous works, a question comes up. Why did they focus upon only autonomy of the robots? Almost all the previous works for the autonomous robots did not deal with infrastructures. As humans need roads, traffic signs and so on, robots also need infrastructures for behaving in dynamically changing worlds. Further, as humans need, the robots also need helps from others (humans and other robots). That is, the previous works lack view points of the infrastructures.

For developing robust robot systems, it is strongly required to consider the infrastructures and interactions with humans. The basic idea of the research approach discussed in this paper is to design the robot systems without loosing important elements of robots which works in real worlds. The elements are robots themselves, infrastructures supporting the robot functions and human interfaces for communicating with humans.

In the following sections, first the PI^2 which is the most important aspect of the research approach is discussed, and then, a distributed vision system for navigating robots is shown as an example of the PI^2. After discussing the PI^2, other elements are discussed and a new research direction of robotics is proposed.

2 Perceptual Information Infrastructure

2.1 From an autonomous robot to distributed agents

As discussed in technical papers on *Active Vision* [2], the most difficult and important issue is attention control to select viewing points according to various events relating to the robot. Two kinds of the attention control exist; one is *Temporal Attention Control* and the other is *Spatial Attention Control*. If the robot has a single vision, it needs to change its gazing direction in a time slicing manner to simultaneously execute several vision tasks. The control of gazing direction is *Temporal Attention Control*. For example, the robot has to detect free regions even while gazing at the targets. We, human, solve this complex temporal attention control with sophisticated mechanisms of memory and prediction. The vision fixed on the robot body often cannot provide proper information for the vision tasks. For example, when a robot estimates collisions with a moving obstacle, the side view in which both the robot itself and the obstacle are observed may be more proper than the view from the robot. This view point selection is called *Spatial Attention Control*.

To realize the attention control is difficult with current technologies for autonomous robots. The following reasons can be considered.

- An active vision system needs a flexible body for acquiring proper visual information like a human. However, vision systems of previous mobile robots are fixed on the mobile platforms and it is generally difficult to build mobile robots which can acquire visual information from arbitrary viewing points in a 3D space.
- An ideal robot builds environment models by itself and uses them for executing commands from humans operators. However, to build a consistent model for a wide dynamic environment and maintain it is basically difficult for a single robot. We, humans, often need helps of other persons to acquire information on the environment.

An idea to solve the problems is to use many perceptual agents embedded in the environment and connected with a computer network. Each perceptual agent independently observes events

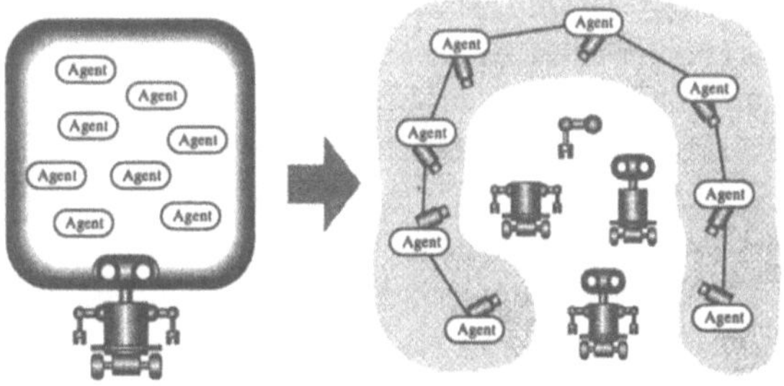

Figure 2: From an autonomous robot to robots integrated with environments

with sensors, such as vision sensors and audio sensors, in the local environment and communicates with other perceptual agents through the computer network. We call the perceptual agent network which supports robots *Perceptual Informative Infrastructure* (PI^2). Fig. 2 shows a conceptual change from an autonomous robot to robots supported by the PI^2. Although almost all of previous autonomous robots consist of software agents, the PI^2 consists of perceptual agents each of which has an original hardware for adapting to the environment. Since the perceptual agents do not have any constraints in the mechanisms, we can install a sufficient number of the perceptual agents according to tasks and the robots can acquire necessary information from various viewing points.

With the PI^2, the attention control is changed to attention selection. As shown in Fig. 1, the robots can obtain proper visual information by selecting sensors and communicating with them. The PI^2 solves the attention control problems with a different but simple manner.

The infrastructure in this paper differs from the infrastructure for mobile robots which move in factories. The purpose is not to develop systems which support individual functions of the robots such as guide lines and landmarks for locomotion, but to develop a flexible and general infrastructure which actively provides various information for real world agents, such as robots and humans. That is,

The PI^2 monitors the environment, maintains the dynamic environment models, and provides information for the real world agents

2.2 Fundamental issues

For realizing the PI^2, the following fundamental issues should be considered.

1. In addition to communication with the computer network, perceptual agents communicate by observing common events. It is an important research issue how to establish sophisticated and flexible communication links through the two types of communication.
2. Static environment models should be generated from dynamic environment models representing the dynamic events. The robust detection of dynamic events of the perceptual agents enables to hierarchically represent the environment.
3. The perceptual agents should be locally and globally organized in order to provide proper information for the robots. The organization of the perceptual agents through the communication links for acquiring and maintaining the dynamic environment models is a key issue of the PI^2.
4. The dynamic environment models are not shared by all perceptual agents, but distributed over perceptual agents which keep relations with related perceptual agents. To distribute the models in the perceptual agent network is important for realizing a flexible and robust PI^2.

2.3 Related works

Recently novel research approaches using distributed sensors and robots have been proposed in robotics. For example, the *Robotic Room* proposed by Mizoguchi and others [11] supports human activities with sensors and robots embedded in a room. Their interests are to design mechanisms and develop sensor systems for executing well-defined local tasks. On the other hand, the purpose of this paper is to propose a flexible sensor system utilized by various kinds of robotics systems as an information infrastructure.

Several vision systems which utilize multiple cameras has been reported, especially, in multimedia. Moezzi and others [12] proposed the concept of *Immersive Video* and developed a vision system using precisely calibrated cameras for building a precise geometrical model of an outdoor environment. Pinhanez and Bobick [13] developed a system which dynamically selects cameras providing proper views for broadcasting a

TV show. The author, however, considers a demerit of the systems is to use calibrated cameras and geometrical models of the world. Geometrical representations of environments obtained by the calibrated cameras lack robustness and flexibility of the systems. In order to solve the problems, the PI^2 employs an alternative approach for modeling dynamic environments, which dynamically and locally estimates the camera parameters and directly represents robot tasks in the images.

3 A PI^2 for mobile robot navigation

This section discusses an example of the PI^2 for mobile robot navigation which is called *Distributed Vision System* (DVS) [9]. The DVS consists of *Vision Agents* (VA) connected with a computer network and observes external worlds.

3.1 Design policies

The VAs are designed based on the following idea:

> **Tasks of robots are closely related to local environments.**

For example, when a mobile robot executes a task of approaching a target, the task is closely related to a local area where the target locates. This idea allows to give VAs specific knowledge for recognizing the local environment, therefore each VA can have simple but robust information processing capabilities.

More concretely, the VAs can easily detect dynamic events since they are fixed in the environment. A vision-guided mobile robot of which camera is fixed on the body has to move for exploring the environment, therefore there exists a difficult problem to recognize the environment through the moving camera. On the other hand, the VA in the DVS easily analyzes the image data and detects moving objects by constructing the background image for the fixed viewing point.

The DVS, which does not keep the precise camera positions for the robustness and flexibility, autonomously and locally calibrates the camera parameters with local coordinate systems according to demand. That is, the VAs iterate to establish representation frames for communicating with other agents. Further, the DVS dose not have a shared clock. The VAs synchronize by observing common events and broadcast messages.

The VAs identify objects with the motions observed in the images in addition to the visual features since they can provide reliable motion information from the fixed viewing points. The author considers the DVS can solve the correspondence problem more robustly and flexibly than the previous vision systems.

The DVS organizes communication between VAs in order to execute given tasks. The design policy that a VA executes particular subtasks in the local environment allows to solve the organization problem in a hierarchical manner. That is, global tasks given to the DVS, generally, can be decomposed into the subtasks and the VAs execute them. However, the subtasks often need to be simultaneously executed and the combinations often change according to various situations. Therefore, the VAs should be globally and locally organized to execute the global tasks.

3.2 Mobile robot navigation

The outline for mobile robot navigation by the DVS is as follows. First, a human operator teaches tasks by manually controlling a robot. The human operator does not directly give task models or behavior models of the robot, but shows examples to the DVS. While the robot moves, each VA tracks it within the visual field with simple image processing functions. Then the DVS decomposes the given example paths into several components which can be maintained by each VA and memorizes then by organizing the VAs. After organizing the VAs, the DVS autonomously navigates the mobile robot while the VAs communicate each other. All of the VAs monitor the robot motions and send messages to other VAs.

3.2.1 *The architecture*

A VA consists basic modules and memory modules as shown in Fig. 3. For the basic modules, the VA has *Image processor*, *Estimator* (Estimator of camera parameters), *Planner*, *Communicator*, and *Controller* (Communication controller). For the memory modules, it has a knowledge database for image processing, memories to memorize global and local tasks, and memories to maintain relations with other VAs for executing the global and local tasks. In the experimentation shown in this section, the global task is to navigate toward goals and the local task is to avoid obstacles and other robots.

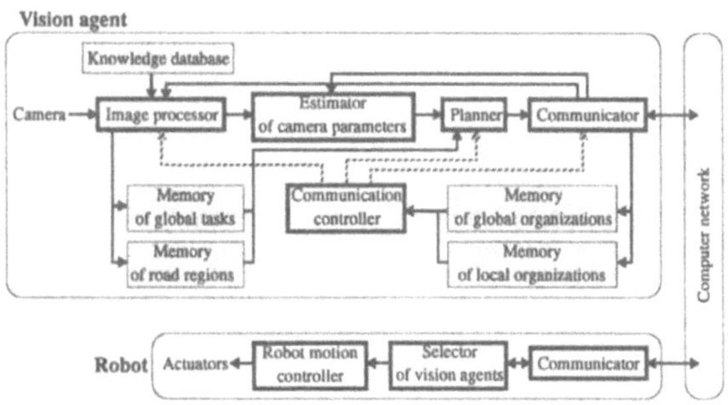

Figure 3: The architecture of the DVS

Figure 4: Model town and robots

Image processor detects moving robots and tracks them by referring to the knowledge database which stores visual features of robots. *Estimator* receives the results and estimates camera parameters for establishing representation frames for sharing robot motion plans with other VAs. As a sophisticated method, it is possible to use an automatic calibration method proposed by Hosoda and others [8]. However, it is often difficult to precisely estimate the camera parameters because of noise in the image processing for detecting the robots. The approach is, rather, to utilize qualitative motion plans generated by many VAs. Therefore, only two camera parameters, pan and tilt, are estimated. For estimating the parameters, the camera projection is assumed as orthogonal so that they can be easily estimated by observing a robot motion. *Planner* plans robot actions based on the estimated camera parameters in order to navigate along the given examples and sends them to the robot through *Communicator*. The robot corrects the plans, selects proper plans with the error estimations, and executes the plans. The error of the motion plan can be considered to be in inverse proportion to the size of the robot projected on the image. If the error is lager than a constant value, the robot ignores the plan, otherwise the robot generates a new plans by computing a weighted sum of the motion plans. The weight is in proportion to the size of the robot projection. The selected plans are sent back to the VAs and memorized. The memorized plans are directly applied in the same situations by *Controller*.

3.2.2 *Experimental setup*

Fig. 4 shows a model town and two mobile robots used in the experimentation. The model town, of which reduced scale is 1/12, has been made for representing enough realities of an outdoor environment, such as shadows, textures of trees, lawns and houses. Sixteen VAs have been established in the model town and used for navigating the mobile robots.

Images taken by the VAs are sent to an image encoder which integrates sixteen images into one image (the size of each image is reduced from $512 \times 512 pixels$ to $128 \times 128 pixels$). Then, the image is sent to a color frame grabber. The main computer, Sun Spark Station 10, executes the vision functions by using data from the color frame grabber in real time. Implementation into parallel computers is a feature work.

3.2.3 *Experimental results*

In the experimentation, first, a human operator taught example paths and goals. The goals are intersections in the model town and two intersections exist in it. Robots randomly select the goals and go toward them by moving along the taught paths. If a robot encounters another robot, VAs observing them generate motion plans to avoid each other. The robot basically moves with only supports from the VAs. However, the VAs sometimes cannot observe the robots from

Figure 5: Robot trajectories navigated by the DVS

proper viewing points. In such cases, the robots avoid obstacles by using their touch sensors. Fig. 5 shows robot trajectories autonomously navigated by the DVS. Because of simplicity of the image processing, the DVS could robustly navigate the mobile robots in a complex environment.

Fig. 6 shows images taken by VAs in the autonomous navigation phase. The vertical axis and horizontal axis indicate the time and ID numbers of the VAs, respectively. The black boxes and dotted boxes indicate selected VAs for navigating the black and gray (the actual color is red) robots shown in Fig. 4, respectively. As shown in Fig. 6, the VAs have been dynamically selected according to the situations, and the robots have received multiple motion plans from the several VAs.

The experimentation has been continued while three days in the AI exhibition held with the last International Joint Conference on Artificial Intelligence (August 26th-28th, 1997 in Nagoya, Japan). While three days, the DVS has never halted and the robots have moved around in the model town. The three days experimentation clearly shows the robustness of the DVS.

In addition to the practical aspect of the DVS, the experimentation shows another important aspects of the DVS. The DVS can memorizes the tasks for navigating the robot along a path by organizing the VAs and iterate to select proper VAs for robustly executing the tasks in a complex environment. That is, the DVS solves the attention control problems for the autonomous robots discussed in Section 2 with a different but more robust manner.

4 Toward robots integrated with environments

As discussed in section 2 and 3, the PI^2 solves attention control problems and realize robust and flexible systems for robot navigation. Further, the computer network connecting the perceptual agents can be used for communication between robots and human operators. That is, the PI^2 is an infrastructure which robustly navigates robots in a real world. With the PI^2, we can proceed to the next step of robotics research.

Figure 7(a) indicates a previous research approach in robotics. Robot tasks are determined based on requirements from humans and human societies, and then robots are designed for the tasks. The robots designed for the specified tasks work with limited human-robot interactions. For example, industrial robots which execute human tasks in factories yield limited relations between humans and robots; the robots do heavy labor and humans monitor the robot behaviors.

However, we cannot determine tasks of robots behaving in open worlds such as a town. The tasks should be rather emerged as results of human robot interactions. Let us imagine a robot which has several basic functions such as carrying baggage, guiding humans, communicating with humans, and so on. When we, humans, meet a robot, we think what we want to do with the robot and what the robot can do, and ask various tasks to the robot. That is, the tasks are developed through human-robot interactions and the emerged tasks influence human societies. In other words, human societies uses robots as a new information infrastructure. Figure 7(b) shows a new research direction in robotics in which robots exist in a human society, humans find tasks through interactions with the robots, and the human society changes with the robots.

The research direction shown in Figure 7(b) is not only for robots but also for various information systems, such as the Internet. The Internet has been designed for just data transmission. However, it is essential for human societies today and we are developing various applications of the Internet. WWW is one of the examples. WWW influences and changes human societies as everybody knows. We consider the robot systems should be similar to the Internet.

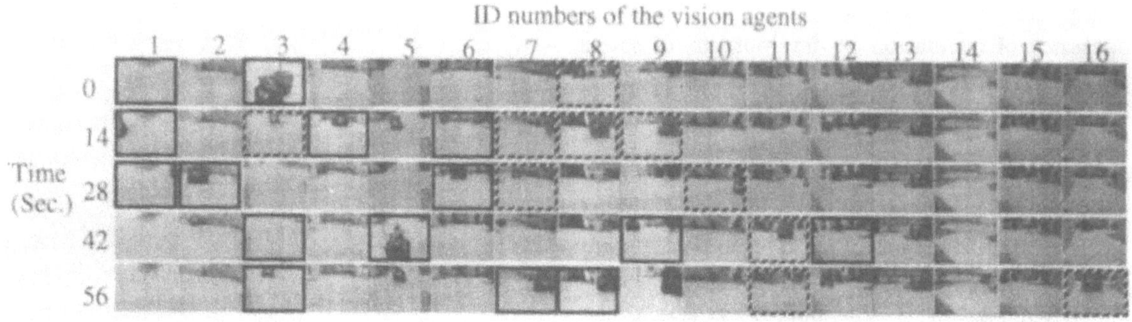

Figure 6: Images taken by the VAs

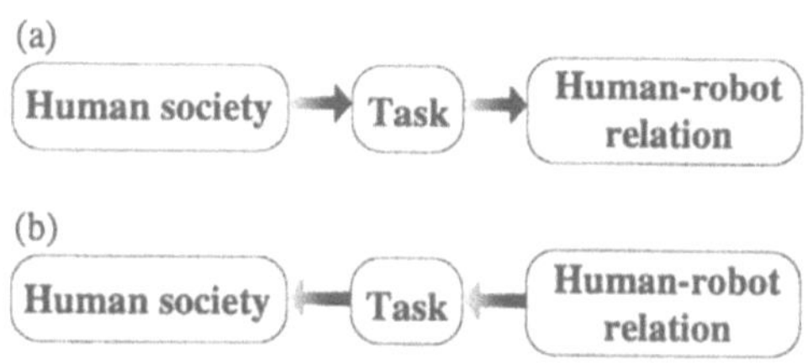

Figure 7: Human society and robot task

The new robot systems used in human societies need to satisfy the following requirements:

- *The system never halt*: For keeping interactions with humans, the system totally keeps on working even if several parts of the system sometime halt.
- *The system emerges tasks*: The system should be flexible and open as humans can find tasks though interactions with it.
- *The system keeps on growing up*: The system should estimate relations with humans and develop new functions by itself while working.

For realizing the above-mentioned functions, new technologies should be developed for dealing with the following three issues in addition to previous robot technologies. The issues have never seriously discussed in previous works.

4.1 Three important issues

4.1.1 *Infrastructure*

For keeping on working, robots need supports from the environment. With current technologies for autonomous robots, it is almost impossible to develop robust robot systems used in a complex and dynamically changing real world. The robots need a support system. As discussed in this paper, the PI^2 is such an infrastructure.

The roles of the infrastructure is not only to provide information, but also to be enable on-line debugging. A robust system used in a real world sometimes needs helps of human operators. Human operators can monitor robot behaviors and send commands to the robots through the computer network of the infrastructure.

4.1.2 *Evaluation*

The robot system also needs to evaluate multiple tasks emerged through interactions with humans. If the system has a single purpose, the evaluation criteria can be easily determined. However, it is basically difficult to evaluate a system which has multiple purpose. For the evaluations, the author considers the system is required the following two functions:

1. Recognizing tasks emerged between the system and humans.
2. Recording and reporting statistical data for task evaluation.

These functions enable to closely relate robot tasks with human behaviors interacting with the robots and to evaluate the complex robot tasks. The results of the evaluation is fed back to the on-line system improvement.

In previous works for intelligent autonomous robots, we have never discussed on evaluation of robot systems. On the other hand, there exist many reports on evaluation of information systems in computer science. For example, we can find such reports in proceedings of Int. Conf.

Computer-Human Interaction. This means importance of evaluation in development of social systems.

Another important issue in the evaluation is that we can find relations between humans and robots and influence of the robots to the human societies. The evaluation results represent human-robot relations and give important information for understanding human societies in which robots work. The author expects to be able to indicate cognitive maps on robot-human interactions.

4.1.3 *Human interface*

The robot system must be a system which never halt so that the system can be used in human societies. In order to support robots, human operators sometimes need to intervene into the system and help the robots. Therefore, the function of operator's intervention should be embedded in the robot system. And further, the robot may ask helps to humans walking in a street. The function of cooperation with the humans is also necessary. In traditional systems, an interface has been to help human's access to system functions. In the robot system, the interface is to help the operator to access to the robot functions and humans who meet the robots. That is, two types of interface are needed: (1) Operator-robot interface and (2) Robot-human interface.

4.2 Development of a new robot system

Figure 8 shows a conceptual figure of a new robot system which has an infrastructure. Although the robots have autonomous behaviors, they can be controlled and maintained their functions by the operators through the infrastructure. The infrastructure works as both extended sensors of the robots and a computer network to support communication between the robots and the operators.

Based on the idea, the author is developing a real robot system which works in university campuses and streets [10]. The robot has sensors, actuators, and energy sources. The system hardware is shown in Figure 9. The size is about 1.35 meters tall including the sensors and the wheels. The width is about 0.6 meters diameter.

Three cameras are attached on the top of the robot as main sensors. A pair of cameras work as a stereo vision. The third camera is specially designed to acquire omni-directional visual information. These vision sensors can rotate with three degrees of freedom, roll, pitch, and yaw by a camera controller. With the vision sensors, the robot can actively obtain visual information which is necessary for visual navigation and interactions with humans. And, the camera controller with three degrees of freedom generates various gestures for communicating with humans. In addition to the vision sensors, sixteen touch sensors are attached around the robot.

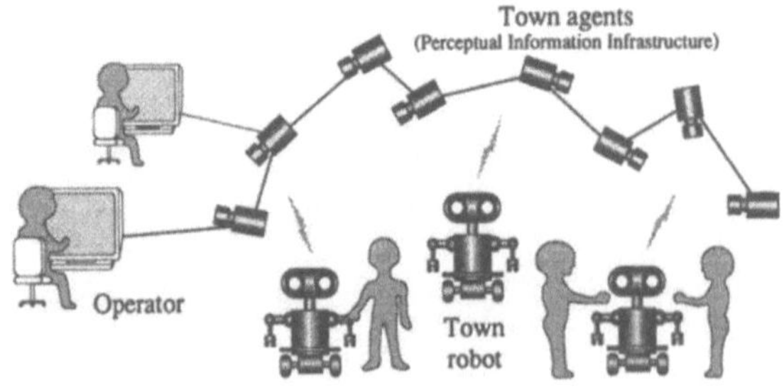

Figure 8: Conceptual figure of a new robot system

The robot has two pairs of wheels in both right and left sides. Two wheels of a pair are chained and driven by one motor. The wheel consisting of eight ellipsoid cylinders specially designed for turn motion. That is, the robot can turn by giving different velocities to the motors. The robot has a special mechanism for posture stabilization. The differential gear mechanism is employed for mounting wheels so that the robot can stand almost upright even when the robot moves across small steps or on irregular terrain.

In addition to the on-board sensors, the robot can utilize sensory information provided by perceptual agents of the infrastructure which support the robots and system operators. The robot equips a wireless network facility, which is used to communicate with the perceptual agents. Figure 10 shows the perceptual agents which are specially designed for the robot system. Each perceptual agents has a camera with a pan-tilt camera controller, computing resources and a wireless communication modem.

The system development employs a methodology which iterates experimentation, evaluation and development in a short cycle for developing the robot system (See Figure 11). Experimentation indicates system's behaviors in a real world and evaluation gives the right directions of the system improvement. While iterating the process, the system can adapt to human societies.

Figure 9: Autonomous mobile robot

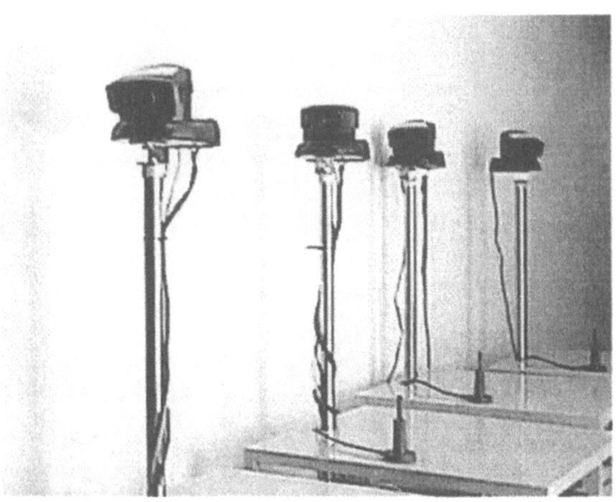

Figure 10: Perceptual agents

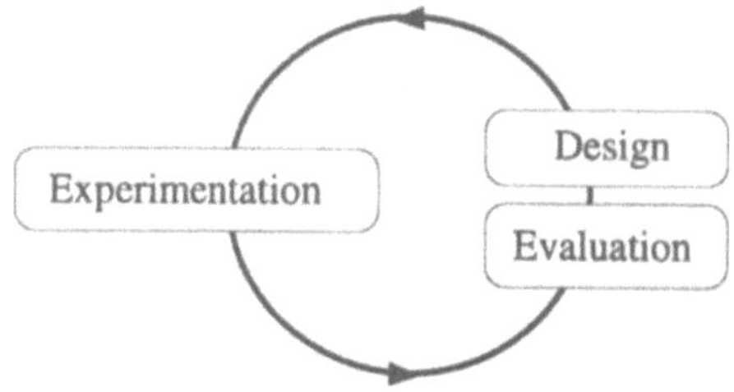

Figure 11: Methodology for developing the robot system

5 Conclusion

This paper has discussed an infrastructure which support robot behaviors in a real world, and shown an example of the infrastructure using vision sensors for robot navigation. With the infrastructure, a new research direction can be considered. The author calls the research direction with the infrastructure *Social Robotics*. In previous works, robots has been designed according to tasks. However, if the tasks are complex, it is hard to design proper mechanisms and sensors. In such cases, robots are required functions to execute possible tasks with the infrastructure.

Acknowledgment

The author would like to thank Prof. Toru Ishida for his advice and support of this research, Dr. Katsumi Kimoto for their stimulating discussions and constructive criticism, and Mr. Takushi Sogo for his programming work for the DVS.

References

[1] R. C. Arkin and others, Active avoidance: Escape and dodging behaviors for reactive control, Active robot vision, World Scientific Publishing, pp. 175-192, 1993.

[2] D. H. Ballard, Reference frames for animate vision, Proc. Int. Joint Conf. Artificial Intelligence, pp. 1635-1641, 1989.

[3] The COG shop, http: //www. ai. mit. edu /projects /cog/.

[4] R.A. Brooks, A Robust Layered Control system for a Mobile Robot, IEEE J. Robotics and Automation, RA-2, pp. 14-23, 1986.

[5] R. Chatila and others, From planning to execution monitoring control for indoor mobile

robots, Proc. Int. Symposium Experimental Robotics, pp. 207-221, 1991.

[6] Dickmanns and others, The seeing passenger car 'VaMoRs-P', Proc. IEEE Symposium on Intelligent Vehicles, pp. 68-73, 1994.

[7] HONDA introduces "Human" robot, http: //www. honda. co. jp /home /hpr /e_news /robot/.

[8] K. Hosoda and M. Asada, Versatile Visual Servoing without Knowledge of True Jacobian, Proc. Int. Conf. Intelligent Robots and Systems, pp. 186-193, 1994.

[9] H. Ishiguro, Distributed vision system: A perceptual information infrastructure for robot navigation, Proc. Int. Joint Conf. Artificial Intelligence, pp. 36-41, 1997.

[10] H. Ishiguro and K. Kimoto, Town robot - Toward social interaction technologies of robot sytems -, to appear in Proc. Int. Conf. Field and Service Robotics, 1997.

[11] H. Mizoguchi, T. Sato and T. Ishikawa, Robotic office room to support office work by human behavior understanding function with networked machines, Proc. Int. Conf. Robotics and Automation, pp. 2968-2975, 1996.

[12] S. Moezzi, An emerging Medium: Interactive three-dimensional digital video, Proc. Int. Conf. Multimedia, pp. 358-361, 1996.

[13] C. S. Pinhanez and A. F. Bobick, Approximate world models: Incorporating qualitative and linguistic information into vision systems, Proc. AAAI, pp. 1116-1123, 1996.

[14] C. E. Thorpe and others, Vision and Navigation for the Carnegie Mellon Navlab, IEEE Trans. PAMI, Vol.10, No.3, pp. 362-373, 1988.

[15] S. Yuta and J. Iijima, State information panel for inter-processor communication for an autonomous mobile robot controller, Proc. Int. Workshop on Intelligent Robots and Systems, 1990.

Bio-robotic Systems Based on Insect Fixed Behavior by Artificial Stimulation

Raphael Holzer, Isao Shimoyama
The University of Tokyo, Mechano-Informatics
7-3-1 Hongo Bunkyo-ku, Tokyo 113, JAPAN
r.holzer@ieee.org, isao@leopard.t.u-tokyo.ac.jp
http://www.leopard.t.u-tokyo.ac.jp

Abstract

Artificial electrical stimulation is one of the tools of neuroethology to investigate and probe a living neural system. By applying artificial inputs to the system, specific reactions can be observed and a model or hypothesis of the inner workings can often be deduced. The escape turn of the American cockroach, Periplaneta Americana (L.), is a well-known fixed reaction pattern in response to the appearance of a predator. Video analysis is one method used to collect quantitative data on the locomotion behavior, but here we have built a trackball device to measure the locomotion behavior mechanically. The major advantage of this method is to obtain locomotion parameters (angle, distance) easily and instantaneously, and to keep the insect immobile for easy access of stimulation and monitoring.

We analyze the locomotory reaction of the Periplaneta Americana to various electrical stimuli. The insect is placed on a light-weight styro-foam trackball which is connected to a computer. This allows to record the turning rate and the forward movement of the insect in response to electric antennal stimulation. Based on this data a simple mathematical model is established.

1 Introduction

Legged locomotion is still a difficult problem in robotics. The complexity of the sensory-motor integration of internal (proprioceptor) and external (receptor) stimuli to generate stable and robust locomotion make legged locomotion a difficult-to-solve problem. Insects have mastered the locomotion problem quite formidably although they have only 10^5-10^6 neurons. This is many orders of magnitudes less than the number of neurons in mammals (which is about 10^{10}-10^{12}). While there are still many neurons involved in visual processing, a single leg is controlled by a population of only about 70 motor neurons which all have been identified [1]. With this small number of motorneurons involved in locomotion, insects seem promising objects for studying the nervous control algorithms of locomotion. The insect locomotory system can then serve as a model and an inspiration for robotic locomotion control.

Kuwana *et al.* [2] used the antennae of a silk moth (*bombyx mori*) as a pheromone sensor on a small mobile robot. Using a recurrent neural network the robot was capable to recreate the naturally occurring zig-zag pattern of the real insect when following a pheromone plume in the air.

Crary *et al.* [3] placed four electrodes on a Madagascar cockroach and experimented with electric stimulation. They achieved locomotion guidance by way of four wires over which stimulation signals were supplied from a handheld unit.

Comer and Dowd [5] investigated the contralateral escape response triggered by wind and touch stimuli on the cerci and the antennae. They established a simple model of the nervous circuit to account for the probabilistic behavior of the experimental outcome.

The present and previous [4] work demonstrate that artificial electrical stimulation can lead to similar results as are found with natural stimuli.

2 Materials and methods

2.1 The measurement set-up

The set-up shown in Fig. 1 was used to measure and record the reaction of tethered adult American cockroaches to electric stimulation. The cockroach is placed on top of a light-weight styrofoam ball. After removal of the wings a small holding rod made of styrofoam is glued to its back. The ball is in mechanical contact with two rotary encoders. The encoder directly below the ball records the forward movement. The encoder on the side records the turning rate. The ball cage is slightly inclined to allow a better mechanical contact between the ball and the encoder on the side. The signals of the two encoders are decoded by a small interface containing an 8-bit PIC microcontroller. 2-byte-sized position values are sent over a serial link to the host computer which runs a data collection and display program written in LabVIEW. The same 8-bit microcontroller also produces the electric stimulation signals. Electrodes have been placed in the antennae and the cercal area.

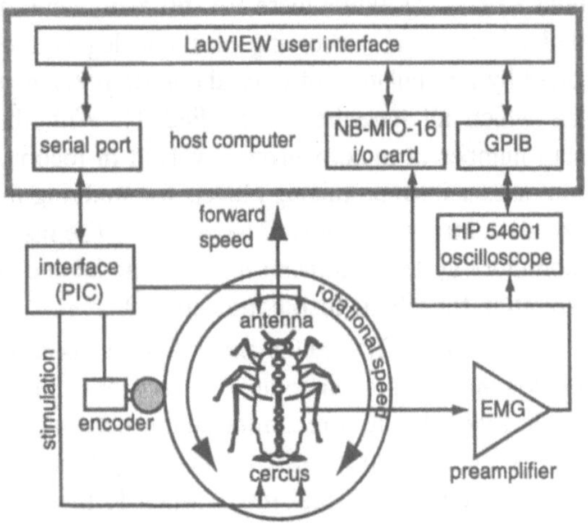

Figure 1: Overview of the locomotion trackball and recording system.

The EMG during walking is taken from muscles on the thorax, preamplified and transferred to the host computer. For high-sampling-rate but short-time acquisitions the HP-54601 digital oscilloscope with a GPIB interface was used. For continuous, but low-sampling-rate acquisitions a DA acquisition board was used.

2.2 Calculation of the inertial forces on the trackball

The inertial forces due to the trackball are slightly different from the inertial forces experienced for free walking. For a simple calculation we approximate the cockroach body as having the dimensions: length 40 mm, width 10 mm, and height 3 mm (see Fig. 2). With a density of about 1 g/cm^3 it weighs 1.2 g. During free walk the force of the legs acts upon the body mass, whereas during walk on the trackball the force acts upon the inertia of the ball. For the forward movement the ratio of these two inertias is:

$$\frac{I_{\text{ball}}}{I_{\text{insect}}} = \frac{(2/5)(2m_{\text{ball}})}{m_{\text{insect}}} = 0.5 \qquad (1)$$

For the turning movement, the inertia of the ball must be compared with the inertia of the cockroach around its vertical axis.

$$\frac{I_{\text{ball}}}{I_{\text{insect}}} = \frac{(2/5)(m_{\text{ball}}R^2)}{(1/12)m_{\text{insect}}(a^2+b^2)} = 2.2 \qquad (2)$$

It is impossible to match the inertia of the trackball for both forward movement and rotational movement simultaneously. The trackball was chosen to approximate the inertial forces as good as possible.

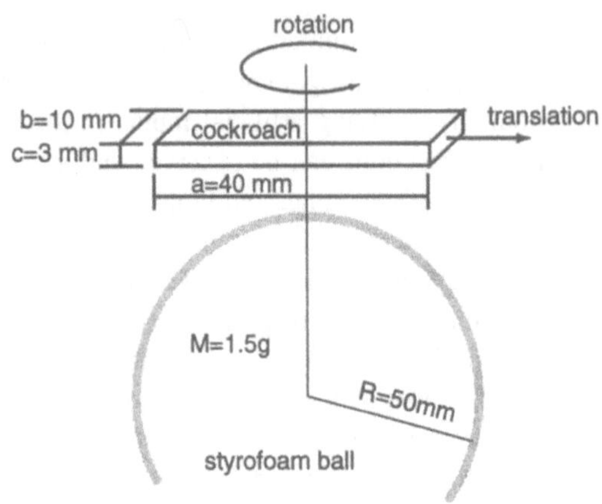

Figure 2: Inertia calculation for the cockroach-trackball system.

2.3 The electric stimulation signals

To stimulate the insect we used two different signal shapes: bursts of unipolar pulses and bursts of bipolar pulses. The current is supplied directly by the digital output of the microcontroller (PIC16C71) which has a CMOS output stage capable of sourcing up to 20 mA. This is largely sufficient. A fixed amplitude has been used, and the stimulation strength was varied by using different pulse widths and different numbers of pulse repetitions. The first series of experiments were done with unipolar signals. This is simpler and allows to have one common ground reference for multiple stimulation sites. However due to polarization of the electrodes, the electrode impedance goes up and the effective stimulation current decreases. Later bipolar pulses were used instead, generated by alternatingly connecting one electrode to the high potential and the other electrode to the low potential (see Fig. 3). In that case there is no constant ground level for the reference electrode and the electrodes must be switched into high-impedance mode (input mode) when not used.

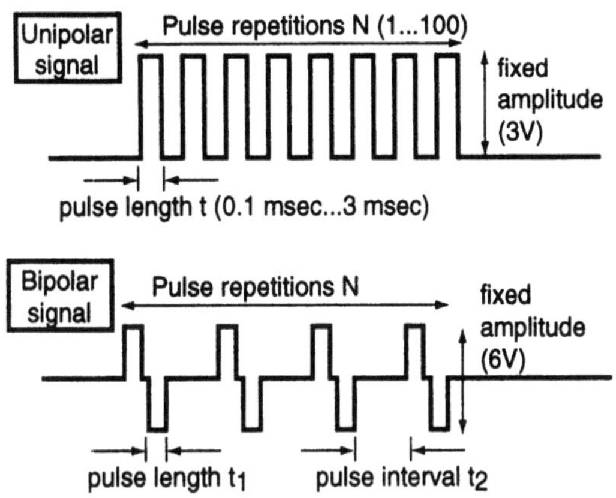

Figure 3: Unipolar and bipolar stimulation signals with their parameters.

100 µm-thin Platinum wires have been used as electrodes. The antennae were cut at about 3 mm distal of the pedicel and the electrodes were inserted inside.

It turned out that fatigue and habituation occur rather quickly and tended to interfere with the experiments. There is a large variance in reaction over time. Only stimulation events which are close in time can reliably be compared. A strategy of repeating the same stimulation parameters for a certain number of times and then vary that parameter is bound to fail due to the above stated reason. A better strategy is to combine the values to investigate in a set, and then repeat the sets for a certain number of timse as shown in Fig. 4. The overall appearance of the reaction will vary from the begin to the end of the experiment, but the distribution of the reaction amplitude within a set will still be valid. The first method clusters equal probing parameters all together in time, whereas the second method of forming sets distributes the probing parameters evenly over time. The stimulation amplitudes within a set have been chosen increasing exponentially. It is assumed that according to Weber's law the sensation of the electrical stimulation is logarithmically related to the strength (energy) of the stimulus.

3 Results

In the following experiments we investigated two input variables: the stimulation strength (number of pulse repetitions), and the location (left or right antenna). The observed output variables are the for-

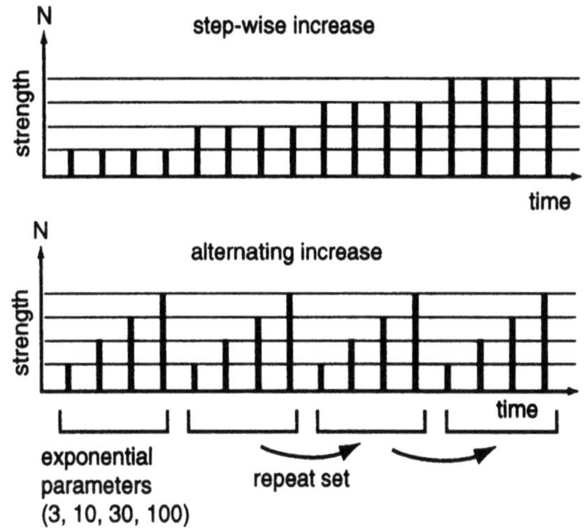

Figure 4: Stimulation in sets, with logarithmically increasing stimulus strength.

ward locomotion distance and the body rotation angle.

3.1 The influence of stimulus strength on forward locomotion distance

A number of experiments have been conducted to isolate the influence of the stimulation strength, i.e. the number of pulse repetitions at constant pulse width (and constant amplitude). The recorded output variable was the forward locomotion distance. Fig. 5 shows the results obtained from six series of experiments. Each series consists of ten sets of stimuli. One set of stimuli comprises four different stimulation strengths. In the present case these values are N=3, 10, 30 and 100. The stimulations within a set occur at 15 second intervals. The results show a forward locomotion distance which has a high stochastic variation but which shows an underlying pattern relating input and output. The higher the pulse repetition number, the larger the distance walked in average, but there are many overcrossings in the graph. If the six series are averaged, the ordering becomes evident. Fig. 5 shows the result: the responses are now in order of increasing stimulation strength N. The graph shows also the decrease of the reaction over time when the series are repeated at 15-minute intervals. If the insect is given time to recover, such as the 2-hour interval between series 4 and 5, the initial performance is partially restituted.

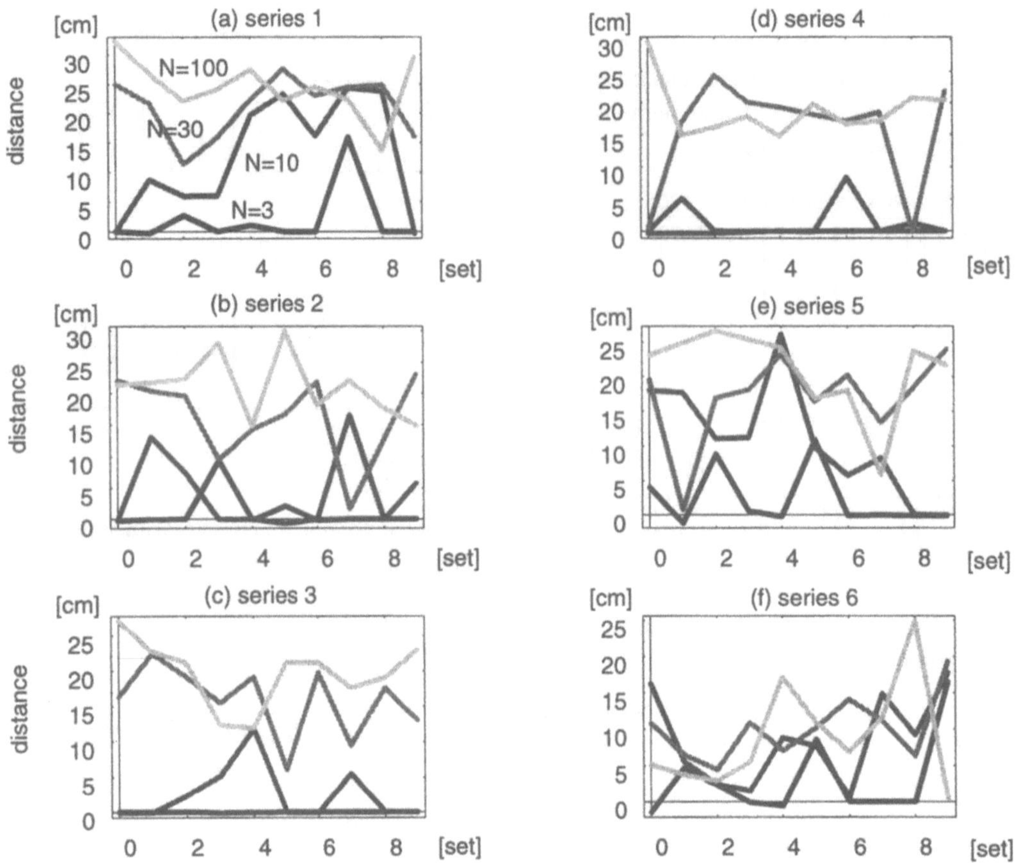

Figure 5: Walked distance for six different series of stimulation experiments with the same insect.

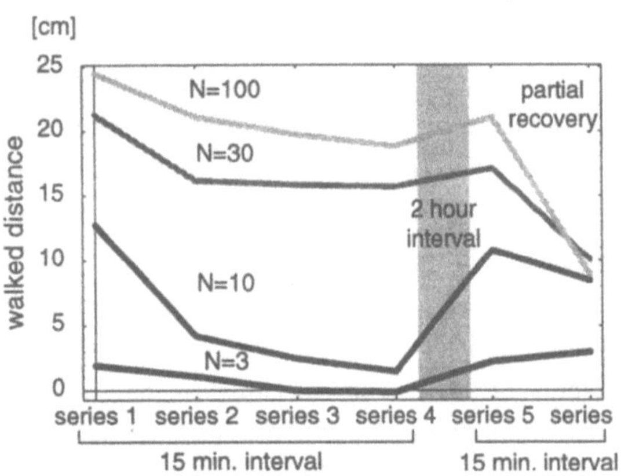

Figure 6: Walking distanced average for 6 series.

3.2 The influence of stimulus location on directionality

The second parameter we investigated is the influence of the electrode location (left or right antenna). In a similar setup to the previous experiment we applied sets consisting of a stimulation once to the left and once to the right antenna. The interval between stimulations was 30 seconds. Twenty sets were executed in a series and the results are graphed in Fig. 7. First we see that the direction of rotation is contralateral to the stimulation site: stimulation of the left antenna results in a turn to the right side, stimulation of the right antenna results in a turn to the left side.

Again we see a similar stochastic distribution pattern, however for most of the sets directionality to the location site is well expressed (few overcrossings). We also find a decrease in rotation angle over time in the fashion of a negative exponential. This is what we would expect, due to fatigue and habituation to the stimulus. The three graphs on the left side use pulses with a pulse length of 1 msec, whereas the three graphs on the right side use a pulse width of 3 msec. From top to bottom the pulse repetition number was increased from 10 to 30 to 100. The turning angle increases with the number of repetitions. There are few overcrossings for the left side. The average and the standard deviation have been calculated and are shown in Fig. 8. The mean turning angle increases

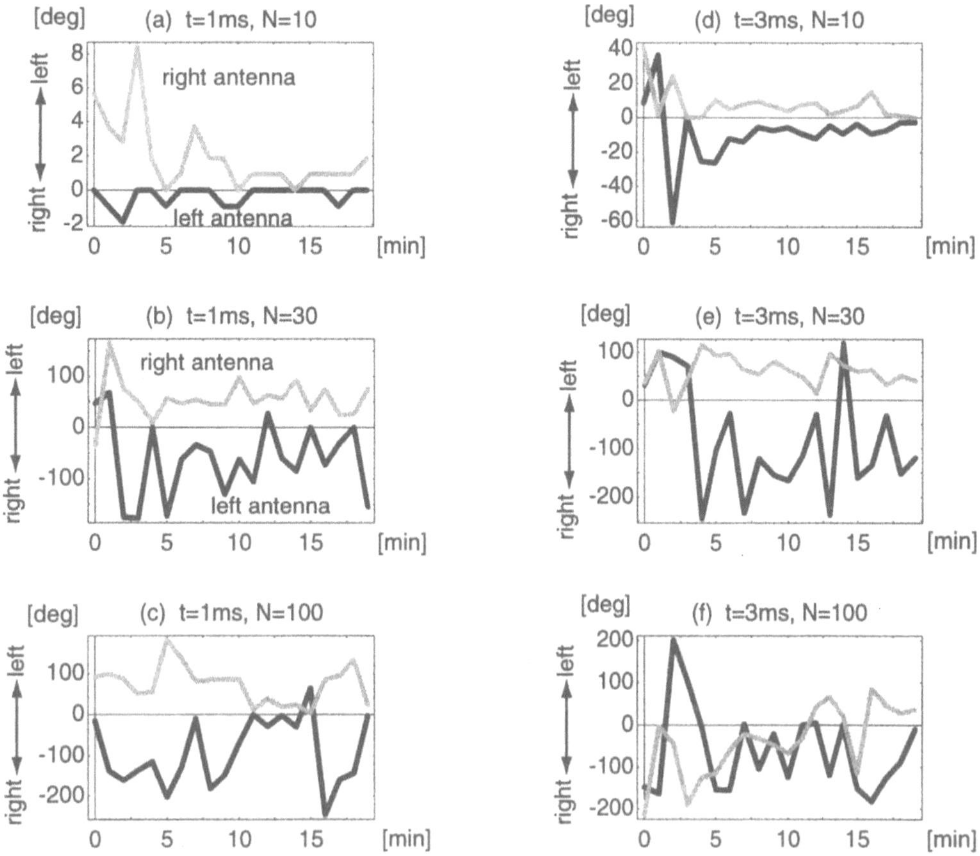

Figure 7: Turning angle recorded for alternating left/right stimulation.

with the number of stimulations, and the standard deviation is slightly smaller than the mean ($\sigma < \mu$).

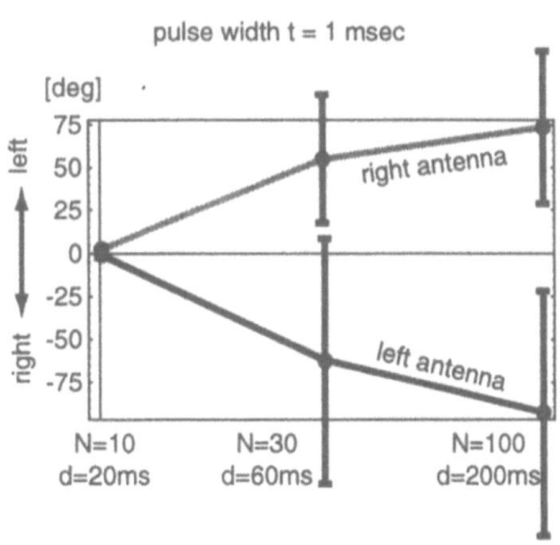

Figure 8: Mean and standard deviation (1ms pulse width).

Fig. 9 shows the case for the longer pulses of 3 ms pulse width. At first the turning angle shows directionality, but for large stimulation strengths, the directionality is lost. The mean turning angles for left and right stimulations now lie close together. The standard deviation is larger than the mean in this case ($\sigma > \mu$). If the total duration of the stimulation becomes longer than 200 msec, the directionality decreases.

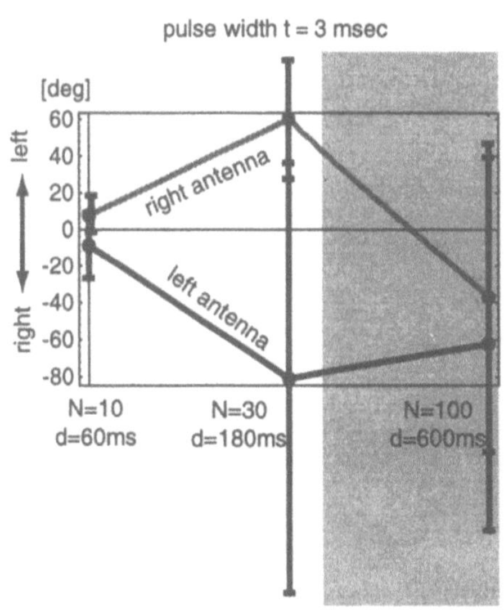

Figure 9: Mean and standard deviation (3ms pulse width).

3.3 Time-dependence

If the same stimulation is repeated continuously the reaction decreases with time. This is shown Fig. 10 where a unilateral stimulus to the left antenna is repeated continuously. The initially strong turning reaction to the right side decreases gradually in a negative exponential fashion.

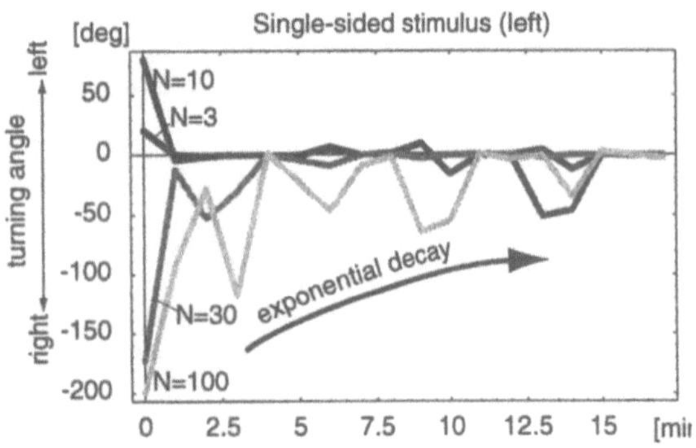

Figure 10: Exponential decay of the reaction to a repeated stimulus.

3.4 Logarithmic scaling

The mean turning angle and the standard deviation were calculated and plotted on a logarithmic-linear graph. The result for the mean is shown in Fig. 11. The experiment was done for pulse repetitions of N=3, 10, 30, and 100. There is a reaction threshold at about 10 below which the mean of the turning angle is zero. Above this value the turning angle is related to the logarithm of the stimulation strength.

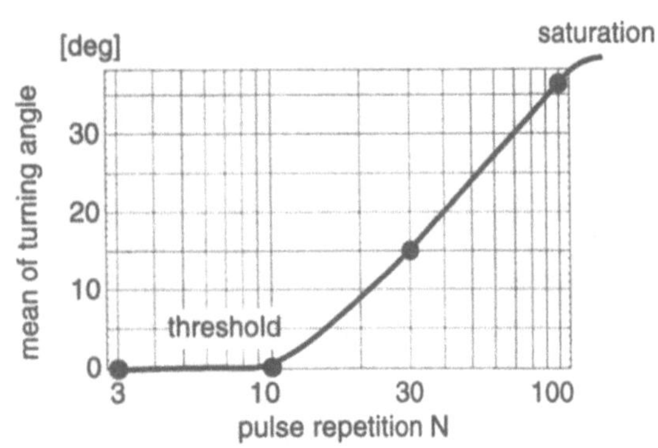

Figure 11: The mean turning angle as a logarithmic function of *N*.

On a second graph (Fig. 12) we plotted the standard deviation of the turning angle versus the logarithm of the stimulation strength. Again we obtain a linear relationship. It is linear even in the area where the stimulation strength is below the reaction threshold for the mean turning angle.

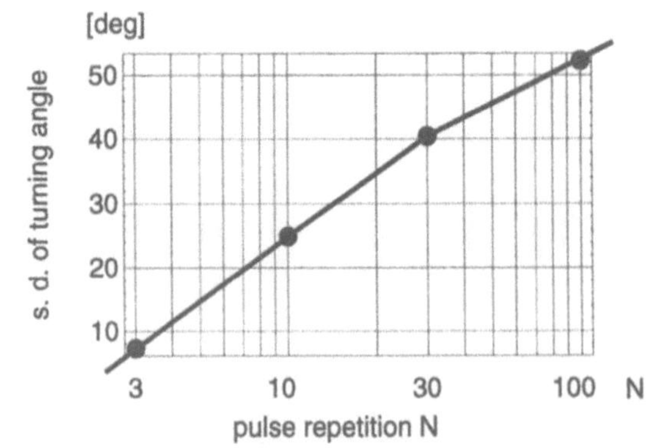

Figure 12: The standard deviation of the turning angle vs. the logarithm of stimulation strength.

4 Mathematical Model

From the experimental results in the previous section we propose the following "black-box" model composed of 3 stages which accounts for the observed input-output relationship (Fig. 13). This simple model relates the input variable N, the number of pulse repetitions, to the output variables s and θ, the forward distance and the turning angle respectively.

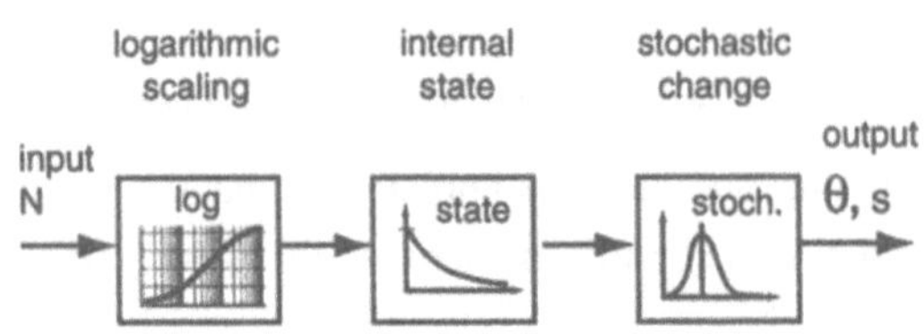

Figure 13: Simple 3-stage mathematical model.

The first block in the model accounts for logarithmic scaling of the inputs according to Weber's law. The equation includes a lower threshold value and an upper saturation value which delimit the valid range of this scaling function. Within the limits of threshold and saturation, the output is equal to the input's logarithm multiplied by a suitable proportionality constant.

$$f(N) = k_1 \cdot log\left(\frac{N_{sat} \cdot (N_{thresh} + N)}{N_{thresh} \cdot (N_{sat} + N)}\right) \tag{3}$$

The second stage of the model provides a simple mechanism for fatigue and habituation. The reasoning is that the performance decrease shows the characteristics of a negative exponential. The exponent is the accumulation of the previous stimuli. It's the logarithms that are accumulated to account for the effect the stimulation had.

$$f(i) = e^{-a(i)} \qquad a(i) = k_2 \cdot \sum_{j=0}^{i} log(N_j) \qquad (4)$$

The final and last stage is the one which most influences the output result. It is taken to be a Gaussian distribution with the following parameters:

$$[\mu = 1 \qquad \sigma = 0.5\ldots1] \qquad (5)$$

The last stage overshadows most of the previous parts and makes the determination of the exact parameters difficult.

5 Conclusions

A mechanical trackball can be used to measure the locomotion behavior of medium or large-sized insects. The main advantage of a mechanical trackball is to keep the insect fixed in one place. This allows to measure the locomotion reaction to a variety of stimuli quite easily. As stimulation source could be used: mechanical touch, wind movement, temperature, humidity, smell, or the optical field recreated artificially with a projection device.

Here we measured the reaction of the insect to artificial electric stimulation. The result turns out to be similar to the known escape behavior of the cockroach due to tactile and wind stimuli on cerci and antennae.

A simple mathematical model was proposed which accounts for the locomotion characteristics obtained in the trackball experiments. The stochastic factor is very large and shadows the other more subtle stages. Averaging over many stimulation experiments allows to estimate the parameters of these other stages.

Acknowledgment

This research was supported by PROBRAIN (Program for Promotion of Basic Research Activities for Innovative Biosciences) Japan.

References

[1] M. Burrows, "The Neurobiology of an Insect Brain," Oxford University Press, 1996.

[2] Y. Kuwana, I. Shimoyama, H. Miura, "Steering Control of a Mobile Robot Using Insect Antennae," proceedings of the IEEE IROS'95 conference in Pittsburgh, 1995.

[3] S. B. Crary, T. E. Moore, T. A. Conklin, F. Sukardi, D. E. Koditschek, "Insect Biobot: Electro-Neural Control of Cockroach Walking," 1996 IEEE International Conference on Robotics and Automation, Workshop WT#, Bio-Mechatronics, Minneapolis, Minnesota, USA, pp. 42-54, April 22-28, 1996.

[4] R. Holzer, I. Shimoyama, "Locomotion Control of a Bio-Robotic System via Electric Stimulation," proceedings of the IEEE/RSJ Conference on Intelligent Robotics and Systems, IROS'97, Grenoble, France, 8-12 September, 1997.

[5] C. M. Comer, J. P. Dowd, "Multisensory Processing for Movement: Antennal and Cercal Mediation of Escape Turning in the Cockroach," in "Biological Neural Networks in Invertebrate Neuroethology and Robotics", Academic Press Inc., 1993.

PART 12

PROJECTS IN JAPAN

SESSION SUMMARY

Hirochika Inoue
Dept. of Mechano-Informatics, University of Tokyo
7-3-1 Hongo, Bunkyo-ku, Tokyo 113 (Japan)

1 Introduction

This session discusses project oriented research activities in Japan. It covers three topics, and includes following six papers :

Arimoto: Physical Understanding of Manual Dexterity

Shirai: Tightly Coupled Sensor and Behavior for Real World Recognition

Sato: Intelligence and Autonomy for Human-machine Cooperative System

Yuta: Biologically Inspired Approaches to Autonomous Systems

Tanie: FNR: Toward a Platform Based Humanoid Project

Hirai: Current and Future Perspective of Honda Humanoid Robot

The first four papers introduce the inter-university research program on intelligent robotics. In the fifth paper, Tanie presents a brief introduction to the preliminary study of the MITI humanoid project. The last presentation introduces the surprising results of Honda Humanoid Robot.

At present, robotics research in Japan seems well funded, and some of the activities are effectively organized toward the next generation robotics. For instance, Japanese Ministry of Education chose "Intelligent Robotics" as one of the priority areas of scientific research, and funded the three-years research program as an inter-university research project. This program mainly supports research on intelligence and autonomy for advanced robots. The tightly coupled perception-action approach is given a focus to create robust systems which behave in real world. The first four papers, Arimoto, Shirai, Sato and Yuta describes four major issues in this research program.

Besides the above university program, the Ministry of International Trade and Industry is now preparing new national R&D project of Humanoid Robot. A two years' preliminary study for materializing the project has been carried out and determined the technical specification of humanoid platform. Currently, it is on the way to determine particular and promising applications that could be implemented on the humanoid. Tanie introduces what is going on about the really challenging R&D project on Humanoid robots.

In December 1996, Honda R&D Co. Ltd. announced their success of humanoid robot development. The prototype "P2", a battery powered astronaut shaped humanoid robot of 180cm height and 210Kg weight, walked around on the floor and climbed up and down the regular stairs. The shock wave of this news hit researchers and raised serious discussions about the future directions of robotics research. Hirai introduces a brief technical description on Honda humanoid prototype P2.

2 Inter-university Research Program on Intelligent Robotics

The first four papers introduce a research program on intelligent robotics which is a three-years program started in 1995 and funded by the Japanese Ministry of Education. The title of the program is "Research on Emergent Mechanisms for Machine Intelligence A Tightly-Coupled Perception-Motion Behavior Approach" (Principal Investigator H. Inoue, Univ. of Tokyo). This program aims to study mechanisms for the emergence of robotic intelligence and autonomy. Intelligence which emerges from physical interaction between a machine and the world is essential for real robots. Adopting a tightly coupled sensor-action approach, and being based on the theoretical foundations of machine intelligence, and this project experimentally studies autonomous sys-

Adopting a tightly coupled sensor-action approach, and being based on the theoretical foundations of machine intelligence, and this project experimentally studies autonomous systems that recognize and behave in the real world.

The research program described here was selected as one of the priority areas of Grant-in-Aid Program for Scientific Research, and given a mission to focus efforts on the challenges of intelligence and autonomy for advanced mechanical systems. It is structured as an inter-university research program, consisting of 21 prescribed research themes and 27 subscribed topics. Since 1995, 50 laboratories from 25 universities have participated in this program.

2.1 Research Issues

Four major issues are chosen to structure the research. They are; (A) physical understanding of dexterity, (B) real world understanding through tightly-coupled sensing and action, (C) human-robot cooperative systems, and (D) biologically inspired autonomous systems. All of these issues require a tight coupling of perception with action, and each issue closely relates with the others. Issues A and B are fundamentals for machine intelligence that should be coupled each other, and issues C and D are integration oriented. The research program consists framework research topics, which defines the domain and approach, and the supplemented research topics which enhance the planned approach.

(A) Physical understanding of dexterity

This issue aims to explore the physical principles underlying the dexterity of human/robot motions in execution of sophisticated tasks from the viewpoint of dynamics and control theory. It also aims to discover implementable algorithms based on the physical principles, which can function in robotic arms and hands and execute tasks with sufficient dexterity even if they are physically interacting with objects in the environment.

(B) Tightly coupled sensing and behavior for real world recognition

So far, the phases of recognize-plan-act in robotic behavior are conventionally connected in a simple loop, if at all. However, recent advances in computing performance provide new possibilities for constructing perception-action coupling. As the speed of visual processing becomes higher and the cycles for perception and sensing action become very fast, the recognition mechanism can be configured into more sophisticated constructs for fast, flexible, efficient and robust perception. This group studies various aspects of tight coupling of sensors and action so that behaviors are improved by sensing and sensing is better performed by proper behaviors.

(C) Intelligence and autonomy for human-machine cooperative system

Modern human life is supported by various kinds of advanced machines. However the relationship between those machines and human users is not natural enough. Machines often force that humans adapt to machines rather than the machines adapting themselves to humans. The human centered design of human-machine process is thus strongly required. The purpose of this issue is the study of intelligence and autonomy that is required to provide new cooperative relationships between human and machines. The role of intelligence and autonomy of advanced robotic systems is studied for making systems more human-centered and friendly.

(D) Biologically inspired approaches to autonomous systems

Biological systems embody a sort of intelligence that enables flexible adaptation to environmental changes and robust pursuit of goals. Thus,they keep themselves alive. In this topic, through the study of biological intelligence, we explore to understand the mechanism of intelligence and autonomy of the living things and attempt to build real mechanical systems of such nature. This approach is expected to uncover the ways to create reliable, robust and flexible machines that behave intelligently in the real environment.

The purpose, approach, and details for each issue are to be introduced in the following four papers by the project leaders (A: S.Arimoto, B: S. Shirai, C: T. Sato, D: S.Yuta, respectively).

2.2 Approach and Guiding Principles

The aim of this program is summed up by the two keywords: perception-motion coupling, and emergent machine intelligence. The objective of this program is to investigate the mechanism or algorithm that gives rise to machine intelligence. For that purpose, we chose a tightly-coupled perception-motion behavior approach. A machine is a physical entity that behaves in the real world. The research interest is focused onto some sort of intelligence, which is inherently required for robust, efficient, and quick interactions between the machine and the real environment. This project aims to find the principles of such mechanical intelligence and to demonstrate them by building physical entities. One of the important motivations for forming this program was to refresh the field and the approach of robotics research. Currently, the program seems to encourage approaches based upon a tight coupling of sensing and action. We expect that the program will open up new topics, new attempts, and new approaches for machine intelligence in the robot-world interaction process.

2.3 Infrastructures for experimental robotics

In robotics research, the development of experimental devices or tools are very important, because the process of building robots is also an effective approach for understanding the robotic principles. However, as proper platforms for robotic experiment are not available, many researchers build similar but not the same experimental devices. If a standard platform like computer hardware were available, software libraries would be accumulated and help to accelerate the speed of research. For this purpose, in the research program, we develop common research tools for distribution such as a real time tracking vision system based upon correlation technique, a high performance general color vision system in DSP based multi-processor architecture, and a compact low cost legged robot platform. This research program supports the initial cost for development of the legged robot and the vision units, and opens the way to further development of general-purpose, common research tools for the sensing-action coupling approach.

3 FNR: Friendly Network Robotics

A preliminary study for MITI R&D project draws new technical perspectives for human friendly robots which could work cooperatively with human in the ordinary environment for human life. The purpose of this project is to develop humanoid technology as an infrastructure that supports future elderly society. The key features should include advanced collaborative manipulation with human, stable mobility on rough terrain, highly safe and reliable physical interaction with human, remote operation through computer network, and advanced autonomy for human-robot symbiosis. Application fields of robots will be expanded from factory floors to construction, public service, social maintenance, service industry, and home. Generally, such robots need not be human shaped, this project, however focuses to develop humanoid robot as a symbolic target for technical integration, and to look for possible applications of humanoids.

Usually, Ri&D projects are formed to begin with elementary technologies. Then, the outcome is integrated toward a target system as bottom-up approach. But, this humanoid project takes in contrasts a new approach referred to as the platform based technology development. A humanoid, if available, is considered as a general purpose worker machine, and various applications can expand using such general humanoid as platform. In order to materialize this scheme, the project first develops a humanoid platform of open architecture, and then research goes in two directions. One direction is to implement new real applications by using those platforms. Practical studies often reveal real technical problems and triggers new development, improvement, and replacement of platform elements. In such away, the platform will shape up its technology. The second direction is to accelerate R&D activities by integrating new research results on theories, software and knowledge since the platform provides the common R&D tool and the base of abstraction. From such viewpoint, the committee of the preliminary study defines a specification of a humanoid platform. Tanie's presentation introduces the proposal of the platform based approach for humanoid R&D project.

4 Honda Humanoid Robot

After ten years unopened research, Honda R&D Co. Ltd. announced their success of humanoid robot. The performance shown by the video tape and real experiments was quite impressive, and gave a considerable shock to researchers. The prototype P2 walks and climbs up and down staircases in a very stable manner, as if a human were inside. Technically, P2 is a biped walking machine that carries an upper body with a head and a couple of arms. The design with enough payload and sufficient stability control made the system an epock-making humanoid.

A complete humanoid robot is considered very difficult to build. Honda first gave a solution to the challenging problem. Honda's success does not terminate the challenge, rather it proved the correctness of direction and provided the starting point for paradigm change of the challenge. In fact, it opens very fertile research topics in reality. Although the humanoid had been a target to build, it also became a general entity to employ. Not only the technical issues for building humanoid, but also issues of theory and implementation of whole body motion of human and humanoid become important: Motions must comply with the complex physical constraints with environment. For instance, the motion under multiple contacts with the outer world by body and hands together with feet is an important issue. Mobility becomes a 3D or 6D problem from a 2D problem of floor constrained mobile robots. Real time vision as well as complex motion planning are more demanded than ever. The method of autonomous and/or remote control is another issue. I consider humanoid research was reborn with reality, and the Honda's constructive proof provided us with the starting point for the Friendly Network Robotics Project stated above.

Physical Understanding of Manual Dexterity

Suguru Arimoto
Department of Robotics, Ritsumeikan University
Nojihigashi 1-1-1, Kusatsu, Shiga,525-8577 Japan
e-mail: arimoto@bkc.ritsumei.ac.jp

One of the great challenges to robotics research in the next century is to design mechanical robots that can imitate human behavior. Manual dexterity is one of intelligent human behaviors generated unconsciously with sensory-motor organization in the central nervous system. Such intelligence is hidden from unconsciousness but it is an outcome of biological evolution that spent billions of years and eased skillful manipulation. It can therefore be believed that there are physical principles underlay the process of such biological evolution toward perfecting human dexterity.

The research of the A-group aims to make robots equipped with manual dexterity by unveiling such hidden physical principles and understanding manual dexterity from the viewpoint of physics and mechanics. In other words, it aims to discover simple but refined motor control algorithms based upon the physical principles, which can function in robotic arms or mechanical hands and execute robotic tasks with sufficient dexterity even if they are physically interacting with objects in environment.

In the third(last) year after the project started, the following research topics are included in three categories.

a) Analysis and Control of Robot Tasks under Physical or Geometrical Constrains

(1) S. Arimoto, K Nagai, and T. Naniwa: Physical understanding of adaptation and learning for dexterous robot motions.

(2) M. Uchiyama, D. Nenchev, and Y. Tsumaki: Mechanics and control of skilled manipulation process.

(3) Yoshida: Control of moving base manipulators for impedance operations.

b) Modeling and Planning of Grasping, Assembling, and Insertion.

(4) T. Yoshikawa, Y. Yokooji, and Y. Yu: Kinetics and control in emerging mechanism of manipulation skill.

(5) Y. Nakamura: Generating smooth behaviors from sensory integration.

(6) K. Kosuge: Decentralized control of multiple robots in coordination.

(7) S. Hirai: Research on machine skills in distributed manipulation of deformable objects.

c) Analysis of Dexterity of Living Things

(8) M. Kaneko, T. Tsuji, K. Harada, and M. Svinin: Analysis of dexterity of living things and its application to robots.

(9) S. Kawamura: Measurement and analysis on task-ability of human multifinger.

Arimoto's group clarified in this research a crucial role of "passivity" in adaptability and learnability of robots that are subject to external constraints or physical interactions with their environments. The group proposed last year "Inertia-only(Function/Gravity-free) Robots" which is realized by incorporating circuit-theoric modules in nonlinear position-dependent circuits expressing robot dynamics. In the third year they are going to propose a new extended concept of "impedance matching" in order to evaluate dexterity in execution of robot tasks.

Uchiyama's group presented a unified motion, force and compliance control scheme for execution of very fast and complicated tasks with using a high speed and accurate parallel robot HEXA. They showed from experimental data the effectiveness of their proposed method of compensation for static friction consisting of two different modes (one is for motion starting from the steady-state and the other is for motion at transient state crossing at zero velocity).

Yoshida developed a computer simulation program for dynamic simulation of moving-base robots, including free-flying robots and flexible structure mounted robots. The program can handle a broad class of robots as far as the robot is composed by open-tree configuration of arm(s) mounted on a single moving base.

Yoshikawa's group presented a simulator of the 2-dimensional peg-in-hole assembly operation in the virtual environment, considering the dynamics and the frictional force. An operator can operate the virtual object by two link mechanisms which display the operating feeling.

Nakamura is going to establish a learning method for a 3-fingered robot hand that grasps an object by integration of reactive behaviors, where inner parameters between sensors and behaviors are self-organized through iteration of trials such that the sensor signals converge to the given teacher signals respectively. A skill abstraction model is also proposed, which accumulates skills of plural tasks into a set of inner parameters.

Kosuge aims to develop a new decentralized control algorithm for multiple non-holonomic mobile robots handling single object in coordination. A leader-follower type control algorithm proposed for holonomic robots last year is now extended to non-holonomic mobile robots.

Hirai is going to investigate machine skills in distributed manipulation of deformable objects by analyzing a relationship between actuation and sensory information including force sensing and visual sensing. In practice, human motion in the insertion of an object into a deformable tube is analyzed to derive functions to detect the transition among process states by force sensor. In the last year of the project dynamic modeling of deformable strings and control of extensional deformation of textile fabrics are to be presented.

Kaneko's group first presented an observation that in grasping human changes the strategy according to the size of objects, even though they have similar geometry, and hence the grasping strategies are roughly classified into three groups, depending on the size of objects. Then they explored a strategy finally grasping column objects placed on a table by a multi-fingered robot hand. In the strategy the most important is to relax the grasping force after lifting up the object so that the object may be moved by slipping.

Kawamura is going to analyze force and friction on human fingers in order to design artificial fingers and measure mechanical impedance on human fingers. In the experiment, much attention is paid to the four levels of impedance: (1) fingertip tissue, (2) finger joint (muscle impedance), (3) object, and (4) sensory feedback impedance. Tsuji in Kaneko's group also investigates adaptability of human operator in a human-robot system by an experiment based on manual tracking tests.

Tightly Coupled Sensor and Behavior for Real World Recognition

Yoshiaki Shirai
Department of Computer-Controlled Machine Systems
Faculty of Engineering, Osaka University
2-1, Yamadaoka, Suita, 565 (Japan)
email: shirai@mech.eng.osaka-u.ac.jp
jointly assigned to Department of Mechano-Informatics, The University of Tokyo

Abstract

"Tightly Coupled Sensor and Behavior for Real World Recognition" is the name of a B Group which is one of four groups in the project "Research on Emerging Mechanism of Machine Intelligence: A Tightly-Coupled Perception-Motion Behavior Approach" supported by Grant-In-Aid for Scientific Research on Priority Area from the Ministry of Education, Science, and Culture, Japanese Government.

The aims of the B Group is to study recognition of a real world by interaction the world so that intelligent systems may efficiently perform given tasks. The Group consists of five planned research teams and eleven teams (in this fiscal year) which are selected according to the submitted proposals. This paper introduces the outline of the research project and the research of author's team as an example.

1 Introduction

Sensing and action is important for many intelligent systems to performs given tasks in the real world. Such systems (robots) should deal with various uncertainties caused by physical constraints. Typical uncertainties are given below.

- uncertainty of sensor information: originated from the sensor constraints.
- uncertainty of environment: only partial information is available.
- uncertainty of motion: originated from the actuator constraints.
- uncertainty of a priori knowledge: for example, a given map.
- uncertainty of models: impossible to have models of all real world objects

One of the promising way of dealing with the uncertainties for a robot is to interact with the environment. Many uncertainties are reduced by repetitive cycles of suitable sensing and actions.

With the development of hardware, this interaction cycle has been greatly reduced. Now is the time to consider how to perform the interaction in order to achieve given goals.

While understanding the environment is necessary for a robot to make actions in the environment, actions are often necessary to understand the environment. If the goal of the robot is not only to understand the environment but to make some actions such as reach a goal position or manipulate an object in the physical environment, we have to consider the trade-off between the cost of actions and that of sensing.

2 Overview of the Research Themes of the Project

The Group consists of five planned research teams and eleven teams (in this fiscal year) which are selected according to the submitted proposals. The research themes of the planned research cover a main field of the B Group. Viewed from the the levels of sensing and actions, they correspond those levels as shown in Figure1.

- Interpretation of sensor information: real time recognition of environments from various sensor information using sensing-action rules. This is useful for immediate actions.
- Understanding of the environment: dealing with uncertainty by sensing and action cycles

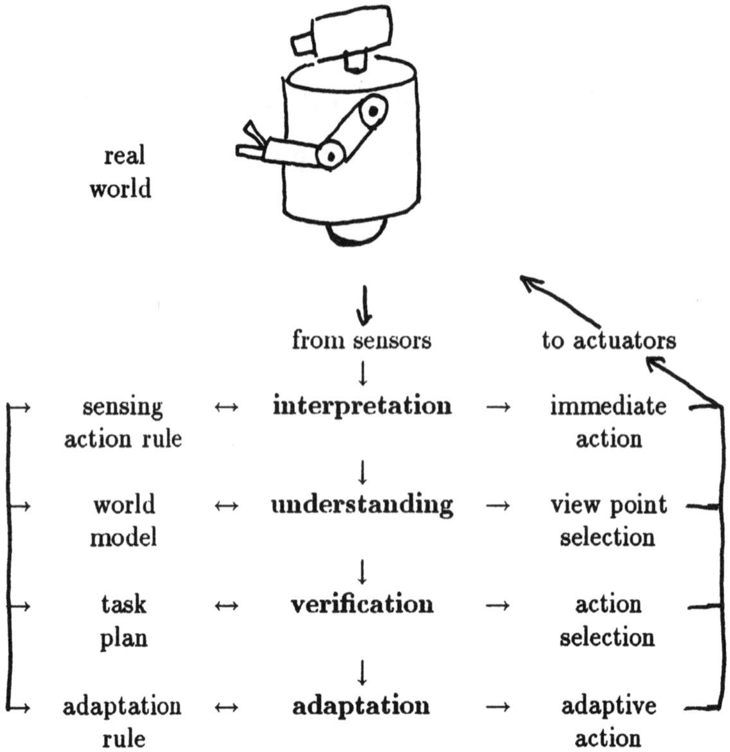

Figure 1: Various level of sensing action

using models of objects and proper viewpoint selection.

- Verification of the environment: flexible sensing action control in the course of the task to observe the state of the environment and the robot itself. This is useful to accomplish a task which consists of a known action sequence.
- Adaptation to the environment: architectures for flexible sensing action control by some adaptation rules.

Actually, however, the sensing and action scheme is not simple as shown here. Many levels of knowledge, information processing, and actions are combined for complicated tasks in the real world.

In the following subsections, this year's research themes of the projects are briefly introduced here by referring to the survey report [1].

2.1 Planned Research Projects

The four planned projects have been studied by five research institutes three years. The outline of the research themes are shown below.

Interpretation of Sensor Information for Real World Understanding

by Y. Shirai (Osaka University)

This research aims to study mechanisms of understanding of real world environments by fusing various sensor data. Based on the result of rough understanding of the environment, decision is made on what information is necessary for decreasing the uncertainty of the interpretation and what is the next action for better understanding of the environment or reaching a given goal. This feedback loop of sensing, interpretation, and action is studied. The features of the research are

1. the reliability measure of interpretation is considered
2. the optimal plan is made by considering the cost of motions and that of sensor information interpretation.

This year, we first study fusion of sensor information such as brightness, range, and motion, to make a real time recognition system using DSP boards. And study optimal planning of sensing and motions under time pressure by considering uncertainty of sensor information and the planning cost. Finally, we integrate real time recognition system and optimal planning into an autonomous mobile system for real environments.

Sensing Mechanism for Behaviors Adaptive to the Real World

by H. Mori (Yamanashi University)

This year we integrate the sign pattern-based sensing system and the rule-based navigation system to make a tightly coupled sensor-behavior system. We evaluate the system through field tests in the open outdoor environment which is free from the traffic accident.

Through the tests the sign pattern based-sensing system is improved to understand the road environment in the counter-sunlight by applying the active sensing.

We equip a real time stereo vision system developed by SUBARU Co.. to another mobile robots. The stereo vision system gets the depth image every 100ms, and detects obstacles and road boundaries in the sidewalk.

Behavior Control Based on Environment Model and Sensory Information

by M. Kakikura (Tokyo Denki University)

Behaviors are improved by sensing and better sensing is performed by behaviors. According to this conceptual guide line, we develop a strategic algorithms which decide the most suitable observation points by multi-sensor system, and proper subtask execution series for final task accomplishment.

The goal is to develop a robot which can handle the non-solid form objects, especially clothes. The detection of error and recovery are studied for handling of clothes. Error avoidance is also studied by inspecting th history of sensing and actions. The real time control of behavior is experimented.

Behavior Control Based on Environment Model and Sensory Information

by T. Hasegawa (Kyushu University)

We study tightly couple sensing and behavior so that behaviors are improved by sensing and better sensing is performed by behaviors. As an actual application, our research focuses on an action control for a given manipulation task based on sensing and a world model. Tha main thrust of the research consists in developing a method of reasoning: what kind of sensor information is required to complete the task and how to obtain it; how to identify the state of the task using sensor information.

The plan of this year is to integrate different sensing devises and actions for sensing to implement assembly motion primitives. We also evaluate developed methods of determining the most adequate sensing devices based on the task sequences and context.

Adaptive Functions in Perception-Action Systems

by H. Inoue (University of Tokyo)

This research aims to study adaptive functions in perception-action systems that behave in the real environments. In order to perform adaptive behaviors in the real world, the autonomous systems should have control mechanisms to select sensor processing and reactive motions to suit the environments. This research remarks the parallel process network system to not only action selection but also automatic acquisition of reactive behaviors. In order to evaluate the architecture, the behavior network system will be implemented on a real quadruped robot connected with wireless distributed network computers. It will show adaptive behaviors with experiments in the real world.

2.2 Selected Research Projects

Ten research projects have been selected this year. Although selected teams changes every year, most of them have contiued the same research three years.

Active Sensory-Motor Integration System Using Vision and Touch

by M. Ishikawa (University of Tokyo)

Our final target of this project is to integrate sensory information such as vision and touch and control information of motor system. In other words, it is a realization of vision-touch-motor integration using active sensing behavior.

We have already realized an architecture and recognition algorithms based on active sensing, developed 1ms sensory-motor fusion system based on them, and evaluated its performance by implementing algorithms for grasping on it.

In this year, we will carry out the following tasks; 1) improvement of 1ms sensory-motor fusion system (expansion of DSPs system), 2) development and evaluation of a new sensing architecture and recognition algorithm, 3) clarification of the functionality of information processing for sensory-motor system considering active sensing.

An Intelligent Mobile Robot Localizing Sound Sources and Characters

by N. Ohnishi (Nagoya University)

Sound and characters are very important information for humans to understand the environment and move in it. We aim to develop an intelligent mobile robot that can localize sound sources such as humans, and character strings such as sign boards in the real scene. To realize such intelligent robot, we study 1) a method for localizing an invisible sound source behind objects and 2) a method for detecting character strings which are observed either orthogonally or obliquely. We are going to implement the developed method into the robot and conduct experiment in the real world to evaluate its effectiveness.

Acquisition of Dynamic Environment Models by a Distributed Vision System

by H. Ishiguro (Kyoto University)

We have proposed the concept of *Distributed Vision* and developed a distributed vision system (DVS) which supports humans and robots with vision agents embedded in the environment. The DVS has the following three features: (1) The DVS solves the attention control problems by dynamically selecting the vision agents. (2) The DVS acquires dynamic environment models and maintains in the vision agent network. (3) The vision agents are organized through communication with the computer network and observations with the vision sensor.

We especially focus upon the acquisition of dynamic environment models this year and try to develop original technologies of the DVS for acquiring the models. Since the vision agents are fixed in the environment, the DVS can robustly detect dynamic events through the vision sensors. The obtained information of the dynamic events are exchanged among vision agents through the computer network and memorized in the vision agents as the dynamic environment models.

Learning Methodology for Acquiring A Perception-Action Strategy

by F. Miyazaki (Osaka University)

This program aims to study a method of estimating unmodeled factors in dynamic manipulation and utilizing the estimation to enhance the robot's performance. The task we deal with in this program is Ping-Pong. The task for the immobile robot is to return the incoming ball to the human opponent with a racket. This project will tackle the learning methodology for acquiring a strategy which links perception and action, and will try to realize a robotic system which can execute the target task.

Visual Navigation in a Dynamic Environment by Using Multiple Image Sensing System MISS

by M. Yachida (Osaka University)

Multiple visual sensing sensor (MISS) combines with an omnidirectional image sensor and binocular vision. Our aim is to propose an effective method for map generation by integrating global azimuth information from an omnidirectional image sensor and local range information from binocular vision.

Because boundaries between the ground plane and wall appear as a closed loop curve in an omnidirectional image, the robot can move along the route by tracking the closed looped curve with an active contour model. Then the binocular vision can fix the attention and observe the detail of the interesting moving object by using the global azimuth information.

Behavior Acquisition and Environmental Representation of a Vision-Guided Legged Robot

by K. Hosoda (Osaka University)

In this research, we propose a behavior acquisition method of a vision-guided legged robot, and its environmental representation based on the acquired behaviors. First, a swaying motion based on visual servoing, a motion not to fall down, and a walking motion utilizing the lifted leg are embedded in the legged robot. If the human designer implements these kinds of motions into the robot, it may walk by tracking the moving visual targets. Second, a method to collect environmental information is proposed during the visually guided walking of the robot. By collecting information, the robot gets an environmental representation. Finally, by the representation, the robot moves around and correcting the representation by its own experience. These strategies will be tested in the real robot.

Robust Estimation of Moving Objects and Ego-motion Parameters from Moving Camera Images

by N. Yokoya (NAIST)

This study aims to develop robust techniques for extracting moving objects and estimating egomotion parameters of a mobile robot from image sequences captured by a moving camera. In the year, we work on estimation of ego-motion parameters from moving camera images.

The main problem here is how to discriminate the object/camera motion from the apparent motion of the static background. The research will be mainly carried out using a mobile robot platform with cameras, which was constructed in last year. Experiments for real-time execution of the algorithm will also be conducted using DSP boards connected to the workstation.

Cognition of Changes in Structure of Environment

by S. Tsuji (Wakayama University

Conventional methods for making environment models by a robot assume the structure of the environment and implement procedures for estimating the environment geometry based upon the assumption. However, the robot fails in performing reasonable actions if the structure changes. This research aims to establish a new method for finding and accommodating to such changes. The following experiments will be conducted.

1. The robot continuously observes the environment while it moves toward its goal and describes the environment structure.
2. When the robot finds a significant change in the environment structure description, it activates actions of verification based upon the perceived change.

Active Recognition System in an Indoor Environment

by H. Ohki (Ohita University)

In order to recognize various objects, viewers attempt to change their position around unknown object to observe. Our research aim is to develop a robust object recognition system focusing on the following topics:

1. Object recognition with partial features and conflict resolution by changing the viewpoint,
2. Integration of recognition and motion,
C) Reduction of image processing computation using previous recognition.

Visual Servoing Combining Environment Prediction and Motion Control

by K. Hashimoto (Okayama University)

This research aims to study the visual feedback system as a dynamical system with nonlinear constraints. The visual cues in the sensory output are considered as controlled variables and the visual servo system is described by nonlinear differential equations.

Nonlinear constraints exist in this differential equation because the visual cues are fixed on the rigid object. The object motion estimation and the tracking are reduced to the control theoretical problems of designing a nonlinear observer and a nonlinear observer-based controller.

We model the object motion on a manifold and study the geometrical structures of the manifold and nonlinear differential equations describing the object motions. Finally, real time experiments on visual tracking will be carried out on industrial manipulators and mobile robots to show the effectiveness of the observer-based controller.

Realtime Stereovision System and Environment Understanding for an Autonomous Field Vehicle

by J. Takeno (Meiji University)

The objective of the study is to develop hardware for the stereovi sion system for mobile robots which the author has been studying. The development of the hardware is almost complete. The hardware constructed with analog/ digital transformation devices with FIFO memories and parallel processing units with many risc tips to analyze each pair of scanning lines of two CCD cameras. The developed system by this study will be adopted as a stereovision system for a real mobile robot called AFV-II currently being studied at my university, and evaluation will be made using an actual robot.

3 Dealing with Uncertainty by Sensing and Action Cycles

This section explains our project "interpretation of sensor information for real world understanding" introduced in the previous section as an example of the research in B Group. We represent the behavior of an intelligent robot as a cycle of sensing, interpretation, planning and actions(including observation) as shown in figure ??.

Uncertainty of sensor information and that of the object model are considered to obtain the interpretation result with uncertainty measures. From the interpretation result, an actio plan is made for motion or observation. Because the future observation result can not be obtained in the planning stage, it is predicted probablistically,

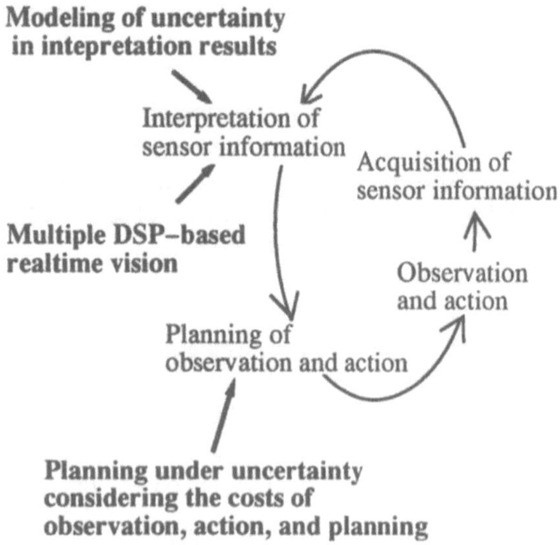

Figure 2: Our approach to sensing and action cycle

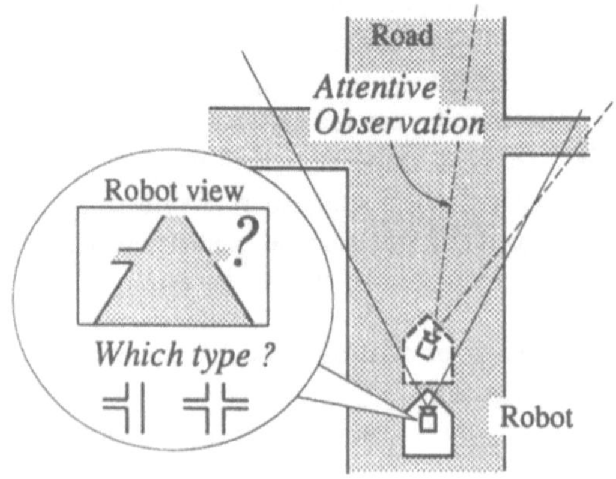

Figure 3: A robot determining type of intersection.

and the planning is made to get the best expectation of the criterion. For quick interaction with the environment, a realtime vision system is constructed with multiple DSPs.

In the next subsections, planning for recognition of road intersections and a realtime vision system are described.

4 Recognition of Intersection Scenes Considering the Uncertainty of Recognition

Suppose that a mobile robot is going to a specied place in a qualitative map and is to use intersections as landmarks. This involves the following problems:

1. Intersection scenes include various objects whose color, size and shape are similar to one another.

2. The objects are often not clearly identified.

To overcome these problems, attentive observation considering the uncertainty of recognition is a promising approach [2].

Figure 3 shows a typical situation, where the robot observes the intersection scene at the first viewpoint and tries to determine the type of the current intersection. If the the recognition result is umbiguous, the robot moves to the second viewpoint and observes the part of the road attentively.

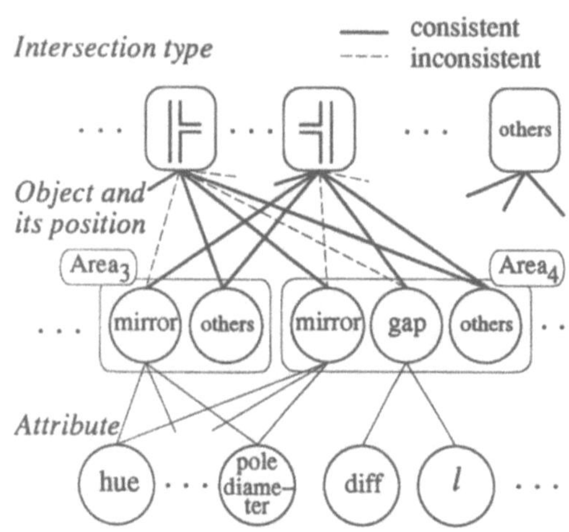

Figure 4: Model of intersection scene.

4.1 Model of Intersections

The types of intersections are characterized by the number and direction of the branching roads. The intersections also have particular objects such as curve mirrors. These intersection objects are described by several attributes such as color, shape and size. The relationship between the intersection types, the intersection objects and their attributes are represented as shown in Figure 4.

The attribute of intersection objects are represented by a probability distribution function. The position of objects in each intersection type also varies within certain ranges. The position is represented by normal distributions.

4.2 Recognition of Intersection Object

The intersection objects are found in an color image as regions with several attributes such as color, shape and size. To extract the regions, the robot uses *finders* specializing in each intersection object. The extraction procedure is as follows:

1. From the distribution of the attributes, the candidate region is extracted.
2. Let $a_1, a_2, \ldots$ denote these attributes of the region. Since $a_1, a_2, \ldots$ are assumed to be independent, the probability of the region r coming from an object o is calculated as follows:

$$P(r|o) = P(a_1, a_2, \ldots|o) = \prod_k P(a_k|o). \quad (1)$$

$P(a_k|o)$ is calculated by using the distribution represented by the knowledge.

The probabilities of the regions selected from the overlapping regions and the isolated regions are used to determine the intersection type.

Figure 5: A sample scene and results of object recognition. The ellipsoids correspond to a gap, a curve mirror, and a white mark.

Figure 5 shows an example scene.

4.3 Recognition of Intersection Type

If the finders extract the regions, the robot uses them as evidence for recognition of the intersection type. By using these pieces of evidence $\mathbf{e} = (e_1, e_2, \ldots)$ and the current observation parameters ω(assumed to be known), the robot calculates the conditional probability of the hypothesis that the current intersection type is $I_i \in \mathcal{I}$ as follows:

$$P(I_i|\mathbf{e}, \omega) = \alpha P(I_i|\omega) P(\mathbf{e}|I_i, \omega), \quad (2)$$

where $\alpha = [P(\mathbf{e}|\omega)]^{-1}$. Since the intersection type is independent of the observation parameters, $P(I_i|\omega) = P(I_i)$. Initially $P(I_i)$ is set to a prior probability. Later it is updated by Eq. (2) at the previous stage.

Let $o_{p,t}$ denotes an intersection object of type t at position p. By observing objects with observation parameters ω, various combinations of objects $\mathbf{o} = (o_{1,t1}, o_{2,t2}, \ldots)$ are observed. From the possible combinations, we make an exclusive and exhaustive set $\mathcal{O}(\omega)$.

By using $\mathcal{O}(\omega)$, $P(\mathbf{e}|I_i, \omega)$ is expressed as:

$$P(\mathbf{e}|I_i, \omega) = \sum_{\mathbf{o} \in \mathcal{O}(\omega)} P(\mathbf{e}|\mathbf{o}, I_i, \omega) P(\mathbf{o}|I_i, \omega). \quad (3)$$

Since the evidence is assumed to be independent of the intersection type and the observation parameters, and each evidence is assumed to be independent of the others,

$$P(\mathbf{e}|\mathbf{o}, I_i, \omega) = P(\mathbf{e}|\mathbf{o}) = \prod_p P(e_p|o_{p,t}). \quad (4)$$

Next, because objects are independent of the observation,

$$P(\mathbf{o}|I_i, \omega) = P(\mathbf{o}|I_i) \quad (5)$$

This can be computed from the consistency of the object with the intersection type. Here, $P(o_{p,t}|I_i)$ is defined as:

$$P(o_{p,t}|I_i) = \begin{cases} 1/N & \text{(if } o_{p,t} \text{ is consistent with } I_i\text{.)} \\ 0 & \text{(otherwise.)} \end{cases} \quad (6)$$

where N indicates the number of the consistent objects existing at the position.

To evaluate the abmiguity, the following entropy is used:

$$H(\mathbf{e}, \omega) = \sum_i -P(I_i|\mathbf{e}, \omega) \log P(I_i|\mathbf{e}, \omega). \quad (7)$$

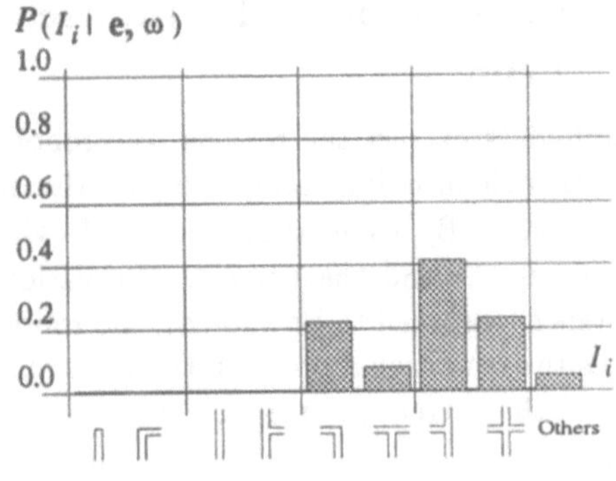

Figure 6: Probability distribution of the intersection type calculated from evidence as shown in Figure 5.

The robot adopts the best hypothesis if the entropy is lower than a certain threshold.

Figure 6 shows the distribution calculated from the evidence shown in Figure 5. The entropy of the distribution is 1.93.

4.4 Attentive Observation to Resolve Ambiguity of Recognition

If the recognition result is ambiguous (the entropy is larger than the threshold), the robot select a informative part of the road to observe attentively. Informativeness is defined by the expectation of the predicted entropy:

$$\bar{H}(\omega') = \int_{\hat{e}\in\hat{\mathcal{E}}(\omega')} H(\hat{e},\omega')P(\hat{e}|\omega')d\hat{e}, \qquad (8)$$

where ω' denotes the new observation parameters achieved by applying an action and $\hat{\mathcal{E}}(\omega')$ denotes a set of possible evidence predicted to be obtained with the observation parameters ω'. $H(\hat{e},\omega')$ is the entropy calculated from Eq. (7).

The robot searches for the ω' which minimizes Eq. (8) and applies the action to achieve the ω'.

In the case of the scene as shown in Figure 5, the best action is to observe a gap of the road boundary at right side(the expectation of the entropy is 1.66). Figure 7 shows a newly obtained image and the result of object recognition. From the probability for the gap, the robot calculates the distribution(see Figure 8). The entropy of the distribution is 1.32. Since the new entropy is lower than the threshold(described in the previous subsection, the intersection type is determined to be the cross type.

Figure 7: A new image obtained by attentive observation. The white circle indicates a newly obtained gap.

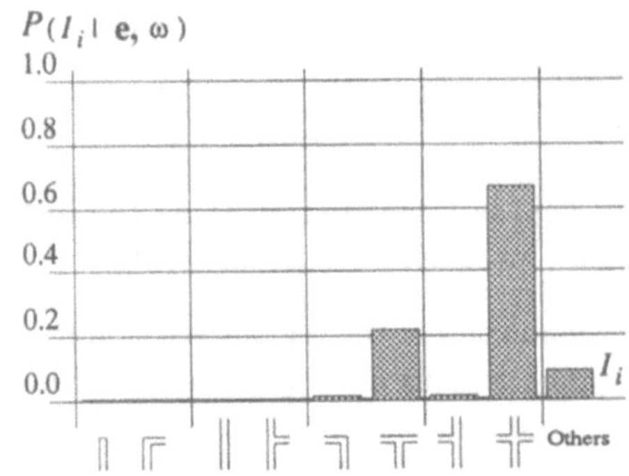

Figure 8: A new probability distribution calculated from new evidence as shown in Figure 7.

Other intersection scenes were also successfully recognized which include road lines, road marks, curve mirrors and so on.

5 Realtime Vision System

For smooth intereaction with the environment, a realtime vision system is useful which cover a variety of visual information processing. Let us estimate the required computation for such a system. Assuming that an image with resolution of 256 *times* 256 pixels is fed at 30 frames/sec, that the operation for each pixel is performed by multiplication and addition of the neiboring 10 pixels, and that 10 stages are required to obtaine the

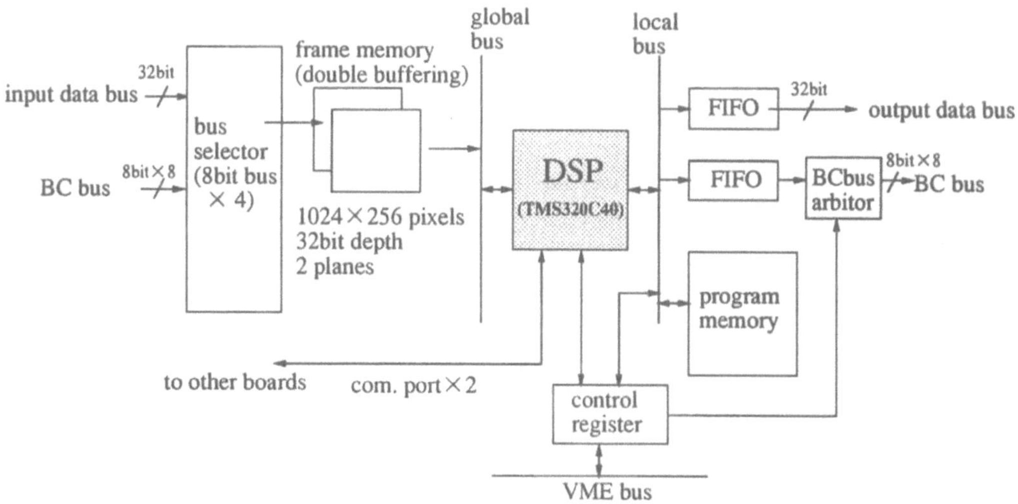

Figure 9: Block diagram of DSP board

result for the pixel, then it results in about 200 milion multiplications and additions per second. Although this may be realized by a supercomputer, it costs much.

In the pase many special devices for image processing were developed for accelirating image processing. However, they are applicable to only a limited kind of processing. If the remaining processing is performed by a conventional computer, theis process becomes a bottleneck. That is, only processes excutable on existing devices have been processed in realtime.

Because the remaining processing is usually not fixed and requires improvement throuth experiments, a special deviece is not suitable for it. Currently, an effective way is to use multiple Digital Signal Processors(DSPs). It is obvious that by programming, each DSP can perform any processing. One difficulty for using DSPs is to design a cuicuit with memories or signal controllers for a given task. One way to solve the problem is to design a board with DSPs, memories, and controllers which is suitable for various image processing.

In the beginning of this project, we designed the outline of the DSP board and asked Fujitsu company for detailed design and manufacturing. The block diagram of the board is shown in Figure 9. The board receives image data (monochrome or color) and store in fouble buffering memory. The result of processing is sent to the next board by output data bus. These buses suppor pepeline processing by multiple DSPs. It can also communicate with other boards by communication ports, and by broadcast bus (BC bus), which support communication between DSPs. Each board is connected to a host computer by VME bus.

5.1 Tracking Objects by Optical Flow

As an example of realtime processing, tracking of moving objects is desceibed here. There are many simple methods for tracking objects in a stationary background. However, if an moving object is tracked by a camera with a pan an tilt mechanism, the background also moves in the image. In this case, the observed motion in the image (optical flow) is a useful feature for identifying the target object. Among varous ways of computing the optical flow, the generalized gradient method[3] is convenient for many applications because of it requires less computation.

The basic method of optical flow computation and object tracking is described in the previous book [?]. For realtime tracking, DSP boards were used each of which contain two DSPs. While three boards were used in the old system, the new system requires five new boards (each of which contains one DSO) to compute the optical flow. More reliable flow can be obtained by applying four filters to an input image sequence [5]. Tracking of one moving object requires one more board and that of two objects requires two more boards.

Figure 5.1 shows an example of tracking two

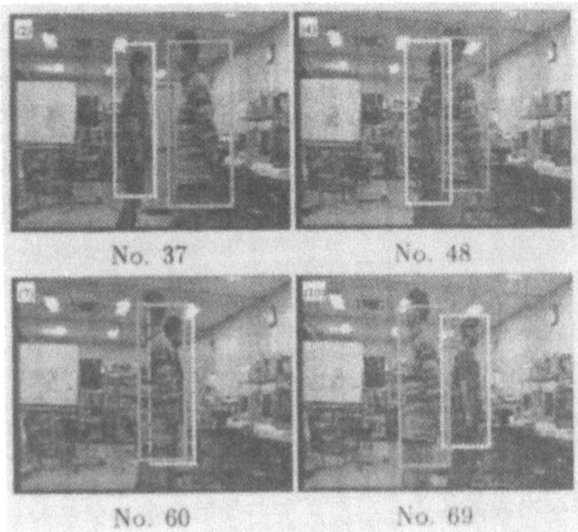

Figure 10: Result of tracking two persons

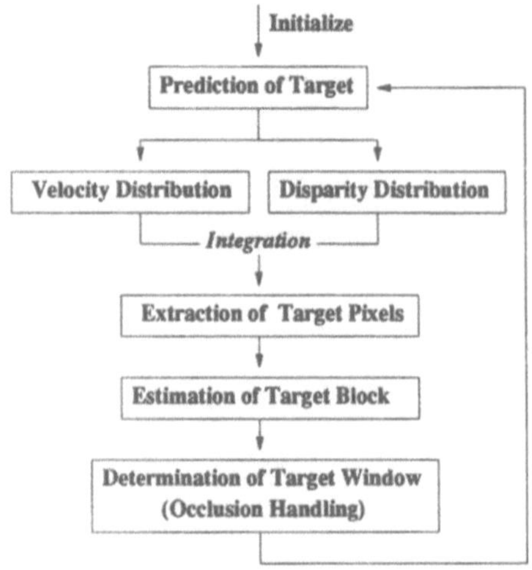

Figure 11: Outline of tracking

persons based on the optical flow obtained by four filters. The system contains 15 DSPs; 11 for optical flow computation by the least squared method, 2 for tracking, and 2 for making output images.

5.2 Tracking Based on Optical Flow and Depth

Tracking with velocity data alone has difficulty if the velocity of other objects are similar to the target. On the other hand, tracking with depth data alone [6][?] has alos difficulty if the target and other objects have similar depth data.

This problem is solve by using two sensor data: optical flow and disparity of a stereo pair of images (the depth is derived from the disparity) [7]. At each frame, the optical flow and the disparity are calculated. The target velocity and the disparity are estimated using the optical flow and the disparity inside the predicted target region. The region which has the similar velocity and disparity to the predicted ones is extracted as the target region. Occlusion of the target is detected from the abrupt disparity change in the target region and the motion is estimated based on the past record.

Initially a target object is detected as a moving region, or it is specified by a user. The region is approximated by a rectangle called *target window*.

As shown in Figure 11, a candidate region of the target is predicted from the target window of the previous frame (initially, the candidate region is the initial target window itself).

We approximate the distribution of the velocity and disparity in the target region with Gaussian distribution. The probability of the pixel belonging to the target is calculated based on the distributions of the target velocity, that of the target disparity and the pixel attributes in the predicted candidate region of the target. The pixels with high probability are extracted and are called *target pixels*. The rectangle circumscribing the target pixels is called a *temporal target window* (see Fig. 12).

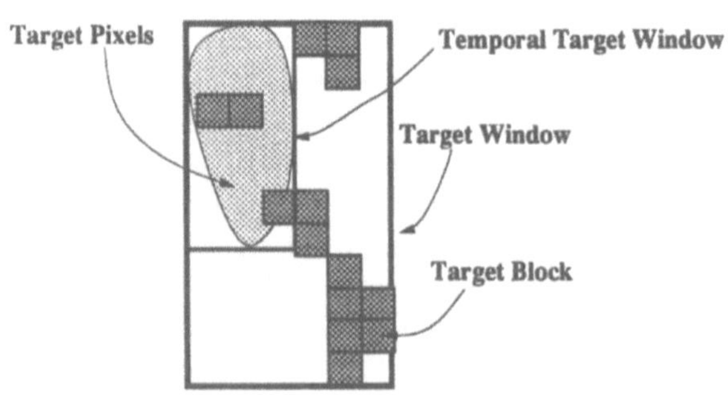

Figure 12: Terminolgies for tracking

Because the extracted optical flow and disparity are noisy, the final decision is made with a sequence of images. When a target region is found, the target window is divided into small fix-sized rectangles (called *block*). The block moves at the same velocity as that of the target window. If enough target pixels are extracted in a block in many frames, the block is determined to be a part of the target and called *target block*.

If the mean disparity in the target block becomes abruptly large, the region is determined to be occluded. Then, the position of the target region is estimated by shifting the previous position by the mean target velocity. This block is called an *occluded block*. The next *target window* is a rectangle circumscribing the target pixels, the target blocks, and the occluded blocks. In the next frame, the target region is predicted as the target window shifted by the mean target velocity.

If the target is completely occluded, the target window is estimated assuming the target moves with a constant velocity so that the target may be tracked again when it comes out of the occluding object.

Figure 13 shows the experimental result. In this experiment, we used 100 images (in about 6.7 seconds). The resolution of each image is 160 × 120 pixels. Persons walk from right to left. In image No. 1, the left person is the target person and the right one is nearer to the camera than the target person. While the right person overtakes the target person (image No. 25 to No. 40), tracking using only the optical flow might fail because the optical flow is similar.

After the target object is overtaken, he passes by the standing person. Because the distance of the standing person is similar to the target person, the tracking using only depth data might also fail.

The target person was successfully tracked in our method. Note that the feet of the target person are not extracted because they violated the assumption that the target velocity is uniform in the target region.

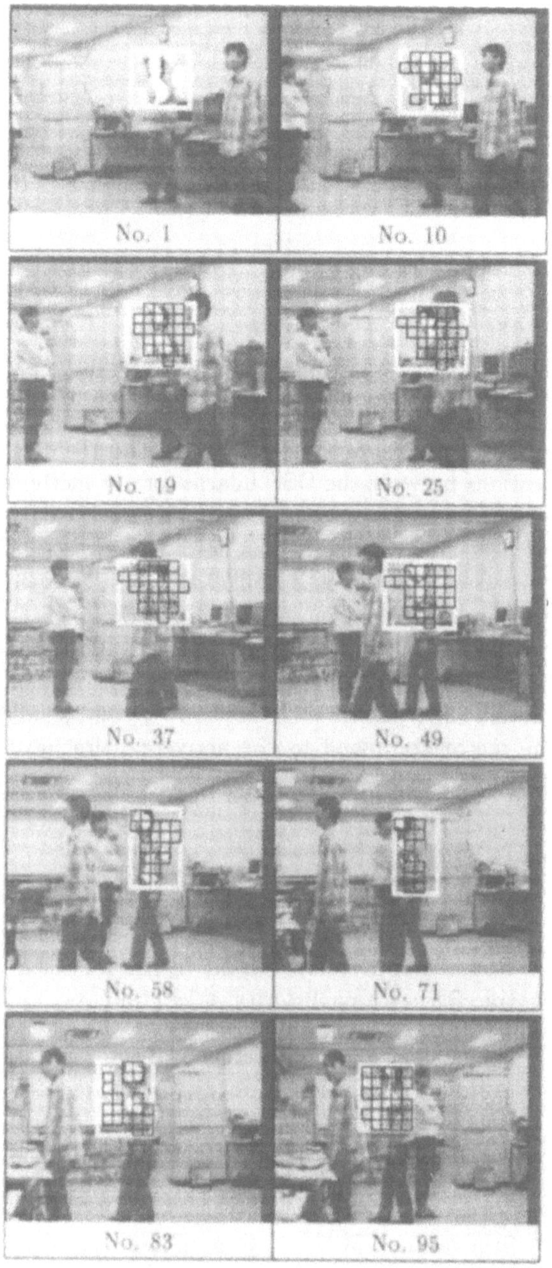

Figure 13: Experimental result (white region: target pixels, black rectangle: target blocks and occluded blocks, and white rectangle: target window)

5.3 Realtime tracking

The hardware configuration of our system is illustrated in Figure14. The image processor processes the sequence of stereo images, and put the result of tracking out to the monitor. The image processor also sends the motor commands to the active camera head controller.

The image processor has many DSPs (Digital Signal Processor). Each DSP is mounted on a DSP board which has memories and interface for data transfer. The connections between DSP boards can be changed according to the image processing algorithm.

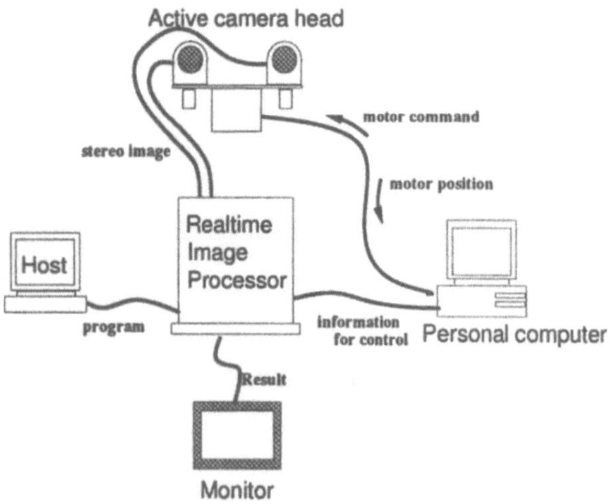

Figure 14: Realtime tracking system

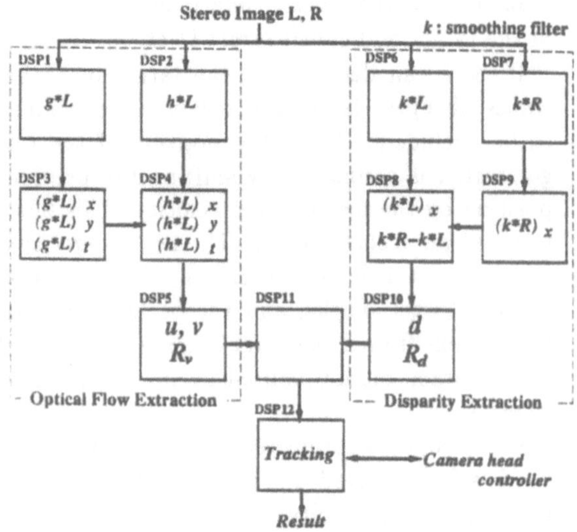

Figure 15: Functions of DSP-boards

The functions of the DSP boards and the connections between the DSP boards for our method are shown in Figure 15. The optical flow is calculated in DSP1, 2, 3, 4, 5 as described in section ??. The disparity is calculated in DSP6, 7, 8, 9, 10 as described in section ??. DSP11 gathers the optical flow and the disparity, and sends them to the next DSP board. DSP12 carries out the tracking described in section ??, and sends the motor command to the active camera head controller.

References

[1] Survey of Research Themes in "Intelligent Robot" supported by Grant-In-Aid for Scientific Research on Priority Area (1997.

[2] H. Takizawa, Y. Shirai, Y. Kuno and J. Miura: Recognition of Intersection Scene by Attentive Observation for a Mobile Robot, IEEE/RSJ Int. Conf. on Intelligent Robots and Systems, pp. 1648-1654 (1996).

[3] Srinivasan, M.V.: Generalized gradient schemes for the measurement of two-dimensional image motion, Biol. Cybern., Vol.63, pp.421-431 (1990)

[4] Shirai, Y., Mae, Y., and Yamamoto, S.: Object Tracking by Using Optical Flows and Edges, G.Giralt and G. Hirzinger Eds., The 7th Int. Symp. on Robotics Research, Springer, pp.440-447 (1996)

[5] Chen, H. -J., Shirai, Y., and Asada, M.:Obtaining optical flow with multi-orientation filters, Proc. CVPR'93, pp.736-737 (1993).

[6] Coombs, D. and Brown, C.: Real-time Smooth Pursuit Tracking for a Moving Binocular Robot, Proc. CVPR'92, pp.23-28 (1992).

[7] Okada, R., Shirai, Y., and Miura, J.: Object Tracking Based on Optical Flow and Disparity, Proc. of IEEE/SICE/RSJ Int. Conf. on Multisensor Fusion and Integration for Intelligent Systems, pp. 565-571, 1996.

Intelligence and Autonomy for Human-machine Cooperative System

Tomomasa SATO*
RCAST, University of Tokyo
4-6-1, Komaba, Meguraku, Tokyo, 153-0041, Japan 305
tomo@lssl.rcast.u-tokyo.ac.jp

1 Introduction

The project of "Emerging Mechanism of Machine Intelligence: A Tightly-Coupled Perception-Motion Behavior Approach" supported by Japanese Ministry of Education has been conducted for these tree years. The main goal of the this project is to make clear and realize the mechanism of the intelligence which emerges from the physical interaction between a machine and the world. "Human-machine Cooperative System" group member of this project , Group C hereafter, are doing researches to realize novel human-machine cooperation systems in this project.

Main goal of the Group C is to realize the intelligence which emerges from the physical interaction between machines and the humans. In another words, the intelligence of Group C is an intelligence for cooperation between machines and humans. The following three fields are dealt with by the Group C. 1) Machine intelligence to support human activities, 2) Machine intelligence to support human cooperative works, and 3) Machine intelligence to realize cooperative teleoperation. Figure 1 summarizes the research coverage of the Group C in terms of cooperative intelligence.

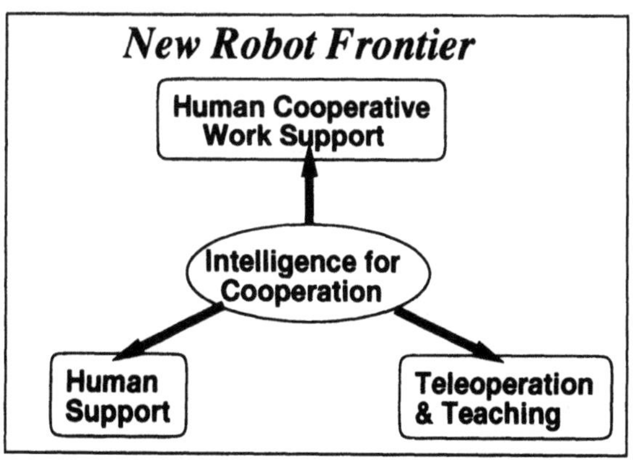

Figure 1: Intelligence for Human-Machine Cooperative System

This paper introduces the research target and results of the Group C of Human-machine Cooperative System research group of the project.

2 Machine Intelligence for Human Support System

2.1 Research on Human Support System

Machine intelligence to support human activities has been investigating by the group of Tomomasa SATO as follows. After the necessity of the robot environing system that supports human activities are clarified such as human care, the concept of a "Robotic Room" is proposed. The Robotic Room is a robot system in the configuration of a room to monitor and offer necessary supports to humans living in that room. The Robotic Sick Room is realized as a first version of Robotic Room. It understands such human behaviors as pointing the object, respiration and turn in sleep[1] (Behavior Understanding). It uses behavior as information expression means (Behavior Expression) to the human[2], i.e. to express the normal operation of the system. The Robotic Room also interacts together with the human(Behavior Interaction) to bring the pointed object to the sick people by long reach manipulator.

2.2 Research on Mechanical Intelligence for Emotional Human-Machine Communication

To open up a new field: "emotional communicable robots," the research conducted by Shigeki Sugano aims to realize the human-assistance of robots based on a natural and smooth communication by the implementation of an "emotion model" into a robot based on the reference of the human brain[3]. Until last year, they have developed a behavior based autonomous mobile robot named WAMOEBA-2 (Waseda Artificial Mind On Emotion BAse) which has a model of the 'internal secretion system' using an original algorithm; "the evaluation function for self-preservation"[4]. In the current year, the effects which the presented emotion model causes on the behaviors of WAMOEBA-2 will be quantitatively investigated through real experiments. Furthermore, the emotional impressions which observers receive from the behaviors of WAMOEBA-2 will be qualitatively investigated by subjective measurements. Finally, by comparing the quantitative data with the qualitative data, the methodology of "emotional communication between robots and human" will be made clear.

2.3 Vision-Based Human-Interface System Understanding Human Intention in Gestures

The research conducted by Yoshinori KUNO aims to realize a human-robot interface system that enables us to communicate with the robot by hand gestures[5]. Conventional gesture interface systems classify every human action into one of the learned categories, issuing the corresponding robot control command. However, we might move our hands without any intention of controlling the robot. Also, we might use the same hand gesture for different meanings in various situations. Thus, a user-friendly interface should pick out the gestures made intentionally and understand their intended meanings. They have been investigating the use of the following three items to realize such a human-robot interface system.

A) Eye contact information: We usually make eye contact with the person when we ask him/her something. We will apply this to human-robot communication.

B) Contextual information[6]: The user and the robot are cooperating to complete a given task with the system. In this context, we can obtain some constraints aiding gesture recognition.

C) 3D human hand model[7]: In principle, we expect the robot to move in the same way as we move our hand. However, the physical structure of the hand cannot allow certain motions. The knowledge about the hand helps to understand intended motions from apparent motions.

2.4 Merging Human/Robot Cooperative Behavior in the IMI Space Based on Human Safety and Intention

In the research project conducted by YAMADA Yoji et al., they propose to provide a robot with the IMI (Interactive Meaning Inferable) space in the Human/Robot (abbreviated as H/R) coexistence. The space is characterized as the intersection of two parametric subspaces representing human safety and intention. Their proposal is aimed to realize H/R cooperativeness which emerges from their interactive motions generated in the IMI space. They have proceeded with Their works in the following ways:

After designing a safe robot based on mechanical risk analyses in order not to cause any physical hazard to a human in contact [8], they propose a biopsychological method of using a gaze point detector for identifying where on the robot the human is watching and when he feels fear. We evaluated it in the multi-modal H/R relative motion parameters in order to ensure human mental safety and stress that unexpectedness, or a large and sudden deviations of the robot motion from the predetermined, is one of the major causative factors to invoke human fear. They show that it is possible to discriminate predeter mined motion trajectories by defining that the initial stage of human behaviors reflects human intention and by detecting it with the application of the HMM (Hidden Markov Model) technique. We further discuss a control method for tracing the desired operational force pattern under the predetermined motion trajectories. Finally, they demonstrate direct teaching operations with the proposed control method installed to learn the usefulness of our approach.

3 Machine Intelligence to support Human Cooperative Work

3.1 Research on Machine Intelligence for Human Cooperative Work

The research of the group of Yu-itsiro ANZAI aims to construct a robot network system that supports various kinds of cooperative work between humans. To realize the system, we need to clarify how communication between human and robots should be and the role that each should have in cooperative work. Therefore we investigate cooperation among autonomous agents (human and robots) in the robot networks, based on a multi-agent computational model. The research topics are as follows. (1)Agent network environment: developing the agent-programming environment and designing the adaptive(planning) mechanism for an agent on the robot network. (2)Human robot interface: designing the human robot interface system based on a multi-agent model with distributed terminals and sensors. Finally, using the agent network environment and the human robot interface, we construct the robot network system with multiple robots and multiple computers for cooperative works between humans and robots.

3.2 Machine Intelligence on cooperating motion of men and intelligent mobile robot

The group of Eiji NAKANO aims at studying and developing an "office messenger robots" working in an indoor environment. One of the missions of these robots is to deliver some documents to someone. When working, the robots' destinations are sometimes quite uncertain and the situations are always changing. Therefore, the main contemporary purpose of this study is to clarify how the robots must act under these uncertain circumstances with the aid of the cooperation and their dialogue with men. In this research, we are specially interested in studying some methodologies for the robots to find their addressees or alternative people/means to deliver the documents to them, while coping with uncertainty in position of the people and the dynamic changes in the environment.

In the last year, two main research works: (1) developing the base parts of the robot hardware and software[9,10] and, (2) studying the possibility of localizing a moving addressee by the messenger robot and using the attained information to decide its action, are performed. Based on these works, they plan to develop an information system for storing and managing the environment information, required by the messenger robot. Also, they will integrate the developed sensor system, including visual and ultrasonic range sensors, and use a fusion of the sensory data to control the robot. Moreover, the robot arm will be improved. In addition, the path planing algorithm of the mobile base will be modified and the overall system will become operational. The target mission of the robot is to move from one room to another autonomously and to deliver a letter to its addressee.

3.3 Research on Visual Attention Control for Human Behavior Understanding by Utilizing Speech Analysis

The group of Yuichi NAKAMURA has been investigating the utilization of spoken language to realize visual attention control for human behavior understanding. For the perception of human activities and behaviors, visual sensing is considered as the most effective way. Human behavior understanding, however, is extremely difficult because human movements are heavily depending on environments and situations. In this sense, we have to deal with the contexts of human behaviors such as the speech given simultaneously with the human movements. For this purpose, they impose two research topics as follows.

A) Cooperative use of spoken language for human behavior understanding. The intention of the human movement can be often clarified by the cooperative use of one or two simple words. To realize this mechanism on a computer, they first built multi-modal human behavior database which can be the basis for examining the varieties, frequency, or possibilities of human behaviors in a certain environment. Then, they are building a system to recognize the intention of human movements by using simple keywords or sentences detected by speech recognition system[11, 12].

B) A method to track human motion. When a human is moving quickly or the environment is quickly changing, a quick motion of a camera aiming at human hands or other portions is often hard to see. To get relevant information in such situations, a system with multiple cameras with constrained motion are developed.

3.4 Motion Planning of Mobile Robots based on Prediction of Human Actions

The group of Toshi TAKAMORI aims to realize home robots which cooperate with persons and can be friends of them. The robots must understand the person's intentions, take appropriate actions to support them, and generate actions with fertile emotions. To approach these goals, the followings are studied :

A person's actions are modeled as a stochastic discrete event system which reflects his habits. The home robot observes person by many sensors and predicts person's next action based on that model. This predicting system will be checked experimentally and must be improved. [13,14]

The Bunraku puppet's actions with plural functions and plural emotions are generated by experts of puppet manipulation and these action time series are observed. The functional factors, the emotional factors and the stochastic fluctuation factors are extracted based on a rigorous statistical method. The laws which generate actions with fertile emotions are considered by generating puppet actions on computer graphic simulations based on the extracted functional, emotional and stochastic factors.

4 Machine Intelligence for Tele-Operation

4.1 Research on Machine Intelligence for Human Skill Acquisition

The purpose of Hideki HASHIMOTO's research group is to study the intelligence of humans, by realizing advanced physical skill, which is a fundamental ability of humans, on the machine. By understanding human-skill, we can construct a robotic system which is based on the advanced skills. Primitive intelligence afford the direction of research to get higher intelligence. For example, in order to make machine which can be used easily, new cooperation between human and machine is required. In this research, they develop Intelligent Cooperative Manipulating System (ICMS) based on human knowledge. ICMS is composed of two parts: 1) Dynamic Force Simulator (DFS)[15] , 2) Robot Hand System. Each other are connected through B-ISDN. They are now making SG (Sensor Globe:20 DOF[16]) and SA (Sensor Arm:7 DOF) as a human-machine interface between real world and computer world. We will measure the data of human movement from SA and SG to achieve the algorithm of primitive intelligence, which is a reflective behavior as a low-level motion skill. In near future, they will do tele-teaching and tele-programing experiments and demonstrate the system.

4.2 Operational Environment Transmission for Tele-Micro-Surgery

The research of Mamoru MITSUISHI concerns the realization of a tele-micro-surgery system with which a human operator can easily execute a surgical operation by reducing surgical motions, conventionally executed by a medical doctor's hands in the normal human world scale, into equivalent motions under the microscope. Among many obstacles to realizing such a system, this research aims at the following items: (a) realization of visual information acquisition and display systems which are controlled in accordance with the inferred intention of the operator[17], and (b) augmented presentation of sensed information in the micro world by an information transformation method which enhances the operability of the system by providing the surgeon with reality sensation, where he or she can operate as if using his or her own hands[18]. Using these functions a tele-micro-surgical experiment was successfully executed. The following items are currently focused on in this research: (1) a trial of infection-free slave manipulators, (2) analysis of the relation between integrated information presentation and the operability of the system, and (3) evaluation of the total system.

4.3 Observation on emergence of intentions in everyday action

The group of Masato SASAKI had two observation. First is a longitudinal observation of a infant having a meals. An infant's having meals were videotaped once a month. His actions were described with various objects and any events in his surrounding environment; such as conversations with parents and meals spatial arrangements. Second is a observation of the adults preparing a cup of tea and some cookies. Their actions were videotaped and described with some small movements in their actions. For both observations , organization and development of every day actions and emergence of intentions are made clear.

4.4 Mechanism for Generating Sensing/Manipulating Behavior based on Visual-Tactual Aspect Graph

For flexibly manipulating an object by robot hands or fingers, not only recognition of shape of the handled object but also geometrical relationship between the handled object and its environment are needed. In this research, a system for recognition of both the object shape and the geometrical relationship between the object and environment is developed by integrating sensing behavior and manipulating behavior. First, the object shape is defined by a tactual aspect graph[19] derived from tactual images of 2D tactile sensors. The way to grasp the object is determined based on the tactual aspect graph. Next, the behavior for obtaining the relationship between the grasped object and its environment is determined based on a visual-tactual aspect graph which is obtained by both a CCD camera attached to the robot hand and force data derived from tactual images[20]. Automatic assembly of objects into uncertain environment by using a parallel jaw gripper with tactile sensors will be done based on the visual-tactual aspect graph.

5 Conclusions

This paper introduces the research scope and results of research group (Group C) of "Human Cooperative System" in the project of "Research on Emerging Mechanism of Machine Intelligence: A Tightly-Coupled Perception-Motion Behavior

Approach" supported by Japanese Ministry of Education.

The intelligence for cooperation between human and machine in the following research areas are revealed in the project. 1) Machine intelligence to support human activities: A function to grasp most of the objects in a room by tracking and updating them as well as a function to express emotional information based on gesture of the robot are realized. 2) Machine intelligence to support in cooperating among humans: Multiple robot environment based on agent programming is realized to supports cooperative works between humans. Also mechanism, sensor, controller and man-machine interface for mobile robot with TV camera are constructed. 3)Machine intelligence to realize cooperative maneuvering: After modeling of human skill, a cooperative maneuvering system are designed.

While the conventional robotics is dealing with the physical object, this Group C deals with the human as a target. This is the most important difference of the research direction from the other groups of this project. Figure 2 summarizes the differences of research topics between conventional robotics and that of our Group C.

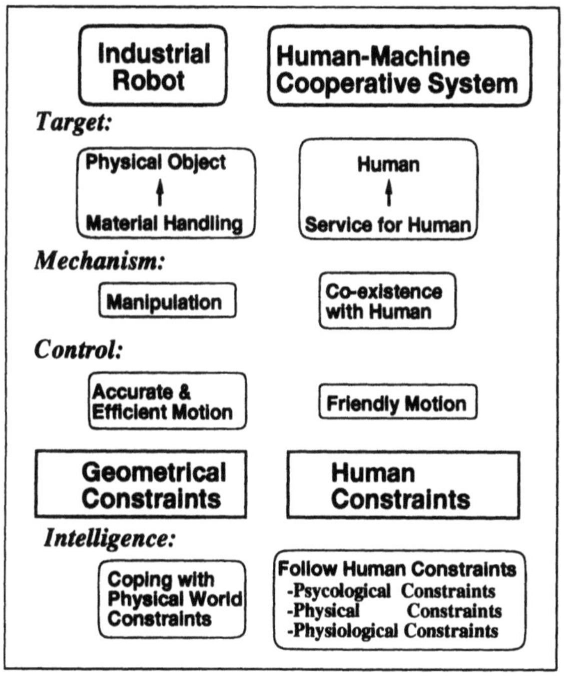

Figure 2: Essential difference of Research Direction

Since the human has physiological, physical, psychological and social aspects, the robotic system in our group should be realized as the integrated human system to satisfy the stated human aspects as a whole. Especially, the physiological, physical, psychological and social aspects opens new aspects of the robotics in the future. Figure 3 illustrates the application areas of such research results. There should be a lot of research efforts in this field.

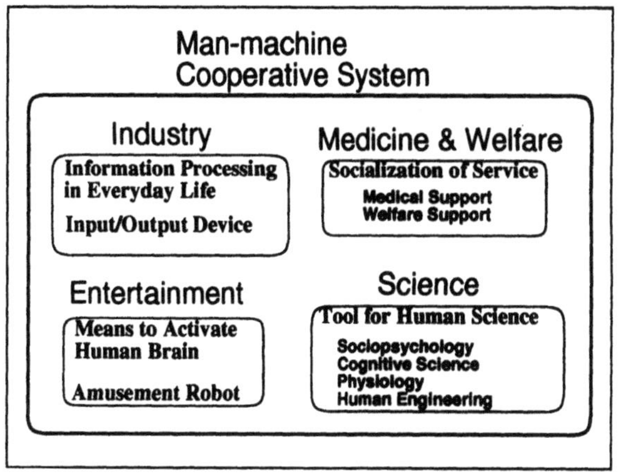

Figure 3: Application Areas of Human-Machine cooperative System

Acknowledgments

This research is supported by the Japanese Ministry of Education.

References

[1] Y. Nishida, M. Takeda, T. Mori, H. Mizoguchi and T.Sato, "Monitoring Patient Respiration and Posture using Human Symbiosis System," Proc. of Int. Conf. on Intelligent Robot, pp.632-639(1997).

[2] T. Nakata, T. Sato, H. Mizoguchi and T.Mori, "Synthesis of Robot-to-Human Expressive Behavior for Human-Robot Symbiosis, " Proc. of the IEEE/RSJ Int. Conf. on Intelligent Robots and System, pp.1608-1613(1996).

[3]S. SUGANO and T. OGATA, "Emergence of Mind in Robots for Human Interface - Research Methodology and Robot Model, " Proc. of IEEE Int. Conf. on Robotics and Automation, pp.1191-1198(1996).

[4]T. OGATA and S. SUGANO, "Mechanical System for Autonomic Nervous System in Robots, " Proc. of IEEE/ASME Int. Conf. on Advanced Intelligent Mechatronics (AIM'97).

[5] K. Hayashi, Y. Kuno, and Y. Shirai , "Pointing gesture recognition system permitting user's freedom of movement", Proc. Workshop on Perceptual User Interfaces, pp.16-19(1997).

[6] K.H. Jo, Y. Kuno, and Y. Shirai, "Context-based recognition of manipulative hand gestures for human computer interaction", Proc. Third Asian Conference on Computer Vision, (1998).

[7]N. Shimada, Y. Shirai, Y. Kuno, and J. Miura, "3-D pose estimation and model refinement of an articulated object from a monocular image sequence", Proc. Third Asian Conf. on Computer Vision, (1998).

[8]Y. Yamada, Y. Hirasawa, S. Y. Huang, Y. Umetani, K. Suita, "Human-Robot Contact in the

Safeguading Space", IEEE/ASME Trans. on Mechatronics, Vol.2, No. (1997).
[9]Y.MORI, E.NAKANO, T.TAKAHASHI and K.TAKAYAMA, "A study on the Mechanism and Control of Omni-Directional Vehicle", IEEE/RSJ Proc. of Int. Conf. on Intelligent Robots and Systems, pp.52-59, Osaka(1996.)
[10]K. HANADA, T. TAKAHASHI and E. NAKANO, "Software architecture for constructing a mobile robot with highly independent modules", Int. Conf. on Control, Automation , Robotics and Vision (ICARCV'96), pp.858-62, Philipine(1996).
[11] Y. Kimura, Y.Yu, Y. Nakamura, Y.Ohta, "Multimodal Presentation Database for Human Behavior Understanding (in Japanese)", Intelligent Information Media Symposium, December (1997).
[12] Y. Kimura, Y.Yu, Y. Nakamura, Y. Ohta, "Presentation Summarization by the Integrated use of Motion and Language Analysis (in Japanese)", IEICE SIG-PRMU97 (1998).
[13] Y. Manabe, M. Hattori, S. Tadokoro and T. Takamori, "A model of human actions by a Petri net and prediction of human acts -For generation of home robots movements based on prediction of human actions - (in Japanese)", Tran. of the Japan society of mechanical engineers, series C, vol.63, no.609, pp.1693-1700(1997).
[14] Y. Manabe, M. Hattori, S. Tadokoro and T. Takamori, "Generation of home robots movement based on prediction of human actions - A model of human actions by a Petri net and prediction of human acts - Proc. of Symposium on robotics and cybernetics, CESA'96 IMACS multiconference, pp.210-215(1996).
[15] Y. Kunii, H. Hashimoto, "Dynamic Force Simulator for Multi-Finger Force Display, " Trans. on IEEE Trans. on Industrial Electronics, Vol.43, No.1, pp.74-80(1996).
[16] Y. Kunii, Y. Nishino, T. Kitada, H. Hashimoto: "Development of 20DOF Glove Type Haptic Interface Device -Sensor Glove II-, " Proc. of Advanced Intelligent Mechatronics AIM'97(1997).
[17] M. Mitsuishi, H. Watanabe, H. Nakanishi, H. Kubota and Y. Iizuka, "Dexterity Enhancement for a Tele-micro-surgery System with Multiple Macro-micro Co-located Operation Point Manipulators and Understanding of the Operator's Intention, " 3rd Int. Symp. on Medical Robotics and Computer-Assisted Surgery (MRCAS'97), pp.821-830, Grenoble(1997).
[18] M. Mitsuishi, K. Kobayashi, T. Watanabe, H. Nakanishi, H. Watanabe and B. Kramer, "Development of an Inter-World Tele-Micro-Surgery System with Operational Environment Information Transmission Capability, " Proc. IEEE Int. Conf. on Robotics and Automation, pp.3081-3088, Nagoya(1995).
[19]G. Kinoshita, E. Mutoh and K. Tanie, "Haptic Aspect Graph Representation of 3-D Object Shapes, " Proc. of the 1992 IEEE Int. Conf. on Robotics and Automation, 1648-1653(1992)
[20]K. Umeda, G. Kinoshita and H. Osumi, "Tactual Servoing on an Object's Surface with the Visual Information, " Proc. 1997 IEEE/ASME Int. Conf. on Advanced Intelligent Mechatronics, (1997)

Biologically Inspired Approach to Autonomous Systems

Shin'ichi YUTA
University of Tsukuba,
Tsukuba 305-8573, Japan.
yuta@roboken.esys.tsukuba.ac.jp

1 Introduction

The historical intelligent robots tended to try to model humans' thought and imitate humans' intelligent action or behavior. However, it has been known that, traditional logical treatment or classical AI techniques do not have enough power to realize such a behavior on the robots. And also, it has been found that the most important point of the robots is the ability to cope with various situations in the usual real environment, which seems easily realized in human or animals. In this sense, the basic necessary function of the robot which we have to study is not a high level judgement of the human, but the ability to survive in the real environment. The necessary intelligence of the robot is also located at here.

So, the model to be approached in robotics may not be a human, but may be a mammal, bird, amphibian or other more low level animals. In this case, the definition of the word "intelligence" is not clear any more and the word "autonomy" may be more suitable to use.

Group-D of the priority research program on intelligent robotics is focusing the research theme as "biologically inspired approach to autonomous system". We are approaching to the robots' intelligence and autonomy of non-human or animals, and its ability to survive in the given various environment or situation.

In some case of animals, the group or the society if formed by many individuals. In such a group, even when each one seems insufficient in ability and acting only simple behavioral patterns, the resultant group looks to perform more complex ability. The mechanism of realizing such a society is also interesting theme on the intelligence or the autonomy of the animals.

2 Autonomy of the robotic systems

Each individual of animal keeps independence in itself, both physically and informatically. And also, they can survive in the given environment. We use the word "autonomous system" for one with such type of independence. We are studying both types of autonomy in group D. The physically independent system is sometimes called self-contained.

The informatically autonomous robotic system will be considered to have the abilities as :

1. The real-time control ability based on both environmental informations taken by itself and plans decided by itself, ie. a robot should controls itself by an embedded algorithm without remote control.

2. The learning ability to obtain the above-mentioned control algorithm by itself, ie. it should be able to learn such algorithms from the result of its own motion.

The first type ability is evaluated by width and variety of situations, which the robot can cope with and perform a certain task in. Generally, the second type ability is considered as higher-level comparing with the first one. But only a simple model of robot functions can be considered for the second type ability in current technology. So, we have to limit the environmental condition strictly when treating the second problem.

The physical independence, which is called self-containedness, is also important issue of autonomy. Self-contained robotic system is one which brings all necessary functions such as a computer for information processing and a power source in the inside of its own body. Furthermore, when we consider the self-contained system, it has to act under the physical constraint such as weight and gravity force. This is basic condition of the robotic system, but it is not easy to fulfill when we consider about the informatic independence.

The biological being is the system which has all these autonomy. And the purpose of group D is, to find the methodology to fulfill these conditions on the real robotic system and to realize high-level autonomous and self-contained artificial machine which can work in the real physical world.

3 Research subjects and members

This group has been organized by 10 or 11 members in each year from 1995 to 1997, which includes 5 programmed members with designated subjects and 5 or 6 subscribed ones. The research on each subject is conducted independently by a subject leader and it is done at the laboratory in his University. We are having several meetings a year to discuss the direction and the results of each research subject.

3.1 Programed research

We have five programmed research in group D. Each subject has three years (two years for a part), and orients to build-up a real concrete robotic system to demonstrate the research results. By developing a concrete system which works in a real environment, we can know the real critical problems and develop the necessary elemental technologies, – this is our standing point.

The subjects and project members are listed below.

3.1.1 *Robust navigation of autonomous robots in real environment*

Researcher: Shin'ichi Yuta (Professor, University of Tsukuba)

This is the author's subject. He is considering the wheel-type mobile robot and investigating the autonomous navigation. He defined the task and environment as : the robot should reach the given destination robustly in indoor or outdoor environment such like his University building or the pedestrian street. In this subjects, the design methodology of sensor system, control system, action algorithm and environment model for such a robot has been studied. The key issues of the research is reliability and robustness of the navigation in real environment. For this purpose, the reliable position estimation technique is developed and the robot is controlled based on the estimated position. He has demonstrated the integrated autonomous mobile robot's navigation in the University building. In which, the wheeled mobile robot with manipulator opens/closes the door of the laboratory and passes through it to the another room. Fig.1 is a photograph of the robot which has used in this experiment. This is an example of realization of the physical autonomy and the first type of informational autonomy in the real robot system.

Figure 1: Autonomous mobile robot 'Yamabico' is now opening the door in the real environment.

3.1.2 *Study of Autonomous Robot with Biological Behavior*

Researcher: Shigeo Hirose (Tokyo Institute of Technology)

He has been developing the four legged walking machine to work at the rough terrain especially on the steep slope. For this purpose, the new mechanism, sensors and control algorithm are investigated as well as the wire supporting system for very steep slope. This research includes the consideration and development of stable gait planning and control in the slope. As a result, the robots' ability to cope with more hazardous environment has been realized. The robot was demonstrated at the real construction site as a future working machine for the construction. This is an excellent example of realizing physical autonomy with the first type of informational autonomy.

3.1.3 *Study on insect-model-based autonomous system*

Researcher: Hirofumi Miura (Professor, University of Tokyo)

He is trying to implement a small size robot inspired by the insect, using micro mechanism technology. Generally, the insects have a very simple behavior algorithm based on reflection, but they seem very vivid and intelligent. So that the artificial insects with such a simple algorithm are considered. As an example, he has tried to build pheromone-tracing robot with a size of real silkworm moth. He is using a real antenna of moth as the pheromone sensor, and installed very simple but well organized algorithm for motion control. As the result, this robot demonstrated a similar motion with real silkworm moth when the pheromone stream is given.

3.1.4 *Autonomous system with ability of learning self-evolution and cooperation (Learning of brachiation motion by monkey robot)*

Researcher: Toshio Fukuda (Professor, Nagoya University)

He is studying the learning methodology for the dynamic mechanical system. For this purpose, the monkey styled robot with many degrees of freedom are developed, and learning algorithm of dynamical motion is studied to realize a brachiation. This research attempt the second type of informational autonomy under the real physical conditions. The learning algorithm named target dynamics method is proposed and applied several brachation motion, such as ladder problem, swing-up problem and rope problem. To obtain the algorithm of such a dynamic motion, the numerical simulation is done before making experiments by real brachation robots. But, learning with real experimental system was also necessary to cope with real problems.

3.1.5 *Implicit communication based cooperative transportation by two 4-legged robots*

Researcher: Tamio Arai (Professor, University of Tokyo)

He is trying to realize a cooperative work of the four legged robots. He defined the robot task as the cooperative transportation of a long object on the flat or sloped floor by two four-legged robots. He is using two TITAN VIII robots which is developed by Prof.Hirose for this priority research program, as platforms, and realizing cooperative motion without using the explicit communication. The relationship between object and the robot is important information for the cooperation, and it realizes the implicit communication between two robots for the cooperation. He showed the necessity of synchronization of the gait of both robots and demonstrated the algorithm to perform it.

3.2 Subscribed research

Besides the programmed subjects, we have six subscribed research subjects in 1997, which are selected among the applications. The approaching method expands widely in this category. The research subjects and chief researcher is listed below.

3.2.1 *Development of autonomous intelligent behaviors based on ecological computations*

Researcher: Keiji Suzuki (Assoc.Professor, Hokkaido University)

The purpose of this research is to find the mechanism to obtain a group behavior algorithm, as a problem of the multi-agent system. He defined a agent model for cooperative monkey-banana problem, and is considering the effective strategy of cooperation using pheromon type communication. This research is performed by simulation on the model.

3.2.2 *Studies on the morphology and functional development of bipedal and precursor activities in ape-like robots*

Researcher: Masayuki Inaba (Assoc.Professor, University of Tokyo)

He is researching the relationship between mechanism and development of behavior in the biped locomotion of ape-like robots. Based on the remote-brain approach, he has developed a framework of a research environment in which both the simulator and real robots can execute a same motion plan. He is observing the development of motion plan on the simulated model with respect to the given behavior index, and testing the plan on the real robot to verify them.

3.2.3 *A study on sensory-behavioral information processing models to develop divergent co-operation between human and robots*

Researcher: Mitsuo Wada (Professor, University of Hokkaido)

He is researching on the base of the autonomy of the robot to cooperate with human. For this purpose, a time series signal from human gotten by several sensors is analyzed to find human's dynamics. And he is considering to give the ability to understand the human's condition to the robot using this dynamics.

3.2.4 *Sensory integration of the human posture control system*

Researcher: Akimitsu Ishida (Professor, Tokyo Medical and Dental University)

He is trying to find a humans' sensor-fusion mechanism in the posture control, by observing human's test motion.

3.2.5 *Study on flight mechanics of insects*

Researcher: Keiji Kawachi (Professor, University of Tokyo)

His hypothesis is that the physical mechanism especially the shape of flying insects has been optimized just for flying, since flight is physically hard task for such animals. For an example of the result of hypothesis, the insects can fly just after its birth. In this point of view, he is observing the shape and motion of the wing and body of insects, as well as analyzing the control framework of their flight.

3.2.6 *Studies on autonomous and cooperative multi-agent type robot systems with life-like behaviors*

Researcher: H.Zha (Assoc.Professor, University of Kyusyu)

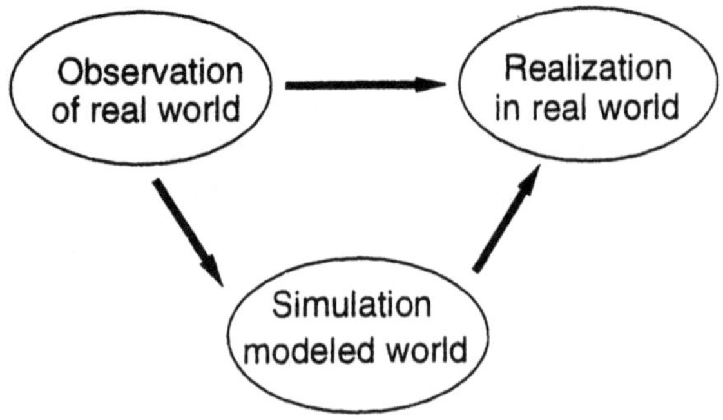

Figure 2: Approaching methodology

He is researching the control strategy of multiple robots using a multi-agent scheme. He is analyzing and implementing the behavior of pushing-up the box by several independently controlled manipulators. In his experiments, each manipulator detects its torque and force from its sensor, and decides a optimum pushing point independently to perform such a cooperative task.

4 Map of approaches

When we consider a biologically-inspired machine, the observation and analysis of animals will be a necessary step of research. On the other hand, to realize a real machine, the procedure of design is also very important. It should include the design of mechanism and information processing. The modeling and simulation is sometimes a very powerful bridge between these analysis and design (Fig.2). Each research of group D has independent standing point among analysis, simulation and design.

Furthermore, the concept of autonomous system has very wide range as well as the biologically inspired one. So, the research interest also ranges widely and may be categorized and characterized by axis as follows :

1. observation and understanding animals, vs. design and realization of the machine. (biology oriented, vs. engineering oriented)

2. individual autonomous system, vs. group behavior of many individuals.

3. autonomy in physical sense, vs. informational sense.

4. level of animal which is considered, observed or imitated.

I tried to categorize the research subjects of group D in these axis in Fig.3

5 What we are learning from animals

5.1 Task or function design for robot

Each animal has many functions and abilities. But, it is impossible and meaningless to consider a man-made machine with all functions of any animal. Instead, when we design the robot, we have to define a concrete function or task performed by it. The concrete function or task of the robot may be found or inspired from ones of animals. It will be a good and reasonable way to define a research subject or development goal. Autonomous navigation in real environment, walk and work in the steep slope, or learning of brachiation are the examples of defined functions or tasks inspired from animals, in group D.

5.2 Mechanism and control

Human or the animals have dextrous mechanism and sophisticated control method for it. All of them are very good goals of the artificial system. Mechanism and structure of animals are worth itself to imitate, since the mechanism and its function are sometimes the results of the evolution and optimization after long history .

The parameters in animals should also be the result of optimization for some criteria. Referring to the animals, we may find the suitable size or speed of the artificial machine respect to the required function to work.

5.3 Sensing and perception

To realize a robust reliable behavior of the robot or artificial machine in the real environment, the recognition and understanding of the environment is most important. However, current existing sensor technology are not good enough for robot's robust behavior. On the other hand, such a frog or a insect seems to have very simple sensors and limited information processing ability, but it can survive in various environment. So we may have to study animal's ability of environment recognition, to find the solution of reliable sensing methodology of the machine.

5.4 Information processing – Architecture and algorithm

In human or monkeys can think logically but lower level animals may be not able to do so. Instead, the behavior of lower level animals are considered as the combination of simple reflective action. Even in the high level animals, a kind of purposive behavior is constructed based of such reflections as basic functions. Can the gathering of such reflective motions perform more complex and meaningful behavior? How to organize these basic functions? These are the interesting subject and worth to apply to the robotics systems.

Further, human and animals can learn much from their experience and can adapt their motion algorithm. We have to study this adaptability to realize a real robust and reliable autonomous system.

5.5 Realization of group behavior

The combination of simple behaviors can realize more complex and sophisticated ones in the group of individual or in the animal society. This is very interesting phenomena. However, in the real world, the multiple autonomous robot systems which work in the same environment, is not so easy to realize. Sometimes we find the necessity of very high level and difficult performance in each individual when we implement a robot society which is requested to achieve some meaningful functions. To avoid such confusion of multiple robot problem, we have to learn more from animal's robustness and flexibility.

6 Conclusion

We can not cover all subjects to research in the field of biologically inspired autonomous system. Rather, the research conducted in group D are only a very small range of this interesting field. It is not possible to find a simple and general principle to design such good system. And we need to approach from many ways to such a problem. Fortunately the back ground and interests of the researchers in group D are ranging widely, and this is very good opportunity to collaborate on this particular topic. I believe our group can

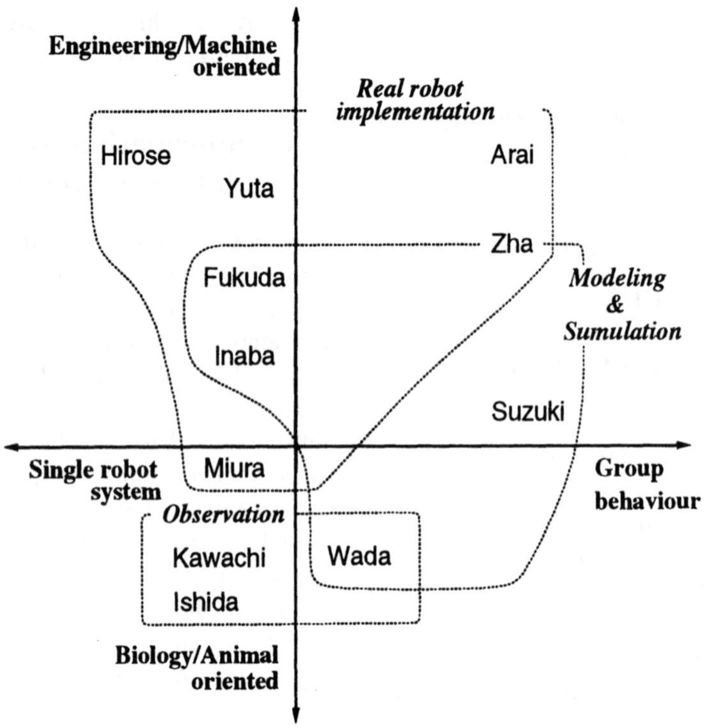

Figure 3: A map of research subjects.

make a big advancement and find a good direction of future research in "biologically-inspired autonomous system".

Acknowledgement

Not only the members but also many co-workers of each subject are working together for this priority research program. I wish to acknowledge all these researches for their contribution. Also, I thank Prof.Inoue, the general leader of this program, for giving me a chance to work in this program. At last, I acknowledge Ms.Teruko Yata for her assistance to prepare this material.

References

[1] S. Yuta : "Biologically inspired autonomous system" *Journal of Society of Instrumentation and Control Engineers*, vol.35 no.4 pp.268-273. 1997. (in Japanese)

FNR: Toward a Platform Based Humanoid Project

Kazuo Tanie
Biorobotics Division, Robotics Department
Mechanical Engineering Laboratory, AIST-MITI
1-2 Namiki, Tsukuba, Ibaraki, 305 Japan
email: tanie@mel.go.jp

1 Introduction

The Ministry of International Trade and Industry, Japan(MITI) has several programs for national projects. In a program, MITI so far promoted one national project in robotics area, which was called "Advanced Robotics". The objectives were to research on the technologies for the robots which can work at unstructured environments. Major robotic application areas considered in the project were relating to nuclear power plant maintenance, undersea oil plant maintenance and fire fighting and rescue operations. The project was carried out from 1982 to 1990. The budget spent during the period was about 180 billion yen. 17 companies, 2 nonprofit organizations, and 2 MITI's national laboratories (Mechanical Engineering Laboratory and Electrotechnical Laboratory) worked in the project[1]. After the end of the project, MITI did not have robotics projects promoted till now. However, three years ago, MITI started the first action for preparing for a new robotics project, considering what kinds of technology our society needs and also new advancement of the technologies. Since last year, the committee on Human Friendly Network Robotics (FNR), including professionals in robotics from universities, national laboratories and companies has been established for the project preparation, which is now promoted as an activity under a MITI's research program which call Industrial Science and Technology Frontier Development Program. With the intermediate report made by the committee in mind, MITI has a plan to launch a project relating to the development of humanoid robot and its applications, next year. The paper describes the outline of this project plan.

2 MITI's Project

A major role of the MITI is to encourage industrial activities. When a research project will be planned, some effects of creating new industrial activities, encouraging current industrial activities, and so on are always expected from the results or through doing research. Of course, the topics which will be pursued must be new. However, the project which is important only from the scientific point of view will not be suitable for MITI's project. Also, the project has to have some kinds of social impact which can improve the quality of our life. It means the project must be promoted to provide us with some solutions to the problems our society has or some means to satisfy social needs. To propose a new project, with the above matters in mind, the research plan must be prepared.

3 Social Contributions

Satisfying the requirements in the MITI's project, the following two social contributions are considered.

1) *Contribution to Improving the Quality of Life in Future Aged Society*
At the beginning of the 21st century, elderly person's population is expected to increase in many countries including Japan. The statistics tells us that the number of people who are older than 65 years will be more than 25 percent of total population in 2010. Such a society is supposed to cause several problems in our daily life. One of the most serious problems will be relating to the difficulty of supplying enough labor power in several areas. The other is to increase people who need some kinds of help in their daily life because of the increase of the aged people

population who needs health care. One of ways to solve these problems will be to develop some kinds of automated machine like robots which can work in lieu of human and also provide some services to human. Since the 1960's, many robots have been developed and introduced especially in the manufacturing industry. They were the machines which work to release human from dirty, tedious, and dangerous jobs. In the 21st century, this kinds of robot will continue to be important, because the lack of labor power in industry will be more serious. On the other hand, even in non-industrial areas in which robots have not been so far used, like home, office and so on, the robots will be expected to be introduced. The patient care robots, home or office use robots which can deal with various tasks in home or office will be such typical examples.

2) *Contribution to Stimulating Economy through Encouraging New Robot Business*
About 35 years have passed since the robot application started. Many robotics researches have been carried out for 35 years. However, limited kinds of robot are practically used in limited areas, though the number of robots practically used increases year by year and they contribute to improving the efficiency of productivity especially in manufacturing industry. Since the beginning of the history of industrial robots, major practical robots used in the factory are position control based robot, like teaching playback robot. Because of only the robots with limited functions and applications practically available, the robot market is getting saturated. When we consider to expand the new robot market and encourage the robot future industries, how to increase robot variations and also how to expand new robot application areas is indispensable problems to be solved. Especially, if new robots will be developed, which are applicable to the non-industrial areas where few robots have been so far used, like home, office, and so on, the robot market will tremendously increase. This also contributes to encouraging economy. Currently, the economical condition is not well in several countries. One of the reasons is relating to that no new products which stimulate customers' interests exist. In the past 40 years, there were several important products, like cars, TV sets, and so on, which could stimulate people's purchasing desires and in turn encourage our economy. However, those products have been already well distributed and reached to the limit of market. New products will be expected. Non-industrial use robots usable in wider application areas is expected to be a candidate of such a product.

4 Key Considerations for Efficient Project Promotion

4.1 Human Robot Coexistence Technology

To obtain the robots which satisfy the above requirements within the limited project period, what kinds of technology should be pursued and how the project should be organized have been deeply considered. Regarding the key technologies in the project to be pursued, two key issues are identified. The first is "human robot coexistence technology".

When we pay attention to the robots which works for human in our daily life, we have to consider the robots which work in environments which are very different from the industrial robot's ones. The first, the robot has to work often with human in same environment. This means that the environment includes human. This requests the robots to have the safe structure for human. The industrial robots in factories are considered to be dangerous. Therefore, they have been used separately from human workers, because of keeping human safety. Generally, it is strongly inhibited for human to enter the working space when the robot is moving. Several precautions to maintain human security have been made before the industrial robot will be introduced. For non-industrial robots considered in the project, we need new technologies for human robot coexistence to assure the safety of human in the environment. The other important feature which the non-industrial robots have is that the users will be non-professional. The industrial robot was the machine which can be used only by well-educated persons who have learned knowledge about the robot itself. In non-industrial robots which will be used in home, office and so on, we can not expect this situation. We have to consider the robots which everybody can use without complex instructions, like home electric products commercially available. This produces new issues for the development of non-industrial robot, which will be relating to the human friendly user interface. The human robot coexistence technologies which cover these issues are considered as the first key topic in the project.

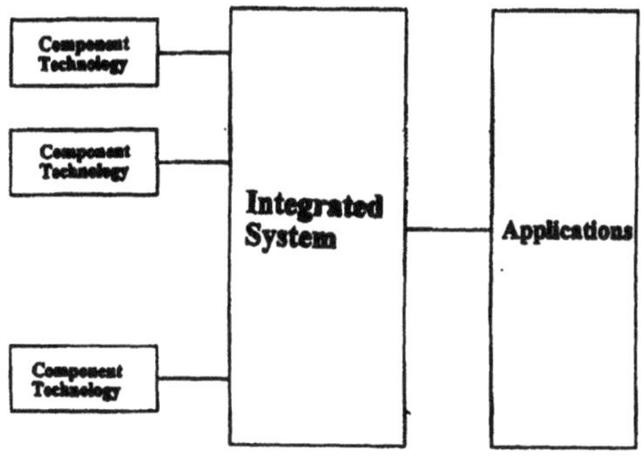

Fig. 1: Fundamental Research Based Project.

4.2 Communication Network Based Robotics

From the observation of history of robotics, it can be recognized that the robot technology is influenced from relating peripheral technologies. For example, the market of teaching playback robots began to increase in 1980. This is closely relating to the advancement of micro computer technologies and reduction of the computer prices. From the middle of the 1970's to 1980, the micro computer technology had remarkable advancement. This caused the increase of computational power and the reduction of price. Since the high quality computer becomes available in almost industrial robot with low price, the cost/performance of the industrial robots was very much improved, and in turn, the product sales were expanded. This story tells us that some advancement of the relating technology plays an important role to expand the robot applications.

One of the most important technical trends in near future to extend the robot applications may be the high speed/high capacity communication networks using optical fiber. The network tele-robotics will be an attractive area which will be made by the combination of robotics and the communication network technology. Suppose to connect the robots on the communication network and control them by joysticks through the network. The high speed/high capacity network technology will enables to remotely control the robots on real time. Using this technology, therefore, we can transmit the "actions" from one site to the other very efficiently through the robot. This will produce several new robot applications, like tele-surgery, tele-assistance and so on. With this situation in mind, how to utilize the communication network technology advancement for robot control is considered as the second key issue in the project.

4.3 Platform Based/Application Oriented Approach

In almost project researches so far promoted, first, we try to identify several fundamental issues to be solved to achieve the objectives and at the beginning period of the project, the fundamental researches used to be separately carried out to pursue each identified fundamental issue. After the fundamental researches will be completed, the technologies obtained from them were integrated to complete the final research target(Fig.1). This popular way has a drawback in that we can not spend enough time for solving practical issues within the limited project period. Generally, we used to spend too much time for fundamental researches and the system integration. However, not enough time was spent to investigate how to apply the technologies for real tasks. This resulted in producing many academic results with less practical applications. With this experiences in mind and in order to remove the drawback the previous project researches had, we decided to introduce "Platform Based/Application Oriented Approach". In this approach, first, we provide an integrated robot system to researchers in the project, which will be called as "Platform". Secondly, the researchers will be asked to propose what kind of practical applications they hope for the platform. In this procedure, the proposal will be carefully evaluated with the criterion which puts high score on the proposal which proposes the application the platform is the best way for and including technical issues solvable in the practically applicable way within the project period. Because of this criterion, for examples, the remote control approach may be preferable to the autonomous control approach using artificial intelligence and/or learning control techniques when the control problem will be considered. The development of the robot with autonomous functions is an interesting research topic. However, we are thinking it includes too

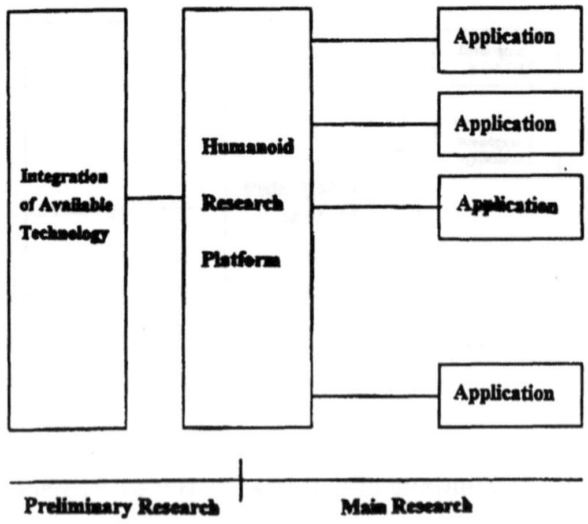

Fig. 2: Platform Based/Application Oriented Project.

many difficult problems to be solved to obtain the practically applicable results within the project period. In the case this kinds of issue will be proposed, the strong explanation from the practical point of view will be requested. In this efforts, anyhow, we are expecting that several applications suitable for the platform will be newly found out and established which will expand the robot application. After selecting the suitable applications, the researches are executed through improving the platform to meet the application requirements(Fig.2). Using this approach, we suppose that the technologies can be always pursued with the practical applications in mind and it contributes to producing practically useful results. Also, the platform as a common facility among researchers will encourage the efficient communications among researchers who are working on different sub-programs. This is the other important effect to the efficient project execution we expected by the platform based approach.

In the platform based approach, what kinds of the platform should be selected is very important. The platform must be the machine(robot) which can be made with the integration of the technologies currently available. Also, in order to collect many noble ideas from various different application areas and develop new applications where the robots have not yet been used, it has to be the noble robot which is different from currently available robots, and also attractive enough to stimulate the various people's attentions. The applicability to the human-robot coexistence system is the other important concern. With the above matters in mind, we selected the humanoid robot as the platform. The reasons are summarized as follows:

1) It is attractive enough to draw various people's attentions.

2) It has a human friendly structure.

3) It has many characteristics which we can not see in usual industrial robots and which may provide a possibility of developing new robot application areas.

4) It will be possible to construct the practical humanoid for the research uses by integrating the current available technologies without deep research efforts.

5 Outline of the Project

As mentioned in the previous sections, at the first period of the project, the humanoid platform will be developed. The rest of the period will be used to research on how the platform should be improved to make it applicable to the proposed specific applications. To collect the application ideas suitable for the research, what kind of platform will be provided will be announced at the beginning of the project. Also, several examples of application ideas will be shown, considering the proposer's convenience. The participants of the project will be the research groups from the private industries which propose the interesting applications selected based on the criterion mentioned in the previous section.

5.1 Humanoid Platform

In order to provide the platform, two kinds of humanoid robot were considered for the development in the project, which are adult type and child type. As an example, the specifications for the adult type are shown in Fig.3. For child type, lighter and smaller humanoid is considered. The height and the weight will be about 135cm and 50kg, respectively. For several different uses' conveniences, the open architecture of the computer system and the humanoid hardware itself with the modular structure will be introduced. Also, the standard software which can execute the fundamental functions of the humanoid, like stable walking on usual surface, basic manipulation functions and so on will be implemented. Constructing the humanoid platform equipped with

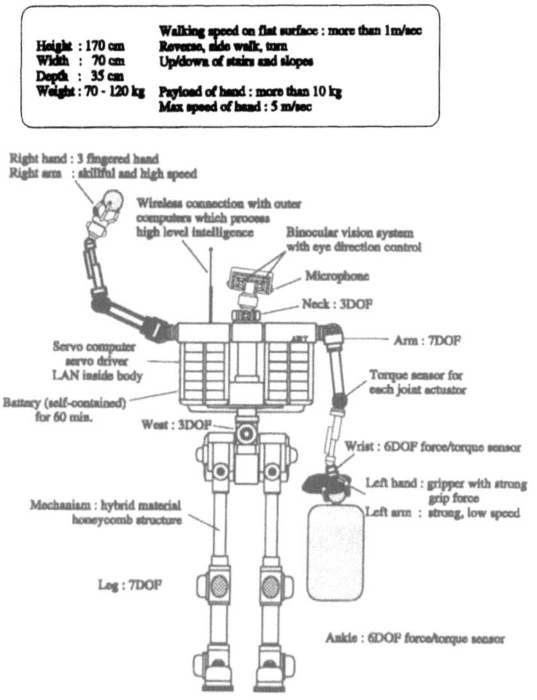

Fig. 3: Humanoid Platform (Adult Type).

Fig. 4: Plant Maintenance Application.

Fig. 5: Tele-Assistance Application.

the open control architecture and standard software will be the major researches at the platform development period.

5.2 Example Applications

To achieve the purposes of the application oriented approach, the application tasks which the humanoid robot can do in the most efficient way are carefully identified in the project. These application tasks will be found out based on the proposals collected from the project participants which want to use the humanoid platform. Before collecting the proposals, the four application tasks are indicated as examples, considering the proposers' convenience. In each application example, human-robot coexistence technology and the networked telerobotic technology will be considered together with the issues mentioned independently in the applications.

1) *Plant Maintenance*
The humanoid has a thinner width and taller body, comparing to the usual industrial robots. This structure will provide a benefit of working with human and also executing several kinds of tasks in the way that human is doing. In the plant maintenance tasks, we are often required to reach to the task location, moving our body in some obstacles, like pipes, valves and so on. The human body has an excellent structure to do such a behavior efficiently. This situation is shown in Fig.4. With this benefits the humanoid has in mind, the software which enables to execute the various skilled motions using humanoid body will be developed. As the final target, a networked tele-operated robot including shared autonomy supported by the body skilled motion software will be developed.

2) *Human Assistance*
The humanoid robot tele-operated through the network will be effectively used for assisting the aged persons in their daily life. The use of the communication network for the robot tele-operation will remove the limitation of the working areas of the current telerobotic system with the locally connected master and slave. Fig.5 shows the application, in which the humanoid remotely controlled through the network is assisting an

aged person walking outside. To realize this application, the safety robot structure for human should be investigated as a key technology, like collision tolerant body motion control and soft skin surface which reduces the impact force caused by unpredictable collision of human and robot. We call this technology as soft robotics technology. Also, the human friendly appearance and the friendly interfaces for human robot communication should be implemented. These research issues will be dealt with in this application to improve the platform.

3) *Disaster Environment Operation*
In disaster environments, there are several needs for the robot. One of such tasks will be to inspect the situation soon after the disaster has happened. In the case some plant will be accidentally wrecked by an explosion, it will be urgently requested to inspect what kinds of situation have happened by the accident. The robot will be an useful tool to achieve such a task because in many cases the situation will be very dangerous for the human to do it. There are many tasks expected to be done by the robot according to the kinds of disaster. The task of finding and saving the wounded persons in disaster environments will be one of the most important tasks. However, to achieve such tasks, a lot of complex and dexterous functions will be needed in the robot. It will include too many difficult problems to be solved within the project period. To select the application with the solvable problems within the period, the tele-observation robot for the disaster environments will be considered as an example. This robot is tele-operated through the network and can move dexterously in the environment destroyed by the disaster. Also, it can do some simple manipulation tasks with the remotely controlled arms and hands to help the observation tasks, like removing obstacles to disclose the surface to be observed. The environment will be detected by the vision, force and tactile sensors which will be installed on the humanoid platform and the detected information will be presented to the operator with tele-presence/tele-existence technology. The ground surface on which the robot will move is assumed to be very rough. Therefore, in this application, the robust biped locomotion which enables to move stably even on the rough surface ground will be identified as a key issue(Fig.6).

Fig. 6: Disaster Environment Application.

Fig. 7: Tele-Inspection Application.

4) *Tele-inspection*
Humanoid or human-like shape robots has a possibility of making the tele-operation easier. Because of this reason, several anthropomorphic robots have been so far researched. In order for the operator to remotely control the robot with feeling as if he/she were the robot, it is important for the tele-robotic system to have a humanoid robot which imitates human body motion and human sensory functions as the slave and the well-designed master devices with proper sensory feedback as the master. This is a well-known concept in designing tele-robotic system using tele-presence/tele-existence technology. The humanoid will provide a key component for the tele-robotic system which realizes the remote presence. Looking at this benefits,

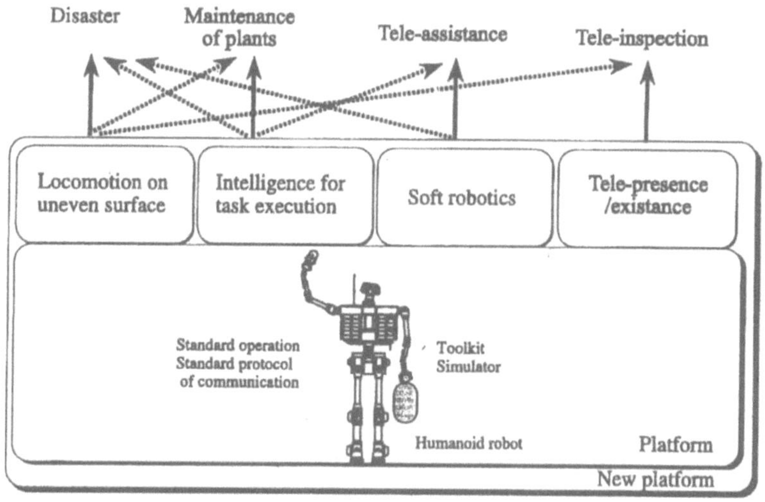

Fig. 8: Key Issues and Platform Technology.

the tele-experience application will be considered as an example application(Fig.7). In this application, a tele-robotic system which enables the operator to feel he/she is presenting remotely at the slave environment will be developed for the application to tele-inspection, tele-experience system in several amusement applications and so on. The key technologies considered will be the design of whole body master device, real-time dynamic image communication technology, design of efficient haptic interface and so on. These technology will be applied to the humanoid platform to obtain the robot applicable to the intended application.

The relations among each example application, humanoid platform and each key issue included in each application are summarized in Fig.8. This shows each application will be realized through adding the key technologies developed in each application research to the humanoid platform.

5.3 Budget and Schedule

The project plan has been already submitted to the MITI through the division in charge of the project. It is now under consideration. If it will be accepted, it will start on April in 1998. The project duration will be 5 years and the expected total budget will be about 5 billion yen (1 billion yen/year). As mentioned in the above section, the first 2 years will be spent for development of the humanoid platform, and the rest of the period will be used for the application research of the humanoid platform.

6 Conclusions and Acknowledgements

The project described in the paper has been discussed in the committee including MITI's officers, researchers of MITI's laboratories(MEL, ETL and NIBHT), university professors, researchers from companies. Especially, the author would like to note that the following members in the committee have played important roles to establish the research plan, Mr. K. Makiuchi(MITI), Mr. N. Ozawa(MITI), Prof. Y. Umetani(Toyota Tech. Inst.), Prof. H. Inoue(Univ. of Tokyo), Prof. S. Tachi(Univ. of Tokyo), Dr. T. Nozaki(MEL), and Dr. H. Tsukune(ETL). We believe that the proposed project includes several new concepts which will be important to expand the future robot applications. We are hoping that the project will be successfully approved and contribute to encouraging the future robot industries through this project.

References

[1] M. Konaka (1991, March), National Project on Advanced Robot Technology in Japan, in Proc. of International Symposium on Advanced Robot Technology, Tokyo, Japan, pp. 9-14.

Current and Future Perspective of Honda Humanoid Robot

Kazuo HIRAI
HONDA R&D Co. Ltd. Wako Research Center
1-4-1 Chuo Wako-shi Saitama 351-0193 JAPAN

Abstract

Honda revealed a humanoid robot with two legs and two arms in December of 1996. The robot walks not only forward and backward but also diagonally either to the right or left and turns into any direction as well.

The robot also can steadily walk up and down staircase without missing a step and push a cart with coordinated movement of its legs and arms. This robot with its original posture stability control can keep its balance against such unexpected disturbances as irregularities and unevenness on the floor surface. The paper introduces an outline of the structure and functions of the robot along with the development history. As to the controls, the basic principle of the robot's posture recovery control is also briefly explained.

1 Start of Humanoid Robot Research

Honda publicly presented a humanoid robot having two legs and two arms in December of 1996. The research and development of this robot was initiated 11 years ago of 1986.

The key words were "intelligence" and "mobility", and our direction and thoughts were "to coexist and collaborate with human, to perform things that human is unable to do and to create a mobility which brings additional value to the human society." That is to say, we aimed at developing a new type of robot to meet the consumer needs but not a robot for special limited purpose.

We first planed a practical wheeled robot having two arms and a video camera installed on the upper body for recognition research which we thought would be very convenient to study such intelligence as judgment and recognition research. However, as we gave a careful thought as to the meaning of a consumer type robot which we initially intended to develop, we came to a conclusion that it does not meet one of our key words "mobility". We then looked into a type of consumer robot which will better meet our initial objective.

If we were to look at a "Domestic Robot," for an example, as a type of robot that consumers may use, it will be necessary for a robot to walk around the furnitures and walk up and down the staircase inside of a house. We found that human with two legs is best suited for such movements. At the same time, if we were able to develop a two legged (biped) robot technology, we believed that the robot should be able to move around the majority of earth environment including rough terrains.

Consequently, we reached to a conclusion that the configuration of lower part of the robot would be better if it has two legged mobile mechanism which can walk like human than wheeled type and therefore decided to develop a robot by concentrating our effort on that objective. Once we established our direction, the next step was how to realize it.

We then began to conduct a study on two legged walking mechanism by first analyzing actual human walking by taking ourselves as a model.

2 Study on Robot's Leg Mechanism

Following 7 subjects were selected to study the leg mechanism.

1). Effectiveness of leg joints relating to the walking.
2). Locations of leg joints.
3). Movable extent of leg joints.
4). Dimension, weight and center of gravity of a leg.
5). Torque placed on leg joints during the walking.
6). Sensors relating to the walking.
7). Grounding impact on leg joints during the walking.

In the following, we would like to explain the result of our initial experiment.

Regarding the subject 1) of the leg joints above,

we found that the walking was not affected even there were no fingers and that the roots of the fingers and heel are more important for supporting the body weight.

As far as the ankle joint and walking function are concerned, if the ankle joints were fixed;

i). there will be a lack of contact feeling with ground surface and the fore-and-aft stability will be weak,
ii). standing still is difficult if eyes were closed, and
iii). when side crossing a sloped surface, the feeling of contact with ground surface and stability are weak.

As far as the knee joint, we found that if the knee joints were fixed, walking up and down staircase is not possible.

From the above observation, we decided to have joints equivalent to such joints as hip joints, knee joints and ankle joints but not finger joints.

In order to study the roles of hip and back bones, those areas were fixed and restrained from outside and we observed how these factors affect such walking patterns as S-curves, straight and turning. As a result, we found that the robot was able to walk at a speed of 8km/h even it was restrained.

Regarding the subject 2) of the locations of leg joint, we carefully observed human bone frames.

Regarding of the subject 3) of the movable extent of leg joints, we experimented the walking on flat surface and up and downs of staircase, measured the movement of joints and the extent of the movement of each joint of the robot.

Regarding the subject 4) of the center of gravity of each leg, we set our objective by referring to "the center of gravity of actual human body."

Regarding the subject 5) of the torque placed on the joints, we the torque of joints by measuring the ground reaction force and the movement of leg joints of human walking.

And regarding the subject 6) of the sensor, we made a careful study on types of sensors to be required for the robot.

Human has three senses for sensing the equilibrium. One is the sensor to sense acceleration by ear drum, the second one is the sensor to sense the tipping rate by semicircular canals and the third one is the sensor to sense the angles of joints movement, angle acceleration, muscular strength, pressure feeling of foot sole and skin.

We also studied the visual sensor which complements and alternates the sense of equilibrium mentioned above and also manages the walking information. Basing on those informations, we concluded that a robot in its system needs a G-sensor, 6 axis force sensor and gyrometer to sense its own posture, and joint angle sensor in order to grasp the leg movement when walking.

Next subject discussed was the ground reaction force; that is, impact force imposed on the foot during the walking. Human body is so designed, as an example, to absorb the impact force with his soft skin tissues surrounding the foot, arch frame of bones forming the foot, and the roots of finger joints, and flexible movement of knee joint as the foot land the ground.

As the walking speed increases, the reaction force becomes larger even with the human's impact

Degree of Freedom (DOF)
Legs' DOF : 6 x 2 = 12

Actuators: DC servo motors
(+ Harmonic drives)

Sensors: Vision cameras
Gyrometers
G-sensors
Six axis force sensors

Weight: 150 kg

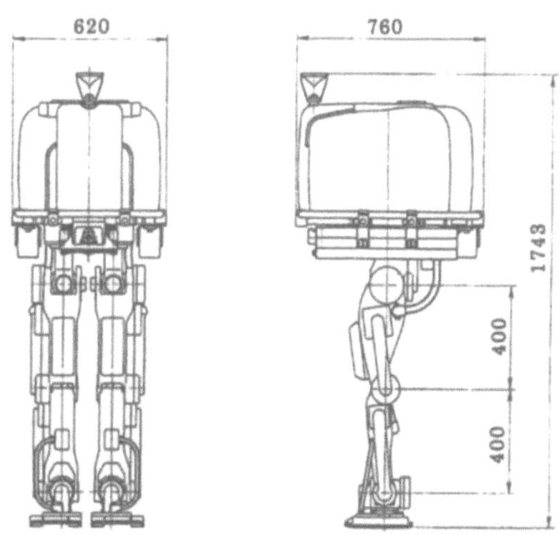

Figure 1: Biped Walking Robot

damper mechanism. We found that at the walking speed of 2 to 4 km/h, the load forced on the foot is about 1.2 to 1.4 times that of the body weight and at the walking speed of 8 km/h, it is about 1.8 times. Basing on these results, we made an initial specifications for our robot.

3 Development of Biped Walking Robot

Basing on a biped walking robot fabricated as a preliminary experimental robot, we then refined the specifications which meet the following functions.

1). To realize the walking speed of 3 km/h.
2). For attaching arms and hands of robot to the upper body.
3). Walking up and down a normal staircase.

For a research purpose, we began with static walk. However, since human walking is mainly of dynamic walking, we thought that the robot must also have dynamic walking feature and therefore our research effort have emphasized more on the dynamic walking. As a result, our walking program is a dynamic walking program based on the human walking data and we conducted our experiment.

Basing on our continuos research effort, we were able to consolidate those specifications mentioned above for a biped robot and complete a wireless robot movable with electric batteries mounted as shown in the figure 1.

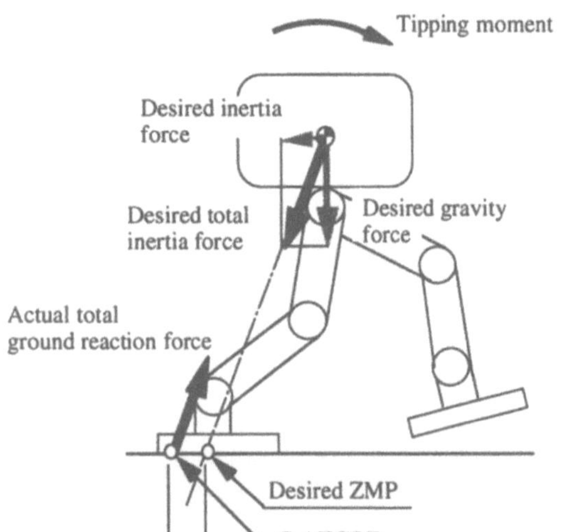

Figure 2: Dynamic Balance while Walking

4 Start of Flexible Walking Robot

Up to this point, the robot we developed was just able to walk on a straight line only. For the next stage, however, we began a research and development of a biped robot which can steadily move around in human life environment without tipover or falling down and is especially maneuverable on different road surfaces, undulation, slope and steps. As a preparation for further study, we developed a walking simulator taking the mechanical characteristics into account.

5 Technical Points for Realizing Stable Walking

When human is about to fall down while walking or standing straight, he pushes ground hard with a part of his foot sole to resist the falling. But when he is no longer able to resist, then he tries to recover his posture by changing his body movement or by taking an extra step. We have tried to achieve a high posture stability by adopting a similar recovery ability to the robot. The figure 2 explains the principle of the robot posture recovering.

Basically, a robot is controlled to follow the angles of leg joints of ideal walking pattern. A combined force of inertia force and gravity force of desired walking pattern is called "desired total inertia force." As it is known, the point on the ground surface where the moment of the desired total inertia force becomes zero except the vertical element is called "Desired Zero Moment Point or Desired ZMP."

The ground reaction force has an affect on each leg of an actual robot of which combined force is called "Actual Total Ground Reaction Force or ATGRF." The point on the ground surface where the moment of ATGRF becomes zero except the vertical element is called "the Center of ATGRF or C-ATGRF."

If the actual robot walks in ideal condition, ZMP and C-ATGRF will be at a same point. In actuality, however, even though the body posture is in accord with the desired posture and the joint angles are following the desired joint angles, C-ATGRF is off the desired ZMP as it is shown in the figure because of the irregular terrain. In this state, since the action lines of ATGRF and the desired total inertia force do not agree, a couple force produced by these forces acts upon the robot and the robot's entire posture tends to tipover. This couple force is called "Tipping Moment" which is calculated

Degree of Freedom (DOF)
Leg's DOF : 6 x 2 = 12
Arm's DOF : 7 x 2 = 14
Hand's DOF : 2 x 2 = 4

Actuators: DC servo motors
(+ Harmonic drives)

Sensors: Vision cameras
Gyrometers
G-sensors
Six axis force sensors
on wrists and feet

Weight: 210 kg

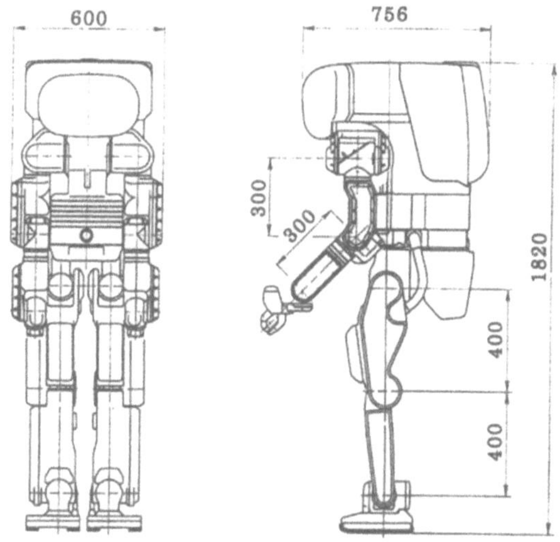

Figure 3: Humanoid Robot

with the following equation.

$$\textit{Tipping moment} = (\textit{Desired ZMP} - \textit{C-ATGRF}) * \textit{Vertical element of desired total inertia force} \quad (1)$$

By observing this equation, we reached to an idea that if the distance between the desired ZMP and C-ATGRF were actively controlled, the tipover posture of a robot can be recovered by conversely utilizing the tipping moment. We believed that this is a basic principle for recovering the robot's tipover posture. The control to operate C-ATGRF is called "Ground Reaction Force Control" and the control to operate the desire ZMP is called "Model ZMP Control."

The ground reaction force control controls C-ATGRF by sensing with 6 axis force sensors and by modifying the desired position and posture of feet.

The model ZMP control is to control the shifting of the desired ZMP to an appropriate position in order to recover the robot posture by changing the ideal body trajectory when the robot is about to tip over (that is, when the difference between the inclination of the body of actual robot and that of the desired body).

For an example, when the body of a robot tips forward, the model ZMP control system increases the acceleration of the desired body. As a result, the magnitude of the desired inertia force changes and the desired ZMP shifts its position to rearward of the original desired ZMP to recover the robot posture.

If the desired body position of the model changes, the spatial configuration of the desired body and feet will be off the ideal state. In order to bring this back to the ideal state gradually, the landing position of the feet is changed.

By having the controls described above to work simultaneously, we were able to achieve the robot to have a posture stabilizing control like human does.

6 Progress toward Humanizing

The next step is to realize a humanoid robot. We defined the functions of this humanoid robot as follows. The robot should be of such a type that he can automatically perform a certain type of works under the known environment and perform an uncertain type of works with assistance from a human operator under unknown environment. The first experimental humanoid robot had an overall length of 1,915mm and weight of 185 kgf.

We had first concentrated our study on how to realize the coordinated movement of legs and arms and therefore the computers for image processing and action plan, electric power supply, etc. were not installed on the first robot. Through the experiment with this robot, we studied a coordinated movement of the robot to perform such tasks as turning switch on and off, grasping and turning door knob and holding and carrying an object.

In the next stage, we developed a wireless humanoid robot as shown in the figure 3 which was publicly revealed by Honda as mentioned earlier. The overall length was 1,820mm with weight of 210 kgf. Computers, motor driver, batteries as a power source and transmitter were installed inside of the robot. The main functional specifications are listed below.

Mobility Performance:

1). Be able to move around on normal flat surfaces.
Example: Plastic tiles, paved road, grazing, etc.
2). Be able to pass through a narrow opening: Width of the opening of 850 mm.
3). Be able to step over and cross over steps and mounds.
Example: Step over steps with a height of about 200 mm. Cross over steps of 150 mm height and 150 mm length.
4). Be able to walk up and down the staircase of general buildings at a normal human speed.
Example: Staircase with 200 mm height and 220 mm depth of each step.
5). Be able to walk on a known slope of about 10%.

Working Ability:

1). Be able to grasp and hold an object with weight of about 5kgf
2). Be able to perform a light work using such a tool as wrench by a remote control.

7 Future Plan

The future development will be divided into short term and long term plan.

For short term, emphasis will focused on the hardware improvement, which are;

1). Smaller, more compact robot
2). Mobility performance improvement
3). Operability performance improvement

For long term, we believe that increasing the physical versatility by way of mobility improvement and environmental adaptability, made possible by hardware and software technology advancement, as well as improving the autonomous mobility without detailed human instructions are important.

On the contrary, we also hope to develop technologies so that the humanoid robot can function not only as a machine, but blend in our social environment and interact with people, and play more important roles in our society.

List of Participants (The 8th ISRR - Hayama, Japan -)

Arimoto, Suguru

Dept. of Robotics
Ritsumeikan Univ.
Kusatsu, Shiga, 525-79, Japan
arimoto@bkc.ritsumei.ac.jp

Bergamasco, Massimo

PERCRO
Scuola Superiore S. Anna
via Carducci, 40, 56100 Pisa, Italy
bergamasco@percro.sssup.it

Asada, Minoru

Dept. of Adaptive Machine Systems, Graduate School of Eng., Osaka Univ.
Suita, Osaka, 565, Japan
asada@ams.eng.osaka-u.ac.jp

Bolles, Bob

AI Center, M/S EK290,
SRI International
333 Ravenswood Ave., Menlo Park, CA 94025, USA
bolles@sunset.ai.sri.com

Ayache, Nicholas

Epidaure Project,
INRIA
2004 Route des Lucioles, 06902 Sophia-Antipolis, France
ayache@sophia.inria.fr

Brooks, Rodney

Artificial Intelligence Lab.
MIT
545 Technology Square
Cambridge, MA 02139, USA
brooks@ai.mit.edu

Bruyninckx, Herman

Dept. of Mechanical Eng.,
Katholieke Universiteit Leuven
Celestijnenlaan 300B, B-3001 Leuven, Belgium
herman.bruyninckx@mech.kuleuven.ac.be

Burdick, Joel

Dept. of Mechanical Eng.,
California Inst. of Tech.
Mail Code 104-44, Pasadena,
CA 91125, USA
jwb@robby.caltech.edu

Carlisle, Brian

Adept Technology, Inc.
150 Rose Orchard Way, San Jose California 95134, USA
brian.carlisle@adept.com

Chatila, Raja

LASS-CNRS
7, Ave. du Colonel Roche
31077 Toulouse Cedex 4, France
raja@laas.fr

Daniel, Ron,W.

Dept. of Eng. Science
Oxford Univ.
Parks Road, Oxford OX1 3PJ, UK
rwd@robots.ox.ac.uk

Dario, Paolo

ARTS Lab
Scuola Superiore Sant' Anna
Via Carducci 40 56100 Pisa, Italy
dario@arts.sssup.it

Dillmann, Ruediger

Fakultat fur Informatik, Institut fur Prozessrechentechnik und Robotik, Universitat Karlsruhe, Postfach 6980, Kaiserstr. 12/Geb. 40.28, D-76128, Karlsruhe, Germany
dillmann@ira.uka.de

Durrant-Whyte, Hugh

Dept. of Mechanical and Mechatronic Eng.
Univ. of Sydney
NSW, 2006, Australia
hugh@tiny.me.su.oz.au

Feiten, Wendelin

Siemens AG, ZFE
Otto-Hahn-Ring 6, D-81730
Munchen, Germany
wendelin.feiten@mchp.siemens.de

Hirzinger, Gerhard

DLR Oberpfaffenhofen
D-82234 Wessling, Germany
gerd.hirzinger@dlr.de

Giralt, Georges

LAAS-CNRS
7 avenue du Colonel Roche
31077 Toulouse Cedex 4,
France
giralt@laas.fr

Hodgins, Jessica

College of Computing
Georgia Institute of Technology
Atlanta, GA 30332-0280,
USA
jkh@cc.gatech.edu

Hirai, Kazuo

HONDA R&D Co. Ltd.
kazuo.hirai@f.rd.honda.co.jp

Hollerbach, John

Univ. of Utah
3190 Merrill Eng. Building,
Salt Lake City UT 84112,
USA
jmh@veto.cs.utah.edu

Hirose, Shigeo

Dept. of Mechano-Aerospace Eng.
Tokyo Inst. of Tech.
2-12-1 O-Okayama,
Meguro-ku, Tokyo, 152,
Japan
hirose@mes.titech.ac.jp

Hollis, Ralph L.

The Robotics Institute
Carnegie Mellon Univ.
5000 Forbes Ave., Pittsburgh,
PA 15213, USA
rhollis@cs.cmu.edu

Howe, Robert

Divi. of Engineering and Applied Sciences
Harvard Univ.
29 Oxford St., Cambridge, MA 02138, USA
howe@arcadia.harvard.edu

Inoue, Hirochika

Dept. of Mechano-Infomatics, Univ. of Tokyo
7-3-1 Hongo, Bunkyo-ku, Tokyo, 113, Japan
inoue@jsk.t.u-tokyo.ac.jp

Ikeuchi, Katsushi

IIS
Univ. of Tokyo
7-22-1, Roppongi, Minato-ku, Tokyo, Japan
ki@iis.u-tokyo.ac.jp

Ishiguro, Hiroshi

Dept. of Information Science
Kyoto Univ.
Sakyo-ku, Kyoto 606-01, Japan
ishiguro@kuis.kyoto-u.ac.jp

Ikuta, Koji

Dept. of Micro System Eng.
Nagoya Univ.
Furocho, Chikusa-ku, Nagoya, 464-01, Japan
ikuta@everest.mech.nagoya-u.ac.jp

Jarvis, Ray A.

Dept. of Electrical and Computer Systems Eng.
Monash Univ.
Wellington Road Clayton, Vic. 3168, Australia
ray.jarvis@eng.monash.edu.au

Inaba, Masayuki

Dept. of Mechano-Infomatics, Univ. of Tokyo
7-3-1 Hongo, Bunkyo-ku, Tokyo, 113, Japan
inaba@jsk.t.u-tokyo.ac.jp

Kaneko, Makoto

Industrial and Systems Eng
Hiroshima Univ.
Higashi-Hiroshima, 739, Japan
kaneko@huis.hiroshima-u.ac.jp

Khatib, Oussama

Robotics Lab., Dept. of Computer Science
Stanford Univ.
Stanford, California 94305, USA
khatib@CS.Stanford.EDU

Latombe, Jean-Claude

Dept. of Computer Science
Stanford Univ.
Stanford, CA 94305, USA
latombe@cs.stanford.edu

Koditschek, Dan

EECS Dept. College of Eng
Univ. of Michigan
1101 Beal Avenue, USA
kod@eecs.umich.edu

Laugier, Chrischan

INRIA Rgone-Alpes
Grenoble, France
Christian.Laugier@inrialpes.fr

Konolige, Kurt

Artificial Intelligence Center
SRI International
333 Ravenswood Ave., Menlo Park, CA 94025, USA
konolige@ai.sri.com

Miura, Hirofumi

Dept. of Mechano-Infomatics, Univ. of Tokyo
7-3-1 Hongo, Bunkyo-ku, Tokyo, Japan
miura@leopard.t.u-tokyo.ac.jp

Kumar, Vijay

General Robotics and Active Sensory Perception Lab.
Univ. of Pennsylvania, 3401 Walnut St., Philadelphia, PA 19104-6228, USA
kumar@central.cis.upenn.edu

Mori, Hideo

Dept. of EE and CS.
Yamanashi Univ.
Takeda-4, Kohu, 400, Japan
forest@kki.esi.yamanashi.ac.jp

Mosemann, Heiko

Institute for Robotics and Process Control
Technical Univ. of Braunsch - weig
Hamburger Str. 267, 38114 Braunschweig, Germany
H.Mosemann@tu-bs.de

Nakamura, Yoshihiko

Dept. of Mechano-Infomatics
Univ. of Tokyo
7-3-1 Hongo, Bunkyo-ku, Tokyo, 113, Japan
nakamura@ynl.t.u-tokyo.ac.jp

Nayar, Shree K.

Dept. of Computer Science
Columbia Univ.
New York City, New York 10027, USA
nayar@cs.columbia.edu

Omichi, Takeo

Takasago R&D Center
Mitsubishi Heavy Industry
2-1-1 Shinhama, Arai-cho, Takasago-sgi, Hyogo, Japan
oomichi@trdc.tksg.mhi.co.jp

Perez, Philippe

Ingenieur des Telecommui-cations, Attache pour la Science et la Technologie
Ambassade de France
4-11-44, Minami-Azabu, Minato-ku, Tokyo, 106, Japan
perez@sst.paradigm.co.jp

Rimon, Elon

Mechanical Engineering Dept.
Technion, Israel Ins.of Trch.
ME Dept. Technion, Haifa 32000, Israel
elon@robby.technion.ac.il

Roth, Bernie

Mechanical Engineering Dept.
Stanford Univ.
Stanford, CA 94305, USA
roth@flamingo.stanford.edu

Sato, Tomomasa

RCAS
Univ. of Tokyo
4-6-1, Komaba, Meguro-ku, Tokyo, Japan
tomo@lssl.rcast.u-tokyo.ac.jp

Shimoyama, Isao

Dept. of Mechano-Infomatics, Univ. of Tokyo
7-3-1 Hongo, Bunkyo-ku,
Tokyo, 113, Japan
isao@leopard.t.u-tokyo.ac.jp

Shirai, Yoshiaki

Dept. of Computer-Controlled Machine Systems
Osaka Univ.
2-1 Yamadaoka, Suita, Osaka,
565, Japan
shirai@mech.eng.osaka-u.ac.jp

Tanie, Kazuo

Biorobotics Divi., Robotics Dept.
MEL, AIST-MITI
1-2 Namiki, Tsukuba, Ibaraki,
305, Japan
m1750@mel.go.jp

Thorpe, Chuck

Robotics Institute
Carnegie Mellon Univ.
Smith Hall 5000 Forbes Ave.
Pittsburgh PA 15213-3890,
USA
thorpe@ri.cmu.edu

Trevelyan, James

Univ. of Western Australia
Nedlands 6907, Australia
jamest@mech.uwa.edu.au

Troccaz, Jocelyne

TIMC/IMAG Lab., Faculte de Medecine de Grenoble
Domaine de la Merci
38706 La Tronche cedex,
France
Jocelyne.Troccaz@imag.fr

Uchiyama, Masaru

Dept. of Aeronautics and Space Eng., Graduate School of Eng., Tohoku Univ.
Aramaki-aza-Aoba, Aoba-ku,
Sendai 980-77, Japan
uchiyama@space.mech.tohoku.ac.jp

Uchiyama, Takashi

M-Project
Fujitsu Lab. Limited
10-1, Wakamiya, Morinosato,
Atsugi, Japan
uchiyama@flab.fujitsu.co.jp

Whitcomb, Louis

Dept. of Mechanical Eng.
Johns Hopkins Univ.
123 Latrobe Hall, 3400 North Charles Street, Baltimore, Maryland, 21218, USA
llw@jhu.edu

Yuta, Shin-ichi

Institue of Information Science and Electronics
Univ. of Tsukuba
Tsukuba, 305, Japan
yuta@roboken.esys..tsukuba.ac.jp

Yang, Hyun S.

Dept. of Computer Science, Center for Artificial Intellige-nce Research, KAIST
373-1 Kusong-dong, Yusong-ku, Taejon, 305-701, Korea
hsyang@cs.kaist.ac.kr

Yoshida, Kazuya

Dept. of Aeronautics and Space Eng.
Tohoku Univ.
Aobayama campus, Sendai 980-77, Japan
yoshida@astro.mech.tohoku.ac.jp

Yoshikawa, Tsuneo

Dept. of Mechanical Eng.
Kyoto Univ.
Yoshida, Sakyo-ku, Kyoto, 606-01, Japan
yoshi@mech.kyoto-u.ac.jp

Professor Yoshiaki Shirai
Department of Compter-Controlled Machine Systems, Faculty of Engineering, Osaka University 2-1, Yamadaoka, Suita, Japan 565

Professor Shigeo Hirose
Mechano-Aerospace, Faculty of Engineering, Tokyo Institute of Technology, 2-12-1 Ohokayama, Meguro-ku, Tokyo, Japan 152

ISBN-13: 978-1-4471-1582-3 e-ISBN-13: 978-1-4471-1580-9
DOI: 10.1007/ 978-1-4471-1580-9

British Library Cataloguing in Publication Data
Robotics research : the eighth international symposium
1.Robotics - Congresses
I.Shirai, Yoshiaki II.Hirose, Shigeo III.International Symposium on Robotics Research (8th : 1998 : Shonan, Japan)
629.8'92

Library of Congress Cataloging-in-Publication Data
Robotics research : the eighth international symposium / Yoshiaki Shirai and Shigeo Hirose, eds.
p. cm.
Includes bibliographical references.

1. Robotics--Research--Congresses. I. Shirai, Yoshiaki.
II. Hirose, Shigeo, 1947-
TJ211.3.R64 1998
629.8'92'072--dc21 98-16079

Softcover reprint of the hardcover 1st edition 1998

Typesetting: Camera ready by editors
69/3830-543210 Printed on acid-free paper

Yoshiaki Shirai and Shigeo Hirose (Eds)

Robotics Research

The Eighth International Symposium

Springer

Springer
London
Berlin
Heidelberg
New York
Barcelona
Budapest
Hong Kong
Milan
Paris
Santa Clara
Singapore
Tokyo